T0135739

TECHNISCHE UNIVERSITÄT MÜNCHEN

Lehrstuhl für Anorganische Chemie

η^2-Complexes of group 10 with cinnamic acid derivatives; spectroscopic properties, stability, reactivity and application potential as photo-switchable catalysts

Magnus Richard Buchner

Vollständiger Abdruck der von der Fakultät für Chemie der Technischen Universität München zur Erlangung des akademischen Grades eines

Doktors der Naturwissenschaften

genehmigten Dissertation.

Vorsitzende: Univ.-Prof. Dr. S. Weinkauf

Prüfer der Dissertation:

        1. Univ.-Prof. Dr. K. D. Ruhland, Universität Augsburg

        2. Univ.-Prof. Dr. Th. Bach

        3. Univ.-Prof. Dr. K. Köhler

Die Dissertation wurde am 26. Januar 2011 bei der Technischen Universität München eingereicht und durch die Fakultät für Chemie am 21. Februar 2011 angenommen.

Bibliografische Information der Deutschen Nationalbibliothek

Die Deutsche Nationalbibliothek verzeichnet diese Publikation in der
Deutschen Nationalbibliografie; detaillierte bibliografische Daten sind
im Internet über http://dnb.d-nb.de abrufbar.

©Copyright Logos Verlag Berlin GmbH 2011
Alle Rechte vorbehalten.

ISBN 978-3-8325-2847-8

Logos Verlag Berlin GmbH
Comeniushof, Gubener Str. 47,
10243 Berlin
Tel.: +49 (0)30 42 85 10 90
Fax: +49 (0)30 42 85 10 92
INTERNET: http://www.logos-verlag.de

The presented work was prepared from January 2008 to January 2011 at the chair for Inorganic Chemistry of the Technische Universität München in the laboratories of the *Wacker* Institute for Silicon Chemistry.

I want to particularly thank my admired academic teacher
Prof. Dr. rer. nat. habil. Klaus Ruhland
for the opportunity to work on this interesting and challenging topic, his confidence, unlimited support and personal interest in the success of this project.

Furthermore my special thanks are dedicated to the *Wacker Chemie AG.* for the financial support of my work. My personal contact persons, *Dr. Jürgen Stohrer* and *Dr. Florian Hoffmann*, deserve particular mentioning.

I want to thank *Prof. Dr. Thorsten Bach* for the unbureaucratic access to the photoreactors of his group and *Christiane* for her exceptional helpfulness and support concerning the irradiations. The whole work on photo-chemistry in this thesis would not have been possible without these two persons.

Daniel deserves thank for his support during the last three years and together with *Lars-Arne* and *Sebastian* for their help in completing this thesis. *Matthias* and *Manuel* need to be thanked for their assistance with software issues. Furthermore I would like to thank *Bettina* and *Bernhard* for the massive amount of X-ray structure determinations. The whole technical staff needs to be thanked for the huge amount of samples they processed for me. Thanks to *Ferenc E. van der Sluijs' Cybernetic Broadcasting System* and its successor, *Intergalactic FM*, for expanding my musical horizon and the high quality music played in my laboratory.

I sincerely thank my family for their unlimited support during the last 28 years and *Mareike* for her understanding and loving support.

Finally I thank all persons involved in the success of this work, who are not explicitly mentioned here.

for

Hilde & Richard

Si vis pacem, para bellum.

Contents

Nomenclature

ATR	attenuated total reflection
boc	*tert*-butyl oxycarbonyl
CCD	charged coupled device
CI	chemical ionisation
cod	1,5-cyclooctadiene
coe	*cis*-cyclooctene
COSY	COrrelation SpectroscopY
d	doublet
dcc	dicyclohexylcarbodiimide
dippp	1,3-bis(di-*iso*-propylphosphino)propane
dmap	N,N-dimethyl-4-aminopyridine
dmso	dimethylsulfoxide
dppb	1,4-bis(diphenylphosphino)butane
dppe	1,2-bis(diphenylphosphino)ethane
dppm	1,1-bis(diphenylphosphino)methane
dppp	1,3-bis(diphenylphosphino)propane
eq.	equivalent(s)
ESI	electro spray ionisation
EXSY	EXchange SpectroscopY
FAB	fast atom bombardment
FT-IR	fourier transform infrared spectroscopy
h	hexet
HMBC	Heteronuclear Multiple Bond Correlation

HMQC	Heteronuclear Multiple Quantum Coherence
hp	heptet
LED	light emitting diode
m	multiplet
MS	mass spectrometry
nbe	bicyclo[2.2.1]heptene
NMR	nuclear magnetic resonance
OEDP	olefin electron density parameter
ORTEP	oak ridge thermal ellipsoid plot
pd	pseudo doublet
pq	pseudo quartet
ps	pseudo singlet
pt	pseudo triplet
q	quartet
qu	quintet
s	singlet
t	triplet
thf	tetrahydrofurane
tms	trimethylsilyl
UV	ultra violet

1. Introduction

1.1. Organosilicon polymers

Organosilicon polymers are one of the most important classes of polymers due to their high thermal and chemical resistance. Therefore these materials are widely applied in industrial processes and highly stressed building parts, e.g. in automotive manufacturing. Another advantage of silicon polymers is their wide scope of application owed to their flexibility in ductility and elasticity, which may easily be adjusted from oils via soft silicon rubbers to extremely hard polymers.

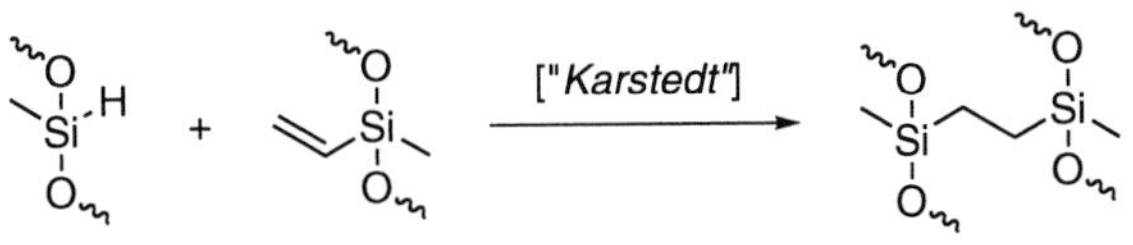

Scheme 1.1: Silicone curing via hydrosilylation.

The polysiloxanes are most commonly synthesised by the addition of polyfunctional silicon hydrides to poly vinyl organo siloxanes, which is exemplified in scheme 1.1.[1] The hydrosilylative curing of silicon rubbers proceeds without side reactions and results generally in high conversion and the products exhibit exceptionally high tensile strength, toughness and low porosity. The most commonly used hydrosilylation catalyst for the afore described polymerisation is the *Karstedt* catalyst, depicted in figure 1.1, which is most commonly synthesised by reaction of hexachloro platinate with divinyl tetramethyl disiloxane.[2–4] However this catalyst is active at ambient temperature, leading to unwanted polymerisation causing low shelf live when the catalyst is mixed with the substrate.

1

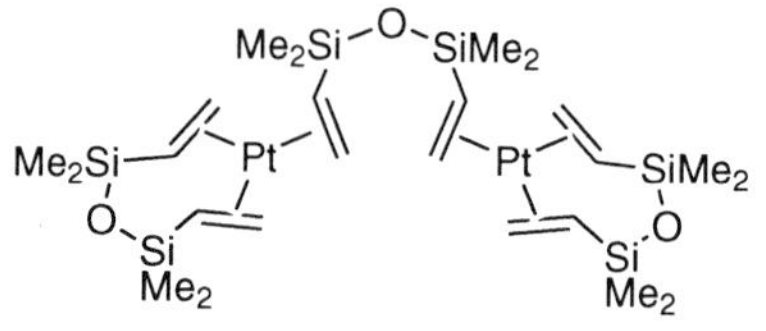

Figure 1.1: *Karstedt's* catalyst.

To achieve reaction mixtures which are polymerisable on demand, substances are added which temporarily inhibit the catalyst. Commonly the inhibitors used are derivatives of electron deficient olefins, like maleates or fumarates, which block polymerisation by coordination to the metal center. [5,6] The catalyst activity is regained by rapid heating of the reaction mixture, which initiates the exothermic polymerisation. The release in reaction energy leads to a self accelerating effect driving the polymerisation to completion. However this approach leads to high costs necessary for the heating. Polysiloxanes are, due to their high resistance towards heat and chemicals as well as their excellent optic properties concerning emission and aging, the ideal material for the encapsulation of LEDs and LED lenses. [7,8] However the electronic parts of the LED unit do not exhibit the heat resistance, necessary to endure the temperatures needed for the heat induced polymerisation described above. Therefore the ideal hydrosilylation catalyst should be light activateable at ambient temperature and exhibit high activities at low temperatures as well.

1.2. Photo-activation of transition metal complexes

Numerous reactions are known where light induces the dissociation of a ligand and subsequent coordination of another ligand. [9,10]

One of the most common class of photo-substitutions is the removal of carbonyl ligands, by promotion of a d_π electron into a d_σ level, which is M−L σ-antibonding, and displacement with another ligand as shown in scheme 1.2. [11] However metal carbonyl complexes are generally toxic and the evolution of carbon monoxide is undesirable in an industrial scale. Also is the CO still present in solution, and therefore capable of re-coordination to

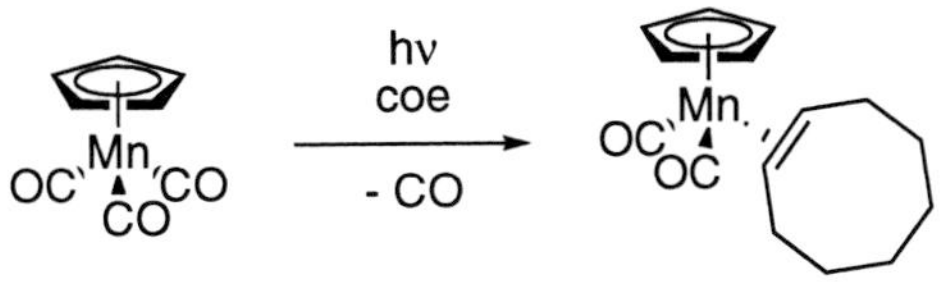

Scheme 1.2: Carbon monoxide dissociation caused by irradiation.

the metal, regenerating the starting complex.

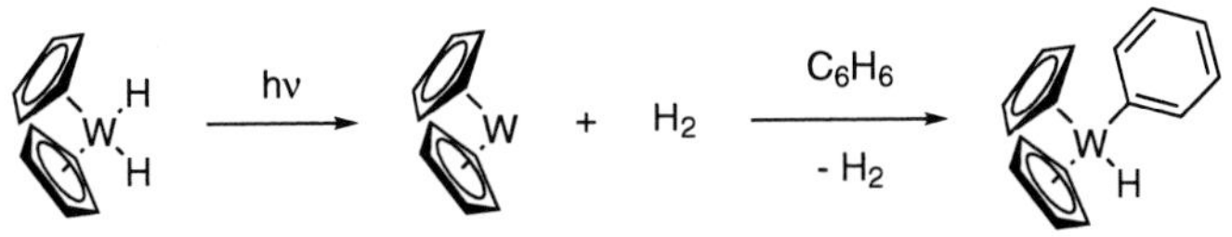

Scheme 1.3: Irradiative reductive hydride elimination.

Another well described reaction is the extrusion of hydrogen from metal hydrides (scheme 1.3).[12,13] The generated reductive elimination products are often highly unstable and undergo fast subsequent reactions.

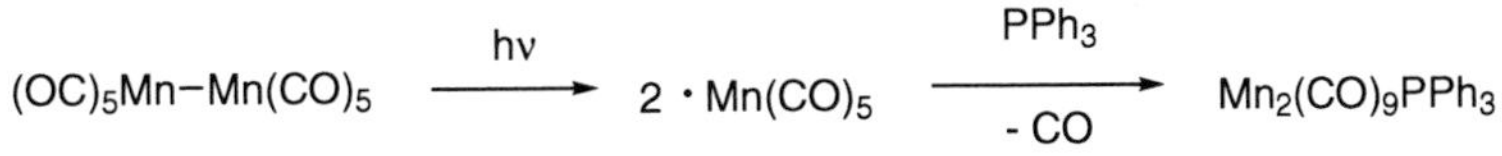

$$(OC)_5Mn-Mn(CO)_5 \xrightarrow{h\nu} 2 \cdot Mn(CO)_5 \xrightarrow[-CO]{PPh_3} Mn_2(CO)_9PPh_3$$

Scheme 1.4: Photo-cleavage of a metal-metal bond.

The third important photo-process in transition metal chemistry is the homolysis of metal-metal bonds, leading to ligand substitution as shown in scheme 1.4.[14,15] However the metal radical species generated upon photolysis are highly reactive and therefore react unselectively.

In more recent years the photo-dissociation of olefins bound to transition metals and subsequent ligand exchange was described.[16–19] This reactions always involve high energy laser pulses and are normally solely observed in matrix or femto second spectroscopy, though.

These principles have been applied in the design of photo-switchable hydrosilylation catalysts of which two examples are depicted in figure 1.2.[20–22] However these designs show lower activities than the *Karstedt* system, are prone to side reactions due to the

Figure 1.2: Photo-activateable platinum(IV) and platinum(II) hydrosilylation catalysts.

radical intermediates formed during the UV-initiation and are hard to modify, since the photo-process causing ligand dissociation involves directly the metal-ligand bond.

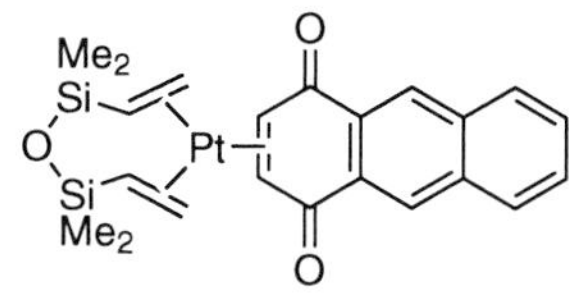

Figure 1.3: Photo-switchable platinum(0) catalyst for hydrosilylation.

Another approach that was undertaken, is the addition of chromophoric olefin inhibitors to technical *Karstedt* solution.[23] The presumably formed inactive species is shown in figure 1.3 and the electron deficient olefin dissociates upon irradiation. However the reaction mechanism is not clarified and the inhibitor stays unchanged in solution, resulting in lower activities of the catalyst than the non inhibited *Karstedt* system.

1.3. Diazo olefins as photo-switchable ligands

To achieve a photo-switchable hydrosilylation catalysis system the concept shown in scheme 1.5 was anticipated. The platinum would be encapsulated by the coordinating ligands, in a fashion preventing any reaction with the substrates. This is achieved through steric shielding of the metal center. Upon irradiation a photo-sensitive group in the outer ligand sphere reacts with parts of the inner ligand sphere. Therefore a free coordination site is generated since some of the coordinating groups were removed through the reaction with the photo-activated group. The generated, due to under coordination, highly reactive metal center would then undergo immediate oxidative insertion into a Si−H bond or coordination of a substrate olefin and therefore start the catalytic cycle.

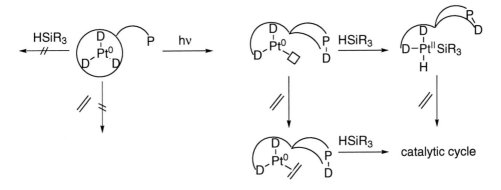

Scheme 1.5: Generic concept of photo-switchable catalysis.

For industrial application of a photo-triggered hydrosilylation catalyst the following points are mandatory:

1. no activity before irradiation

2. stability at ambient temperature without self activation or decomposition

3. defined absorption wavelength, ideally below 500 nm to provide handleability of the catalyst mixture, without activation, under red light

4. selective ligand cleavage, without side reactions, to start catalysis

5. post-irradiation activity comparable to *Karstedt´s* catalyst

These demands were assigned with the system depicted on the bottom of scheme 1.6. The anticipated photo-switchable ligand system is an olefin functionalised with a diazo group. Upon irradiation nitrogen is cleaved from the diazo group and a carbene is formed, which undergoes intramolecular cyclopropanation with the double bond. The ligand therefore loses the coordination ability and dissociates from the platinum center, which is then capable of catalytic hydrosilylation. The photo-dissociation of the ligand is similar to the dissociation of the bridging divinyl tetramethyl disiloxane in the *Karstedt* catalyst, which is prior to the catalytic hydrosilylation cycle (scheme 1.6 top). This concept has the advantage of a permanent removal of the ligand, so no interference is caused by the ligand

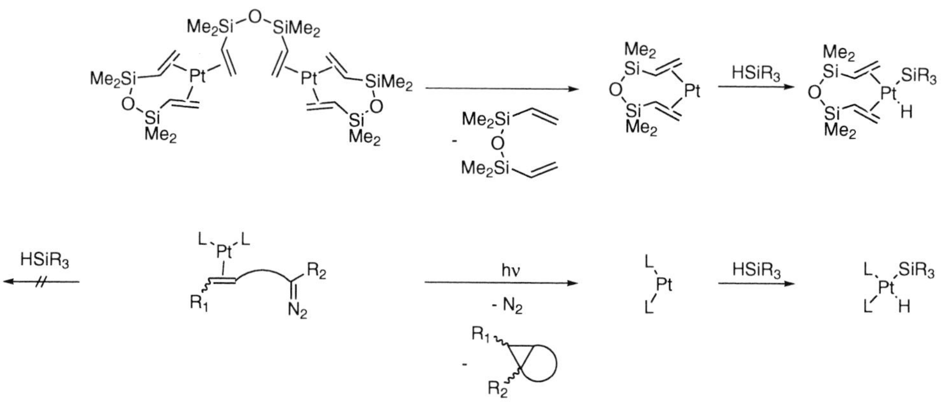

Scheme 1.6: Use of diazo olefins as photo-switch.

concerning post-irradiation activity. Due to the light triggered intra-ligand cyclisation, the properties of the metal are of no significance to the photo-activation and may therefore be optimised concerning the catalytic properties. Also do the other ligands provide further variation options concerning the catalyst's properties without interference with the photo-switch.

The results concerning photo-triggered catalysis with diazo olefins are described in chapter 2, while the synthesis of the respective ligands and their irradiation behaviour is given in chapter 3. Chapter 4 deals with the effects exhibited by connatural ligands on the spectroscopic properties, stability and reactivity of group 10 complexes. Photoswitchability may also be achieved by the addition of photo-inactivatable inhibitors to technical *Karstedt* solution. This approach is dealt with in chapter 5, while chapter 6 addresses the synthesis and properties of highly under coordinated palladium systems which arose from the synthesis of the palladium complexes in chapter 4.

2. Photo-switchable catalysts

To achieve photo-induced catalytic reactivity a selective cleavage of an olefin ligand from a bis(triphenylphosphine)platinum olefin complex was anticipated. This concept is depicted in scheme 2.1.

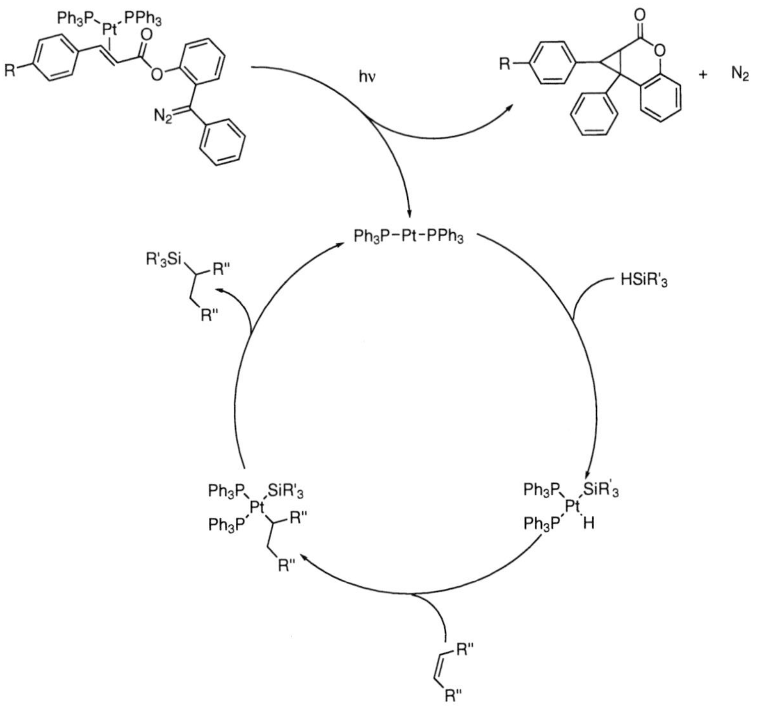

Scheme 2.1: Anticipated photo-induced hydrosilylation cycle.

Upon irradiation an *in situ* generated carbene, derived from nitrogen cleavage of the diazo group, would react in an intramolecular cyclopropanation with the olefin tethered to the metal. The loss of the double bond leads to immediate dissociation of the ligand leaving an under-coordinated and therefore highly reactive 14 e$^-$ metal fragment which would undergo immediate oxidative insertion into Si−H bonds, impossible before removal of the olefin. This oxidative addition is the first step in the *Chalk-Harrod* mechanism and would start the catalytic cycle.[1]

Bis(triphenylphosphine)platinum cinnamates were chosen due to their straightforward synthesis[1] and the electronic tuneability of the olefin via the *para*-substituent of the cinnamic acid´s phenyl group. Due to the phosphines a highly sensitive NMR probe is present as well, which allows reaction monitoring in non-deuterated solvents. The diazo ester was chosen due to the relatively high stability and reasonable accessibility of the compound. Furthermore a six membered ring would be formed upon cyclopropanation, which provides a thermodynamic driving force for the latter reaction.

Since the diazo complexes were estimated to be light- and temperature-sensitive the corresponding benzophenone cinnamate complexes were synthesised first, as reference compounds. The NMR spectra give no evidence for an interaction of the carbonyl group with the metal, which is important since the proximity of metal and carbonyl, detectable by coupling or shift variations, would imply the same for the diazo group and therefore the *in situ* generated carbene. The latter could then react with the metal directly further inactivating the complex.

Crystals suitable for X-ray diffraction analysis could be grown by slow diffusion of pentane into a saturated toluene solution of **PtPhNO$_2$−bp**. The complex is trigonal planar coordinated with the olefin twisted 5.9° out of the coordination plane. The C37-C38 distance is with 1.45 Å between a typical C−C single (1.48 Å) and a C=C double bond (1.33 Å), as expected for coordinated olefins. The C37-Pt, C38-Pt, P1-Pt and P2-Pt distances are with 2.12 Å, 2.14 Å, 2.30 Å and 2.28 Å in accordance to the equivalent bond lengths in simpler cinnamate complexes, which are discussed in more detail in section 4.2.3. All trends concerning their relation with each other are also in agreement with the findings of the other complexes, *vide infra*. The C37-Pt-C38, P1-Pt-P2, C37-Pt-P2 and

---

[1]See chapter 4 for a detailed and more general discussion of synthesis, stability, spectroscopic properties and reactivity of bisphosphine olefin complexes of group 10.

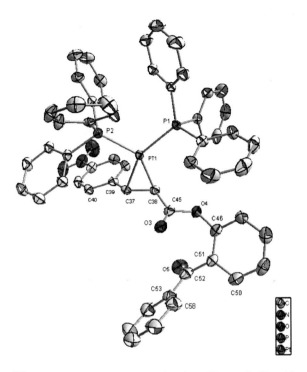

Figure 2.1: ORTEP[24] representation of **PtPhNO$_2$–bp**. Thermal ellipsoids are given at 50% probability level. Hydrogen atoms are omitted for clarity.

C38-Pt-P1 angles are 39.8°, 108.4°, 102.8° and 109.0° respectively and therefore also in consistence with the other cinnamic acid ester complexes of platinum discussed later. The same thing applies to the C39-C37-C38-C45 torsion angle of 150.2°.

Table 2.1.: Selected bond lengths and angles of **PtPhNO$_2$–bp**:

d [Å]		∠ [°]	
C1–P1	1.841(3)	C1–P1–Pt1	121.0(1)
C7–P1	1.830(3)	C1–P1–C13	102.0(1)
C13–P1	1.827(3)	C1–P1–C7	102.1(1)
C19–P2	1.827(3)	C7–P1–Pt1	113.6(1)
C25–P2	1.840(3)	C7–P1–C13	104.0(2)

continued on next page

Table 2.1 – continued from previous page

d [Å]		∠ [°]	
C31–P2	1.840(3)	C13–P1–Pt1	112.1(1)
C37–C38	1.449(5)	C37–C38–C45	119.7(3))
C37–Pt1	2.119(3)	C37–Pt1–P2	102.8(1)
C37–C39	1.478(4)	C37–Pt1–C38	39.8(1)
C38–Pt1	2.139(3)	C38–Pt1–P1	109.0(1)
C38–C45	1.459(5)	C38–C37–C39	120.2(3)
C45–O4	1.387(4)	C38–C45–O3	128.9(3)
C45–O3	1.207(5)	C38–C45–O4	109.9(2)
C46–O4	1.403(4)	C45–O4–C46	118.8(2)
C46–C51	1.390(5)	C46–C51–C52	122.2(3)
C51–C52	1.503(5)	C51–C52–C53	119.7(4)
C52–O5	1.218(5)	C51–C46–O4	123.1(3)
C52–C53	1.496(7)	C51–C52–O5	119.2(3)
P1–Pt1	2.296(1)	C52–C53–C58	123.7(4)
P2–Pt1	2.284(1)	C53–C52–O5	121.1(4)
		P1–Pt1–P2	108.4(0)

The C52-O5 distance is with 1.22 Å comparable to the one in free **PhOMe−bp**
(1.22 Å) and gives no indication of any interaction between the metal and the carbonyl
bond. This is confirmed by a C52-Pt distance of more than 6 Å and is in agreement
with the NMR data. The sector of the double bond opposite to the metal is not sterically
shielded and the carbonyl group points towards this side of the olefin. This should render
the carbene attack, necessary for cyclopropanation, and therefore light induced ligand
cleavage possible. The other structural parameters, apart from the olefin´s bond length
and torsion angle, both distorted due to metal coordination (ref. section 4.2.3.), are in
agreement with the free **PhOMe−bp** ligand.

Table 2.2.: Selected torsion angles of **PtPhNO$_2$−bp**[°]:

C37–C38–Pt1–P1	173.8(2)
C38–C37–Pt1–P2	177.4(2)
C39–C37–C38–C45	150.2(3)
C46–C51–C52–O5	51.2(5)
C50–C51–C52–C53	54.3(5)
C51–C52–C53–C58	8.7(6)
C58–C53–C52–O5	171.5(4)

2.1. Light induced ligand exchange

Since the structural data of the benzophenone cinnamates indicated sufficient distance of metal and carbonyl but unhindered excess to the olefin, the most simple diphenyl diazomethane functionalised cinnamic acid ester complex was synthesised, **PtPh−bpN$_2$**.[2]

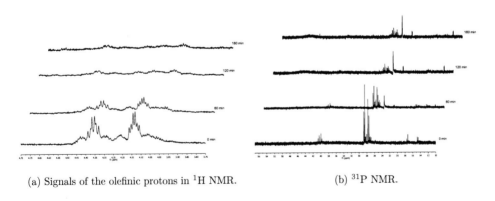

(a) Signals of the olefinic protons in ^{1}H NMR.   (b) ^{31}P NMR.

Figure 2.2: NMR spectra in dependence of irradiation time.

Upon irradiation of a mixture of **PtPh−bpN$_2$** and one equivalent of triphenylphos-phine at 300 nm or 419 nm in benzene-d$_6$ or toluene-d$_8$ both signals of the olefin, which

---

[2]The synthetic routes to the complex and the properties are discussed in chapter 4, while the synthesis and behaviour of the diazo ligand are given in chapter 3.

split into a doublet of doublets of doublets with platinum satellites due to coupling with both phosphine phosphorus, [195]Pt and each other, are depleted in the ^{1}H NMR spectra as shown in figure 2.2 (a) in the course of three hours. This is due to removal of the double bond since no signals appear for non coordinated olefins. In the ^{31}P NMR spectra the two doublets, typical for trigonal diphosphine complexes with unsymmetrical alkenes, decrease while a broad signal at ca. 51 ppm appears, which belongs to $Pt(PPh_3)_3$ (b). The broadening is caused by fast exchange of coordinated and free triphenylphosphine.

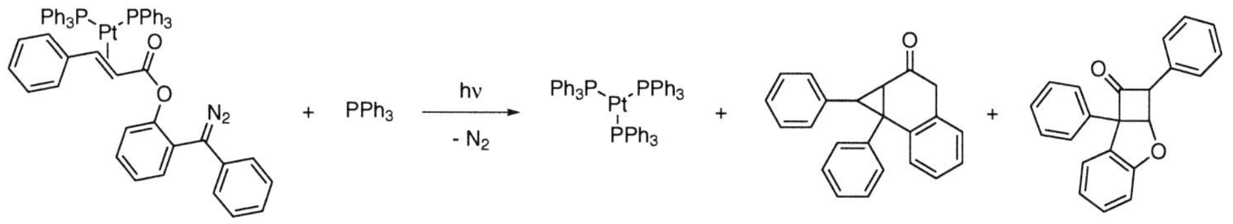

Scheme 2.2: Photo-induced olefin cleavage and subsequent triphenylphosphine coordination.

The *in situ* generated carbene reacts therefore with the coordinated double bond under the formation of cyclopropane **3.1** and cyclobutanone **3.12**.[3] After removal of the olefin ligand the free triphenylphosphine immediately binds to the vacant coordination site (scheme 2.2). If the same reaction conditions without irradiation are applied and the reaction mixture is kept under the strict exclusion of light, the signals were retained during the same time period, in which the signals completely disappeared when irradiation was performed. However the educt is converted to $Pt(PPh_3)_3$ in a substantially quantity over a period of 48 hours due to slow decomposition of the diazo compound at room temperature, which also leads to olefin cleavage. This process is shown in figures 2.3 (a) and (b).

To exclude other light induced reactions, which could also lead to ligand exchange the corresponding carbonyl compound, **PtPh−bp** was subjected to the same reaction and irradiation conditions. As expected no reaction was observed and the complex remained stable in solution over the whole examination period, as evident from unaltered ^{1}H and

---

[3]A detailed study concerning the behaviour of the diazo olefin esters upon irradiation including all products and side products is presented in chapter 3.

^{31}P NMR spectra.

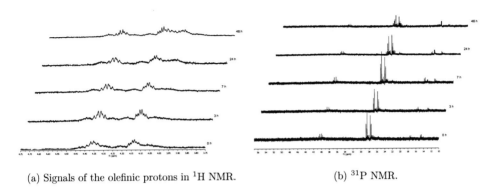

(a) Signals of the olefinic protons in ^{1}H NMR.   (b) ^{31}P NMR.

Figure 2.3: NMR spectra of the reaction mixture in the dark over time.

If norbornene and **PtPh−bpN$_2$** are dissolved in toluene or benzene an equilibrium establishes between coordination of norbornene and cinnamate, which is almost completely on the **PtPh−bpN$_2$** side. After exposure to UV light the only observed compound is $(PPh_3)_2Pt(nbe)$ though. When **PtPh−bpN$_2$** was irradiated without the addition of one equivalent of additional ligand the complex decomposed rapidly, while the complex is still detectable, even though some thermal decomposition takes place, in solution after two days, if stored in the dark.

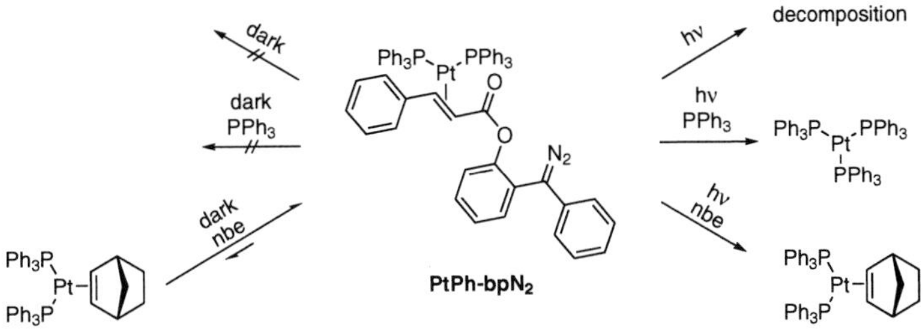

Scheme 2.3: Reactivity of **PtPh−bpN$_2$** in the dark & with UV/vis irradiation.

These findings, which are summarised in scheme 2.3, prove that the cleavage of olefins is possible in the anticipated mild way by selective intra-ligand cyclisation.

2.2. Photo-activated oxidative addition

To apply the selective ligand cleavage for photo-induced hydrosilylation, oxidative addition
of silanes must not occur before irradiation.

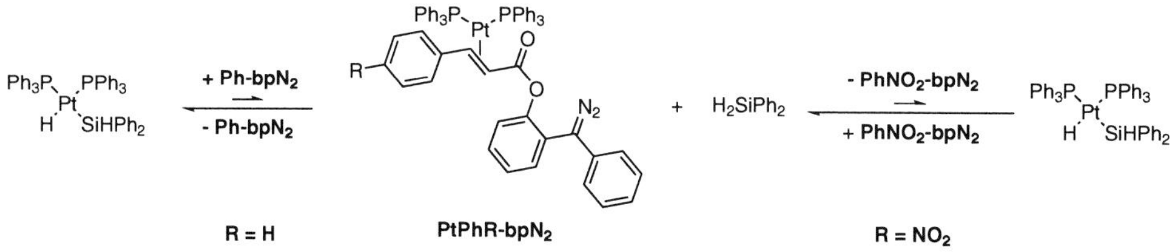

Scheme 2.4: Reactions of **PtPh−bpN$_2$** & **PtPhNO$_2$−bpN$_2$** with H$_2$SiPh$_2$ in the absence of
light.

However cinnamic acid esters undergo rapid oxidative insertion into one Si−H bond
of diphenyl silane due to easy dissociation of the olefin. This establishes an equilibrium
between the platinum(II) silane addition and the platinum(0) olefin complex, which is
considerably on the platinum(II) side. Therefore the ligand had to be modified in order
to enforce the metal olefin bond strength, increasing complex stability. This was easiest
established by the introduction of a nitro group in 4 position of the cinnamic acid´s
phenyl ring. Through the introduction of this electron withdrawing substituent, the overall
electron density of the double bond was decreased, which leads to stronger back bonding
of platinum into the olefinic π^* orbital. With this small modification the equilibrium could
be shifted almost totally onto the cinnamate complex side (scheme 2.4).[4]

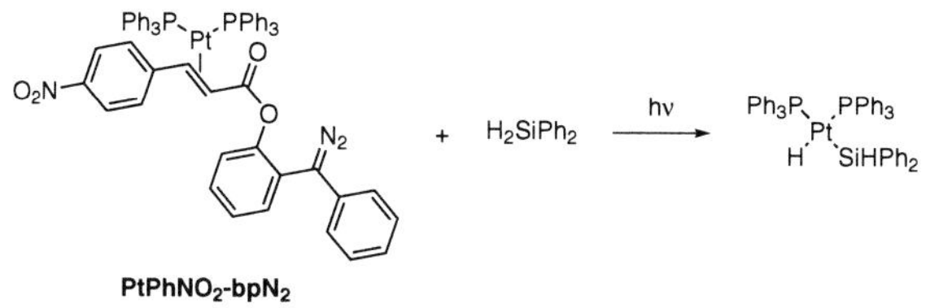

Scheme 2.5: Platinum insertion into an Si−H bond upon UV/vis irradiation.

When H$_2$SiPh$_2$ is added to a toluene or benzene solution of **PtPhNO$_2$−bpN$_2$** the

---
[4]For a comprehensive analysis of the equilibrium refer to section 4.3.

equilibrium remained on the olefin complex side as long as the reaction solution was kept in the dark. When irradiation was commenced, the colour of the solution changed from deep orange, typical for diazo nitocinnamic acid platinum complexes, to yellow. In the ^{31}P NMR spectrum the two doublets of the inequivalent phosphines changed to one broad signal at ca. 36 ppm. This is the signal of the two inequivalent phosphorus atoms in $(PPh_3)Pt(H)SiHPh_2$, broadened by their fast exchange in position.[25-28] The formation of the oxidative addition product is also evidenced through a broad hydride signal at ca. 1.2 ppm, with a $^2J_{HP}$ coupling constant of ca. 130 Hz to the phosphorus in *trans*-position. The *cis*-coupling is not resolved due to the aforementioned dynamic behaviour, as are the platinum satellites. The reaction was complete after three hours and no starting material was detectable after irradiation.

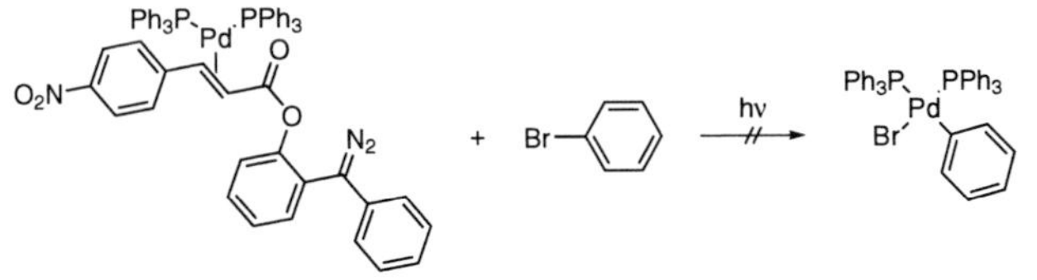

Scheme 2.6: Reaction of **PdPhNO$_2$−bpN$_2$** with bromo benzene upon UV/vis irradiation.

Even though photo-triggered oxidative addition of Si−H bonds to platinum could be achieved with 4-nitro cinnamates the transfer of this concept to other group 10 metals failed. While no diazo functionalised olefin complexes of nickel were accessible due to oxidation of Ni(0) to Ni(II) by the diazo group, if nitro groups are present, these oxidise the metal even faster, the palladium complex **PdNO$_2$−bpN$_2$** is synthesisable via standard procedures in good yields, *vide infra*. However the nitrocinnamate complexes of palladium are of deep red colour. This leads to absorption of the light necessary for the nitrogen cleavage from the diazo group. Therefore no photo-induced ligand removal was achieved with palladium and the anticipated light induced oxidative insertion into bromobenzene could not be performed (scheme 2.6).

PtNO$_2$−bpN$_2$ was tested in the hydrosilylation of *para*-fluoro acetophenone with diphenyl silane, which is shown in scheme 2.7. This model reaction was chosen due

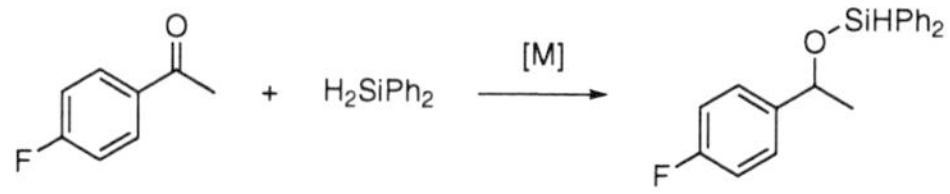

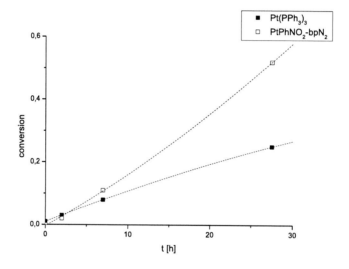

Scheme 2.7: Hydrosilylation of 4-fluoro acetophenone.

to the excellent monitorability of the reaction process via ^{19}F NMR and the expected fast reaction progress, necessary because of the limited stability of the diazo group at ambient temperature.

Graph 2.1: Time conversion plot of 4-fluoro acetophenone hydrosilylation in the absence of light.

However the reaction proceeded extremely slow. This led to thermal cleavage of the olefin causing acceleration of the hydrosilylation rate with time. In graph 2.1 it is clearly observable, that the reaction proceeds slower than the $Pt(PPh_3)_3$ reference in the first two hours of the catalysis run. Afterwards **$PtNO_2-bpN_2$** becomes increasingly faster in the hydrosilylation due to olefin cleavage. Despite the unwanted decomposition of **$PtNO_2-bpN_2$**, this shows the potential of the concept, since the activity of the complex after olefin elimination is significantly faster due to the absence of free coordinating ligand in solution. The free triphenylphosphine, formed upon oxidative addition of H_2SiPh_2 to

Pt(PPh$_3$)$_3$, inhibits Pt(PPh$_3$)$_2$, which is the active species in both cases.

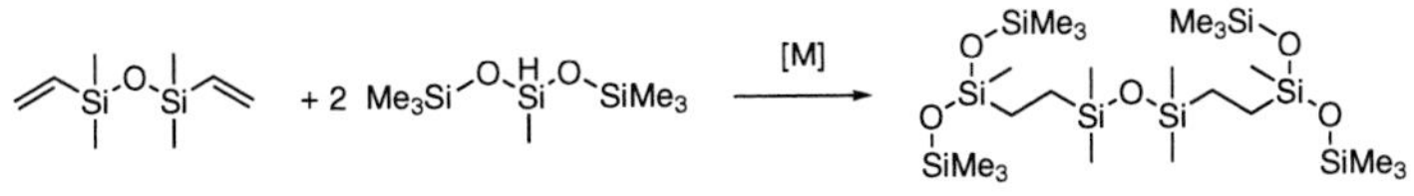

Scheme 2.8: Hydrosilylation of C=C double bonds with 1,1,1,3,5,5,5-heptamethyl trisiloxane.

The catalytic performance of platinum phosphine complexes is poor in hydrosilyla-tion. This is especially the case for synthetically more interesting substrates, like bridged vinyl siloxanes and, compared to diphenyl silane, less active silanes (e.g. heptamethyl trisiloxane). The first approach to achieve higher reactivities was the substitution of the phosphines by olefins. However these complexes were too reactive and no inhibition of the oxidative addition of silanes could be achieved (ref. section 4.4.3.).

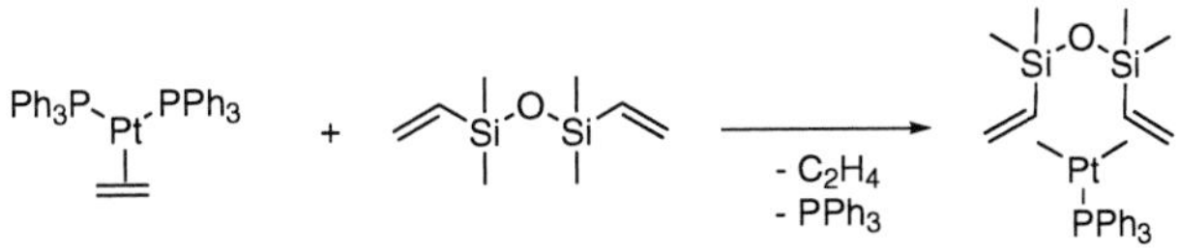

Scheme 2.9: Formation of η^4-divinyltetramethydisiloxane triphenylphosphine platinum(0).

One reason for the low activity of bis(triphenylphosphine)platinum systems was pre-sumed to be the dissociation of one phosphine and the coordination of the chelating divinyl tetramethyl silane, described by *Beuter et al.*, as shown in scheme 2.9.[29] Therefore anal-ogous cinnamate complexes with chelating phosphines (e.g. dppe, dppp) were synthesised to disable phosphine dissociation. Alas no increase in reactivity could be achieved with this approach, *vide infra*. Furthermore is the solubility of dppe and dppp complexes sig-nificantly lower than their triphenylphosphine analogues, decreasing reactivity actually.

2.3. Stability enhancement

Since the enhancement of catalytic performance would involve significant changes in the ligand design, it had to be evaluated, whether the stability of the diazo moiety could be enhanced. Anticipated was a complex stable at ambient temperature for an unlimited

amount of time and handleable at temperatures above 100 °C for at least one hour without
significant decomposition.

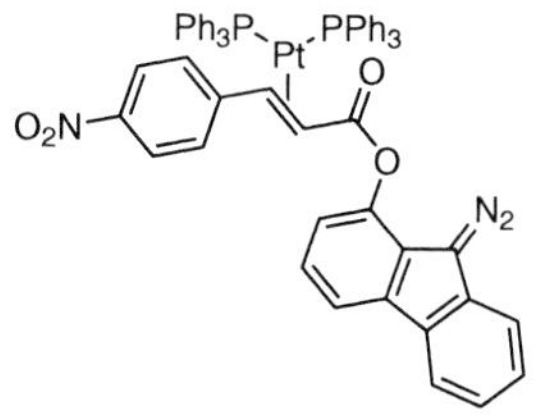

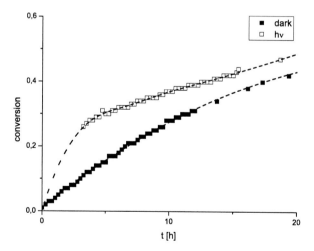

PtPhNO$_2$-flN$_2$

Figure 2.4: Ambient temperature stable diazofluorene ester complex.

A significant increase in the stability of the complexes could be achieved by the substitution of the benzophenone by a fluorene skeleton. The synthesis and reactivity of these diazofluorene olefin esters is discussed in section 3.2. Synthesis of catalyst **PtPhNO$_2$-flN$_2$**
(figure 2.4) from the diazo olefin followed the same procedure applicable to diphenyl diazomethanes.

Graph 2.2: **PtPhNO$_2$-flN$_2$** catalysed hydrosilylation of benzophenone in absence of light and
after UV/vis irradiation.

The catalyst was tested with the hydrosilylation of acetophenone with diphenyl silane. Even though the catalyst is stable in solution at ambient temperature, the reaction rate is above zero before irradiation. This is caused by the equilibrium between coordinated olefin and oxidative addition of silane (scheme 2.10). Although this equilibrium is far on the olefin complex side if only one equivalent of silane is present, excess of the latter upon catalytic conditions leads to significant olefin dissociation and therefore catalytic activity. However the reaction velocity is observably increased after irradiation as shown in graph 2.2.

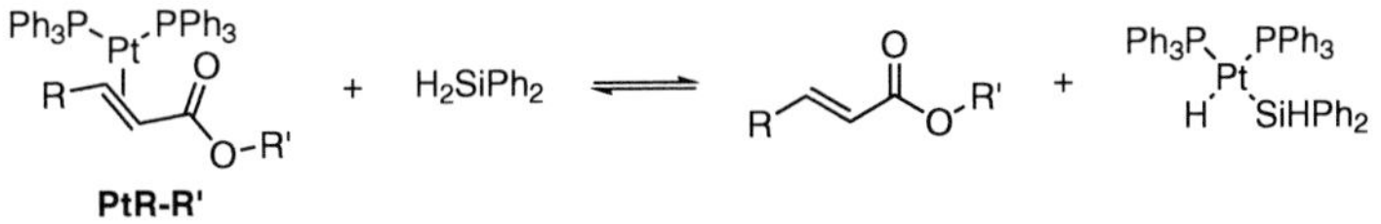

Scheme 2.10: Equilibrium between Pt(0) olefin complexes and the oxidative addition product of silane.

2.4. Conclusions & outlook

It could be shown, that light induced cleavage of olefin ligands is possible under mild conditions, via the *in situ* generation of a carbene through photo-cleavage of nitrogen from a diazo compound and subsequent intramolecular cyclopropanation. The induced free coordination site inserts into Si−H bonds and therefore starts the catalytic cycle as depicted in scheme 2.11. The stability of the catalyst could be raised to ambient temperature. However due to low catalytic activity of platinum phosphine complexes only C=O double bonds could be hydrosilylated. Since the metal olefin bond is too weak in the tested systems, the catalysts show significant pre-irradiation activities, though.

To solve the latter problem olefin ligands are needed which are even more electron deficient than *para*-nitro cinnamic acid esters. The most promising class of ligands are unsaturated diacids, like fumarates and maleates, that are already applied as inhibitors added to *Karstedt* solution to control reactivity (figure 2.5 (a)).[3,5,30] Tests with 2-hydroxybenzophenone esters of these already show promising trends concerning reac-

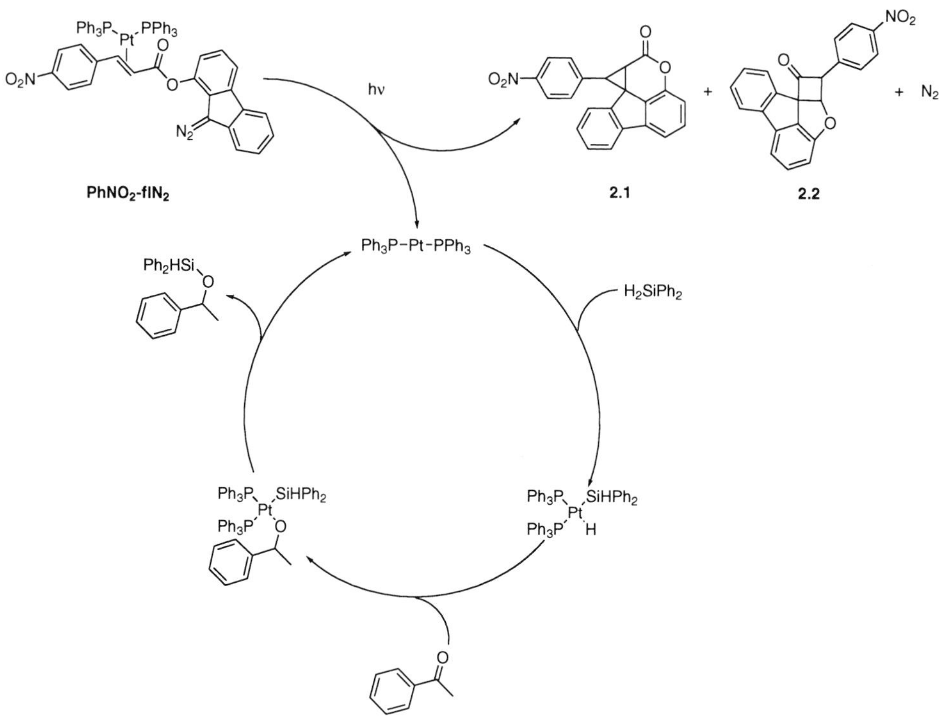

Scheme 2.11: Photo-induced hydrosilylation of benzophenone.

tivity (section 4.4.1.). Another advantage of these diacid systems is the transparency of these complexes concerning UV light. Unlike the nitro cinnamates, which are intensely coloured when coordinated to metals, leading to partial and in case of palladium total absorption of the irradiated light, lowering the quantum yield of the diazo group and therefore slow the activation times down. The same should also apply for the irradiation products, since the cyclisation products derived from nitro cinnamic acid fluorene esters do also absorb in the same wavelength range as the diazo group does.

To improve the catalytic performance after irradiation the phosphine content needs to be reduced. This might be achieved by the utilisation of chelating diacid ligands, as shown in figure 2.5 (b). The coordinating group depicted with D may either be an olefin, in which case L probably needs to be a phosphine or *vice versa*. In both cases the phosphine content is halved and should increase the turn over frequency significantly. However the faster

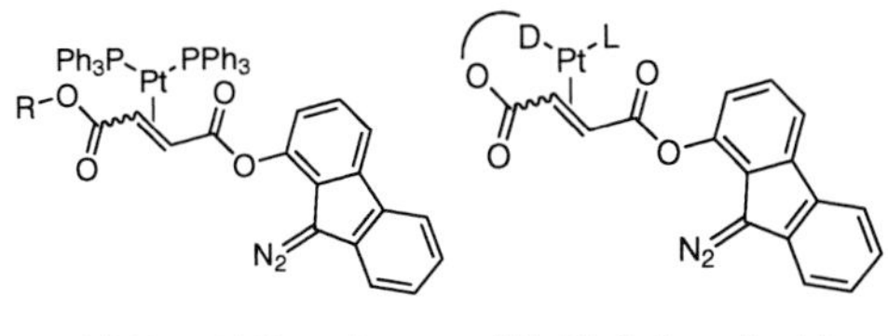

(a) Diacid ligands.   (b) Chelating diacids.

Figure 2.5: Improved complex designs.

approach to high activities after irradiation is presumably the use of diazo olefin esters as inhibitors in technical *Karstedt* solution. These should be photo-inactivatable which would lead to a restoration of uninhibited *Karstedt* activity.[23] This approach is examined in chapter 5.

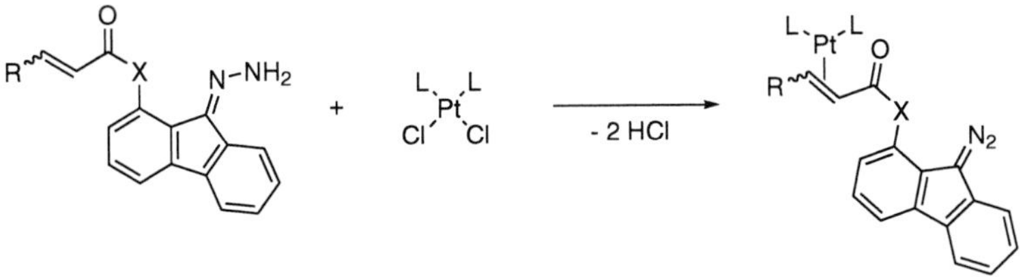

Scheme 2.12: Formation of diazo olefin complexes via redox chemistry.

A defined catalyst however has the advantage, that a combined redox synthesis of the compound might be possible. A possibility is depicted in scheme 2.12 where a hydrazone is used to reduce platinum(II) to platinum(0) and is oxidised in the same step to the anticipated diazo compound. This would have the advantage of air and light stable pre-catalysts and the photo-switchability is unlocked only shortly before it is needed.

3. Photo-reactivity of diazo olefin esters

In order to photo-chemically cleave the olefin from the complex a removal of the double bond is necessary, *vide supra*. Therefore the intramolecular cyclopropanation depicted in scheme 3.1 was anticipated.

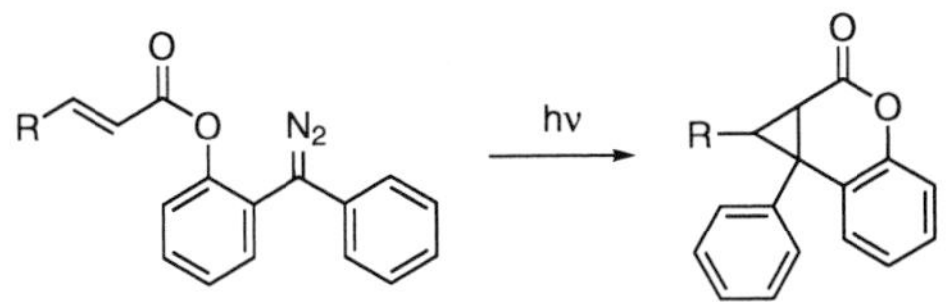

Scheme 3.1: Anticipated intramolecular cyclopropanation.

This reaction seemed applicable since the light-induced photo-cleavage of nitrogen to generate the carbene from diphenyl diazomethane and diazo-9-fluorene is possible at an irradiation wavelength of ca. 250–550 nm and thus no expensive quartz equipment is needed. [31] The absence of absorption bands beyond 550 nm facilitates handling under red light without accidentally triggering the reaction. [32] Furthermore a six membered ring would be formed, providing a thermodynamic driving force.

3.1. Diphenyl diazo esters

To investigate the dependence of the reaction on the electron density of the olefin diazo ester, derivatives of cinnamic acid were chosen, since the *para*-substituent provides an

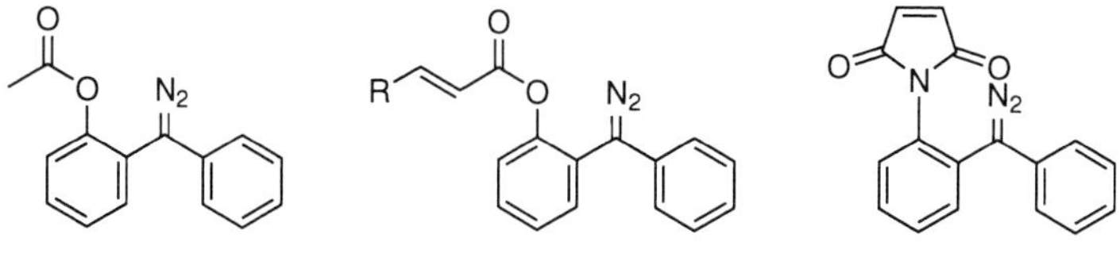

Figure 3.1: Targeted diazo compounds. R $\equiv$ Me (**Me**), p-C$_6$H$_4$OMe (**PhOMe**), p-C$_6$H$_4$Me
(**PhMe**), Ph (**Ph**), p-C$_6$H$_4$NO$_2$ (**PhNO$_2$**), C(O)OEt (**fu**).

excellent possibility of electronic control, together with the less bulky and more electron-
rich crotonic acid. For the influence of the geometry of the double bond, fumaric and
maleic acid systems were included as well. In order to examine possible side reactions,
the acetate derivative (**Ac-bpN$_2$**) was also examined, due to the lack of a double bond.
These systems are summarised in Figure 3.1.

3.1.1. Synthesis

The first route anticipated, started from the benzophenyl ester with subsequent hydrazonation followed by oxidation to the diazo ester. This synthetic approach was chosen
since the hydrazone ester **bpNNH$_2$** was one possible precursor for direct redox complex
synthesis (chapter 2). Also oxidation of hydrazones to the diazo compound is the most
common route to diphenyl diazomethane.[33]

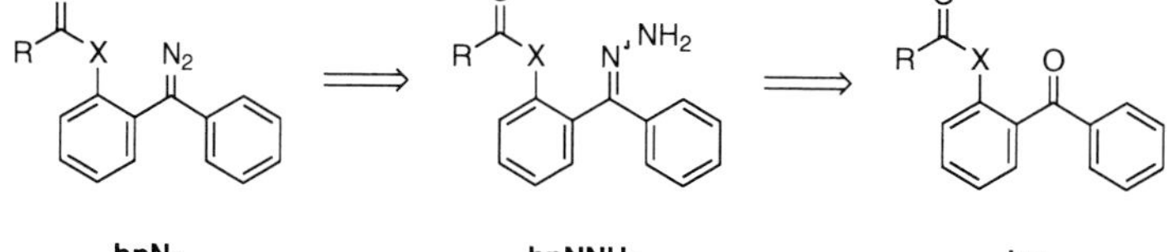

Scheme 3.2: Direct hydrazonation of carbonyl esters. X $\equiv$ O, NH, NR.

Synthesis of 2-hydroxy benzophenyl esters from the alcohol and the adequate acid chloride following standard literature procedures[34,35] gave the anticipated esters in good
to excellent yields. N(2-Benzoylphenyl)maleimide was synthesised by nucleophilic ring

opening of maleic anhydride with 2-amino benzophenone and subsequent recyclisation by dehydration.[36–38]

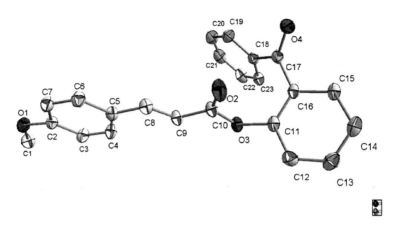

Figure 3.2: ORTEP[24] representation of **PhOMe−bp**. Thermal ellipsoids are given at 50% probability level. Hydrogen atoms are omitted for clarity.

Single crystals of **PhOMe-bp** suitable for X-ray diffraction analysis could be grown by slow solvent evaporation from a saturated ethanol solution. The C8-C9 distance of 1.34 Å is in the range of a C=C double bond and the C5-C8 and C9-C10 with 1.46 Å have typical C−C single bond distances as expected. O2-C10 and O4-C17 are with 1.20 Å and 1.22 Å in the typical range of a C=O double bond while O3-C10 with 1.38 Å has the normal C−O ester single bond distance. The C8-C9-C10, O2-C10-O3 and O4-C17-C18 angles are approximately 120° as estimated for sp²-hybridised carbon atoms whilst C5-C8-C9 and O2-C10-C9 angles are with 126.6° and 127.6° slightly larger. The olefin is almost planar with torsion angles of 179.8° and 177.6°. The C16-C17 and C17-C18 distances are with 1.50 Å and 1.49 Å single bonds and therefore show no evidence of a conjugated π-system via the carbonyl carbon. This is also supported by the torsion angles C11-C16-C17-C18 (44.3°) and C16-C17-C18-C19 (145.9°) indicating an out-of-plane angle of 45° between the two phenyl rings. All the structure parameters are in accordance with literature.[39,40]

Table 3.1.: Selected bond length, angles and torsion angles of **PhOMe−bp**:

d [Å]		∠ [°]		torsion ∠ [°]	
C5–C8	1.460(2)	C5–C8–C9	126.6(1)	C9–C8–C5–C6	179.8(1)
C8–C9	1.341(1)	C8–C9–C10	120.3(1)	C11–C16–C17–C18	44.3(1)
C9–C10	1.464(2)	C11–C16–C17	124.0(1)	C16–C17–C18–C19	145.9(1)
C16–C17	1.500(1)	C16–C17–C18	120.3(1)	O3–C10–C9–C8	177.6(1)
C17–C18	1.492(1)	O2–C10–C9	127.6(1)		
O1–C1	1.435(1)	O2–C10–O3	121.8(1)		
O1–C2	1.363(2)	O3–C11–C16	122.1(1)		
O2–C10	1.203(1)	O4–C17–C18	120.1(1)		
O3–C10	1.37671)				
O4–C17	1.217(1)				

However reaction of the keto-ester with hydrazine monohydrate in refluxing ethanol in the presence of acid, the standard hydrazonation conditions,[41,42] led to the cleavage of the ester bond and the formation of free acid and 2-hydroxy benzophenone hydrazone (**3.2-OH**). Even stirring at room temperature in ethanol or the non-protic solvent toluene leads to ester cleavage without observation of hydrazone formation. Reflux in toluene gave a mixture of free acid, acid amide and 2-hydroxy benzophenone (scheme 3.3). To exclude the stoichiometric amounts of water introduced through the hydrazine monohydrate as the reason for the cleavage, water scavengers were added. Nonetheless no improvement was observed. Obviously the high basicity of the hydrazine results in nucleophilic attack at the carboxyl group prior to addition to the carbonyl group, causing an ester opening in all cases.

Amides were tested since these are less prone towards basic cleavage. However under the reaction conditions in which the amide was stable, no conversion could be obtained.

Therefore an alternative route was chosen, starting from the hydrazone **3.2**, followed by esterification and oxidation (scheme 3.4).

2-Amino benzophenone hydrazone (**3.2-NH₂**) was obtained almost quantitatively by

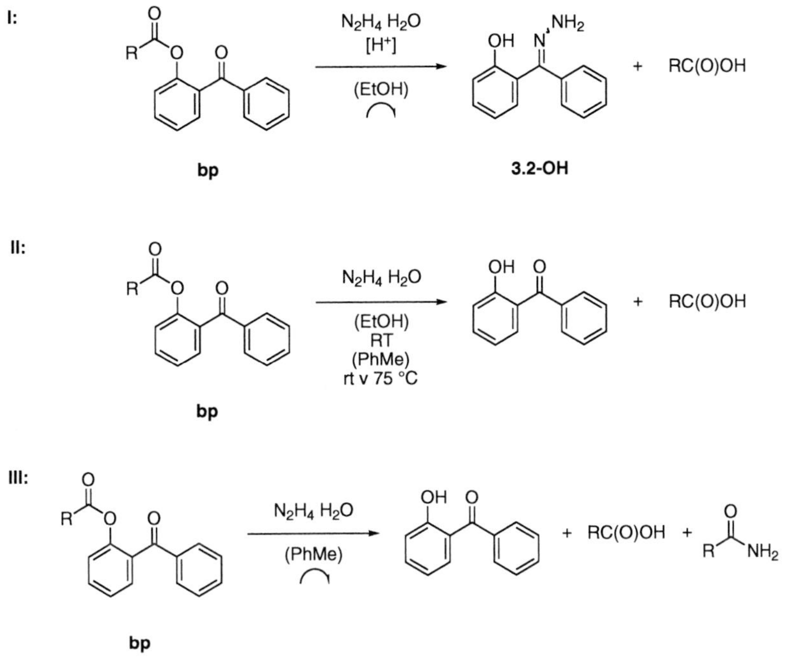

Scheme 3.3: Reactions of hydrazine monohydrate with 2-hydroxybenzophenone esters.

reflux of 2-amino benzophenone and hydrazine hydrate in ethanol in the presence of acetic acid, following the route of *Adger et al.*[43] This route could also be transferred to the synthesis of **3.2-OH** directly with similar yields.

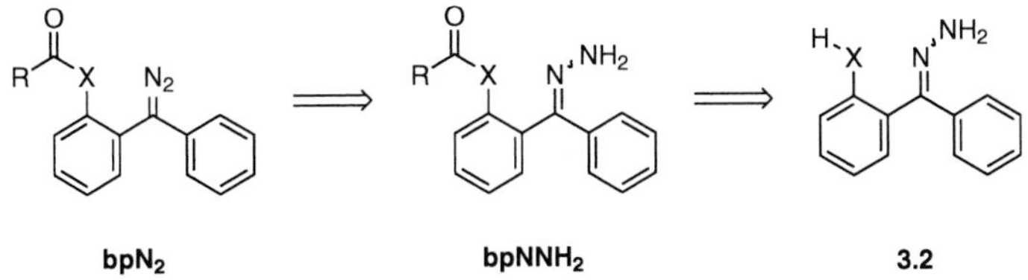

Scheme 3.4: Esterification of hydroxy or amino functionalised hydrazones. $X \equiv O$, NH.

However reaction of **3.2-OH** with acid chlorides in pyridine at 100 °C led exclusively to the formation of the hydrazide **3.3-OH**. This is due to the formation of the thermodynamic product which is driven by the formation of the, in comparison to the ester, more stable amide bond.

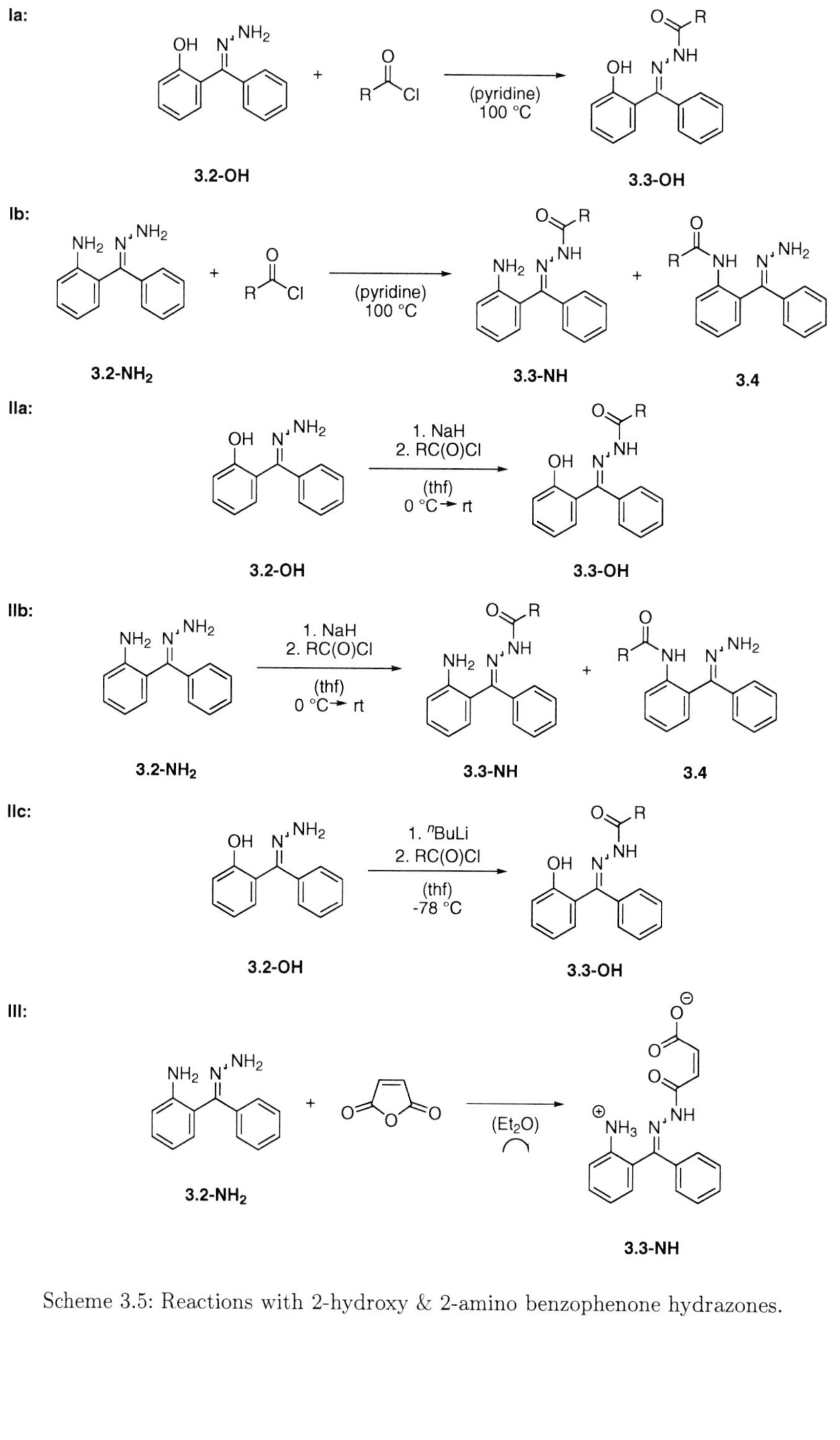
Ia:
(pyridine)
100 °C
3.2-OH
3.3-OH
Ib:
(pyridine)
100 °C
3.2-NH2
3.3-NH
3.4
IIa:
1. NaH
2. RC(O)Cl
(thf)
0 °C→ rt
3.2-OH
3.3-OH
IIb:
1. NaH
2. RC(O)Cl
(thf)
0 °C→ rt
3.2-NH2
3.3-NH
3.4
IIc:
1. ⁿBuLi
2. RC(O)Cl
(thf)
-78 °C
3.2-OH
3.3-OH
III:
(Et2O)
3.2-NH2
3.3-NH

Scheme 3.5: Reactions with 2-hydroxy & 2-amino benzophenone hydrazones.

Therefore the reaction temperature was reduced and the alcohol deprotonated with sodium hydride to enforce ester formation. Nevertheless, only amide was obtained. Even at −78 °C and deprotonation with ^{n}BuLi only a mixture of amide and ester could be isolated, in which the amide was the predominant product still. To exclude the unlike possibility that deprotonation of the hydrazone is favored towards phenol deprotonation, the deprotonated 2-hydroxy benzophenone hydrazone was quenched with tms chloride. This gave exclusively the expected tms protected alcohol **3.5** as shown in scheme 3.6. In order to circumvent the high difference in bond strength reactions with 2-amino benzophenone were executed (scheme 3.5). At 100 °C in pyridine only mixtures were isolated however. Upon the move to ambient temperature and deprotonation of the more acidic phenylic amine, product distribution did only improve slightly whereas reaction times increased to an extent that rendered the reaction inapplicable. Even the nucleophilic opening of maleic anhydride gave the hydrazide exclusively.

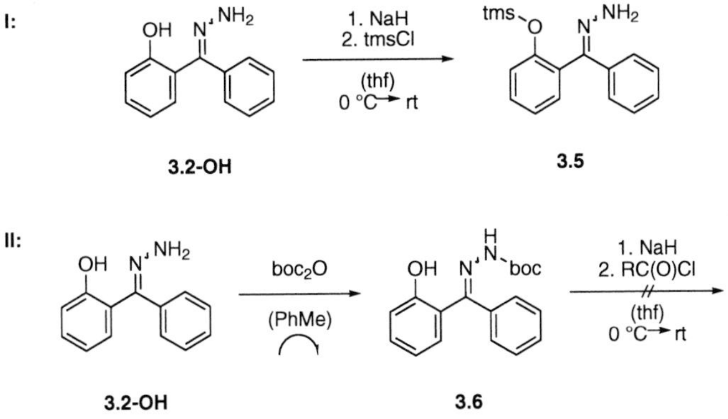

Scheme 3.6: Protecting group approach.

To reduce the high basicity and the resulting nucleophilicity of the hydrazone, attempts were made to protect the nitrogen with a boc group. However the reaction did not proceed in the necessary yield and selectivity to the desired product, apart from the copious and lengthy approach.

(2-Hydroxy phenyl)phenyl diazomethane synthesised from **3.2-OH** via oxidation with yellow mercury oxide following a procedure described by *Staudinger* [33] proved too unsta-

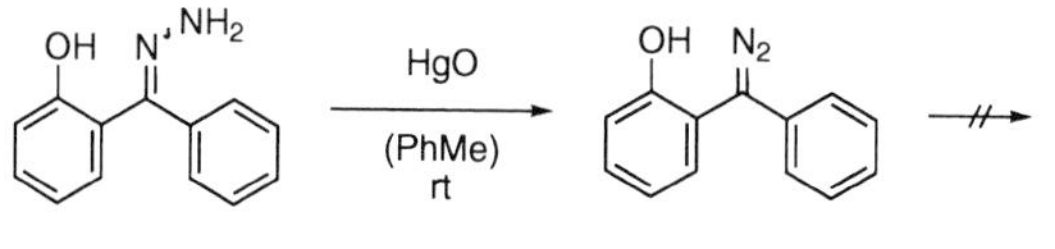

3.2-OH   **3.7**

Scheme 3.7: Synthesis of 2-hydroxyphenyl phenyl diazomethan and further conversion.

ble for further synthesis. This fact in conjunction with the low yields in the first step, rendered the direct esterification of diphenyl diazomethanes unfeasible as well (scheme 3.7).

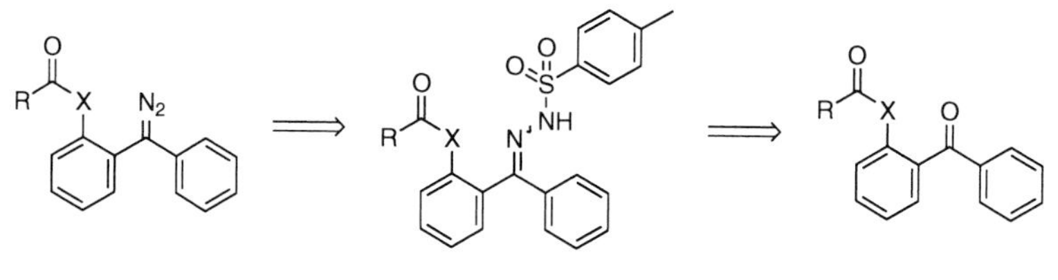

bpN$_2$ **bpNNHTs**   **bp**

Scheme 3.8: Direct tosyl hydrazonation of carbonyl esters and amides. X $\equiv$ O, NH, NR.

Due to these basicity issues the first aspired route was modified by the use of tosyl-hydrazone instead of hydrazine. This utilizes the electron withdrawing properties of the sulfonic acid moiety and further removes the water as a source of error. The generation of the diazo compound would then follow thermal rearrangement of the deprotonated hydrazone.[44]

In contrast to the common procedures[45–51] no alcohols could be used due to imminent ester cleavage. A procedure by *Jones et al.*[52] using CH$_2$Cl$_2$ as solvent resulted in no conversion, probably due to the lower electron density of the benzophenone caused by the electron withdrawing ester substituent, and therefore reduced carbonyl reactivity. However no degradation of the ester was observed. Therefore a synthesis by *Guldi et al.*[53] was modified and the ester and tosylhydrazide were refluxed in toluene over night in the presence of catalytic amounts of toluene sulfonic acid to give the desired tosylhydrazones **bpNNHTs** in good yields.

This route failed for the maleimide **mi-bp** and only starting material was isolated.

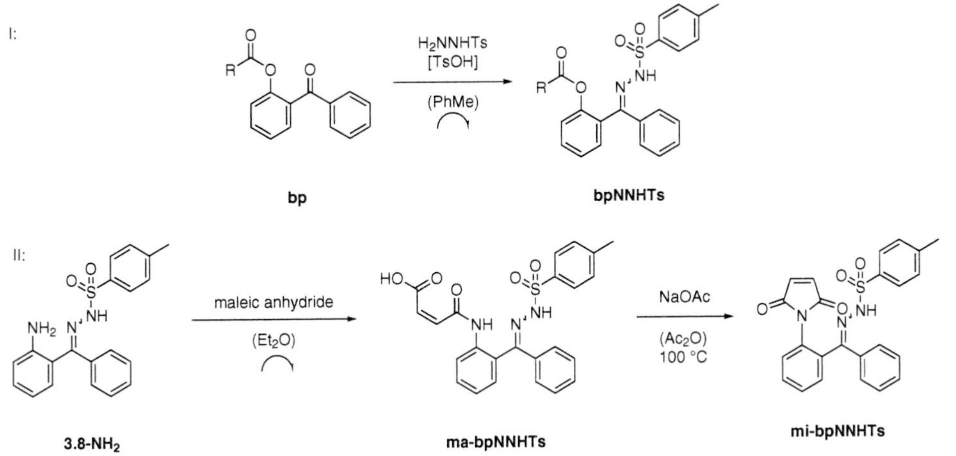

Scheme 3.9: Synthesis of tosyl hydrazone esters.

A rise of reaction temperature by the substitution of toluene with mesitylene resulted in decomposition. Reaction of 2-amino benzophenone tosylhydrazone **3.8**, which is easily synthesised by the same procedure as the other tosylhydrazones, with maleic anhydride followed by subsequent dehydrative recyclisation led to the desired product, however (scheme 3.9).

Single crystals of **Me-bpNNHTs** suitable for X-ray diffraction analysis could be obtained by slow evaporation of the solvent from a concentrated ethanol solution. The C25-C25 distance is with 1.32 Å comparable to the olefin bond distance in **PhOMe-bp**, as are the other bond lengths and angles of the acid moiety. The C3-C22 and C16-C22 distances are also in the range of a single bond with 1.50 Å and 1.48 Å. This also applies for the N1-N2 (1.40 Å) and N1-S1 (1.65 Å) distances which are typical N−N and N−S single bond distances. The C22-N2 distance is with 1.28 Å in the normal range of a C=N double bond. Characteristic for a sp^2 carbon atom are the bond angles C3-C22-C16, N2-C22-C3 and N2-C22-C16 with 119.5°, 123.8° and 116.6°. The N1-N2-C22 and N2-N1-S1 angles are 115.9° and 111.7°. The two phenyl rings are almost perpendicular to each other with C3-C22-C16-C15 and C4-C3-C22-C16 torsion angles of 161.3° and 80.4°, whereas the C22-N2-N1-S1 torsion is with 176.2° close to 180°. All structural parameters are in

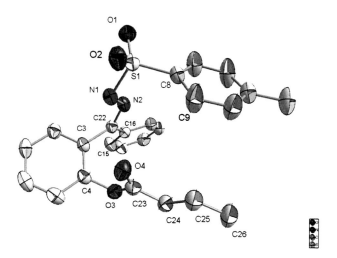

Figure 3.3: ORTEP[24] representation of **Me−bpNNHTs**. Thermal ellipsoids are given at 50% probability level. Hydrogen atoms are omitted for clarity.

accordance with literature examples.[49]

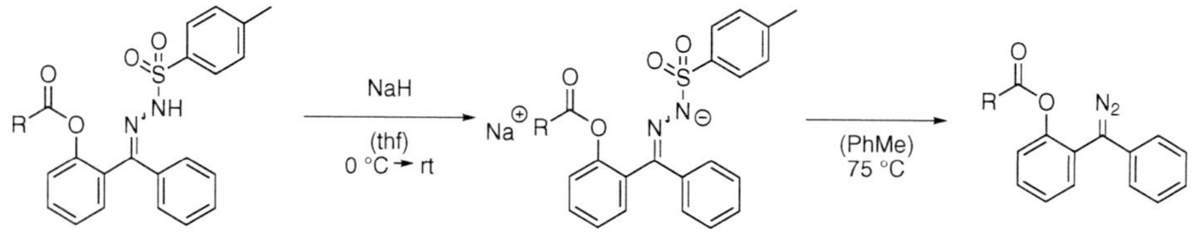

Scheme 3.10: Diazo ester synthesis.

Synthesis of the diazo compounds was achieved by deprotonation of the hydrazone with sodium hydride over night and subsequent thermal rearrangement by a modified procedure of *Tomioka et al.*[54] Due to the low volatility of the diazo esters these could not be extracted by distillation and therefore heating had to be performed in toluene, not neat, and extraction with pentane or diethylether depending on the solubility of the ester had to be performed. The hydrazide salt **3.9** (scheme 3.10) is isolable and may be stored under an argon atmosphere, however direct conversion of the residue of the deprotonation after removal of the solvent in vacuo, proved successful without loss of yields.

Table 3.2.: Selected bond length, angles and torsion angles of **Me−bpNNHTs**:

d [Å]		∠ [°]		torsion ∠ [°]	
C3–C22	1.500(1)	C3–C22–C16	119.5(2)	C3–C22–C16–C17	161.3(2)
C16–C22	1.485(0)	C23–C24–C25	121.8(2)	C4–C3–C22–C16	80.4(3)
C23–C24	1.458(0)	C24–C25–C26	124.1(2)	C22–N2–N1–S1	176.2(2)
C24–C25	1.318(0)	N1–S1–C8	108.3(2)	C23–C24–C25–C26	177.2(2)
C25–C26	1.490(0)	N1–N2–C22	116.0(2)	C25–C24–C23–O3	172.5(2)
N1–S1	1.654(0)	N2–N1–S1	111.7(2)		
N1–N2	1.404(0)	N2–C22–C3	123.8(2)		
N2–C22	1.285(0)	N2–C22–C16	116.6(2)		
O1–S1	1.424(0)	O1–S1–C8	107.8(2)		
O2–S1	1.431(0)	O2–S1–N1	103.8(2)		
O3–C23	1.370(0)	O2–S1–O1	120.1(2)		
O4–C23	1.208(0)	O3–C23–O4	122.3(2)		
S1–C8	1.755(0)	O4–C23–C24	127.4(2)		

Despite the expected similarities of the acids this approach failed with maleimides and the fumaric acid esters. Even though deprotonation was successful, no rearrangement could be achieved. Even a rise in temperature to 120 °C did not prove to be effective. This might be due to coordination of the sodium by not only the deprotonated nitrogen but also by the carboxylic oxygen. This is supported by the N1-O4 distance in the crystal structure of **Me-bpNNHTs** which is with 2.94 Å below the combined *van der Waals* radius[55] of Na (2.27 Å) and the semi-*van der Waals* radii of N (0.78 Å) and O (0.76 Å). This hypothesis is supported by the fact, that in FAB-MS not only the anions are detected but also their sodium adducts. The ratio of the adduct rises from **3.9-Me** to **3.9-fu** as depicted in table 3.3. Due to this fact the cation was exchanged with lithium by the use of ^{n}BuLi as base giving the desired **fu-bpN$_2$** and **mi-bpN$_2$** after thermal rearrangement at 75 °C in toluene.

Table 3.3.: Ratio of **3.9** and its sodium adduct in MS [%]:

	Me	Ph	fu
3.9	70	64	53
3.9 + Na	30	36	47

The received diazo esters proved to be fairly stable in air. They may be handled in air at ambient temperature and stored at -35 °C in air without any decomposition observed for months. However at ambient temperature they decompose slowly in substance and within days in solution. In the FT-IR spectra two prominent bands are observed, the one around 1720 cm^{-1} being assigned to the C=O stretch frequency of the carboxyl C=O double bond and the one at around 2050 cm^{-1} to the typical N≡N resonance. [31]

Table 3.4.: Selected IR and UV/vis bands of diazo esters:

	N=N [cm^{-1}]	C=O [cm^{-1}]	N_2 [nm]
Ac−bpN$_2$	2045	1766	285
Me−bpN$_2$	2046	1737	286
PhOMe−bpN$_2$	2046	1726	293
PhMe−bpN$_2$	2043	1730	286
Ph−bpN$_2$	2045	1732	282
Ph−flN$_2$	2065	1733	290
PhNO$_2$−bpN$_2$	2046	1734	291
PhNO$_2$−flN$_2$	2066	1720	285
fl−bpN$_2$	2048	1736	–
mi−bpN$_2$	2046	1716	–

The ^{1}H NMR spectra show the expected signals without any unusual shifts. In ^{13}C NMR spectroscopy all signals are observed except for the diazo carbon, due to the low amount of substance and the limited recording time enforced by the instability of the compounds in CDCl$_3$. [56] All substances where isolated as red solids or oils showing absorption in the aromatic region of the UV/vis range as well as at around 290 nm. The latter absorption band is caused by the diazo group.

3.1.2. Irradiation

Upon irradiation at 300 nm, which was chosen due to its proximity to the absorption frequency of the diazo group, for one hour four different products were obtained, whereof three are shown in the detail of the ^{1}H NMR spectrum in figure 3.4. The irradiation was performed in toluene to minimize activation via excitation of the phenyl rings of the substrate.

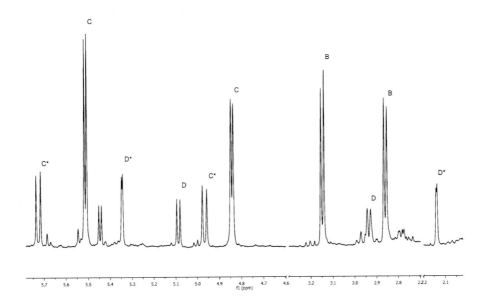

Figure 3.4: ^{1}H NMR after irradiation of **Ph−bpN$_2$** in toluene. Product assignment: B ≡ **3.1-Ph**, C & C* ≡ **3.12-Ph** (both diastereoisomers), D & D* ≡ **3.13-Ph** (both diastereoisomers).

The signals marked B were assigned to the anticipated cyclopropane **3.1** which is formed by the *in situ* generation of the carbene and its subsequent attack of the double bond forming the product (scheme 3.11).

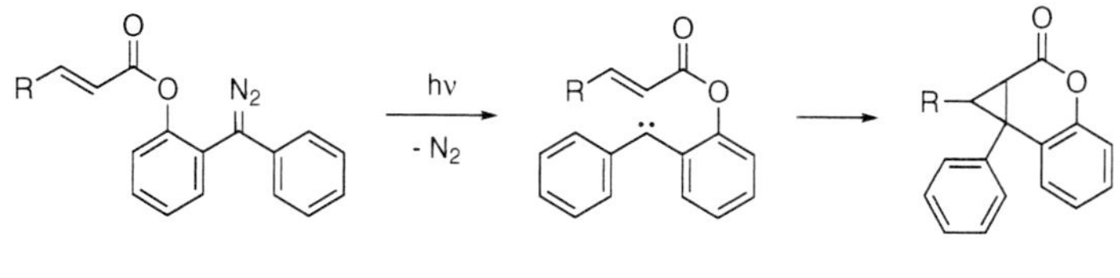

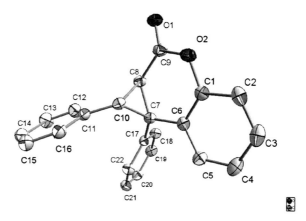

Scheme 3.11: *In situ* carbene generation and subsequent cyclopropanation.

The chemical shifts of ca. 2.9 ppm and 3.1 ppm in the ^{1}H NMR spectra are typical for ester substituted cyclopropanes as is the shift of ca. 31 ppm, 38 ppm and 42 ppm in ^{13}C NMR spectroscopy.[57,58] The $^2J_{HH}$ coupling constants of about 5 Hz however are smaller than in literature and as expected from the torsion angle observed in the single crystal structure. This might be caused by distortion of the three-membered ring in solution caused by the anellated lactone. The carbonyl stretching frequencies at ca. 1750 cm^{-1} are in the typical range of six membered lactones.[59,60] An overview of selected IR bands is given in table 3.6. All cyclopropanes were isolated as colourless solids.

Figure 3.5: ORTEP[24] representation of **3.1-Ph**. Thermal ellipsoids are given at 50% probability level. Hydrogen atoms are omitted for clarity.

Single crystals of **3.1-Ph** could be obtained by slow evaporation of a concentrated chloroform solution. The distances in the cyclopropane ring C7-C8, C7-C10 and C8-C10 are with 1.52 Å , 1.53 Å and 1.52 Å slightly longer than typical C−C single bonds caused by sp^2-hybridisation of the carbon atoms in the three membered ring. The lactone ring shows no extraordinary features with all bond lengths and angles matching the expected values of the assigned single and double bonds. The angles of the cyclopropane carbon atoms differ strongly from the tetrahedral angle due to the ring strain and sp^2-hybridisation and are in agreement with the literature.[61] The torsion angle of the hydrogen atoms tethered to the three-membered ring is with 146° consistent with literature but too large concerning the torsion angle calculated from the $^2J_{HH}$ coupling constants via the *Karplus* equation of 129°.[62] This is expected because the *Karplus* relation is only directly applicable for sp^3-hybridised carbon atoms. The lactone ring is almost planar, only distorted by O2 and C9 slightly bent out of plane. The two non-anellated phenyl rings are directed almost perpendicular to the plane defined by the lactone, while the cyclopropane plane is pointing towards the lactone ring at an angle of roughly 105°.

Table 3.5.: Selected bond length, angles and torsion angles of **3.1-Ph**:

d [Å]		∠ [°]		torsion ∠ [°]	
C1–C6	1.388(2)	C1–C6–C7	120.1(1)	C1–O2–C9–O1	167.4(1)
C6–C7	1.490(2)	C6–C7–C8	114.0(1)	C6–C7–C8–C9	0.2(2)
C7–C8	1.517(2)	C6–C7–C10	114.5(1)	C7–C8–C9–O1	171.7(2)
C7–C10	1.531(2)	C6–C7–C17	116.8(1)	C17–C7–C10–C11	0.9(2)
C7–C17	1.499(2)	C7–C8–C9	121.1(1)		
C8–C9	1.472(3)	C7–C8–C10	60.5(1)		
C8–C10	1.523(2)	C7–C10–C8	59.6(1)		
C10–C11	1.488(2)	C7–C10–C11	123.0(1)		
O1–C9	1.201(2)	C8–C7–C10	59.9(1)		
O2–C1	1.400(2)	C8–C7–C17	119.7(1)		
O2–C9	1.365(2)	C8–C10–C11	121.6(1)		

continued on next page

Table 3.5 – continued from previous page

d [Å]	∠ [°]		torsion ∠ [°]
	C9–C8–C10	116.9(1)	
	C10–C7–C17	119.7(1)	
	O2–C9–C8	117.9(1)	
	O2–C1–C6	123.2(1)	

The second main product is cyclobutanone **3.12** with the signals C for the major diastereoisomer and C* for the minor diastereoisomer in figure 3.4.

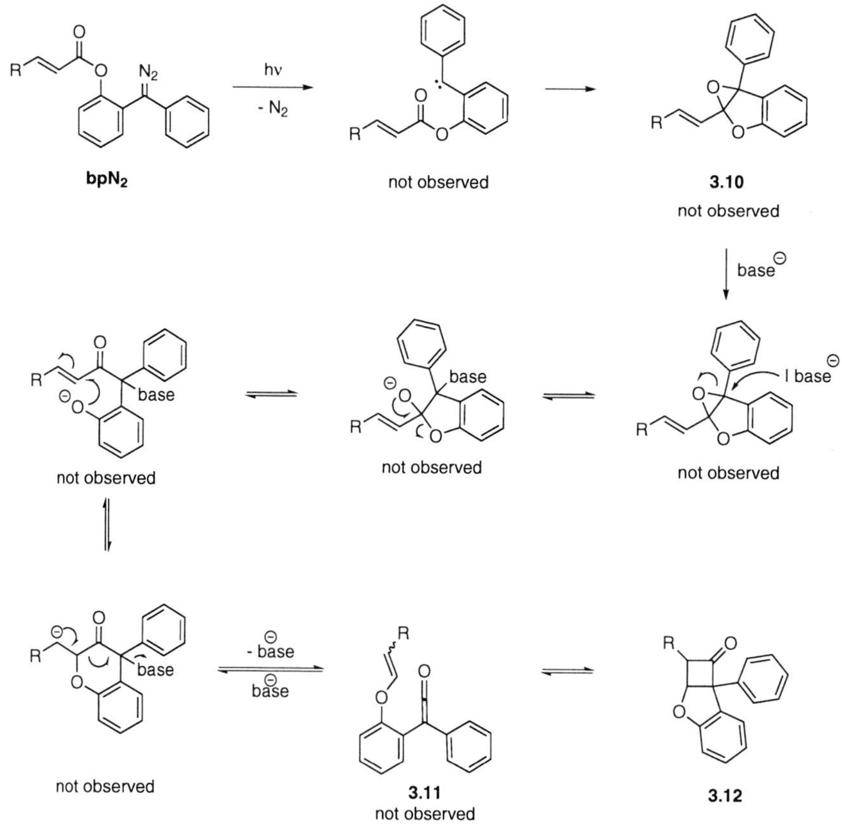

Scheme 3.12: Epoxidation and subsequent rearrangement to cyclobutanone via a ketene intermediate.

This cyclobutanone is presumably formed via the cascade depicted in scheme 3.12. In the first step the *in situ* generated carbene attacks the C=O double bond of the carbonyl forming epoxide **3.10** in analogy to the diazo derived epoxidations described by *Padwa et al.* and *Suga et al.*[63,64] This reaction should be kinetically favored towards cyclopropanation due to the formation of a five- instead of a six-membered ring. **3.10** is subsequently opened to release ring strain by attack of base, which is presumably sodium toluene sulfinate which was not totally removed from the prior stage or sodium hydroxide originating from reaction of the sulfinate with water condensed into the flask while stored in the freezer. Attack should predominantly take place at the more electrophilic position, the diphenylmethylen carbon. The formed dihydrofuranolate then undergoes rearrangement to the phenolate that subsequently attacks the cinnamic acid's double bond in α-position. The formed carbanion may now rotate freely around the former C=C double bond, which accounts for the two diastereoisomers formed consequently. Reformation of the conjugated π-system leads to a rearrangement forming ketene **3.11** which undergoes thermal [2+2]-cyclisation to cyclobutanone **3.12** and regenerates the base which then undergoes the next rearrangement cycle.[65] This anionic mechanism is supported by traces of alcohols found in the ^{1}H NMR spectrum at shift regions typical for phenols and non aromatic alcohols.

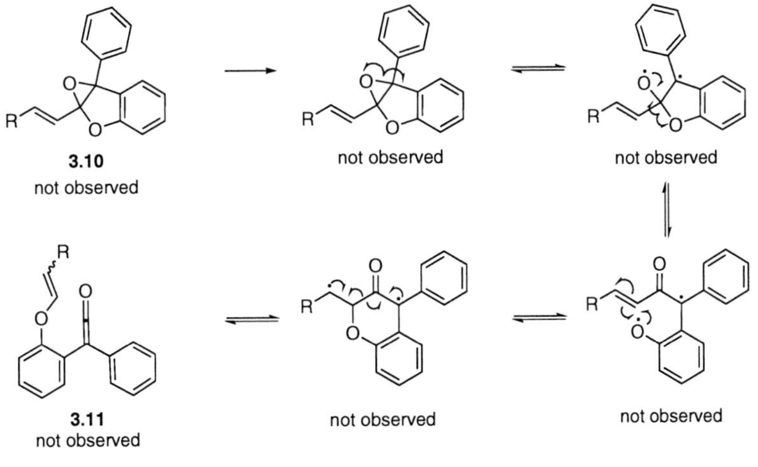

3.10
not observed

not observed

not observed

3.11
not observed

not observed

not observed

Scheme 3.13: Alternative mechanism for the ketene formation.

The major drawback of the base catalysed mechanism is however the α-attack at the double bond, which is unlikely. Because of this the alternative radical cascade depicted in scheme 3.13 has to be considered as well, even though radical cleavage of the epoxide is necessary. The ring opening should occur at the bond between the oxygen and the bencylic carbon atom since the formed radical in the benzylic position of the two phenyl rings is stabilised through delocalisation. The more reactive oxygen centered radical rearranges to the more stable radical centered at the phenolic oxygen and reforms the carbonyl group. The latter radical may now attack in α-position of the olefinic acid. In the case of cinnamic acids, the generated radical is again in a benzylic position and therefore stabilised. Free rotation around the former double bond is here also possible. Finally ketene **3.11** is formed via the reformation of the double bond, cleaving the bond between the α- and the carbonyl-carbon, and recombination of the two radicals.

Table 3.6.: Selected IR bands of the irradiation products $[cm^{-1}]$:

	C=O
3.1-PhOMe	1749
3.1-Ph	1757
3.1-PhNO$_2$	1760
3.12-Me	1802
3.12-PhOMe	1779
3.12-Ph	1780
3.12-PhNO$_2$	1802

These cyclobutanones show extreme low field shifts for the two hydrogen atoms tethered to the four membered ring as expected for phenyl substituted anellated keto ethers. The major isomer shows $^2J_{HH}$ coupling constants of 3.3–3.7 Hz accounting for a torsion angle of 118°–120° according to the *Karplus* equation.[62] However this value may only serve as a rough trend since the *Karplus* correlation in expected to be inaccurate for ring systems. With $^2J_{HH}$ coupling constants of 7.8–8.0 Hz the minor isomer has a hydrogen torsion angle of 23°–25°. Therefore the major diastereoisomer is substituted in a *trans*-fashion concerning the hydrogen, while the minor one is *cis*-substituted. This finding means that the major isomer is derived from the *E*-olefin while the minor one originates from the *Z*-olefin, which is conclusive since the carbanionic and radical intermediate respectively responsible for *E-Z*-isomerisation is expected to be short lived and therefore rearranges fast to the ketene without much time to isomerise. The *E*-isomer is also the more stable one. The structure assignment is also supported by COSY, HMQC and HMBC spectroscopy. The C=O stretching vibration of 1779–1802 cm^{-1} is in the range expected for four-membered cyclic ketones.[59,60]

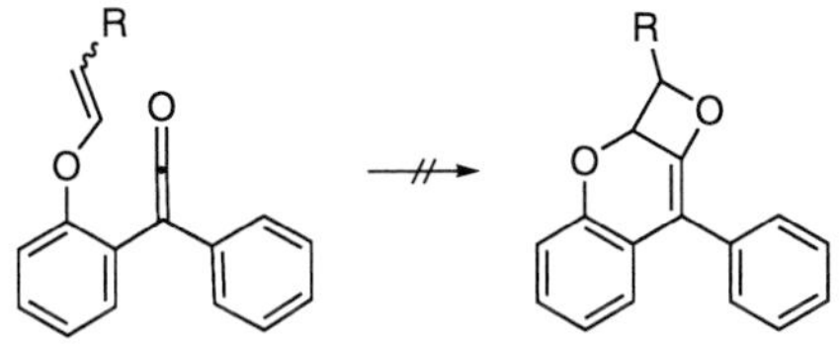

3.11
not observed

Scheme 3.14: [2+2]-cyclisation to cyclobutanone.

The *Paterno Büchi* analogue [2+2]-cyclisation depicted in scheme 3.14 seemed likely since a six-membered ring is formed, which would provide a thermodynamic driving force. However no evidence for this cyclisation product was found in the 1D- and 2D-NMR spectra.

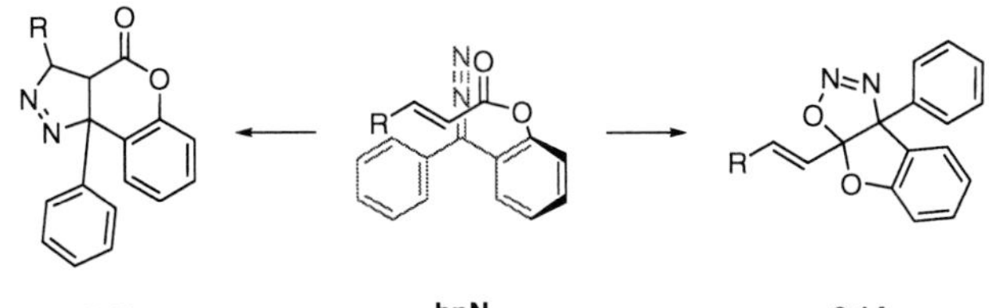

3.13   **bpN$_2$**   **3.14**

Scheme 3.15: 1,3-Dipolar cycloaddition of the diazo group to the olefin & carbonyl.

Besides the two products derived from photo-cyclisation, two byproducts were observed in the NMR spectra. These are presumably pyrazole **3.13** which also shows signals in the ^{1}H NMR spectrum depicted in figure 3.4 (signals D & D*) formed by 1,3-dipolar cycloaddition to the olefin and oxadiazole **3.14**[1] derived from 1,3-dipolar cycloaddition to the C=O double bond.[65,66] However only small amounts of impure samples could be isolated via *flash*-column-chromatography. This is the reason, why no certain structure assignment is possible.

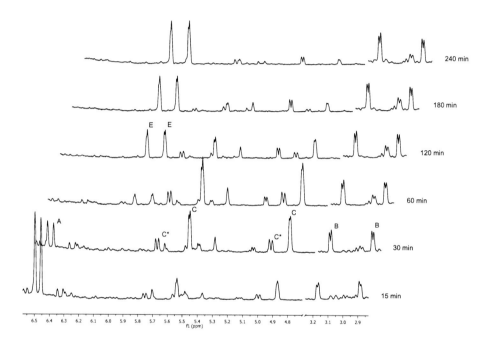

Figure 3.6: ^{1}H NMR of the reaction progress. Product assignment: A ≡ **Ph−bpN$_2$** B ≡ **3.1-Ph**, C & C* ≡ **3.12-Ph** (both diastereoisomers), E ≡ **3.15-Ph**.

Upon further irradiation at 300 nm the signals accounted for the cyclobutanone are depleted and two new signals formed (E in figure 3.6). The other irradiation products are not effected by longer irradiation times however. The newly formed product is the oxetane **3.15** which is generated besides 3-phenyl benzofurane **3.16**.

The generation of **3.15** presumably follows the pathway shown in scheme 3.16. In

---

[1]No quantification could be performed in the crude product mixture by NMR spectroscopy, due to the overlap of the olefin signals with aromatic signals of the other products. Therefore **3.14** is omitted in the following tables.

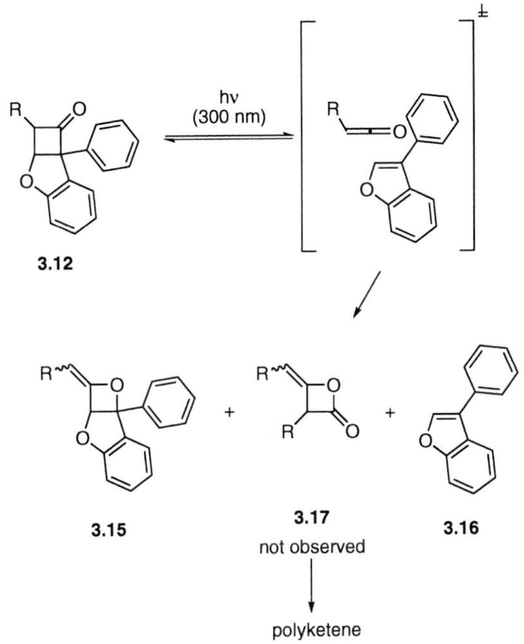

Scheme 3.16: Photo-chemical ring opening of the cyclobutanone and subsequent *Paterno Büchi* like recyclisation and decomposition of the ketene.

the first step cyclobutanone **3.12** undergoes a [2+2]-photo-cycloreversion. This is in analogy to the opening of cyclobutanones, generating a ketene and olefins described by *Staudinger*[67] thermally, by *Koda et al.*[68] laser induced and was achieved by *Majima et al.*[69] via photo-sensitizers. The formed ketene may then undergo either thermal [2+2]-photo-cycloaddition reforming the starting material or undergo a *Paterno Büchi*[70,71] analogous [2+2]-photo-cycloaddition generating oxetane **3.15**. Due to the radical intermediate of the *Paterno Büchi* reaction only the thermodynamically more stable conformation of the four membered ring is formed and therefore only one configuration of the double bond is observed. In competition to the intermolecular [2+2] *Paterno Büchi* reaction the ketene is also able to react with another equivalent of itself to generate diketene **3.17**, which then undergoes spontaneous polymerisation to polyketene leaving 3-phenyl benzofurane **3.16** unreacted in solution.[72–75]

The two non-aromatic protons of **3.15** show no coupling in ^{1}H NMR and correlation spectroscopy as expected. The shifts are with 5.73 and 5.85 ppm in the range expected for olefins or electron deficient, strongly deshielded aliphatic hydrogen atoms. The proposed structure is supported by 2D NMR spectroscopy experiments. No significant absorption besides aromatic C–H stretching bands at 3080 and 3050 cm^{-1} were observed in IR. The spectroscopic properties of **3.16** match the ones given in literature.[76,77]

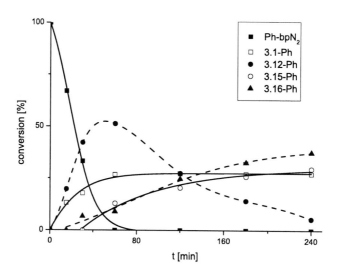

Graph 3.1: Time conversion plot for **Ph−bpN$_2$** irradiated at 300 nm in toluene.

The time conversion plot in graph 3.1 shows clearly that the diazo compound is completely converted after one hour and the formation of oxetane **3.15** and benzofurane **3.16** begins after a reasonable amount of cyclobutanone **3.12** has been generated. The concentration of cyclopropane **3.1** remains unchanged upon complete conversion of **bpN$_2$**. Slightly more benzofurane is formed than oxetane.

To examine the product distribution and the reaction pathways in more detail irradiation was conducted in different solvents and at other wavelengths. Since the cyclobutanone opening was considered to be photo-induced by excitation via the π-systems of the phenyl rings, non aromatic solvents were included to enforce this pathway. This had the expected effect that the oxetane ratio was significantly higher in pentane and Et_2O in comparison to toluene. The content of **3.15** was highest in Et_2O probably prone to stabilisation of the ketene by the ether. Diethylether also reduces 1,3-dipolar cycloaddition. Upon the move to longer irradiation wavelengths, the cyclobutanone opening was suppressed completely. However the product ratio was shifted towards **3.10**.

Table 3.7.: Solvent dependent irradiation-product distribution[2]:

	λ [nm]	3.1	3.12	3.15	3.13
pentane	300	1	1.56	0.3	0.78
pentane	419	1	4.42	–	–
PhMe	300	1	1.84	0.19	0.90
PhMe	419	1	6.54	–	–
Et_2O	300	1	0.43	0.63	0.46
CH_2Cl_2	419	1	4.42	–	–

Interestingly no pyrazole **3.13** was found at 419 nm indicating, that the 1,3-dipolar cycloaddition is photo-induced as well. To our surprise the reaction did proceed in CH_2Cl_2 as it did in pentane without any side reactions caused by the halogenated solvent. At both wavelengths toluene favoured the cyclobutanone formation more strongly than non-aromatic solvents.

Table 3.8.: Primary irradiation product distribution:

	σ	3.10^3/ 3.1
PhOMe	-0.27	3.43
PhMe	-0.17	1.90
Ph	0	1.70
PhNO$_2$	0.78	0.44

To examine the influence of the electron density of the olefinic double bond the

---

[2]Normalised on cyclopropane.

[3]Epoxide ratio determined by the daughter products.

para-substituent of the cinnamic acid was varied. By this method the relative electron density may be simply assigned through the *Hammett* parameter σ of the substituent.[78] Therefore the cinnamic acid esters are directly comparable.

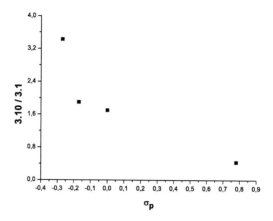

Graph 3.2: Dependence of the primary product distribution on the olefin electron density.

The relative product ratio for the different cinnamic acid diazo ester is summarized in table 3.8. While the concentration of cyclopropane **3.1** could directly be appointed by the integral value of the aliphatic protons in the ^{1}H NMR, the content of epoxide **3.10** had to be calculated from its direct daughter product cyclobutanone **3.11** and its daughter products oxetane **3.15** and benzofuran **3.16**. It was observed, that the preference for epoxidation rises with the electron density of the olefin. This is explained by the nucleophilic character of the *in situ* generated carbene. Therefore it attacks preferably electron deficient double bonds. This relation is outlined in graph 3.2.

Similar reactivity was also expected for the crotonic acid derivative **Me-bpN$_2$**. However cyclobutanone was only observed in low yields and the reaction did not proceed as cleanly as for the cinnamic acids. No indications of cyclopropane formation were found, which is consistent with the even higher electron density of the olefin. From the reaction mixture only diketone **3.18-Me** could be isolated purely. Most of the product mixture is composed of a substance containing many inequivalent but similar methyl groups, which could not be isolated by *flash*-chromatography. **3.18-Me** might be produced via the mechanism shown in scheme 3.17. The first steps follow the reaction pathways discussed earlier for the formation of cyclobutanone **3.12**. However the α-attack would not yield a stabilised product since the phenyl group is missing. Therefore an alternative rearrangement of the π-system takes place, generating **3.18-Me**.

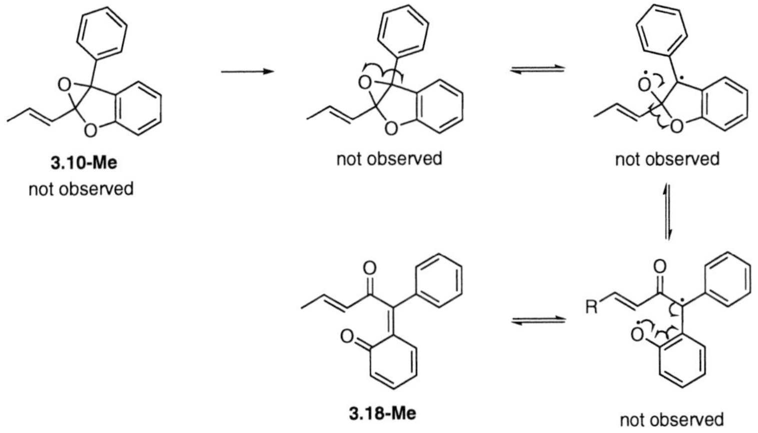

(a) Radical mechanism.

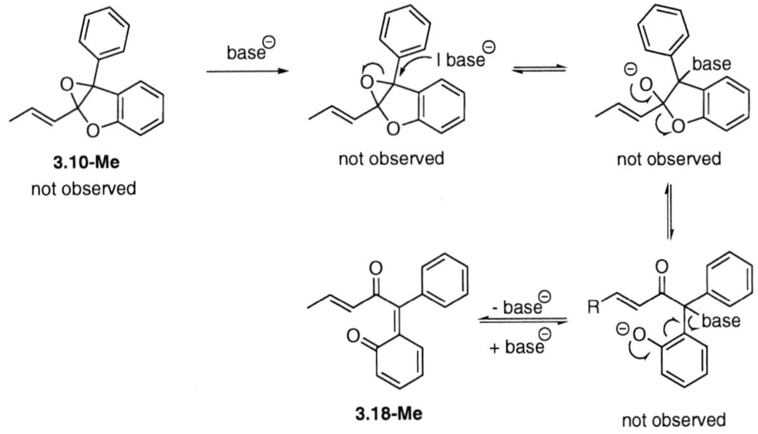

(b) Base catalysed mechanism.

Scheme 3.17: Alternative rearrangement pathway of the epoxide.

Irradiation of the double-bond-free derivative **Ac-bpN₂** did not give defined product mixtures. Although C−H-insertion as depicted in scheme 3.18 was anticipated, only traces could be observed in the ^{1}H NMR spectrum.

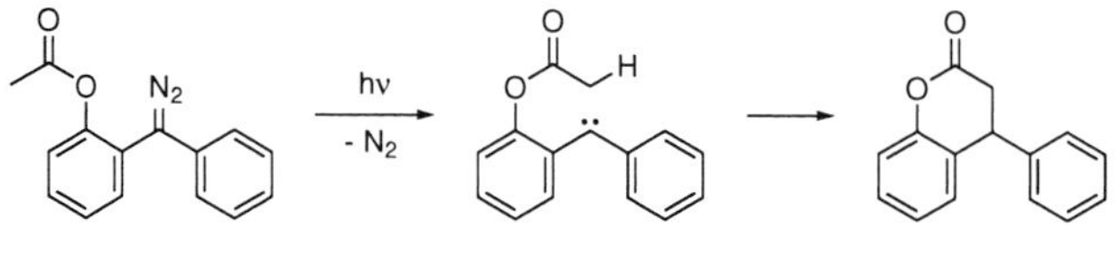

Scheme 3.18: C−H insertion in the absence of a double bond.

Thermal decomposition or rhodium catalysis only gave traces of cyclopropane and no indications for the formation of cyclobutanone were found. Besides some evidence for C−H-insertion, based on the observation of CH−CH$_2$-fragments in the ^{1}H and 2D (COSY, HMBC, HMQC) NMR spectra, no further information could be gathered due to unselective reaction proceeding and ill defined reaction products.

3.2. 9-Diazofluorene esters

For application of diazoesters as switches in catalysts, stability of the compounds in solution at room temperature is mandatory. This is not realizable with diphenyl diazomethanes though. Therefore transfer of the established ester systems to the more stable 9-diazo fluorene moiety was aspired.[79]

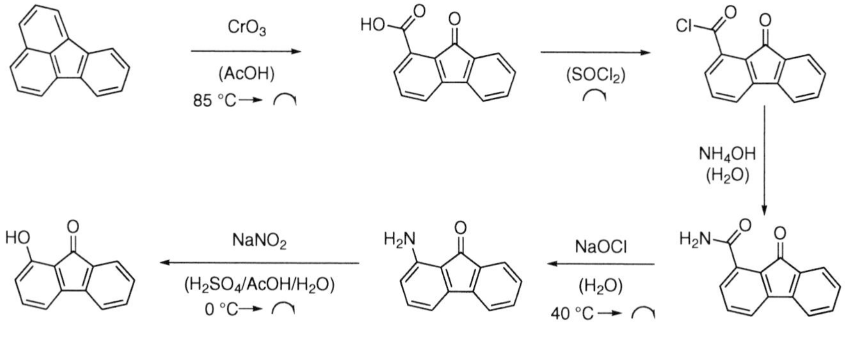

Scheme 3.19: Synthesis of 1-amino & 1-hydroxy fluorenone.

Although 1-hydroxy and 1-amino functionalised 9-fluorenones are literature known, little research has been done upon their reactivity. This is most likely prone to the lengthy synthesis as depicted in scheme 3.19.[80] Fluoranthene, the only substance available commercially in scales applicable for multi step reactions and at a reasonable price, is oxidized with seven equivalents of $Cr(VI)O_3$ in acetic acid to give 9-fluorenone-1-carboxylic acid.[81–83] The carboxylic acid is refluxed in $SOCl_2$ to yield the acid chloride which is then treated with concentrated ammonium hydroxide solution to give 9-fluorenone-1-carboxamide.[84] Via *Hofmann* rearrangement the carboxamide is converted to 1-amino-9-fluorenone by treatment with NaOCl.[85] Finally diazotation of the amine and subsequent heating to 80 °C gives the desired 1-hydroxy-9-fluorenone.[86]

Application of the reaction conditions successful in the diphenyl diazomethane ester synthesis failed however for the fluorenones. Even though the 9-fluorenone esters and amides could be prepared in comparable yields to 2-hydroxy and 2-amino benzophenone the subsequent hydrazonation by reflux of the ester and tosylhydrazone in toluene in the presence of *para*-toluene sulfonic acid did not give any conversion and only starting material was isolated. Even in boiling xylene no formation of hydrazone was observed. Since all literature procedures known used alcohols as solvent[50,87,88] which are not applicable to our system due to the reasons discussed in section 3.1.1, a new synthetic approach had to be developed.

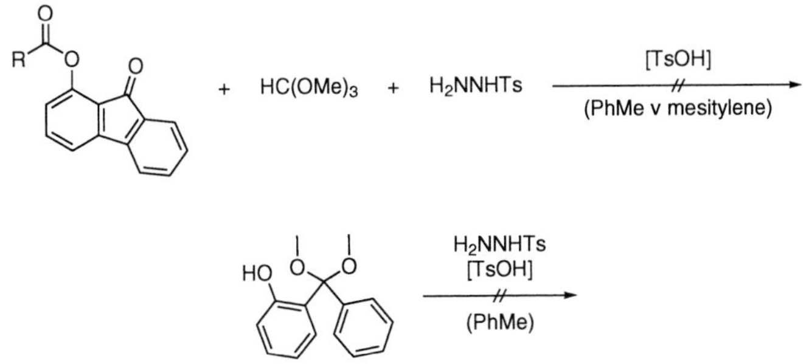

Scheme 3.20: Activation via acetal formation.

It was presumed that the low reactivity of the fluorenone carbonyl towards nucleophilic

attack, liable for hydrazonation, derives from the integration into the aromatic π-system, unlike in the benzophenone esters, where the phenyl rings are twisted against each other by approximately 45°. Therefore the ketone was converted into an acetal to break the conjugated π-system at the carbonyl open to enhance reactivity and also provide driving force for hydrazonation through rearomatisation.[89,90] First experiments of an *in situ* acetal formation as described in scheme 3.20 proved unsuccessful. Therefore tests were performed with the dimethoxy acetal derived from 2-hydroxy benzophenone showing no conversion at all and rendering this attempt a dead end.

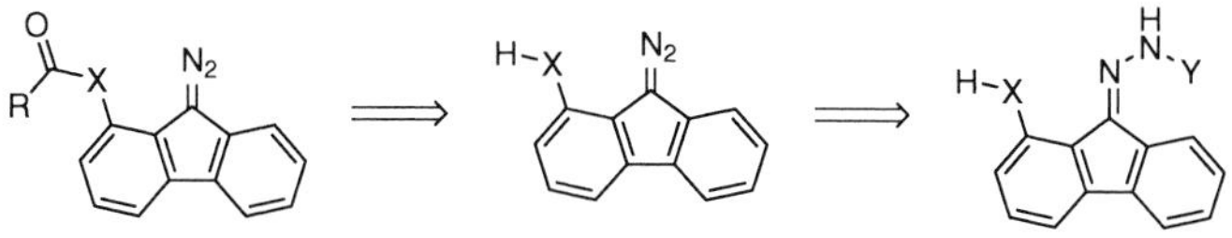

Scheme 3.21: Synthesis of the ester from diazo alcohols and amides. X ≡ O, NH; Y ≡ H, $C_7H_7SO_2$.

Since the diazofluorenes were expected to be sufficiently more stable than their benzophenone analogues 1-hydroxy and 1-amino 9-diazofluorene was anticipated as precursor to the diazo esters as outlined in scheme 3.21.

9-Fluorenone hydrazone is prepared easily by refluxing 9-fluorenone and hydrazine hydrate in ethanol in the presence of acid.[91] However the same reaction conditions led to an insoluble red solid for 1-hydroxy-9-fluorenone. Even at ambient temperature the insoluble substance is formed which is probably coordination polymer caused by deprotonation of the alcohol by the formed hydrazone or hydrazine hydrate. To circumvent this problem the alcohol was tms protected. This led to no improvement however since the tms group is obviously cleaved by the basic hydrazine. Upon reaction of 1-amino-9-fluorenone with hydrazine in ethanol in the presence of acetic acid the hydrazono acetamide **3.22** was formed, in which the acetate groups blocked further functionalisation. The substitution with other acids led to no conversion at all. However the use of tosyl hydrazone proved successful again. 1-Amino and 1-hydroxy tosylhydrazone-9-fluorenone monohydrate were obtained in reasonable yields by reflux of the adequate fluorenone and tosylhydrazone in ethanol, catalysed by hydrochloric acid.

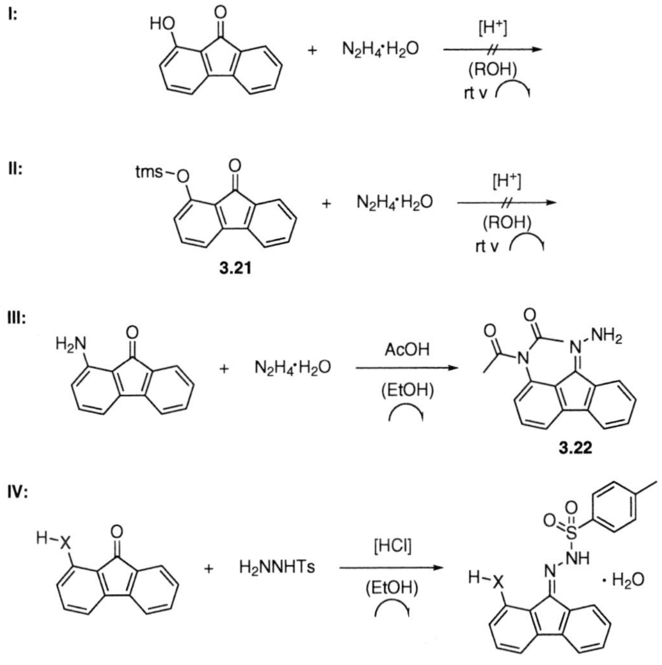

Scheme 3.22: Reactions of 1-hydroxy & 1-amino fluorenone with hydrazones. $X \equiv O$, NH.

Since the single known literature procedure[92] to 1-amino-9-diazofluorene emerged non-reproducible in reasonable yields, the high acidity of the phenylic OH group was harnessed. Deprotonation of the alcohol with sodium hydride with an extra equivalent to remove the hydrate at 0 °C in thf and salt metathesis with the appropriate acid chloride led to the desired tosylhydrazone acid esters. These could be deprotonated and thermally rearranged to the related diazofluorene esters following the procedure for the diphenyl diazo esters. However the rearrangement failed for diacids again (scheme 3.24) and no reactions with lithium bases could be performed up until to date.

The 9-diazofluorene acid esters are obtained as orange air stable solids and are in contrast to the diphenyl diazomethanes insoluble in pentane. They may be stored in air and at ambient temperature in the dark without decomposition. The NMR spectra show no unexpected features. In comparison to the diphenyl diazomethanes the $N{\equiv}N$

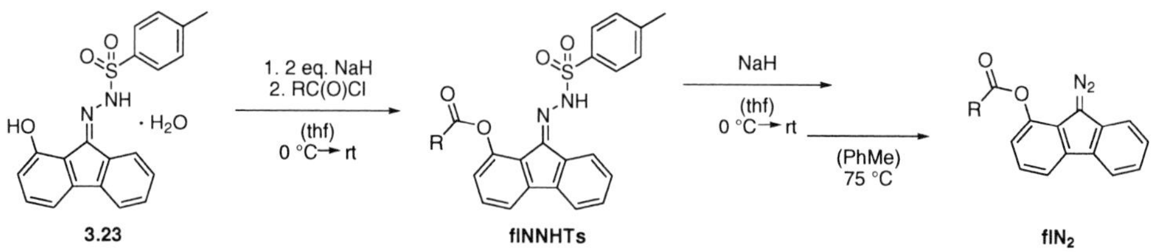

Scheme 3.23: Synthesis of olefin tosyl hydrazone fluorenone ester, deprotonation and subsequent thermal rearrangement to the diazo compound.

stretching frequency in IR is slightly shifted to larger wavenumbers accounting for the higher stability. UV/vis absorption is in the same region as the diphenyl diazomethanes. A collection of the IR and UV/vis bands is given in table 3.4.

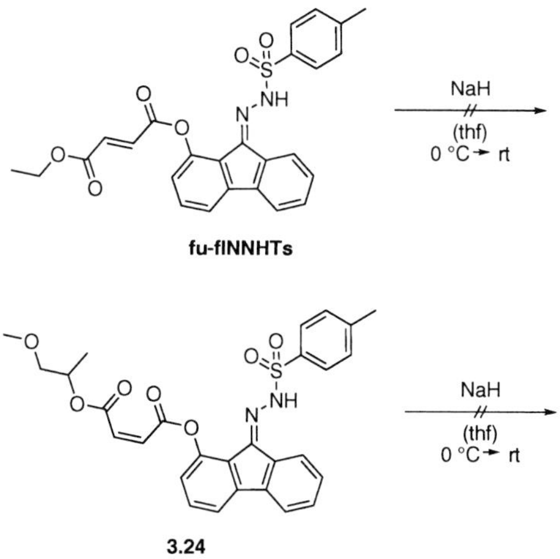

Scheme 3.24: Thermal rearrangement fails with diacids.

First irradiation tests with **Ph-flN$_2$** suggest similar reactivity to the diphenyl di-azomethanes. Irradiation in toluene leads to a complete removal of the double bond and traces of cyclopropane were observed in ^{1}H NMR. However, due to the low solubility of the irradiation products a definitive conclusion may not be given to date.

3.3. Conclusions & outlook

The synthesis of diphenyl diazomethane acid esters is possible by reaction of benzophenyl esters with tosylhydrazone, as shown in scheme 3.9 and subsequent deprotonation and thermal rearrangement (scheme 3.10). The electron withdrawing ester substituent leads to a reduction of the carbonyl group reactivity hence the catalytic protonation, necessary for nucleophilic attack of the hydrazone, is hindered. Therefore the reaction temperature had to be elevated to 110 °C. Due to its basicity hydrazine may not be used. The use of alcohols as solvent causes ester cleavage and has to be avoided as well. The integration of the carbonyl group into the conjugated π-system and steric rigidity of the fluorenone induces a decrease of the carbonyl reactivity. This leads in conjunction with the electron withdrawing ester group to a massive reduction of reactivity. Hence no hydrazonation of the fluorenone esters was possible. However tosylhydrazones of 1-hydroxy and 1-amino-9-fluorenone were accessible and reaction with acid chlorides gave the desired hydrazone esters, which could be thermally rearranged to diazofluorenes. Rearrangement of the sodium salts of the tosyl hydrazones failed for diacids and imides, probably caused by chelate complexation, though. So lithium salts should be used.

Upon irradiation of the diazo olefin esters at 300 or 419 nm the reaction cascade depicted in scheme 3.25 is triggered. The *in situ* generated carbene attacks the olefinic and the carboxylic double bond forming cyclopropane **3.1** and epoxide **3.10**. The more electron deficient the olefin is the more **3.1** is formed. This is owed to the nucleophilic character of the generated carbene. UV light of 300 nm apparently also provides activation energy for the 1,3-dipolar cycloaddition of the diazo group to both double bonds generating pyrazole **3.13** and oxadiazole **3.14** as byproducts. No side reactions occur at 419 nm however. The highly strained epoxide **3.10** is unstable at the conditions applied and rearranges to ketene **3.11** which undergoes instantaneous thermal [2+2]-cyclisation to cyclobutanone **3.12**. The cyclobutanone [2+2]-photo-cyclo-reverts to a second ketene and 3-phenyl benzofurane **3.16** if irradiation is continued at 300 nm. Via a *Paterno Büchi* analogous [2+2]-photo-cyclisation the ketene reacts with benzofurane **3.16** to oxetane **3.15** or reacts with another equivalent of ketene to form diketene which then undergoes spontaneous polymerisation.

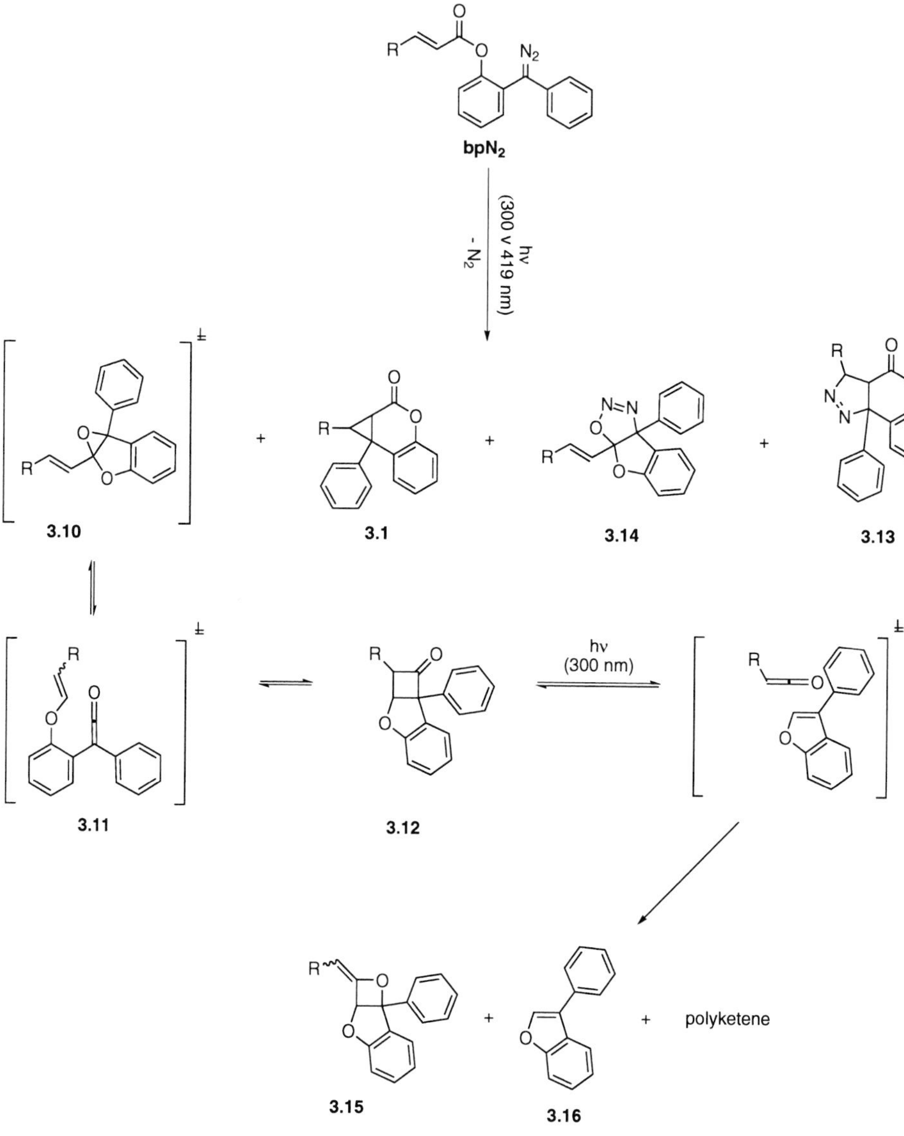

Scheme 3.25: Overview of the photo-cyclisation and decomposition pathways.

However the energy provided by light at 419 nm is insufficient to re-open cyclobutanone **3.12**. Therefore it is assumed that the 1,3-dipolar cycloaddition and the [2+2]-photo-cyclo-reversion, which need 300 nm, involve sensitisation via the phenyl rings.

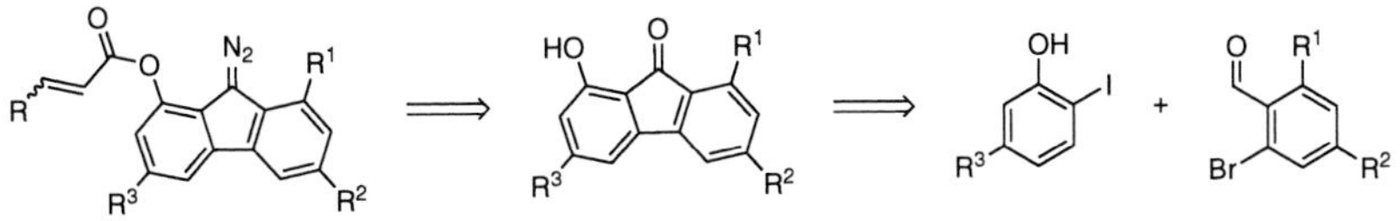

Scheme 3.26: Functionalised 1-hydroxy fluorenone ester.

In recent years several studies on palladium catalyzed cross coupling cyclisations to fluorenones were published. [93–100] With the application of transition metal mediated reactions a reduction in synthetic steps for the diazofluorenes should be possible. This should also improve the accessibility of functionalised 1-hydroxy and 1-amino-9-fluorenones, giving access to stability and reactivity tuning of the diazofluorenes through steric and electronic control. This in combination with a non-toxic and short, therefore cheap route is mandatory for industrial application of the diazo esters in photo-switching. Possible building blocks are given in the retrosynthesis of scheme 3.26.

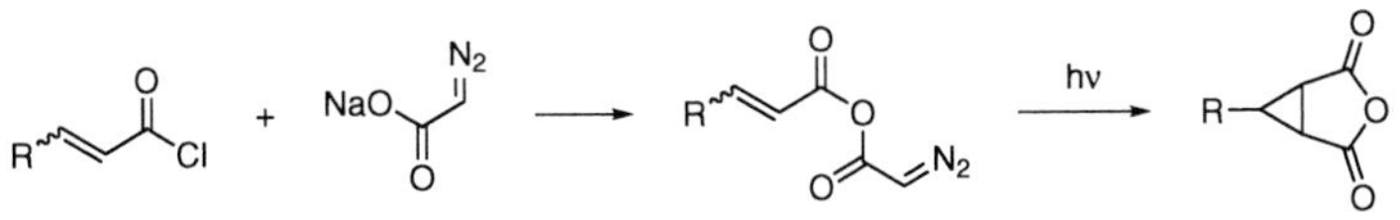

Scheme 3.27: Potential synthesis and photo-cyclisation of diazo acetate anhydrides.

Also the influence of the nucleo- or electrophilicity of the carbene upon intramolecular cyclisation needs to be studied in more detail. In combination with functionalised diazofluorenes the diazo acetate in scheme 3.27 is promising for this examination due to easy accessibility and high electrophilicity of the generated carbene.

4. Bis(triphenylphosphine) cinnamic acid ester complexes of group 10

To obtain a detailed insight into the influence of the different substituents of the photoswitchable ligands onto the complex stability and reactivity a detailed investigation of bis(triphenylphosphine) cinnamic acid ester complexes of group 10 was executed. The focus was laid upon the correlation of directly measurable spectroscopic properties with the electronics and sterics of the ligands and the reactivity. Even though detailed studies concerning the electronic properties of distant ligand systems have been performed, no examinations were conducted on subtle changes in the electronic and steric properties of the ligands to date.

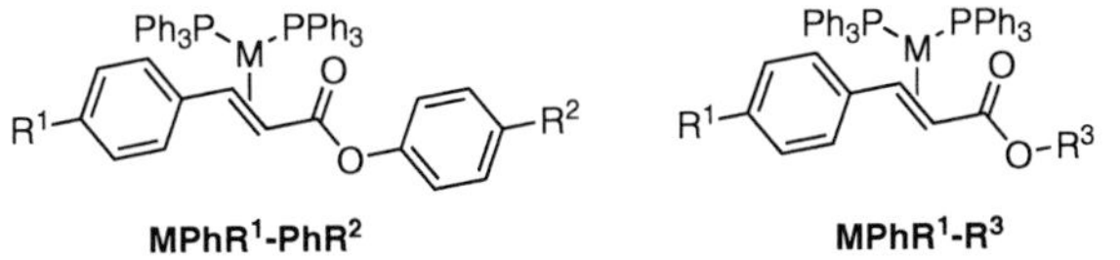

Figure 4.1: Investigated cinnamic acid complexes. $R^1 \cup R^2 \equiv$ OMe, Me, H, Cl, CF$_3$, NO$_2$; $R^3 \equiv$ Me, Et, ^{i}Pr, ^{t}Bu

The analysed systems, summarised in figure 4.1, were chosen due to their high flexibility. In these complexes electronic properties of the double bond may be roughly adjusted by the substituent in para-position[78] to the olefin ($\mathbf{R^1}$). By variation of the ester group a fine tuning on the electronics ($\mathbf{R^2}$) of the olefin and the steric demand[101] ($\mathbf{R^3}$) of the ligand can be performed. The electronic influence was assigned via the *Hammett* parameter (σ_p) while the steric demand was determined via the A values, which are compiled in table 4.1.

Table 4.1.: *Hammett & A values:*

	σ_p		A
OMe	-0.27	Me	1.70
Me	-0.17	Et	1.75
H	0.00	^{i}Pr	2.15
Cl	0.23	^{t}Bu	4.50
CF$_3$	0.54		
NO$_2$	0.78		

4.1. Synthesis

The cinnamic acid ester ligands used, were in case of phenole esters and the *tert*-butyl ester [102,103] either synthesised by reaction of the appropriate acid chloride with the alcohol in pyridine at 100 °C [34,35] or by deprotonation of the alcohol with sodium hydride and subsequent salt metathesis with the acid chloride in nearly quantitative yield. The other alkyl esters were obtained by refluxing the cinnamic acid in the corresponding alcohol in the presence of catalytic amounts of H$_2$SO$_4$ in quantitative yield. [104]

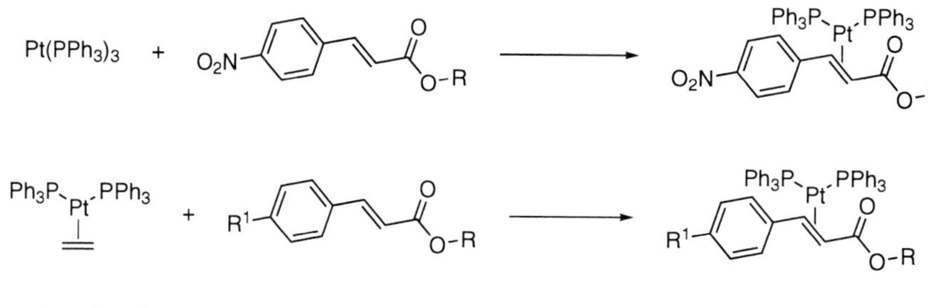

Scheme 4.1: Synthetic routes to bis(triphenylphosphine)platinum cinnamates. R^1 ≡ OMe, Me, H, Cl, CF$_3$, NO$_2$

Reaction of Pt(PPh$_3$)$_3$ [105] with one equivalent 4-nitrocinnamic acid esters in toluene led to the desired bistriphenylphosphine olefin complexes (**PtPhNO$_2$–PhR^2**) in quantitative yields. [106] Triphenylphosphine could be easily removed by washing with pentane. The same procedure was successful with 4-chloro and 4-trifluoromethyl cinnamic acid phenol esters as well as with cinnamic acid phenol ligands even though yields were lower caused by comparably higher solubility of the product complexes in pentane. Due to the higher electron density at the double bond of 4-methylcinnamic acid 10 eq. of ligand were needed to shift the equilibrium to the product side. The extensive amounts of pentane needed to remove the excess ligand and the increased solubility of the product led to extremely low yields via this route. However this synthetic approach failed when 4-methoxycinnamic acid was used and only mixtures of $Pt(PPh_3)_3$ and product could be isolated. *In situ* NMR experiments showed the same trends for the alkyl esters. However only the nitro complexes could be isolated via this route. With all other ligands, even though 100% conversion was observed by NMR for 4-chloro, 4-tifluoro and cinnamic acid alkyl esters, only $Pt(PPh_3)_3$ could be obtained. This is due to the high solubility of the non nitrosubstituted alkyl ester complexes. Upon washing to remove triphenylphosphine, $Pt(PPh_3)_3$ precipitated while the alkyl ester complexes stayed in solution, shifting the equilibrium to the educt side. Another reason accountable is the unexpectedly high electron density of the double bond in alkyl esters, *vide infra*. Therefore all non nitrosubstituted complexes where synthesised from $(PPh_3)_2PtC_2H_4$ [107] in benzene, which gave the desired products in good to excellent yields. All platinum complexes were isolated as white or cream, in case of nitro cinnamic acid esters yellow, coloured solids and may be handled in air for a short period of time, but should be stored in an inert gas atmosphere. However only the nitro cinnamic acid derivatives are air stable in solution, whereas all other complexes decompose slowly under the formation of platinum black. All complexes are stable in solution in the absence of oxygen.

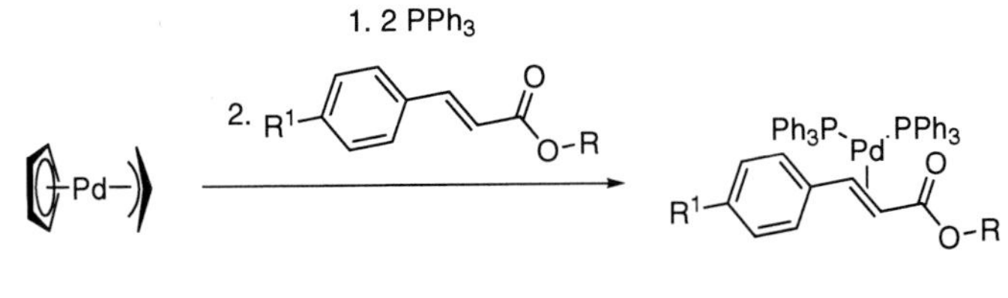

Scheme 4.2: Synthetic routes to bis(triphenylphosphine)palladium cinnamates. $R^1 \equiv$ OMe, Me, H, Cl, CF_3, NO_2

The corresponding palladium complexes were all synthesised by reaction of Pd(C$_3$H$_5$)Cp[108] with 2 eq. triphenylphosphine to generate a bis(triphenylphosphine)-palladium(0) fragment[1], *in situ*, via reductive elimination, in analogy to the synthesis of bis(tri-*tert*-butylphosphine)palladium(0).[109] Subsequent addition of olefin gives the desired complexes in quantitative yield in this simple one pot reaction. The nitro cinnamic acid complexes were isolated as deep red, air stable solids while all other palladium complexes are orange to yellow solids which decompose in air within minutes. All complexes with the exception of nitro and trifluoromethyl cinnamic acid esters decompose slowly, in the presence of oxygen within seconds, in solution. The sterically less shielded methyl ester complexes decompose within minutes in solution at room temperature. However the degradation is slow enough at -25 °C for sufficient spectroscopic analysis. Despite the lability in solution all complexes are stable at ambient temperature in the solid state.

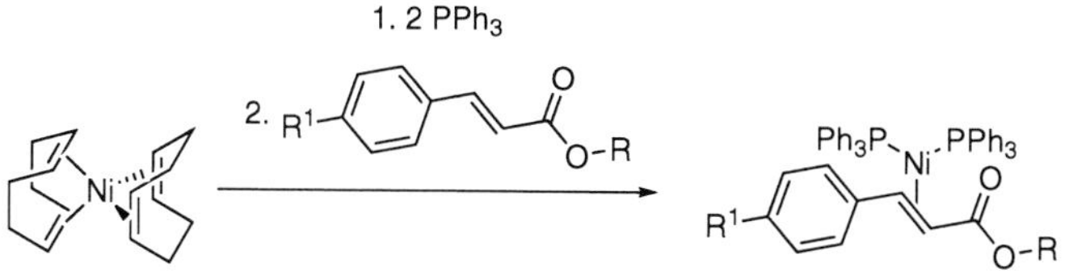

Scheme 4.3: Synthetic routes to bis(triphenylphosphine)nickel cinnamates. R^1 ≡ OMe, Me, H, Cl, CF$_3$

Following a procedure described by *Maciejewski et al.*, the nickel analogues could be obtained quantitatively by *in situ* generation of η^4-cycloocta-1,5-diene bis(triphenylphosphine)nickel(0), from Ni(cod)$_2$ and PPh$_3$, followed by addition of ligand.[110] However no nitro substituted ligands could be used due to the oxidation of Ni(0) to Ni(II) by the nitro group and subsequent decomposition of the complex. All nickel compounds were isolated as deep red, pyrophoric solids which decompose immediately in the presence of traces of oxygen. Despite the air and moisture sensitivity all complexes are stable in substance and solution for months in an atmosphere of argon.

---

[1]For a detailed investigation of the intermediate refer to chapter 6.

4.2. **Spectroscopic properties**

4.2.1. **NMR**

All platinum complexes show static behaviour in ^{1}H, ^{13}C, ^{31}P and ^{195}Pt NMR spectroscopy on the spectral time scale.

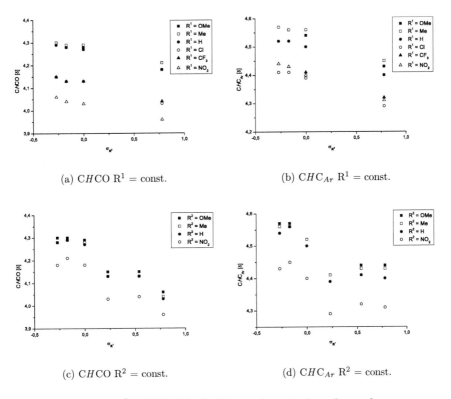

(a) $CHCO$ $R^1 = $ const.

(b) CHC_{Ar} $R^1 = $ const.

(c) $CHCO$ $R^2 = $ const.

(d) CHC_{Ar} $R^2 = $ const.

Graph 4.1: ^{1}H NMR shifts for Pt complexes in dependence of σ.

All phenol cinnamate complexes of platinum show a characteristic high field shift of the olefinic protons compared to the free ligand. The observed shifts of ca. 4.0 - 4.3 ppm for the olefinic proton closer to the carboxyl group $(CHCO)$ and ca. 4.3 - 4.6 ppm for the one next to the phenyl ring (CHC_{Ar}) are caused by back bonding of the metal into the olefin's

π^* orbital.[2] Therefore the σ character of the olefinic C-C bond is more pronounced and the signal is shifted upfield. Both signals show coupling to each other with coupling constants of ca. 8 - 9 Hz which are significantly smaller than the ones observed for E-substituted protons of non coordinated olefins. This is due the coordination of the platinum which distorts the planar geometry of the double bond by forcing all substituents of the olefin to move away from the metal into the opposite side of the plane spanned by the double bond. The coupling constants indicate a $H-C=C-H$ torsion angle of ca. 150°, determined via the *Karplus* equation.[62] Coupling is also observed to both phosphorus nuclei via the platinum with a *cis*-coupling constant of ca. 4 Hz and a *trans*-coupling constant of ca. 8 - 9 Hz. Therefore both olefinic protons show a multiplicity of a doublet of pseudo triplets (dpt) and not the expected doublet of doublets of doublets (ddd). Both protons also couple to ^{195}Pt which is shown through the presence of Pt satellites. The electron density on the double bond is the main parameter determining the strength of the metal-olefin bond. The electron poorer the olefin is the stronger is the back bonding into the olefin's π^* orbital. Therefore the σ character of the olefinic C-C bond is more pronounced. This leads to a high field shift of the olefinic proton signals in NMR spectroscopy. Due to the dependence of σ bonding from the olefin to the metal and the π back-bonding to the olefin, this high field shift should be a direct indicator of the organometallic bond strength. The simplest way to describe the electron density at the double bond is to use the *Hammett* parameters of the substituents $\mathbf{R}^1$ (σ_{R1}) and $\mathbf{R}^2$(σ_{R2}). As expected a linear correlation of σ_{R2} and the upfield shift of both olefin signals is observed in the ^{1}H NMR spectra (graph 4.1 (a) & (b)). Since the influence on the double bond electron density of R^1 is stronger than of R^2 due to the direct incorporation into the cinnamic acid´s π-system a more pronounced trend was expected for σ_{R1}. Even though this trend is observed in general, no linear correlation is possible and obviously two distinct regimes are present (graph 4.1 (c) & (d)). All other signals in the proton NMR show no characteristic change in shift or additional coupling due to metal coordination. All relevant NMR shifts are summarised in table 4.2.

The signals of both olefinic carbon atoms are, as expected, shifted to higher field in ^{13}C

---

[2]Signal assignment is based on comparison with the free ligand and related complexes.

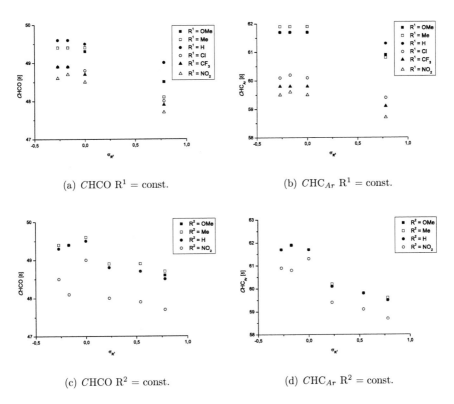

(a) $CHCO$ R^1 = const.

(b) CHC_{Ar} R^1 = const.

(c) $CHCO$ R^2 = const.

(d) CHC_{Ar} R^2 = const.

Graph 4.2: ^{13}C NMR shifts for Pt complexes in dependence of σ.

NMR spectroscopy, due to platinum coordination. The signals of the carbon ($CHCO$) next to the carboxyl carbon is observed at ca. 48 - 49 ppm while the one closer to the aryl group (CHC_{Ar}) is at ca. 59 - 62 ppm. Both carbon atoms show coupling to the two phosphorus atoms with a *cis*-coupling constant of ca. 5 Hz and a *trans*-coupling constant of ca. 30 Hz. Therefore both signals are observed as doublet of doublets. Both signals exhibit ^{195}Pt satellites indicating additional coupling to the metal center. In analogy to the ^{1}H NMR shift of the olefinic protons a dependence of the upfield shift of the double bond's carbon atoms in dependence of σ was expected in the ^{13}C NMR spectra. A linear correlation could be determined for both signals and σ_{R^2} (graph 4.2 (a) & (b)) but no trend was observed for σ_{R^1}. However the same two regimes are formed in analogy to the ^{1}H

NMR spectra (graph 4.2 (c) & (d)). The signals of the carboxylic carbon and the aromatic carbon next to the double bond are both observed as doublet of doublets with platinum satellites due to coupling via the olefin to the metal and via the olefin and platinum to both phosphorus nuclei but no significant change in NMR shift is observed. Coupling to only one ^{31}P atom is observed for the carbon atom in *ipso*-position to R^1 when R^1 is OMe or NO_2. Due to overlap of the aromatic carbon signals of the triphenylphosphine with the *ipso*-carbon to $R^1 \equiv$ Me, H, Cl and CF_3 no statement may be made whether any interaction occurs between phosphorus and the aforementioned carbon nuclei. An assignment of the PPh_3 carbon atoms and the remaining aromatic ligand ^{13}C nuclei is not possible due to mutual overlap. The methyl groups, if present, show no characteristic change in shift or additional coupling.

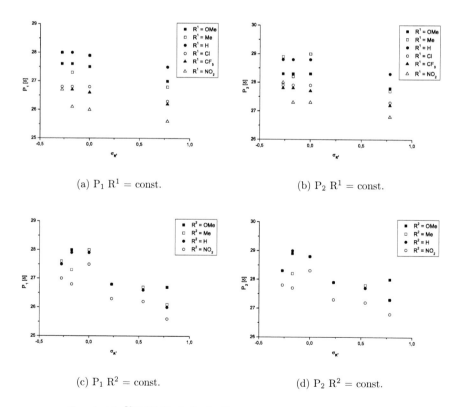

(a) P_1 R^1 = const.   (b) P_2 R^1 = const.

(c) P_1 R^2 = const.   (d) P_2 R^2 = const.

Graph 4.3: ^{31}P NMR shifts for Pt complexes in dependence of σ.

The two phosphorus nuclei can be observed as two distinct signals in the ^{31}P NMR spectra. Both signals are split to doublets due to $^2J_{PP}$ coupling. The coupling constants are with ca. 30 - 45 Hz within the typical range of *cis*-couplings. Both signals show platinum satellites due to direct interaction with the metal. The more down field signal (P_2) is observed as a doublet of doublets with Pt satellites, if R^1 is CF_3. Due to the direct bond to the metal center it was expected, that the ^{31}P NMR shift would be dependent on the electron density of platinum and therefore on the olefin electronics, in analogy to ^{1}H and ^{13}C NMR spectroscopy. Yet only a slight trend could be observed for the dependence onto σ_{R^2} (graph 4.3 (a) & (b)) but no trend was observed for σ_{R^1}. However the same two regimes are formed in analogy to the two nuclei discussed above (graph 4.3 (c) & (d)).

In ^{19}F NMR only one singlet appears without any indication of coupling, neither to ^{31}P nor to ^{195}Pt. The shift also only changes slightly upon coordination, probably caused by solvent effects. Therefore no signs of interaction of the CF_3 group and the metal could be observed.

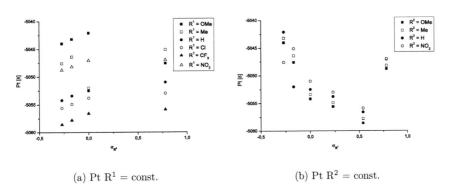

(a) Pt R^1 = const. (b) Pt R^2 = const.

Graph 4.4: ^{195}Pt NMR shifts in dependence of σ.

The signals in the ^{195}Pt NMR spectra are with ca -5050 ppm shifted high field in comparison to known complexes of similar geometry.[111] Due to the direct influence of the electron density of a nucleus upon its shift in NMR spectroscopy a direct correlation of the electron density of the olefin and ^{195}Pt NMR shift was expected. However no such trends could be observed concerning σ_{R^1} and σ_{R^2}. While graph 4.4 (a) shows the expected low field shift upon increasing electron deficiency upon the move from $R^2 \equiv OMe$ to $R^2 \equiv H$ the trend is reversed when R^2 is NO_2. If R^2 is kept constant a minimum between -5055 ppm and -5060 ppm can be observed at $\sigma_{R^1} \approx 0.5$ (graph 4.4 (b)).

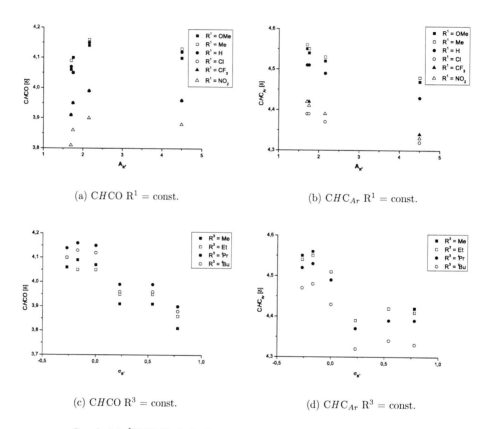

(a) $CHCO$ R^1 = const.   (b) CHC_{Ar} R^1 = const.

(c) $CHCO$ R^3 = const.   (d) CHC_{Ar} R^3 = const.

Graph 4.5: ^{1}H NMR shifts for Pt complexes in dependence of σ & A.

All alkyl cinnamate complexes show the same characteristic high field shift of the olefinic protons in ^{1}H NMR spectroscopy, with signals at ca. 3.8 - 4.1 ppm ($CHCO$) and ca. 4.3 - 4.6 ppm (CHC_{Ar}), analogous to the phenol esters. $^3J_{HH}$ coupling constants of ca. 8 - 9 Hz for both signals indicate a $H-C=C-H$ torsion angle of ca. 150°, comparable to the systems discussed earlier. *Karplus cis*-coupling constants of ca. 4 Hz and *trans*-coupling constants of ca. 8 - 9 Hz to phosphorus are also observed for both signals. Therefore both olefinic protons show a multiplicity of a doublet of pseudo triplets with platinum satellites, which indicates a similar situation as observed for the aryl esters. In contrast to the later mentioned the ethyl and *iso*-propyl esters possess diastereotopic groups which give two distinct signals upon coordination respectively. For the ethyl esters two doublets of quartets are observed for the methylene moiety, one of which is shifted relatively high upfield to ca. 3.7 ppm, while two separated signals are observed for the two methyl groups of the *iso*-propyl group. Coupling constants match the expected values.

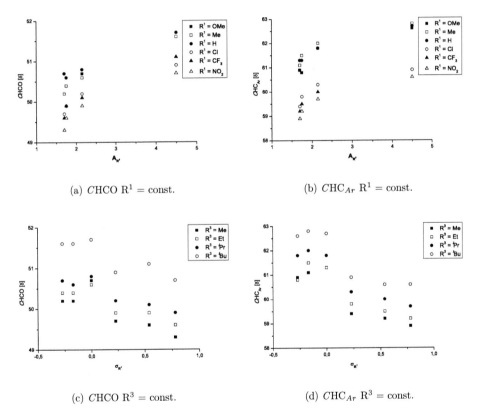

(a) *CHCO* R^1 = const.

(b) *CHC$_{Ar}$* R^1 = const.

(c) *CHCO* R^3 = const.

(d) *CHC$_{Ar}$* R^3 = const.

Graph 4.6: ^{13}C NMR shifts for Pt complexes in dependence of σ & A.

All remaining signals are in analogy to the systems discussed above. Upon the increase of the steric demand of the ester moiety the signal of *CH*CO shifts to slightly lower field while the opposite is the case for *CHC$_{Ar}$* (graph 4.5 (a) & (b)). The dependence of the olefinic proton shifts upon σ_{R^1} shows the same trend as the aforementioned systems (graph 4.5 (c) & (d)).

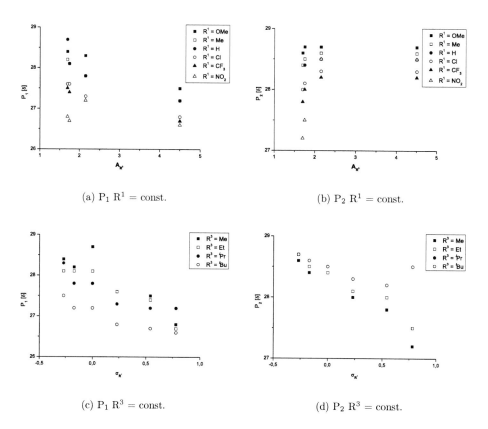

(a) P_1 R^1 = const. (b) P_2 R^1 = const.

(c) P_1 R^3 = const. (d) P_2 R^3 = const.

Graph 4.7: ^{31}P NMR shifts for Pt complexes in dependence of σ & A.

The ^{13}C NMR signals of $CHCO$ at ca. 50 - 52 ppm and CHC_{Ar} at ca. 59 - 61 ppm are shifted to high field in the same extent as the phenol analogues. Both signals show a similar coupling to phosphorus and platinum as for the ligands mentioned before. Therefore two doublets of doublets with platinum satellites are observed. Two separate signals are given by the methyl groups of the *iso*-propyl ester due to their diastereotopicity. The residual features of the spectra are in accordance with the aryl ester complexes discussed prior like the two regimes concerning the shift dependence on σ_{R^1}. However, in contrast to the ^{1}H NMR spectra both signals of the olefin´s carbon nuclei shift to lower field when the steric demand of R^3 is increased (graph 4.6 (a) & (b)). This is probably caused by an overall increased olefin platinum distance leading to less orbital overlap and therefore higher sp^2 character. This is also supported by the larger distance between the olefin and the platinum in the crystal structure of **PtPhNO$_2$–^{i}Pr** in comparison to the other 4-nitro cinnamate platinum complexes of platinum.

The two phosphine phosphorus nuclei show similar multiplicities, coupling constants and shifts to the aryl ester complexes. Unlike the other nuclei no trends show up in ^{31}P NMR spectroscopy concerning the steric influence of the alkyl group (graph 4.7 (a) & (b)), whereas the two regimes in dependence of σ_{R^1} are observed again (graph 4.7 (c) & (d)).

The ^{19}F NMR spectra give no indication of interaction of the CF$_3$ group and the metal, *vide supra*.

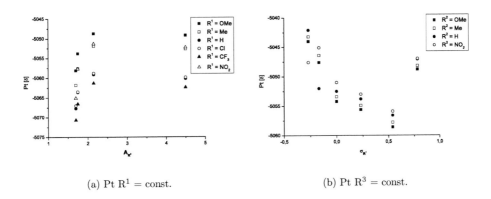

(a) Pt R^1 = const. (b) Pt R^3 = const.

Graph 4.8: ^{195}Pt NMR shifts in dependence of σ & A.

The ^{195}Pt shifts are in the same region as described above with matching signal multiplicities. While no correlation could be observed for the influence of A$_{R^3}$ (graph 4.8 (a)), the same minimum is observed at $\sigma_{R^1} \approx 0.5$ as for the aryl esters (graph 4.8 (b)).

Table 4.2.: Selected NMR shifts for triphenylphosphine Pt complexes [δ]:

	$CHCO$	CHC_{Ar}	$CHCO$	CHC_{Ar}	P_1	P_2	Pt
PtPhOMe–Me	4.06	4.55	50.2	60.9	28.4	28.6	-5058.1
PtPhOMe–Et	4.10	4.54	50.4	60.8	28.1	28.7	-5053.8
PtPhOMe–^{i}Pr	4.14	4.52	50.7	61.8	28.3	28.7	-5048.7
PtPhOMe–^{t}Bu	4.10	4.47	51.6	62.6	27.5	28.7	-5049.2
PtPhMe–Me	4.09	4.56	50.2	61.1	28.2	28.4	-5061.8
PtPhMe–Et	4.05	4.55	50.4	61.5	28.1	28.5	-5057.6
PtPhMe–^{i}Pr	4.16	4.53	50.6	62.0	27.8	28.6	-5051.9
PtPhMe–^{t}Bu	4.13	4.48	51.6	62.8	27.2	28.6	-5052.7
PtPh–Me	4.07	4.51	50.7	61.3	28.7	29.1	-5067.7
PtPh–Et	4.05	4.51	50.6	61.3	28.1	28.4	-5063.6
PtPh–^{i}Pr	4.15	4.49	50.8	61.8	27.8	28.5	-5059.1
PtPh–^{t}Bu	4.12	4.43	51.7	62.7	27.2	28.5	-5060.2
PtPhCl–Me	3.91	4.39	49.7	59.4	27.6	28.0	-5067.0
PtPhCl–Et	3.95	4.39	49.9	59.8	27.6	28.1	-5063.5
PtPhCl–^{i}Pr	3.99	4.37	50.2	60.3	27.3	28.3	-5058.8
PtPhCl–^{t}Bu	3.96	4.32	50.9	60.9	26.8	28.3	-5059.9
PtPhCF$_3$–Me	3.91	4.42	49.6	59.2	27.5	27.8	-5070.7
PtPhCF$_3$–Et	3.95	4.42	49.9	59.5	27.4	28.0	-5066.6
PtPhCF$_3$–^{i}Pr	3.99	4.39	50.1	60.0	27.2	28.2	-5061.3
PtPhCF$_3$–^{t}Bu	3.96	4.34	51.1	60.6	26.7	28.2	-5062.4
PtPhNO$_2$–Me	3.81	4.42	49.3	58.9	26.8	27.2	-5065.1
PtPhNO$_2$–Et	3.86	4.41	49.6	59.2	26.7	27.5	-5057.3
PtPhNO$_2$–^{i}Pr	3.90	4.39	49.9	59.7	27.2	28.5	-5051.3
PtPhNO$_2$–^{t}Bu	3.88	4.33	50.7	60.6	26.6	28.5	-5052.3
PtPhOMe–PhOMe	4.30	4.57	49.4	61.7	27.6	28.3	-5044.0
PtPhOMe–PhMe	4.28	4.56	49.4	61.7	27.6	28.3	-5043.2

continued on next page

Table 4.2 – continued from previous page

	$CHCO$	CHC_{Ar}	$CHCO$	CHC_{Ar}	P_1	P_2	Pt
PtPhOMe–Ph	4.28	4.54	49.3	61.7	27.5	28.3	-5042.1
PtPhOMe–PhNO$_2$	4.18	4.43	48.5	60.9	27.0	27.8	-5047.6
PtPhMe–PhOMe	4.30	4.57	49.4	61.9	28.0	28.9	-5047.6
PtPhMe–PhMe	4.29	4.56	49.4	61.9	27.3	28.2	-5046.4
PtPhMe–Ph	4.29	4.56	49.4	61.9	27.9	29.0	-5052.0
PtPhMe–PhNO$_2$	4.21	4.45	48.1	60.8	26.8	27.7	-5045.1
PtPh–PhOMe	4.29	4.52	49.6	61.7	28.0	28.8	-5054.2
PtPh–PhMe	4.28	4.52	49.6	61.7	28.0	28.8	-5053.4
PtPh–Ph	4.27	4.50	49.5	61.7	27.9	28.8	-5052.5
PtPh–PhNO$_2$	4.18	4.40	49.0	61.3	27.5	28.3	-5051.0
PtPhCl–PhOMe	4.15	4.41	48.9	60.1	26.8	27.9	-5055.6
PtPhCl–PhMe	4.13	4.41	48.9	60.2	26.8	27.9	-5054.9
PtPhCl–Ph	4.13	4.39	48.8	60.1	26.8	27.9	-5053.8
PtPhCl–PhNO$_2$	4.03	4.29	48.0	59.4	26.3	27.3	-5053.0
PtPhCF$_3$–PhOMe	4.15	4.44	48.9	59.8	26.7	27.8	-5058.6
PtPhCF$_3$–PhMe	4.13	4.43	48.9	59.8	26.7	27.8	-5057.8
PtPhCF$_3$–Ph	4.13	4.41	48.7	59.8	26.6	27.7	-5056.6
PtPhCF$_3$–PhNO$_2$	4.04	4.32	47.9	59.1	26.2	27.2	-5055.9
PtPhNO$_2$–PhOMe	4.06	4.44	48.6	59.5	26.7	28.0	-5048.8
PtPhNO$_2$–PhMe	4.04	4.43	48.7	59.6	26.1	27.3	-5048.2
PtPhNO$_2$–Ph	4.03	4.40	48.5	59.5	26.0	27.3	-5047.1
PtPhNO$_2$–PhNO$_2$	3.96	4.31	47.7	58.7	25.6	26.8	-5047.0

Since the coupling in NMR spectroscopy is strongly dependent on the bond properties between the two coupling nuclei it was assumed, that the $^1J_{CPt}$ coupling constants would be a good indicator for the platinum carbon bond strength and subsequently the olefinic electron density estimated through σ. However as shown in graph 4.9 only a random distribution around 200 Hz is present without any indication of a steric or electronic

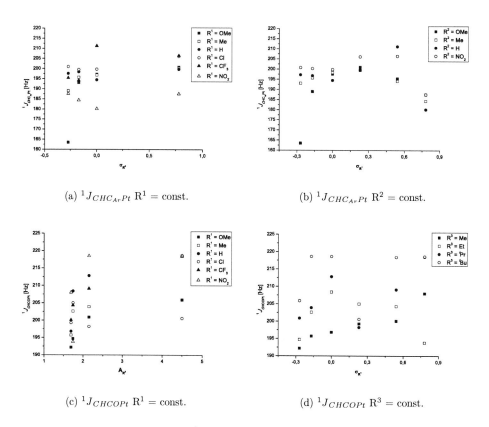

(a) $^1J_{CHC_{Ar}Pt}$ R^1 = const.

(b) $^1J_{CHC_{Ar}Pt}$ R^2 = const.

(c) $^1J_{CHCOPt}$ R^1 = const.

(d) $^1J_{CHCOPt}$ R^3 = const.

Graph 4.9: $^1J_{CPt}$ in dependence of σ & A.

influence raised by the olefin. This is presumably caused by additional effects onto the coupling from the olefin´s geometry and more dominantly by interaction of the metal with either the oxygen or the C=O bond of the carboxyl group or the phenyl ring of the cinnamic acid. Even interaction with the substituents R^1 might be possible. All relevant coupling constants are summarised in table 4.3.

However probably due to the further distance from the olefin, the $^2J_{PP}$ coupling constant proved as an excellent indicator for the electron density of the olefin and therefore the metal olefin bond strength. As shown in graph 4.10 (a) upon an increase of σ_{R2} the $^2J_{PP}$ coupling constant decreases slightly. This is caused by a decrease in electron density on platinum caused by the increasing electron deficiency of the double bond. This finally

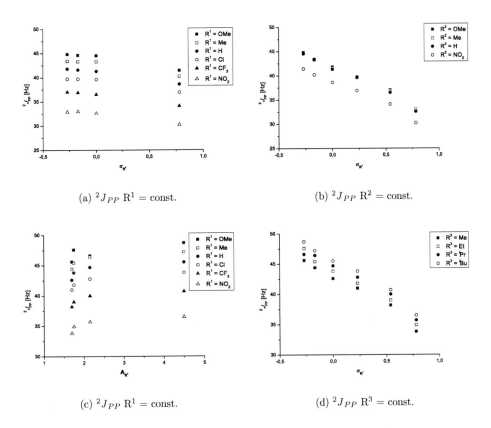

(a) $^2J_{PP}$ R^1 = const.

(b) $^2J_{PP}$ R^2 = const.

(c) $^2J_{PP}$ R^1 = const.

(d) $^2J_{PP}$ R^3 = const.

Graph 4.10: $^2J_{PP}$ for Pt complexes in dependence of σ & A.

results in the reduction of the coupling constant. A similar, but significantly stronger linear correlation can be observed in dependence of σ_{R^1} as depicted in graph 4.10 (b) & (d). The higher impact of σ_{R^1} is logical, since the electronic influence of the substituent on the phenyl ring, which is directly incorporated in the cinnamic acid´s π-system, is expected to be significantly higher than the electronic coupling via the ester bond. However no linear correlation is observed for the steric demand of R^3 (graph 4.10 (c)). Here the coupling constant increases with the sterical demand of the ester group which is expected since upon the move from R$^3 \equiv$ Me to R$^3 \equiv$ ^{t}Bu the electron donation increases significantly as well.

Since the $^1J_{PPt}$ coupling constant is also sensitive towards the oxidation state of the

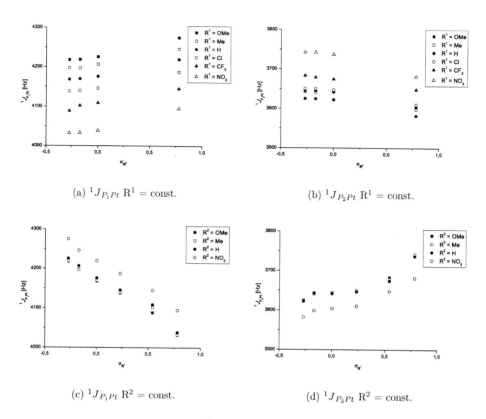

(a) $^1J_{P_1Pt}$ $R^1 = $ const.   (b) $^1J_{P_2Pt}$ $R^1 = $ const.

(c) $^1J_{P_1Pt}$ $R^2 = $ const.   (d) $^1J_{P_2Pt}$ $R^2 = $ const.

Graph 4.11: $^1J_{PPt}$ in dependence of σ.

metal and the *trans*-influence of other ligands, a correlation to the electronic and steric properties of the olefin was expected. As shown in graphs 4.11 (c) and 4.12 (c) the coupling constant of P_1 does decrease with increasing electron density of the olefin. This is caused by stronger influence of R^1 upon the vinylogue carbon ($CHCO$). Therefore the *trans*-influence is increased and the $Pt-P_1$ bond is weakened. However due to the decrease in this bond strength the $Pt-P_2$ bond is enforced leading to a higher coupling constant as depicted in graphs 4.11 (d) and 4.12 (d). The exact opposite trends, though smaller because of weaker electronic coupling via the ester bond, is observed for the dependence upon σ_{R^2} (graph 4.11 (a) & (b)). This is caused by the higher impact of the electronic coupling onto the vinylogue carbonyl carbon CHC_{Ar}. Again the steric demand of the

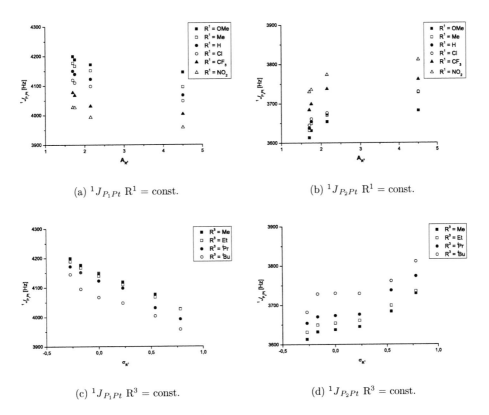

(a) $^1J_{P_1Pt}$ R^1 = const.

(b) $^1J_{P_2Pt}$ R^1 = const.

(c) $^1J_{P_1Pt}$ R^3 = const.

(d) $^1J_{P_2Pt}$ R^3 = const.

Graph 4.12: $^1J_{PPt}$ in dependence of σ & A.

ester is compensated by the electron donation properties of the alkyl moieties leading to the same correlation as R^2.

Table 4.3.: Selected NMR coupling constants for triphenylphosphine Pt complexes [Hz]:

	$\lvert^1J_{CHCOPt}\rvert$	$\lvert^1J_{CHC_{Ar}Pt}\rvert$	$\lvert^2J_{PP}\rvert$	$\lvert^1J_{P_2Pt}\rvert$	$\lvert^1J_{P_1Pt}\rvert$
PtPhOMe–Me	192.2	197.4	45.6	3613.5	4199.4
PtPhOMe–Et	194.7	194.8	47.6	3631.5	4188.7
PtPhOMe–^{i}Pr	200.9	194.5	46.6	3654.1	4171.7
PtPhOMe–^{t}Bu	205.9	196.0	48.7	3682.4	4144.9
PtPhMe–Me	195.7	197.3	44.4	3632.2	4176.9

continued on next page

Table 4.3 – continued from previous page

	$|^1J_{CHCOPt}|$	$|^1J_{CHC_{Ar}Pt}|$	$|^2J_{PP}|$	$|^1J_{P_2Pt}|$	$|^1J_{P_1Pt}|$
PtPhMe–Et	202.6	195.2	45.4	3649.7	4167.1
PtPhMe–^{i}Pr	203.9	195.2	46.4	3670.8	4151.7
PtPhMe–^{t}Bu	218.6	192.2	47.2	3728.6	4095.2
PtPh–Me	196.8	199.1	42.6	3637.8	4149.9
PtPh–Et	208.4	205.4	43.8	3654.3	4139.6
PtPh–^{i}Pr	212.8	196.6	44.7	3673.8	4122.8
PtPh–^{t}Bu	218.6	195.2	45.5	3730.3	4067.2
PtPhCl–Me	199.3	197.2	41.0	3645.1	4119.5
PtPhCl–Et	205.0	196.2	41.8	3661.1	4110.3
PtPhCl–^{i}Pr	198.2	186.4	42.8	3676.7	4098.5
PtPhCl–^{t}Bu	200.6	190.2	43.8	3729.6	4047.8
PtPhCF$_3$–Me	200.1	197.1	38.2	3684.4	4077.2
PtPhCF$_3$–Et	204.3	199.9	39.0	3699.4	4067.6
PtPhCF$_3$–^{i}Pr	209.1	186.9	40.0	3737.7	4031.3
PtPhCF$_3$–^{t}Bu	218.4	199.9	40.7	3761.4	4003.4
PtPhNO$_2$–Me	208.0	191.6	33.8	3730.2	4027.3
PtPhNO$_2$–Et	193.8	190.7	34.9	3735.7	4026.5
PtPhNO$_2$–^{i}Pr	218.5	188.1	35.7	3773.7	3992.9
PtPhNO$_2$–^{t}Bu	218.6	175.8	36.5	3811.1	3957.7
PtPhOMe–PhOMe	193.7	163.5	44.8	3625.4	4218.0
PtPhOMe–PhMe	188.9	193.1	44.6	3624.9	4218.8
PtPhOMe–Ph	193.2	197.2	44.5	3622.5	4225.6
PtPhOMe–PhNO$_2$	157.0	200.8	41.4	3582.5	4275.2
PtPhMe–PhOMe	186.0	189.0	43.4	3642.2	4197.8
PtPhMe–PhMe	198.9	195.7	43.3	3641.1	4197.0
PtPhMe–Ph	197.6	196.8	43.3	3643.7	4206.7
PtPhMe–PhNO$_2$	186.0	200.3	40.2	3598.7	4246.2

continued on next page

Table 4.3 – continued from previous page

	$|^1J_{CHCOPt}|$	$|^1J_{CHC_{Ar}Pt}|$	$|^2J_{PP}|$	$|^1J_{P_2Pt}|$	$|^1J_{P_1Pt}|$
PtPh–PhOMe	198.3	197.6	41.8	3645.0	4168.4
PtPh–PhMe	199.1	198.4	41.6	3645.1	4169.8
PtPh–Ph	196.0	194.5	41.3	3642.0	4176.1
PtPh–PhNO$_2$	188.3	199.8	38.6	3604.4	4219.6
PtPhCl–PhOMe	206.3	201.0	39.7	3650.2	4138.6
PtPhCl–PhMe	205.1	199.5	39.7	3650.2	4139.8
PtPhCl–Ph	199.9	199.7	39.6	3647.0	4146.1
PtPhCl–PhNO$_2$	175.0	206.2	36.9	3610.7	4187.5
PtPhCF$_3$–PhOMe	197.5	195.4	37.0	3683.4	4088.8
PtPhCF$_3$–PhMe	201.0	194.2	36.9	3679.0	4102.1
PtPhCF$_3$–Ph	209.8	211.3	36.5	3674.5	4109.1
PtPhCF$_3$–PhNO$_2$	193.8	206.5	34.1	3648.5	4145.4
PtPhNO$_2$–PhOMe	206.0	187.6	32.9	3741.8	4032.3
PtPhNO$_2$–PhMe	209.0	184.4	33.0	3741.9	4033.1
PtPhNO$_2$–Ph	201.0	180.2	32.6	3737.1	4038.7
PtPhNO$_2$–PhNO$_2$	201.4	187.6	30.2	3681.7	4095.1

All palladium cinnamic acid ester complexes except for the 4-nitro cinnamic acids show line broadening of the signals in ^{1}H, ^{13}C and ^{31}P NMR spectroscopy due to rotation of the olefin about the π-bond axis. The olefinic protons are shifted high field in NMR spectroscopy, characteristic for the metal coordination. However the shift is not as pronounced as for the platinum complexes, indicating a weaker metal olefin bond in the palladium systems. Despite the line broadening, *vide infra*, and the aforementioned smaller high field shift, the ^{1}H NMR spectra show features similar to the Pt-complexes. The shift of the olefinic protons shows the same correlation to the electronic influence of R^1 and R^2 assigned by σ_{R^1} and σ_{R^2} and steric demand of R^3 as the analogous platinum cinnamic acid complexes, exemplified in graph 4.13 (a) and (b).

The same trends concerning shifts are present in ^{13}C NMR spectroscopy. The signals

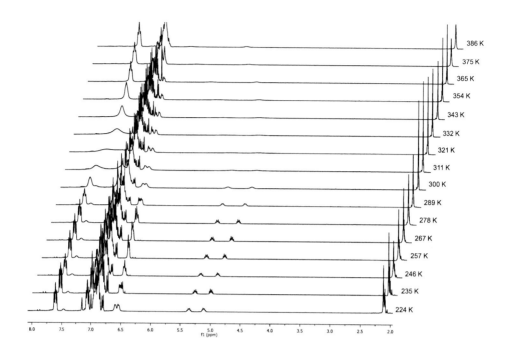

Figure 4.2: ^{1}H NMR spectra of **PdPh−PhMe** in dependence of the temperature.

arising from the olefinic carbon nuclei are also shifted upfield with a less pronounced shift difference to the non coordinated ligand in comparison to the platinum complexes. Due to the dynamic behaviour no coupling constants could be determined, except for the 4-nitro cinnamates, and in some cases the olefinic carbon signals were not observable at ambient temperature at all. In contrast to the Pt-complexes a splitting of all methoxy carbon signals into doublets occurs in the 4-methoxy cinnamic acid systems. Surprisingly, this interaction is observed even for the R^2 methoxy moiety. All other features are similar to the ^{13}C NMR spectra of the platinum compounds. This accounts also for the dependence of the shift towards electronic (R^1 & R^2) and steric (R^3) influences (graph 4.13 (c) & (d)). Two doublets are present in the ^{31}P NMR spectra which couple to each other. However coupling is once again not observable in the highly dynamic systems at ambient temperature. The chemical shift is comparable to the ones observed for the Pt systems, as is the coupling. Even the additional coupling of P_2 is observed if a CF_3 group is present.

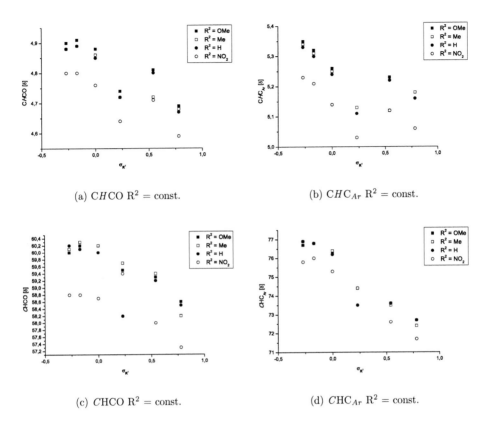

(a) $CHCO$ R^2 = const.

(b) CHC_{Ar} R^2 = const.

(c) $CHCO$ R^2 = const.

(d) CHC_{Ar} R^2 = const.

Graph 4.13: ^{1}H & ^{13}C NMR shifts for Pd complexes in dependence of σ.

The ^{19}F NMR spectra show the same features as for platinum.

In coherence with platinum the $^2J_{PP}$ coupling constant, if determinable at ambient temperature, is dependent on the electron density of the double bond. Therefore the same linear correlations depending on σ_{R1}, σ_{R2} and R^3 are observed as depicted in graph 4.14. The same processes already discussed for the Pt-complexes are accountable for this behaviour. The deviation from linearity at low values of σ_{R1} is caused by the highly dynamic behaviour of the 4-methoxy cinnamate systems leading to a decrease of the coupling constant due to severe line broadening. However the $^2J_{PP}$ coupling constants are overall smaller than the ones observed in the corresponding platinum compounds.

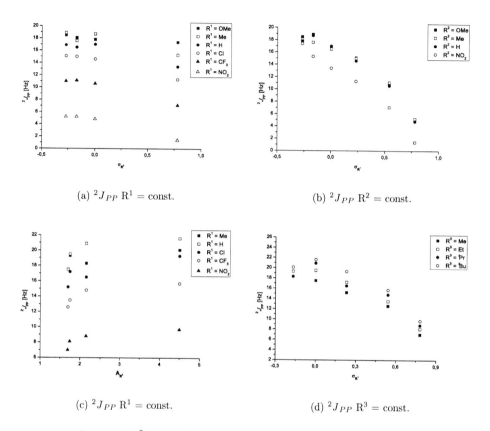

(a) $^2J_{PP}$ R^1 = const.

(b) $^2J_{PP}$ R^2 = const.

(c) $^2J_{PP}$ R^1 = const.

(d) $^2J_{PP}$ R^3 = const.

Graph 4.14: $^2J_{PP}$ for Pd complexes in dependence of σ & A at rt.

Table 4.4.: Selected NMR data for triphenylphosphine Pd complexes:

	$CHCO$	CHC_{Ar}	$CHCO$	CHC_{Ar}	P$_1$	P$_2$	$|^2J_{PP}|$
	[δ]	[δ]	[δ]	[δ]	[δ]	[δ]	[Hz]
PdPhOMe−Me	4.77	5.39	−	−	24.8	25.9	−
PdPhOMe−Et	4.78	5.37	−	−	24.5	25.7	−
PdPhOMe−^{i}Pr	4.79	5.35	61.7	77.1	23.4	24.7	−
PdPhOMe−^{t}Bu	4.75	5.29	63.2	77.7	22.9	24.3	−
PdPhMe−Me	4.75	5.33	−	−	23.8	25.4	−

continued on next page

Table 4.4 – continued from previous page

	CHCO [δ]	CHC$_{Ar}$ [δ]	CHCO [δ]	CHC$_{Ar}$ [δ]	P$_1$ [δ]	P$_2$ [δ]	$|^2 J_{PP}|$ [Hz]
PdPhMe–Et	4.78	5.33	61.5	76.7	24.3	26.1	19.3
PdPhMe–^{i}Pr	4.77	5.30	61.7	77.0	24.0	25.8	18.3
PdPhMe–^{t}Bu	4.73	5.26	63.0	77.7	22.8	24.6	20.1
PdPh–Me	4.72	5.26	61.2	75.8	24.9	26.3	17.5
PdPh–Et	4.75	5.25	61.2	76.2	24.7	26.1	19.5
PdPh–^{i}Pr	4.76	5.25	61.8	76.8	24.4	25.9	20.9
PdPh–^{t}Bu	4.71	5.20	63.0	77.8	24.0	25.5	21.6
PdPhCl–Me	4.55	5.13	61.6	74.0	24.8	26.3	15.2
PdPhCl–Et	4.58	5.13	60.5	74.4	23.9	25.5	17.2
PdPhCl–^{i}Pr	4.58	5.11	61.4	74.8	23.4	25.3	16.5
PdPhCl–^{t}Bu	4.56	5.07	62.4	75.3	23.0	24.9	19.3
PdPhCF$_3$–Me	4.64	5.23	60.3	73.1	25.1	26.7	12.6
PdPhCF$_3$–Et	4.68	5.24	60.6	73.5	24.9	26.6	13.5
PdPhCF$_3$–^{i}Pr	4.70	5.19	60.9	74.0	24.6	26.5	14.8
PdPhCF$_3$–^{t}Bu	4.63	5.17	62.0	74.6	24.1	26.1	15.7
PdPhNO$_2$–Me	4.51	5.19	59.5	72.3	25.0	27.0	7.0
PdPhNO$_2$–Et	4.53	5.18	59.9	72.6	24.9	27.1	8.1
PdPhNO$_2$–^{i}Pr	4.56	5.15	60.1	73.2	24.5	27.0	8.8
PdPhNO$_2$–^{t}Bu	4.53	5.10	61.2	73.8	24.1	26.7	9.7
PdPhOMe–PhOMe	4.90	5.35	60.0	76.7	24.4	25.5	18.5
PdPhOMe–PhMe	4.88	5.34	60.1	76.9	24.4	25.5	18.1
PdPhOMe–Ph	4.88	5.33	60.2	76.9	24.4	25.5	17.8
PdPhOMe–PhNO$_2$	4.80	5.23	58.8	75.8	24.6	25.2	17.4
PdPhMe–PhOMe	4.91	5.32	60.2	76.8	24.3	25.8	18.9
PdPhMe–PhMe	4.89	5.31	60.3	76.8	24.4	26.0	17.6
PdPhMe–Ph	4.89	5.30	60.1	76.8	24.4	26.0	18.7

continued on next page

Table 4.4 – continued from previous page

	$CHCO$	CHC_{Ar}	$CHCO$	CHC_{Ar}	P_1	P_2	$|{}^2J_{PP}|$
	$[\delta]$	$[\delta]$	$[\delta]$	$[\delta]$	$[\delta]$	$[\delta]$	[Hz]
PdPhMe–PhNO$_2$	4.80	5.21	58.8	76.0	24.5	25.5	15.3
PdPh–PhOMe	4.88	5.26	60.2	76.3	24.8	25.9	16.9
PdPh–PhMe	4.86	5.25	60.2	76.4	24.8	25.9	16.5
PdPh–Ph	4.85	5.24	60.0	76.2	24.8	25.9	17.0
PdPh–PhNO$_2$	4.76	5.14	58.7	75.3	25.0	25.6	13.4
PdPhCl–PhOMe	4.74	5.13	59.5	74.4	23.8	25.3	15.1
PdPhCl–PhMe	4.72	5.13	59.7	74.4	23.8	25.3	15.0
PdPhCl–Ph	4.72	5.11	58.2	73.5	23.7	25.2	14.6
PdPhCl–PhNO$_2$	4.64	5.03	59.4	74.4	24.0	25.0	11.3
PdPhCF$_3$–PhOMe	4.81	5.23	59.3	73.6	25.0	26.4	11.0
PdPhCF$_3$–PhMe	4.72	5.12	59.4	73.5	25.0	26.4	11.1
PdPhCF$_3$–Ph	4.80	5.22	59.2	73.6	25.0	26.4	10.6
PdPhCF$_3$–PhNO$_2$	4.71	5.12	58.0	72.6	25.1	26.1	7.1
PdPhNO$_2$–PhOMe	4.69	5.18	58.6	72.7	25.0	26.9	5.2
PdPhNO$_2$–PhMe	4.68	5.18	58.2	72.4	25.0	26.9	5.2
PdPhNO$_2$–Ph	4.67	5.16	58.5	72.7	25.0	26.9	4.8
PdPhNO$_2$–PhNO$_2$	4.59	5.06	57.3	71.7	25.2	26.5	1.4

To determine exact coupling constants and observe all signals the rotation about the olefin palladium bond had to be frozen. Therefore spectra of all dynamic complexes were recorded at 246 K, in case of $R^1 \equiv CF_3$ at 278 K. Due to cooling the dynamic behaviour was inhibited and all signals could be observed. At low temperature ${}^2J_{PP}$ coupling shows the expected monotone slope towards the influence of R^1, R^2 and R^3 as shown in graph 4.15.

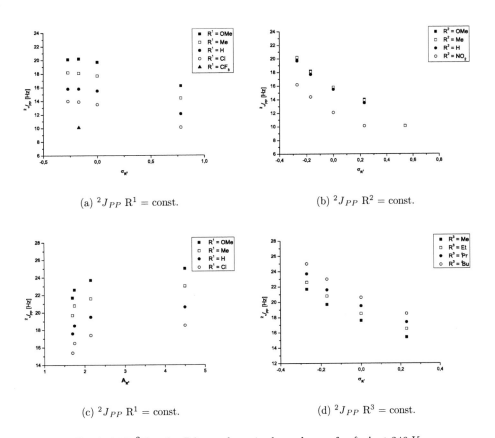

(a) $^2J_{PP}$ R^1 = const.

(b) $^2J_{PP}$ R^2 = const.

(c) $^2J_{PP}$ R^1 = const.

(d) $^2J_{PP}$ R^3 = const.

Graph 4.15: $^2J_{PP}$ for Pd complexes in dependence of σ & A at 246 K.

Table 4.5.: Selected NMR data for triphenylphosphine Pd complexes at 246 K:

	$CHCO$	CHC_{Ar}	$CHCO$	CHC_{Ar}	P$_1$	P$_2$	$\lvert^2J_{PP}\rvert$
	[δ]	[δ]	[δ]	[δ]	[δ]	[δ]	[Hz]
PdPhOMe–Me	4.86	5.40	60.7	76.0	24.1	26.0	21.7
PdPhOMe–Et	4.90	5.40	60.9	76.3	23.9	25.6	22.6
PdPhOMe–^{i}Pr	4.91	5.39	61.1	76.9	23.5	25.3	23.7
PdPhOMe–^{t}Bu	4.83	5.33	62.2	77.3	23.1	24.9	25.0
PdPhMe–Me	4.86	5.39	60.7	76.0	24.1	26.2	19.7

continued on next page

Table 4.5 – continued from previous page

	$CHCO$	CHC_{Ar}	$CHCO$	CHC_{Ar}	P_1	P_2	$|^2J_{PP}|$
	$[\delta]$	$[\delta]$	$[\delta]$	$[\delta]$	$[\delta]$	$[\delta]$	$[Hz]$
PdPhMe−Et	4.90	5.39	60.9	76.3	23.8	25.9	20.8
PdPhMe−^{i}Pr	4.91	5.37	61.1	76.9	23.5	25.6	21.6
PdPhMe−^{t}Bu	4.85	5.32	62.2	77.5	23.0	25.2	23.0
PdPh−Me	4.84	5.33	60.7	75.4	24.4	26.2	17.6
PdPh−Et	4.89	5.32	60.8	75.8	24.1	26.0	18.5
PdPh−^{i}Pr	4.89	5.31	61.0	76.3	23.8	25.7	19.5
PdPh−^{t}Bu	4.83	5.25	62.1	76.8	23.5	25.2	20.6
PdPhCl−Me	4.69	5.19	59.9	73.5	24.2	26.1	15.4
PdPhCl−Et	4.74	5.19	60.1	73.9	23.9	26.0	16.5
PdPhCl−^{i}Pr	4.75	5.17	60.3	74.5	23.5	25.7	17.4
PdPhCl−^{t}Bu	4.69	5.11	61.4	75.2	23.1	25.3	18.5
PdPhOMe−PhOMe	5.09	5.44	59.4	76.5	23.8	25.2	20.1
PdPhOMe−PhMe	5.07	5.43	59.5	76.6	23.8	25.3	20.2
PdPhOMe−Ph	5.03	5.37	59.3	76.7	23.8	25.2	19.7
PdPhOMe−PhNO$_2$	4.94	5.26	58.1	75.6	23.9	25.0	16.2
PdPhMe−PhOMe	5.06	5.38	59.5	76.7	23.7	25.6	18.2
PdPhMe−PhMe	5.04	5.37	59.5	76.6	23.7	25.6	18.1
PdPhMe−Ph	5.03	5.36	59.3	76.6	23.7	25.6	17.7
PdPhMe−PhNO$_2$	4.95	5.26	58.2	75.6	23.9	25.3	14.4
PdPh−PhOMe	5.04	5.32	59.4	76.1	24.1	25.6	15.8
PdPh−PhMe	5.03	5.32	59.5	76.2	24.1	25.6	15.8
PdPh−Ph	5.02	5.30	59.3	76.1	24.1	25.6	15.5
PdPh−PhNO$_2$	4.93	5.19	58.1	75.1	24.2	25.3	12.1
PdPhCl−PhOMe	4.90	5.18	58.7	74.3	23.8	25.7	14.0
PdPhCl−PhMe	4.88	5.17	55.8	74.3	23.8	25.7	13.9
PdPhCl−Ph	4.87	5.16	58.6	74.3	23.8	25.7	13.5

continued on next page

Table 4.5 – continued from previous page

	$CHCO$	CHC_{Ar}	$CHCO$	CHC_{Ar}	P_1	P_2	$|^2J_{PP}|$
	$[\delta]$	$[\delta]$	$[\delta]$	$[\delta]$	$[\delta]$	$[\delta]$	[Hz]
PdPhCl–PhNO$_2$	4.79	5.05	57.5	73.3	24.0	25.4	10.1
PdPhCF$_3$–PhMe[3]	4.88	5.16	–	–	24.3	26.1	10.1

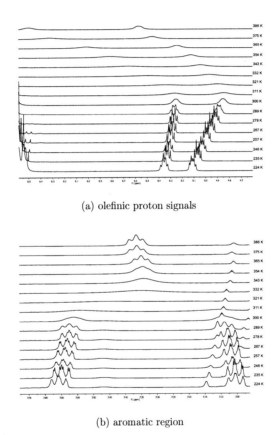

(a) olefinic proton signals

(b) aromatic region

Figure 4.3: Magnifications of ^{1}H NMR spectra of **PdPh–PhMe** in dependence of the temperature.

The olefinic proton signals are only resolved properly below 270 K and show a splitting

---

[3]No ^{13}C NMR was recorded at 246 K.

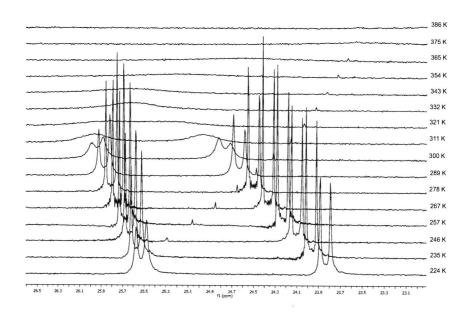

Figure 4.4: ^{31}P NMR spectra of **PdPh−PhMe** in dependence of the temperature.

into a doublet of pseudo triplets with $^3J_{HH}$ coupling constants of ca. 10 Hz which indicates a smaller distortion from 180° than in the platinum complexes but is still significantly smaller than in non coordinated olefins. Upon warming, the signals broaden due to rotation around the π-bond. However no sharp signals are obtained upon heating beyond the coalescence point of ca. 320 K. This is caused by the partial dissociation of the ligand, which leads to an additional equilibrium causing a shift to lower field upon higher temperature and a second line broadening effect. Above ca. 360 K the compounds start to decompose slowly. In contrast the two sets of aromatic protons observed for the triphenylphosphines show coalescence at ca. 320 K and combine to a new signal upon higher temperature, which supports the assumption that only the olefin dissociates. The two signals of the ethyl´s methylene protons as well as the ones for the *iso*-propyl´s methyl groups also show dynamic behaviour and merge to one signal due to rotation of the alkyl group.

Table 4.6.: Selected NMR data for triphenylphosphine Pd complexes at 278 K:

	$CHCO$	CHC_{Ar}	P_1	P_2	$|^2J_{PP}|$
	$[\delta]$	$[\delta]$	$[\delta]$	$[\delta]$	[Hz]
PdPhCF$_3$–Me	4.69	5.26	24.9	26.6	12.5
PdPhCF$_3$–Et	4.74	5.26	24.6	26.5	13.4
PdPhCF$_3$–^{i}Pr	4.75	5.23	24.3	26.3	14.3
PdPhCF$_3$–^{t}Bu	4.70	5.18	23.9	26.0	15.4
PdPhCF$_3$–PhOMe	4.89	5.25	24.7	26.3	10.6
PdPhCF$_3$–PhMe	4.77	5.14	24.7	26.3	10.8
PdPhCF$_3$–Ph	4.86	5.22	24.7	26.3	10.1
PdPhCF$_3$–PhNO$_2$	4.77	5.12	24.8	26.0	6.8

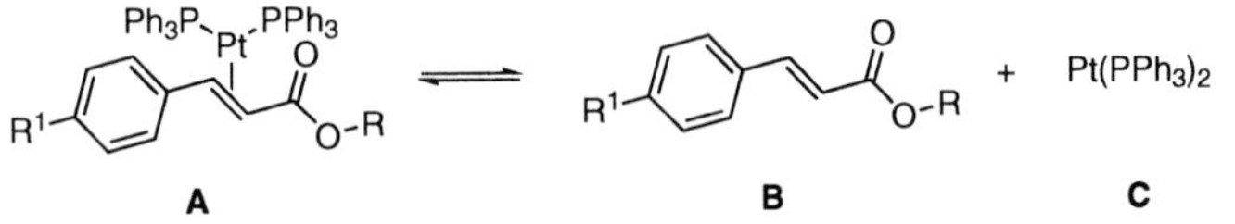

Scheme 4.4: Partial dissociation of the ligand at elevated temperature.

The equilibrium constant for the observed dissociation/coordination equilibrium depicted in scheme 4.4 is given by:

$$K = \frac{[B][C]}{[A]} \tag{4.1}$$

with $n_B = n_C$ and the volume of the solvent follows:

$$K = \frac{[B]^2}{[A]} = \frac{X_B^2(n_A + 2n_B)}{X_A V} \tag{4.2}$$

Since one equivalent of B is formed from one equivalent of A, $n_B = n_{A_0} - n_A$. With the initial amount of complex n_{A_0}. From these assumptions follows:

$$n_A + 2n_B = n_A + 2(n_{A_0} - n_A) = 2n_{A_0} - n_A = 2n_{A_0} - X_A(n_A + 2n_B) = \frac{2n_{A_0}}{1 + X_A} \tag{4.3}$$

The definition of the mole fraction:

$$1 = X_A + X_B + X_C \tag{4.4}$$

implies with $X_B = X_C$:

$$X_B = \frac{1 - X_A}{2} \tag{4.5}$$

When 4.3 and 4.5 are inserted into 4.2 we receive the following expression for K:

$$K = \frac{(1 - X_A)^2 [A_0]}{2X_A(1 + X_A)} \tag{4.6}$$

The average position of the signal[112] is described through:

$$\delta \approx \frac{n_A \delta_A + n_B \delta_B}{n_A + n_B} \tag{4.7}$$

with $n_B = n_{A_0} - n_A$ follows:

$$n_A \approx \frac{n_{A_0}(\delta - \delta_B)}{\delta_A - \delta_B} \tag{4.8}$$

The mole fraction may be written as:

$$X_A = \frac{n_A}{n_A + 2n_B} = \frac{n_A}{2n_{A_0} - n_A} \tag{4.9}$$

The combination of 4.8 and 4.9 and some mathematical transformations give:

$$X_A = \frac{\frac{\delta - \delta_B}{\delta_A - \delta_B}}{2 - \frac{\delta - \delta_B}{\delta_A - \delta_B}} \tag{4.10}$$

with the chemical shift of the olefin in the pure complex δ_A and of the non coordinated ligand δ_B at ambient temperature. Because of the temperature dependence of the shift the chemical shift of the equilibrium signal needs to be corrected with:

$$\delta = \delta_{obs} + \Delta\delta(T) \tag{4.11}$$

where δ_{obs} is the observed signal and

$$\Delta\delta(T) = \delta_{A;rt} - \delta_{A;theo} \qquad (4.12)$$

with $\delta_{A;rt} \equiv$ chemical shift of the olefin in the complex at ambient temperature. The theoretical shift of the olefinic protons in the absence of dissociation may be calculated in dependence of the temperature via:

$$\delta_{A;theo} = aT + b \qquad (4.13)$$

The factors a and b were calculated from the chemical shift of the olefinic protons in the complex in dependence of the temperature via linear regression at low temperatures, to disable dissociation. The quality of the fit R^2 and the factors a and b are summarized in table 4.7.

Table 4.7.: Linear regression coefficients for $\delta_{A;theo}$:

		R^2	a	b [ppm]
PdPhOMe–PhOMe	$CHCO$	1.00000	-0.00298(0)	5.74016(14)
	$CHCAr$	0.83621	-0.00070(17)	5.54303(4597)
PdPhMe–PhOMe	$CHCO$	0.98519	-0.00249(15)	5.61105(4136)
	$CHCAr$	0.69100	-0.00129(40)	5.69105(11147)
PdPh–PhOMe	$CHCO$	0.95871	-0.00349(32)	5.89144(8949)
	$CHCAr$	0.92711	-0.00116(14)	5.59715(4016)
PdPh–PhMe	$CHCO$	0.96192	-0.00150(11)	5.68848(2711)
	$CHCAr$	0.99247	-0.00343(10)	5.87716(2625)
PdPh–Ph	$CHCO$	0.98427	-0.00347(19)	5.86527(5421)
	$CHCAr$	0.91142	-0.00099(14)	5.53333(3805)
PdPh–PhNO$_2$	$CHCO$	0.99412	-0.00361(12)	5.82289(3436)
	$CHCAr$	0.97443	-0.00114(10)	5.47096(2275)
PdPh–Me	$CHCO$	0.92151	-0.00239(34)	5.39405(9371)

continued on next page

Table 4.7 – continued from previous page

		R^2	a	b [ppm]
	$CHCAr$	0.88884	-0.00109(19)	5.57302(5179)
PdPh−Et	$CHCO$	0.95395	-0.00278(30)	5.54205(8267)
	$CHCAr$	0.91844	-0.00129(19)	5.62102(5180)
PdPh−^{i}Pr	$CHCO$	0.91763	-0.00338(50)	5.71404(13619)
	$CHCAr$	0.86514	-0.00129(25)	5.60902(6808)
PdPh−^{t}Bu	$CHCO$	0.95370	-0.00229(25)	5.36904(6810)
	$CHCAr$	0.92708	-0.00109(15)	5.50503(4130)
PdPhCl−PhOMe	$CHCO$	0.98869	-0.00263(11)	5.50285(3237)
	$CHCAr$	0.90237	-0.00060(10)	5.31667(2406)
PdPhCF$_3$−PhOMe	$CHCO$	0.98110	-0.00264(14)	5.51534(3976)
	$CHCAr$	0.94075	-0.00060(10)	5.30857(1737)
PdPhNO$_2$−PhOMe	$CHCO$	0.98530	-0.00227(10)	5.29103(2868)
	$CHCAr$	0.79047	-0.00023(4)	5.16766(1217)

From 4.6 and 4.10 the temperature dependent equilibrium constants could be calculated. With:

$$\Delta G = \Delta H - T\Delta S = -RT lnK \qquad (4.14)$$

the dissociation enthalpy and entropy can be calculated by plotting lnK against T^{-1} which are given in table 4.8 together with the quality of the linearity R^2.

Table 4.8.: Calculated dissociation enthalpies and entropies:

	R^2	$\Delta H [kJ/mol]$	$\Delta S[J/Kmol]$
PdPhOMe−PhOMe	0.95171	56.3(56)	111.5(168)
PdPhMe−PhOMe	0.99105	62.9(27)	125.1(79)
PdPh−PhOMe	0.94407	66.6(72)	131.8(214)
PdPh−PhMe	0.97669	83.0(52)	186.8(149)

continued on next page

Table 4.8 – continued from previous page

	R^2	$\Delta H[kJ/mol]$	$\Delta S[J/Kmol]$
PdPh−Ph	0.99181	71.3(32)	152.6(95)
PdPh−PhNO$_2$	0.94585	66.6(71)	127.3(211)
PdPh−Me	0.99154	77.2(32)	173.6(95)
PdPh−Et	0.98920	85.3(40)	193.7(119)
PdPh−^{i}Pr	0.99844	72.1(13)	162.1(38)
PdPh−^{t}Bu	0.99777	80.6(17)	189.0(51)
PdPhCl−PhOMe	0.77180	92.2(242)	205.8(707)
PdPhCF$_3$−PhOMe	0.93683	101.6(131)	208.3(384)
PdPhNO$_2$−PhOMe	0.95768	183.8(172)	435.9(512)

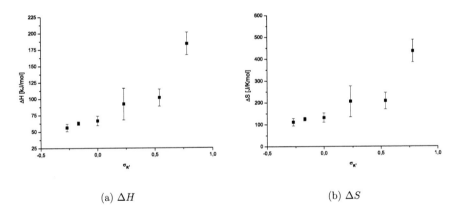

(a) ΔH (b) ΔS

Graph 4.16: Dissociation enthalpies and entropies in dependence of σ_{R^1}. $R^2 \equiv$ OMe

The dissociation enthalpy increases with electron deficiency of the double bond, as expected, since the metal-olefin bond strength is increased (graph 4.16 (a)). Surprisingly the dissociation entropy increases as well as shown in graph 4.16 (b). This is was unexpected because the entropy should be of comparable value for all complexes caused by the similarity of the dissociation reactions.

The two signals in ^{31}P NMR start to broaden at ca. 270 K then transit through

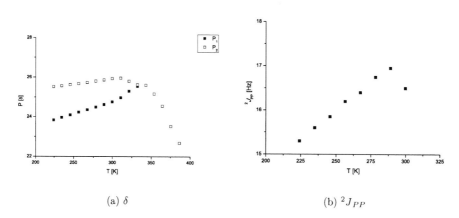

Graph 4.17: ^{31}P NMR shifts and $^{2}J_{PP}$ of **PdPh−PhMe** in dependence of the temperature.

coalescence at ca. 320 K and start to reform a new sharp signal at ca. 330 K. However this signal broadens again at higher temperature and massively shifts to the high field region of the spectrum.

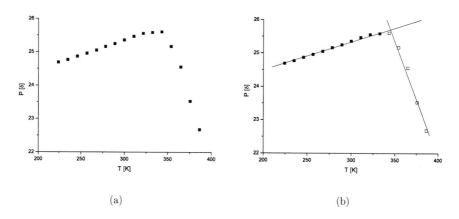

Graph 4.18: Average ^{31}P NMR shifts of **PdPh−PhMe** in dependence of the temperature.

This is due to the formation of an equilibrium with bis(triphenylphosphine)palladium, formed through dissociation of olefin. The aforementioned decomposition of the complexes also leads to the formation of free triphenylphosphine which results in the new equilibrium of $Pd(PPh_3)_2$, triphenylphosphine and tris(triphenylphosphine)palladium, intensifying the upfield shift. These processes are shown in graphs 4.17 (a) and 4.18. Before line broadening occurs, the signals in the ^{31}P NMR spectra are shifted to lower field with increasing temperature and the $^2J_{HH}$ coupling constants increase. From the point of line broadening the signals move towards each other and the coupling constant decreases as expected (graph 4.17).

Via line shape analysis[112] the rotational barriers of the cinnamates about the olefin palladium axis and the rotational barriers of the ethyl and *iso*-propyl groups around the ester bond could be determined and are compiled in table 4.9. The rotational barrier increases with the electron deficience of the olefin. As expected a linear trend is observed for σ_{R^1} and a similar, less pronounced behaviour is also observed in dependence of σ_{R^2}, with the exception of **PdPh−PhMe** which shows extremely large values, as may be seen in graph 4.19. However, unlike with $^2J_{PP}$ coupling, no direct correlation of the electronic properties of R^3 and the double bond rotation is present. This is most likely caused by steric interactions of the alkyl moieties with the phosphines, hindering the rotation. Therefore the methyl substituted systems rotate easier than the other three alkyl esters, which should also account for the low stability of these systems in solution.

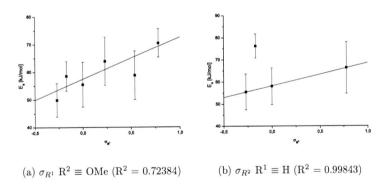

(a) σ_{R^1} $R^2 \equiv OMe$ ($R^2 = 0.72384$)   (b) σ_{R^2} $R^1 \equiv H$ ($R^2 = 0.99843$)

Graph 4.19: Rotational energies in dependence of σ.

While the barrier is similar for ethyl and *tert*-butyl it is highest for *iso*-propyl. This is supposedly caused through the competition of higher steric demand raising the rotation barrier and increasing electron donor properties with the opposite effect. Apparently the latter overrules in *tert*-butyl systems. The rotation barrier for the ester group is significantly lower in the *iso*-propyl system (**PdPh–^{i}Pr**) than in the ethyl ester complex (**PdPh–Et**).

Table 4.9.: Rotational barriers determined by line shape analysis:

	R^2	E_a(olefin)$[kJ/mol]$	$E_a(R^3)[kJ/mol]$
PdPhOMe–PhOMe	0.96727	49.9(61)	–
PdPhMe–PhOMe	0.97552	58.6(53)	–
PdPh–PhOMe	0.95901	55.5(80)	–
PdPh–PhMe	0.97942	76.2(55)	–
PdPh–Ph	0.92309	58.1(83)	–
PdPh–PhNO$_2$	0.91235	66.4(117)	–
PdPh–Me	0.95882	52.5(49)	–
PdPh–Et	0.98551	58.8(36)	48.6(44)
PdPh–^{i}Pr	0.99075	68.6(38)	28.6(50)
PdPh–^{t}Bu	0.98971	62.0(44)	–
PdPhCl–PhOMe	0.92933	63.9(87)	–
PdPhCF$_3$–PhOMe	0.91643	58.8(88)	–
PdPhNO$_2$–PhOMe	0.97872	70.4(52)	–

The ^{1}H, ^{13}C and ^{31}P NMR spectra of the nickel complexes are similar to platinum and literature known nickel complexes[113] concerning shift, coupling and multiplicities. The trends concerning the shifts described for platinum also apply here. The $^2J_{PP}$ coupling constants are a direct indicator for the electron density of the double bond and bond strength, *vide supra*. This clearly shows in graph 4.20.

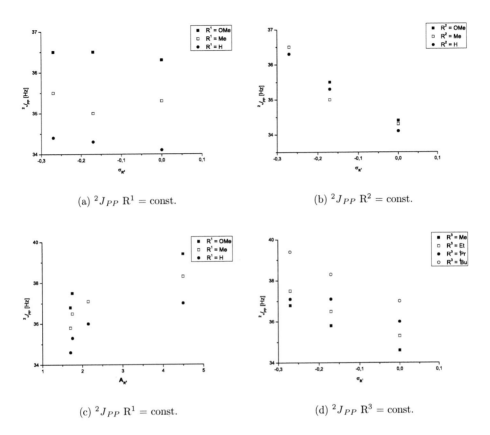

(a) $^2J_{PP}$ R^1 = const.

(b) $^2J_{PP}$ R^2 = const.

(c) $^2J_{PP}$ R^1 = const.

(d) $^2J_{PP}$ R^3 = const.

Graph 4.20: $^2J_{PP}$ for Ni complexes in dependence of σ & A.

Table 4.10.: Selected NMR data for triphenylphosphine Ni complexes:

	$CHCO$	CHC_{Ar}	$CHCO$	CHC_{Ar}	P$_1$	P$_2$	$|^2J_{PP}|$
	[δ]	[δ]	[δ]	[δ]	[δ]	[δ]	[Hz]
NiPhOMe−Me	3.60	4.35	53.6	65.2	31.8	33.7	36.8
NiPhOMe−Et	3.60	4.36	53.8	65.4	31.7	33.6	37.5
NiPhOMe−^{i}Pr	3.55	4.37	54.0	65.0	31.5	33.5	37.1
NiPhOMe−^{t}Bu	3.52	4.35	55.0	66.2	30.8	32.8	39.4
NiPhMe−Me	3.58	4.33	53.4	65.5	31.6	34.2	35.8

continued on next page

Table 4.10 – continued from previous page

	$CHCO$	CHC_{Ar}	$CHCO$	CHC_{Ar}	P_1	P_2	$|^2J_{PP}|$
	$[\delta]$	$[\delta]$	$[\delta]$	$[\delta]$	$[\delta]$	$[\delta]$	$[Hz]$
NiPhMe−Et	3.58	4.35	53.7	65.7	31.5	34.1	36.5
NiPhMe−^{i}Pr	3.54	4.35	54.0	66.1	31.3	34.0	37.1
NiPhMe−^{t}Bu	3.51	4.33	54.9	66.5	30.6	33.3	38.3
NiPh−Me	3.58	4.30	53.4	65.2	32.1	34.2	34.6
NiPh−Et	3.55	4.32	53.6	65.4	31.9	34.1	35.3
NiPh−^{i}Pr	3.53	4.32	53.9	65.8	31.5	33.7	36.0
NiPh−^{t}Bu	3.50	4.30	54.8	66.2	31.1	33.3	37.0
NiPhOMe−PhOMe	3.71	4.36	52.5	65.2	31.7	33.3	36.5
NiPhOMe−PhMe	3.70	4.36	52.5	65.1	32.0	33.6	36.5
NiPhOMe−Ph	3.70	4.35	52.4	65.0	32.1	33.6	36.3
NiPhMe−PhOMe	3.73	4.38	52.4	65.5	30.6	32.9	35.5
NiPhMe−PhMe	3.70	4.36	52.4	65.5	30.6	32.9	35.0
NiPhMe−Ph	3.69	4.34	52.3	65.3	31.6	33.8	35.3
NiPh−PhOMe	3.72	4.34	52.3	65.3	31.5	33.3	34.4
NiPh−PhMe	3.71	4.34	52.4	65.2	32.0	33.8	34.3
NiPh−Ph	3.71	4.32	52.2	65.1	31.2	32.9	34.1

Since only ligands could be compared using the *Hammett* parameters when one of the substituents was kept constant another parameter had to be used to assign the overall electronic properties of the olefin. Since the $^2J_{PP}$ coupling constants proved sensitive towards the metal-olefin bond strength and therefore olefin electronics as well, a new parameter may be calculated from it, the olefin electron density parameter (OEDP) ω. This parameter is a sum-characteristic of the electron density on the metal and since the other substituents remain unchanged also for the olefinic double bond. Therefore ω is defined 0 for **Ph−Ph** due to $\sigma_{R^1} \equiv 0$ and $\sigma_{R^2} \equiv 0$ and $\sigma_{R^1} \equiv \omega$ for **PhOMe−Ph**, **PhMe−Ph** and **PhNO$_2$−Ph**. With these points a linear regression is possible linking $^2J_{PP}$ couping constants with the OEDP for all three metals:

$$\omega_{Pt} = \frac{^2J_{PP} - 41.62Hz \pm 0.25Hz}{-10.74Hz \pm 0.59Hz} \tag{4.15}$$

$$\omega_{Pd} = \frac{^2J_{PP} - 16.23Hz \pm 0.70Hz}{-12.49Hz \pm 1.68Hz} \tag{4.16}$$

$$\omega_{Ni} = \frac{^2J_{PP} - 34.06Hz \pm 0.14Hz}{-8.03Hz \pm 0.77Hz} \tag{4.17}$$

with regression coefficients of $R^2_{Pt} = 0.98486$, $R^2_{Pd} = 0.91555$ and $R^2_{Ni} = 0.98157$. With these three equations ω of the other ligands can be calculated from the $^2J_{PP}$ coupling constants for each metal separately. The received OEDPs are corrected afterwards by subtraction of the calculated ω_{Ph-Ph}. Finally the mean value is calculated for each ligand from the values received of the three metals respectively:

$$\omega_{PhR^1-R} = \frac{\omega_{PhR^1-R;Pt} + \omega_{PhR^1-R;Pd} + \omega_{PhR^1-R;Ni}}{3} \tag{4.18}$$

The olefin electron density parameters ω calculated via this route are summarised in table 4.11.

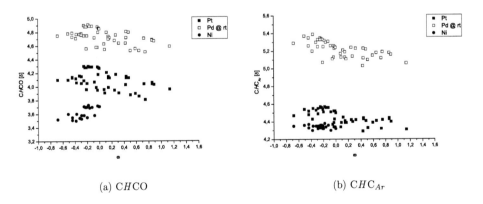

(a) $CHCO$ (b) CHC_{Ar}

Graph 4.21: ^{1}H NMR shifts in dependence of ω.

No correlation could be observed for the dependence of the shifts in ^{1}H and ^{13} C NMR for the olefin upon the *Hammett* parameters of R^1 and R^2 or the steric influence of R^3

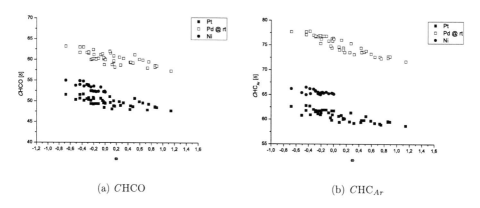

(a) $CHCO$ (b) CHC_{Ar}

Graph 4.22: ^{13}C NMR shifts in dependence of ω.

due to additional complex influence of back bonding and the olefins. However when these shifts are plotted against ω a clear trend is observed, showing that with increasing OEDP, therefore decreasing electron density of the olefin, the signals shift to higher field. This is caused by the stronger back bonding into the π^*-orbital of the double bond increasing the sp^3-character of the olefinic carbon atoms (graphs 4.21 & 4.22).

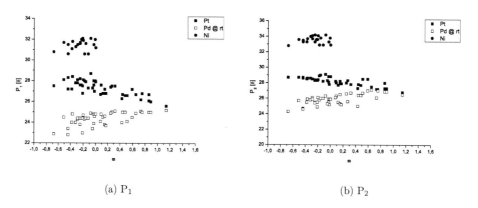

(a) P$_1$ (b) P$_2$

Graph 4.23: ^{31}P NMR shifts in dependence of ω.

The signals in ^{31}P NMR move to lower field when the electron density is decreased at the double bond for the platinum complexes. This was expected since the increasing π back-

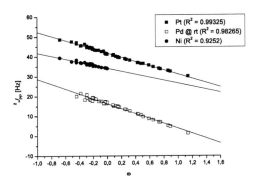

Graph 4.24: $^2J_{PP}$ in dependence of ω.

bonding needs to be compensated by stronger σ bonding from the phosphorus to the metal and therefore decreasing electron density there. However the opposite trend is observed for palladium, indicating a decrease in phosphine bond strength upon stronger coordination of the olefin. This might be caused by reduced back bonding to the phosphorus. No trend may be assigned to nickel due to a lack of data points. These findings are shown in graph 4.23. The linearity of $^2J_{PP}$ coupling constants and ω is clear due to the calculation of the OEDP from the coupling. However the accuracy of the mean values shows through the good linearity (graph 4.24).

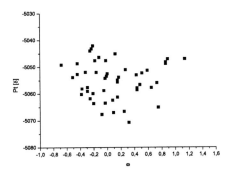

Graph 4.25: ^{195}Pt NMR shifts in dependence of ω.

Even though some trends could be observed concerning the ^{195}Pt NMR shifts and the

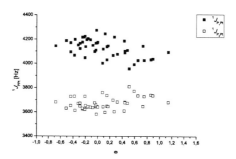

Graph 4.26: $^1J_{PPt}$ in dependence of ω.

Hammett parameters of the substituents no correlation is observed for ω (graph 4.25). When the $^1J_{PPt}$ coupling constants are observed, the two sets obtained for both coupling constants similarise with increasing ω as shown in graph 4.26, which is explained by the increasing similarity of the two olefinic carbon atoms leading to closer *trans*-effects upon the two phosphines.

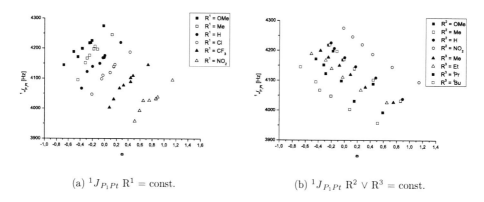

(a) $^1J_{P_1Pt}$ R^1 = const.

(b) $^1J_{P_1Pt}$ $R^2 \vee R^3$ = const.

Graph 4.27: $^1J_{P_1Pt}$ in dependence of ω.

Upon closer examination two linear correlations are observable (graph 4.27) if either R^1 (a) or R^2/R^3 (b) are kept constant. This indicates, that in general the coupling constants are dependent on the olefinic electron density and ω may be applied on a broader range. However *trans*-influence dominates for $^1J_{PPt}$ coupling, which is not described correctly with the OEDP, because ω is only a sum-characteristic of the two carbon atoms and does not account for the individual properties of the carbon atoms. Therefore the linearisation with ω fails, without keeping one substituent unchanged.

Table 4.11.: Olefin electron density parameters ω of the ligands:

PhOMe–PhOME	-0.25(2)	PhOMe–^{t}Bu	-0.68(3)
PhOMe–PhMe	-0.23(2)	PhOMe–^{i}Pr	-0.44(1)
PhOMe–Ph	-0.21(2)	PhOMe–Et	-0.51(2)
PhOMe–PhNO$_2$	-0.02(4)	PhOMe–Me	-0.37(1)
PhMe–PhOME	-0.18(2)	PhMe–^{t}Bu	-0.44(2)
PhMe–PhMe	-0.12(2)	PhMe–^{i}Pr	-0.32(2)
PhMe–Ph	-0.16(2)	PhMe–Et	-0.29(1)
PhMe–PhNO$_2$	0.12(5)	PhMe–Me	-0.25(1)
Ph–PhOMe	-0.03(3)	Ph–^{t}Bu	-0.38(1)
Ph–PhMe	-0.01(3)	Ph–^{i}Pr	-0.29(1)
Ph–Ph	0.00(3)	Ph–Et	-0.20(1)
Ph–PhNO$_2$	0.27(7)	Ph–Me	-0.08(3)
PhCl–PhOMe	0.15(6)	PhCl–^{t}Bu	-0.21(2)
PhCl–PhMe	0.15(6)	PhCl–^{i}Pr	-0.05(4)
PhCl–Ph	0.17(6)	PhCl–Et	-0.03(4)
PhCl–PhNO$_2$	0.43(9)	PhCl–Me	0.09(5)
PhCF$_3$–PhOMe	0.44(9)	PhCF$_3$–^{t}Bu	0.08(5)
PhCF$_3$–PhMe	0.44(9)	PhCF$_3$–^{i}Pr	0.15(6)
PhCF$_3$–Ph	0.48(9)	PhCF$_3$–Et	0.25(7)
PhCF$_3$–PhNO$_2$	0.73(12)	PhCF$_3$–Me	0.32(8)
PhNO$_2$–PhOMe	0.86(14)	PhNO$_2$–^{t}Bu	0.51(10)
PhNO$_2$–PhMe	0.86(14)	PhNO$_2$–^{i}Pr	0.59(11)
PhNO$_2$–Ph	0.89(14)	PhNO$_2$–Et	0.65(11)
PhNO$_2$–PhNO$_2$	1.14(17)	PhNO$_2$–Me	0.75(11)

4.2.2. IR

The C=O stretching frequency of carbonyl complexes is strongly dependent upon π back-bonding from the metal and therefore electron density at the carbonyl. Therefore a similar, though weaker trend, was expected for the carboxylic C=O. However no correlations between ω, and the FT-IR data was observed as shown in graph 4.28 and no assignment of the electronic properties is possible with FT-IR spectroscopy.

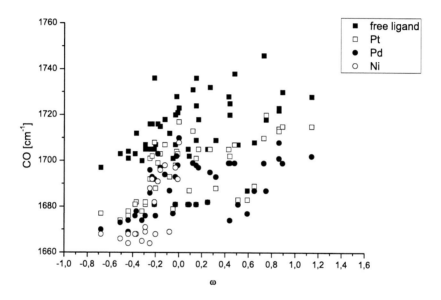

Graph 4.28: C=O stretching frequencies in dependence of ω.

Despite the lack of correlation to the olefin's electron density, the C=O stretching frequency shifts to lower wave numbers upon coordination to the metals. This shift is strongest in the nickel compounds, followed by palladium and platinum shows the smallest changes. This is an indication for interaction between the metal and the carboxyl, which is strongest for the oxophilic nickel. The stronger interaction of palladium in comparison to platinum might be caused by the high dynamics of the complex system.

Table 4.12.: Ester C=O stretching frequency of the free ligand and in the complexes [cm^{-1}]:

	free ligand	Pt	Pd	Ni
PhOMe–Me	1712	1682	1678	1668
PhOMe–Et	1703	1674	1673	1666
PhOMe–^{i}Pr	1701	1676	1669	1664
PhOMe–^{t}Bu	1697	1677	1670	1668
PhMe–Me	1705	1692	1686	1664
PhMe–Et	1707	1681	1676	1669
PhMe–^{i}Pr	1700	1678	1674	1665
PhMe–^{t}Bu	1704	1678	1674	1668
Ph–Me	1712	1693	1687	1669
Ph–Et	1707	1688	1676	1668
Ph–^{i}Pr	1705	1682	1671	1671
Ph–^{t}Bu	1703	1681	1676	1668
PhCl–Me	1702	1687	1681	–
PhCl–Et	1707	1681	1681	–
PhCl–^{i}Pr	1701	1679	1677	–
PhCl–^{t}Bu	1705	1681	1681	–
PhCF$_3$–Me	1709	1688	1693	–
PhCF$_3$–Et	1705	1682	1682	–
PhCF$_3$–^{i}Pr	1709	1681	1681	–
PhCF$_3$–^{t}Bu	1705	1681	1681	–
PhNO$_2$–Me	1718	1720	1687	–
PhNO$_2$–Et	1708	1689	1687	–
PhNO$_2$–^{i}Pr	1687	1683	1677	–
PhNO$_2$–^{t}Bu	1707	1683	1681	–
PhOMe–PhOMe	1716	1701	1696	1688
PhOMe–PhMe	1716	1702	1693	1692

continued on next page

Table 4.12 – continued from previous page

	free ligand	Pt	Pd	Ni
PhOMe–Ph	1736	1708	1692	1682
PhOMe–PhNO$_2$	1728	1704	1702	–
PhMe–PhOMe	1716	1699	1703	1691
PhMe–PhMe	1718	1707	1694	1698
PhMe–Ph	1715	1703	1697	1696
PhMe–PhNO$_2$	1731	1713	1699	–
Ph–PhOMe	1720	1698	1693	1697
Ph–PhMe	1721	1703	1698	1692
Ph–Ph	1723	1717	1710	1708
Ph–PhNO$_2$	1732	1705	1695	–
PhCl–PhOMe	1724	1701	1698	–
PhCl–PhMe	1736	1701	1698	–
PhCl–Ph	1718	1705	1697	–
PhCl–PhNO$_2$	1728	1705	1699	–
PhCF$_3$–PhOMe	1725	1702	1699	–
PhCF$_3$–PhMe	1720	1702	1674	–
PhCF$_3$–Ph	1738	1707	1699	–
PhCF$_3$–PhNO$_2$	1746	1710	1699	–
PhNO$_2$–PhOMe	1722	1714	1708	–
PhNO$_2$–PhMe	1723	1713	1701	–
PhNO$_2$–Ph	1730	1715	1699	–
PhNO$_2$–PhNO$_2$	1728	1715	1702	–

4.2.3. X-Ray

Single crystals of **PtPhOMe–PhOMe, PtPhMe–Ph, PtPh–PhMe, PtPh–Ph,**
PtPhNO$_2$–PhOMe, PtPhNO$_2$–PhMe, PtPhNO$_2$–Ph, PtPhNO$_2$–PhNO$_2$,
PtPhNO$_2$–^{i}Pr, PdPhCl–PhNO$_2$, PdPhNO$_2$–PhOMe, PdPhNO$_2$–Ph,

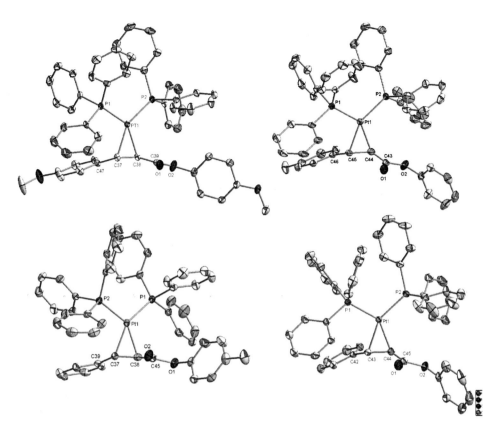

Figure 4.5: ORTEP[24] representation of **PtPhOMe–PhOMe** (top left), **PtPhMe–Ph** (top right), **PtPh–PhMe** (bottom left) and **PtPh–Ph** (bottom right). Thermal ellipsoids are given at 50% probability level. Hydrogen atoms are omitted for clarity.

PdPhNO$_2$–PhNO$_2$ and **NiPh–Me** suitable for X-ray diffraction analysis could be grown by slow diffusion of pentane into a toluene solution of the appropriate complex. Due to incoherencies in the atom labeling the nomenclature defined in figure 4.6 is used to refer to the relevant atoms in all structures.

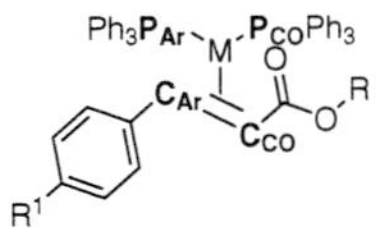

Figure 4.6: Atom nomenclature assignment for X-ray structures of M(PPh$_3$)$_2$olefin.

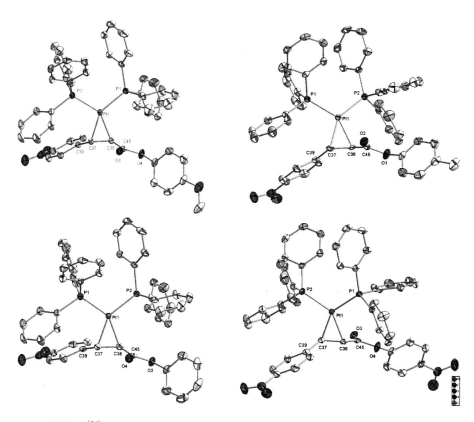

Figure 4.7: ORTEP [24] representation of **PtPhNO₂−PhOMe** (top left), **PtPhNO₂−PhMe** (top right), **PtPhNO₂−Ph** (bottom left) and **PtPhNO₂−PhNO₂** (bottom right). Thermal ellipsoids are given at 50% probability level. Hydrogen atoms are omitted for clarity.

All complexes show trigonal planar geometry, with the olefin slightly twisted out of the plane. Except of the olefin carbon atoms (C_{Ar} & C_{CO}) no interaction is observable between metal and cinnamate. The distances of phosphines and cinnamate give also no indication of contact. The length of the olefin bond in the platinum complexes lies within 1.43 - 1.46 Å and is therefore consistent with the olefin in $Pt(PPh_3)_2C_2H_4$.[114] The C_{Ar}-C_{CO} distances in the palladium compounds is with 1.41 - 1.43 Å slightly shorter as in **NiPh−Me** with 1.42 Å. This indicates a slightly stronger back bonding of platinum into the π^*-orbital of the double bond, decreasing the bond order of the olefin, than palladium

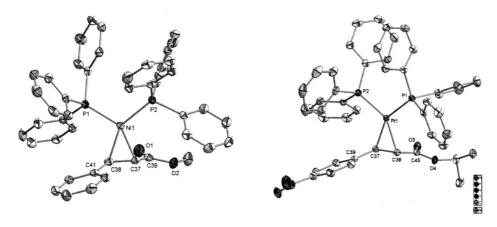

Figure 4.8: ORTEP[24] representation of **NiPh−Me** (left) and **PtPhNO$_2$−^{i}Pr** (right). Thermal ellipsoids are given at 50% probability level. Hydrogen atoms are omitted for clarity.

or nickel. Therefore the metal olefin bond is stronger in the palladium compounds which is consistent with the higher complex stability. All C_{Ar}-C_{CO} distances are between the typical bond lengths of C−C single and C=C double bonds, which is expected due to coordination and the resulting reduction of bond order. The C_{Ar}-Pt bond is, with exception of **PtPhMe−Ph**, **PhNO$_2$−PhMe** and **PtPhNO$_2$−^{i}Pr**, slightly shorter than the C_{CO}-Pt bond due to the lower electron density at C_{Ar}. Both bond lengths are in the range of 2.11 - 2.14 Å and comparable to other platinum complexes like 2.12 and 2.11 Å in Pt(PPh$_3$)$_2$C$_2$H$_4$ and 2.10 and 2.12 Å in Pt(PPh$_3$)$_2$C$_2$(CN)$_4$. The same trend is observed for all palladium complexes with comparable bond lengths of 2.13 - 2.14 Å (C_{Ar}-Pd) and 2.14 - 2.17 Å (C_{CO}-Pd). However in **NiPh−Me** both bond lengths are of similar length with 1.99 Å. The shorter distance in the nickel compound is caused by the significantly smaller *van der Waals* radius of nickel compared to palladium and platinum.[55] The P_{CO}-Pt distance is in all complexes longer than the P_{Ar}-Pt distance, caused by the *trans*-influence of C_{Ar} (lower electron density compared to C_{CO}). The same is also observed for palladium and nickel. The P-Pt distances are with 2.28 - 2.30 Å for P_{CO} and 2.27 - 2.28 Å for P_{Ar} slightly longer than 2.27 Å in Pt(PPh$_3$)$_2$C$_2$H$_4$ and sightly shorter than 2.29 Å in Pt(PPh$_3$)$_2$C$_2$(CN)$_4$ and 2.28 and 2.29 Å in Pt(PPh$_3$)$_2$C$_2$Cl$_4$. The P_{CO}-Pd

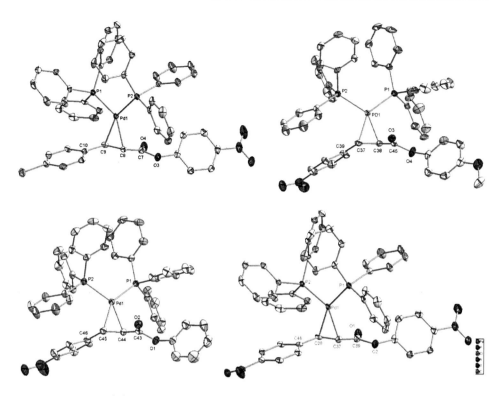

Figure 4.9: ORTEP[24] representation of **PdPhCl–PhNO$_2$** (top left), **PdPhNO$_2$–PhOMe** (top right), **PdPhNO$_2$–Ph** (bottom left) and **PdPhNO$_2$–PhNO$_2$** (bottom right). Thermal ellipsoids are given at 50% probability level. Hydrogen atoms are omitted for clarity.

and P$_{Ar}$-Pd distances are with 2.32 - 2.34 Å and 2.31 - 2.33 Å significantly longer, a sign for the over all reduced stability of these complexes due to weaker bonds. In contrast to this the P$_{CO}$-Ni (2.17 Å) and P$_{Ar}$-Ni (2.18 Å) bonds are approximately 0.1 Å shorter than in the Pt-systems due to the smaller size of nickel. All other bond lengths are consistent with literature and show no additional relevant parameters.

The length of the olefinic double bond increases with decreasing electron density of the double bond due to enhanced π-back bonding resulting in a reduction of bond order, as depicted in graph 4.29. For the palladium and nickel systems not enough single crystal structures were obtained to analyse tendencies in the bond length behaviour.

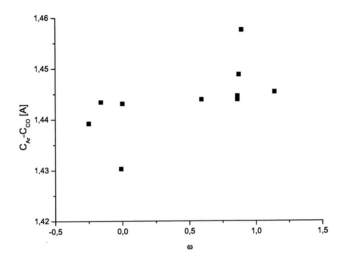

Graph 4.29: C_{Ar}-C_{CO} distance of platinum complexes in dependence of ω.

Table 4.13.: Selected bond lenghts [Å]:

	C_{Ar}-C_{CO}	C_{Ar}-M	C_{CO}-M	P_{CO}-M	P_{Ar}-M
PtPhOMe–PhOMe	1.439(0)	2.132(0)	2.140(0)	2.291(0)	2.279(0)
PtPhMe–Ph	1.443(4)	2.121(3)	2.133(3)	2.292(1)	2.278(1)
PtPh–PhMe	1.430(0)	2.128(0)	2.127(0)	2.283(0)	2.278(0)
PtPh–Ph	1.443(3)	2.108(2)	2.136(2)	2.284(1)	2.272(1)
PtPhNO$_2$–PhOMe	1.445(0)	2.110(0)	2.117(0)	2.293(0)	2.274(0)
PtPhNO$_2$–PhMe	1.444(0)	2.122(0)	2.115(0)	2.290(0)	2.273(0)
PtPhNO$_2$–Ph	1.458(0)	2.116(0)	2.133(0)	2.292(0)	2.273(0)
PtPhNO$_2$–PhNO$_2$	1.445(0)	2.106(0)	2.116(0)	2.292(0)	2.274(0)
PtPhNO$_2$–^{i}Pr	1.444(3)	2.123(2)	2.122(2)	2.288(1)	2.284(1)
PtPhNO$_2$–bp	1.449(0)	2.119(0)	2.138(0)	2.296(0)	2.284(0)
PdPhCl–PhNO$_2$	1.427(4)	2.137(2)	2.166(2)	2.345(1)	2.325(1)
PdPhNO$_2$–PhOMe	1.412(5)	2.132(3)	2.140(3)	2.333(1)	2.322(1)

continued on next page

Table 4.13 – continued from previous page

	C_{Ar}-C_{CO}	C_{Ar}-M	C_{CO}-M	P_{CO}-M	P_{Ar}-M
PdPhNO$_2$–Ph	1.422(2)	2.136(2)	2.151(2)	2.331(0)	2.318(0)
PdPhNO$_2$–PhNO$_2$	1.411(5)	2.125(3)	2.157(3)	2.342(1)	2.329(1)
PdPhNO$_2$–^{i}PrOMe	1.417(5)	2.131(4)	2.138(4)	2.319(1)	2.310(1)
NiPh–Me	1.417(2)	1.994(2)	1.994(2)	2.171(1)	2.181(1)

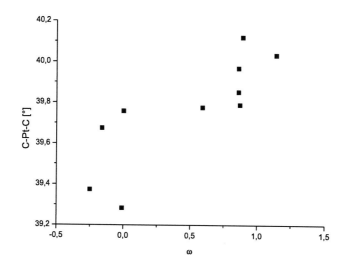

Graph 4.30: C-Pt-C angle of platinum complexes in dependence of ω.

The C_{Ar}-Pt-C_{CO} angle is close to 40° for all compounds, whereas the C_{Ar}-Pd-C_{CO} angle is with 38.5° - 38.8° smaller due to a larger metal olefin distance. With 41.6° the C_{Ar}-Ni-C_{CO} is significantly wider, since the aforementioned distance is smaller. The P_{CO}-Pt-P_{Ar} angle is in the range of 107.7° to 110.8°, with exception of **PtPhMe–Ph** with 115.0°. The deviation from the ideal 120° angle is caused by the higher steric demand of the olefin. The same trend is also observable for palladium with P_{CO}-Pd-P_{Ar} between 106.8° and 112.3° and nickel (P_{CO}-Ni-P_{Ar} = 101.0°). In all structures P_{Ar}-M-C_{Ar} angles are smaller than the C_{CO}-M-P_{CO} angle due to the higher steric demand of the carboxyl group.

With increasing electron deficiency of the double bond the C_{Ar}-Pt-C_{CO} angle widens. This indicates a decrease in the olefin platinum distance and therefore a rise in bond strength, as expected (graph 4.30). Again not enough data points were available for the palladium and nickel complexes to give any trends.

Table 4.14.: Selected bond angles [°]:

	C-M-C	C_{CO}-M-P_{CO}	P-M-P	P_{Ar}-M-C_{Ar}
PtPhOMe–PhOMe	39.4(0)	109.5(0)	110.5(0)	100.7(0)
PtPhMe–Ph	39.7(1)	105.6(1)	115.0(0)	100.2(1)
PtPh–PhMe	39.3(0)	112.5(0)	107.7(0)	101.7(0)
PtPh–Ph	39.8(1)	109.0(1)	109.0(0)	102.6(1)
PtPhNO$_2$–PhOMe	40.0(0)	106.0(0)	110.5(0)	103.5(0)
PtPhNO$_2$–PhMe	39.9(0)	105.6(0)	110.4(0)	104.2(0)
PtPhNO$_2$–Ph	40.1(0)	106.3(0)	109.3(0)	104.3(0)
PtPhNO$_2$–PhNO$_2$	40.0(0)	106.1(0)	110.8(0)	103.1(0)
PtPhNO$_2$–^{i}Pr	39.8(1)	109.7(1)	107.7(0)	102.8(1)
PtPhNO$_2$–bp	39.8(0)	109.0(0)	108.4(0)	102.8(0)
PdPhCl–PhNO$_2$	38.7(1)	114.8(1)	107.5(0)	99.3(1)
PdPhNO$_2$–PhOMe	38.6(1)	105.9(1)	112.3(0)	103.1(1)
PdPhNO$_2$–Ph	38.7(1)	106.2(0)	111.3(0)	103.9(0)
PdPhNO$_2$–PhNO$_2$	38.5(1)	114.5(1)	106.8(0)	100.7(1)
PdPhNO$_2$–^{i}PrOMe	38.8(1)	104.9(1)	113.0(0)	103.7(1)
NiPh–Me	41.6(1)	106.0(1)	111.9(0)	101.0(1)

The C-C_{Ar}-C_{CO}-C torsion angles of the olefins are with 145.9° to 155.7° significantly smaller than the ideal 180° for non coordinated E-olefins due to the hybridization change of the olefin carbon atoms caused by the π back-bonding of platinum. This is also in consistence with the torsion angles calculated from the $^3J_{HH}$ coupling constants of the olefinic protons in ^{1}H NMR spectra, *vide supra*. The same deviation of 180° is observed for palladium and nickel, however not as pronounced. This is caused in the Pd-systems by the weaker metal olefin bond, resulting in a larger bond distance. Therefore the steric pressure upon the olefin substituents is smaller. The higher sp^2-character of the olefinic carbon atoms, caused by lower back bonding from palladium, enhances this as well. The wider torsion angle in **NiPh−Me** compared to palladium complexes, despite the smaller Ni olefin distance, is caused by the smaller size of nickel, which results in lower steric pressure as well.

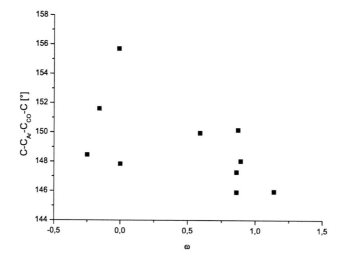

Graph 4.31: C-C$_{Ar}$-C$_{CO}$-C torsion angle of platinum complexes in dependence of ω.

The torsion angle of the olefins in the platinum compounds decrease with increasing ω due to the decreasing metal olefin distance and the resulting stronger sterical interaction, depicted in graph 4.31. As mentioned above not enough crystal structures were available for nickel and palladium to give any correlations.

Table 4.15.: Selected torsion angles [°]:

	C-C_{Ar}-C_{CO}-C
PtPhOMe–PhOMe	148.5(0)
PtPhMe–Ph	151.6(3)
PtPh–PhMe	155.7(0)
PtPh–Ph	147.9(2)
PtPhNO$_2$–PhOMe	145.9(0)
PtPhNO$_2$–PhMe	147.3(0)
PtPhNO$_2$–Ph	148.0(0)
PtPhNO$_2$–PhNO$_2$	146.0(0)
PtPhNO$_2$–^{i}Pr	150.0(2)
PtPhNO$_2$–bp	150.2(0)
PdPhCl–PhNO$_2$	154.7(2)
PdPhNO$_2$–PhOMe	149.6(3)
PdPhNO$_2$–Ph	154.9(1)
PdPhNO$_2$–PhNO$_2$	156.8(3)
PdPhNO$_2$–^{i}PrOMe	156.5(4)
NiPh–Me	159.2(2)

4.3. Reactivity

To determine the reactivity of the complexes, simple ligand substitution and oxidative addition reactions were performed.

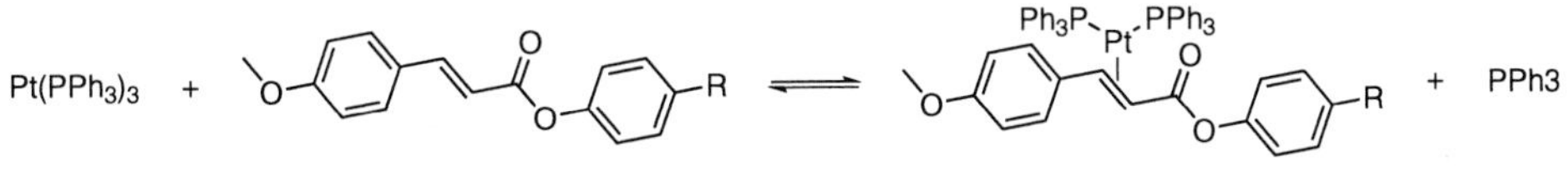

Scheme 4.5: Equilibrium between coordination of triphenylphosphine and cinnamate.

The synthesis of **PtPhMe–PhR** from Pt(PPh$_3$)$_3$ required 10 eq. of ligand and

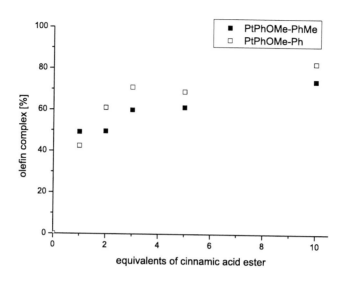

Graph 4.32: NMR titration of $Pt(PPh_3)_3$ with cinnamate ligands.

this route failed for **PhOMe–PhR** and all alkyl complexes with the exception of 4-nitro cinnamates, *vide supra*. Therefore a NMR titration of a $Pt(PPh_3)_3$ solution in benzene-d_6 was performed with **PhOMe–PhOMe, PhOMe–PhMe, PhOMe–Ph** and **PhOMe–PhNO$_2$** (scheme 4.5). However due to overlap of the methoxyl signals in **PtPhOMe–PhOMe** with the free ligand this titration could not be analysed quantitatively. The same accounts for **PtPhOMe–PhNO$_2$**, due to the low solubility of the ligand, resulting in coordination ratios of over 100%. Therefore only the systems **PtPhOMe–PhMe, PtPhOMe–Ph** could be used for quantitative NMR titration. The results are shown in graph 4.32. As expected the cinnamate complex concentration increases with the ligand. However even with 10 eq. only conversions of ca. 80% could be obtained. Higher concentrations of free ligand could not be investigated due to saturation of the solution with cinnamate. No significant differences are observable for the two ligands as expected from the similar OEDPs $\omega_{PhOMe-PhMe} = $ -0.23 and $\omega_{PhOMe-Ph} = $ -0.21. The failure of the route for the alkyl compounds may be explained by the even higher olefin electron density parameters (table 4.11).

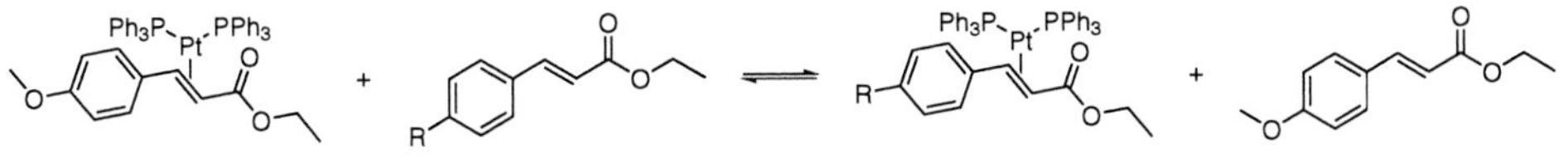

Scheme 4.6: Cinnamate exchange equilibrium.

To examine the difference in ligand properties in more detail a substitution experiment of **PtPhOMe−Et** with **PhR−Et** was performed (scheme 4.6). However even with **PhMe−Et** the ligand substitution was to fast for NMR spectroscopy. For this reason only the complex bearing the ligand with larger ω, therefore lower electron density, could be observed. Also no equilibria between the two different ligands present could be observed, caused by the high impact of the electron density upon complex stability.

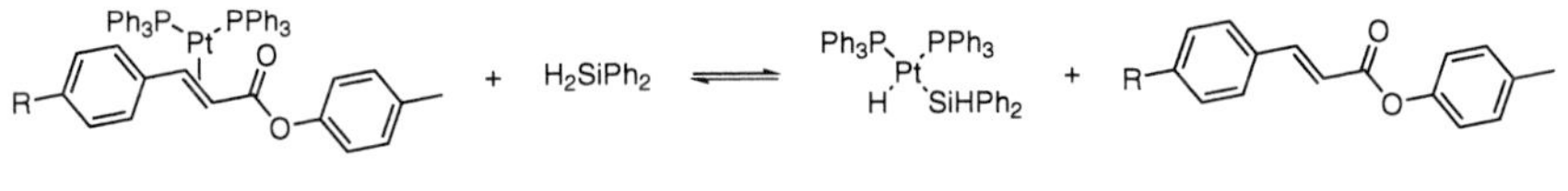

Scheme 4.7: Equilibrium between oxidative addition of Ph_2SiH_2 and cinnamate coordination.

Since the oxidative addition of silane is one of the key steps in hydrosilylation, the dependence of the insertion of platinum into the Si−H bond onto the properties of the cinnamate ligand was investigated.[25–28] Upon the addition of 1.1 eq. Ph_2SiH_2 to the olefin complexes the olefin dissociates and an equilibrium between $(PPh_3)_2PtH(SiHPh_2)$ and the educt forms (scheme 4.7). Surprisingly no hydrosilylation of the cinnamate was observed, not even with the most electron rich ligands. However hydrogen evolution and some evidence for homo-couping of the silane was observed in ^{29}Si NMR spectroscopy when an excess of silane was used or the complexes were added to the neat silane.

The concentration of coordinated cinnamate is directly correlated with the olefin´s electron density, as shown in graph 4.33. From this plot a function may be derived to calculate the percentage of remaining cinnamate complex, utilizing linear regression:

$$\frac{[\text{PtPhR}-\text{R}']}{[\text{PtPhR}-\text{R}'] + [(PPh_3)_2PtH(SiHPh_2)]} = 0.75\omega + 0.25 \tag{4.19}$$

This implies, that with $\omega > 1.0$ no oxidative addition should occur with 1.1 eq. H_2SiPh_2.

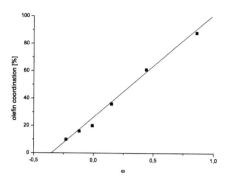

Graph 4.33: Percentage of coordinated olefin upon addition of Ph_2SiH_2 in dependence of the
olefin´s electron density. ($R^2 = 0.99075$)

From the signals in the ^1H NMR spectrum the equilibrium constants could be determined
via:

$$K = \frac{[(PPh_3)_2PtH(SiHPh_2)][PhR-R']}{[PtPhR-R'][H_2SiPh_2]} \quad (4.20)$$

The equilibrium constant K decays exponentially with increasing electron density as expected and may be linearised as shown in graph 4.34.

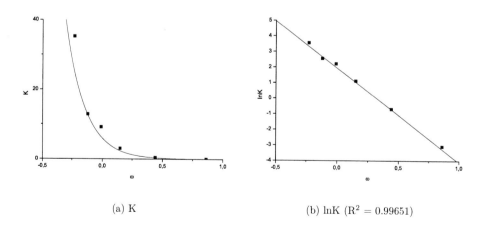

(a) K   (b) lnK ($R^2 = 0.99651$)

Graph 4.34: Equilibrium constant K in dependence of the electronic properties of the ligand.

The transition point of the equilibrium ($K = 1$, $lnK = 0$) could be determined at $\omega_0 = 0.34 \pm 0.02$.

Table 4.16.: Equilibrium constants for the oxidative addition of H_2SiPh_2:

	K	lnK
PtPhOMe–PhMe	32.25	3.56
PtPhMe–PhMe	12.89	2.56
PtPh–PhMe	9.19	2.22
PtPhCl–PhMe	3.05	1.12
PtPhCF$_3$–PhOMe	0.50	-0.69
PtPhNO$_2$–PhMe	0.05	-3.08

The palladium complexes are more reactive and insert into aryl bromine and iodine bonds readily.[115–117] The reaction rate roughly correlates with the olefins electron density. Therefore all complexes immediately insert into the C–I bond at ambient temperature under the formation of $(PPh_3)_2PdI(Ph)$. However only **PdPhOMe–R**, **PdPhMe–R** and **PdPh–R** are capable of C–Br bond activation, whereof only **PdPhOMe–R** is totally converted within 48 hours. The other two compounds require heating to 75 °C to achieve sufficient reaction rates. **PdPhNO$_2$–R** does not insert at all into aryl bromides since the complex decomposition is faster than insertion at the temperatures needed (> 100 °C) to achieve conversion. **PdPhOMe–R** is capable of oxidative addition of chloro benzene quantitatively, however elevated temperatures are needed. Only traces of $(PPh_3)_2PdCl(Ph)$ were observed for **PdPhMe–R**. Fluoro aromatics are not activated at all and only decomposition of the complexes to palladium black occurred.[118]

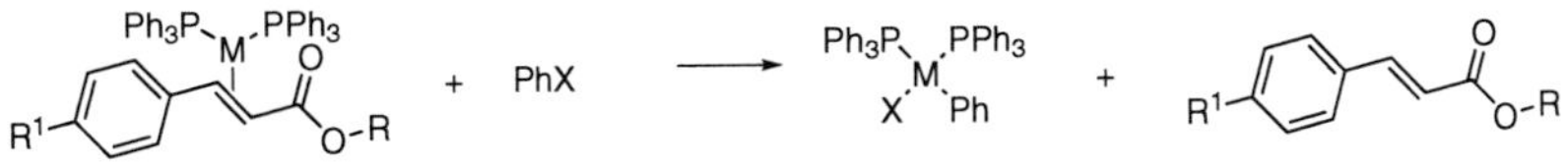

Scheme 4.8: Oxidative addition of arylhalogenides. M $\equiv$ Pd, Ni; X $\equiv$ Cl, Br, I

All nickel complexes synthesised insert immediately into aryl bromine and iodine bonds and slowly into aryl chlorine bonds. Yet again more electron rich olefins promote the reactivity. C−F bonds did not react as for palladium.

4.4. Connatural complexes systems

4.4.1. Derivatives of more complex esters

To investigate the effects of more complexly substituted olefins a series of complexes was synthesised with esters relevant for possible photo-switchable catalysis (figure 4.10). The detailed reactivity of the complexes and the reason for their design is discussed in detail in chapter 2 for all benzophenone and fluorenone derived systems whereas 1-methoxy-iso-propanol ester complexes are discussed in chapter 5. Therefore only the synthesis, spectroscopic properties and reactivities concerning the aforementioned cinnamate ester complexes are described here.

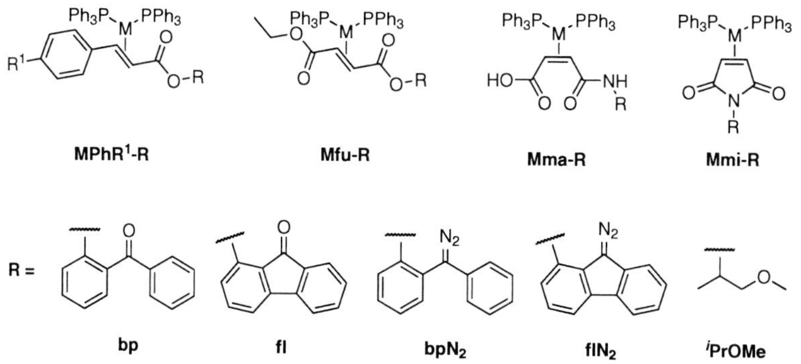

Figure 4.10: Derivatives of more complex esters. $M \equiv Pt, Pd, Ni$; $R^1 \equiv H, NO_2$

The platinum complexes could be synthesised via both routes described in section 4.1 (scheme 4.1) in good to excellent yields. However the reaction of olefin and $(PPh_3)_2PtC_2H_4$ has the advantage that no free triphenylphosphine is formed. Therefore highly lipophilic systems like **Ptfu-R** are obtained in higher yields since no phosphine has to be removed and so less pentane is necessary to wash the product. Also reaction times are shorter in this route, which is of advantage for the thermo labile **bpN$_2$** systems. All platinum complexes were obtained as white, in case of **PtPhNO$_2$−R** yellow and diazo ester complexes orange, solids, which are all stable at ambient temperature, except for **bpN$_2$** compounds, and may all be handled in air for a short period of time. The palladium complexes were synthesised utilizing the route described for the afore mentioned Pd-complexes (scheme 4.2) in quantitative yields. All complexes were obtained as orange, in case of **PdPhNO$_2$−R** red, solids, stable under an argon atmosphere at ambient temperature, with exception of **bpN$_2$** esters. The corresponding nickel systems could be synthesised from Ni(cod)$_2$ following the route already described (scheme 4.3), in almost quantitative yield as red solids. However no **PhNO$_2$−R** esters could be used due to oxidation of Ni(0) to Ni(II), *vide supra*. The same oxidation, even though slower, did also occur with diazo esters, rendering these complexes unaccessible. All synthesised nickel compounds are stable at ambient temperature and highly sensitive towards traces of oxygen.

Table 4.17.: Selected NMR shifts for bis(triphenylphosphine) group 10 complexes [δ]:

	$CHCO$	CHC	$CHCO$	CHC	P$_1$	P$_2$	Pt
PtPh−bp	3.57	3.81	49.2	61.9	27.6	28.5	-5038.1
PtPhNO$_2$−bp	3.33	3.69	–	–	26.5	27.7	–
Ptfu−bp	3.45	3.68	47.8	49.9	27.0	27.7	-5107.4
Ptma−bp	2.89	3.59	48.8	51.3	26.3	27.9	–
Ptmi−bp	3.47	3.47	48.6	48.6	27.8	27.8	-5036.37
PtPh−fl	4.49	4.74	49.4	61.9	28.0	28.9	-5047.6
Ptfu−fl	4.43	4.58	48.4	50.2	27.3	28.0	-5111.1
PtPh−bpN$_2$	3.91	4.23	–	–	27.6	28.6	–
PtPhNO$_2$−bpN$_2$	3.87	4.23	–	–	25.8	27.1	–
PtPhNO$_2$−flN$_2$	3.94	4.45	–	–	26.1	26.7	–
PtPhNO$_2$−^{i}PrOMe$_{ma}$	3.93	4.35	–	–	26.2	27.7	–
PtPhNO$_2$−^{i}PrOMe$_{mi}$	3.92	4.36	–	–	26.2	27.8	–
PdPh−bp	4.34	4.69	–	–	25.1	26.2	–
Pdfu−bp	4.11	4.33	58.1	60.3	26.7	27.2	–
PdPhNO$_2$−bpN$_2$	4.52	4.98	–	–	24.7	26.6	–

continued on next page

Table 4.17 – continued from previous page

	$CHCO$	CHC	$CHCO$	CHC	P_1	P_2	Pt
PdPhNO$_2$–^{i}PrOMe$_{ma}$	4.56	5.09	59.4	72.9	24.3	26.9	–
PdPhNO$_2$–^{i}PrOMe$_{mi}$	4.56	5.09	59.2	72.8	24.4	27.0	–
NiPh–bp	3.18	3.76	51.4	63.8	32.4	34.0	–
Nifu–bp	3.20	3.37	49.8	50.9	33.4	34.0	–

The NMR spectra of the cinnamic acid ester complexes are similar to the ones discussed earlier. The shift regions of the signals in ^{1}H NMR spectroscopy are in the same region as in the simpler cinnamic acid complexes with corresponding R^1 with exception of **MPhR–bp** systems. Here the olefinic signals are shifted significantly upfield. This is surprising since the electronic and steric properties are expected to be similar to the corresponding diazo ester or the fluorene ligands. This indicates some kind of interaction of the benzophenone´s carbonyl group with either the double bond or the metal. However no other evidence for this, i.e. coupling to platinum or phosphorus, is present in the NMR spectra of the compounds or the crystal structure of **PtPhNO$_2$–bp** (chapter 2). No abnormal behaviour concerning the shift was observed in ^{13}C, ^{31}P and ^{195}Pt NMR spectra. The fumaric acid ester (**Mfu-R**) and maleamide (**Mma-R**) complexes show similar behaviour, with shifts corresponding to **PtPh–R** for the prior, while maleic acid derivatives show a pronounced high field shift of the double bond in the ^{1}H NMR spectra. The methylene protons of the fumaric acid´s ethyl group show the expected splitting into two signals, *vide supra*. The spectra of the maleimide derivative **Ptmi-bp** are comparable to, even though more simple due to the symmetry of the ligand, **Ptma-bp** and the shifts are comparable to **Ptfu-bp**.

Table 4.18.: Selected NMR coupling constants of bis(triphenylphosphine) complexes [Hz]:

	$|^1J_{CHCOPt}|$	$|^1J_{CHCPt}|$	$|^2J_{PP}|$	$|^1J_{P_2Pt}|$	$|^1J_{P_1Pt}|$
PtPh–bp	190.7	195.7	40.6	3630.4	4195.8
PtPhNO$_2$–bp	–	–	32.0	3703.3	4071.8

continued on next page

Table 4.18 – continued from previous page

	$|^1J_{CHCOPt}|$	$|^1J_{CHCPt}|$	$|^2J_{PP}|$	$|^1J_{P_2Pt}|$	$|^1J_{P_1Pt}|$
Ptfu–bp	209.1	213.7	26.8	3826.1	3895.7
PtPh–fl	203.5	220.0	41.6	3630.8	4212.6
Ptfu–fl	209.2	213.8	27.2	3826.9	3914.2
Ptma–bp	210.0	220.1	24.1	3706.9	3883.1
Ptmi–bp	223.7	223.7	–	3770.1	3770.1
PtPh–bpN$_2$	–	–	40.6	3634.4	4185.3
PtPhNO$_2$–bpN$_2$	–	–	32.4	3723.2	4045.2
PtPhNO$_2$–flN$_2$	–	–	30.7	3701.8	4073.5
PtPhNO$_2$–^{i}PrOMe$_{ma}$	–	–	35.2	3769.0	4003.5
PtPhNO$_2$–^{i}PrOMe$_{mi}$	–	–	35.2	3768.9	3996.3
PdPh–bp	–	–	14.9	–	–
Pdfu–bp	–	–	0.0	–	–
PdPhNO$_2$–bpN$_2$	–	–	4.7	–	–
PdPhNO$_2$–^{i}PrOMe$_{ma}$	–	–	8.1	–	–
PdPhNO$_2$–^{i}PrOMe$_{mi}$	–	–	8.4	–	–
NiPh–bp	–	–	33.7	–	–
Nifu–bp	–	–	22.3	–	–

The coupling constants to ^{195}Pt observed for the platinum systems are in accordance to the ones obtained for the simpler cinnamic acid ligands with the same R^1. To all three metals applies, that the $^2J_{PP}$ coupling constants are similar to the prior described complexes. Since the coupling constant proved extremely sensitive towards the electron density of the olefin a grouping of the ester groups is possible. The electronic influence exhibited by the **bp**, the **bpN$_2$** and the **flN$_2$** moieties lies between the one of phenyl and 4-nitrophenyl while the one of **fl** is similar to 4-methylphenyl and ^{i}**PrOMe** to *iso*-propanol. Via the latter coupling constant a calculation of the OEDP is possible, *vide supra*, which assigns the electron density of the ligand´s double bond (compiled in table 4.19). The obtained ω shows that the electron density of the fumarates is significantly lower than of the most electron deficient cinnamic acids, the 4-nitro derivatives. This indicates, that sufficiently higher complex stabilities and therefore lower reactivities can be achieved with diacid esters in general, since ω of **ma-bp** is even higher. This is in accordance with the observed reactivities described in chapter 2. However due to the lack of $^2J_{PP}$ coupling no classification of **mi-bp** is possible with this method.

Table 4.19.: OEDP (ω) of 2-Hydroxybenzophenone & related olefin ester:

Ph−bp	0.09(5)	**ma−bp**	1.60(12)
PhNO$_2$−bp	0.87(8)	**Ph−bpN$_2$**	0.06(3)
fu−bp	1.39(18)	**PhNO$_2$−bpN$_2$**	0.91(14)
Ph−fl	-0.03(2)	**PhNO$_2$−flN$_2$**	0.99(8)
fu−fl	1.31(10)	**PhNO$_2$−^{i}PrOMe**	0.63(11)

4.4.2. Chelating phosphines

To investigate the influence of the cinnamic acid esters more general and to generalise the applicability of ω, the triphenylphosphine was substituted by the chelating phosphines: dppm, dppe, dppp, dippp and dppb.

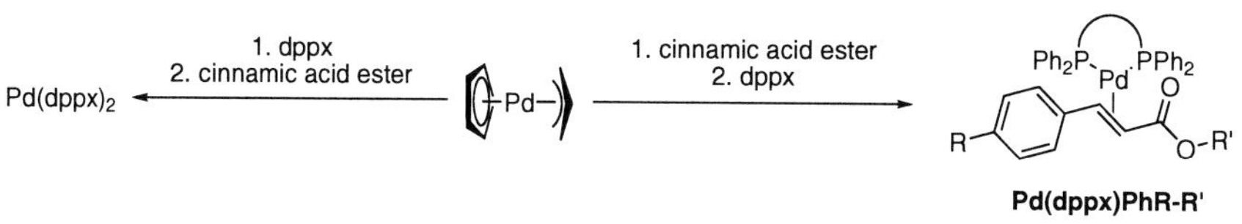

Scheme 4.9: Reaction of palladium with dppe and dppp.

While the dppb and dippp complexes could be synthesised via the route established for the bis(triphenylphosphine)palladium cinnamates, this failed for dppm, dppe and dppp. Upon reaction of Pd(C$_3$H$_5$)Cp with the latter two phosphines the bis-diphosphines Pd(dppe)$_2$ and Pd(dppp)$_2$ formed which do not react with the cinnamic acid ligands.[119] No bis(diphosphine)Pd complex is formed in the case of dippp due to the sterically demanding iso-propyl groups, whereas Pd(dppb)$_2$ might be formed but one ligand is easily substituted by the cinnamic acids due to the virtually non-existent chelate effect of seven membered rings. With dppm the homo bimetallic complex $Pd_2(dppm)_3$ is formed which

is as inert as the aforementioned dppp and dppe complexes.[120–122] Therefore the dppe

and dppp complexes had to be synthesised by slow addition of phosphine to a mixture of

$Pd(C_3H_5)Cp$ and cinnamate (scheme 4.9). The ^{31}P NMR shifts and coupling constants

suggest the formation of mixtures of undefined dppm bridged multi metal compounds; no

defined bis(triphenylphosphine)palladium olefin-analogue dppm complexes could be iso-

lated. All other complexes were isolated as yellow, in case of $R^1 \equiv NO_2$ red, solids, with

similar properties to bis(triphenylphosphine)palladium cinnamates, with the exception of

significantly enhanced stability in solution.

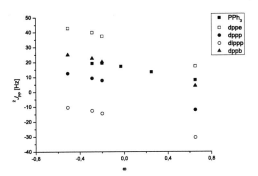

Graph 4.35: $^2J_{PP}$ for Pd complexes in dependence of ω. $R^1 \equiv$ OMe, Me, H, Cl, CF_3, NO_2;
$R^3 \equiv$ Et

Unlike their triphenylphosphine analogues none of the complexes exhibits dynamic be-

haviour at ambient temperature in NMR spectroscopy. The olefin signals in the ^1H NMR

spectra are comparable to the triphenylphosphine systems, shifted by ca. 0.2 ppm down

field for dppe and ca. 0.2 ppm upfield for dppb. This is caused by the different donor

properties of the phosphines.[123] However it does not effect the other signals in the proton

NMR spectrum and in ^{13}C NMR spectroscopy no deviations are observed. The signals

in the ^{31}P NMR spectra are compared to triphenylphosphine shifted ca. 10 ppm down

field for dppe complexes while dppp is shifted ca. 16 ppm and dppb ca. 7 ppm upfield,

typical for phosphorus in five, six and seven membered rings.[124,125] The signal in the dippp complexes is shifted slightly down field by ca. 3 ppm.

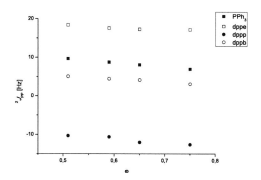

Graph 4.36: $^2J_{PP}$ for Pd complexes in dependence of ω. ; $R^1 \equiv NO_2$; $R^3 \equiv$ Me, Et, ^iPr, ^tBu

The $^2J_{PP}$ coupling constants are ca. 18 Hz larger for dppe, ca. 12 Hz and 34 Hz smaller for dppp and dippp and almost equal for dppb, in comparison to PPh_3. The coupling constants show the same linear correlation to ω observed for the triphenylphosphine complexes, proving the generality of the OEDP, as depicted in graphs 4.35 and 4.36.

Table 4.20.: Selected NMR data for Pd complexes with chelating phosphines:

| | $CHCO$ | CHC_{Ar} | $CHCO$ | CHC_{Ar} | P_1 | P_2 | $\lvert^2J_{PP}\rvert$ |
	$[\delta]$	$[\delta]$	$[\delta]$	$[\delta]$	$[\delta]$	$[\delta]$	[Hz]
Pd(dppe)PhOMe−Et	5.05	6.11	54.8	66.9	32.2	32.6	42.9
Pd(dppe)PhOMe−^{i}Pr	–	–	–	–	31.9	32.3	42.9
Pd(dppe)PhMe−Et	–	–	–	–	32.3	32.7	39.9
Pd(dppe)Ph−Et	5.06	6.03	54.7	66.9	33.0	33.5	37.4
Pd(dppe)PhNO$_2$−Me	4.75	5.75	–	–	36.8	36.5	17.2
Pd(dppe)PhNO$_2$−Et	4.77	5.77	53.4	65.2	36.3	36.6	17.3
Pd(dppe)PhNO$_2$−^{i}Pr	4.73	5.79	54.0	65.4	36.5	36.1	17.6
Pd(dppe)PhNO$_2$−^{t}Bu	4.69	5.75	–	–	35.3	36.4	18.4
Pd(dppe)PhNO$_2$−PhOMe	4.71	5.44	–	–	37.8	37.9	18.2
Pd(dppp)PhOMe−Et	4.80	5.69	54.7	67.2	7.6	10.4	12.6

continued on next page

Table 4.20 – continued from previous page

	$CHCO$	CHC_{Ar}	$CHCO$	CHC_{Ar}	P_1	P_2	$|^2J_{PP}|$
	$[\delta]$	$[\delta]$	$[\delta]$	$[\delta]$	$[\delta]$	$[\delta]$	[Hz]
Pd(dppp)PhOMe$-^i$Pr	–	–	–	–	7.1	11.0	13.7
Pd(dppp)PhMe$-$Et	–	–	–	–	8.0	10.0	9.3
Pd(dppp)Ph$-$Et	4.81	5.63	54.8	67.3	8.1	10.5	7.8
Pd(dppp)PhNO$_2$$-$Me	4.51	5.37	53.1	65.4	9.8	11.5	12.6
Pd(dppp)PhNO$_2$$-$Et	4.50	5.39	53.4	65.4	9.2	12.0	12.0
Pd(dppp)PhNO$_2$$-^i$Pr	4.50	5.40	53.8	65.5	8.9	12.5	10.6
Pd(dppp)PhNO$_2$$-^t$Bu	4.42	5.34	54.8	66.0	8.5	11.9	10.3
Pd(dppp)PhNO$_2$$-$PhOMe	4.68	5.40	–	–	8.8	12.7	16.4
Pd(dippp)PhOMe$-$Et	–	–	–	–	26.8	27.9	10.4
Pd(dippp)PhMe$-$Et	–	–	–	–	26.8	28.2	12.5
Pd(dippp)Ph$-$Et	–	–	–	–	26.4	28.5	14.4
Pd(dippp)PhNO$_2$$-$Et	–	–	–	–	26.2	30.7	30.5
Pd(dppb)PhOMe$-$Et	4.60	5.43	56.9	69.3	17.3	18.2	25.0
Pd(dppb)PhMe$-$Et	4.63	5.41	56.9	69.5	17.7	17.9	22.6
Pd(dppb)Ph$-$Et	4.60	5.35	57.2	69.2	17.8	18.5	20.2
Pd(dppb)PhNO$_2$$-$Me	4.28	5.19	55.0	66.6	19.7	20.2	3.1
Pd(dppb)PhNO$_2$$-$Et	4.28	5.21	55.5	66.7	19.5	20.2	4.2
Pd(dppb)PhNO$_2$$-^i$Pr	4.25	5.20	56.0	66.9	19.3	21.1	4.5
Pd(dppb)PhNO$_2$$-^t$Bu	4.16	5.17	57.1	67.5	19.1	19.6	5.1
Pd(dppb)PhNO$_2$$-$PhOMe	4.41	5.22	54.6	66.6	19.3	20.7	0.0

Single crystals suitable for X-ray diffraction analysis could be grown by slow diffusion of pentane into a saturated toluene solution of **Pd(dppe)PhNO$_2$$-^i$Pr**. The complex is trigonal planar coordinated with the plane defined by the olefin and palladium twisted by 9.9° against the plane defined by both phosphorus atoms and palladium. The C5-C6 distance is with 1.41 Å comparable to the olefin bond length in the bis(triphenylphosphine)palladium cinnamates and lies between the distance of un-

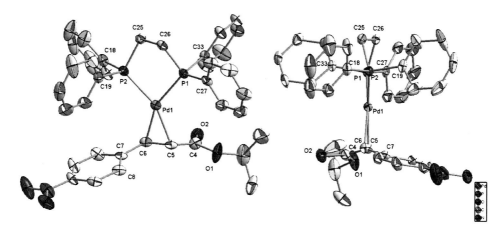

Figure 4.11: ORTEP[24] representation of **Pd(dppe)PhNO₂−ⁱPr** in the front and side on view. Thermal ellipsoids are given at 50% probability level. Hydrogen atoms are omitted for clarity.

coordinated olefins and C-C single bonds as expected. The palladium olefin distance is 2.00 Å and like the Pd-C6 distance with 2.13 Å in accordance with the triphenylphosphine complexes. However the Pd-C5 distance of 2.10 Å is significantly shorter than with non chelating phosphines and is also shorter than the Pd-C6 distance. The Pd-P distances are 2.31 Å (P1) and 2.29 Å (P2) and therefore slightly shorter than for non-chelating phosphines. The Pd-P2 distance is shorter than the Pd-P1 distance due to the smaller *trans*-effect of C5 compared to C6, *vide supra*. The C5-Pd-C6 angle of 35.9° is slightly wider than in the triphenylphosphine complexes due to the slightly smaller palladium olefin distance. The P1-Pd-P2 angle is significantly smaller due to the ring strain with 85.8°. This results in the widening of the C5-Pd-P1 and C6-Pd-P2 angles to 118.2° and 117.5°. The deviation between the latter two angles is less pronounced than in the PPh₃ complexes due to the lower steric demand of the phosphine. Therefore the higher steric demand of the alkyl group is less structure determining. The C4-C5-C6-C7 torsion angle is 148.8°, which is smaller than in the non chelated complexes. This deviation may be caused by stronger back donation into the π^* orbital, caused by the better σ donor properties of the dppe.

The P1-C26 and P2-C25 distances are with 1.83 Å slightly longer than the other P-C distances due to the strain enforced by the five membered ring. Due to this reason the C25-C26 bond length (1.54 Å) is also slightly elongated. However the P1-C26-C25 (108.3°) and P2-C25-C26 (110.4°) angles do not deviate strongly from the ideal 109.5°. C25 is bent to one side of the pallada-phospha-cycle, while C26 is displaced onto the other side. This is expressed in the torsion angles P1-C26-C25-P2, Pd-P1-C26-C25 and C26-C25-P2-Pd of 51.8°, 39.5° and 40.6°. The plane spanned by the phosphorus atoms and the *ipso*-carbon atoms of their phenyl substituents stands perpendicular to the coordination plane of the palladium. The bond lengths and angles of the cinnamate ligand are in accordance with the structures described above.

Table 4.21.: Selected bond length, angles and torsion angles of **Pd(dppe)PhNO$_2$-^{i}Pr**:

d [Å]		∠ [°]		torsion ∠ [°]	
C5–C6	1.411(5)	C5–Pd1–C6	38.9(2)	C5–Pd1–P1–C26	165.2(2)
C6–C7	1.465(6)	C18–P2–C19	102.9(2)	C7–C6–C5–C4	148.8(4)
C25–C26	1.538(5)	C18–P2–C25	103.6(2)	C18–P2–Pd1–P1	130.5(2)
O1–C4	1.354(5)	C26–P1–C33	102.8(2)	C19–P2–Pd1–P1	104.6(1)
O2–C4	1.220(5)	C27–P1–C33	103.3(2)	C25–P2–Pd1–C6	163.1(2)
P1–C26	1.834(4)	O1–C4–C5	110.8(3)	C26–C25–P2–Pd1	39.5(3)
P1–C33	1.823(4)	O2–C4–C5	127.0(4)	C27–P1–Pd1–P2	136.4(2)
P1–C27	1.820(1)	P1–Pd1–C5	118.2(1)	C33–P1–Pd1–P2	97.9(1)
P2–C19	1.817(4)	P1–Pd1–P2	85.8(0)	Pd1–P1–C26–C25	40.6(3)
P2–C18	1.811(4)	P1–C26–C25	108.3(2)	P1–C26–C25–P2	51.8(0)
P2–C25	1.835(4)	P2–Pd1–C6	117.5(1)		
Pd1–C5	2.108(4)	P2–C25–C26	110.1(3)		
Pd1–C6	2.131(4)	Pd1–P2–C19	113.4(1)		
Pd1–P1	2.308(1)	Pd1–P1–C27	126.5(1)		
Pd1–P2	2.293(1)				

Crystals of **Pd(dppp)PhNO$_2$-Et** which were suitable for X-ray diffraction analysis

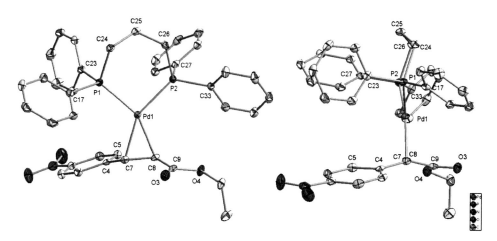

Figure 4.12: ORTEP[24] representation of **Pd(dppp)PhNO$_2$–Et** in the front and side on view. Thermal ellipsoids are given at 50% probability level. Hydrogen atoms are omitted for clarity.

could also be grown by slow diffusion of pentane into a saturated toluene solution. In accordance with the other examples the complex is coordinated in a trigonal planar fashion and the P1-Pd-P2 and C7-Pd-C8 planes are tilted against each other by only 7.2°. The C7-C8 distance of 1.46 Å and the C7-Pd-C8 angle of 38.9° indicate slightly stronger metal olefin bonding than in the previously discussed complex. However the C4-C7-C8-C9 torsion angle is slightly larger with 151.6°. The C7-Pd and C8-Pd distances are slightly longer than in the structure of **Pd(dppe)PhNO$_2$–^{i}Pr**, while the same trend is observed concerning C8-Pd being shorter than C7-Pd, this also accounts for the length of the Pd-P bonds and their relationship. The P1-Pd-P2 angle is with 96.4° wider than in the dppe complex due to reduced ring strain. Therefore the P1-Pd-C7 and P2-Pd-C8 angles are smaller and their difference is more pronounced, P1-Pd-C7 is wider, due to increasing steric repulsion of the ester group and the phosphine, *vide supra*.

The phospha-pallada-cycle deviates from planarity by tilting C25 out of the plane, defined through C23, C24, C26 and C27, by 65.1° and Pd is displaced by 25.8°. The plane spanned by the phosphorus atoms and the *ipso*-carbon atoms of their phenyl substituents is now tilted by 72.0° towards the coordination plane of the palladium and no longer perpendicular to it. The cinnamates structural properties are similar to the ones discussed above.

Table 4.22.: Selected bond length, angles and torsion angles of **Pd(dppp)PhNO₂−Et**:

d [Å]		∠ [°]		torsion ∠ [°]	
C4–C7	1.459(4)	C7–Pd1–C8	38.9(1)	C4–C7–C8–C9	151.6(2)
C7–C8	1.423(3)	C17–P1–C23	102.5(1)	C8–Pd1–P2–C26	156.2(1)
C8–C9	1.469(3)	C23–P1–C24	104.0(1)	C17–P1–Pd1–P2	144.3(1)
C24–C25	1.532(3)	C24–C25–C26	114.5(2)	C23–P1–Pd1–P2	92.3(1)
C25–C26	1.538(3)	C26–P2–C27	101.2(1)	C24–P1–Pd1–C7	160.3(1)
O3–C9	1.216(3)	C27–P2–C33	101.3(1)	C25–C24–P1–Pd1	51.3(2)
O4–C9	1.362(3)	O3–C9–C8	127.3(2)	C26–C25–C24–P1	76.8(2)
P1–C23	1.835(2)	O4–C9–C8	109.9(2)	C27–P2–Pd1–P1	85.2(1)
P1–C17	1.834(2)	P1–Pd1–P2	96.4(0)	C33–P2–Pd1–P1	155.6(1)
P1–C24	1.842(3)	P1–Pd1–C7	113.8(1)	P2–C26–C25–C24	75.4(2)
P2–C27	1.836(2)	P1–C24–C25	115.8(2)	Pd1–P2–C26–C25	51.2(2)
P2–C26	1.837(3)	P2–Pd1–C8	110.6(1)		
P2–C33	1.836(2)	P2–C26–C25	113.6(2)		
Pd1–C8	2.127(3)	Pd1–P1–C24	113.8(1)		
Pd1–C7	2.148(3)	Pd1–P2–C26	115.2(1)		
Pd1–P2	2.302(1)	Pd1–P2–C33	123.7(1)		
Pd1–P1	2.312(1)	Pd1–P1–C17	120.0(1)		

Single crystal of **Pd(dppb)PhNO₂−PhOMe** suitable for X-ray diffraction analysis could be obtained by over-saturation of a benzene solution and crystalisation at ambient temperature. Like in all other palladium cinnamate complexes the palladium is trigonal planar coordinated in **Pd(dppb)PhNO₂−PhOMe**. The C7-Pd-C8 and P1-Pd-P2 planes are twisted against each other even less than in **Pd(dppp)PhNO₂−Et** with 3.7°. The C7-C8 distance of 1.42 Å together with the C7-Pd-C8 angle of 39.1° and the C7-Pd and C8-Pd distances of 2.12 Å and 2.13 Å is comparable to the dppp complex discussed

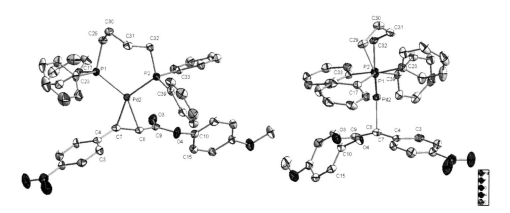

Figure 4.13: ORTEP[24] representation of **Pd(dppb)PhNO$_2$–PhOMe** in the front and side on view. Thermal ellipsoids are given at 50% probability level. Hydrogen atoms are omitted for clarity.

above. This applies for both P-Pd bond lengths as well. The P1-Pd-P2 angle is wider than in the dppe and dppp structures, but with 104.6° still significantly smaller the in the triphenylphosphine complexes, due to the ring strain. This results in a decrease in the P1-Pd-C7 and P2-Pd-C8 angles to 107.3° and 108.8° respectively. Interestingly the difference between the two angles is less pronounced than in **Pd(dppp)PhNO$_2$–Et**. The C4-C7-C8-C9 torsion angle is with 151.7° almost identical to the prior discussed structure. The phospha-pallada-cycle deviates strongly from planarity as expectable for seven membered rings.

Table 4.23.: Selected bond length, angles and torsion angles of **Pd(dppb)PhNO$_2$–PhOMe**:

d [Å]		∠ [°]		torsion ∠ [°]	
C4–C7	1.465(3)	C7–Pd2–C8	39.1(1)	C17–P1–Pd2–P2	138.0(1)
C7–C8	1.422(4)	C17–P1–C29	100.8(1)	C23–P1–Pd2–P2	101.4(1)
C8–C9	1.458(3)	C17–P1–C23	103.8(1)	C29–P1–Pd2–C7	154.6(1)
C29–C30	1.529(4)	C29–C30–C31	116.2(2)	C30–C29–P1–Pd2	79.9(2)
C30–C31	1.525(4)	C30–C31–C32	113.6(2)	C31–C30–C29–P1	51.5(3)
C31–C32	1.534(4)	C32–P2–C33	101.0(1)	C32–C31–C30–C29	55.7(3)

continued on next page

Table 4.23 – continued from previous page

d [Å]		∠ [°]		torsion ∠ [°]	
O3–C9	1.205(3)	C33–P2–C39	103.2(1)	C33–P2–Pd2–P1	114.9(1)
O4–C9	1.381(3)	O3–C9–C8	128.1(2)	C39–P2–Pd2–P1	125.7(1)
P1–C23	1.826(2)	O4–C9–C8	110.1(2)	C4–C7–C8–C9	151.7(2)
P1–C17	1.827(3)	P1–Pd2–C7	107.3(1)	C8–Pd2–P2–C32	179.1(1)
P1–C29	1.846(2)	P1–Pd2–P2	104.6(0)	P2–C32–C31–C30	109.7(2)
P2–C39	1.827(2)	P1–C29–C30	115.7(1)	Pd2–P2–C32–C31	42.5(2)
P2–C32	1.835(2)	P2–C32–C31	114.5(1)		
P2–C33	1.823(2)	P2–Pd2–C8	108.8(1)		
Pd2–C7	2.123(2)	Pd2–P2–C32	117.1(1)		
Pd2–C8	2.128(2)	Pd2–P2–C39	115.7(1)		
Pd2–P1	2.306(0)	Pd2–P1–C23	116.7(1)		
Pd2–P2	2.295(1)				

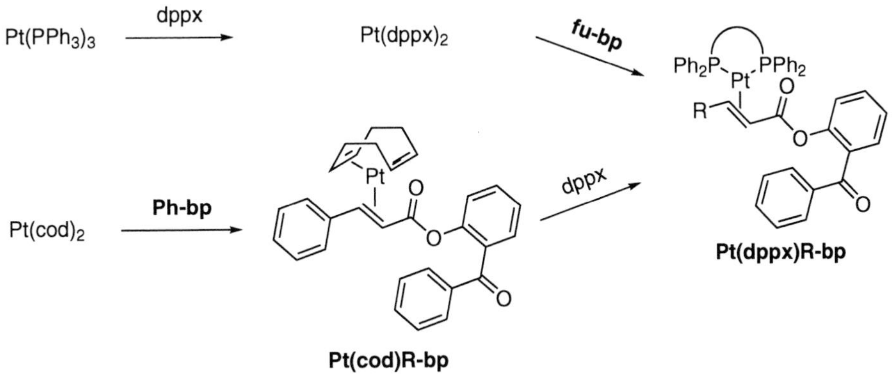

Scheme 4.10: Synthetic routes to dppe & dppp platinum olefin complexes.

Due to the high relevance of platinum benzophenone ester complexes for photo-switchable catalysis some complexes were synthesised with chelating phosphines. In coherence with the findings for the palladium route, all dppb and dippp complexes could be obtained by simple addition of the chelating phosphine to Pt(PPh$_3$)$_3$ and subsequent addition of cinnamate ligand. With dppe and dppp, bi(diphosphine) complexes formed, that did not react with cinnamates.[126] However these complexes could be opened with fumarates under the formation of the desired products (scheme 4.10). The standard route to Pt(dppx) olefin complexes, described in literature, via reduction of Pt(dppx)Cl$_2$ in the presence of olefin failed.[127–129] Even though Pt(dppx)Cl$_2$[130–133] could be synthesised for all phosphines in quantitative yield, by stirring Pt(cod)Cl$_2$[134,135] with one equivalent of phosphine over night in dichloromethane, the major products isolated after reduction were the bi(diphosphine)platinum complexes. To obtain complexes of less electron deficient olefins the most convenient route found was to generate the corresponding cyclooctadiene platinum cinnamate *in situ* from Pt(cod)$_2$ and react it with one equivalent of phosphine. With dppm no defined platinum complexes were obtained. All complexes were isolated as off-white solids, with properties similar to their triphenylphosphine analogues. Their spectroscopic properties match the ones of the bis(triphenylphosphine) systems described in section 4.4.1 concerning the ester ligand and above regarding the chelating phosphines. The reactivity was already discussed in chapter 2.

4.4.3. Pure olefin complexes

Due to the bad performance of phosphine complexes and the high relevance of the *Karstedt* catalyst[3,5,30] in hydrosilylation, a number of cinnamate complexes were synthesised where only olefin ligands are present.

 All the cod and nbe complexes could be obtained by simple ligand substitution starting either from Pt(cod)$_2$[136] or Pt(nbe)$_3$[137] and addition of one equivalent of olefin ester.[138–141] The liberated ligand could then easily be removed by crystallisation of the complex from pentane or diethylether. This in combination with the low stability in solution resulted in only moderate yield though. Both complex systems were isolated, as off-white solids. These are stable at ambient temperature in the solid state if they are stored under the strict exclusion of oxygen. In solution however they decompose within hours to platinum black and free ligands. The siloxane complexes **Pt(dvtmsi)R-bp**, analogue to the *Karstedt* catalyst, were synthesised from Pt(cod)$_2$ via *in situ* generated

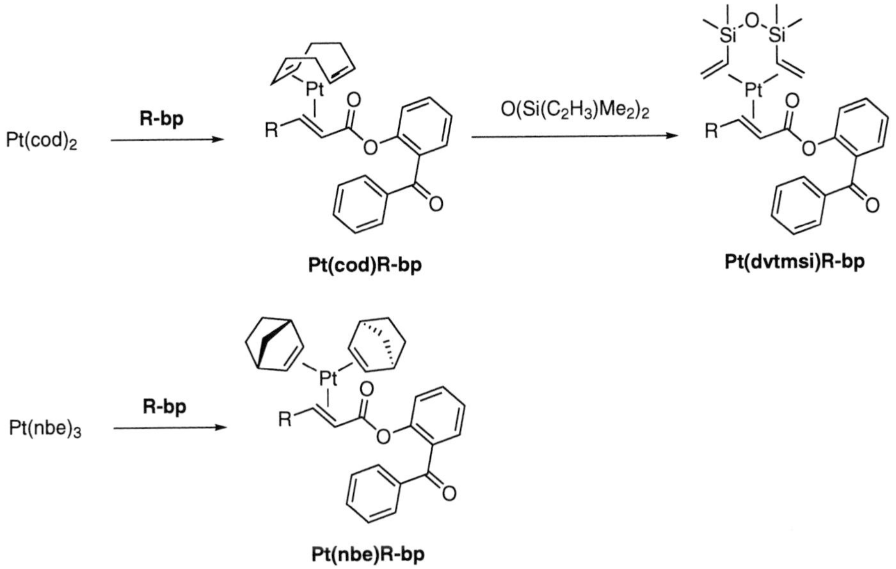

Scheme 4.11: Synthesis of pure olefin complexes.

Pt(cod)R-bp and subsequent ligand exchange with divinyl tetramethyl disiloxane and isolated as a white solid by crystallisation from pentane. These routes are summarised in scheme 4.11. In contrast to the cod and nbe complex systems, **Pt(dvtmsi)R-bp** is stable in solution at ambient temperature. All complexes are superbly soluble in non polar organic solvents like pentane and react with silanes under immediate loss of one of the non cinnamate ligands and oxidative addition of silane. If an excess of silane is employed the electron rich double bonds of the ligands are hydrosilylated, rendering these complex types too reactive for application in photo-induced hydrosilylation.

4.5. Conclusions

The electronic properties of cinnamic acid ligands may be tuned as proposed, with the R^1 substituent being more influential. The electron density of the olefin has direct influence upon the spectroscopic and structural parameters of the complexes as well as onto their reactivity. The electron density at the metal center and therefore also of the double bond

can directly be calculated from the $^2J_{PP}$ coupling constants and the obtained olefin electron density parameter (OEDP) w may also be applied to ligands, where easy assignment
via the *Hammett* parameters is not possible. The OEDP concept is also transferable
to other phosphines with the preservation of all trends observed for triphenylphosphine.
Since the coupling constants may be determined with all phosphines this is an advantage
to the assignment via the C=O stretching frequency in carbonyl compounds.[142] The
electron density of the olefinic double bond is also directly correlated to reactivity and
therefore equilibria may be predicted with w. It is also evident, that the ester moiety´s
influence is minor and therefore may be varied without altering the over all reactivity of
the complexes.

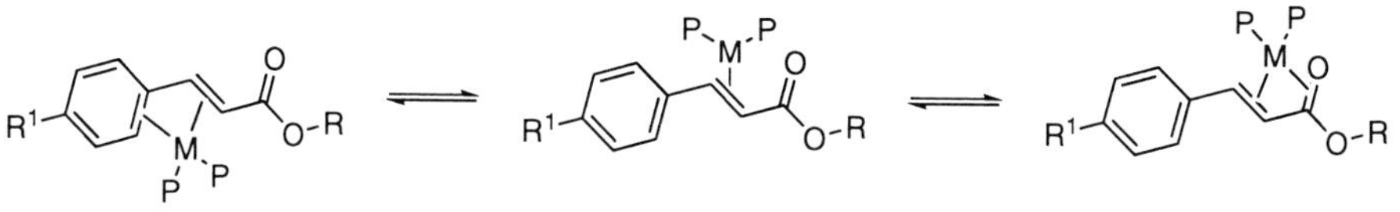

Scheme 4.12: Proposed dynamic coordination modes.

The observed minima in the ^{195}Pt NMR spectra could be explained by the dynamic
behaviour depicted in scheme 4.12,[143] since a higher bond strength of the metal olefin
bond would reduce the interaction with the carboxyl group. This should overcompensate
the electron withdrawing effects of the nitro group and therefore re-increase the relative
electron density on the central atom.

5. *Karstedt* catalyst inhibition studies

Tetramethyl di-η^2-vinyl disiloxane(triphenylphosphine)platinum is formed from tri-phenylphosphine containing platinum complexes in the presence of divinyl tetramethyl disiloxane.[29] This compound shows extremely low reactivity in hydrosilylation. It was also found in general, that phosphines inhibit platinum based hydrosilylation catalysts to an extent that rendered them non-applicable in industrial catalysis. Further it was noticed, that pure olefin complexes are too reactive independent of electron deficient ligands, because of the hydrosilylation of the most electron rich olefin ligand, *vide supra*. Therefore an alternative approach had to be examined in order to yield photo-switchable hydrosilylation catalysts with a post-irradiation activity suitable for large scale application.

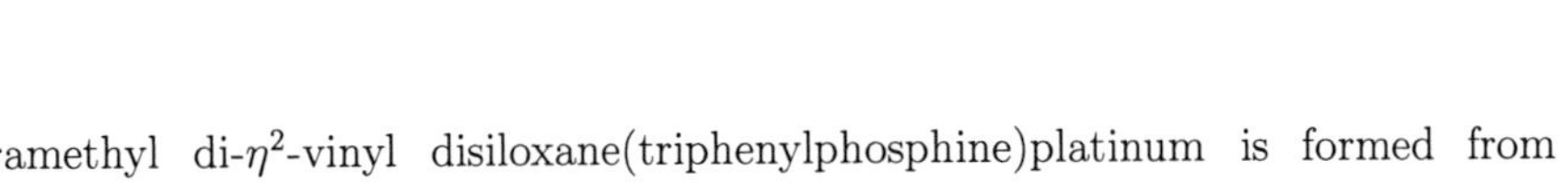

Scheme 5.1: Assay to study reaction deceleration through addition of models for photo-switches.

There are numerous examples for the inhibition capabilities of electron deficient olefins on the *Karstedt* catalyst.[3,6,23] Particularly low reactivity could be achieved with maleic and fumaric acid derivatives.[5,30] To apply the concept of photo-activatable catalysts onto *Karstedt* systems it was decided to perform reactivity studies with technical *Karstedt* solution in the presence of olefins, with model designs easily converted into photo-switchable systems. The hydrosilylation of divinyl tetramethyl disiloxane with heptamethyl tri-siloxane was chosen due to air stability of the substrates and product (scheme 5.1).

Besides maleic and fumaric acid derivatives, which were chosen due to the reasons

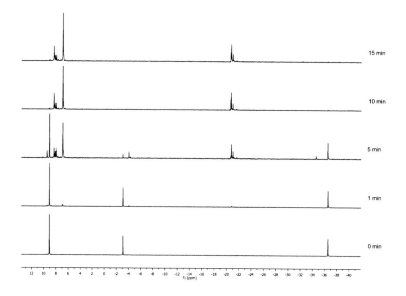

Figure 5.1: Models for photo-switchable ligands.

mentioned above, cinnamic acid systems were included as well, for comparison with the defined cinnamic acid systems investigated in chapters 2 and 4. Different substituents were chosen in regard of their sterical and electronical similarity to the diazo moieties solubility effects in siloxanes and chelating properties. An overview of the different inhibitor building blocks is given in figure 5.1.

Figure 5.2: Hydrosilylation kinetics without inhibitor, monitored by ^{29}Si NMR.

Catalytic test reactions were conducted neat in a 1:2 mixture of divinyl tetramethyl disiloxane and heptamethyl trisiloxane in the presence of 100 ppm catalyst and 1000 ppm inhibitor at 100 °C. The reaction progress was monitored via inverse gated ^{29}Si NMR spectroscopy with a relaxation delay of 60 s to ensure the return of all nuclei into thermo-dynamic equilibrium, guaranteeing quantitative signals in the spectra.[144,145] The reaction proceeding without inhibitor is shown in figure 5.2.

5.1. Synthesis

The synthesis of the inhibitors followed standard organic synthesis procedures. Esters were synthesised by refluxing in the appropriate acid in alcohol, in the presence of catalytic amounts of sulfuric acid, if simple acids and alcohols were used. More complex alcohols were reacted with acid chlorides either in pyridine at 100 °C or first deprotonated with sodium hydride followed by salt metathesis with the acid chloride. Even more complex systems, where the acid was too sensitive for conversion into acid chlorides, had to be synthesised via *Steglich* esterification.[146]

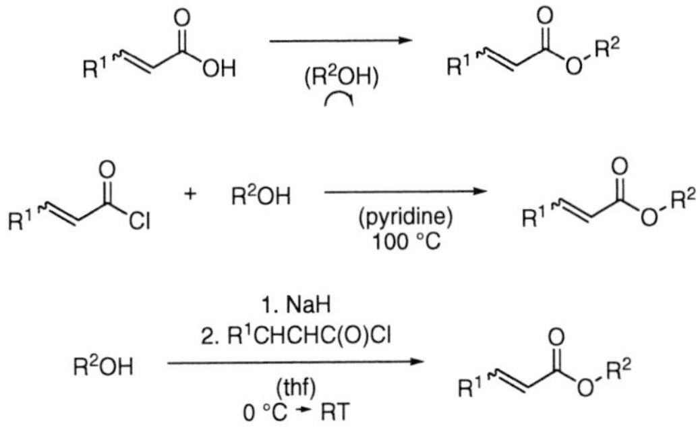

Scheme 5.2: Synthetic routes to the esters.

Monoesters of maleic acid were obtained by opening of maleic anhydride with the corresponding alcohol in substance.[147,148] The same route could be applied for maleic monoamides using amines. Other amides were synthesised by reaction of acid chlorides with amines at 100 °C in pyridine. Maleimids could be achieved by dehydrative ring closure of maleic monoamides in acetic anhydride (scheme 5.3).[36]

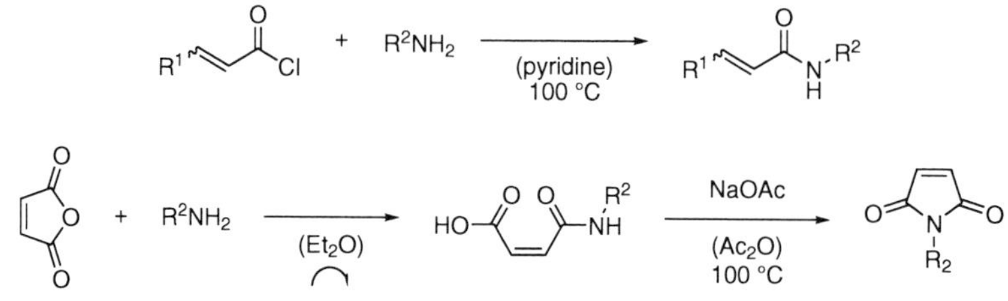

Scheme 5.3: Amide & imide synthesis.

Silylester were synthesised by reaction of acid with vinyl dimethyl chlorosilane in refluxing toluene in the presence of an excess of triethylamine as a HCl scavenger as shown in scheme 5.4 and subsequent extraction with pentane.[149–152]

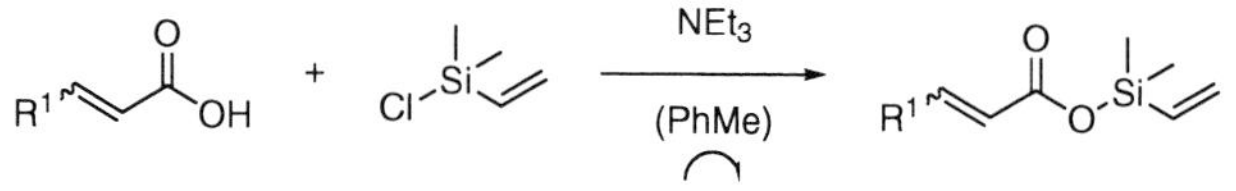

Scheme 5.4: Reaction yielding silyl esters.

All compounds were obtained in nearly quantitative yield. All substances are air and moisture stable, with exception of silylesters which hydrolyse within hours. They may be handled in air however. All inhibitors are superbly soluble in organic solvents and well soluble in non polar, non coordinating solvents like pentane, apart from the nitro substituted compound and the maleimides.

5.2. Inhibition assays

The various experiments are grouped by building blocks to ensure comparability.

Three different cinnamic acid derivatives were investigated to determine the influence of the electron density of the olefin; unsubstituted cinnamic acid, the more electron rich *p*-methoxy and the electron deficient *p*-nitro cinnamic acid. Various alcohols were esterificated with the three acids. 2-Hydroxybenzophenone was chosen due to its electronic and especially steric resemblance of the anticipated diazo groups (**PhOMe–bp**, **Ph–bp**, **PhNO$_2$–bp**). 1-Methoxy-*iso*-propanol esters are described to be among the best *Karstedt* inhibitors and were therefore included (**PhOMe–^{i}PrOMe**, **Ph–^{i}PrOMe, PhNO$_2$–^{i}PrOMe**). As reference compounds cinnamic acid (**5.4**) together with ethylesters (**PhOMe–Et, Ph–Et, PhNO$_2$–Et**) were employed. The latter were selected since no interaction of the ethyl moiety with platinum was expected and the alkyl group enhances solubility. To examine the effect of potentially chelating systems vinyl dimethyl silylesters (**5.1-5.3**) and the triolefin **5.5** were included. Both are theoretically capable to form 6-membered rings upon coordination to platinum and mimic divinyl tetramethyl disiloxane. The cinnamic acid inhibitors are shown in figure 5.3.

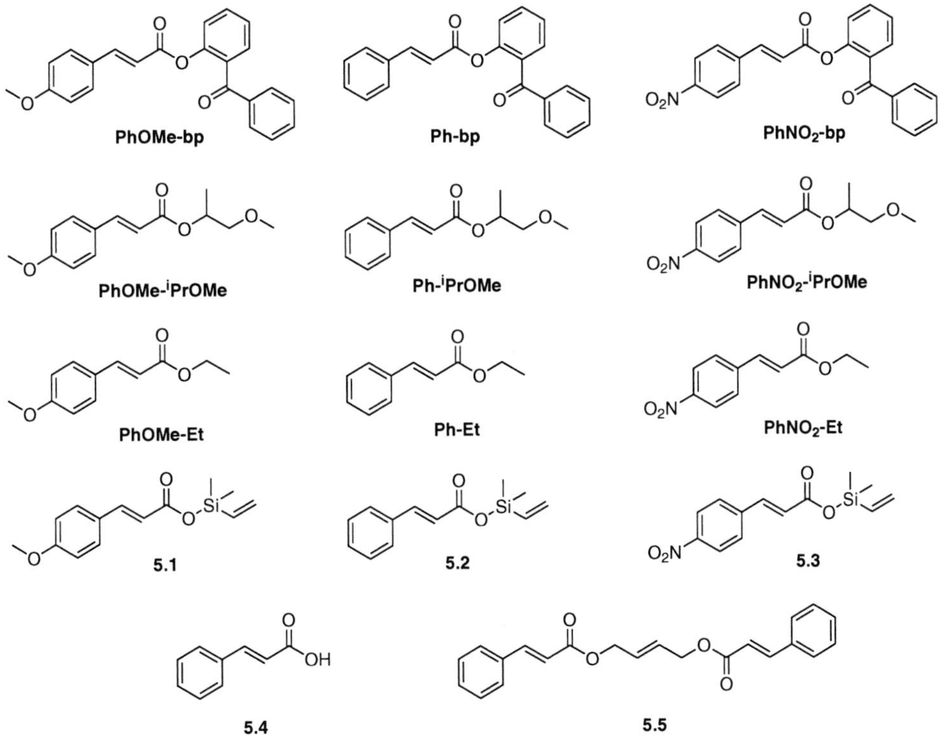

Figure 5.3: Inhibitors containing cinnamic acids.

Unexpectedly, the most electron deficient systems, the nitro substituted, did not show superior inhibition of the catalyst. Only in the cases of methoxy-*iso*-propanol and vinyl dimethyl silyl substitution did they perform better than *p*-methoxy and unsubstituted cinnamic acid. The latter two gave similar results with cinnamic acid, performing better for the 1-hydroxybenzophenone esters and *p*-methoxy cinnamic acid for ethyl esters. This is supposedly caused by poor solubility of the nitro compounds, leading to extremely low concentrations in solution. This is supported by the fact that nitro cinnamic acid performed best for the two substance classes that had the best solubility, methoxy-*iso*-propanol and vinyl dimethyl silyl respectively. Surprisingly cinnamic acid performed even better, which is probably due to partial protonation of the catalyst. This poisoning would be irreversible however, rendering free acids unsuitable for photo-activation. The best inhibitor in the cinnamic acid class proved to be the triolefin **5.5**. This is either because of chelatisation of platinum and therefore higher complex stability, or the close proximity of two relatively electron deficient olefins to the third coordinated one increases the chance of recoordination after dissociation, as long as the inhibitor is close to the metal. The inferior performance of the vinyl dimethyl silyl systems, which should also show chelate properties, is presumably caused by the high electron density provided by the silicon. This leads to fast hydrosilylation of the inhibitors double bonds, rendering it inactive. The conversions for these catalytic systems are summarised in graph 5.1.

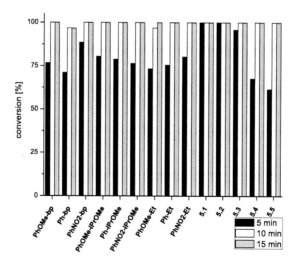

Graph 5.1: Conversions in the presence of cinnamic acid inhibitors.

Fumaric acid shows two advantages compared to cinnamic acid, higher electron deficiency and substitution flexibility due to two acid groups. Again systems comparable to
photo-switchable compounds were investigated, namely 2-hydroxybenzophenone- (**fu-bp**),
1-hydroxyfluorenoneesters (**fu-fl**) and the amide of 2-aminobenzophenone (**fu-Nbp**). For
comparability an innocent second substituent, ethyl, was used for all three systems. The
ethyl group should also increase the solubility. As reference compounds free fumaric acid
(**5.6**) and monoethyl fumarate (**5.7**) were used in analogy to the cinnamic acids. Two
methoxy-*iso*-propanol derivatives were included, the mixed ethyl methoxy-*iso*-propanol
ester **5.8** and the dimethoxy-*iso*-propanol system **5.9**. To investigate chelating effects
triolefin **5.10** and vinyl dimethyl silyl esters (**5.11, 5.12**) were used.

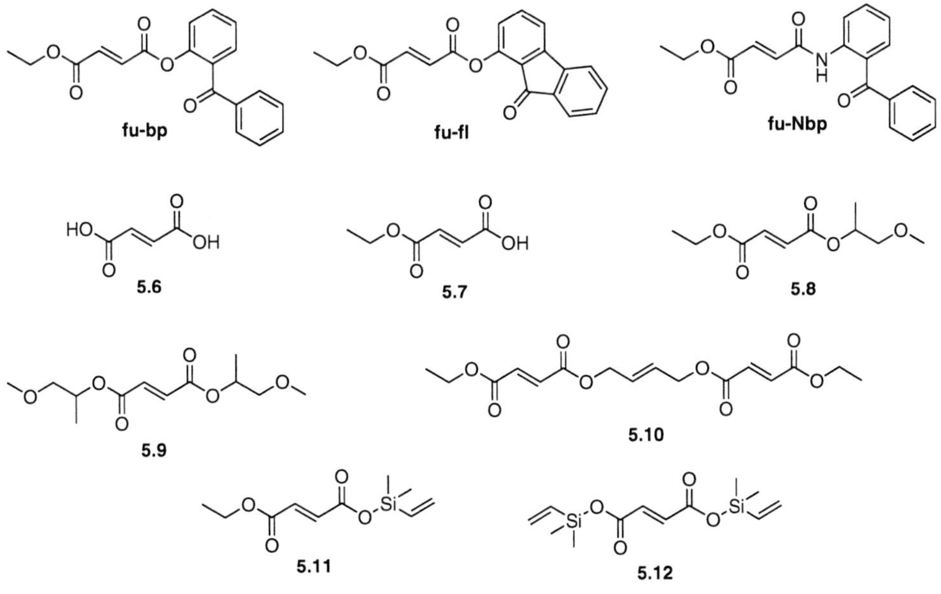

Figure 5.4: Inhibitors containing fumaric acid.

Solubility proved to be the main factor upon inhibition properties, again. For this reason the benzophenone derivative **fu-bp** shows better performance than the significantly
less soluble, fluorenone **fu-fl**. The same fact applies for the amide **fu-Nbp**. The monoethyl ester **5.7** shows good inhibition properties, probably due to catalyst poisoning.

However, free fumaric acid has a comparably low inhibition ability, supposedly caused by extreme low solubility in siloxanes. Methoxy-*iso*-propanol esters inhibit the reaction rate reasonably, whereof the dimethoxy-*iso*-propanol ester **5.9** is superior to mixed ester **5.8**. This indicates additional effects from the methoxy-*iso*-propanol since both systems are completely soluble in the reaction mixture. The triolefin system **5.10** shows the best performance, presumably for the same reasons discussed for cinnamic acid derivative **5.5**. Vinyl dimethyl silyl esters show poor performance again, *vide supra*.

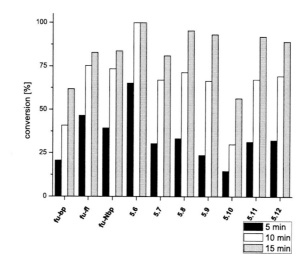

Graph 5.2: Conversions in the presence of fumaric acid inhibitors.

Maleic acids have the same properties as fumaric acid. However the Z-substitution leads to higher solubility in non polar solvents in general. In analogy to fumaric acid, amides of 2-aminobenzophenone (**ma-bp**) and 1-aminofluorenone (**ma-fl**) were investigated, as well as the vinyl dimethyl silyl functionalised amide (**5.13**) and the mixed ethyl benzophenone and methoxy-*iso*-propanol benzophenone esters **5.14** and **5.18**. Derived from **ma-bp** and **ma-fl**, maleimides **mi-bp** and **mi-fl** were investigated as well. As reference compounds maleic acid (**5.23**), maleic anhydride (**5.19**) and maleic acid monoethyl ester (**5.22**) were chosen. For further insight into the maleamide-maleimide relation, the simpler derivatives

5.20 and **5.21** were examined. Mono- (**5.15**) and dimethoxy-*iso*-propanol (**5.24**) esters were included together with the vinyl dimethyl silyl systems **5.16**, **5.17** and **5.25** for the considerations mentioned above.

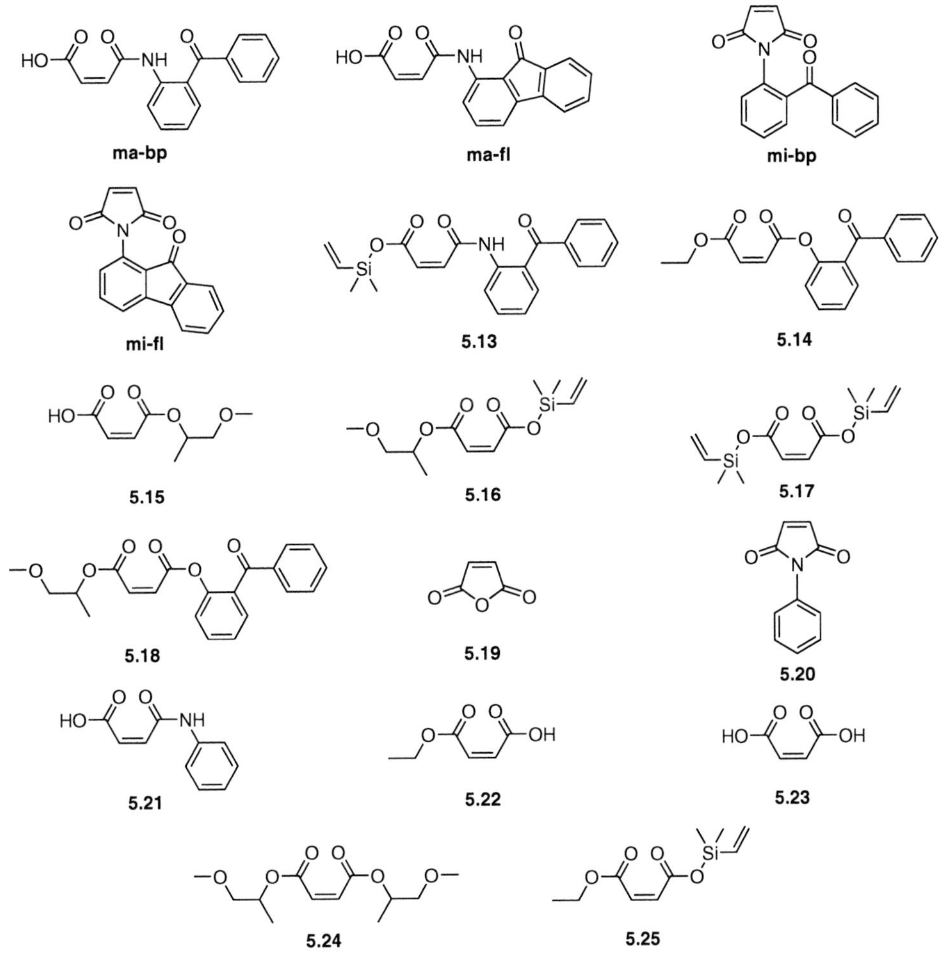

Figure 5.5: Inhibitors containing maleic acid.

Solubility proved to be the main factor again. Therefore benzophenone systems (**ma-bp, mi-bp**) showed better performance than fluorenone systems (**ma-fl,mi-fl**), while the introduction of groups improving solubility led to sufficiently higher inhibition ratios (**5.13**). Esters were better inhibitors than amides due to solubility as well (**5.14**).

Methoxy-*iso*-propanol enhances the performance of the inhibitors if the other acid group bears an ester moiety (**5.18, 5.24**). Even the mono ester of methoxy-*iso*-propanol (**5.15**) is superior to the monoethyl maleate (**5.22**), in contrast to fumaric acid. Free maleic acid shows little inhibition, due to low solubility, *vide supra*, whereas maleic anhydride shows excellent reaction rates, which are caused by poisoning though. Simple maleimide **5.20** is superior to the corresponding maleamide **5.21** due to its better solubility. However more complex maleimides show significantly lower solubility. The accelerating effect of vinyl dimethyl silyl is less pronounced (**5.16, 5.17**) and even improves the performance of the ethyl systems (**5.25**).

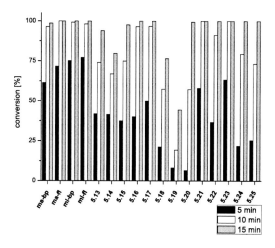

Graph 5.3: Conversions in the presence of maleic acid inhibitors.

To compare the different olefins all inhibitors bearing ethyl groups were compared (figure 5.6). The cinnamic acids perform inferior to all other systems, probably due to electronics and sterics. As described above, the solubility of the inhibitor is the main issue, therefore the fluorenone system **fu-fl** and amide **fu-Nbp** are inferior to **fu-bp**. However the analogue maleic acid system **5.14** performs significantly worse, *vide infra*. Due to the good solubility all other fumaric acid systems exhibit a comparable inhibition potential, with the free acid being slightly better. The maleic acid systems surprisingly are inferior to

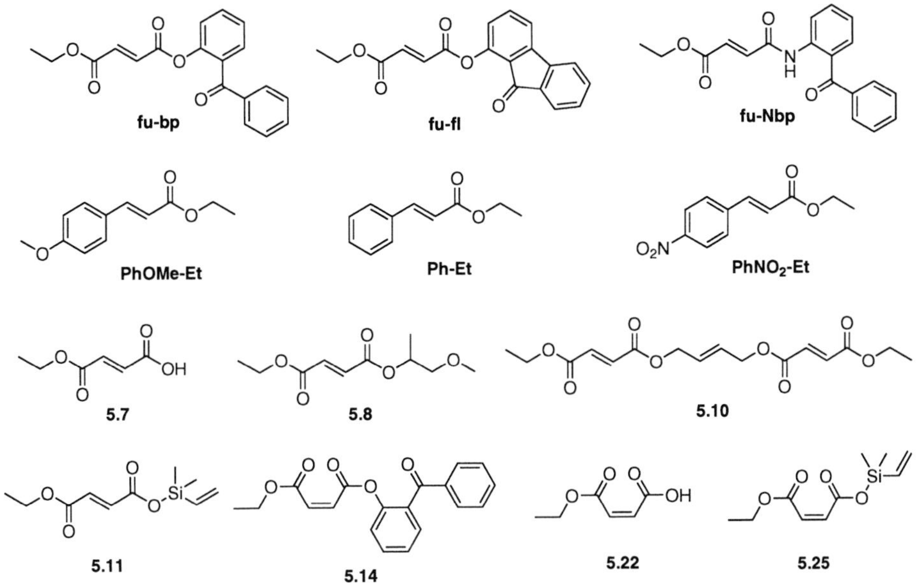

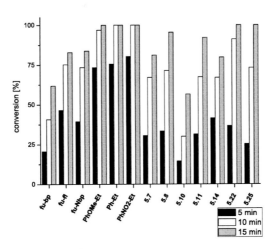

Figure 5.6: Inhibitors containing ethyl groups.

fumaric acid, probably due to the easier attack on the Z-substituted double bond.

Graph 5.4: Conversions in the presence of ethyl bearing inhibitors.

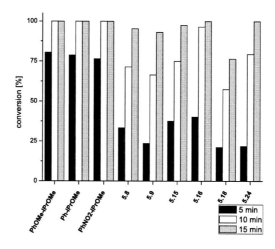

Figure 5.7: Inhibitors containing methoxy-*iso*-propanol groups.

Graph 5.5: Conversions in the presence of methoxy-*iso*-propanol bearing inhibitors.

Upon comparison of the methoxy-*iso*-propanol substituted inhibitors (figure 5.7 & graph 5.5), the cinnamic acids proved to be the least applicable, *vide supra*. Systems bearing two methoxy-*iso*-propanol groups show better inhibition capabilities than the ones with only one. **5.18** has similar inhibition rates however, probably caused by the electron withdrawing effect of benzophenone in addition to the excellent solubility provided by the methoxy-*iso*-propanol. This indicates that methoxy-*iso*-propanol moieties exhibit, besides the increase in solubility, an additional inhibiting effect, which had to be investigated further (section 5.3). Maleic and fumaric acid inhibitors show comparable properties.

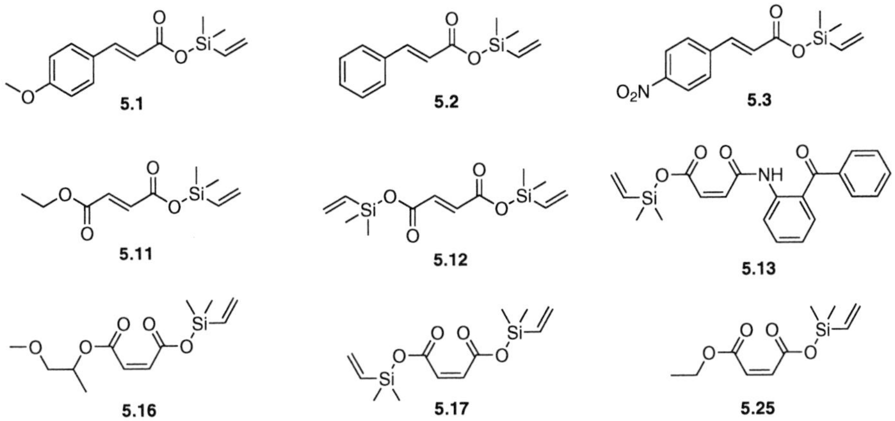

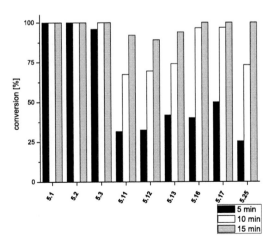

Figure 5.8: Inhibitors containing vinyl dimethyl silyl groups.

Graph 5.6: Conversions in the presence of vinyl dimethyl silyl bearing inhibitors.

The vinyl dimethyl silyl inhibitors show worse performance over all in comparison to the respective alkyl or aryl systems (figure 5.8 & graph 5.6). Maleic and fumaric acid have similar inhibition capabilities whereas cinnamic acids do not reduce reaction velocity at all. Even though all silylesters are highly soluble in siloxanes no advantages to non-substituted compounds are observed. In some cases the inhibition capability is even reduced. However at long reaction times the vinyl dimethyl silyl derivatives prevented the formation of colloidal platinum and therefore the reaction solution stayed colourless.

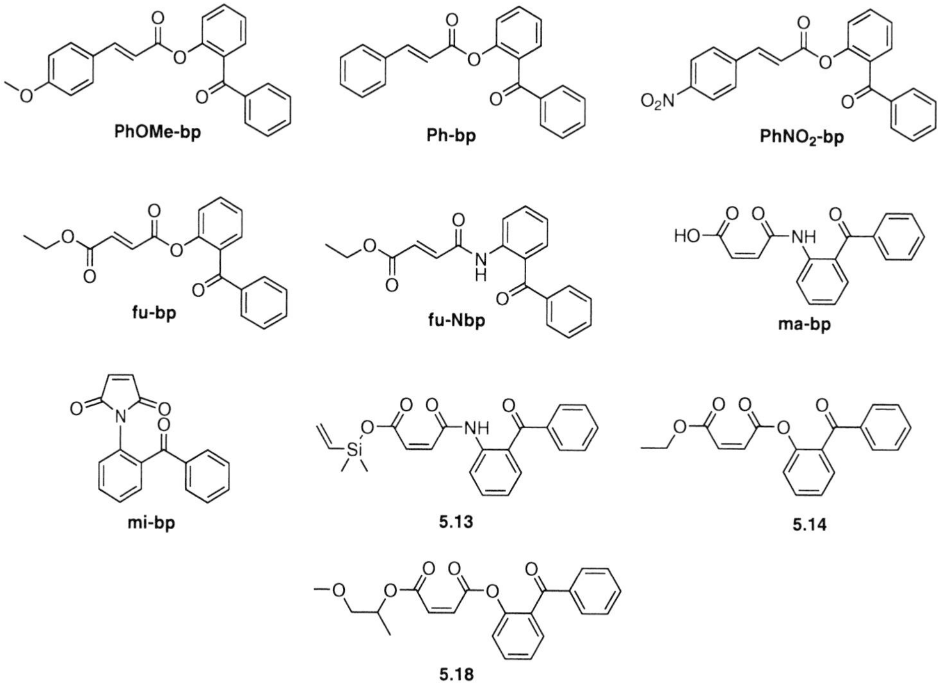

Figure 5.9: Inhibitors containing benzophenones.

Comparison of all 1-hydroxybenzophenone esters supports, that solubility is the main contributor to the *Karstedt* inhibition ability of olefins. Therefore **fu-bp** and **5.18** exhibit the best characteristics, followed by **fu-Nbp**, **5.13** and **5.14** which are all well soluble in the reaction mixture. The cinnamic acid derivatives **PhOMe−bp**, **Ph−bp** and **PhNO$_2$−bp** again give the worst results.

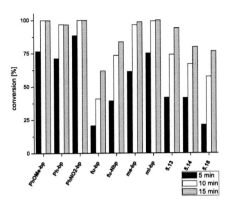

Graph 5.7: Conversions in the presence of benzophenone bearing inhibitors.

Since diazofluorene derivatives are needed to obtain photo-switches stable at ambient temperature similar benzophenone and fluorenone systems were compared (figure 5.10).

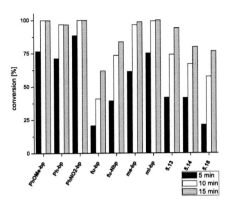

Figure 5.10: Comparison of benzophenone to fluorenone.

All benzophenone inhibitors show superior performance to fluorenone derivatives (graph 5.8). However this is only caused by the different solution characteristics of the compounds. Therefore the difference is less pronounced for poorly soluble substances like

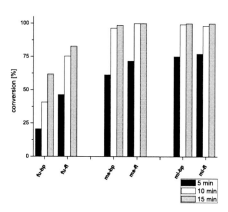

Graph 5.8: Conversions dependent of benzophenone and fluorenone inhibitors.

ma-bp and **ma-fl**. Since diazofluorenes are far more soluble than 9-fluorenones they are expected to react comparable to the benzophenone systems.

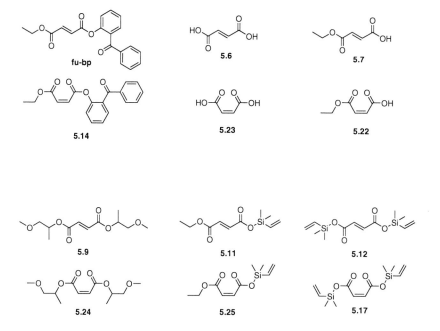

Figure 5.11: Comparison of *E* vs. *Z* substitution.

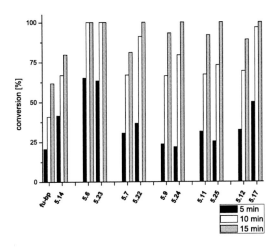

Graph 5.9: Conversions in the presence of E and Z substituted diacids.

Finally the dependence of E- and Z-substitution was investigated. Therefore similarly substituted maleic and fumaric acids esters were compared. The fumaric acid derivative proves superior for benzophenone ethyl diesters (**fu-bp, 5.14**), ethyl mono esters (**5.7, 5.22**) and vinyl dimethyl silane diesters (**5.12, 5.17**). The free acids (**5.6, 5.23**) show equal performance, which supports the theory, that catalyst poisoning via protonation is the main inhibition mechanism for these two compounds. For methoxy-*iso*-propanol diesters (**5.9, 5.24**) and ethyl vinyl dimethyl silyl diesters (**5.11, 5.25**) maleic acid is superior in the beginning of the catalysis however overall faster than fumarates. This might be caused by the easier accessibility of the Z-substituted double bond, which coordinates better to platinum but is also hydrosilylated faster. All catalytic conversions are summarised in table 5.1.

Table 5.1.: Conversions with different *Karstedt* inhibitors [%]:

	5 min	10 min	15 min
PhOMe−bp	76.69	100.00	100.00
Ph−bp	71.18	96.81	96.69

continued on next page

Table 5.1 – continued from previous page

	5 min	10 min	15 min
PhNO$_2$–bp	88.56	100.00	100.00
PhOMe–^{i}PrOMe	80.47	100.00	100.00
Ph–^{i}PrOMe	78.81	100.00	100.00
PhNO$_2$–^{i}PrOMe	76.47	100.00	100.00
PhOMe–Et	73.40	96.86	100.00
Ph–Et	75.55	100.00	100.00
PhNO$_2$–Et	80.35	100.00	100.00
fu–bp	20.63	40.83	61.83
fu–fl	46.52	75.31	82.88
fu–Nbp	39.39	73.47	83.79
ma–bp	61.39	96.43	98.61
ma–fl	71.67	100.00	100.00
mi–bp	75.12	99.24	100.00
mi–fl	77.12	98.13	100.00
5.1	100.00	100.00	100.00
5.2	100.00	100.00	100.00
5.3	95.84	100.00	100.00
5.4	67.74	100.00	100.00
5.5	61.54	100.00	100.00
5.6	65.16	100.00	100.00
5.7	30.56	67.00	81.13
5.8	33.33	71.43	95.45
5.9	23.66	66.44	93.22
5.10	14.53	30.07	56.52
5.11	31.51	67.32	92.08
5.12	32.43	69.42	89.17
5.13	41.86	74.03	94.01

continued on next page

Table 5.1 – continued from previous page

	5 min	10 min	15 min
5.14	41.52	66.89	79.76
5.15	37.50	74.94	97.65
5.16	40.12	96.54	100.00
5.17	50.00	96.74	100.00
5.18	21.26	57.45	76.53
5.19	8.26	19.35	44.44
5.20	6.54	57.26	99.45
5.21	57.98	100.00	100.00
5.22	36.71	91.09	100.00
5.23	63.37	100.00	100.00
5.24	21.88	79.38	100.00
5.25	25.37	73.19	100.00

5.3. Further complexation studies

To investigate the mechanism accountable for the remarkable inhibition enhancement exhibited by the methoxy-*iso*-propanol moiety, further complexation studies were performed. Therefore bis(triphenylphosphine) olefin complexes of palladium and platinum were synthesised with *p*-nitro cinnamic acid methoxy-*iso*-propanol ester as olefin. Potential interaction of the ester´s ether group with the metal, as depicted in figure 5.12, should show up in either coupling of the methyl group via the oxygen to the phosphines, in case of platinum also to ^{195}Pt, or by a change of the $^2J_{PP}$ coupling constant caused by symmetry change, in comparison to the bis(triphenylphosphine) olefin systems discussed in chapter 4. The shift of the signals in ^{31}P NMR should change as well, due to the change in electron density at the metal center.

However no significant difference could be observed in comparison to the simpler cinnamic acid complexes. All NMR data recorded for the palladium complex

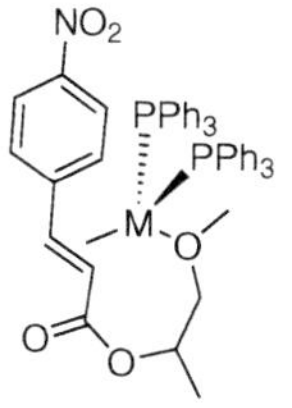

Figure 5.12: Possible coordination mode of methoxy-*iso*-propanol olefin ligands. M ≡ Pd, Pt.

PdPhNO$_2$–^{i}PrOMe and platinum complex **PtPhNO$_2$–^{i}PrOMe** are conform to the signals expected and shift regions are in analogy to the ones discussed in section 4.2.1. The only remarkable feature is the formation of a pair of diastereoisomers due the chirality of the methoxy-*iso*-propanol group. Therefore no further discussion of the NMR spectroscopic data is conducted.

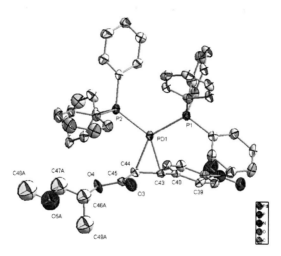

Figure 5.13: ORTEP[24] representation of **PdPhNO$_2$–^{i}PrOMe**. Thermal ellipsoids are given at 50% probability level. Hydrogen atoms are omitted for clarity.

Crystals suitable for X-ray diffraction analysis could be grown by slow diffusion of pentane into a saturated toluene solution of **PdPhNO$_2$–^{i}PrOMe**. The crystal structure shows a disorder of the methoxy-*iso*-propanol group, either standing along the plane spanned by the olefin or perpendicular to it. Only the former is shown in the figures 5.13 and 5.14. The length of the olefin bond lies with 1.46 Å between a C-C single and a

C=C double bond as expected. The Pd-C43 bond is with 2.13 Å slightly shorter than the Pd-C44 bond with 2.14 Å due to the lower electron density at C43. The C43-Pd-C44 angle is 38.8° and the P1-Pd-P2 angle is 113.0°. The deviation from the ideal 120° is caused by the higher steric demand of the olefin. P1-Pd-C43 angle is with 103.7° slightly smaller than the P2-Pd-C44 angle with 104.9° due to the higher steric demand of the carboxyl group. The torsion angles of the olefin is with 156.5° significantly smaller than the ideal 180° for non coordinated E-olefins due to the hybridization change of the olefin carbon atoms caused by the π back-bonding of palladium. The complex shows almost perfect coplanarity concerning the olefin carbon atoms, palladium and both phosphorus atoms.

Table 5.2.: Selected bond length and angles of **PdPhNO$_2$–^{i}PrOMe**:

d [Å]		$\angle$ [°]	
C40–C43	1.464(5)	C40–C43–C44	123.4(4)
C43–C44	1.417(5)	C43–C44–C45	117.6(4)
C44–C45	1.463(6)	C43–Pd1–C44	38.8(1)
C46A–C49B	1.484(8)	C46A–C47A–O5A	91.6(6)
C46A–C47B	1.506(10)	C46A–C49A–O5B	96.9(7)
C47B–O5B	1.517(12)	C47A–O5A–C48A	105.3(9)
C48A–O5A	1.425(14)	C47A–C46A–C49A	113.7(5)
C48B–O5B	1.399(18)	C49A–O5B–C48B	118.9(10)
O3–C45	1.205(6)	O3–C45–C44	126.3(4)
O4–C45	1.362(6)	O4–C45–C44	111.4(4)
O4–C46A	1.464(7)	O4–C46A–C47A	106.6(5)
O5A–C49B	1.563(11)	O4–C46A–C49A	107.6(5)
Pd1–P2	2.319(1)	P1–Pd1–C43	103.7(1)
Pd1–P1	2.310(1)	P2–Pd1–C44	104.9(1)
Pd1–C44	2.138(4)	P2–Pd1–P1	113.0(0)
Pd1–C43	2.131(4)		

Table 5.3.: Selected torsion angles of **PdPhNO$_2$**–^{i}**PrOMe** [°]:

C40–C43–C44–C45	156.5(4)
C43–C44–Pd1–P2	172.5(2)
C44–C43–Pd1–P1	173.5(2)

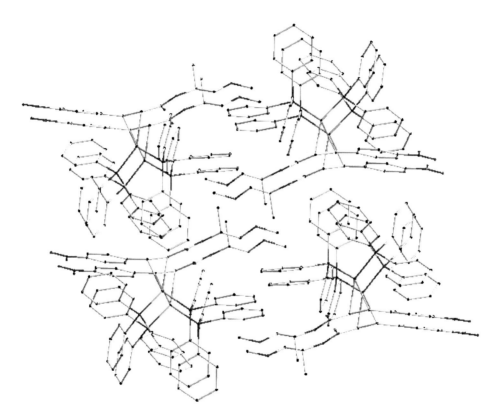

Figure 5.14: Wire frame image showing the packing mode of **PdPhNO$_2$**–^{i}**PrOMe** in the crystal lattice.

The Pd-O5A distance of 6.99 Å gives no indication of interaction nor does the distance to the neighboring molecule in the crystal lattice as shown in figure 5.14. Therefore it is assumed, that the inhibition properties of methoxy-*iso*-propanol are caused by simple steric shielding in combination with the excellent solubility evoked by this group.

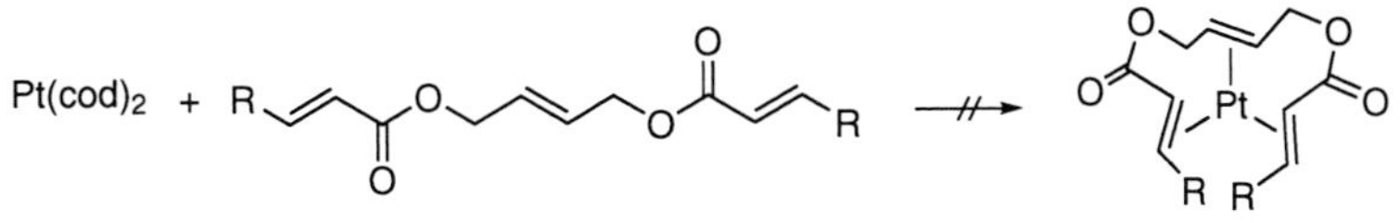

Scheme 5.5: Diol diesters as chelating ligands. R ≡ Ph, C(O)OEt

Since the triolefins **5.5** and **5.10** exhibited also good inhibition properties, complexation studies were conducted with Pt(cod)$_2$. Even though the 1,6-olefin-sequence should be ideal for chelate complexes and the similarity of the ligand to the one employed by *Roglans et al.*[153–156] and *Moreno-Mañas et al.*,[157–159] no coordination was observed in NMR experiments (scheme 5.5). However the precursor did decompose rapidly in solution without any stabilisation by the triolefin. Therefore further experiments were suspended.

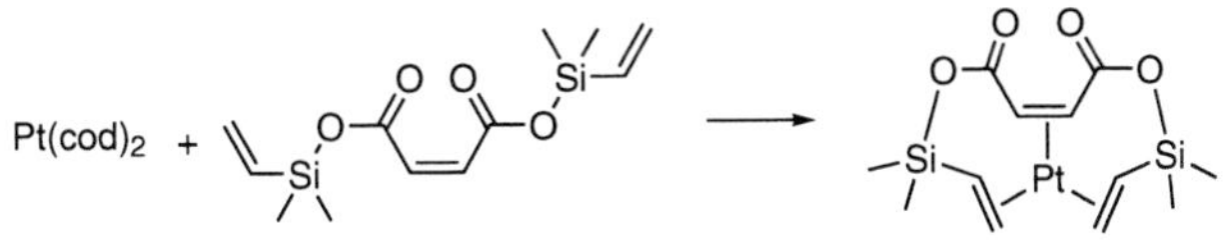

Scheme 5.6: Vinyl silyl diesters as chelating ligands.

The behaviour of vinyl dimethyl silyl diester **5.17** with Pt(cod)$_2$ was examined as well. Even though coordination to all three olefins could be observed in ^{1}H NMR no substance could be isolated because of its good solubility in all tested organic solvents. Therefore no purification could be performed. Due to the disappointing performance and the similarity to known complex systems[160–162] work on this system was discontinued.

5.4. Conclusions & outlook

It was found that solubility is the main problem concerning inhibition of *Karstedt* type catalysts. Cinnamic acid derivatives did not fulfill the demand due to low inhibition ability, caused by low solubility and inadequate electron deficiency. The mimic of *Karstedt* analogous ligand systems via the introduction of vinyl dimethyl silyl groups did not prove successful and no substantial decrease in reaction rate was observed, even though chelating potential was observed in NMR studies for this ligand class. This is supposedly caused by fast hydrosilylation of the electron rich double bonds of the inhibitor, rendering it ineffective. However these systems are capable to suppress catalyst decomposition and therefore keep the reaction solution colourless. Maleic and fumaric acid had the best inhibition properties of the simple modifiable compounds with **fu-bp** and **5.18** being the most potent. Amides and imides rendered unfeasible because of their low solubility. Free carboxylic groups exhibit good inhibition rates, however solubility issues arise and the inhibition is presumably caused by protonation of the catalyst. This renders this substance class non-applicable in photo-activated catalysis.

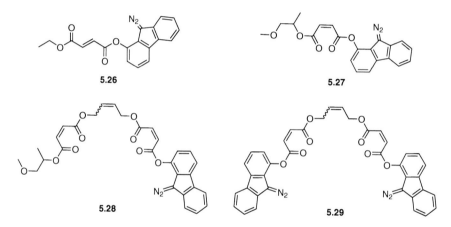

Figure 5.15: Proposed photo-switchable *Karstedt* inhibitors.

Methoxy-*iso*-propanol enhances the inhibition effects, likely due to the increased solubility of the compound and increased sterical shielding of the catalytic center is likely when coordinated. The best results were obtained for the triolefins **5.5** and **5.10**. One reason may be chelatisation of platinum and therefore higher complex stability, yet no evidence in NMR spectroscopy could be observed. An other reason may be the close proximity of two relatively electron deficient olefins to the third coordinated one, which increases the chance of recoordination after dissociation as long as the inhibitor is close to the metal.

Therefore four different systems for photo-switchable *Karstedt* inhibitors are proposed. Systems **5.26** and **5.27** derive directly from the two good inhibitors **fu-bp** and **5.18** and should be easily accessible via the routes established in this thesis. Whereas **5.28** and **5.29** should show superior inhibition capability, as expected from the results with the triolefins **5.5** and **5.10**. However the latter systems pose a higher synthetic challenge and should be medium term targets. In all systems a photo-induced partial or complete removal of the double bonds should be possible. This would then trigger hydrosilylation or at least increase the activity of the catalyst substantially in comparison to the pre-irradiation activity.

6. Bis(triphenylphosphine)palladium(0)

During the synthesis of the bis(triphenylphosphine)palladium olefin systems (scheme 6.1) it was observed, that upon addition of triphenylphosphine to a $Pd(C_3H_5)Cp$ solution, immediate colour change from deep red to yellow occurred. In concentrated solution even precipitation of a yellow solid took place.

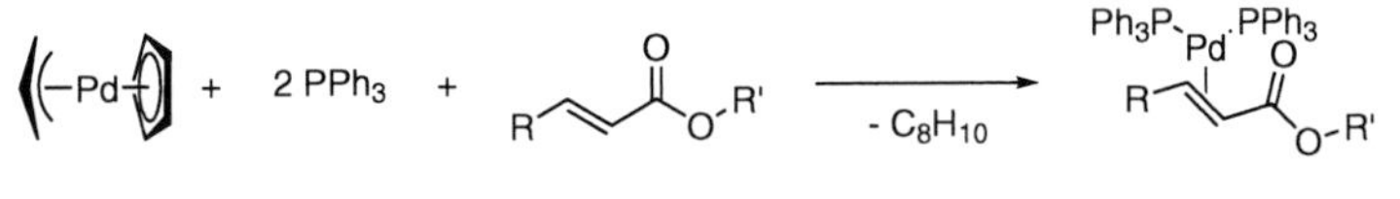

Scheme 6.1: General synthesis of bis triphenylphosphine palladium olefin complexes.

Therefore an investigation of the reactivity of $Pd(C_3H_5)Cp$ with triphenylphosphine was performed with the focus on stable reaction products. These could serve as alternative precursors to bis(triphenylphosphine)palladium(0) olefin complexes.

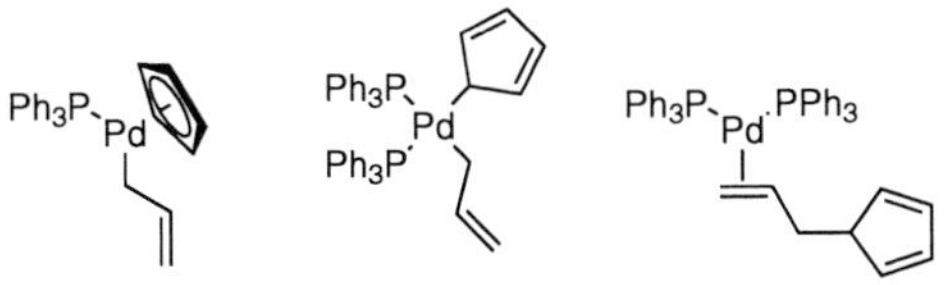

Figure 6.1: Possibly isolable intermediates.

It was expected, that upon addition of triphenylphosphine either a η^3-η^1 shift of the allyl moiety or a η^5-η^3 shift of the Cp-ring yielding 18 electron complexes would occur. Alternatively a η^5-η^1 shift of Cp to a 16 electron complex would be possible as well. After the addition of a second equivalent of triphenylphosphine either a Pd(II) complex with two alkyl substituents or already a bis(triphenylphosphine)Pd(0) olefin system should be generated. In figure 6.1 possible addition products of triphenylphosphine to $Pd(C_3H_5)Cp$ are depicted.

6.1. Synthesis

At first reactivity studies were conducted on NMR spectroscopy scale. Upon addition of one equivalent of triphenylphosphine to a toluene or benzene solution of $Pd(C_3H_5)Cp$ the colour changes immediately from deep red to orange. In the ^{31}P NMR spectrum one signal at 24.7 ppm is observed. The ^{1}H NMR spectrum shows one multiplet at 3.4 ppm with an integral value of one, two doublets at 2.15 ppm and 2.63 ppm with coupling constants of $^2J_{HH} = 12.7$ Hz and an integral value of two and a singlet at 6.07 ppm with an integral value of five. In addition two aromatic multiplets and signals of one equivalent of allyl-cyclopentadiene (C_8H_{10}) are observed. These signals indicate, that exclusively the homo-bimetallic complex **6.1** is formed (scheme 6.2). Similar bridged complexes were obtained by *Werner et al.* and *Norton et al.*[163,164]

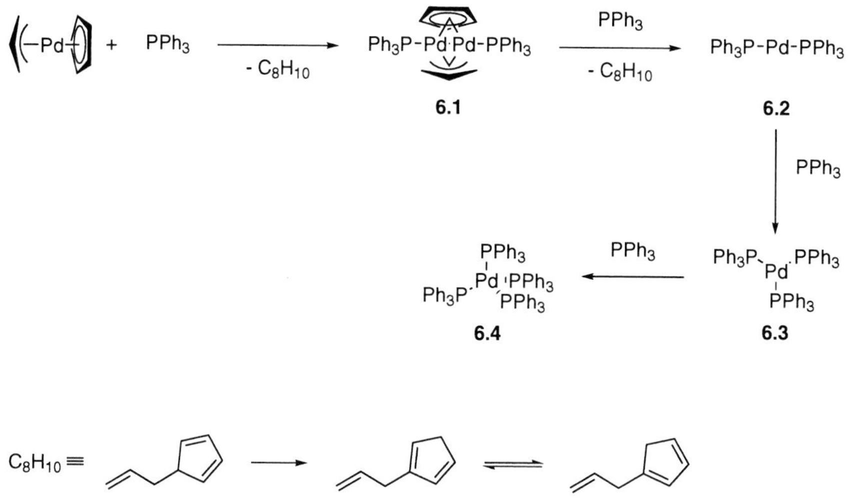

Scheme 6.2: Reaction cascade of $(\eta^3$-allyl$)(\eta^5$-cyclopentadienyl)palladium(II) with triphenylphosphine.

After the addition of a second equivalent of PPh_3 the colour of the reaction solution turns bright yellow and the signal in the ^{31}P NMR spectrum shifts to 22.7 ppm. The ^{1}H NMR spectrum shows exclusively signals of free C_8H_{10} in the non aromatic region of the spectrum without any indication of coordination. This is due to the formation of bis(triphenylphosphine)palladium(0) (**6.2**).[165] This reaction is in analogy to the synthesis of bis(tri-*o*-tolylphosphine)palladium(0) by *Böhm* and *Herrmann*[116] and bis(trialkylphosphine)palladium(0) for bulky alkyl substituents by *Yoshida* and *Otsuka*.[109] Surprisingly this 14 electron species is stable in solution if free allyl-cyclopentadiene is present. Therefore it must be assumed, that partial coordination of the olefin stabilises the system, even though no direct evidence of coordination could be observed in NMR spectroscopy.

At the addition of the third equivalent triphenylphosphine, a light yellow precipitate forms and the ^{31}P NMR spectroscopy signal shifts to 22.6 ppm. This is in accordance with the signal estimated for tris(triphenylphosphine)palladium(0) (**6.3**).[166] However, the broadness of the signal in the ^{31}P NMR spectrum indicates the presence of an equilibrium between twofold, threefold and fourfold coordinated palladium species caused by dissociation of phosphine as described by *Fischer* and *Werner*.[167]

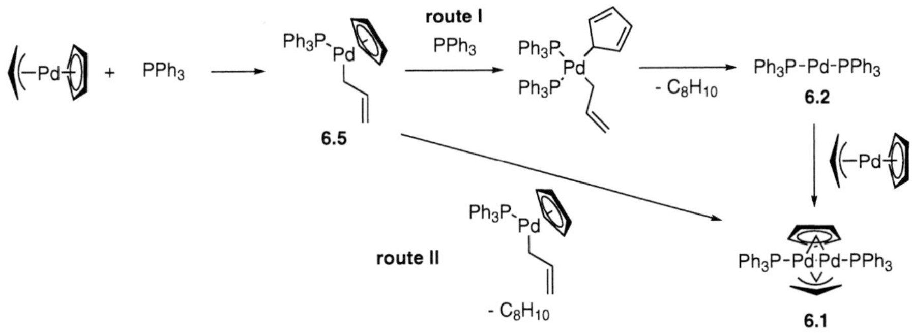

Scheme 6.3: Pathways to **6.1** proposed by literature.[164,168]

Upon the presence of four equivalents PPh$_3$ a shift to 1.5 ppm occurs in ^{31}P NMR spectroscopy. Even though the chemical shift of tetrakis(triphenylphosphine)palladium(0) (**6.4**) in ^{31}P NMR spectra is reported at 29.8 ppm,[169] it is assumed that Pd(PPh$_3$)$_4$ is formed as described by *Harder* and *Werner*.[170] The non matching chemical shift could be explained by an equilibrium similar to the one mentioned above. Due to the low solubility of $Pd(PPh_3)_3$ and $Pd(PPh_3)_4$ a large excess of free triphenylphosphine is present in solution and therefore the signal is observed close to free triphenylphosphine.[171]

Two ways were proposed for the generation of **6.1** from $Pd(C_3H_5)Cp$ (scheme 6.3). *Werner et al.* suggest, that $Pd(PPh_3)_2$ is generated first and reacts subsequently with one equivalent $Pd(C_3H_5)Cp$ to **6.1**.[163,168,172] *Norton et al.* however favour the direct conversion of two equivalents of a triphenylphosphine palladium(II)-η^1-ally-η^5-cyclopentadienyl (**6.5**) to **6.1** under reductive elimination of C_8H_{10}.[164] Due to the lack of signals for **6.2** in the ^{31}P NMR spectrum we support the latter proposal. Both authors state **6.5** as the first intermediate, even though no spectroscopic evidence is present. So the intermediate must be extremely short lived.

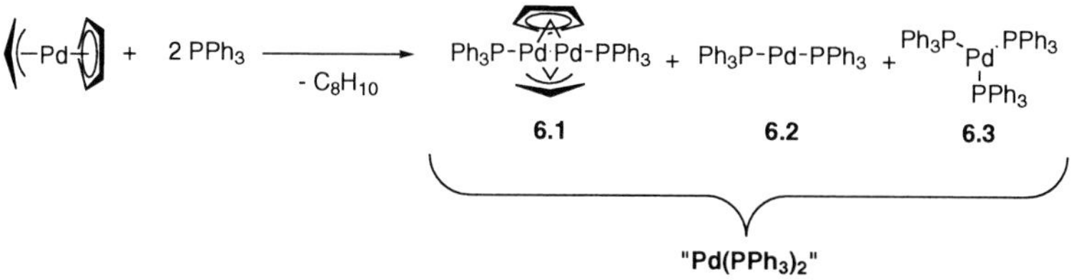

Scheme 6.4: Large scale reactivity of $(\eta^3$-allyl)$(\eta^5$-cyclopentadienyl)palladium(II) with triphenylphosphine.

Albeit the smooth reaction proceeding in NMR spectroscopic scale, attempts to synthesise **6.2** in a large scale by simple reaction of triphenylphosphine with $Pd(C_3H_5)Cp$ gave only mixtures, as shown in scheme 6.4. The average product distribution was: 7% **6.1**, 79% **6.2** and 14% **6.3**. The main reason for this are due to solubility problems in large scale. This leads to precipitation of **6.1** and the thereby generated excess of PPh_3 reacts with **6.2** to **6.3**.

Another reason is an equilibrium between all three species. EXSY spectroscopy[173-175] shows extremely strong exchange signals for **6.1** and **6.2** which are even more intensive than the diagonal signal of **6.1** (figure 6.2). The fast equilibrium of **6.2** and **6.3** was already described by *Fischer* and *Werner*.[167]

To obtain pure **6.2** the product mixture had to be separated by inert gas *flash* col-

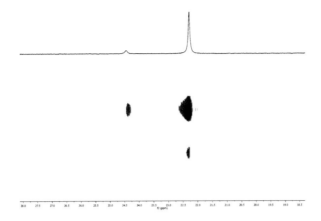

Figure 6.2: ^{31}P-^{31}P EXSY NMR spectrum of "Pd(PPh$_3$)$_2$".

umn chromatography over degased Al$_2$O$_3$. Even though this procedure is laborious it still provides a substantial simplification to the previous known one.[165] Pd(PPh$_3$)$_2$ is a light yellow, extremely air sensitive solid, which decomposes at ambient temperature to Pd(PPh$_3$)$_3$ and Pd black, presumably auto-catalytically according to optical observations. However the crude product mixture of **6.1**, **6.2** and **6.3** ("**Pd(PPh$_3$)$_2$**") proved stable at ambient temperature in solid and solution and could be used as a synthetic equivalent of Pd(PPh$_3$)$_2$ after mass correction for the still present allyl- and cyclopentadienyl-moieties, *vide infra*.

Attempts to crystallise **6.2** failed due to decomposition to **6.3** in solution and the significantly lower solubility of the latter. However crystals of **6.3** suitable for X-ray diffraction analysis could be grown by slow diffusion of pentane into a saturated toluene solution of "**Pd(PPh$_3$)$_2$**". The Palladium is coordinated in a trigonal planar fashion with a displacement from the plane spanned by the three phosphorus atoms of 0.08 Å. This is only half the deviation as in the previous described crystal structure of Pd(PPh$_3$)$_3$[176] with 0.16 Å and is even closer to ideal trigonal planarity than Pt(PPh$_3$)$_3$[177] with 0.1 Å. The Pd-P distances are with 2.32 Å to 2.33 Å slightly shorter than in the sterically more crowded Pd(PPh$_3$)$_4$ (2.43 Å-2.46 Å)[178] but longer than in Pt(PPh$_3$)$_3$ (2.27 Å)[179], as would be expected for the larger and therefore less sterically crowded metal. The

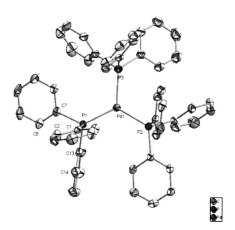

Figure 6.3: ORTEP[24] representation of **6.3**. Thermal ellipsoids are given at 50% probability
level. Hydrogen atoms are omitted for clarity.

P-Pd-P angles are almost ideal for trigonal planar systems with 118.4°-120.2°, comparable to Pt(PPh$_3$)$_3$, which contrasts the Pd(PPh$_3$)$_3$ structure described by *Sergienko* and
Porai-Koshits.[176] The structural parameters for the triphenylphosphine (table 6.1) are
in consistence with the data described for the mentioned complexes, as expected.

Table 6.1.: Selected bond length and angles **6.3**:

d [Å]		∠ [°]	
C1–C2	1.392(5)	C1–C6–C5	120.7(3)
C1–P1	1.840(4)	C1–C2–C3	120.5(3)
C1–C6	1.387(5)	C2–C3–C4	120.1(3)
C2–C3	1.390(5)	C2–C1–C6	118.4(3)
C3–C4	1.377(5)	C3–C4–C5	119.7(4)
C4–C5	1.374(6)	C4–C5–C6	120.5(4)
C5–C6	1.384(5)	C37–P3–C43	103.7(2)
C7–P1	1.845(4)	C37–P3–C49	102.9(2)
C13–P1	1.834(4)	C37–P3–Pd1	105.2(1)
C19–P2	1.848(4)	C43–P3–Pd1	123.4(1)

continued on next page

Table 6.1 – continued from previous page

d [Å]		∠ [°]	
C25–P2	1.851(4)	C43–P3–C49	98.7(2)
C31–P2	1.840(4)	C49–P3–Pd1	120.3(1)
C37–P3	1.843(4)	P1–Pd1–P2	120.2(0)
C43–P3	1.845(4)	P1–Pd1–P3	118.7(0)
C49–P3	1.838(4)	P2–Pd1–P3	118.4(0)
P1–Pd1	2.330(1)		
P2–Pd1	2.323(1)		
P3–Pd1	2.328(1)		

6.2. Reactivity

As already stated the "Pd(PPh$_3$)$_2$" is an ideal precursor to organometallic chemistry containing a Pd(PPh$_3$)$_2$ moiety, due to its relative stability and the easily in vacuo removable C$_8$H$_{10}$, *vide supra*.

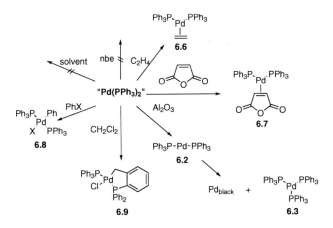

Scheme 6.5: Reactivity of "Pd(PPh$_3$)$_2$". Solvent ≡ acetone, dmso, acetonitrile, pyridine, Et$_2$O, thf. X ≡ Br, I.

When a toluene solution of "Pd(PPh$_3$)$_2$" is saturated with ethylene and stirred for two

days under an ethylene atmosphere the yellow solution decolourises. After removal of the
solvent and crystalisation from pentane $Pd(PPh_3)_2(C_2H_4)$ (**6.6**) is obtained in 79% yield.
This is, to our knowledge, the best yield obtained for **6.6** to date and the described route
renders the previously needed pyrophoric aluminium reagents obsolete.[180–182] Reaction
with maleic anhydride yields the corresponding bis(triphenylphosphine)palladium(0)olefin
6.7[182] complex in quantitative yield. However electron rich olefins like norbornene do
not react at all and only starting material was isolated. No indications for coordination
of norbornene were observed in NMR experiments, either. No reaction with coordinating solvents like acetone, dmso, acetonitrile, pyridine, diethylether or thf did take place
and only precursor or decomposition products were isolated. Due to the high air and
temperature sensitivity no harsher conditions could be applied to enforce reaction. However "**Pd(PPh$_3$)$_2$**" readily inserts into carbon-bromine and carbon-iodine bonds yielding the oxidative addition product in quantitative yields within minutes. "**Pd(PPh$_3$)$_2$**"
interestingly reacts with dichloromethane under C-C bond formation to the phospha-
pallada-cycle **6.9**, via an unknown route. This compound is similar in structure to the
phospha-pallada-cycle described by *Fenton*, which was obtained by C-H-activation of tri-
o-tolylphosphine.[183]

Crystals of **6.9** suitable for X-ray diffraction analysis could be grown by slow diffusion
of dichloromethane into a saturated toluene solution of **6.2**. The complex is of distorted
square planar symmetry with an angle of 9.6° between the P1-Pd-C37 and the P2-Pd-Cl
planes. The C37-Pd-P1 angle is due to ring strain with 82.3° significantly smaller than
90°. The C37-Pd-P2 angle is also smaller than the ideal angle for square planar complexes with 86.2°, whereas the P1-Pd-Cl and P2-Pd-Cl angles are slightly to large with
95.2° and 97.1° respectively. This deviation is mainly caused by the high steric demand induced by chlorine. The Pd-P2 distance is with 2.33 Å in the range of similar
complexes[184] containing triphenylphosphine, whereas the Pd-P1 distance is significantly
shorter (2.29 Å) due to the ring strain.[185] The Pd-C37 distance is comparable to other
trans-chlorine-substituted alkyl complexes[186], with 2.06 Å and in the typical range of
palladium carbon single bonds. The Pd-Cl bond length is with 2.39 Å consistent with
comparable examples.[185–188] The bond lengths and angles of the triphenylphosphine moi-

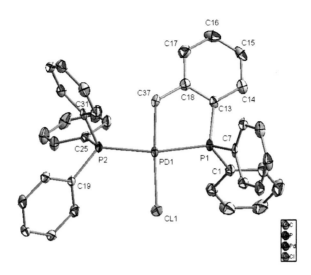

Figure 6.4: ORTEP [24] representation of **6.9**. Thermal ellipsoids are given at 50% probability
level. Hydrogen atoms are omitted for clarity.

ety are in the typical range of triphenylphosphine. The cyclometallated phosphine shows
almost no distortion of bond lengths with similar P1-C distances. The phenyl ring anellated to the pallada-phosphacylce has the expected bond lengths of ca. 1.39 Å, typical for
C=C double bonds, and the angles are close to 120° as expected. The C18-C37 distance
is with 1.15 Å in the typical single bond range and therefore consistent with theory. The
ring strain of the palladacycle slightly distorts the planar geometry of the anellated phenyl
ring, shown by the P1-C13-C18-C37 torsion angle of 7.6°.

Table 6.2.: Selected bond length, angles and torsion angles of **6.9**:

d [Å]		$\angle$ [°]		torsion $\angle$ [°]	
C1–P1	1.818(1)	C1–P1–C13	107.6(2)	C13–C18–C37–Pd1	22.7(5)
C7–P1	1.810(1)	C1–P1–Pd1	118.8(2)	C18–C37–Pd1–P1	21.0(3)
C13–P1	1.819(1)	C1–P1–C7	103.2(2)	C37–Pd1–P1–C13	14.8(2)
C13–C14	1.397(1)	C7–P1–Pd1	116.5(2)	Cl1–P2–Pd1–C37	174.3(2)
C13–C18	1.397(1)	C7–P1–C13	105.7(3)	P1–C13–C18–C37	7.6(5)

continued on next page

Table 6.2 – continued from previous page

d [Å]		∠ [°]		torsion ∠ [°]	
C14–C15	1.385(1)	C13–P1–Pd1	104.1(2)	P2–Cl1–Pd1–P1	173.2(2)
C15–C16	1.382(1)	C13–C14–C15	119.2(4)	Pd1–P1–C13–C18	8.1(4)
C16–C17	1.388(1)	C13–C18–C37	119.8(3)		
C17–C18	1.395(1)	C13–C18–C17	118.3(3)		
C18–C37	1.515(1)	C14–C15–C16	120.6(3)		
C19–C24	1.392(1)	C14–C13–C18	121.0(4)		
C19–P2	1.835(1)	C15–C16–C17	119.9(4)		
C19–C20	1.391(1)	C16–C17–C18	120.9(4)		
C20–C21	1.388(1)	C18–C37–Pd1	116.8(3)		
C21–C22	1.380(1)	C18–C13–P1	112.3(3)		
C22–C23	1.381(1)	C19–P2–C31	101.8(2)		
C23–C24	1.379(1)	C31–P2–Pd1	113.2(2)		
C25–P2	1.829(1)	C37–Pd1–P1	82.3(2)		
C31–P2	1.822(1)	C37–Pd1–P2	86.2(2)		
C37–Pd1	2.063(1)	P1–Pd1–Cl1	95.2(2)		
P1–Pd1	2.292(1)	P2–Pd1–Cl1	97.1(2)		
P2–Pd1	2.325(1)				
Pd1–Cl1	2.392(1)				

Due to the long reaction times and low solubility of **6.9**, no substantial quantities could be synthesised for a complete characterisation. Therefore no statements towards stability, further spectroscopic properties and the reaction pathway can be given to date.

6.3. Conclusions

$Pd(C_3H_5)Cp$ reacts with two equivalents of PPh_3 to "$Pd(PPh_3)_2$" of which pure **6.2** could be isolated. However this mixture may be used without purification as a $Pd(PPh_3)_2$ source and reacts readily with electron deficient or sterically less demanding olefins to bis(triphenylphosphine)palladium(0) olefin complexes (scheme 6.5). It also undergoes fast oxidative addition into C-Br and C-I bonds and is also capable of the insertion into non aromatic C-Cl bonds. However no reaction with various solvents was observed.

7. Summary

7.1. Photo-switchable catalysis

Diphenyl diazo methane esters of cinnamic acid tethered to platinum via a metal olefin bond can be cleaved selectively upon irradiation. The carbene generated *in situ* from the diazo group reacts in an intramolecular cyclopropanation with the olefinic double bond and therefore removes the ligand from the metal. This leads to decomposition of the complex if no additional ligand is present to bind to the free coordination site. Upon irradiation the diazo olefin ester can be substituted with triphenylphosphine and norbornene. This ligand substitution does not occur if the same reaction conditions, apart from the irradiation, are applied and light is strictly excluded form the reaction mixture. These substitution reactions and the fact, that the diazo group is essential for the light triggered ligand exchange, prove the proposed mechanism. These ligand exchange reactions are summarised in scheme 7.1.

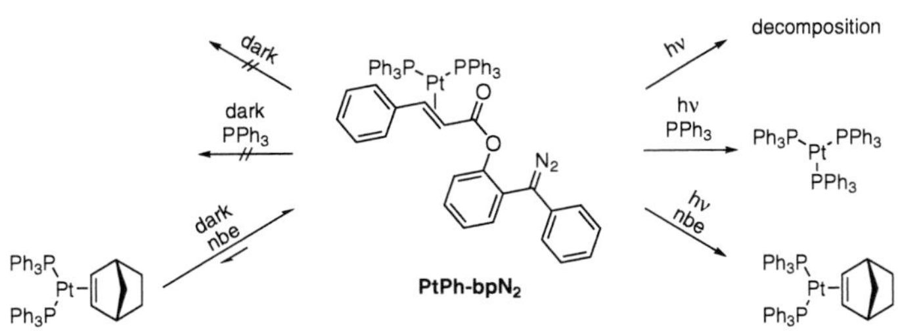

Scheme 7.1: Reactivity of **PtPh−bpN$_2$** in the dark & with UV irradiation.

Since cinnamic acid complexes of platinum are too reactive towards oxidative addition of silanes, the electron density of the double bond was decreased by the introduction of

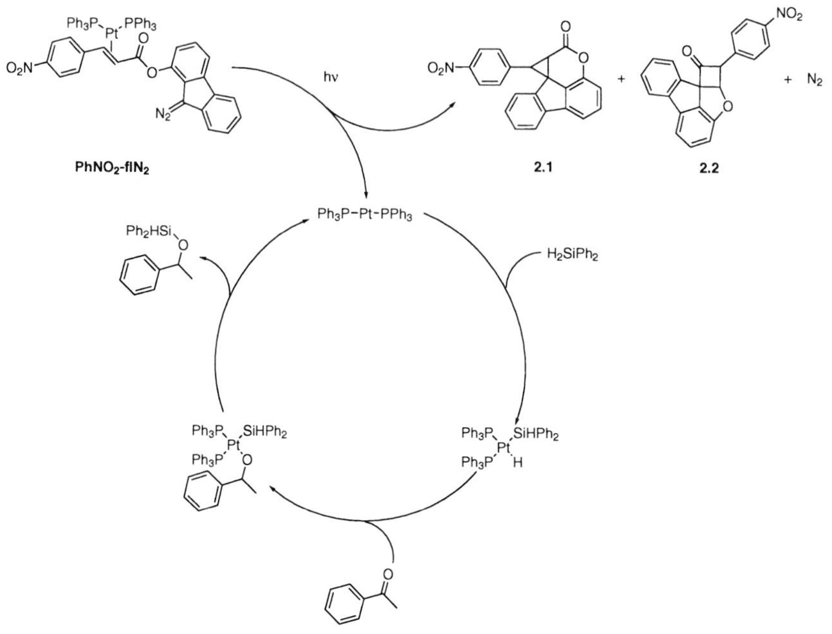

Scheme 7.2: Photo-induced hydrosilylation of benzophenone.

a nitro group in 4-position of the cinnamate. This renderes the oxidative addition of H_2SiPh_2 photo-switchable. However diphenyl diazo methanes decompose slowly at room temperature, which leads to self activation of the catalyst. This stability issue could be solved by the substitution with diazo-9-fluorene systems, which are stable at ambient temperature in solution. Upon irradiation these ligands undergo intramolecular cyclisation and are no longer capable of binding to the metal. The generated highly reactive bis(triphenylphosphine)platinum immediately inserts into the silane´s Si−H bond and therefore starts the catalytic cycle as shown in scheme 7.2. However the hydrosilylation is limited to C=O double bonds due to the poor catalytic performance of these platinum phosphine complexes. A second drawback is the pre-irradiation activity of the catalysts due to partial dissociation of the olefin when a large excess of silane is present. However the latter problem should be easily solvable with unsaturated diacid ligands, like fumarates, of which first reactivity studies show significantly enhanced metal olefin bond strength due to their more pronounced electron deficiency.

7.2. Diazo olefin photo-reactions

The synthesis of diphenyl diazomethane acid esters is possible by reaction of benzophenyl esters with tosylhydrazone and subsequent deprotonation and thermal rearrangement (scheme 7.3 (a)). The electron withdrawing ester substituent leads to a reduction of the carbonyl group reactivity, hence the catalytic protonation necessary for nucleophilic attack of the hydrazone is hindered. Therefore reaction temperature had to be elevated to 110 °C. Due to its basicity hydrazine may not be used. The use of alcohols as solvent causes ester cleavage and has to be avoided as well. The integration of the carbonyl group into the conjugated π-system and steric rigidity of the fluorenone induces a decrease of the carbonyl reactivity. This leads in conjunction with the electron withdrawing ester group to a massive reduction of reactivity. Hence no hydrazonation of the fluorenone esters was possible. However tosylhydrazones of 1-hydroxy and 1-amino-9-fluorenone were accessible and reaction with acid chlorides gave the desired hydrazone esters which could be thermally rearranged to diazofluorenes (scheme 7.3 (b)). Rearrangement of the sodium salts of the tosyl hydrazones failed for diacids and immides, probably caused by chelate complexation, though. So lithium salts should be used.

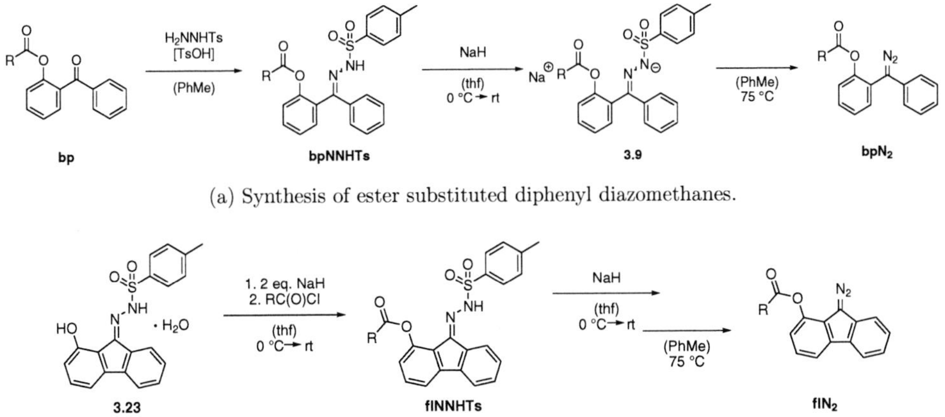

(a) Synthesis of ester substituted diphenyl diazomethanes.

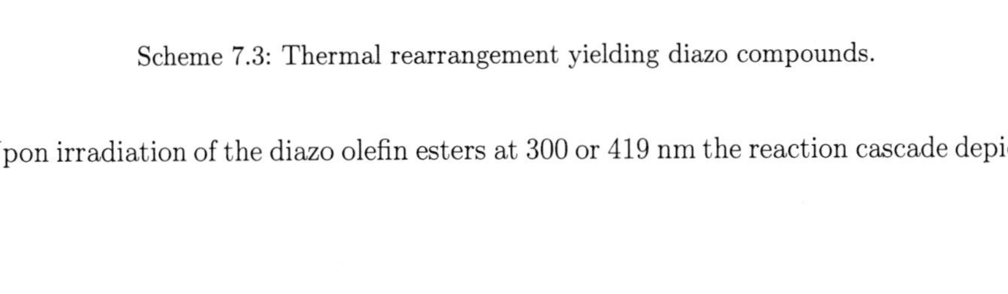

(b) Synthesis of diazo-9-fluorene esters.

Scheme 7.3: Thermal rearrangement yielding diazo compounds.

Upon irradiation of the diazo olefin esters at 300 or 419 nm the reaction cascade depicted

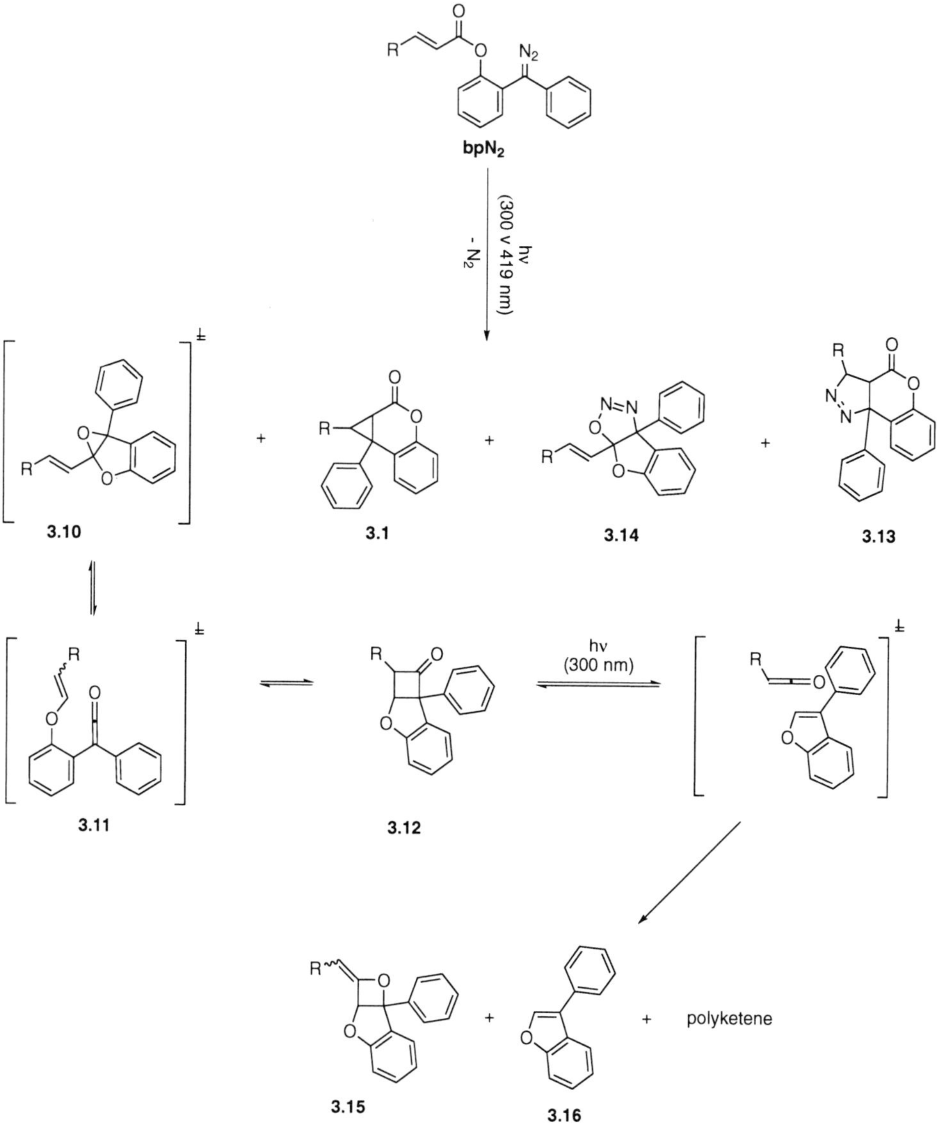

Scheme 7.4: Overview of the photo-cyclisation and decomposition pathways.

in scheme 7.4 is triggered. The *in situ* generated carbene attacks the olefinic and the carboxylic double bond forming cyclopropane **3.1** and epoxide **3.10**. The more electron deficient the olefin is the more **3.1** is formed. This is owed to the nucleophilic character of the generated carbene. UV light of 300 nm apparently also provides activation energy for the 1,3-dipolar cycloaddition of the diazo group to both double bonds generating pyrazole **3.13** and oxadiazole **3.14** as byproducts. No side reactions occur at 419 nm however. The highly strained epoxide **3.10** is unstable at the conditions applied and rearranges to ketene **3.11** which undergoes instantaneous thermal [2+2]-cyclisation to cyclobutanone **3.12**. The cyclobutanone [2+2]-photo-cyclo-reverts to a second ketene and 3-phenyl benzofurane **3.16** if irradiation is continued at 300 nm. Via a *Paterno Büchi* analogous [2+2]-photo-cyclisation the ketene reacts with benzofurane **3.16** to oxetane **3.15** or reacts with another equivalent of ketene to form diketene which then undergoes spontaneous polymerisation. However the energy provided by light at 419 nm is insufficient to re-open cyclobutanone **3.12**. Therefore it is assumed that the 1,3-dipolar cycloaddition and the [2+2]-photo-cyclo-reversion, which need 300 nm, involve sensitisation via the phenyl rings.

7.3. Cinnamic acid ester & related complexes

The electronic and steric properties of cinnamic acid ligands may be tuned via the substituents R^1, R^2 and R^3, with the R^1 substituent being the one with most influence. The electron density of the olefin has direct influence upon the spectroscopic and structural parameters of the complexes as well as onto their reactivity and can be directly calculated from the $^2J_{PP}$ coupling constants.

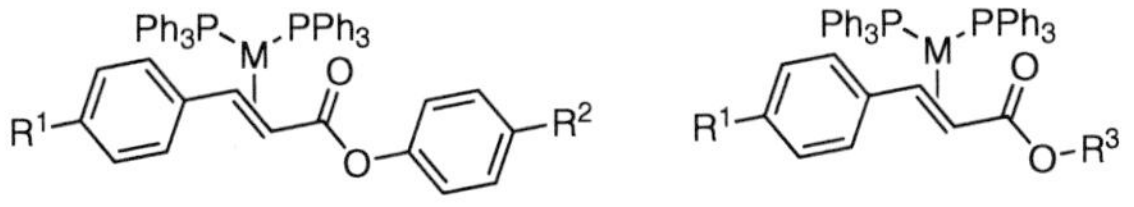

Figure 7.1: Investigated cinnamic acid complexes. $R^1 \cup R^2 \equiv$ OMe, Me, H, Cl, CF$_3$, NO$_2$; $R^3 \equiv$ Me, Et, ^{i}Pr, ^{t}Bu

The obtained olefin electron density parameter (OEDP) ω may also be applied to ligands where easy assignment of the electronics via the *Hammett* parameters is impossible. The OEDP concept is also transferable to other phosphines with the preservation of all trends observed for triphenylphosphine. Since the coupling constants may be determined with all phosphines this is an advantage to the assignment via the C=O stretching frequency in carbonyl compounds. It is also evident, that the ester moiety´s influence is minor and therefore may be varied without altering the over all reactivity of the complexes.

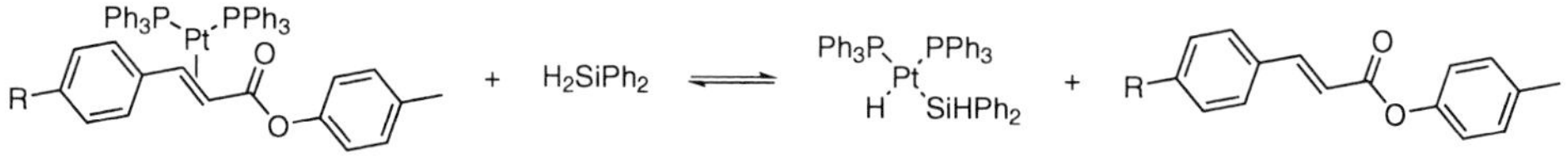

Scheme 7.5: Equilibrium between oxidative addition of Ph_2SiH_2 and cinnamate coordination.

The electron density of the olefinic double bond is also directly correlated to reactivity and therefore equilibria, like the oxidative addition of silane as shown in scheme 7.5, may be predicted with ω. This enables specific ligand design concerning reactivity and stability of the respective complexes. With this knowledge the introduction of functional groups to the ligand is also possible, for example to enhance solubility or incorporation of photo-switchability, without changing the complex reactivity.

7.4. *Karstedt* catalyst inhibition

In order to investigate the potential of photo-triggered catalysis in industrial scale, the inhibition properties of olefin esters were tested, which should easily be modified to systems where the double bond is removable via light induced intramolecular cyclopropanation, *vide supra*. Figure 7.2 gives a summary of the investigated olefins and the used substituents.

It was found that solubility is the main problem concerning inhibition of *Karstedt* type catalysts. Cinnamic acid derivatives did not fulfill the demand due to low inhibition ability, caused by low solubility and inadequate electron deficiency. The mimic of *Karstedt* analogous ligand systems via the introduction of vinyl dimethyl silyl groups did not prove successful and no substantial decrease in reaction rate was observed, even though chelating potential was observed in NMR studies for this ligand class. This is supposedly caused by fast hydrosilylation of the electron rich double bonds of the inhibitor, rendering it ineffective. However these systems are capable to suppress catalyst decomposition

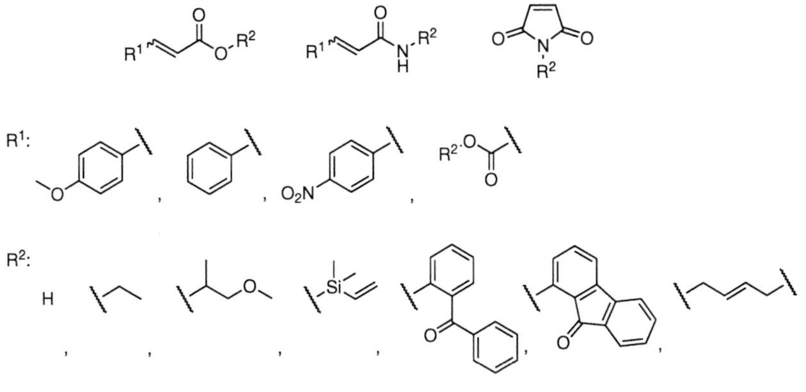

Figure 7.2: Models for photo-switchable ligands.

and therefore keep the reaction solution colourless. Maleic and fumaric acid had the best inhibition properties of the simply modifiable compounds. Amides and imides rendered unfeasible because of their low solubility. Free carboxylic groups show good inhibition rates, however solubility issues arise and the inhibition is presumably caused by protonation of the catalyst. This renders this substance class non applicable in photo-activated catalysis. Methoxy-*iso*-propanol enhances the inhibition effects, due to the increased solubility of the compound and propably increased sterical shielding of the catalytic center upon coordination. The best results were obtained for triolefins however. This is either because of chelatisation of platinum and therefore higher complex stability, although no evidence in NMR could be observed, or the close proximity of two relatively electron deficient olefins to the third coordinated one, which increases the chance of recoordination after dissociation as long as the inhibitor is close to the metal.

Therefore four different systems for photo-switchable *Karstedt* inhibitors are proposed, which are depicted in figure 7.3. Systems **5.26** and **5.27** should be easily accessible via the routes established in this thesis, whereas **5.28** and **5.29** should show superior inhibition capability, as expected from the results with triolefins. However the latter systems pose a higher synthetic challenge and should be medium term targets. In all systems a photo-induced partial or complete removal of the double bonds should be possible. This would then trigger hydrosilylation or at least increase the activity of the catalyst substantially

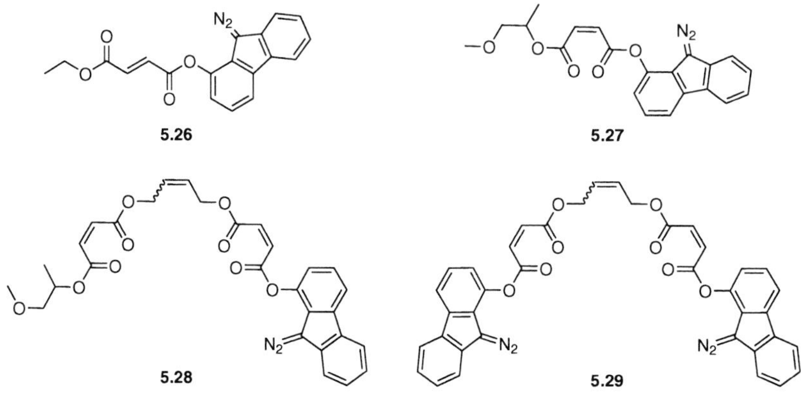

Figure 7.3: Proposed photo-switchable *Karstedt* inhibitors.

in comparison to the pre-irradiation activity.

7.5. Bis(triphenylphosphine)palladium(0)

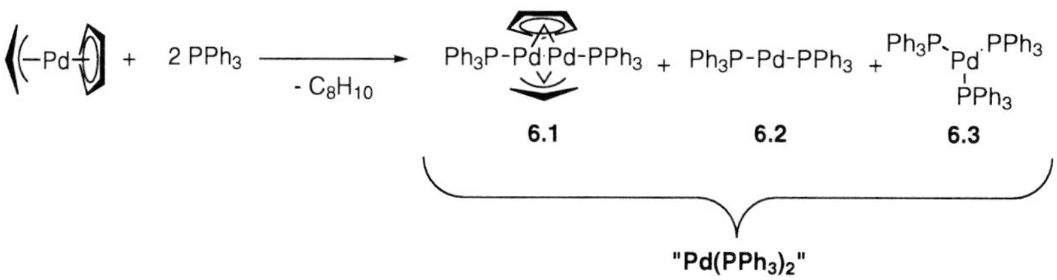

Scheme 7.6: Large scale reactivity of $(\eta^3\text{-allyl})(\eta^5\text{-cyclopentadienyl})$palladium(II) with triphenylphosphine.

Pd(C$_3$H$_5$)Cp reacts with two equivalents of PPh$_3$ to "**Pd(PPh$_3$)$_2$**", a mixture consisting of homo bimetallic complex **6.1**, Pd(PPh$_3$)$_2$ (**6.2**) and Pd(PPh$_3$)$_3$ (**6.3**), of which pure **6.2** could be isolated (scheme 7.6). However this mixture may be used without purification as a Pd(PPh$_3$)$_2$ source and reacts readily with electron deficient or sterically little demanding olefins to bis(triphenylphosphine)palladium(0) olefin complexes (scheme 7.7). It also undergoes fast oxidative addition into C-Br and C-I bonds and is capable of insertion into non aromatic C-Cl bonds. However no reaction with various solvents was observed.

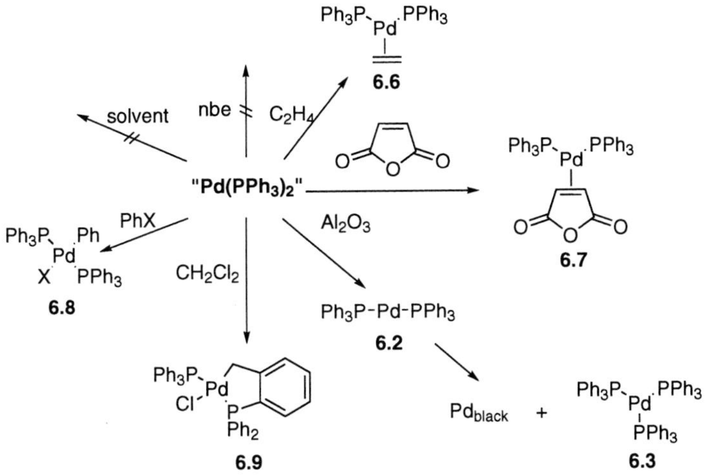

Scheme 7.7: Reactivity of "**Pd(PPh$_3$)$_2$**". Solvent $\equiv$ acetone, dmso, acetonitrile, pyridine, Et$_2$O, thf. X $\equiv$ Br, I.

8. Experimental section

8.1. General procedures

8.1.1. Preparative techniques

All manipulations and experiments were performed under argon using standard *Schlenck* techniques and in a glovebox filled with argon unless otherwise stated. Pentane, dichloromethane and acetonitrile were dried and degassed using a two-column drying system (*MBraun*),[189] diethylether, thf, petroleum ether, benzene and toluene were distilled from sodium. Pyridine was distilled from potassium hydroxide, methanol dried with magnesium and ethanol with sodium, distilled and degased with three freeze pump thaw cycles. All solvents were stored under an argon atmosphere. $CDCl_3$, dmso-d_6, acetone-d_6 and D_2O were used as received from *Deutero GmbH*, benzene-d_6 and toluene-d_8 were dried and degased by stirring over sodium potassium alloy, purified by condensation and stored under argon.[190] Silica with a particle size of 40-63 μm and degased alumina with a *Brockman*-activity[191] of II were used for *flash*-chromatography. Column diameter and filling height were chosen according to *Still et al.*[192]

8.1.2. Analytical methods

Nuclear magnetic resonance (NMR)

^1H NMR, ^{13}C NMR and ^{31}P NMR measurements were performed on a *Bruker* Avance 400, a *Bruker* AMX 400 and a *Jeol* JNM GX 400. ^{195}Pt NMR measurement was performed on a *Bruker* AMX 400. ^1H NMR (400 MHz) and ^{13}C NMR (100 MHz) chemical shifts are given relative to the solvent signal[193,194], ^{31}P NMR (162 MHz) used 85% H_3PO_4, ^{29}Si NMR (79 MHz) used tetramethylsilane in $CDCl_3$, ^{2}D NMR (61 MHz) used neat tetramethylsilane-d_{12} and ^{195}Pt NMR (85 MHz) used Na_2PtCl_6 in D_2O as an external standard. Signal multiplicitys are assigned the following: s - singlet, d - doublet, t - triplet, q - quartet, qu - quintet, h - hexet, hp - heptet, m - multiplet. Pseodo multiplicitys are asigned with the prefix p. Coupling constants are given as average of the experimentally determined.

Infrared spectroscopy (IR)

Transmission FT-IR was carried out on a *Jasco* FT/IR-460 Plus spectrometer and ATR-FT-IR was done with a *Thermo Scientific* Nicolet 380 Smart Orbit.

UV/vis spectroscopy

UV/vis spectra were recorded on a *Jasco* V-550 spectrophotometer using quartz cuvettes.

Mass spectrometry (MS)

FAB-MS analysis was carried out on a *Finnigan* MAT-90 in a *para*-nitrobenzylalcohol matrix under bombardment with ionised xenon. ESI-MS were conducted on a *Finnigan* LCQ in acetonitrile. CI-MS was performed on a *Finnigan* MAT-90 (*iso*-buten 150 eV).

X-ray diffraction

The single crystals were stored under perfluorinated oil, transferred into a *Lindemann* capillary, fixed and sealed. Preliminary examination and data collection were carried out on an area detecting system (APEX II, κ-CCD) at the window of a rotating anode (*Bruker* AXS, FR591) and graphite monochromated MoK_α radiation ($\lambda = 0.71073$ Å). Raw data were corrected for *Lorentz*, polarization, and, arising from scaling procedure, for latent decay and absorption effects. The structures were solved by a combination of direct methods and difference *Fourier* syntheses. All non-hydrogen atoms were refined with anisotropic displacement parameters, whereas all hydrogen atoms were refined with isotropic displacement parameters. Full-matrix least-square refinements were carried out by minimizing $\sum(F_0^2 - F_c^2)^2$ with SHELXL-97 weighting scheme.[195] The final residual electron density maps showed no remarkable features. Neutral atom scattering factors for all atoms and anomalous dispersion corrections for non-hydrogen atoms were taken from *International Tables for Crystallography*.[196]

Elemental analysis

Micro analytical analysis was performed in the micro analytical lab of the Technische Universität München.

8.1.3. Literature known precursors

The following compounds were prepared according to literature procedures:

Organic precursors

2-aminobenzophenone hydrazone,[43] 1-amino-9-fluorenone,[85] 1-amino-9-fluorenone-(p-tosyl)hydrazone,[92] 1-hydroxy-9-fluorenone,[86] cinnamic acid chloride,[104] crotonic acid chloride,[197] 9-fluorenone-1-carboxylic acid,[83] 9-fluorenone-1-carboxylic acid amide,[84] 9-fluorenone-1-carboxylic acid chloride,[86] fumaric monomethylester chloride,[198] maleanilic acid,[36] monoethyl maleate,[148] N-phenylmaleimide.[36]

Organometallic precursors

$(\eta^3$-allyl$)(\eta^5$-cyclopentadienyl)palladium(II),[108] bis$(\eta^4$-1,5-cyclooctadiene)platinum(0),[199] $(\eta^4$-1,5-cyclooctadiene)platinum(II)chloride,[134] $(\eta^2$-ethene)bis(triphenylphosphine)platinum(0),[107] tetrakis(triphenylphosphine)platinum(0),[105] tris$(\eta^2$-bicyclo[2.2.1]heptene)platinum(0),[199] tris(triphenylphosphine)platinum(0).[105]

The following compounds were prepared according to the procedures described below and analytical data matches the literature given one:

Organic precursors

ethyl 4-chlorocinnamate,[200] ethyl cinnamate,[201] ethyl 4-methoxycinnamate,[202] ethyl 4-methylcinnamate,[201] ethyl 4-nitrocinnamate,[201] ethyl 4-trifluorocinnamate,[203] iso-propyl 4-chlorocinnamate,[204] iso-propyl cinnamate,[205] iso-propyl 4-methoxycinnamate,[206] iso-propyl 4-methylcinnamate,[204] iso-propyl 4-nitrocinnamate,[207] iso-propyl 4-trifluorocinnamate,[204] methyl 4-chlorocinnamate,[208] methyl cinnamate,[205] methyl 4-methoxycinnamate,[208] methyl 4-methylcinnamate,[206] methyl 4-nitrocinnamate,[208] methyl 4-trifluorocinnamate;[206] **route:** *Ester-I.*

4-methoxyphenyl cinnamate,[209] 4-methoxyphenyl 4-methoxycinnamate,[210] 4-methoxyphenyl 4-nitrocinnamate,[211] 4-methylphenyl cinnamate,[210] 4-methyl-phenyl 4-methoxycinnamate,[209] 4-nitrophenyl cinnamate,[212] 4-nitrophenyl 4-methoxycinnamate,[213] 4-nitrophenyl 4-methylcinnamate,[214] 4-nitrophenyl 4-nitro-cinnamate,[215] phenyl cinnamate,[216] phenyl 4-methoxycinnamate,[217] phenyl 4-nitrocinnamate,[218] *tert*-butyl cinnamate,[219] *tert*-butyl 4-methoxycinnamate,[206] *tert*-butyl 4-methylcinnamate,[219] *tert*-butyl 4-nitrocinnamate;[219] **route:** *Ester-II.*

2-acetoxybenzophenone,[220] 4-nitrophenyl 4-trifluorocinnamate,[221] phenyl 4-chlorocin-namate,[222] *tert*-butyl 4-chlorocinnamate,[223] *tert*-butyl 4-trifluorocinnamate;[224] **route:** *Ester-III.*

4-chlorocinnamic acid chloride,[225] 4-methoxycinnamic acid chloride,[226] 4-methylcin-namic acid chloride,[227] 4-nitrocinnamic acid chloride,[228] 4-trifluoromethylcinnamic acid chloride;[229] **route:** *Acidchloride-I.*

Organometallic precursors

(1,4-bis(diphenylphosphino)butane)platinum(II)chloride,[132] (1,2-bis(diphenylphosphino)ethane)platinum(II)chloride,[130,132,133]   (bis(diphenylphosphino)methane)platinum(II)chloride,[130,131,133]   (1,3-bis(diphenylphosphino)propane)platinum(II)chloride;[130,132] **route:** *Pt-I.*

bis(triphenylphosphine) maleic anhydride palladium(0);[182] **route:** *Pd-V.*

(η^2-ethene)bis(triphenylphosphine)palladium(0);[182] **route:** *Pd-VI.*

8.2. Organic compounds

8.2.1. General synthetic procedures

Acidchloride-I

The acid was suspended in $SOCl_2$ and refluxed for one hour. The solvent was then removed in vacuo to give the corresponding acid chloride in quantitative yield.

Acidchloride-II

The acid was dissolved in toluene, 1.1 eq. of $SOCl_2$ was added and the reaction mixture refluxed over night. The solvent was removed under reduced pressure to give the corresponding acid chloride.

Amide-I

16.5 mmol amine was dissolved in 5 ml pyridine, 1.6 eq. of acidchloride were added and the reaction mixture was heated to 100 °C for 2 h. The reaction mixture was then poured into a mixture of 200 g of ice and 350 ml of 0.5 [M] hydrochloric acid and extracted three times with Et_2O. The combined organic layers were washed twice with saturated $NaHCO_3$ solution and once with brine, dried over Na_2SO_4, filtered and the solvent removed under reduced pressure to give the desired amide.

Amide-II

Maleic anhydride and 1 eq. of amine were suspended in Et_2O and stirred at ambient temperature for 1 h. The reaction mixture was then refluxed for 2 h and the solvent removed in vacuo afterwards to yield the desired product.

Diazo

The tosylhydrazone was added to 1 eq. of NaH in thf at 0 °C, warmed to ambient temperature and stirred over night. The solvent was removed in vacuo, the residue taken up in toluene and heated to 75 °C for 45 min. Again the solvent was removed in vacuo afterwards, the residue extracted with pentane and filtrated until the filtrate was colourless. The diazocompounds were isolated by removal of the solvent in vacuo.

Ester-I

The acid was dissolved in the adequate alcohol, a few drops of concentrated H_2SO_4 were added and refluxed over night. The reaction mixture was poured into a mixture of ice and water and extracted three times with Et_2O. The combined organic layers were washed twice with saturated $NaHCO_3$ solution and once with brine, dried over Na_2SO_4, filtered and the solvent removed under reduced pressure to yield the desired ester almost quantitatively.

Ester-II

16.5 mmol alcohol was dissolved in 5 ml pyridine, 1.6 eq. of acidchloride were added and the reaction mixture was heated to 100 °C for 2 h. The reaction mixture was poured into a mixture of 200 g of ice and 350 ml of 0.5 [M] hydrochloric acid and extracted three times with Et_2O. The combined organic layers were washed twice with saturated $NaHCO_3$ solution and once with brine, dried over Na_2SO_4, filtered and the solvent removed under reduced pressure to give the product.

Ester-III

The alcohol was added to 1 eq. of NaH in thf at 0 °C and warmed to ambient temperature. After 15 min the appropriate acid chloride was added and stirring continued over night. The reaction was quenched with water and extracted three times with Et_2O. The combined organic layers were washed twice with saturated $NaHCO_3$ solution and once with brine, dried over Na_2SO_4, filtered and the solvent removed under reduced pressure to give the anticipated ester.

Ester-IV

The alcohol was added to 2 eq. of NaH in thf at 0 °C and warmed to ambient temperature. After 15 min the appropriate acidchloride was added and stirring continued for 1 h. The reaction was quenched with water and extracted three times with Et_2O. The combined organic layers were washed twice with saturated $NaHCO_3$ solution and once with brine, dried over Na_2SO_4, filtered and the solvent removed under reduced pressure to give the anticipated ester.

Ester-V

10.0 g (1.02 mol) maleic anhydride was dissolved in 1 eq. of alcohol and heated in a pressure tube to 90 °C for four days. Unreacted reactants were removed in vacuo to yield the product.

Hydrazone

The carbonyl compound was dissolved in toluene, 1 eq. of tosylhydrazine and one crystal of *para*-toluene sulfonic acid monohydrate added and refluxed over night. The solvent was removed under reduced pressure and the residue stirred in pentane and filtered off to give the product.

Imide

Amide and 0.5 eq. sodiumacetate were dissolved in aceticanhydride and heated to 100 °C for one hour. The reaction solution was cooled to ambient temperature, poured into a mixture of ice and water and extracted three times with EtOAc. The combined organic layers were washed with brine, dried over Na_2SO_4, filtered and the solvent removed under reduced pressure to give the imide.

Irradiation

The diazocompound was dissolved in toluene and irradiated at 300 nm for 2 h. After removal of the solvent under reduced pressure the products were separated by *flash* chromatography (SiO_2; 9:1 pentane/Et_2O).

Silylester

Acid was dissolved in toluene, 1.1 eq triethylamine and 1 eq. dimethyl(vinyl)silyl chloride were added and the solution refluxed over night. The solvent was removed in vacuo, the residue extracted with pentane and filtered. The product was received by removal of the pentane in vacuo.

8.2.2. Esters & amides of benzophenone and 9-fluorenone derivatives

2-Benzoylphenyl crotonate

Me-bp: $C_{17}H_{14}O_3$, M = 266.29 g/mol, colourless oil, yield: 86%, **route**: *Ester-III*.

^{1}H-NMR (399.78 MHz; CDCl$_3$): δ = 1.79 (d, $^3J_{HH}$ = 6.9 Hz, 3H, CH_3), 5.72 (d, $^3J_{HH}$ = 14.7 Hz, 1H, CHCHCO), 6.84 (dq, $^3J_{HH}$ = 6.9 Hz, $^3J_{HH}$ = 14.7 Hz, 1H, CH$_3$CHCH), 7.21 - 7.26 (m, 1H, H_{Ar}), 7.28 - 7.35 (m, 1H, H_{Ar}), 7.38 - 7.46 (m, 2H, H_{Ar}), 7.48 - 7.59 (m, 3H, H_{Ar}), 7.73 - 7.81 (m, 2H, H_{Ar}).

^{13}C-NMR (100.53 MHz; CDCl$_3$): δ = 18.3 (CH_3), 121.5 (C$_{Ar}$H), 123.4 (CHCHCO), 125.7 (C$_{Ar}$H), 128.5 (C$_{Ar}$H), 130.0 (C$_{Ar}$H), 130.5 (C$_{Ar}$H), 132.1 (C$_{Ar}$), 132.3 (C$_{Ar}$H), 133.1 (C$_{Ar}$H), 137.8 (C$_{Ar}$), 147.4 (CH$_3$$C$HCH), 148.9 (C$_{Ar}$O), 164.3 (CH$C$OO), 195.1 (C$_{Ar}$$COC_{Ar}$).

ESI-MS: m/z (%): 267.2 (17) [C$_{17}$H$_{14}$O$_3$H$^+$], 289.2 (100) [C$_{17}$H$_{14}$O$_3$Na$^+$], 306.5 (80) [C$_{17}$H$_{14}$O$_3$H$_2$ONa$^+$], 419.1 (47) [(C$_{17}$H$_{14}$O$_3$)$_3$HK^{2+}], 503.1 (28) [C$_{17}$H$_{14}$O$_3$(CH$_3$CN)C$_{13}$H$_9$O$_2$$^+$], 543.5 (17) [C$_{17}H_{14}O_3$(CH$_3$CN)H$_2$ONaC$_{13}H_9O_2$$^+$].

elemental analysis for C$_{17}$H$_{14}$O$_3$ (266.29 g/mol):

calcd.: C 76.68, H 5.30, O 18.02

found: C 76.40, H 5.10

2-Benzoylphenyl 4-methoxycinnamate

PhOMe-bp: C$_{23}$H$_{18}$O$_4$, M = 358.39 g/mol, light yellow solid, yield: 83%, **route**: *Ester-III*, extraction with an Et$_2$O/CH$_2$Cl$_2$ mixture (1:1).

Single crystals suitable for X-ray analysis were obtained by slow evaporation of a saturated ethanol solution.

^{1}H-NMR (400.13 MHz; CDCl$_3$): δ = 3.81 (s, 3H, OCH_3), 6.15 (d, $^3J_{HH}$ = 15.9 Hz, 1H, CHCHCO), 6.87 (d, $^3J_{HH}$ = 8.8 Hz, 2H, H_{Ar}), 7.23 - 7.63 (m, 10H, H_{Ar}, C$_{Ar}$$C$HCH), 7.80 (d, $^3J_{HH}$ = 7.1 Hz, 2H, H_{Ar}).

^{13}C-NMR (100.62 MHz; CDCl$_3$): δ = 55.5 (OCH_3), 113.9 (CHCHCO), 114.4 (C_{Ar}H), 123.3 (C_{Ar}H), 125.6 (C_{Ar}), 126.8 (C_{Ar}H), 128.4 (C_{Ar}H), 129.9 (C_{Ar}H), 130.1 (C_{Ar}H), 130.4 (C_{Ar}H), 132.0 (C_{Ar}), 132.2 (C_{Ar}H), 133.1 (C_{Ar}H), 137.7 (C_{Ar}), 146.5 (C_{Ar}CHCH), 148.9 (C_{Ar}O), 161.8 (C_{Ar}OCH$_3$), 165.1 (CHCOO), 195.1 ($C_{Ar}$$COC_{Ar}$).

ESI-MS: m/z (%): 161.2 (99) [C$_{10}$H$_9$O$_2$$^+$], 381.2 (100) [C$_{23}H_{18}O_4Na^+$], 421.6 (79) [C$_{23}H_{17}O_4H_2ONa_2$$^+$], 739.0 (41) [(C$_{23}H_{18}O_4$)$_2Na^+$], 754.8 (6) [(C$_{23}H_{18}O_4$)$_2K^+$], 869.0 (7) [(C$_{23}H_{18}O_4$)$_2$(HCOOH)(HCOO)NaK$^+$].

elemental analysis for C$_{23}$H$_{18}$O$_4$ (358.39 g/mol):

calcd.: C 77.08, H 5.06, O 17.86

found: C 76.92, H 5.08

2-Benzoylphenyl 4-methylcinnamate

PhMe-bp: C$_{23}$H$_{18}$O$_3$, M = 342.39 g/mol, yellow crystals, yield: 86%, **route**: *Ester-III*.

^{1}H-NMR (400.13 MHz; CDCl$_3$): δ = 2.36 (s, 3H, CH_3), 6.24 (d, $^3J_{HH}$ = 16.0 Hz, 1H, CHCHCO), 7.16 (d, $^3J_{HH}$ = 8.0 Hz, 2H, H_{Ar}), 7.27 - 7.62 (m, 10H, H_{Ar}, C_{Ar}CHCH), 7.80 (d, $^3J_{HH}$ = 7.2 Hz, 2H, H_{Ar}).

^{13}C-NMR (100.62 MHz; CDCl$_3$): δ = 21.6 (CH_3), 115.4 (CHCHCO), 123.4 (C_{Ar}H), 125.7 (C_{Ar}), 128.4 (C_{Ar}H), 128.5 (C_{Ar}H), 129.7 (C_{Ar}H), 129.9 (C_{Ar}H), 130.5 (C_{Ar}H), 131.4 (C_{Ar}H), 132.0 (C_{Ar}), 132.3 (C_{Ar}H), 133.1 (C_{Ar}H), 137.7 (C_{Ar}), 141.3 (C_{Ar}CH$_3$), 146.9 (C_{Ar}CHCH), 148.9 (C_{Ar}O), 165.0 (CHCOO), 195.1 ($C_{Ar}$$COC_{Ar}$).

ESI-MS: m/z (%): 145.2 (85) [C$_{10}$H$_9$O$^+$], 365.2 (98) [C$_{23}$H$_{18}$O$_3$Na$^+$], 405.7 (100) [C$_{23}$H$_{17}$O$_3$H$_2$ONa$_2$$^+$], 475.8 (39) [C$_{23}H_{18}O_3$(HCOOH)(HCOO)NaK$^+$], 717.0 (41) [(C$_{23}H_{18}O_3$)$_2Na^+$], 837.1 (10) [(C$_{23}H_{18}O_3$)$_2$(HCOOH)(HCOO)NaK$^+$].

elemental analysis for $C_{23}H_{18}O_3$ (342.39 g/mol):

calcd.: C 80.68, H 5.30, O 14.02

found: C 79.56, H 5.54

2-Benzoylphenyl cinnamate

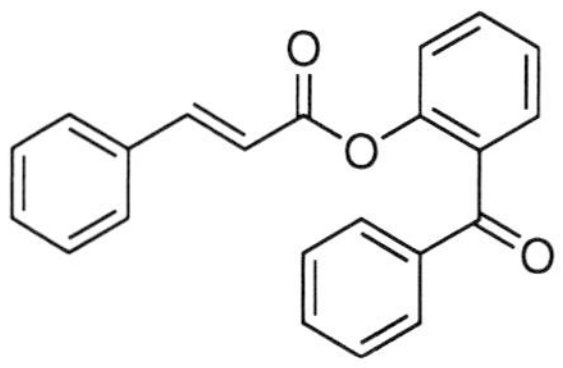

Ph-bp: $C_{22}H_{16}O_3$, M = 328.36 g/mol, off-white solid, yield: >99%, **route**: *Ester-II.*

^{1}H-NMR (399.78 MHz; CDCl$_3$): δ = 6.29 (d, $^3J_{HH}$ = 16.0 Hz, 1H, CH*CH*CO), 7.30 - 7.43 (m, 9H, H_{Ar}), 7.50 - 7.60 (m, 4H, H_{Ar}, $C_{Ar}CH$CH), 7.81 (d, $^3J_{HH}$ = 7.3 Hz, 2H, H_{Ar}).

^{13}C-NMR (100.53 MHz; CDCl$_3$): δ = 116.6 (CH*CH*CO), 123.4 (C_{Ar}H), 125.8 (C_{Ar}H), 128.4 (C_{Ar}H), 128.5 (C_{Ar}H), 129.0 (C_{Ar}H), 130.0 (C_{Ar}H), 130.5 (C_{Ar}H), 130.8 (C_{Ar}H), 132.0 (C_{Ar}), 132.3 (C_{Ar}H), 133.2 (C_{Ar}H), 134.2 (C_{Ar}), 137.8 (C_{Ar}), 146.9 (C_{Ar}CHCH), 148.9 (C_{Ar}O), 164.9 (CH*C*OO), 195.0 ($C_{Ar}COC_{Ar}$).

ESI-MS: m/z (%): 351.2 (100) $[C_{22}H_{16}O_3Na^+]$, 391.8 (50) $[C_{22}H_{16}O_3(CH_3CN)Na^+]$, 475.9 (77) $[C_{22}H_{16}O_3(CH_3CN)_3Na^+]$, 565.1 (57) $[C_{22}H_{16}O_3(CH_3CN)C_{14}H_9O_2^+]$, 605.6 (85) $[C_{22}H_{16}O_3(CH_3CN)C_{14}H_9O_2K^+]$, 678.9 (19). $[(C_{22}H_{16}O_3)_2Na^+]$.

elemental analysis for $C_{22}H_{16}O_3$ (328.36 g/mol):

calcd.: C 80.47, H 4.91, O 14.62

found: C 80.20, H 4.86

9-Fluorenone-1-yl cinnamate

Ph-fl: $C_{22}H_{14}O_3$, M = 326.34 g/mol, orange solid, yield: 85%, **route:** *Ester-II.*

^{1}H-NMR (399.78 MHz; CDCl$_3$): δ = 6.73 (d, $^3J_{HH}$ = 16.0 Hz, 1H, CHCHCO), 7.04 (d, $^3J_{HH}$ = 8.1 Hz, 1H, H_{Ar}), 7.29 (t, $^3J_{HH}$ = 7.4 Hz, 1H, H_{Ar}), 7.37 - 7.65 (m, 9H, H_{Ar}), 7.96 (d, $^3J_{HH}$ = 16.0 Hz, 1H, $C_{Ar}CH$CH).

^{13}C-NMR (100.53 MHz; CDCl$_3$): δ = 116.9 (CHCHCO), 118.2 (C_{Ar}H), 120.7 (C_{Ar}H), 123.7 (C_{Ar}H), 124.4 (C_{Ar}H), 124.8 (C_{Ar}), 128.6 (C_{Ar}H), 129.1 (C_{Ar}H), 129.6 (C_{Ar}H), 130.9 (C_{Ar}H), 134.2 (C_{Ar}), 134.4 (C_{Ar}), 134.7 (C_{Ar}H), 136.3 (C_{Ar}H), 143.7 (C_{Ar}), 146.1 (C_{Ar}), 147.4 ($C_{Ar}CH$CH), 148.3 (C_{Ar}O), 164.6 (CHCOO), 191.1 ($C_{Ar}COC_{Ar}$).

ESI-MS: m/z (%): 326.4 (38.04) [$C_{22}H_{14}O_3{}^+$].

elemental analysis for $C_{22}H_{14}O_3$ (326.34 g/mol):

calcd.: C 80.97, H 4.32, O 14.71

found: C 80.39, H 4.18

2-Benzoylphenyl 4-nitrocinnamate

PhNO$_2$-bp: $C_{22}H_{15}NO_5$, M = 373.36 g/mol, white solid, yield: 85%, **route:** *Ester-III.*

^{1}H-NMR (399.78 MHz; CDCl$_3$): δ = 6.45 (d, $^3J_{HH}$ = 16.0 Hz, 1H, CHCHCO), 7.31 (d, $^3J_{HH}$ = 8.1 Hz, 1H, H_{Ar}), 7.38 (t, $^3J_{HH}$ = 7.5 Hz, 1H, H_{Ar}), 7.43 (t, $^3J_{HH}$ = 7.6 Hz,

2H, H_{Ar}), 7.48 - 7.66 (m, 6H, H_{Ar}, $C_{Ar}CHCH$), 7.80 (d, $^3J_{HH}$ = 7.6 Hz, 2H, H_{Ar}), 8.22 (d, $^3J_{HH}$ = 16.0 Hz, 2H, H_{Ar}).

^{13}C-NMR (100.53 MHz; CDCl$_3$): δ = 120.9 (CHCHCO), 123.2 (C_{Ar}H), 124.6 (C_{Ar}H), 126.0 (C_{Ar}H), 128.5 (C_{Ar}H), 129.0 (C_{Ar}H), 130.0 (C_{Ar}H), 130.7 (C_{Ar}H), 131.7 (C_{Ar}), 132.5 (C_{Ar}H), 133.2 (C_{Ar}H), 137.6 (C_{Ar}), 140.1 (C_{Ar}), 143.6 (C$_{Ar}$$CH$CH), 148.7 ($C_{Ar}$O), 148.8 ($C_{Ar}$N), 164.0 (CH$C$OO), 194.8 (C$_{Ar}$$COC_{Ar}$).

ESI-MS: m/z (%): 176.1 (64) [C$_9$H$_6$NO$_3{}^+$], 768.8 (45) [(C$_{22}$H$_{15}$NO$_5$)$_2$Na$^+$].

elemental analysis for C$_{22}$H$_{15}$NO$_5$ (373.36 g/mol):

calcd.: C 70.77, H 4.05, N 3.75, O 21.43

found: C 70.61, H 4.12, N 3.64

9-Fluorenone-1-yl 4-nitrocinnamate

PhNO$_2$-fl: C$_{22}$H$_{13}$NO$_5$, M = 371.34 g/mol, orange solid, yield: 93%, **route**: *Ester-III*, product is filtered of after quenching and washed with water and ethanol.

^{1}H-NMR (400.13 MHz; CDCl$_3$): δ = 6.84 (d, $^3J_{HH}$ = 16.1 Hz, 1H, CHCHCO), 7.04 (d, $^3J_{HH}$ = 8.1 Hz, 1H, H_{Ar}), 7.30 (t, $^3J_{HH}$ = 7.3 Hz, 1H, H_{Ar}), 7.41 - 7.63 (m, 5H, H_{Ar}), 7.77 (d, $^3J_{HH}$ = 8.4 Hz, 2H, H_{Ar}), 7.98 (d, $^3J_{HH}$ = 16.1 Hz, 1H, C$_{Ar}$CHCH), 8.28 (d, $^3J_{HH}$ = 8.3 Hz, 2H, H_{Ar}).

^{13}C-NMR (100.62 MHz; CDCl$_3$): δ = 118.4 (CHCHCO), 120.8 (C_{Ar}H), 121.3 (C_{Ar}H), 123.4 (C_{Ar}H), 124.4 (C_{Ar}H), 124.5 (C_{Ar}H), 124.6 (C_{Ar}), 129.2 (C_{Ar}H), 129.7 (C_{Ar}H), 134.2 (C_{Ar}), 134.9 (C_{Ar}H), 136.5 (C_{Ar}H), 140.4 (C_{Ar}), 143.7 (C_{Ar}), 144.2 (C$_{Ar}$$CH$CH), 146.2 ($C_{Ar}$), 148.0 ($C_{Ar}$), 148.9 ($C_{Ar}$N), 163.7 (CH$C$OO), 191.1 (C$_{Ar}$$COC_{Ar}$).

N(2-Benzoylphenyl) 4-nitrocinnamide

PhNO$_2$-Nbp: C$_{22}$H$_{16}$N$_2$O$_4$, M = 372.37 g/mol, yellow solid, yield: 92%, **route:** *Amide-I*, product is filtered of after quenching and washed with water and ethanol.

^{1}H-NMR (400.13 MHz; CDCl$_3$): δ = 6.74 (d, $^3J_{HH}$ = 15.6 Hz, 1H, CHC*H*CO), 7.14 (t, $^3J_{HH}$ = 7.6 Hz, 1H, H_{Ar}), 7.50 (t, $^3J_{HH}$ = 7.6 Hz, 2H, H_{Ar}), 7.58 - 7.66 (m, 3H, H_{Ar}), 7.68 - 7.74 (m, 4H, H_{Ar}), 7.78 (d, $^3J_{HH}$ = 15.6 Hz, 1H, C$_{Ar}$C*H*CH), 8.24 (d, $^3J_{HH}$ = 8.6 Hz, 2H, H_{Ar}), 8.80 (d, $^3J_{HH}$ = 8.3 Hz, 1H, H_{Ar}), 11.33 (s, 1H, N*H*).

^{13}C-NMR (100.62 MHz; CDCl$_3$): δ = 121.8 (CH*C*HCO), 122.7 (C_{Ar}H), 123.3 (C_{Ar}), 124.3 (C_{Ar}H), 126.2 (C_{Ar}H), 128.6 (C_{Ar}H), 128.8 (C_{Ar}H), 130.0 (C_{Ar}H), 132.8 (C_{Ar}H), 134.1 (C_{Ar}H), 134.7 (C_{Ar}H), 138.7 (C_{Ar}), 139.7 (C$_{Ar}$*C*HCH), 140.7 (C_{Ar}), 141.0 (C_{Ar}), 148.5 (C_{Ar}N), 163.6 (CH*C*OO), 200.3 (C$_{Ar}$*C*OC$_{Ar}$).

2-Benzoylphenyl ethyl fumarate

fu-bp: C$_{19}$H$_{16}$O$_5$, M = 324.33 g/mol, colourless oil, yield: 93%, **route:** *Ester-II*.

^{1}H-NMR (399.78 MHz; CDCl$_3$): δ = 1.29 (t, $^3J_{HH}$ = 7.1 Hz, 3H, C*H$_3$*), 4.22 (q, $^3J_{HH}$ = 7.1 Hz, 2H, CH$_3$C*H$_2$*), 6.70 (pq, $^3J_{HH}$ = 15.8 Hz, 2H, COC*H*C*H*CO), 7.23 - 7.29 (m, 1H, H_{Ar}), 7.34 - 7.39 (m, 1H, H_{Ar}), 7.40 - 7.47 (m, 2H, H_{Ar}), 7.52 - 7.62 (m, 3H, H_{Ar}), 7.71 - 7.81 (m, 2H, H_{Ar}).

^{13}C-NMR (100.53 MHz; CDCl$_3$): $\delta = 14.3$ (CH_3), 61.6 (CH$_3$$CH_2$), 123.2 ($C_{Ar}$H), 126.3 ($C_{Ar}$H), 128.7 ($C_{Ar}$H), 130.0 ($C_{Ar}$H), 130.9 ($C_{Ar}$H), 131.6 ($C_{Ar}$), 132.0 (CO$CH$CH), 132.5 ($C_{Ar}$H), 133.4 ($C_{Ar}$H), 135.5 (CO$C$HCH), 137.6 ($C_{Ar}$), 148.5 ($C_{Ar}$O), 163.2 (CH$C$OO), 164.7 (CH$C$OO), 194.7 (C$_{Ar}$$COC_{Ar}$).

ESI-MS: m/z (%): 324.9 (64) [C$_{19}$H$_{16}$O$_5$H$^+$], 347.1 (91) [C$_{19}$H$_{16}$O$_5$Na$^+$], 364.3 (97) [C$_{19}$H$_{16}$O$_5$K$^+$], 384.7 (60) [C$_{19}$H$_{16}$O$_5$(CH$_3$CN)Na$^+$], 561.1 (61) [C$_{19}$H$_{16}$O$_5$C$_{14}$H$_9$O$_2$K$^+$], 601.5 (100) [C$_{19}$H$_{16}$O$_5$C$_{17}$H$_{11}$O$_4$$^+$].

elemental analysis for C$_{19}$H$_{16}$O$_5$ (324.33 *g/mol*):

calcd.: C 70.36, H 4.97, O 24.67

found: C 70.31, H 5.33

9-Fluorenone-1-yl ethyl fumarate

fu-fl: C$_{19}$H$_{14}$O$_5$, M = 322.31 *g/mol*, orange solid, yield: 68%, **route:** *Ester-II.*

^{1}H-NMR (399.78 MHz; CDCl$_3$): $\delta = 1.34$ (t, $^3J_{HH} = 7.1$ Hz, 3H, CH_3), 4.30 (q, $^3J_{HH} = 7.1$ Hz, 2H, CH$_3$CH_2), 6.98 (d, $^3J_{HH} = 8.1$ Hz, 1H, H_{Ar}), 7.12 (ps, 2H, COCHCHCO), 7.29 (t, $^3J_{HH} = 7.3$ Hz, 1H, H_{Ar}), 7.42 (d, $^3J_{HH} = 7.4$ Hz, 1H, H_{Ar}) 7.45 - 7.54 (m, 3H, H_{Ar}), 7.58 (d, $^3J_{HH} = 7.4$ Hz, 1H, H_{Ar}).

^{13}C-NMR (100.53 MHz; CDCl$_3$): $\delta = 14.3$ (CH_3), 61.6 (CH$_3$$CH_2$), 118.6 ($C_{Ar}$H), 120.8 ($C_{Ar}$H), 123.2 ($C_{Ar}$H), 124.4 ($C_{Ar}$), 124.5 ($C_{Ar}$H), 129.7 ($C_{Ar}$H), 132.3 (CO$CH$CH), 134.1 ($C_{Ar}$), 134.9 (CO$C$HCH), 135.9 ($C_{Ar}$H), 136.5 ($C_{Ar}$H), 143.6 ($C_{Ar}$), 146.2 ($C_{Ar}$), 147.5 ($C_{Ar}$O), 162.8 (CH$C$OO), 164.8 (CH$C$OO), 190.8 (C$_{Ar}$$COC_{Ar}$).

CI-MS: m/z (%): 196.7 (100) [C$_{13}$H$_8$O$_2$H$^+$], 322.3 (34.01) [C$_{19}$H$_{14}$O$_5$$^+$].

elemental analysis for C$_{19}$H$_{14}$O$_5$ (322.31 *g/mol*):

calcd.: C 70.80, H 4.38, O 24.82

found: C 69.89, H 4.55

N(2-Benzoylphenyl) ethyl fumaramide

fu-Nbp: $C_{19}H_{17}NO_4$, M = 323.34 g/mol, yellow solid, yield: 97%, **route**: *Amide-I.*

^{1}H-NMR (399.78 MHz; CDCl$_3$): δ = 1.32 (t, $^3J_{HH}$ = 7.1 Hz, 3H, CH_3), 4.26 (q, $^3J_{HH}$ = 7.1 Hz,, 2H, CH$_3$CH_2), 6.92 (d, $^3J_{HH}$ = 15.4 Hz, 1H, COCHCHCO), 7.09 (d, $^3J_{HH}$ = 15.4 Hz, 1H, COCHCHCO), 7.14 (dt, $^3J_{HH}$ = 1.2 Hz, $^3J_{HH}$ = 7.7 Hz, 1H, H_{Ar}), 7.46 - 7.53(m, 2H, H_{Ar}), 7.56 - 7.65(m, 3H, H_{Ar}), 7.66 - 7.73 (m, 2H, H_{Ar}), 8.74 (d, $^3J_{HH}$ = 8.7 Hz, 1H, H_{Ar}), 11.33 (bs, 1H, NH).

^{13}C-NMR (100.53 MHz; CDCl$_3$): δ = 14.4 (CH$_3$), 61.5 (CH$_3$$CH_2$), 121.9 ($C_{Ar}$H), 123.1 ($C_{Ar}$H), 123.5 ($C_{Ar}$), 128.6 (CO$C$HCH), 130.1 ($C_{Ar}$H), 131.6 ($C_{Ar}$H), 132.8 ($C_{Ar}$H), 134.0 ($C_{Ar}$H), 134.7 ($C_{Ar}$H), 137.3 (CO$C$HCH), 138.6 ($C_{Ar}$), 140.3 ($C_{Ar}$N), 162.6 (CH$C$OO), 165.4 (CH$C$OO), 200.1 ($C_{Ar}$$COC_{Ar}$).

ESI-MS: m/z (%): 324.1 (68) [$C_{19}H_{17}NO_4H^+$], 363.5 (100) [$C_{19}H_{17}NO_4H_2ONa^+$], 383.7 (76) [$C_{19}H_{16}NO_4NaK^+$], 504.8 (38) [($C_{19}H_{17}NO_4$)$_3$HK$^+$], 669.1 (11) [($C_{19}H_{17}NO_4$)$_2$Na$^+$], 685.1 (44) [($C_{19}H_{17}NO_4$)$_2$K$^+$].

elemental analysis for $C_{19}H_{17}NO_4$ (323.34 g/mol):

calcd.: C 70.58, H 5.30, N 4.33, O 19.79

found: C 70.41, H 5.20, N 4.28

2-Benzoylphenyl ethyl maleate

5.14: $C_{19}H_{16}O_5$, M = 324.33 g/mol, colourless oil, yield: 71%, **route**: *Ester-III*.

^{1}H-NMR (400.13 MHz; CDCl$_3$): $\delta = 1.30$ (t, $^3J_{HH} = 7.0$ Hz, 3H, CH$_3$), 4.23 (q, $^3J_{HH} = 7.0$ Hz, 2H, CH$_3$CH$_2$), 6.68 (d, $^3J_{HH} = 15.8$ Hz, 1H, COCHCHCO), 6.73 (d, $^3J_{HH} = 15.8$ Hz, 1H, COCHCHCO), 7.30 - 7.48 (m, 3H, H_{Ar}), 7.48 - 7.62 (m, 4H, H_{Ar}), 7.79 (d, $^3J_{HH} = 7.4$ Hz, 2H, H_{Ar}).

^{13}C-NMR (100.62 MHz; CDCl$_3$): $\delta = 14.2$ (CH$_3$), 61.6 (CH$_3$CH$_2$), 123.1 (C_{Ar}H), 126.2 (C_{Ar}H), 128.6 (C_{Ar}H), 129.3 (C_{Ar}H), 129.9 (C_{Ar}H), 130.8 (C_{Ar}), 131.9 (CO CHCH), 132.5 (C_{Ar}H), 133.3 (C_{Ar}H), 135.4 (CO CHCH), 136.5 (C_{Ar}), 148.4 (C_{Ar}O), 163.1 (CHCOO), 164.6 (CHCOO), 194.6 (C_{Ar} CO C$_{Ar}$).

2-Benzoylphenyl 1-methoxypropan-2-yl maleate

5.18: $C_{21}H_{20}O_6$, M = 368.38 g/mol.

1.00 g (5.30 mmol) 1-methoxypropan-2-yl maleate, 1.05 g (5.30 mmol) 2-hydroxybenzophenone and 50.0 mg (0.41 mmol) dmap were dissolved in 5 ml of dichloromethane, cooled to 0 °C and 1.10 g (5.30 mmol) dcc was added. After 5 min the solution mixture was warmed to ambient temperature and stirred over night. 10 ml dichloromethane were added, the reaction mixture filtered and the solvent removed in vacuo. The residue was taken up in 10 ml dichloromethane, washed twice with 5 [M] hydrochloric acid, once with saturated NaHCO$_3$ solution, dried over Na$_2$SO$_4$, filtered and the solvent removed under reduced pressure. The crude product was purified by *flash* chromatography (SiO$_2$; CH$_2$Cl$_2$) to yield 0.69 g (1.96 mmol, 37%) 2-benzoylphenyl 1-methoxypropan-2-yl maleate as a colourless oil.[146]

^{1}H-NMR (400.13 MHz; CDCl$_3$): δ = 1.25 (d, $^3J_{HH}$ = 6.5 Hz, 3H, CHCH_3), 3.35 (s, 3H, OCH_3), 3.40 (dd, $^3J_{HH}$ = 3.5 Hz, $^2J_{HH}$ = 10.8 Hz, 1H, CHCH_2O), 3.46 (dd, $^3J_{HH}$ = 6.2 Hz, $^2J_{HH}$ = 10.8 Hz, 1H, CHCH_2O), 5.11 - 5.17 (m, 1H, OCH), 6.68 (d, $^3J_{HH}$ = 15.8 Hz, 1H, COCHCHCO), 6.73 (d, $^3J_{HH}$ = 15.8 Hz, 1H, COCHCHCO), 7.24 (d, $^3J_{HH}$ = 8.2 Hz, 1H, H_{Ar}), 7.35 (t, $^3J_{HH}$ = 7.5 Hz, 1H, H_{Ar}), 7.42 (t, $^3J_{HH}$ = 7.5 Hz, 2H, H_{Ar}), 7.50 - 7.61 (m, 3H, H_{Ar}), 7.74 (d, $^3J_{HH}$ = 7.9 Hz, 2H, H_{Ar}).

^{13}C-NMR (100.62 MHz; CDCl$_3$): δ = 16.6 (CHCH$_3$), 59.3 (OCH$_3$), 70.7 (OCH), 74.9 (CHCH$_2$O), 123.1 (C_{Ar}H), 126.2 (C_{Ar}H), 128.6 (C_{Ar}H), 129.9 (C_{Ar}H), 130.7 (C_{Ar}H), 131.6 (C_{Ar}), 132.0 (COCHCH), 132.4 (C_{Ar}H), 133.3(C_{Ar}H), 135.5 (CHCO), 137.5 (C_{Ar}), 148.4 (C_{Ar}O), 163.1 (CHCOO), 164.1 (CHCOO), 194.5 ($C_{Ar}C$OC$_{Ar}$).

N (2-Benzoylphenyl) maleamide

ma-bp: C$_{17}$H$_{13}$NO$_4$, M = 295.29 *g/mol*, yellow powder, yield: >99%, **route:** *Amide-II.*

^{1}H-NMR (399.78 MHz; CDCl$_3$): δ = 6.45 (d, $^3J_{HH}$ = 17.2 Hz, 1H, COCHCHCO), 6.49 (d, $^3J_{HH}$ = 17.2 Hz, 1H, COCHCHCO), 7.28 (dt, $^3J_{HH}$ = 1.1 Hz, $^3J_{HH}$ = 7.7 Hz, 1H, H_{Ar}), 7.49 - 7.54 (m, 2H, H_{Ar}), 7.60 - 7.75 (m, 5H, H_{Ar}), 7.70 (dd, $^3J_{HH}$ = 1.0 Hz, $^3J_{HH}$ = 8.7 Hz, 1H, H_{Ar}), 11.79 (bs, 1H, NH).

^{13}C-NMR (100.53 MHz; CDCl$_3$): δ = 122.6 (C_{Ar}H), 124.4 (C_{Ar}), 125.0 (C_{Ar}H), 128.7 (C_{Ar}H), 130.2 (C_{Ar}H), 132.0 (C_{Ar}H), 133.3 (C_{Ar}H), 134.3 (COCHCH), 134.8 (C_{Ar}H), 138.0 (C_{Ar}), 138.1 (COCHCH), 138.5 (C_{Ar}N), 164.4 (CHCOO), 165.1 (CHCOO), 200.2 ($C_{Ar}$$COC_{Ar}$).

ESI-MS: m/z (%): 198.2 (28) [C$_{13}$H$_{11}$NOH$^+$], 278.2 (10) [C$_{17}$H$_{12}$NO$_3$$^+$], 296.1 (44) [C$_{17}H_{13}NO_4H^+$], 318.1 (30) [C$_{17}H_{13}NO_4Na^+$], 613.1 (23) [(C$_{17}H_{13}NO_4$)$_2Na^+$], 629.1 (100) [(C$_{17}H_{13}NO_4$)$_2K^+$].

elemental analysis for C$_{17}$H$_{13}$NO$_4$ (295.29 *g/mol*):

calcd.: C 69.15, H 4.44, N 4.74, O 21.67

found: C 69.45, H 4.30, N 4.66

N(2-Benzoylphenyl) dimethyl(vinyl)silyl maleate

5.13: C$_{21}$H$_{21}$NO$_4$Si, M = 379.48 *g/mol*, orange oil, yield: 26%, **route**: *Silylester*.

^{1}H-NMR (400.13 MHz; CDCl$_3$): δ = 0.27 (s, 3H, Si(CH_3)$_2$), 0.37 - 0.44 (m, 3H, Si(CH_3)$_2$), 5.75 (dd, $^3J_{HH}$ = 3.5 Hz, $^2J_{HH}$ = 20.3 Hz, 1H, CHCH_2), 5.92 (dd, $^3J_{HH}$ = 3.5 Hz, $^3J_{HH}$ = 14.8 Hz, 1H, CHCH_2), 6.05 - 6.16 (m, 1H, CHCH$_2$), 6.13 (d, $^3J_{HH}$ = 12.4 Hz, 1H, COCHCHCO), 6.47 (d, $^3J_{HH}$ = 12.4 Hz, 1H, COCHCHCO), 7.11 (t, $^3J_{HH}$ = 7.6 Hz, 1H, H_{Ar}), 7.38 - 7.53 (m, 2H, H_{Ar}), 7.53 - 7.65 (m, 3H, H_{Ar}), 7.69 (d, $^3J_{HH}$ = 7.9 Hz, 2H, H_{Ar}), 8.68 (d, $^3J_{HH}$ = 8.3 Hz, 1H, H_{Ar}), 11.05 (s, 1H, NH).

^{13}C-NMR (100.62 MHz; CDCl$_3$): δ = -2.1 (SiCH$_3$), -2.0 - -1.9 (m, SiCH$_3$), 122.0 (C_{Ar}H), 122.7 (C_{Ar}H), 128.3 (C_{Ar}H), 128.5 (C_{Ar}H), 129.3 (C_{Ar}), 130.1 (C_{Ar}H), 132.7 (C_{Ar}H), 133.5 (C_{Ar}H), 134.3 (CHCH$_2$), 134.6 (COCHCH), 135.1 (COCHCH), 135.3 (CHCH$_2$), 138.6 (C_{Ar}), 140.0 (C_{Ar}N), 164.1 (CHCOO), 165.0 (CHCOO), 199.5 ($C_{Ar}$$COC_{Ar}$).

N(9-Fluorenone-1-yl) maleamide

ma-fl: $C_{17}H_{11}NO_4$, M $= 293.27$ g/mol, red-orange solid, yield: $>99\%$, **route**: *Amide-II*.

^{1}H-NMR (399.78 MHz; CDCl$_3$/dmso-d$_6$): δ $=$ 5.74 (d, $^3J_{HH}$ $=$ 12.1 Hz, 1H, COC*H*CHCO), 5.97 (d, $^3J_{HH}$ $=$ 12.1 Hz, 1H, COCH*C*HCO), 6.65 - 6.80 (m, 2H, H_{Ar}), 6.84 - 7.05 (m, 4H, H_{Ar}), 7.68 (d, $^3J_{HH}$ $=$ 8.4 Hz, 1H, H_{Ar}), 9.72 (s, 1H, N*H*).

^{13}C-NMR (100.53 MHz; CDCl$_3$/dmso-d$_6$): δ $=$ 115.1 (C_{Ar}H), 119.5 (C_{Ar}H), 119.9(C_{Ar}H), 122.8 (C_{Ar}H), 128.2 (C_{Ar}H), 129.2 (C_{Ar}H), 131.6 (C_{Ar}H), 132.4 (C_{Ar}), 133.9 (CO*C*HCH), 135.6 (CO*C*HCH), 135.8 (C_{Ar}N), 136.9 (C_{Ar}), 142.3 (C_{Ar}), 142.8 (C_{Ar}), 163.1 (CH*C*OO), 165.2 (CH*C*OO), 193.5 (C_{Ar}*C*OC_{Ar}).

CI-MS: m/z (%): 194.8 (100) [C$_{13}$H$_8$NO$^+$], 293.7 (26.24) [C$_{17}$H$_{11}$NO$_4{}^+$].

ESI-MS: m/z (%): 196.2 (71) [C$_{13}$H$_9$NOH$^+$], 294.2 (93) [C$_{17}$H$_{11}$NO$_4$H$^+$], 609.3 (61) [(C$_{17}$H$_{11}$NO$_4$)$_2$Na$^+$], 625.2 (100) [(C$_{17}$H$_{11}$NO$_4$)$_2$K$^+$], 902.0 (17) [(C$_{17}$H$_{11}$NO$_4$)$_3$Na$^+$], 918.0 (63) [(C$_{17}$H$_{11}$NO$_4$)$_3$K$^+$].

elemental analysis for C$_{17}$H$_{11}$NO$_4$ (293.27 g/mol):

calcd.: C 69.62, H 3.78, N 4.78, O 21.82

found: C 69.93, H 3.70, N 4.71

N(2-Benzoylphenyl) maleimide

mi-bp: $C_{17}H_{11}NO_3$, M = 277.27, yellow crystals, yield: 98%, **route**: *Imide*.

^{1}H-NMR (399.78 MHz; CDCl$_3$): δ = 6.63 (s, 2H, COC*H*C*H*CO), 7.34 - 7.44 (m, 3H, H_{Ar}), 7.46 - 7.56 (m, 2H, H_{Ar}), 7.60 - 7.68 (m, 2H, H_{Ar}), 7.69 - 7.74 (m, 2H, H_{Ar}).

^{13}C-NMR (100.53 MHz; CDCl$_3$): δ = 128.5 (C_{Ar}H), 129.4 (C_{Ar}H), 130.1 (C_{Ar}H, C_{Ar}), 131.1 (C_{Ar}H), 132.3 (C_{Ar}H), 133.2 (C_{Ar}H), 134.3 (CO*C*H), 136.0 (C_{Ar}), 137.2 (C_{Ar}N), 169.5 (CH*C*ON), 195.4 (C_{Ar}*C*OC_{Ar}).

ESI-MS: m/z (%): 278.2 (100) [$C_{17}H_{11}NO_3H^+$], 317.5 (81) [$C_{17}H_{11}NO_3K^+$], 337.9 (74) [$C_{17}H_{11}NO_3H_2OCH_3CNH^+$].

elemental analysis for $C_{17}H_{11}NO_3$ (277.27 *g/mol*):

calcd.: C 73.64, H 4.00, N 5.05, O 17.31

found: C 73.37, H 3.99, N 5.02

N(9-Fluorenone-1-yl) maleimide

mi-fl: $C_{17}H_9NO_3$, M = 275.26 *g/mol*, yellow solid, yield: >99%, **route**: *Imide*.

^{1}H-NMR (399.78 MHz; CDCl$_3$): δ = 6.92 (s, 2H, COC*H*C*H*CO), 7.09 - 7.18 (m, 1H, H_{Ar}), 7.28 (t, 1H, $^3J_{HH}$ = 7.1 Hz, H_{Ar}), 7.44 - 7.63 (m, 5H, H_{Ar}).

^{13}C-NMR (100.53 MHz; CDCl$_3$): δ = 120.8 (C_{Ar}H), 121.0 (C_{Ar}H), 124.6 (C_{Ar}H), 128.6 (C_{Ar}), 129.3 (C_{Ar}), 129.4 (C_{Ar}H), 129.7 (C_{Ar}H), 134.0 (C_{Ar}), 134.8 (COCH), 135.0 (C_{Ar}H), 135.7 (C_{Ar}H), 143.5 (C_{Ar}), 146.1 (C_{Ar}N), 169.0 (CHCON), 191.2 (C_{Ar}COC_{Ar}).

CI-MS: m/z (%): 275.5 (49.74) [C$_{17}$H$_9$NO$_3$$^+$].

elemental analysis for C$_{17}$H$_9$NO$_3$ (275.26 g/mol):

calcd.: C 74.18, H 3.30, N 5.09, O 17.44

found: C 73.95, H 3.41, N 5.00

8.2.3. Hydrazones

2-(Hydrazono(phenyl)methyl)phenol

3.2-OH: C$_{13}$H$_{12}$N$_2$O, M = 212.25 g/mol.

1.00 g (5.37 mmol) 2-hydroxybenzophenone was dissolved in 20 ml of ethanol, 1.25 ml (1.25 g, 25.0 mmol) hydrazine hydrate and 0.25 ml acetic acid were added and the solution was refluxed over night. The reaction mixture was poured into 100 ml of water and extracted two times with 25 ml CHCl$_3$. The combined organic layers were washed once with water and once with brine, dried over Na$_2$SO$_4$, filtered and the solvent removed under reduced pressure to give 1.02 g (4.81 mmol, 90%) 2-(hydrazono(phenyl)methyl)phenol as a white solid. A microanalytically pure sample was obtained by crystallisation from methanol at -40 °C.

^{1}H-NMR (399.78 MHz; CDCl$_3$): δ = 5.31 (s, 2H, NH_2), 6.64 - 6.71 (m, 2H, H_{Ar}), 6.98 (d, $^3J_{HH}$ = 8.2 Hz, 1H, H_{Ar}), 7.14 - 7.20 (m, 1H, H_{Ar}), 7.25 - 7.35 (m, 2H, H_{Ar}), 7.48 - 7.60 (m, 3H, H_{Ar}), 12.40 (s, 1H, OH).

^{13}C-NMR (100.53 MHz; CDCl$_3$): $\delta = 117.1$ (C_{Ar}H), 118.5 (C_{Ar}H), 128.1 (C_{Ar}), 128.4 (C_{Ar}), 128.6 (C_{Ar}H), 129.4 (C_{Ar}H), 129.7 (C_{Ar}H), 129.8 (C_{Ar}H), 129.9 (C_{Ar}H), 154.4 ($C_{Ar}CNC_{Ar}$), 158.7 (C_{Ar}O).

ESI-MS: m/z (%): 196.2 (49) [C$_{13}$H$_9$NO$^+$], 213.1 (100) [C$_{13}$H$_{12}$N$_2$OH$^+$].

N'-((2-Hydroxyphenyl)(phenyl)methylene)-4-methylbenzenesulfonohydrazide

3.8-OH: C$_{20}$H$_{18}$N$_2$O$_3$S, M = 366.43 *g/mol*, white solid, yield: >99%, **route:** *Hydrazone.*

^{1}H-NMR (399.78 MHz; CDCl$_3$/dmso-d$_6$): $\delta = 2.30$ (s, 3H, C$_{Ar}$CH_3), 6.54 (d, $^3J_{HH} = 4.1$ Hz, 2H, H_{Ar}), 6.83 (d, $^3J_{HH} = 8.3$ Hz, 1H, H_{Ar}), 6.99 - 7.05 (m, 2H, H_{Ar}), 7.06 - 7.14 (m, 1H, H_{Ar}), 7.23 (d, $^3J_{HH} = 8.0$ Hz, 2H, H_{Ar}), 7.39 - 7.46 (m, 3H, H_{Ar}), 7.70 (d, $^3J_{HH} = 8.1$ Hz, 2H, H_{Ar}), 8.52 (bs, 1H, NH), 11.30 (bs, 1H, OH).

^{13}C-NMR (100.53 MHz; CDCl$_3$/dmso-d$_6$): $\delta = 21.5$ ($_{Ar}$CH_3), 117.3 (C_{Ar}H), 118.6 (C_{Ar}H), 118.9 (C_{Ar}), 127.6 (C_{Ar}H), 128.1 (C_{Ar}H), 128.8 (C_{Ar}), 129.5 (C_{Ar}H), 129.9 (C_{Ar}H), 130.1 (C_{Ar}H), 130.2 (C_{Ar}), 130.5 (C_{Ar}H), 131.5 (C_{Ar}H), 134.9 (C_{Ar}), 144.4 ($C_{Ar}CNC_{Ar}$), 158.6 (C_{Ar}O).

ESI-MS: m/z (%): 211.2 (43) [C$_{13}$H$_{11}$N$_2$O$^+$], 367.2 (100) [C$_{20}$H$_{18}$N$_2$O$_3$SH$^+$], 389.2 (3) [C$_{20}$H$_{18}$N$_2$O$_3$SNa$^+$], 755.1 (12) [(C$_{20}$H$_{18}$N$_2$O$_3$S)$_2$Na$^+$], 771.1 (12) [(C$_{20}$H$_{18}$N$_2$O$_3$S)$_2$K$^+$].

N-acetyl-N-(9-hydrazono-9-fluoren-1-yl)acetamide

3.22: $C_{17}H_{15}N_3O_2$, M = 293.32 g/mol.

1.00 g (5.12 mmol) 1-amino-9-fluorenone was suspended in 25 ml of diethylether, 1.25 ml (1.25 g, 25.0 mmol) hydrazine hydrate and 0.25 ml acetic acid were added and the solution was refluxed over night. The reaction mixture was poured into 100 ml of water and extracted three times with 75 ml CHCl$_3$. The combined organic layers were washed once with water and once with brine, dried over Na$_2$SO$_4$, filtered and the solvent removed under reduced pressure. The resulting oil is extracted with diethylether, filtered and the solvent removed from the filtrate under reduced pressure to give 0.588 g (2.00 mmol, 39%) product as a cream coloured solid.

^{1}H-NMR (399.78; CDCl$_3$): δ = 2.05 (s, 3H, CH_3), 2.24 (s, 3H, CH_3), 5.58 (bs, 2H, NH_2), 6.59 (d, $^3J_{HH}$ = 8.1 Hz, 1H, H_{Ar}), 7.00 (d, $^3J_{HH}$ = 7.2 Hz, 1H, H_{Ar}), 7.17 (t, $^3J_{HH}$ = 7.6 Hz, 1H, H_{Ar}), 7.23 - 7.27 (m, 1H, H_{Ar}), 7.38 (d, $^3J_{HH}$ = 7.4 Hz, 1H, H_{Ar}), 7.60 (d, $^3J_{HH}$ = 7.5 Hz, 1H, H_{Ar}), 8.25 (d, $^3J_{HH}$ = 7.6 Hz, 1H, H_{Ar}).

^{13}C-NMR (100.53 MHz; CDCl$_3$): δ = 18.4 (CH_3), 25.4 (CH_3), 109.4 (C_{Ar}H), 115.7 (C_{Ar}H), 117.4 (C_{Ar}), 120.1 (C_{Ar}H), 127.8 (C_{Ar}H), 129.6 (C_{Ar}H), 130.7 (C_{Ar}H), 131.6 (C_{Ar}H), 132.3 (C_{Ar}), 141.8 (C_{Ar}), 142.3 (C_{Ar}), 145.5 (C_{Ar}CNC$_{Ar}$), 158.8 (C_{Ar}O), 162.6 (CH$_3$CON).

N'-(1-Hydroxy-9H-fluoren-9-ylidene)-4-methylbenzenesulfonohydrazide monohydrate

3.23: $C_{20}H_{18}N_2O_4S$, M = 382.43 g/mol.

1.00 g (5.10 mmol) 1-hydroxy-9-fluorenone, 1.41 g (7.57 mmol) tosylhydrazone and 875 μl conc. hydrochloric acid were suspended in 22 ml ethanol and refluxed over night. The reaction mixture was cooled to ambient temperature and the solvent removed under reduced pressure. The product was isolated by *flash* chromatography (SiO$_2$; CH$_2$Cl$_2$) as a white solid (0.82 g, 2.14 mmol, 42%).[92]

^{1}H-NMR (400.13; CDCl$_3$): δ = 1.54 (s, 2H, H_2O), 2.42 (s, 3H, $C_{Ar}CH_3$), 6.78 (d, $^3J_{HH}$ = 8.5 Hz, 1H, H_{Ar}), 7.09 (d, $^3J_{HH}$ = 7.3 Hz, 1H, H_{Ar}), 7.21 - 7.26 (m, 1H, H_{Ar}), 7.29 - 7.40 (m, 3H, H_{Ar}), 7.46 (t, $^3J_{HH}$ = 7.6 Hz, 1H, H_{Ar}), 7.63 (d, $^3J_{HH}$ = 7.3 Hz, 1H, H_{Ar}), 7.86 (d, $^3J_{HH}$ = 7.7 Hz, 1H, H_{Ar}), 7.91 (d, $^3J_{HH}$ = 8.0 Hz, 2H, H_{Ar}), 8.09 (bs, 1H, NH), 8.50 (bs, 1H, OH).

N'-((2-Aminophenyl)(phenyl)methylene)-4-methylbenzenesulfonohydrazide

3.8-NH₂: $C_{20}H_{19}N_3O_2S$, M = 365.45 g/mol, yellow solid, yield: 75%, **route:** *Hydrazone*.

two diastereoisomers:

^{1}H-NMR (399.78 MHz; CDCl$_3$): δ = 2.41 (s, 3H, C$_{Ar}$CH_3), 2.42 (s, 3H, C$_{Ar}$CH_3), 6.38 - 6.47 (m, 1H, H_{Ar}), 6.54 - 6.59 (m, 1H, H_{Ar}), 6.67 (d, $^3J_{HH}$ = 8.2 Hz, 1H, H_{Ar}), 6.75 - 6.86 (m, 4H, H_{Ar}), 6.99 - 7.09 (m, 3H, H_{Ar}), 7.23 - 7.38 (m, 11H, H_{Ar}, NH_2), 7.46 - 7.57 (m, 6H, H_{Ar}, NH), 7.78 (d, $^3J_{HH}$ = 8.3 Hz, 2H, H_{Ar}), 7.82 (bs, 1H, NH), 7.87 (d, $^3J_{HH}$ = 8.3 Hz, 2H, H_{Ar}).

^{13}C-NMR (100.53 MHz; CDCl$_3$): δ = 21.8 ($_{Ar}$CH_3), 116.0 (C_{Ar}), 116.0 (C_{Ar}H), 116.6 (C_{Ar}H), 116.7 (C_{Ar}H), 119.4 (C_{Ar}H), 127.7 (C_{Ar}H), 127.8 (C_{Ar}H), 128.2 (C_{Ar}H), 128.2 (C_{Ar}H), 128.6 (C_{Ar}H), 129.5 (C_{Ar}H), 129.8 (C_{Ar}H), 130.0 (C_{Ar}H), 130.1 (C_{Ar}H), 130.3 (C_{Ar}H), 130.7 (C_{Ar}H), 131.6 (C_{Ar}H), 131.8 (C_{Ar}), 132.1 (C_{Ar}H), 135.7 (C_{Ar}), 136.0 (C_{Ar}), 143.4 (C_{Ar}), 144.3 (C_{Ar}), 144.5 (C_{Ar}N), 147.7 (C_{Ar}N), 152.8 (C$_{Ar}$CNC$_{Ar}$), 158.2 (C$_{Ar}$CNC$_{Ar}$).

ESI-MS: m/z (%): 195.2 (28) [C$_{13}$H$_{11}$N$_2^+$], 210.2 (21) [C$_{13}$H$_{12}$N$_3^+$], 366.1 (100) [C$_{20}$H$_{19}$N$_3$O$_2$SH$^+$], 388.1 (3) [C$_{20}$H$_{19}$N$_3$O$_2$SNa$^+$], 752.8 (2) [(C$_{20}$H$_{19}$N$_3$O$_2$S)$_2$Na$^+$], 769.0 (3) [(C$_{20}$H$_{19}$N$_3$O$_2$S)$_2$K$^+$].

elemental analysis for C$_{20}$H$_{19}$N$_3$O$_2$S (365.45 *g/mol*):

calcd.: C 65.73, H 5.24, N 11.50, O 8.76, S 8.77

found: C 65.59, H 5.30, N 11.40, S 8.64

2-(Phenyl(2-tosylhydrazono)methyl)phenyl acetate

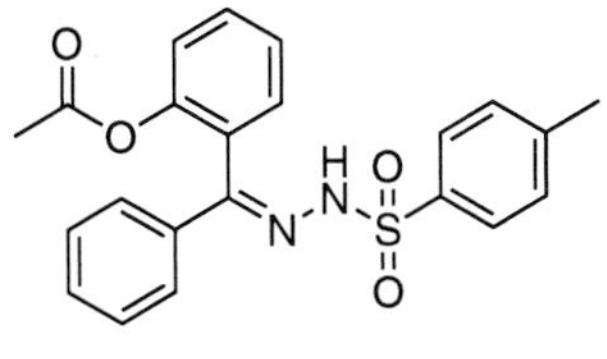

Ac-bpNNHTs: C$_{22}$H$_{20}$N$_2$O$_4$S, M = 408.47 *g/mol*, white solid, yield: 53%, **route:** *Hydrazone.*

^{1}H-NMR (400.13 MHz; CDCl$_3$): $\delta = 1.64$ (s, 3H, CH_3CO), 2.42 (s, 3H, CH_3), 7.13 - 7.35 (m, 6H, H_{Ar}), 7.39 (d, $^3J_{HH} = 7.1$ Hz, 2H, H_{Ar}), 7.52 - 7.57 (m, 2H, H_{Ar}), 7.63 (s, 1H, NH), 7.80 (d, $^3J_{HH} = 8.2$ Hz, 1H, H_{Ar}), 7.91 (d, $^3J_{HH} = 8.2$ Hz, 2H, H_{Ar}).

^{13}C-NMR (100.62 MHz; CDCl$_3$): $\delta = 20.1$ (CH$_3$), 21.7 (CH$_3$), 123.7 (C_{Ar}H), 127.2 (C_{Ar}H), 127.5 (C_{Ar}H), 127.9 (C_{Ar}), 128.5 (C_{Ar}H), 128.6 (C_{Ar}H), 129.4 (C_{Ar}H), 130.1 (C_{Ar}H), 130.2 (C_{Ar}H), 131.7 (C_{Ar}H), 135.5 (C_{Ar}), 135.9 (C_{Ar}), 144.0 (C_{Ar}), 148.1 (C_{Ar}O), 150.8 (C_{Ar}CNC$_{Ar}$), 169.6 (CHCOO).

ESI-MS: m/z (%): 367.2 (97) [C$_{20}$H$_{18}$N$_2$O$_3$SH$^+$], 409.1 (69) [C$_{22}$H$_{20}$N$_2$O$_4$SH$^+$], 431.2 (100) [C$_{22}$H$_{20}$N$_2$O$_4$SNa$^+$], 838.9 (11) [(C$_{22}$H$_{20}$N$_2$O$_4$S)$_2$Na$^+$], 855.0 (13) [(C$_{22}$H$_{20}$N$_2$O$_4$S)$_2$K$^+$].

elemental analysis for C$_{22}$H$_{20}$N$_2$O$_4$S (408.47 g/mol):

calcd.: C 64.69, H 4.94, N 6.86, O 15.67, S 7.85

found: C 62.22, H 4.95, N 7.20, S 9.07

2-(Phenyl(2-tosylhydrazono)methyl)phenyl crotonate

Me-bpNNHTs: C$_{24}$H$_{22}$N$_2$O$_4$S, M = 434.51 g/mol, white solid, yield: 67%, **route**: *Hydrazone.*

Single crystals suitable for X-ray analysis were obtained by slow evaporation of a saturated ethanol solution.

^{1}H-NMR (399.78 MHz; CDCl$_3$): $\delta = 1.70$ (d, $^3J_{HH} = 6.8$ Hz, 3H, CHCH_3), 2.42 (s, 3H, C$_{Ar}$CH_3), 5.40 (d, $^3J_{HH} = 14.3$ Hz, 1H, CHCHCO), 6.52 (dq, $^3J_{HH} = 7.0$ Hz, $^3J_{HH} = 15.4$ Hz, 1H, CH$_3$CHCH), 7.15 - 7.46 (m, 10H, H_{Ar}), 7.50 - 7.60 (m, 1H, H_{Ar}), 7.73 (s, 1H, NH), 7.90 (d, $^3J_{HH} = 8.1$ Hz, 2H, H_{Ar}).

^{13}C-NMR (100.53 MHz; CDCl$_3$): δ = 18.3 (CHCH_3), 21.8 (C$_{Ar}$$CH_3$), 120.7 (C$_{Ar}$H), 123.7 (CH$CH$CO), 125.8 (C$_{Ar}$), 127.3 (C$_{Ar}$H), 127.4 (C$_{Ar}$H), 128.4 (C$_{Ar}$H), 128.7 (C$_{Ar}$H), 129.5 (C$_{Ar}$H), 130.0 (C$_{Ar}$H), 130.2 (C$_{Ar}$H), 131.7 (C$_{Ar}$H), 135.8 (C$_{Ar}$), 136.1 (C$_{Ar}$), 143.7 (C$_{Ar}$), 148.0 (CH$_3$$C$HCH), 148.3 (C$_{Ar}$O), 151.2 (C$_{Ar}$$CNC_{Ar}$), 164.9 (CH$C$OO).

ESI-MS: m/z (%): 367 (21) [C$_{20}$H$_{19}$N$_2$O$_3$S$^+$], 435.1 (22) [C$_{24}$H$_{22}$N$_2$O$_4$SH$^+$], 457.2 (100) [C$_{24}$H$_{22}$N$_2$O$_4$SNa$^+$].

elemental analysis for C$_{24}$H$_{22}$N$_2$O$_4$S (434.51 g/mol):

calcd.: C 66.34, H 5.10, N 6.45, O 14.73, S 7.38

found: C 65.83, H 5.10, N 6.44, S 7.34

2-(Phenyl(2-tosylhydrazono)methyl)phenyl 4-methoxycinnamate

PhOMe-bpNNHTs: C$_{30}$H$_{26}$N$_2$O$_5$S, M = 526.60 g/mol, cream solid, yield: 86%, **route:** *Hydrazone*.

^{1}H-NMR (400.13 MHz; CDCl$_3$): δ = 2.21 (s, 3H, CH_3), 3.85 (s, 3H, OCH_3), 5.87 (d, $^3J_{HH}$ = 15.9 Hz, 1H, CHCHCO), 6.89 (d, $^3J_{HH}$ = 8.8 Hz, 2H, H_{Ar}), 7.19 - 7.4 (m, 9H, H_{Ar}), 7.40 - 7.44 (m, 3H, H_{Ar}, C$_{Ar}$CHCH), 7.53 - 7.59 (m, 2H, H_{Ar}), 7.83 (s, 1H, NH), 7.90 (d, $^3J_{HH}$ = 8.3 Hz, 2H, H_{Ar}).

^{13}C-NMR (100.62 MHz; CDCl$_3$): δ = 21.6 (CH$_3$), 55.6 (OOCH$_3$), 113.1 (CHCHCO), 114.5 (C$_{Ar}$H), 123.7 (C$_{Ar}$H), 125.8 (C$_{Ar}$), 126.7 (C$_{Ar}$), 127.3 (C$_{Ar}$H), 128.0 (C$_{Ar}$), 128.0 (C$_{Ar}$H), 128.4 (C$_{Ar}$H), 128.5 (C$_{Ar}$H), 129.4 (C$_{Ar}$H), 130.0 (C$_{Ar}$H), 130.2 (C$_{Ar}$H), 131.7 (C$_{Ar}$H), 135.7 (C$_{Ar}$), 136.2 (C$_{Ar}$), 143.7 (C$_{Ar}$), 147.1 (C$_{Ar}$$C$HCH), 148.4 (C$_{Ar}$O), 151.1 (C$_{Ar}$$CNC_{Ar}$), 162.1 (C$_{Ar}OCH_3$), 165.8 (CH$C$OO).

ESI-MS: m/z (%): 161.2 (14) $[C_{10}H_9O_2{}^+]$, 367.2 (9) $[C_{20}H_{18}N_2O_3SH^+]$, 527.0 (2) $[C_{30}H_{26}N_2O_5SH^+]$, 549.3 (100) $[C_{30}H_{26}N_2O_5SNa^+]$, 565.1 (2) $[C_{30}H_{26}N_2O_5SK^+]$, 1075.0 (15) $[(C_{30}H_{26}N_2O_5S)_2Na^+]$, 1091.1 (7) $[(C_{30}H_{26}N_2O_5S)_2K^+]$.

elemental analysis for $C_{30}H_{26}N_2O_5S$ (526.60 *g/mol*):

calcd.: C 68.42, H 4.98, N 5.32, O 15.19, S 6.09

found: C 67.27, H 5.06, N 5.18, S 6.24

2-(Phenyl(2-tosylhydrazono)methyl)phenyl 4-methylcinnamate

PhMe-bpNNHTs: $C_{30}H_{26}N_2O_4S$, M = 510.60 *g/mol*, cream solid, yield: 67%, **route**: *Hydrazone*.

^{1}H-NMR (400.13 MHz; CDCl$_3$): δ = 2.41 (s, 3H, CH_3), 2.46 (s, 3H, CH_3), 6.43 (d, $^3J_{HH}$ = 16.0 Hz, 1H, CHCHCO), 6.63 - 6.76 (m, 1H, H_{Ar}), 7.02 (d, $^3J_{HH}$ = 8.1 Hz, 1H, H_{Ar}), 7.09 - 7.16 (m, 1H, H_{Ar}), 7.19 - 7.34 (m, 6H, H_{Ar}, NH), 7.38 (d, $^3J_{HH}$ = 8.1 Hz, 2H, H_{Ar}), 7.48 (d, $^3J_{HH}$ = 8.0 Hz, 3H, H_{Ar}), 7.57 - 7.60 (m, 2H, H_{Ar}), 7.80 (d, $^3J_{HH}$ = 7.2 Hz, 1H, C$_{Ar}$CHCH), 7.84 (d, $^3J_{HH}$ = 8.3 Hz, 2H, H_{Ar}).

^{13}C-NMR (100.62 MHz; CDCl$_3$): δ = 21.5 (CH$_3$), 21.6 (CH$_3$), 115.9 (CHCHCO), 117.6 (C_{Ar}H), 118.7 (C_{Ar}), 118.8 (C_{Ar}H), 127.1 (C_{Ar}H), 127.8 (C_{Ar}H), 127.9 (C_{Ar}H), 128.4 (C_{Ar}H), 129.3 (C_{Ar}H), 129.7 (C_{Ar}H), 129.7 (C_{Ar}H), 130.0 (C_{Ar}H), 130.1 (C_{Ar}H), 130.6 (C_{Ar}H), 130.7 (C_{Ar}H), 131.3 (C_{Ar}), 131.8 (C_{Ar}H), 134.7 (C_{Ar}H), 141.4 (C_{Ar}CNC_{Ar}), 147.5 (C_{Ar}CHCH), 158.8 (C_{Ar}O), 172.2 (CHCOO).

ESI-MS: m/z (%): 145.2 (9) $[C_{10}H_9O^+]$, 367.3 (24) $[C_{20}H_{18}N_2O_3SH^+]$, 511.1 (2) $[C_{30}H_{26}N_2O_4SH^+]$, 533.3 (100) $[C_{30}H_{26}N_2O_4SNa^+]$, 549.2 (3) $[C_{30}H_{26}N_2O_4SK^+]$, 1043.0 (12) $[(C_{30}H_{26}N_2O_4S)_2Na^+]$, 1059.1 (4) $[(C_{30}H_{26}N_2O_4S)_2K^+]$.

2-(Phenyl(2-tosylhydrazono)methyl)phenyl cinnamate

Ph-bpNNHTs: $C_{29}H_{24}N_2O_4S$, M $= 496.58$ g/mol, off-white solid, yield: 53%, **route:** *Hydrazone.*

^{1}H-NMR (399.78 MHz; CDCl$_3$): $\delta = 2.16$ (s, 3H, CH_3), 5.98 (d, $^3J_{HH} = 16.0$ Hz, 1H, CHCHCO), 7.17 - 7.47 (m, 16H, H_{Ar}, C_{Ar}CHCH), 7.54 - 7.61 (m, 1H, H_{Ar}), 7.78 (s, 1H, NH), 7.89 (d, $^3J_{HH} = 8.3$ Hz, 2H, H_{Ar}).

^{13}C-NMR (100.53 MHz; CDCl$_3$): $\delta = 21.6$ (CH$_3$), 115.7 (CHCHCO), 123.7 (C_{Ar}H), 125.8 (C_{Ar}), 127.3 (C_{Ar}H), 127.5 (C_{Ar}H), 128.5 (C_{Ar}H), 128.6 (C_{Ar}H), 129.1 (C_{Ar}H), 129.5 (C_{Ar}H), 130.1 (C_{Ar}H), 130.3 (C_{Ar}H), 131.1 (C_{Ar}H), 131.8 (C_{Ar}H), 133.9 (C_{Ar}), 135.6 (C_{Ar}), 136.1 (C_{Ar}), 143.9 (C_{Ar}), 147.4 (C_{Ar}CHCH), 148.3 (C_{Ar}O), 151.1 (C_{Ar}CNC$_{Ar}$), 165.5 (CHCOO).

ESI-MS: m/z (%): 367.1 (16) [$C_{20}H_{18}N_2O_3SH^+$], 497.1 (21) [$C_{29}H_{24}N_2O_4SH^+$], 519.2 (100) [$C_{29}H_{24}N_2O_4SNa^+$], 535.2 (7) [$C_{29}H_{24}N_2O_4SK^+$], 764.2 (10) [$(C_{29}H_{24}N_2O_4S)_3HK^{2+}$], 1014.9 (19) [$(C_{29}H_{24}N_2O_4S)_2Na^+$], 1031.0 (17) [$(C_{29}H_{24}N_2O_4S)_2K^+$].

elemental analysis for $C_{29}H_{24}N_2O_4S$ (496.58 g/mol):

calcd.: C 70.14, H 4.87, N 5.64, O 12.89, S 6.46

found: C 69.94, H 4.89, N 5.68, S 6.46

9-(2-Tosylhydrazono)fluoren-1-yl cinnamate

Ph-flNNHTs: $C_{29}H_{22}N_2O_4S$, M = 494.56 g/mol, white solid, yield: 88%, **route**: *Ester-IV.*

^{1}H-NMR (400.13 MHz; CDCl$_3$): δ = 2.30 (s, 3H, CH_3), 6.75 (d, $^3J_{HH}$ = 16.0 Hz, 1H, CHCHCO), 7.03 - 7.13 (m, 3H, H_{Ar}), 7.30 - 7.51 (m, 7H, H_{Ar}), 7.65 - 7.76 (m, 5H, H_{Ar}), 7.85 (d, $^3J_{HH}$ = 7.5 Hz, 1H, H_{Ar}), 7.94 (d, $^3J_{HH}$ = 16.0 Hz, 1H, C$_{Ar}$CHCH), 8.50 (s, 1H, NH).

2-(Phenyl(2-tosylhydrazono)methyl)phenyl 4-nitrocinnamate

PhNO$_2$-bpNNHTs: $C_{29}H_{23}N_3O_6S$, M = 541.57 g/mol, cream solid, yield: 61%, **route**: *Hydrazone.*

^{1}H-NMR (399.78 MHz; CDCl$_3$): δ = 2.27 (s, 3H, CH_3), 6.20 (d, $^3J_{HH}$ = 16.1 Hz, 1H, CHCHCO), 7.21 - 7.49 (m, 11H, H_{Ar}, C$_{Ar}$CHCH), 7.53 (d, $^3J_{HH}$ = 8.3 Hz, 2H, H_{Ar}), 7.60 (t, $^3J_{HH}$ = 7.8 Hz, 1H, H_{Ar}), 7.73 (s, 1H, NH), 7.91 (d, $^3J_{HH}$ = 7.7 Hz, 2H, H_{Ar}), 8.24 (d, $^3J_{HH}$ = 8.2 Hz, 2H, H_{Ar}).

^{13}C-NMR (100.53 MHz; CDCl$_3$): δ = 21.7 (*C*H$_3$), 120.2 (CH*C*HCO), 123.6 (*C$_{Ar}$*H), 124.3 (*C$_{Ar}$*H), 125.2 (*C$_{Ar}$*), 127.3 (*C$_{Ar}$*H), 127.7 (*C$_{Ar}$*H), 128.5 (*C$_{Ar}$*H), 128.5 (*C$_{Ar}$*H), 129.0 (*C$_{Ar}$*H), 129.5 (*C$_{Ar}$*H), 130.1 (*C$_{Ar}$*H), 130.3 (*C$_{Ar}$*H), 131.9 (*C$_{Ar}$*H), 135.6 (*C$_{Ar}$*), 136.0 (*C$_{Ar}$*), 139.9 (*C$_{Ar}$*), 143.9 (*C$_{Ar}$*), 144.2 (*C$_{Ar}$*CHCH), 148.0 (*C$_{Ar}$*O), 148.9 (*C$_{Ar}$*N), 150.6 (*C$_{Ar}$C*NC$_{Ar}$), 164.3 (CH*C*OO).

ESI-MS: m/z (%): 542.2 (56) [C$_{29}$H$_{23}$N$_3$O$_6$SH$^+$], 564.2 (100) [C$_{29}$H$_{23}$N$_3$O$_6$SNa$^+$], 1105.0 (47) [(C$_{29}$H$_{23}$N$_3$O$_6$S)$_2$Na$^+$], 1121.1 (41) [(C$_{29}$H$_{23}$N$_3$O$_6$S)$_2$K$^+$].

elemental analysis for C$_{29}$H$_{23}$N$_3$O$_6$S (541.57 *g/mol*):

calcd.: C 64.31, H 4.28, N 7.76, O 17.73, S 5.92

found: C 63.97, H 4.33, N 7.64, S 5.73

9-(2-Tosylhydrazono)fluoren-1-yl 4-nitrocinnamate

PhNO$_2$-flNNHTs: C$_{29}$H$_{21}$N$_3$O$_6$S, M = 539.56 *g/mol*, white solid, yield: 74%, **route:** *Ester-IV*, extraction with EtOAc instead of Et$_2$O.

^{1}H-NMR (400.13 MHz; CDCl$_3$/dmso-d$_6$): δ = 2.23 (s, 3H, C*H$_3$*), 6.69 (d, $^3J_{HH}$ = 16.1 Hz, 1H, CH*CH*CO), 6.91 - 7.06 (m, 3H, *H$_{Ar}$*), 7.09 - 7.21 (m, 2H, *H$_{Ar}$*), 7.29 - 7.49 (m, 3H, *H$_{Ar}$*), 7.56 - 7.65 (m, 2H, *H$_{Ar}$*), 7.71 (d, $^3J_{HH}$ = 7.0 Hz, 2H, *H$_{Ar}$*), 7.81 (d, $^3J_{HH}$ = 16.1 Hz, 1H, C$_{Ar}$*CH*CH), 8.00 (d, $^3J_{HH}$ = 7.0 Hz, 1H, *H$_{Ar}$*), 8.19 (d, $^3J_{HH}$ = 8.5 Hz, 2H, *H$_{Ar}$*), 8.67 (s, 1H, N*H*).

Ethyl 2-(phenyl(2-tosylhydrazono)methyl)phenyl fumarate

fu-bpNNHTS: $C_{26}H_{24}N_2O_6S$, M = 492.54 g/mol, colourless oil, yield: 88%, **route**: *Hydrazone*.

^{1}H-NMR (399.78 MHz; CDCl$_3$): δ = 1.30 (t, $^3J_{HH}$ = 7.1 Hz, 3H, CH_3CH$_2$), 2.41 (s, 3H, $C_{Ar}CH_3$), 4.22 (q, $^3J_{HH}$ = 7.1 Hz, 2H, CH$_3$CH_2), 6.35 (pq, $^3J_{HH}$ = 15.8 Hz, 2H, COCHCHCO), 7.19 (d, $^3J_{HH}$ = 8.2 Hz, 1H, H_{Ar}), 7.24 - 7.38 (m, 8H, H_{Ar}), 7.46 (t, $^3J_{HH}$ = 7.6 Hz, 1H, H_{Ar}), 7.54 - 7.60 (m, 1H, H_{Ar}), 7.65 (s, 1H, NH), 7.88 (d, $^3J_{HH}$ = 7.9 Hz, 2H, H_{Ar}).

^{13}C-NMR (100.53 MHz; CDCl$_3$): δ = 14.3 (CH_3CH$_2$), 21.8 ($C_{Ar}CH_3$), 61.8 (CH$_3$$CH_2$), 123.4 ($C_{Ar}$H), 125.6 ($C_{Ar}$), 127.1 ($C_{Ar}$H), 128.1 ($C_{Ar}$H), 128.5 ($C_{Ar}$H), 128.6 ($C_{Ar}$H), 129.6 ($C_{Ar}$H), 130.3 (CO$C$HCH), 130.5 ($C_{Ar}$H), 131.1 ($C_{Ar}$H), 132.0 ($C_{Ar}$H), 135.4 ($C_{Ar}$), 135.7 ($C_{Ar}$), 135.9 (CO$C$HCH), 144.3 ($C_{Ar}$), 147.7 ($C_{Ar}$O), 150.6 ($C_{Ar}CNC_{Ar}$), 163.6 (CH$C$OO), 164.3 (CH$C$OO).

ESI-MS: m/z (%): 493.2 (100) [$C_{26}H_{24}N_2O_6$SH$^+$], 515.2 (74) [$C_{26}H_{24}N_2O_6$SNa$^+$], 531.2 (7) [$C_{26}H_{24}N_2O_6$SK$^+$], 758.2 (27) [($C_{26}H_{24}N_2O_6$S)$_3$HK^{2+}], 1006.8 (19) [($C_{26}H_{24}N_2O_6$S)$_2$Na$^+$], 1022.9 (11) [($C_{26}H_{24}N_2O_6$S)$_2$K$^+$].

9-(2-Tosylhydrazono)fluoren-1-yl ethyl fumarate

fu-flNNHTs: $C_{26}H_{22}N_2O_6S$, M = 490.53 g/mol, white solid, yield: 83%, **route**: *Ester-IV*.

^{1}H-NMR (400.13 MHz; CDCl$_3$): δ = 1.37 (t, $^3J_{HH}$ = 6.9 Hz, 3H, CH_3), 2.39 (s, 3H, CH_3), 4.34 (q, $^3J_{HH}$ = 6.9 Hz, 2H, CH$_3$CH_2), 7.09 (s, 2H, COCHCHCO), 7.26 - 7.41 (m, 4H, H_{Ar}), 7.43 - 7.55 (m, 2H, H_{Ar}), 7.78 (d, $^3J_{HH}$ = 7.4 Hz, 2H, H_{Ar}), 7.82 - 7.94 (m, 3H, H_{Ar}), 8.50 (s, 1H, NH).

N(2-(Phenyl(2-tosylhydrazono)methyl)phenyl) maleamide

ma-bpNNHTs: $C_{24}H_{21}N_3O_5S$, M = 463.51 g/mol, light yellow solid, yield: 97%, **route**: *Amide-II*.

^{1}H-NMR (399.78 MHz; CDCl$_3$): δ = 2.44 (s, 3H, C$_{Ar}$CH_3), 6.53 (d, $^3J_{HH}$ = 12.8 Hz, 1H, COCHCHCO), 6.82 (d, $^3J_{HH}$ = 7.8 Hz, 1H, H_{Ar}), 6.92 (d, $^3J_{HH}$ = 12.8 Hz, 1H, COCHCHCO), 6.99 - 7.11 (m, 3H, H_{Ar}), 7.32 - 7.45 (m, 3H, H_{Ar}), 7.55 - 7.63 (m, 3H,

H_{Ar}, SNH), 7.67 - 7.76 (m, 3H, H_{Ar}), 8.68 (d, $^3J_{HH}$ = 8.3 Hz, 1H, H_{Ar}), 12.50 (bs, 1H, CONH).

^{13}C-NMR (100.53 MHz; CDCl$_3$): δ = 21.9 (C$_{Ar}$CH$_3$), 122.2 (C$_{Ar}$H), 123.3 (C$_{Ar}$), 125.2 (C$_{Ar}$H), 127.3 (C$_{Ar}$H), 128.0 (C$_{Ar}$H), 130.3 (C$_{Ar}$), 130.5 (C$_{Ar}$H), 130.6 (C$_{Ar}$H), 131.1 (C$_{Ar}$H), 131.3 (C$_{Ar}$H), 132.2 (C$_{Ar}$H), 132.9 (COCHCH), 135.3 (C$_{Ar}$), 136.7 (C$_{Ar}$), 137.8 (COCHCH), 145.4 (C$_{Ar}$N), 156.5 (C$_{Ar}$CNC$_{Ar}$), 164.9 (CHCOO), 165.6 (CHCOO).

ESI-MS: m/z (%): 366.2 (40) [C$_{20}$H$_{19}$N$_3$O$_2$SH$^+$], 446.2 (23) [C$_{24}$H$_{20}$N$_3$O$_4$S$^+$], 464.1 (100) [C$_{24}$H$_{21}$N$_3$O$_5$SH$^+$], 486.1 (66) [C$_{24}$H$_{21}$N$_3$O$_5$SNa$^+$], 502.0 (6) [C$_{24}$H$_{21}$N$_3$O$_5$SK$^+$], 926.7 (12) [(C$_{24}$H$_{21}$N$_3$O$_5$S)$_2$H$^+$], 948.9 (62) [(C$_{24}$H$_{21}$N$_3$O$_5$S)$_2$Na$^+$], 965.0 (77) [(C$_{24}$H$_{21}$N$_3$O$_5$S)$_2$K$^+$], 1177.9 (8) [(C$_{24}$H$_{21}$N$_3$O$_5$S)$_5$HK^{2+}], 1408.9 (4) [(C$_{24}$H$_{21}$N$_3$O$_5$S)$_3$Na$^+$], 1427.5 (4) [(C$_{24}$H$_{21}$N$_3$O$_5$S)$_3$K$^+$].

elemental analysis for C$_{24}$H$_{21}$N$_3$O$_5$S (463.51 *g/mol*):

calcd.: C 62.19, H 4.57, N 9.07, O 17.26, S 6.92

found: C 61.87, H 4.63, N 8.58, S 6.81

1-Methoxypropan-2-yl 9-(2-tosylhydrazono)-9-fluoren-1-yl maleate

3.24: C$_{28}$H$_{26}$N$_2$O$_7$S, M = 534.58 *g/mol*, white solid, yield: >99%, **route**: *Ester-IV*.

^{1}H-NMR (400.13 MHz; CDCl$_3$): δ = 1.34 (d, $^3J_{HH}$ = 6.8 Hz, 3H, CHCH_3), 2.39 (s, 3H, CH_3), 3.37 (s, 3H, OCH_3), 3.41 - 3.59 (m, 2H, CHCH_2O), 5.23 - 5.32 (m, 1H, OCH), 7.10 (s, 2H, COCHCHCO), 7.27 - 7.41 (m, 5H, H_{Ar}), 7.43 - 7.54 (m, 2H, H_{Ar}), 7.70 (d,

$^3J_{HH}$ = 7.2 Hz, 1H, H_{Ar}), 7.78 (d, $^3J_{HH}$ = 7.5 Hz, 2H, H_{Ar}), 7.85 (d, $^3J_{HH}$ = 8.3 Hz, 1H, H_{Ar}), 8.50 (s, 1H, NH).

N(2-(Phenyl(2-tosylhydrazono)methyl)phenyl) maleimide

mi-bpNNHTs: $C_{24}H_{19}N_3O_4S$, M = 445.49 g/mol, white solid, yield: 14%, **route:** *Imide*, purification by *flash* chromatography (SiO$_2$; 36-100% Et$_2$O/pentane).

^{1}H-NMR (399.78 MHz; CDCl$_3$): δ = 2.44 (s, 3H, C_{Ar}CH_3), 6.36 (s, 2H, COCHCHCO), 7.16 - 7.37 (m, 8H, H_{Ar}), 7.43 - 7.49 (m, 1H, H_{Ar}), 7.57 - 7.65 (m, 2H, H_{Ar}), 7.82 (s, 1H, NH), 7.94 (d, $^3J_{HH}$ = 8.3 Hz, 2H, H_{Ar}).

^{13}C-NMR (100.53 MHz; CDCl$_3$): δ = 21.9 (C_{Ar}CH$_3$), 127.3 (C_{Ar}H), 128.4 (C_{Ar}H), 128.9 (C_{Ar}H), 129.4 (C_{Ar}H), 130.0 (C_{Ar}H), 130.1 (C_{Ar}), 130.3 (COCH), 130.9 (C_{Ar}H), 131.5 (C_{Ar}H), 131.6 (C_{Ar}), 134.0 (C_{Ar}H), 135.5 (C_{Ar}), 135.6 (C_{Ar}), 144.0 (C_{Ar}N), 152.1 (C_{Ar}CNC_{Ar}).

ESI-MS: m/z (%): 292.3 (68) [$C_{17}H_{13}N_3O_2H^+$], 446.2 (100) [$C_{24}H_{19}N_3O_4SH^+$], 468.2 (89) [$C_{24}H_{19}N_3O_4SNa^+$], 912.8 (15) [($C_{24}H_{19}N_3O_4S)_2Na^+$], 928.9 (16) [($C_{24}H_{19}N_3O_4S)_2K^+$].

***N'*-((2-hydroxyphenyl)(phenyl)methylene)cinnamohydrazide**

3.3-OH-Ph: $C_{22}H_{18}N_2O_2$, M = 342.39 g/mol, off-white solid, yield: 19%, **route**: *Amide-I.*

^{1}H-NMR (399.78 MHz; CDCl$_3$): $\delta = 6.19$ (d, $^3J_{HH} = 16.0$ Hz, 1H, CHC*H*CO), 7.10 - 7.33 (m, 10H, H_{Ar}), 7.41 (d, $^3J_{HH} = 16.0$ Hz, 2H, H_{Ar}), 7.46 - 7.52 (m, 3H, H_{Ar}, N*H*), 7.54 (d, $^3J_{HH} = 16.0$ Hz, 1H, C$_{Ar}$C*H*CH), 8.42 (s, 1H, O*H*).

^{13}C-NMR (100.53 MHz; CDCl$_3$): $\delta = 116.3$ (CH*C*HCO), 116.7 (C_{Ar}H), 124.3 (C_{Ar}H), 127.4 (C_{Ar}H), 128.5 (C_{Ar}H), 128.6 (C_{Ar}H), 128.9 (C_{Ar}H), 129.0 (C_{Ar}H), 129.9 (C_{Ar}), 130.2 (C_{Ar}H), 130.3 (C_{Ar}), 130.9 (C_{Ar}), 131.7 (C_{Ar}H), 144.1 (C_{Ar}H), 147.3 (C_{Ar}*C*HCH), 148.6 (C_{Ar}*C*NC_{Ar}), 164.7 (C_{Ar}O), 167.1 (CH$_3$*C*ON).

(E)-ethyl 4-(2-((2-hydroxyphenyl)(phenyl)methylene)hydrazinyl)-4-oxobut-2-enoate

3.3-OH-fu: $C_{19}H_{18}N_2O_4$, M = 338.36 g/mol, off-white solid, yield: 25%, **route**: *Amide-I*.

^{1}H-NMR (399.78 MHz; CDCl$_3$): δ = 1.29 (t, $^3J_{HH}$ = 7.1 Hz, 3H, CH_3), 4.22 (q, $^3J_{HH}$ = 7.1 Hz, 2H, CH$_3$CH_2), 6.67 (d, $^3J_{HH}$ = 15.3 Hz, 1H, COCHCHCO), 6.71 - 6.79 (m, 2H, H_{Ar}), 6.95 (d, $^3J_{HH}$ = 15.3 Hz, 1H, COCHCHCO), 7.06 (d, $^3J_{HH}$ = 7.9 Hz, 1H, H_{Ar}), 7.25 - 7.36 (m, 3H, H_{Ar}, NH), 7.56 - 7.70 (m, 3H, H_{Ar}), 8.42 (s, 1H, OH).

(Z)-4-(2-((2-ammoniophenyl)(phenyl)methylene)hydrazinyl)-4-oxobut-2-enoate

3.3-NH: $C_{17}H_{15}N_3O_3$, M = 309.32 g/mol, cream solid, yield: >99%, **route**: *Amide-II*.

^{1}H-NMR (399.78 MHz; CDCl$_3$): δ = 6.34 (d, $^3J_{HH}$ = 12.7 Hz, 1H, COCHCHCO), 6.70 - 6.79 (m, 2H, H_{Ar}), 6.80 - 6.92 (m, 1H, H_{Ar}), 7.18 (d, $^3J_{HH}$ = 12.7 Hz, 1H, COCHCHCO),

7.22 - 7.35 (m, 4H, H_{Ar}), 7.45 - 7.56 (m, 2H, H_{Ar}, NH), 7.58 (d, $^3J_{HH}$ = 7.2 Hz, 1H, H_{Ar}), 8.81 (bs, 3H, NH_3).

8.2.4. Diazocompounds

2-(Diazo(phenyl)methyl)phenyl acetate

Ac-bpN$_2$: $C_{15}H_{12}N_2O_2$, M = 252.27 *g/mol*, red oil, yield: 43%, **route**: *Diazo*.

FT-IR (KBr plates, cm^{-1}): 3062 (w), 3028 (w), 2924 (m), 2854 (w), 2045 (s, N$_2$), 1766 (s, CO), 1673 (m), 1597 (s), 1495 (w), 1370 (w), 1319 (w), 1192 (m), 1087 (w), 1040 (w), 911 (w), 882 (w), 817 (w), 755 (w), 696 (m), 665 (m), 578 (w), 561 (w),

UV/vis (pentane, nm): 207 (s, Ar), 285 (s, N$_2$).

2-(Diazo(phenyl)methyl)phenyl crotonate

Me-bpN$_2$: $C_{17}H_{14}N_2O_2$, M = 278.31 *g/mol*, red oil, yield: 43%, **route**: *Diazo*.

^{1}H-NMR (399.78 MHz; CDCl$_3$): δ = 1.87 (d, $^3J_{HH}$ = 6.9 Hz, 3H, CH_3), 5.89 (dd, $^3J_{HH}$ = 1.7 Hz, $^3J_{HH}$ = 15.5 Hz, 1H, CHCHCO), 7.00 (dq, $^3J_{HH}$ = 6.9 Hz, $^3J_{HH}$ = 15.5 Hz, 1H, CH$_3$CHCH), 7.07 - 7.13 (m, 2H, H_{Ar}), 7.20 - 7.36 (m, 6H, H_{Ar}), 7.42 - 7.46 (m, 1H, H_{Ar}).

^{13}C-NMR (100.53 MHz; CDCl$_3$): $\delta = 18.4$ (CH_3), 121.5 (C_{Ar}H), 122.3 ($C_{Ar}CNC_{Ar}$), 123.8 (C_{Ar}H), 123.9 (CHCHCO), 124.9 (C_{Ar}H), 126.5 (C_{Ar}H), 128.5 (C_{Ar}H), 129.1 (C_{Ar}H), 130.0 (C_{Ar}H), 130.3 (C_{Ar}), 130.5 (C_{Ar}), 147.5 (CH$_3$CHCH), 148.5 (C_{Ar}O), 164.3 (CHCOO).

FT-IR (KBr plates, cm^{-1}): 3089 (w), 3060 (m), 3030 (w), 2925 (m), 2851 (m), 2046 (s, N$_2$), 1970 (w), 1737 (s, CO), 1654 (s), 1597 (m), 1500 (w), 1494 (s), 1440 (s), 1381 (w), 1299 (s), 1254 (m), 1194 (m), 1149 (m), 1090 (m), 1030 (w), 976 (s), 910 (w), 828 (w), 816 (m), 814 (w), 753 (m), 693 (m), 647 (m), 579 (m), 523 (w), 477 (m).

UV/vis (pentane, nm): 208 (s, Ar), 286 (s, N$_2$).

2-(Diazo(phenyl)methyl)phenyl 4-methoxycinnamate

PhOMe-bpN$_2$: C$_{23}$H$_{18}$N$_2$O$_3$, M $= 370.40$ *g/mol*, red oil, yield: 49%, **route**: *Diazo*.

^{1}H-NMR (400.13 MHz; CDCl$_3$): $\delta = 3.85$ (s, 3H, OCH_3), 6.32 (d, $^3J_{HH} = 16.0$ Hz, 1H, CHCHCO), 6.91 (d, $^3J_{HH} = 8.7$ Hz, 2H, H_{Ar}), 7.06 - 7.14 (m, 3H, H_{Ar}), 7.24 - 7.39 (m, 6H, H_{Ar}), 7.41 - 7.49 (m, 2H, H_{Ar}), 7.63 (d, $^3J_{HH} = 16.0$ Hz, 1H, $C_{Ar}CH$CH).

^{13}C-NMR (100.62 MHz; CDCl$_3$): $\delta = 55.6$ (OCH_3), 114.1 (CHCHCO), 114.5 (C_{Ar}H), 122.2 (C_{Ar}), 123.8 (C_{Ar}H), 123.9 (C_{Ar}H), 124.9 (C_{Ar}H), 126.5 (C_{Ar}H), 127.0 (C_{Ar}), 128.4 (C_{Ar}), 128.5 (C_{Ar}H), 129.1 (C_{Ar}H), 129.9 (C_{Ar}H), 130.2 (C_{Ar}H), 146.6 ($C_{Ar}CH$CH), 148.6 (C_{Ar}O), 161.9 (C_{Ar}OCH$_3$), 165.2 (CHCOO)

FT-IR (KBr plates, cm^{-1}): 3066 (m), 3036 (m), 2954 (m), 2931 (m), 2842 (m), 2046 (s, N$_2$), 1726 (s, CO), 1636 (m), 1602 (s), 1568 (w), 1512 (m), 1486 (m), 1442 (m), 1419 (w), 1307 (m), 1255 (m), 1195 (m), 1173 (w), 1128 (s), 1030 (m), 982 (m), 919 (w), 852 (w), 828 (m), 777 (m), 694 (m), 665 (w), 636 (w), 568 (w), 553 (m), 509 (m).

UV/vis (pentane, nm): 208 (s, Ar), 227 (s, Ar), 293 (s, N_2).

2-(Diazo(phenyl)methyl)phenyl 4-methylcinnamate

PhMe-bpN$_2$: $C_{23}H_{18}N_2O_2$, M = 354.40 g/mol, red oil, yield: 25%, **route:** *Diazo*.

^{1}H-NMR (400.13 MHz; CDCl$_3$): δ = 2.38 (s, 3H, CH_3), 6.40 (d, $^3J_{HH}$ = 16.0 Hz, 1H, CHCHCO), 7.07 - 7.13 (m, 3H, H_{Ar}), 7.20 (d, $^3J_{HH}$ = 7.8 Hz, 2H, H_{Ar}), 7.27 - 7.42 (m, 7H, H_{Ar}), 7.47 (d, $^3J_{HH}$ = 7.6 Hz, 2H, H_{Ar}), 7.64 (d, $^3J_{HH}$ = 16.0 Hz, 1H, C$_{Ar}$CHCH).

^{13}C-NMR (100.62 MHz; CDCl$_3$): δ = 21.7 (CH_3), 155.6 (CHCHCO), 122.3 (C_{Ar}), 123.8 (C_{Ar}H), 123.9 (C_{Ar}H), 124.9 (C_{Ar}H), 126.5 (C_{Ar}H), 128.5 (C_{Ar}H), 129.1 (C_{Ar}H), 129.8 (C_{Ar}H), 130.0 (C_{Ar}H), 130.5 (C_{Ar}), 131.6 (C_{Ar}), 141.4 (C_{Ar}), 147.0 (C$_{Ar}$$C$HCH), 148.6 ($C_{Ar}$O), 165.0 (CH$C$OO).

FT-IR (KBr plates, cm^{-1}): 3029 (w), 2925 (s), 2850 (m), 2043 (s, N_2), 1730 (s, CO), 1687 (w), 1628 (m), 1599 (m), 1496 (m), 1449 (m), 1411 (w), 1374 (w), 1314 (w), 1269 (w), 1240 (s), 1195 (s), 1180 (s), 1142 (w), 1098 (w), 1031 (w), 987 (m), 946 (m), 919 (m), 866 (w), 815 (m), 753 (m), 697 (m), 628 (w), 499 (m), 473 (w).

UV/vis (pentane, nm): 208 (m, Ar), 222 (m, Ar), 286 (s, N_2).

2-(Diazo(phenyl)methyl)phenyl cinnamate

Ph-bpN$_2$: C$_{22}$H$_{16}$N$_2$O$_2$, M = 340.37 g/mol, red oil, yield: 34%, **route**: *Diazo*.

^{1}H-NMR (399.78 MHz; CDCl$_3$): δ = 6.44 (d, $^3J_{HH}$ = 16.0 Hz, 1H, CHC*H*CO), 7.04 - 7.17 (m, 2H, H_{Ar}), 7.25 - 7.42 (m, 5H, H_{Ar}), 7.44 - 7.52 (m, 2H, H_{Ar}), 7.66 (d, $^3J_{HH}$ = 16.0 Hz, 1H, C$_{Ar}$C*H*CH).

^{13}C-NMR (100.53 MHz; CDCl$_3$): δ = 116.8 (*C*HCHCO), 122.3 (C$_{Ar}$*C*NC$_{Ar}$), 123.8 (*C*$_{Ar}$H), 123.9 (*C*$_{Ar}$H), 125.0 (*C*$_{Ar}$H), 126.7 (*C*$_{Ar}$H), 128.5 (*C*$_{Ar}$H), 128.6 (*C*$_{Ar}$H), 128.9 (*C*$_{Ar}$), 129.1 (*C*$_{Ar}$H), 129.2 (*C*$_{Ar}$H), 130.0 (*C*$_{Ar}$H), 130.5 (*C*$_{Ar}$), 130.9 (*C*$_{Ar}$H), 134.3 (*C*$_{Ar}$), 147.0 (C$_{Ar}$*C*HCH), 148.5 (*C*$_{Ar}$O), 164.9 (CH*C*OO).

FT-IR (KBr plates, cm^{-1}): (m), 3029 (m), (w), 2925 (m), 2856 (w), 2045 (vs, N$_2$), 1957 (w), 1732 (vs, CO), 1666 (w), 1634 (s), 1597 (m), 1577 (w), 1496 (s), 1449 (s), 1308 (s), 1235 (s), 1193 (s), 1133 (s), 1100 (w), 1029 (w), 979 (s), 949 (w), 945 (w), 916 (w), 860 (m), 751 (s), 693 (s), 619 (m), 577 (w), 561 (w), 473 (m).

UV/vis (pentane, nm): 208 (s, Ar), 282 (s, N$_2$).

9-Diazofluoren-1-yl cinnamate

Ph-flN$_2$: C$_{22}$H$_{14}$N$_2$O$_2$, M = 338.36 g/mol, orange solid, yield: 29%, **route**: *Diazo*, extraction with Et$_2$O instead of pentane.

^{1}H-NMR (400.13 MHz; CDCl$_3$): δ = 6.62 (d, $^3J_{HH}$ = 16.0 Hz, 1H, CHC*H*CO), 7.29 - 7.49 (m, 8H, H_{Ar}), 7.58 - 7.67 (m, 2H, H_{Ar}), 7.82 (d, $^3J_{HH}$ = 7.5 Hz, 1H, H_{Ar}), 7.90 (d, $^3J_{HH}$ = 16.0 Hz, 1H, C$_{Ar}$C*H*CH), 7.94 (d, $^3J_{HH}$ = 7.6 Hz, 1H, H_{Ar}).

FT-IR (KBr plates, cm^{-1}): 3065 (w), 2961 (w), 2931 (m), 2856 (w), 2065 (vs, N$_2$), 1733 (s, CO), 1633 (m), 1572 (w), 1453 (m), 1425 (m), 1333 (w), 1303 (w), 1236 (w), 1199 (w), 1139 (m), 1124 (m), 1020 (w), 960 (w), 856 (w), 788 (w), 751 (m), 542 (w).

UV/vis (pentane, nm): 235 (s, Ar), 290 (s, N$_2$).

2-(Diazo(phenyl)methyl)phenyl 4-nitrocinnamate

PhNO$_2$-bpN$_2$: C$_{22}$H$_{15}$N$_3$O$_4$, M = 385.37 g/mol, orange solid, yield: 39%, **route:** *Diazo*, extraction with Et$_2$O instead of pentane.

^{1}H-NMR (399.78 MHz; CDCl$_3$): δ = 6.54 (d, $^3J_{HH}$ = 16.0 Hz, 1H, CHC*H*CO), 7.05 - 7.13 (m, 3H, H_{Ar}, C$_{Ar}$C*H*CH), 7.27 - 7.40 (m, 6H, H_{Ar}), 7.48 (d, $^3J_{HH}$ = 7.5 Hz, 1H, H_{Ar}), 7.60 - 7.69 (m, 2H, H_{Ar}), 8.25 (d, $^3J_{HH}$ = 8.7 Hz, 2H, H_{Ar}).

^{13}C-NMR (100.53 MHz; CDCl$_3$): δ = 121.0 (C*H*COO), 122.1 (C_{Ar}), 123.7 (C_{Ar}H), 123.9 (C_{Ar}H), 124.3 (C_{Ar}H), 125.1 (C_{Ar}H), 126.9 (C_{Ar}H), 128.5 (C$_{Ar}$CNC$_{Ar}$), 128.6 (C_{Ar}H), 129.0 (C_{Ar}H), 129.1 (C_{Ar}H), 130.1 (C_{Ar}H), 130.3 (C_{Ar}), 140.2 (C_{Ar}), 143.7 (C$_{Ar}$CHCH), 148.2 (C_{Ar}O), 148.8 (C_{Ar}N), 163.8 (CH*C*OO).

FT-IR (KBr plates, cm^{-1}): 3091 (m), 2942 (m), 2852 (m), 2046 (s, N$_2$), 1734 (s, CO), 1643 (w), 1598 (m), 1520 (s), 1486 (m), 1449 (w), 1404 (w), 1345 (s), 1237 (w), 1199 (m), 1140 (m), 976 (w), 844 (w), 755 (m), 699 (m), 662 (w), 579 (w).

UV/vis (pentane, nm): 208 (s, Ar), 291 (s, N$_2$).

9-Diazofluoren-1-yl 4-nitrocinnamate

PhNO$_2$-flN$_2$: C$_{22}$H$_{13}$N$_3$O$_4$, M = 383.36 g/mol, orange solid, yield: 25%, **route:** *Diazo,* extraction with Et$_2$O instead of pentane.

^{1}H-NMR (400.13 MHz; CDCl$_3$): δ = 6.71 (d, $^3J_{HH}$ = 16.1 Hz, 1H, CHCHCO), 7.54 - 7.26 (m, 7H, H_{Ar}), 7.73 (d, $^3J_{HH}$ = 8.7 Hz, 2H, H_{Ar}), 7.89 (d, $^3J_{HH}$ = 16.1 Hz, 1H, C$_{Ar}$CHCH), 8.26 (d, $^3J_{HH}$ = 8.6 Hz, 2H, H_{Ar}).

FT-IR (KBr plates, cm^{-1}): 2962 (m), 2925 (s), 2850 (w), 2066 (s, N$_2$), 1720 (s, CO), 1644 (m), 1599 (m), 1521 (m), 1450 (w), 1420 (w), 1345 (m), 1300 (w), 1270 (w), 1144 (m), 897 (w), 748 (m), 644 (w).

UV/vis (pentane, nm): 210 (s, Ar), 260 (s, Ar), 285 (m, N$_2$).

2-(Diazo(phenyl)methyl)phenyl ethyl fumarate

fu-bpN$_2$: C$_{19}$H$_{16}$N$_2$O$_4$, M = 336.34 g/mol, red oil, yield: 29%, **route:** *Diazo,* deprotonation with 1 eq. of ^{n}BuLi instead of NaH.

^{1}H-NMR (400.13 MHz; CDCl$_3$): δ = 1.21 (t, $^3J_{HH}$ = 7.0 Hz, 3H, CH_3), 4.09 (q, $^3J_{HH}$ = 7.0 Hz, 2H, CH$_3$CH_2), 7.01 (d, $^3J_{HH}$ = 7.9 Hz, 1H, COCHCHCO), 7.04 - 7.11 (m, 1H, H_{Ar}), 7.20 - 7.40 (m, 8H, H_{Ar}), 7.44 (d, $^3J_{HH}$ = 7.9 Hz, 1H, COCHCHCO).

FT-IR (KBr plates, cm^{-1}): 2955 (s), 2925 (s), 2851 (s), 2048 (s, N$_2$), 1795 (w), 1765 (m), 1736 (s, CO), 1598 (m), 1463 (m), 1377 (w), 1262 (w), 1176 (w), 1117 1027 (w), 859 (w), 814 (m), 740 (m), 701 (w).

N(2-(Diazo(phenyl)methyl)phenyl) maleimide

mi-bpN$_2$: C$_{17}$H$_{11}$N$_3$O$_2$, M = 277.27 *g/mol*, red oil, yield: 24%, **route:** *Diazo*, deprotonation with 1.1 eq. of ^{n}BuLi instead of NaH.

^{1}H-NMR (400.13 MHz; CDCl$_3$): δ = 6.25 (s, 2H, COC*H*C*H*CO), 7.16 - 7.21 (m, 2H, H_{Ar}), 7.27 - 7.40 (m, 7H, H_{Ar}),

^{13}C-NMR (100.53 MHz; CDCl$_3$): δ = 123.2 (C_{Ar}), 124.2 (C_{Ar}H), 125.4 (C_{Ar}H), 126.4 (C_{Ar}H), 127.9 (C_{Ar}H), 128.9 (CO*C*H), 129.0 (C_{Ar}H), 129.2 (C_{Ar}H), 138.8 (C_{Ar}), 140.2 (C_{Ar}N), 160.2 CH*C*ON).

FT-IR (KBr plates, cm^{-1}): 3062 (m), 3032 (m), 2957 (s), 2927 (s), 2868 (m), 2046 (s, N$_2$), 1716 (s, CO), 1640 (s), 1605 (m), 1486 (m), 1456 (m), 1378 (m), 1322 (w), 1245 (m), 1185 (m), 1021 (m), 991 (m), 909 (w), 857 (w), 805 (w), 765 (s), 698 (m), (w), 803 (w).

8.2.5. Irradiation products

1-(4-Methoxyphenyl)-7b-phenyl-1,1a-dihydrocyclopropa[c]chromen-2(7bH)-one

3.1-PhOMe: $C_{23}H_{18}O_3$, M = 342.39 g/mol, white solid, R_f = 0.35, **route**: *Irradiation.*

^{1}H-NMR (400.13 MHz; CDCl$_3$): δ = 2.80 (d, $^3J_{HH}$ = 4.8 Hz, 1H, CHCHCO), 3.04 (d, $^3J_{HH}$ = 4.8 Hz, 1H, C$_{Ar}$CHCH), 3.73 (s, 3H, OCH_3), 6.82 - 7.22 (m, 9H, H_{Ar}), 7.29 - 7.55 (m, 6H, H_{Ar}).

^{13}C-NMR (100.53 MHz; CDCl$_3$): δ = 31.9 (CHCOO), 38.3 (C$_{Ar}$$C$HCH), 41.8 (C$_{Ar}$$C$), 55.4 (O$CH_3$), 113.8 ($C_{Ar}$H), 114.6 ($C_{Ar}$), 117.8 ($C_{Ar}$H), 124.4 ($C_{Ar}$H), 126.2 ($C_{Ar}$), 127.6 ($C_{Ar}$H), 128.2 ($C_{Ar}$H), 128.7 ($C_{Ar}$H), 128.9 ($C_{Ar}$H), 130.0 ($C_{Ar}$), 132.1 ($C_{Ar}$H), 135.0 ($C_{Ar}$H), 149.5 ($C_{Ar}$O), 158.8 ($C_{Ar}$O), 166.1 (CH$C$OO).

ESI-MS: m/z (%): 313.3 (100) [$C_{21}H_{17}O_2H^+$], 343.3 (27) [$C_{23}H_{18}O_3H^+$].

FT-IR (KBr plates, cm^{-1}): 2953 (m), 2924 (m), 2834 (w), 1749 (s, CO), 1606 (m), 1515 (s), 1455 (m), 1249 (s), 1032 (m), 829 (w), 757 (m), 702 (m).

1,7b-Diphenyl-1,1a-dihydrocyclopropa[c]chromen-2(7bH)-one

3.1-Ph: $C_{22}H_{16}O_2$, M = 312.36 g/mol, white solid, R_f = 0.60, **route**: *Irradiation.*

Single crystals suitable for X-ray analysis were obtained by slow evaporation of a saturated CHCl$_3$ solution.

^{1}H-NMR (400.13 MHz; CDCl$_3$): $\delta = 2.85$ (d, $^3J_{HH} = 5.4$ Hz, 1H, CHCHCO), 3.14 (d, $^3J_{HH} = 5.4$ Hz, 1H, C$_{Ar}$CHCH), 6.70 - 6.80 (m, 2H, H_{Ar}), 2.80 - 2.91 (m, 2H, H_{Ar}), 7.07 - 7.19 (m, 6H, H_{Ar}), 7.20 - 7.30 (m, 4H, H_{Ar}).

^{13}C-NMR (100.53 MHz; CDCl$_3$): $\delta = 31.8$ (CHCOO), 38.7 (C$_{Ar}$$C$HCH), 42.1 (C$_{Ar}$$C$), 117.8 ($C_{Ar}$H), 124.4 ($C_{Ar}$H), 125.6 ($C_{Ar}$), 127.2 ($C_{Ar}$H), 127.7 ($C_{Ar}$H), 127.9 ($C_{Ar}$H), 128.2 ($C_{Ar}$H), 128.3 ($C_{Ar}$H), 128.3 ($C_{Ar}$H), 128.7 ($C_{Ar}$H), 132.0 ($C_{Ar}$H), 134.4 ($C_{Ar}$H), 134.9 ($C_{Ar}$H), 149.5 ($C_{Ar}$H), 166.0 (CH$C$OO).

ESI-MS: m/z (%): 283.3 (100) [C$_{21}$H$_{14}$OH$^+$], 313.2 (38) [C$_{22}$H$_{16}$O$_2$H$^+$], 378.3 (28) [C$_{24}$H$_{19}$NO$_2$Na$^+$].

FT-IR (KBr plates, cm^{-1}): 3061 (m), 2923 (m), 1757 (vs, CO), 1604 (w), 1486 (m), 1453 (m), 1211 (s), 1112 (w), 957 (w), 750 (s), 699 (s), 571 (w).

1-(4-Nitrophenyl)-7b-phenyl-1,1a-dihydrocyclopropa[*c*]chromen-2(7b*H*)-one

3.1-PhNO$_2$: C$_{22}$H$_{15}$NO$_4$, M $= 357.36$ *g/mol*, yellow solid, R$_f = 0.24$, **route:** *Irradiation*, eluent (9:2 pentane/Et$_2$O).

^{1}H-NMR (400.13 MHz; CDCl$_3$): $\delta = 2.92$ (d, $^3J_{HH} = 5.2$ Hz, 1H, CHCHCO), 3.21 (d, $^3J_{HH} = 5.2$ Hz, 1H, C$_{Ar}$CHCH), 6.87 (d, $^3J_{HH} = 8.5$ Hz, 2H, H_{Ar}), 6.97 - 7.21 (m, 5H, H_{Ar}), 7.27 - 7.54 (m, 4H, H_{Ar}), 7.99 (d, $^3J_{HH} = 8.5$ Hz, 2H, H_{Ar}).

^{13}C-NMR (100.53 MHz; CDCl$_3$): $\delta = 29.9$ (CHCOO), 38.0 (C$_{Ar}$$C$HCH), 43.1 (C$_{Ar}$$C$), 118.0 ($C_{Ar}$H), 123.4 ($C_{Ar}$H), 124.7 ($C_{Ar}$H), 127.7 ($C_{Ar}$H), 128.5 ($C_{Ar}$H), 128.8 ($C_{Ar}$H),

128.9 ($C_{Ar}H$), 129.1 ($C_{Ar}H$), 129.2 (C_{Ar}), 131.8 ($C_{Ar}H$), 133.8 (C_{Ar}), 142.5 (C_{Ar}), 147.0 ($C_{Ar}N$), 149.4 ($C_{Ar}O$), 165.0 (CHCOO).

ESI-MS: m/z (%): 282.3 (100) [$C_{21}H_{13}OH^+$], 328.2 (29) [$C_{21}H_{13}NO_3H^+$], 358.2 (10) [$C_{22}H_{15}NO_4H^+$].

FT-IR (KBr plates, cm^{-1}): 2925 (s), 1760 (s, CO), 1602 (m), 1520 (s), 1455 (w), 1345 (s), 1213 (m), 1110 (w), 853 (w), 760 (s), 703 (s).

2-(4-Methoxyphenyl)-7b-phenyl-2,2a-dihydrobenzo[*d*]cyclobuta[*b*]furan-1(7b*H*)-one

3.12-PhOMe: $C_{23}H_{18}O_3$, M = 342.39 *g/mol*, white solid, R$_f$ = 0.65, **route:** *Irradiation.*

^{1}H-NMR (400.13 MHz; CDCl$_3$): δ = 3.78 (s, 3H, OCH_3), 4.76 (d, $^3J_{HH}$ = 3.7 Hz, 1H, CHCHO), 5.43 (d, $^3J_{HH}$ = 3.7 Hz, 1H, CHCHO), 6.87 (d, $^3J_{HH}$ = 8.4 Hz, 2H, H_{Ar}), 6.96 (t, $^3J_{HH}$ = 7.4 Hz, 1H, H_{Ar}), 7.04 (d, $^3J_{HH}$ = 8.2 Hz, 1H, H_{Ar}), 7.15 - 7.23 (m, 3H, H_{Ar}), 7.26 - 7.31 (m, 2H, H_{Ar}), 7.32 - 7.43 (m, 4H, H_{Ar}).

^{13}C-NMR (100.53 MHz; CDCl$_3$): δ = 55.5 (OCH$_3$), 69.7 (CHCHO), 80.4 (CHCCO), 85.5 (CHCHO), 111.8 ($C_{Ar}H$), 114.6 ($C_{Ar}H$), 122.4 ($C_{Ar}H$), 126.0 ($C_{Ar}H$), 126.1 (C_{Ar}), 126.3 ($C_{Ar}H$), 127.6 (C_{Ar}), 128.0 ($C_{Ar}H$), 128.6 ($C_{Ar}H$), 129.1 ($C_{Ar}H$), 130.4 ($C_{Ar}H$), 137.3 (C_{Ar}), 159.2 ($C_{Ar}O$), 159.6 ($C_{Ar}O$), 203.0 (CCOCH).

ESI-MS: m/z (%): 313.3 (100) [$C_{21}H_{17}O_2H^+$].

FT-IR (KBr plates, cm^{-1}): 2957 (w), 2927 (m), 2830 (w), 1779 (s, CO), 1607 (s), 1513 (s), 1461 (m), 1251 (s), 1179 (w), 1068 (m), 1031 (m), 831 (w), 753 (m), 699 (w).

2,7b-Diphenyl-2,2a-dihydrobenzo[*d*]cyclobuta[*b*]furan-1(7b*H*)-one

3.12-Ph: $C_{22}H_{16}O_2$, 312.36 *g/mol*, white solid, $R_f = 0.86$, **route**: *Irradiation*.

^{1}H-NMR (400.13 MHz; CDCl$_3$): δ = 4.69 (d, $^3J_{HH}$ = 3.3 Hz, 1H, C*H*CHO), 5.35 (d, $^3J_{HH}$ = 3.3 Hz, 1H, CHC*H*O), 6.83 (t, $^3J_{HH}$ = 7.5 Hz, 1H, H_{Ar}), 6.90 (d, $^3J_{HH}$ = 8.2 Hz, 1H, H_{Ar}), 7.06 - 7.29 (m, 12H, H_{Ar}).

^{13}C-NMR (100.53 MHz; CDCl$_3$): δ = 70.2 (*C*HCHO), 80.5 (*C*CO), 85.1 (CH*C*HO), 111.8 (C_{Ar}H), 122.4 (C_{Ar}H), 126.0 (C_{Ar}H), 126.3 (C_{Ar}H), 127.5 (C_{Ar}H), 127.6 (C_{Ar}), 127.8 (C_{Ar}H), 128.0 (C_{Ar}H), 129.1 (C_{Ar}H), 129.2 (C_{Ar}H), 130.4 (C_{Ar}H), 134.0 (C_{Ar}), 137.2 (C_{Ar}), 159.6 (C_{Ar}O), 202.5 (C*C*OCH).

ESI-MS: m/z (%): 283.4 (100) [$C_{21}H_{14}OH^+$].

FT-IR (KBr plates, cm^{-1}): 3064 (w), 2929 (w), 1780 (s, CO), 1604 (m), 1495 (w), 1472 (w), 1461 (w), 1223 (m), 1063 (w), 752 (s), 697 (s).

2-Benzylidene-7b-phenyl-2a,7b-dihydro-2*H*-benzo[*d*]oxeto[3,2-*b*]furan

3.15-Ph: $C_{22}H_{16}O_2$, M = 312.36 *g/mol*, white solid, **route**: *Irradiation*, 4h irradiation time.

^{1}H-NMR (400.13 MHz; CDCl$_3$): δ = 5.73 (s, 1H, CHO), 5.85 (s, 1H, CHC$_{Ar}$), 6.71 - 6.84 (m, 2H, H_{Ar}), 6.89 - 7.03 (m, 3H, H_{Ar}), 7.21 - 7.34 (m, 6H, H_{Ar}), 7.37 - 7.47 (m, 3H, H_{Ar}).

^{13}C-NMR (100.53 MHz; CDCl$_3$): δ = 77.1 (CHO), 109.6 (COCH), 116.7 (CHC$_{Ar}$), 121.2 (C_{Ar}H), 123.4 (C_{Ar}H), 126.0 (C_{Ar}H), 127.4 (C_{Ar}H), 128.1 (C_{Ar}H), 128.5 (C_{Ar}H), 128.8 (C_{Ar}H), 128.9 (C_{Ar}H), 129.4 (C_{Ar}), 129.7 (C_{Ar}H), 131.1 (C_{Ar}H), 136.9 (C_{Ar}), 138.2 (C_{Ar}), 140.7 (CHC$_{Ar}$CH), 153.9 (C_{Ar}O).

ESI-MS: m/z (%): 283.3 (100) [C$_{21}$H$_{14}$OH$^+$].

FT-IR (KBr plates, cm^{-1}): 3080 (m), 3050 (m), 2931 (m), 1736 (m), 1602 (m), 1479 (s), 1449 (s), 1263 (m), 1196 (s), 1061 (m), 750 (s), 698 (s).

(*Z*)-6-((*E*)-2-oxo-1-phenylpent-3-enylidene)cyclohexa-2,4-dienone

3.18-Me: C$_{17}$H$_{14}$O$_2$, M = 250.29 g/mol, white solid, R$_f$ = 0.2, **route**: *Irradiation*.

^{1}H-NMR (400.13 MHz; CDCl$_3$): δ = 1.87 (d, $^3J_{HH}$ = 7.2 Hz, 3H, CH_3), 6.53 (d, $^3J_{HH}$ = 15.2 Hz, 1H, CHCHCO), 6.77 (dq, $^3J_{HH}$ = 7.2 Hz, $^3J_{HH}$ = 15.2 Hz, 1H, CH$_3$CHCH), 7.27 - 7.61 (m, 7H, H_{Ar}), 7.75 (d, $^3J_{HH}$ = 7.8 Hz, 2H, H_{Ar}).

^{13}C-NMR (100.53 MHz; CDCl$_3$): δ = 18.7 (CH$_3$), 128.5 (C_{Ar}H), 128.9 (C_{Ar}H), 129.0 (C_{Ar}H), 129.9 (C_{Ar}H), 130.0 (C_{Ar}H), 130.1 (C_{Ar}H), 131.0 (C_{Ar}H), 133.1 (CHCHCO), 137.4 (C_{Ar}), 139.5 (C_{Ar}), 140.6 (C_{Ar}), 146.7 (CH$_3$$C$HCH), 192.9 ($C$O), 197.4 ($C$O).

ESI-MS: m/z (%): 209.2 (50) [C$_{14}$H$_9$O$_2$$^+$], 233.2 (53) [C$_{17}H_{13}O^+$], 251.1 (100) [C$_{17}H_{14}O_2H^+$].

FT-IR (KBr plates, cm^{-1}): 3070 (w), 3040 (w), 2980 (w), 2936 (w), 1667 (s, CO), 1621 (m), 1448 (m), 1284 (s), 1147 (w), 969 (w), 931 (m), 760 (m), 704 (m), 637 (w).

8.2.6. Other cinnamic acid esters

4-Methoxyphenyl 4-methylcinnamate

PhMe-PhOMe: $C_{17}H_{16}O_3$, M = 268.31 g/mol, white solid, yield: 87%, **route**: *Ester-II.*

^{1}H-NMR (399.78 MHz; CDCl$_3$): δ = 2.39 (s, 3H, CH_3), 3.81 (s, 3H, CH_3O), 6.58 (d, $^3J_{HH}$ = 16.0 Hz, 1H, CHCHCO), 6.92 (d, $^3J_{HH}$ = 9.2 Hz, 2H, H_{Ar}), 7.09 (d, $^3J_{HH}$ = 9.2 Hz, 2H, H_{Ar}), 7.22 (d, $^3J_{HH}$ = 8.1 Hz, 2H, H_{Ar}), 7.48 (d, $^3J_{HH}$ = 8.1 Hz, 2H, H_{Ar}), 7.84 (d, $^3J_{HH}$ = 16.0 Hz, 1H, C$_{Ar}$CHCH).

^{13}C-NMR (100.53 MHz; CDCl$_3$): δ = 21.7 (CH_3), 55.8 (CH_3O), 114.6 (C$_{Ar}$H), 116.4 (CHCHCO), 122.6 (C$_{Ar}$H), 128.5 (C$_{Ar}$H), 129.9 (C$_{Ar}$H), 131.7 (C$_{Ar}$), 141.3 (C$_{Ar}$), 144.5 (C$_{Ar}$O), 146.6 (C$_{Ar}$CHCH), 157.4 (C$_{Ar}$OCH$_3$), 166.1 (CHCOO).

ESI-MS: m/z (%): 145.1 (38) [C$_{10}$H$_9$O$^+$], 269.2 (100) [C$_{17}$H$_{16}$O$_3$H$^+$].

FT-IR (ATR, cm^{-1}): 2973 (w), 1716 (s, CO), 1622 (m), 1603 (m), 1443 (w), 1307 (m), 1244 (w), 1186 (m), 1140 (s), 1100 (m), 1034 (w), 1015 (m), 976 (w), 849 (w), 817 (s), 765 (w), 520 (m), 502 (s), 485 (w), 425 (m).

elemental analysis for $C_{17}H_{16}O_3$ (268.31 g/mol):

calcd.: C 76.10, H 6.01, O 17.89

found: C 76.24, H 6.04

4-Methylphenyl 4-methylcinnamate

PhMe-PhMe: $C_{17}H_{16}O_2$, M = 252.31 g/mol, white solid, yield: 87%, **route**: *Ester-II*.

^{1}H-NMR (399.78 MHz; CDCl$_3$): δ = 2.36 (s, 3H, CH_3), 2.39 (s, 3H, CH_3), 6.58 (d, $^3J_{HH}$ = 16.0 Hz, 1H, CHCHCO), 7.05 (d, $^3J_{HH}$ = 8.4 Hz, 2H, H_{Ar}), 7.17 - 7.24 (m, 4H, H_{Ar}), 7.48 (d, $^3J_{HH}$ = 8.0 Hz, 2H, H_{Ar}), 7.84 (d, $^3J_{HH}$ = 16.0 Hz, 1H, C$_{Ar}$CHCH).

^{13}C-NMR (100.53 MHz; CDCl$_3$): δ = 21.1 (CH$_3$), 21.7 (CH$_3$), 116.5 (CHCHCO), 121.5 (C_{Ar}H), 128.5 (C_{Ar}H), 129.9 (C_{Ar}H), 130.1 (C_{Ar}H), 131.7 (C_{Ar}), 135.5 (C_{Ar}), 141.3 (C_{Ar}), 146.6 (C_{Ar}CHCH), 148.8 (C_{Ar}O), 166.0 (CHCOO).

ESI-MS: m/z (%): 145.1 (65) [C$_{10}$H$_9$O$^+$], 253.2 (100) [C$_{17}$H$_{16}$O$_2$H$^+$].

FT-IR (ATR, cm^{-1}): 2975 (w), 1718 (s, CO), 1629 (m), 1603 (m), 1505 (m), 1309 (m), 1195 (m), 1164 (w), 1138 (s), 1018 (w), 992 (m), 973 (w), 849 (w), 839 (w), 812 (s), 758 (w), 683 (w), 499 (s), 440 (w).

elemental analysis for $C_{17}H_{16}O_2$ (252.31 g/mol):

calcd.: C 80.93, H 6.39, O 12.68

found: C 79.66, H 6.18

Phenyl 4-methylcinnamate

PhMe-Ph: $C_{16}H_{14}O_2$, M = 238.28 g/mol, white solid, yield: 89%, **route**: *Ester-II*.

^{1}H-NMR (399.78 MHz; CDCl$_3$): δ = 2.41 (s, 3H, CH_3), 6.61 (d, $^3J_{HH}$ = 16.0 Hz, 1H, CHCHCO), 7.17 - 7.30 (m, 5H, H_{Ar}), 7.42 (t, $^3J_{HH}$ = 7.8 Hz, 2H, H_{Ar}), 7.50 (d, $^3J_{HH}$ = 7.9 Hz, 2H, H_{Ar}), 7.87 (d, $^3J_{HH}$ = 16.0 Hz, 1H, C$_{Ar}$CHCH).

^{13}C-NMR (100.53 MHz; CDCl$_3$): δ = 21.7 (CH$_3$), 116.4 (CHCHCO), 121.8 (C_{Ar}H), 125.9(C_{Ar}H), 128.5 (C_{Ar}H), 129.6 (C_{Ar}H), 129.9 (C_{Ar}H), 131.6 (C_{Ar}), 141.4 (C_{Ar}), 146.7 (C_{Ar}CHCH), 151.0 (C_{Ar}O), 165.7 (CHCOO).

FAB-MS: m/z (%): 144.8 (100) [C$_{10}$H$_9$O$^+$], 238.8 (53.04) [C$_{16}$H$_{14}$O$_2{}^+$].

FT-IR (ATR, cm^{-1}): 2973 (w), 1715 (s, CO), 1633 (m), 1614 (w), 1481 (w), 1309 (m), 1194 (m), 1144 (s), 1069 (w), 989 (s), 819 (s), 791 (w), 751 (w), 711 (m), 686 (w), 496 (s), 447 (m).

elemental analysis for C$_{16}$H$_{14}$O$_2$ (238.28 *g/mol*):

calcd.: C 80.65, H 5.92, O 13.43

found: C 80.38, H 5.93

4-Methoxyphenyl 4-chlorocinnamate

PhCl-PhOMe: C$_{16}$H$_{14}$O$_2$, M = 288.73 *g/mol*, white solid, yield: 75%, **route**: *Ester-III*.

^{1}H-NMR (399.78 MHz; CDCl$_3$): δ = 3.80 (s, 3H, CH_3O), 6.58 (d, $^3J_{HH}$ = 16.0 Hz, 1H, CHCHCO), 6.91 (d, $^3J_{HH}$ = 8.8 Hz, 2H, H_{Ar}), 7.07 (d, $^3J_{HH}$ = 8.8 Hz, 2H, H_{Ar}), 7.38 (d, $^3J_{HH}$ = 8.2 Hz, 2H, H_{Ar}), 7.50 (d, $^3J_{HH}$ = 7.5 Hz, 2H, H_{Ar}), 7.79 (d, $^3J_{HH}$ = 16.0 Hz, 1H, C$_{Ar}$CHCH).

^{13}C-NMR (100.53 MHz; CDCl$_3$): δ = 55.7 (CH$_3$O), 114.6 (C_{Ar}H), 118.1 (CHCHCO), 122.5 (C_{Ar}H), 129.4 (C_{Ar}H), 128.6 (C_{Ar}H), 132.9 (C_{Ar}), 136.7 (C_{Ar}), 144.4 (C_{Ar}), 145.0 (C_{Ar}CHCH), 157.4 (C_{Ar}O), 165.7 (CHCOO).

FT-IR (ATR, cm^{-1}): 2982 (w), 1724 (s, CO), 1634 (m), 1585 (w), 1500 (m), 1403 (w), 1318 (m), 1237 (w), 1193 (m), 1142 (s), 1086 (w), 1029 (w), 988 (w), 843 (w), 812 (s), 767 (w), 555 (w), 525 (m), 501 (s), 403 (m).

4-Methylphenyl 4-chlorocinnamate

PhCl-PhMe: $C_{16}H_{13}ClO_2$, M $= 272.73$ g/mol, white solid, yield: 85%, **route**: *Ester-III.*

^{1}H-NMR (399.78 MHz; CDCl$_3$): $\delta = 2.36$ (s, 3H, CH_3), 6.59 (d, $^3J_{HH} = 16.0$ Hz, 1H, CHCHCO), 7.04 (d, $^3J_{HH} = 7.4$ Hz, 2H, H_{Ar}), 7.19 (d, $^3J_{HH} = 7.7$ Hz, 2H, H_{Ar}), 7.38 (d, $^3J_{HH} = 7.6$ Hz, 2H, H_{Ar}), 7.50 (d, $^3J_{HH} = 7.4$ Hz, 2H, H_{Ar}), 7.80 (d, $^3J_{HH} = 16.0$ Hz, 1H, C$_{Ar}$CHCH).

^{13}C-NMR (100.53 MHz; CDCl$_3$): $\delta = 21.0$ (CH$_3$), 118.2 (CHCHCO), 121.4 (C_{Ar}H), 129.4 (C_{Ar}H), 129.6 (C_{Ar}H), 130.1 (C_{Ar}H), 132.9 (C_{Ar}), 135.6 (C_{Ar}), 136.7 (C_{Ar}), 145.0 (C_{Ar}CHCH), 148.6 (C_{Ar}O), 165.5 (CHCOO).

FT-IR (ATR, cm^{-1}): 2949 (w), 1736 (s, CO), 1635 (m), 1580 (w), 1504 (m), 1489 (m), 1398 (w), 1315 (m), 1192 (m), 1163 (w), 1141 (s), 1087 (s), 1108 (w), 981 (s), 848 (m), 822 (s), 762 (w), 537 (w), 494 (s), 462 (m).

4-Nitrophenyl 4-chlorocinnamate

PhCl-PhNO$_2$: $C_{15}H_{10}ClNO_4$, M $= 303.70$ g/mol, white solid, yield: 46%, **route**: *Ester-III.*

^{1}H-NMR (399.78 MHz; CDCl$_3$): δ = 6.59 (d, $^3J_{HH}$ = 16.0 Hz, 1H, CH*CH*CO), 7.36 (d, $^3J_{HH}$ = 7.9 Hz, 2H, H_{Ar}), 7.41 (d, $^3J_{HH}$ = 7.9 Hz, 2H, H_{Ar}), 7.53 (d, $^3J_{HH}$ = 7.6 Hz, 2H, H_{Ar}), 7.85 (d, $^3J_{HH}$ = 16.0 Hz, 1H, C$_{Ar}$*CH*CH), 8.29 (d, $^3J_{HH}$ = 7.9 Hz, 2H, H_{Ar}).

^{13}C-NMR (100.53 MHz; CDCl$_3$): δ = 116.9 (CH*CH*CO), 122.6 (C_{Ar}H), 125.4 (C_{Ar}H), 129.6 (C_{Ar}H), 129.7 (C_{Ar}H), 132.4 (C_{Ar}), 137.4 (C_{Ar}), 145.5 (C_{Ar}), 146.6 (C$_{Ar}$*CH*CH), 155.7 (C_{Ar}O), 162.2 (CH*C*OO).

FT-IR (ATR, cm^{-1}): 3122 (w), 1728 (s, CO), 1628 (m), 1593 (m), 1515 (s), 1488 (m), 1394 (w), 1350 (m), 1305 (m), 1206 (s), 1140 (s), 1091 (s), 989 (s), 970 (m), 866 (s), 816 (m), 806 (s), 728 (m), 682 (w), 490 (s), 409 (m).

4-Methoxyphenyl 4-trifluoromethylcinnamate

PhCF$_3$-PhOMe: C$_{17}$H$_{13}$F$_3$O$_3$, M = 322.28 *g/mol*, white solid, yield: 82%, **route**: *Ester-III*.

^{1}H-NMR (399.78 MHz; CDCl$_3$): δ = 3.81 (s, 3H, C*H$_3$*O), 6.69 (d, $^3J_{HH}$ = 16.1 Hz, 1H, CH*CH*CO), 6.92 (d, $^3J_{HH}$ = 7.5 Hz, 2H, H_{Ar}), 7.08 (d, $^3J_{HH}$ = 7.5 Hz, 2H, H_{Ar}), 7.68 (s, 4H, H_{Ar}), 7.86 (d, $^3J_{HH}$ = 16.1 Hz, 1H, C$_{Ar}$*CH*CH).

^{13}C-NMR (100.53 MHz; CDCl$_3$): δ = 55.8 (*C*H$_3$O), 114.7 (C_{Ar}H), 120.2 (CH*CH*CO), 122.4 (C_{Ar}H), 126.1 (q, $^3J_{CF}$ = 3.7 Hz, C_{Ar}H), 128.5 (C_{Ar}H), 132.2 (q, $^2J_{CF}$ = 32.6 Hz, C_{Ar}), 137.7 (C_{Ar}), 144.3 (C_{Ar}), 144.6 (C$_{Ar}$*CH*CH), 157.5 (C_{Ar}O), 135.4 (CH*C*OO).

FT-IR (ATR, cm^{-1}): 2937 (w), 2846 (w), 1725 (s, CO), 1637 (m), 1611 (w), 1505 (s), 1459 (w), 1315 (s), 1283 (w), 1250 (w), 1201 (w), 1126 (s), 1064 (s), 1020 (m), 989 (m), 976 (m), 833 (s), 761 (w), 720 (w), 622 (w), 593 (w), 548 (w), 523 (w), 503 (w).

4-Methylphenyl 4-trifluoromethylcinnamate

PhCF$_3$-PhMe: C$_{17}$H$_{13}$F$_3$O$_2$, M = 306.28 g/mol, white solid, yield: 77%, **route:** *Ester-III.*

^{1}H-NMR (399.78 MHz; CDCl$_3$): δ = 2.61 (s, 3H, CH_3), 6.94 (d, $^3J_{HH}$ = 16.0 Hz, 1H, CHCHCO), 7.29 (d, $^3J_{HH}$ = 7.4 Hz, 2H, H_{Ar}), 7.45 (d, $^3J_{HH}$ = 7.7 Hz, 2H, H_{Ar}), 7.93 (s, 4H, H_{Ar}), 8.11 (d, $^3J_{HH}$ = 16.0 Hz, 1H, C$_{Ar}$CHCH).

^{13}C-NMR (100.53 MHz; CDCl$_3$): δ = 21.0 (CH_3), 120.2 (CHCHCO), 121.3 (C$_{Ar}$H), 126.1 (q, $^3J_{CF}$ = 7.3 Hz, C$_{Ar}$H), 128.5 (C$_{Ar}$H), 130.2 (C$_{Ar}$H), 135.8 (C$_{Ar}$), 137.7 (C$_{Ar}$), 144.6 (C$_{Ar}$CHCH), 148.6 (C$_{Ar}$O), 165.2 (CHCOO).

FT-IR (ATR, cm^{-1}): 3053 (w), 2932 (w), 1720 (s, CO), 1637 (m), 1505 (m), 1355 (w), 1307 (s), 1211 (m), 1153 (s), 1067 (s), 1022 (s), 1007 (s), 849 (w), 833 (s), 762 (w), 719 (w), 593 (w), 534 (w), 502 (w), 458 (w).

Phenyl 4-trifluoromethylcinnamate

PhCF$_3$-Ph: C$_{16}$H$_{11}$F$_3$O$_2$, M = 292.25 g/mol, white solid, yield: 46%, **route:** *Ester-III.*

^{1}H-NMR (399.78 MHz; CDCl$_3$): δ = 6.70 (d, $^3J_{HH}$ = 16.1 Hz, 1H, CHCHCO), 6.92 (t, $^3J_{HH}$ = 7.4 Hz, 1H, H_{Ar}), 7.17 (d, $^3J_{HH}$ = 8.0 Hz, 2H, H_{Ar}), 7.41 (t, $^3J_{HH}$ = 7.2 Hz, 2H, H_{Ar}), 7.68 (s, 4H, H_{Ar}), 7.87 (d, $^3J_{HH}$ = 16.1 Hz, 1H, C$_{Ar}$CHCH).

FT-IR (ATR, cm^{-1}): 1738 (s, CO), 1637 (m), 1592 (w), 1485 (m), 1408 (w), 1315 (s), 1205 (m), 1272 (s), 1107 (s), 1066 (s), 989 (m), 831 (s), 782 (w), 741 (s), 723 (m), 688 (m), 635 (w), 593 (w), 500 (m), 467 (w).

4-Methylphenyl 4-nitrocinnamate

PhNO$_2$-PhMe: C$_{16}$H$_{13}$NO$_4$, M = 283.28 g/mol, brown solid, yield: 89%, **route**: *Ester-II.*

^{1}H-NMR (399.78 MHz; CDCl$_3$): δ = 2.36 (s, 3H, CH_3), 6.74 (d, $^3J_{HH}$ = 16.0 Hz, 1H, CHCHCO), 7.04 (d, $^3J_{HH}$ = 8.3 Hz, 2H, H_{Ar}), 7.20 (d, $^3J_{HH}$ = 8.3 Hz, 2H, H_{Ar}), 7.72 (d, $^3J_{HH}$ = 8.6 Hz, 2H, H_{Ar}), 7.87 (d, $^3J_{HH}$ = 16.0 Hz, 1H, C$_{Ar}$CHCH), 8.26 (d, $^3J_{HH}$ = 8.6 Hz, 2H, H_{Ar}).

^{13}C-NMR (100.53 MHz; CDCl$_3$): δ = 21.1 ($C H_3 C_{Ar}$), 121.3 (C_{Ar}H), 121.9 (CHCHCO), 124.4 (C_{Ar}H), 129.0 (C_{Ar}H), 130.2 (C_{Ar}H), 136.0 (C_{Ar}), 140.4 (C_{Ar}), 143.5 (C$_{Ar}$CHCH), 148.5 (C_{Ar}O), 148.9 (C_{Ar}N), 164.9 (CHCOO).

FAB-MS: m/z (%): 175.8 (25.45) [C$_9$H$_6$NO$_3{}^+$], 283.7 (100) [C$_{16}$H$_{13}$NO$_4{}^+$].

FT-IR (ATR, cm^{-1}): 1723 (s, CO), 1626 (m), 1588 (m), 1504 (s), 1337 (s), 1300 (m), 1188 (m), 1143 (s), 1118 (m), 1002 (m), 967 (m), 861 (w), 815 (s), 755 (s), 664 (w), 540 (m), 506 (m).

elemental analysis for C$_{16}$H$_{13}$NO$_4$ (283.28 g/mol):

calcd.: C 67.84, H 4.63, N 4.94, O 22.59

found: C 67.07, H 4.49, N 4.97

9-Fluoren-9-yl cinnamate

Ph-9fl: $C_{22}H_{16}O_2$, M = 312.36 g/mol, orange solid, yield: 86%, **route**: *Ester-III*.

^{1}H-NMR (399.78 MHz; CDCl$_3$): δ = 6.54 (d, $^3J_{HH}$ = 16.0 Hz, 1H, CHCHCO), 6.96 (s, CH) 7.31 (t, $^3J_{HH}$ = 7.5 Hz, 2H, H_{Ar}), 7.35 - 7.47 (m, 5H, H_{Ar}), 7.49 - 7.56 (m, 2H, H_{Ar}), 7.62 (d, $^3J_{HH}$ = 7.5 Hz, 2H, H_{Ar}), 7.69 (d, $^3J_{HH}$ = 7.5 Hz, 2H, H_{Ar}), 7.79 (d, $^3J_{HH}$ = 16.0 Hz, 1H, $C_{Ar}CH$CH).

^{13}C-NMR (100.53 MHz; CDCl$_3$): δ = 75.3 ($C_{Ar}CHC_{Ar}$), 117.9 (CHCHCO), 120.2 (C_{Ar}H), 125.3 (C_{Ar}), 126.1 (C_{Ar}H), 128.0 (C_{Ar}H), 128.3 (C_{Ar}H), 129.0 (C_{Ar}H), 129.6 (C_{Ar}H), 130.6 (C_{Ar}H), 141.2 (C_{Ar}), 142.3 (C_{Ar}), 145.8 ($C_{Ar}CH$CH), 167.8 (CHCOO).

CI-MS: m/z (%): 165.0 (100) [($C_{13}H_9^+$], 181.0 (89.05) [($C_{13}H_9O^+$], 345.1 (13.63) [($C_{26}H_{17}O^+$], 361.1 (8.53) [($C_{26}H_{17}O_2^+$].

ESI-MS: m/z (%): 663.6 (38) [($C_{22}H_{16}O_2)_2K^+$], 726.0 (100) [($C_{22}H_{16}O_2)_2H_2O(CH_3CN)_2H^+$].

elemental analysis for $C_{22}H_{16}O_2$ (312.36 g/mol):

calcd.: C 84.59, H 5.16, O 10.24

found: C 82.94, H 5.23

8.2.7. *Karstedt* inhibitors

1-Methoxypropan-2-yl 4-methoxycinnamate

PhOMe–^{i}PrOMe: $C_{14}H_{18}O_4$, M = 250.29 g/mol, colourless oil, yield: 80%, **route**: *Ester-I*.

^{1}H-NMR (400.13 MHz; CDCl$_3$): δ = 1.29 (d, $^3J_{HH}$ = 6.4 Hz, 3H, CHCH_3), 3.38 (s, 3H, OCH_3), 3.45 (dd, $^2J_{HH}$ = 10.5 Hz, $^3J_{HH}$ = 4.1 Hz, 1H, CHCH_2O), 3.51 (dd, $^2J_{HH}$ = 10.5 Hz, $^3J_{HH}$ = 5.9 Hz, 1H, CHCH_2O), 3.81 (s, 3H, OCH_3), 5.19 (dpq, 1H, $^3J_{HH}$ = 6.4 Hz, $^3J_{HH}$ = 5.9 Hz, OCH), 6.31 (d, $^3J_{HH}$ = 15.9 Hz, 1H, CHCHCO), 6.87 (d, $^3J_{HH}$ = 8.6 Hz, 2H, H_{Ar}), 7.45 (d, $^3J_{HH}$ = 8.6 Hz, 2H, H_{Ar}), 7.63 (d, $^3J_{HH}$ = 15.9 Hz, 1H, C$_{Ar}$CHCH).

^{13}C-NMR (100.62 MHz; CDCl$_3$): δ = 16.8 (CHCH$_3$), 55.4 (OCH$_3$), 59.3 (OCH$_3$), 69.3 (OCH), 75.2 (CHCH$_2$O), 114.4 (C_{Ar}H), 116.0 (CHCHCO), 127.3 (C_{Ar}), 129.8 (C_{Ar}H), 144.5 ($C_{Ar}$$C$HCH), 161.4 ($C_{Ar}$O), 166.9 (CH$C$OO).

Dimethyl(vinyl)silyl 4-methoxycinnamate

5.1: $C_{14}H_{18}O_3$Si, M = 262.38 g/mol, colourless oil, yield: 81%, **route**: *Silylester*.

^{1}H-NMR (400.13 MHz; CDCl$_3$): δ = 0.42 (s, 6H, CH_3), 3.82 (s, 3H, OCH_3), 5.89 (dd, $^2J_{HH}$ = 20.4 Hz, $^3J_{HH}$ = 3.5 Hz, 1H, CHCH_2), 6.09 (dd, $^3J_{HH}$ = 3.5 Hz, $^3J_{HH}$ = 14.8 Hz,

1H, CH*CH*$_2$), 6.24 - 6.34 (m, 2H, CHC*H*$_2$, CHC*H*CO), 6.89 (d, $^3J_{HH}$ = 8.4 Hz, 2H, H_{Ar}), 7.46 (d, $^3J_{HH}$ = 8.4 Hz, 2H, H_{Ar}), 7.59 (d, $^3J_{HH}$ = 15.8 Hz, 1H, C_{Ar}C*H*CH).

^{13}C-NMR (100.62 MHz; CDCl$_3$): δ = -1.8 (*C*H$_3$), 55.5 (O*C*H$_3$), 114.5 (C_{Ar}H), 117.3 (CH*C*H$_2$), 127.3 (C_{Ar}), 129.9 (C_{Ar}H), 134.3 (CH*C*HCO), 135.9 (*C*HCH$_2$), 145.3 (C_{Ar}*C*HCH), 161.5 (C_{Ar}O), 167.4 (CH*C*OO).

1-Methoxypropan-2-yl cinnamate

Ph–^{i}PrOMe: C$_{13}$H$_{16}$O$_3$, M = 220.26 g/mol, colourless oil, yield: 81%, **route**: *Ester-III*, alcohol deprotonation overnight.

^{1}H-NMR (399.78 MHz; CDCl$_3$): δ = 1.31 (d, $^3J_{HH}$ = 6.5 Hz, 3H, CHC*H*$_3$), 3.39 (s, 3H, OC*H*$_3$), 3.48 (dd, $^3J_{HH}$ = 3.9 Hz, $^2J_{HH}$ = 10.6 Hz, 1H, CHC*H*$_2$O), 3.53 (dd, $^3J_{HH}$ = 6.0 Hz, $^2J_{HH}$ = 10.6 Hz, 1H, CHC*H*$_2$O), 5.21 (dpq, 1H, $^3J_{HH}$ = 3.9 Hz, $^3J_{HH}$ = 6.5 Hz, OC*H*), 6.46 (d, $^3J_{HH}$ = 16.0 Hz, 1H, CHC*H*CO), 7.34 - 7.47 (m, 3H, H_{Ar}), 7.48 - 7.61 (m, 2H, H_{Ar}), 7.68 (d, $^3J_{HH}$ = 16.0 Hz, 1H, C_{Ar}C*H*CH).

^{13}C-NMR (100.53 MHz; CDCl$_3$): δ = 16.8 (CH*C*H$_3$), 59.4 (O*C*H$_3$), 69.5 (O*C*H), 75.2 (CH*C*H$_2$O), 118.6 (CH*C*HCO), 128.2 (C_{Ar}H), 129.0 (C_{Ar}H), 130.4 (C_{Ar}H), 134.6 (C_{Ar}), 144.9 (C_{Ar}*C*HCH), 166.7 (CH*C*OO).

Dimethyl(vinyl)silyl cinnamate

5.2: C$_{13}$H$_{16}$O$_2$Si, M = 232.35 g/mol, colourless oil, yield: 76%, **route**: *Silylester*.

^{1}H-NMR (400.13 MHz; CDCl$_3$): δ = 0.43 (s, 6H, C*H*$_3$), 5.90 (dd, $^2J_{HH}$ = 20.4 Hz, $^3J_{HH}$ = 3.6 Hz, 1H, CHC*H*$_2$), 6.10 (dd, $^3J_{HH}$ = 3.6 Hz, $^3J_{HH}$ = 14.9 Hz, 1H, C*H*CH$_2$), 6.30 (dd, $^2J_{HH}$ = 20.4 Hz, $^3J_{HH}$ = 14.9 Hz, 1H, CHC*H*$_2$), 6.41(d, $^3J_{HH}$ = 15.9 Hz, 1H, CHC*H*CO), 7.34 - 7.41 (m, 3H, *H*$_{Ar}$), 7.48 - 7.55 (m, 2H, *H*$_{Ar}$), 7.64 (d, $^3J_{HH}$ = 15.9 Hz, 1H, C$_{Ar}$C*H*CH).

^{13}C-NMR (100.62 MHz; CDCl$_3$): δ = -1.8 (*C*H$_3$), 119.8 (*C*HCH$_2$), 128.3 (*C*$_{Ar}$H), 129.0 (*C*$_{Ar}$H), 130.4 (CH*C*HCO), 134.4 (*C*$_{Ar}$H), 134.6 (*C*$_{Ar}$), 135.7 (CH*C*H$_2$), 145.6 (*C*$_{Ar}$CHCH), 167.1 (CH*C*OO).

1-Methoxypropan-2-yl 4-nitrocinnamate

PhNO$_2$–^{i}PrOMe: C$_{13}$H$_{15}$NO$_5$, M = 265.26 *g/mol*, yellow oil, yield: 63%, **route:** *Ester-III*, alcohol deprotonation overnight.

^{1}H-NMR (400.13 MHz; CDCl$_3$): δ = 1.30 (d, $^3J_{HH}$ = 6.5 Hz, 3H, CHC*H*$_3$), 3.38 (s, 3H, OC*H*$_3$), 3.47 (dd, $^2J_{HH}$ = 10.6 Hz, $^3J_{HH}$ = 3.9 Hz, 1H, CHC*H*$_2$O), 3.52 (dd, $^2J_{HH}$ = 10.6 Hz, $^3J_{HH}$ = 6.0 Hz, 1H, CHC*H*$_2$O), 5.22 (dpq, 1H, $^3J_{HH}$ = 3.9 Hz, $^3J_{HH}$ = 6.5 Hz, OC*H*), 6.56 (d, $^3J_{HH}$ = 16.0 Hz, 1H, CHC*H*CO), 7.65 (d, $^3J_{HH}$ = 8.6 Hz, 2H, *H*$_{Ar}$), 7.69 (d, $^3J_{HH}$ = 16.0 Hz, 1H, C$_{Ar}$C*H*CH), 8.22 (d, $^3J_{HH}$ = 8.6 Hz, 2H, *H*$_{Ar}$).

^{13}C-NMR (100.62 MHz; CDCl$_3$): δ = 16.7 (CH*C*H$_3$), 59.3 (O*C*H$_3$), 70.1 (O*C*H), 75.1 (CH*C*H$_2$O), 122.9 (CH*C*HCO), 124.3 (*C*$_{Ar}$H), 128.7 (*C*$_{Ar}$H), 140.7 (*C*$_{Ar}$), 141.9 (*C*$_{Ar}$CHCH), 148.6 (*C*$_{Ar}$N), 165.7 (CH*C*OO).

Dimethyl(vinyl)silyl 4-nitrocinnamate

5.3: $C_{13}H_{15}NO_4Si$, M $= 277.35$ g/mol, yellow solid, yield: 7.0%, **route**: *Silylester*.

^{1}H-NMR (400.13 MHz; CDCl$_3$): $\delta = 0.43$ (s, 6H, CH_3), 5.91 (dd, $^2J_{HH} = 20.3$ Hz, $^3J_{HH} = 3.6$ Hz, 1H, CHCH_2), 6.11 (dd, $^3J_{HH} = 3.6$ Hz, $^3J_{HH} = 14.9$ Hz, 1H, CHCH$_2$), 6.28 (dd, $^2J_{HH} = 20.3$ Hz, $^3J_{HH} = 14.9$ Hz, 1H, CHCH_2), 6.52 (d, $^3J_{HH} = 16.0$ Hz, 1H, CHCHCO), 7.60 - 7.69 (m, 3H, H_{Ar}, C_{Ar}CHCH), 8.23 (d, $^3J_{HH} = 8.7$ Hz, 2H, H_{Ar}).

^{13}C-NMR (100.62 MHz; CDCl$_3$): $\delta = -1.9$ (CH_3), 124.2 (CHCH$_2$), 124.3 (C_{Ar}H), 128.8 (C_{Ar}H), 134.8 (CHCHCO), 135.3 (CHCH$_2$), 140.7 (C_{Ar}), 142.5 (C_{Ar}CHCH), 148.7 (C_{Ar}N), 166.0 (CHCOO).

Di-1-methoxypropan-2-yl fumarate

5.9: $C_{12}H_{20}O_6$, M $= 260.28$ g/mol, colourless oil, yield: 50%, **route**: *Ester-I*.

^{1}H-NMR (400.13 MHz; CDCl$_3$): $\delta = 1.25$ (d, $^3J_{HH} = 6.5$ Hz, 6H, CHCH_3), 3.34 (s, 6H, OCH_3), 3.41 (dd, $^2J_{HH} = 10.6$ Hz, $^3J_{HH} = 3.9$ Hz, 2H, CHCH_2O), 3.46 (dd, $^2J_{HH} = 10.6$ Hz, $^3J_{HH} = 6.2$ Hz, 2H, CHCH_2O), 5.15 (dpq, $^3J_{HH} = 4.0$ Hz, $^3J_{HH} = 6.5$ Hz, 2H, OCH), 6.84 (s, 2H, COCHCHCO).

^{13}C-NMR (100.62 MHz; CDCl$_3$): $\delta = 16.6$ (CHCH$_3$), 59.3 (OCH$_3$), 70.5 (OCH), 75.0 (CHCH$_2$O), 134.0 (COCH), 164.6 (CHCOO).

1-Methoxypropan-2-yl ethyl fumarate

5.8: $C_{10}H_{16}O_5$, M = 216.23 g/mol, brown oil, yield: 67%, **route**: *Ester-III*, alcohol deprotonation overnight.

^{1}H-NMR (400.13 MHz; CDCl$_3$): δ = 1.13 - 1.46(m, 6H, CH$_2$CH_3, CHCH_3), 3.35 (s, 3H, OCH_3), 3.42 (dd, $^3J_{HH}$ = 4.2 Hz, $^2J_{HH}$ = 10.6 Hz, 1H, CHCH_2O), 3.47 (dd, $^3J_{HH}$ = 6.2 Hz, $^2J_{HH}$ = 10.6 Hz, 1H, CHCH_2O), 4.23 (q, $^3J_{HH}$ = 7.1 Hz, 2H, CH_2CH$_3$), 5.17 (dpq, 1H, $^3J_{HH}$ = 4.2 Hz, $^3J_{HH}$ = 6.2 Hz, OCH), 6.84 (d, $^3J_{HH}$ = 5.0 Hz, 2H, COCHCHCO), 6.85 (d, $^3J_{HH}$ = 5.0 Hz, 2H, COCHCHCO).

^{13}C-NMR (100.62 MHz; CDCl$_3$): δ = 14.2 (CH$_2$$CH_3$), 16.6 (CH$CH_3$), 59.3 (O$CH_3$), 61.4 ($CH_2CH_3$), 70.6 (O$C$H), 75.0 (CH$CH_2$O), 133.9 (CO$C$HCH), 164.7 (CH$C$OO), 165.1 (CH$C$OO).

Dimethyl(vinyl)silyl ethyl fumarate

5.11: $C_{10}H_{16}O_4Si$, M = 228.32 g/mol, colourless oil, yield: 82%, **route**: *Silylester*.

^{1}H-NMR (400.13 MHz; CDCl$_3$): δ = 0.39 (s, 6H, Si(CH_3)$_2$), 1.30 (t, $^3J_{HH}$ = 7.2 Hz, 3H, CH$_2$CH_3), 4.24 (q, $^3J_{HH}$ = 7.2 Hz, 2H, CH_2CH$_3$), 5.88 (dd, $^3J_{HH}$ = 3.7 Hz, $^2J_{HH}$ = 20.2 Hz, 1H, CHCH_2), 6.09 (dd, $^3J_{HH}$ = 3.7 Hz, $^3J_{HH}$ = 14.9 Hz, 1H, CHCH$_2$), 6.23 (dd, $^3J_{HH}$ = 14.9 Hz, $^2J_{HH}$ = 20.2 Hz, 1H, CHCH_2), 5.79 (ps, 2H, COCHCHCO).

^{13}C-NMR (100.62 MHz; CDCl$_3$): δ = -2.0 (Si(CH_3)$_2$), 14.3 (CH$_2$$CH_3$), 61.5 ($CH_2CH_3$), 134.3 ($CHCH_2$), 134.9 (CO$C$HCH), 135.0 (CO$C$HCH), 135.1 (CH$CH_2$), 165.0 (CH$C$OO), 165.3 (CH$C$OO).

Bis(dimethyl(vinyl)silyl) fumarate

5.12: C$_{12}$H$_{20}$O$_4$Si$_2$, M = 284.46 g/mol, colourless oil, yield: 84%, **route:** *Silylester*, with 2 eq. dimethyl(vinyl)silylchloride and 2.2 eq. NEt$_3$.

^{1}H-NMR (400.13 MHz; CDCl$_3$): δ = 0.39 (s, 12H, CH_3), 5.87 (dd, $^2J_{HH}$ = 20.2 Hz, $^3J_{HH}$ = 3.6 Hz, 2H, CHCH_2), 6.08 (dd, $^3J_{HH}$ = 3.6 Hz, $^3J_{HH}$ = 14.9 Hz, 2H, CHCH$_2$), 6.22 (dd, $^2J_{HH}$ = 20.2 Hz, $^3J_{HH}$ = 14.9 Hz, 2H, CHCH_2), 6.74 (s, 2H, COCHCHCO).

^{13}C-NMR (100.62 MHz; CDCl$_3$): δ = -2.0 (CH$_3$), 135.0 (CHCH$_2$), 135.0 (COCH), 135.6 (CHCH$_2$), 165.1 (CHCOO),

1-Methoxypropan-2-yl maleate

5.15: C$_8$H$_{12}$O$_5$, M = 188.18 g/mol, colourless oil, yield: 50%, **route:** *Ester-V.*

^{1}H-NMR (400.13 MHz; CDCl$_3$): δ = 1.29 (d, $^3J_{HH}$ = 6.4 Hz, 3H, CHCH_3), 3.35 (s, 3H, OCH_3), 3.44 (dd, $^3J_{HH}$ = 3.8 Hz, $^2J_{HH}$ = 10.8 Hz, 1H, CHCH_2O), 3.49 (dd, $^3J_{HH}$ = 6.4 Hz, $^2J_{HH}$ = 10.8 Hz, 1H, CHCH_2O), 5.21 (dpq, $^3J_{HH}$ = 3.8 Hz, $^3J_{HH}$ = 6.4 Hz, 1H, OCH), 6.36 (s, 2H, COCHCHCO), 11.35 (bs, 1H, OH).

^{13}C-NMR (100.62 MHz; CDCl$_3$): δ = 16.2 (CHCH_3), 59.2 (OCH_3), 72.3 (OCH), 74.5 (CHCH_2O), 130.4 (CHCO), 134.5 (CHCO), 165.9 (CHCOO), 166.9 (CHCOO).

ESI-MS: m/z (%): 187.0 (100) [C$_8$H$_{11}$O$_5{}^-$].

elemental analysis for C$_8$H$_{12}$O$_5$ (188.18 *g/mol*):

calcd.: C 51.06, H 6.43, O 42.51

found: C 51.03, H 6.37

Di-1-methoxypropan-2-yl maleate

5.24: C$_{12}$H$_{20}$O$_6$, M = 260.28 *g/mol*, colourless oil, yield: 59%, **route**: *Ester-I.*

^{1}H-NMR (400.13 MHz; CDCl$_3$): δ = 1.28 (d, $^3J_{HH}$ = 6.5 Hz, 6H, CHCH_3), 3.35 (s, 6H, OCH_3), 3.42 (dd, $^2J_{HH}$ = 10.6 Hz, $^3J_{HH}$ = 4.1 Hz, 2H, CHCH_2O), 3.48 (dd, $^2J_{HH}$ = 10.6 Hz, $^3J_{HH}$ = 5.6 Hz, 2H, CHCH_2O), 5.15 (dpq, $^3J_{HH}$ = 5.7 Hz, $^3J_{HH}$ = 11.1 Hz, 2H, OCH), 6.22 (s, 2H, COCHCHCO).

^{13}C-NMR (100.62 MHz; CDCl$_3$): δ = 16.5 (CHCH_3), 59.3 (OCH_3), 70.5 (OCH), 74.8 (CHCH_2O), 130.1 (COCH), 164.8 (CHCOO).

Ethyl dimethyl(vinyl)silyl maleate

5.25: C$_{10}$H$_{16}$O$_4$Si, M = 228.32 *g/mol*, colourless oil, yield: 72%, **route**: *Silylester.*

^{1}H-NMR (400.13 MHz; CDCl$_3$): δ = 0.39 (s, 6H, CH_3), 1.28 (t, $^3J_{HH}$ = 7.1 Hz, 3H, CH$_2$CH_3), 4.22 (q, $^3J_{HH}$ = 7.1 Hz, 2H, CH_2CH$_3$), 5.87 (dd, $^2J_{HH}$ = 20.3 Hz,

$^3J_{HH}$ = 3.4 Hz, 1H, CHCH_2), 6.07 (dd, $^3J_{HH}$ = 3.4 Hz, $^3J_{HH}$ = 14.8 Hz, 1H, CHCH$_2$), 6.13 - 6.30 (m, 3H, CHCH_2, COCHCHCO).

^{13}C-NMR (100.62 MHz; CDCl$_3$): δ = -2.0 (Si(CH_3)$_2$), 14.2 (CH$_2$$CH_3$), 61.3 ($CH_2CH_3$), 129.8 ($CHCH_2$), 130.8 (CO$C$HCH), 134.8 (CO$C$HCH), 135.2 (CH$CH_2$), 165.1 (CH$C$OO), 165.4 (CH$C$OO).

1-Methoxypropan-2-yl dimethyl(vinyl)silyl maleate

5.16: C$_{12}$H$_{20}$O$_5$Si, M = 272.37 g/mol, colourless oil, yield: 40%, **route**: *Silylester*.

^{1}H-NMR (400.13 MHz; CDCl$_3$): δ = 0.39 (s, 6H, CH_3), 1.27 (d, $^3J_{HH}$ = 6.5 Hz, 3H, CHCH_3), 3.35 (s, 3H, OCH_3), 3.41 (dd, $^2J_{HH}$ = 10.6 Hz, $^3J_{HH}$ = 4.3 Hz, 1H, CHCH_2O), 3.46 (dd, $^2J_{HH}$ = 10.6 Hz, $^3J_{HH}$ = 5.6 Hz, 1H, CHCH_2O), 5.15 (dpq, $^3J_{HH}$ = 6.0 Hz, $^3J_{HH}$ = 10.7 Hz, 1H, OCH), 5.86 (dd, $^2J_{HH}$ = 20.3 Hz, $^3J_{HH}$ = 3.6 Hz, 1H, CHCH_2), 6.17 (dd, $^3J_{HH}$ = 3.6 Hz, $^3J_{HH}$ = 14.9 Hz, 1H, CHCH$_2$), 6.14 - 6.28 (m, 3H, CHCH_2, COCHCHCO).

^{13}C-NMR (100.62 MHz; CDCl$_3$): δ = -2.0 (CH$_3$), 16.5 (CHCH$_3$), 59.3 (OCH$_3$), 70.4 (OCH), 74.8 (CHCH$_2$O), 129.8 (CHCH$_2$), 130.9 (COCH), 134.7(COCH), 135.2 (CHCH$_2$), 164.9 (CHCOO), 165.0 (CHCOO).

Bis(dimethyl(vinyl)silyl) maleate

5.17: $C_{12}H_{20}O_4Si_2$, M = 284.46 g/mol, colourless oil, yield: 67%, **route**: *Silylester*, with 2 eq. dimethyl(vinyl)silylchloride and 2.2 eq. NEt$_3$.

^{1}H-NMR (400.13 MHz; CDCl$_3$): δ = 0.39 (s, 12H, CH_3), 5.86 (dd, $^2J_{HH}$ = 20.3 Hz, $^3J_{HH}$ = 3.6 Hz, 2H, CHCH_2), 6.07 (dd, $^3J_{HH}$ = 3.6 Hz, $^3J_{HH}$ = 14.9 Hz, 2H, CHCH$_2$), 6.15 (s, 2H, COCHCHCO), 6.23 (dd, $^2J_{HH}$ = 20.3 Hz, $^3J_{HH}$ = 14.9 Hz, 2H, CHCH_2).

^{13}C-NMR (100.62 MHz; CDCl$_3$): δ = -2.0 (CH$_3$), 130.6 (CHCH$_2$), 134.7 (COCH), 135.3 (CHCH$_2$), 165.1 (CHCOO).

(*E*)-But-2-ene-1,4-diyl biscinnamate

5.5: $C_{22}H_{20}O_4$, M = 348.39 g/mol, orange solid, yield: 12%, **route**: *Ester-II*, purification by *flash* chromatography (SiO$_2$; 9:1 pentane/Et$_2$O).

^{1}H-NMR (399.78 MHz; CDCl$_3$): δ = 4.72 - 4.77 (m, 4H, OCH_2CH), 5.95 - 6.05 (m, 2H, OCH$_2$CH), 6.46 (d, $^3J_{HH}$ = 16.0 Hz, 2H, CHCHCO), 7.33 - 7.45 (m, 6H, C$_{Ar}$$H$), 7.48 - 7.60 (m, 4H, C$_{Ar}$$H$), 7.72 (d, $^3J_{HH}$ = 16.0 Hz, 2H, C$_{Ar}$CHCH).

^{13}C-NMR (100.53 MHz; CDCl$_3$): δ = 64.2 (OCH$_2$CH), 117.9 (CHCHCO), 128.3 (C_{Ar}H), 128.4 (OCH$_2$$C$H), 129.1 ($C_{Ar}$H), 130.6 ($C_{Ar}$H), 134.5 ($C_{Ar}$), 145.5 (C$_{Ar}$$C$HCH), 166.8 (CH$C$OO).

ESI-MS: m/z (%): 371.1 (100) [$C_{22}H_{20}O_4$Na$^+$], 388.2 (61) [$C_{22}H_{20}O_4$K$^+$], 718.8 (19) [($C_{22}H_{20}O_4$)$_2$Na$^+$].

elemental analysis for $C_{22}H_{20}O_4$ (348.39 g/mol):

calcd.: C 75.84, H 5.79, O 18.37

found: C 75.65, H 5.31

((E)-But-2-ene-1,4-diyl) diethyl difumarate

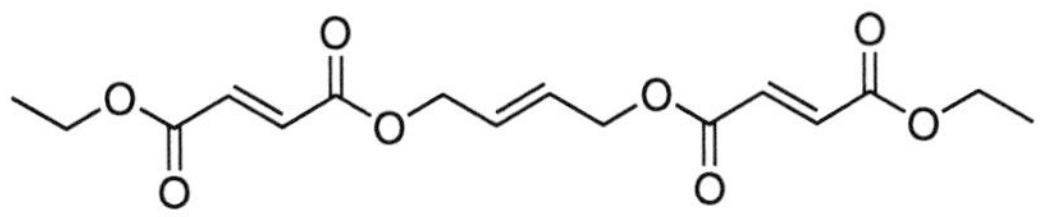

5.10: $C_{16}H_{20}O_8$, M = 340.33 g/mol, orange oil, yield: 63%, **route:** *Ester-II.*

^{1}H-NMR (399.78 MHz; CDCl$_3$): δ = 1.28 (t, $^3J_{HH}$ = 7.1 Hz, 6H, CH_3), 4.23 (q, $^3J_{HH}$ = 7.1 Hz, 4H, CH$_3$CH_2), 4.67 - 4.71 (m, 4H, OCH_2CH), 5.87 - 5.92 (m, 2H, OCH$_2$CH), 6.84 (s, 4H, COCHCHCO).

^{13}C-NMR (100.53 MHz; CDCl$_3$): 14.3 (CH$_3$), 61.5 (CH$_3$$CH_2$), 64.6 (O$CH_2$CH), 128.1 (OCH$_2$$C$H), 133.2 (CO$C$HCH), 134.4 (CO$C$HCH), 164.7 (CH$C$O), 165.0 (CH$C$O).

CI-MS: m/z (%): 196.7 (100) $[C_{10}H_{13}O_4{}^+]$, 340.7 (23.09) $[C_{16}H_{20}O_8{}^+]$.

elemental analysis for $C_{16}H_{20}O_8$ (340.33 g/mol):

calcd.: C 56.47, H 5.92, O 37.61

found: C 56.48, H 6.16

8.2.8. Further organic compounds

Ethyl maleicacid chloride

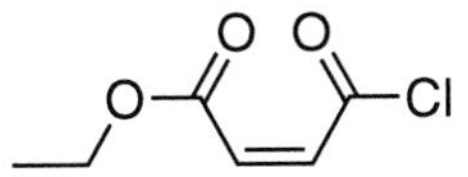

$C_6H_7ClO_3$, M = 162.57 g/mol, brown oil, yield: 87%, **route:** *Acidchloride-II.*

^{1}H-NMR (400.13 MHz; CDCl$_3$): δ = 1.33 (t, $^3J_{HH}$ = 7.2 Hz, 3H, CH_3), 4.29 (q, $^3J_{HH}$ = 7.2 Hz, 2H, CH$_3$CH_2), 6.98 (s, 1H, COCHCHCO), 7.02 (s, 1H, COCHCHCO).

^{13}C-NMR (100.62 MHz; CDCl$_3$): δ = 14.2 (CH$_3$), 62.2 (CH$_3$$CH_2$), 136.9 (CO$C$HCH), 138.0 (CO$C$HCH), 163.9 (CH$C$OO), 165.6 (CH$C$OO).

1-Methoxypropan-2-yl maleicacid chloride

C$_8$H$_{11}$ClO$_4$, M = 206.62 g/mol, brown oil, yield: 84%, **route**: *Acidchloride-II*.

^{1}H-NMR (400.13 MHz; CDCl$_3$): δ = 1.27 (d, $^3J_{HH}$ = 6.4 Hz, 3H, CHCH_3), 3.35 (s, 3H, OCH_3), 3.39 - 3.51 (m, 2H, CHCH_2O), 5.19 (dpq, $^3J_{HH}$ = 3.9 Hz, $^3J_{HH}$ = 6.4 Hz, 1H, OCH), 6.97 (s, 2H, COCHCHCO).

^{13}C-NMR (100.62 MHz; CDCl$_3$): δ = 16.5 (CHCH$_3$), 59.3 (OCH$_3$), 71.5 (OCH), 74.8 (CHCH$_2$O), 136.9 (CHCO), 138.2 (CHCO), 163.4 (CHCOO), 165.5 (CHCOO).

1-(Trimethylsilyloxy)-fluoren-9-one

3.21: C$_{16}$H$_{16}$O$_2$Si, M = 268.38 g/mol.

0.20 g (1.02 mmol) 1-hydroxy-9-fluorenone was added to a suspension of 24.4 mg (1.02 mmol) NaH in 5 ml of thf at 0 °C and warmed to ambient temperature. After 15 min 0.13 ml (0.11 g, 1.02 mmol) tms chloride was added and stirring continued over night. The solvent was removed in vacuo, the residue extracted with dichloromethane, filtered and the solvent removed under reduced pressure to give 0.27 g (1.02 mmol, >99%) 1-(trimethylsilyloxy)-fluoren-9-one as colourless oil.

^{1}H-NMR (400.13 MHz; CDCl$_3$): $\delta = 0.32$ (s, 9H Si(CH_3)$_3$), 6.68 (d, 1H, $^3J_{HH} = 8.3$ Hz, H_{Ar}), 7.14 (d, 1H, $^3J_{HH} = 7.2$ Hz, H_{Ar}), 7.27 - 7.36 (m, 2H, H_{Ar}), 7.41 - 7.53 (m, 2H, H_{Ar}), 7.61 (d, 1H, $^3J_{HH} = 7.3$ Hz, H_{Ar}).

^{13}C-NMR (100.62 MHz; CDCl$_3$): $\delta = 0.44$ (Si(CH_3)$_3$), 114.1 (C_{Ar}H), 120.3 (C_{Ar}H), 123.3 (C_{Ar}H), 123.4 (C_{Ar}), 123.9 (C_{Ar}H), 129.2 (C_{Ar}H), 134.2 (C_{Ar}H), 134.7 (C_{Ar}), 136.2 (C_{Ar}H), 143.6 (C_{Ar}), 146.3 (C_{Ar}), 154.3 (C_{Ar}O), 192.1 (C_{Ar}COC_{Ar}).

2-(Dimethoxy(phenyl)methyl)phenol

3.20: C$_{15}$H$_{16}$O$_3$, M $= 244.26$ g/mol.

0.50 g (2.50 mmol) 2-hydroxybenzophenone, 0.55 ml (0.53 g, 5.00 mmol) trimethoxy orthoformate and one crystal of *para*-toluene sulfonic acid monohydrate were dissolved in 10 ml of methanol and stirred for one week at ambient temperature. The solvent was removed in vacuo to yield 0.52 g (2.13 mmol, 85%) of product as a white solid.

^{1}H-NMR (400.13 MHz; CDCl$_3$): $\delta = 3.22$ (s, 6H, OCH_3), 6.78 (t, $^3J_{HH} = 7.5$ Hz, 1H, H_{Ar}), 6.86 (d, $^3J_{HH} = 8.1$ Hz, 1H, H_{Ar}), 7.08 - 7.18 (m, 2H, H_{Ar}), 7.23 - 7.37 (m, 3H, H_{Ar}), 7.54 (d, $^3J_{HH} = 7.6$ Hz, 2H, H_{Ar}), 8.71 (bs, 1H, OH).

^{13}C-NMR (100.62 MHz; CDCl$_3$): $\delta = 49.6$ (OCH$_3$), 106.5 (C_{Ar}) 117.5 (C_{Ar}H) 119.8 (C_{Ar}H) 124.6 (C(OCH$_3$)$_2$) 126.8 (C_{Ar}H) 128.3 (C_{Ar}H) 128.3 (C_{Ar}H) 129.3 (C_{Ar}H) 129.8 (C_{Ar}H) 141.2 (C_{Ar}) 155.7 (C_{Ar}O).

8.3. Complexes

8.3.1. General synthetic procedures

Ni

40.0 mg (145 μmol) Ni(cod)$_2$ and 76.3 mg (290 μmol) PPh$_3$ were dissolved in toluene and added to 145 μmol of the appropriate ligand and stirred for 1 h. After removal of the solvent in vacuo the residue was stirred with pentane over night. The solvent was decanted off and the residue washed with pentane to give the desired complex.

Pd-I

30.0 mg (141 μmol) (η^3-allyl)(η^5-cyclopentadienyl)palladium(II) and 74.0 mg (282 μmol) triphenylphosphine were dissolved in toluene and added to 141 μmol of ligand and stirred for 1 h. After removal of the solvent in vacuo the residue was stirred with pentane over night. The solvent was decanted off and the residue washed once with pentane to give the product in microanalytical grade.

Pd-II

40.0 mg (0.18 mmol) (η^3-allyl)(η^5-cyclopentadienyl)palladium(II) and 0.18 mmol phosphine were dissolved in toluene and added to 0.18 mmol of olefin and stirred for 1 h. After removal of the solvent in vacuo the residue was stirred with pentane over night. The solvent was decanted off and the residue washed once with pentane to give the complex.

Pd-III

40.0 mg (0.18 mmol) (η^3-allyl)(η^5-cyclopentadienyl)palladium(II), 0.18 mmol phosphine and 0.18 mmol cinnamic acid ester were dissolved in toluene. After 1 h the solvent was removed in vacuo and the residue was stirred with pentane over night. The solvent was decanted off and the residue washed once with pentane to give the desired product.

Pd-IV

40.0 mg (0.18 mmol) (η^3-allyl)(η^5-cyclopentadienyl)palladium(II) and 0.18 mmol of the appropriate olefin were dissolved in benzene and a benzene solution of 0.18 mmol phosphine was added dropwise over night via a pressure equalizing funnel. After lyophilization the residue was stirred in pentane over night. The solvent was decanted off and the residue washed twice with pentane to yield the anticipated complex.

Pd-V

100 mg (154 μmol) crude Pd(PPh$_3$)$_2$ and 154 μmol olefin were dissolved in toluene and stirred for 1 h. After removal of the solvent in vacuo the residue was stirred with pentane over night. The solvent was decanted off and the residue washed once with pentane to give the complex.

Pd-VI

100 mg (154 μmol) crude Pd(PPh$_3$)$_2$ was suspended in toluene, C$_2$H$_4$ bubbled through the suspension for 10 min and stirring continued under an ethene atmosphere for 2 days. The solvent was removed by a fast flow of ethene, the residue dissolved in pentane and cooled to -40 °C retaining the C$_2$H$_4$ atmosphere. The solvent was decanted off and the white solid dried in an ethene flow to yield 87.0 mg (125 μmol, 79%) product.

Pt-I

200 mg (0.53 mmol) (η^4-1,5-cyclooctadiene)platinum(II)chloride and 0.53 mmol phosphine were dissolved in CH$_2$Cl$_2$ and stirred over night. The solvent was removed in vacuo and the residue washed with pentane to obtain the product in quantitative yield.

Pt-II

50.0 mg (0.10 mmol) Pt(nbe)$_3$ and 0.10 mmol of the olefin were dissolved in benzene and stirred for 1 h. After lyophilization the residue was stirred in pentane over night. The solvent was decanted off and the obtained complex dried in vacuo.

Pt-III

100 mg (0.10 mmol) Pt(PPh$_3$)$_3$ and 0.10 mmol of the appropriate ligand were dissolved in toluene and stirred for 1 h. After removal of the solvent in vacuo the residue was stirred with pentane over night. The solvent was decanted off and the residue washed twice with pentane to give the product.

Pt-IV

10.0 mg (24.0 μmol) Pt(cod)$_2$ and 24.0 μmol of olefin ligand were dissolved in benzene, 24.0 μmol phosphine was added after 10 min and stirring continued for 1 h. The solvent was removed in vacuo and the residue washed with pentane to obtain the desired complex.

Pt-V

30.0 mg (31.0 μmol) Pt(PPh$_3$)$_3$ and 31.0 μmol phosphine were dissolved in benzene,

24.0 μmol olefin was added after 15 min and stirring continued for 1 h. The solvent was removed in vacuo and the residue washed twice with pentane to obtain the desired complex.

Pt-VI

75.0 mg (0.10 mmol) Pt(PPh$_3$)$_2$C$_2$H$_4$ were dissolved in 5 ml of benzene, added to 0.10 mmol of the appropriate ligand and stirred for 1 h. After lyophilization the residue was stirred in pentane over night. The solvent was decanted off and the obtained powder dried in vacuo to yield the desired complex in microanalytical grade.

8.3.2. 2-Benzoylphenyl ester complexes

(η^4-Cycloocta-1,5-diene)-(η^2-2-benzoylphenyl cinnamate)platinum(0)

Pt(cod)Ph-bp: C$_{30}$H$_{28}$O$_3$Pt, M = 631.62 g/mol.

50 mg (122 μmol) Pt(cod)$_2$ and 39.9 mg (122 μmol) 2-benzoylphenyl cinnamate were dissolved in toluene and stirred for 2 h. The solvent was removed in vacuo and the residue stirred in pentane over night. The solvent was decanted off and the residue washed with pentane to yield 69.0 mg (109 mmol, 90%) of (η^4-cycloocta-1,5-diene)-(η^2-2-benzoylphenyl cinnamate)platinum(0) as off-white solid.

^{1}H-NMR (399.78 MHz; C$_6$D$_6$): δ = 1.16 - 1.84 (m, 8H, CH_2CH), 3.92 (d, $^3J_{HH}$ = 10.1 Hz, 1H, CHCHCO), 3.92 (dd, $^3J_{HH}$ = 10.1 Hz, $^2J_{HPt}$ = 88.8 Hz, 1H, CHCHCO), 4.35 - 4.46 (m, 2H, C$_{Ar}$CHCH, CH$_2$CH), 4.35 - 4.46 (dm, $^2J_{HPt}$ = 77.2 Hz, 2H, C$_{Ar}$CHCH, CH$_2$CH), 4.88 - 4.97 (m, 1H, CH$_2$CH), 4.88 - 4.97 (dm, $^2J_{HPt}$ = 69.9 Hz, 1H, CH$_2$CH), 5.32 - 5.42 (m, 1H, CH$_2$CH), 5.32 - 5.42 (dm, $^2J_{HPt}$ = 60.1 Hz, 1H, CH$_2$CH),

5.74 - 5.83 (m, 1H, CH_2CH), 5.74 - 5.83 (dm, $^2J_{HPt}$ = 59.5 Hz, 1H, CH_2CH), 6.80 (t, $^3J_{HH}$ = 7.7 Hz, 1H, H_{Ar}), 6.88 - 7.12 (m, 9H, H_{Ar}), 7.21 (d, $^3J_{HH}$ = 7.1 Hz, 1H, H_{Ar}), 7.32 (d, $^3J_{HH}$ = 8.0 Hz, 1H, H_{Ar}), 7.98 (d, $^3J_{HH}$ = 6.0 Hz, 2H, H_{Ar}).

^{13}C-NMR (100.53 MHz; C_6D_6): δ = 29.1 (CH_2CH), 29.5 (CH_2CH), 29.8 (CH_2CH), 30.2 (CH_2CH), 43.1 ($CHCHCO$), 43.1 (d, $^1J_{CPt}$ = 314.4 Hz, $CHCHCO$), 52.2 ($C_{Ar}CHCH$), 52.2 (d, $^1J_{CPt}$ = 353.6 Hz, $C_{Ar}CHCH$), 91.4 (CH_2CH), 91.4 (d, $^1J_{CPt}$ = 111.5 Hz, CH_2CH), 93.3 (CH_2CH), 93.3 (d, $^1J_{CPt}$ = 132.2 Hz, CH_2CH), 96.3 (CH_2CH), 96.3 (d, $^1J_{CPt}$ = 78.5 Hz, CH_2CH), 96.7 (CH_2CH), 96.7 (d, $^1J_{CPt}$ = 76.1 Hz, CH_2CH), 123.5 ($C_{Ar}H$), 124.2 ($C_{Ar}H$), 124.5 ($C_{Ar}H$), 126.2 ($C_{Ar}H$), 129.3 ($C_{Ar}H$), 130.0 ($C_{Ar}H$), 130.2 ($C_{Ar}H$), 130.8 (C_{Ar}), 131.0 ($C_{Ar}H$), 132.5 ($C_{Ar}H$), 133.4 (C_{Ar}), 138.0 ($C_{Ar}H$), 144.0 ($C_{Ar}CH$), 144.0 (d, $^2J_{CPt}$ = 40.0 Hz, $C_{Ar}CH$), 149.9 ($C_{Ar}O$), 169.9 ($CHCOO$), 169.9 (d, $^2J_{CPt}$ = 67.6 Hz, $CHCOO$), 194.4 ($C_{Ar}COC_{Ar}$).

^{195}Pt-NMR (85.48 MHz; C_6D_6): δ = -6178.3.

Di(η^2-bicyclo[2.2.1]hept-2-ene)-(η^2-2-benzoylphenyl cinnamate)platinum(0)

Pt(nbe)Ph-bp: $C_{36}H_{36}O_3Pt$, M = 711.75 g/mol, off-white solid, yield: 92%, **route**: Pt-II.

^{1}H-NMR (399.78 MHz; C_6D_6): δ = -0.06 (d, $^2J_{HH}$ = 8.6 Hz, 2H, $CHCH_2CH$), 0.25 (d, $^2J_{HH}$ = 7.3 Hz, 2H, $CHCH_2CH$), 1.03 (d, $^3J_{HH}$ = 8.7 Hz, 4H, $CHCH_2CH_2CH$), 1.42 (d, $^3J_{HH}$ = 6.9 Hz, 4H, $CHCH_2CH_2CH$), 2.53 (s, 2H, $CHCH_2CH_2CH$), 2.85 (s, 2H, $CHCH_2CH_2CH$), 3.26 (s, 2H, $CHCHCHCH$), 3.26 (d, $^2J_{HPt}$ = 62.6 Hz, 2H, $CHCHCHCH$), 3.55 (s, 2H, $CHCHCHCH$), 3.55 (d, $^2J_{HPt}$ = 61.2 Hz, 2H, $CHCHCHCH$),

4.08 (d, $^3J_{HH}$ = 10.0 Hz, 1H, CHC*H*CO), 4.08 (dd, $^3J_{HH}$ = 10.0 Hz, $^2J_{HPt}$ = 62.8 Hz, 1H, CHC*H*CO), 4.76 (d, $^3J_{HH}$ = 10.0 Hz, 1H, C_{Ar}C*H*CH), 4.76 (dd, $^3J_{HH}$ = 10.0 Hz, $^2J_{HPt}$ = 53.0 Hz, 1H, C_{Ar}C*H*CH), 6.75 - 7.09 (m, 11H, H_{Ar}), 7.24 (d, $^3J_{HH}$ = 8.1 Hz, 2H, H_{Ar}), 7.84 - 7.91 (m, 2H, H_{Ar}).

^{13}C-NMR (100.53 MHz; C_6D_6): δ = 27.8 (CHC*H$_2$*CH), 27.8 (d, $^2J_{CPt}$ = 40.8 Hz, CHC*H$_2$*CH), 28.0 (CHC*H$_2$*CH), 28.0 (d, $^2J_{CPt}$ = 41.5 Hz, C*C*HC*H$_2$*CH), 40.7 (C*H*CH$_2$CH), 40.7 (d, $^3J_{CPt}$ = 39.2 Hz, C*H*CH$_2$CH), 43.0 (CHC*H$_2$*CH$_2$CH), 43.6 (CHC*H$_2$*CH$_2$CH), 49.9 (C*H*CHCO), 49.9 (d, $^1J_{CPt}$ = 163.7 Hz, C*H*CHCO), 63.3 (C_{Ar}*C*HCH), 63.3 (d, $^1J_{CPt}$ = 149.9 Hz, C_{Ar}*C*HCHCH), 68.7 (CH*C*HCHCH), 68.7 (d, $^1J_{CPt}$ = 189.8 Hz, CH*C*HCH), 78.7 (CHC*H*CHCH), 78.7 (d, $^1J_{CPt}$ = 166.0 Hz, CHC*H*CHCH), 79.7 (CHCH*C*HCH), 79.7 (d, $^1J_{CPt}$ = 179.8 Hz, CHC*H*CHCH), 123.7 (C_{Ar}H), 125.3 (C_{Ar}H), 126.4 (C_{Ar}H), 127.0 (C_{Ar}H), 128.6 (C_{Ar}H), 128.8 (C_{Ar}H), 130.2 (C_{Ar}H), 130.6 (C_{Ar}H), 131.6 (C_{Ar}H), 133.1 (C_{Ar}H), 133.7 (C_{Ar}), 138.5 (C_{Ar}), 142.4 (C_{Ar}CH), 142.4 (d, $^2J_{CPt}$ = 32.3 Hz, C_{Ar}CH), 149.9 (C_{Ar}O), 169.1 (CH*C*OO), 169.1 (d, $^2J_{CPt}$ = 47.6 Hz, CH*C*OO), 194.5 (C_{Ar}*C*OC_{Ar}).

^{195}Pt-NMR (85.48 MHz; C_6D_6): δ = -6161.1.

FAB-MS: m/z (%): 522.8 (43.74) [$C_{22}H_{16}O_3Pt^+$], 616.9 (11.10) [$C_{22}H_{16}O_3PtC_7H_{10}^+$].

Bis(triphenylphosphine)-(η^2-2-benzoylphenyl cinnamate)nickel(0)

NiPh-bp: $C_{58}H_{46}NiO_3P_2$, M = 911.62 *g/mol*, light yellow solid, yield: 96%, **route**: *Ni*.

^{1}H-NMR (399.78 MHz; C_6D_6): δ = 3.18 (m, 1H, CHC*H*CO), 3.76 (m, 1H, C_{Ar}C*H*CH), 6.34 (d, $^3J_{HH}$ = 7.5 Hz, 2H, H_{Ar}), (m, 24H, H_{Ar}), 7.53 (t, $^3J_{HH}$ = 8.5 Hz, 6H, H_{Ar}), 7.88 (d, $^3J_{HH}$ = 7.4 Hz, 2H, H_{Ar}).

^{13}C-NMR (100.53 MHz; C$_6$D$_6$): δ = 51.4 (d, $^2J_{CP}$ = 11.9 Hz, CHCHCO), 63.8 (d, $^2J_{CP}$ = 17.2 Hz, C$_{Ar}$$C$HCH), 123.9 ($C_{Ar}$H), 124.3 ($C_{Ar}$H), 124.6 ($C_{Ar}$H), 125.5 ($C_{Ar}$H), 128.7 ($C_{Ar}$H), 129.2 ($C_{Ar}$H), 129.4 ($C_{Ar}$H), 129.8 ($C_{Ar}$H), 130.1 ($C_{Ar}$H), 131.0 ($C_{Ar}$H), 131.6 ($C_{Ar}$H), 132.5 ($C_{Ar}$H), 133.2 ($C_{Ar}$), 134.1 - 134.5 (m, C_{Ar}P), 134.9 (C_{Ar}P), 135.7 (C_{Ar}P), 136.0 (C_{Ar}P), 138.7 (C_{Ar}), 143.4 (C_{Ar}CH), 150.2 (C_{Ar}O), 169.9 (CHCOO), 195.4 (C$_{Ar}$$COC_{Ar}$).

^{31}P-NMR (161.83 MHz; C$_6$D$_6$): δ = 32.4 (d, $^2J_{PP}$ = 33.7 Hz), 34.0 (d, $^2J_{PP}$ = 33.7 Hz).

Bis(triphenylphosphine)-(η^2-2-benzoylphenyl cinnamate)palladium(0)

PdPh-bp: C$_{58}$H$_{46}$O$_3$P$_2$Pd, M = 959.35 g/mol, yellow solid, NMR scale, **route:** Pd-I.

^{1}H-NMR (399.78 MHz; C$_6$D$_6$): δ = 4.34 (bs, 1H, CHCHCO), 4.69 (bs, 1H, C$_{Ar}$CHCH), 6.29 - 6.49 (m, 2H, H_{Ar}), 6.72 - 7.09 (m, 32H, H_{Ar}), 7.41 - 7.67 (m, 8H, H_{Ar}), 7.90 (d, $^3J_{HP}$ = 7.4 Hz, 2H, H_{Ar}).

^{31}P-NMR (161.83 MHz; C$_6$D$_6$): δ = 25.1 (d, $^2J_{PP}$ = 14.9 Hz), 26.2 (d, $^2J_{PP}$ = 14.9 Hz).

Bis(triphenylphosphine)-(η^2-2-benzoylphenyl cinnamate)platinum(0)

PtPh-bp: $C_{58}H_{46}O_3P_2Pt$, M = 1048.01 g/mol, light yellow solid, yield: >99%, **route:** *Pt-III*.

^{1}H-NMR (399.78 MHz; C_6D_6): δ = 3.57 (dpt, $^3J_{HH}$ = 8.7 Hz, $^3J_{HP}$ = 4.4 Hz, $^3J_{HP}$ = 8.7 Hz, 1H, CHCHCO), 3.57 (ddpt, $^3J_{HH}$ = 8.7 Hz, $^3J_{HP}$ = 4.4 Hz, $^3J_{HP}$ = 8.7 Hz, $^2J_{HPt}$ = 59.8 Hz, 1H, CHCHCO), 3.81 (dpt, $^3J_{HH}$ = 8.7 Hz, $^3J_{HP}$ = 5.6 Hz, $^3J_{HP}$ = 8.7 Hz, 1H, $C_{Ar}CH$CH), 3.81 (ddpt, $^3J_{HH}$ = 8.7 Hz, $^3J_{HP}$ = 5.6 Hz, $^3J_{HP}$ = 8.7 Hz, $^2J_{HPt}$ = 51.3 Hz, 1H, $C_{Ar}CH$CH), 6.28 (d, $^3J_{HH}$ = 5.5 Hz, 2H, H_{Ar}), 6.73 - 7.10 (m, 28H, H_{Ar}), 7.42 - 7.58 (m, 12H, H_{Ar}), 7.91 (d, $^3J_{HH}$ = 7.6 Hz, 2H, H_{Ar}).

^{13}C-NMR (100.53 MHz; C_6D_6): δ = 49.2 (dd, $^2J_{CP}$ = 5.4 Hz, $^2J_{CP}$ = 27.7 Hz, CHCHCO), 49.2 (ddd, $^2J_{CP}$ = 5.4 Hz, $^2J_{CP}$ = 27.7 Hz, $^1J_{CPt}$ = 190.7 Hz, CHCHCO), 61.9 (dd, $^2J_{CP}$ = 3.9 Hz, $^2J_{CP}$ = 33.1 Hz, $C_{Ar}CH$CH), 61.9 (ddd, $^2J_{CP}$ = 3.9 Hz, $^2J_{CP}$ = 33.1 Hz, $^1J_{CPt}$ = 195.7 Hz, $C_{Ar}CH$CH), 124.6 (C_{Ar}H), 124.7 (C_{Ar}H), 125.1 (C_{Ar}H), 126.7 (C_{Ar}H), 126.7 (C_{Ar}H), 128.3 (C_{Ar}H), 128.7 (C_{Ar}H), 129.1 (C_{Ar}H), 129.2 (C_{Ar}H), 130.0 (C_{Ar}H), 130.1 (C_{Ar}H), 130.3 (C_{Ar}H), 130.6 (C_{Ar}H), 131.6 (C_{Ar}H), 133.2 (C_{Ar}H), 133.5 (C_{Ar}) 134.6 (C_{Ar}H), 134.8 (C_{Ar}H), 134.8 (C_{Ar}H), 134.9 (C_{Ar}H), 135.0 (C_{Ar}H), 135.7 (d, $^1J_{CP}$ = 3.0 Hz, C_{Ar}P), 135.7 (dd, $^1J_{CP}$ = 3.0 Hz, $^2J_{CPt}$ = 35.4 Hz, C_{Ar}P), 136.1 (d, $^1J_{CP}$ = 2.3 Hz, C_{Ar}P), 136.1 (dd, $^1J_{CP}$ = 2.3 Hz, $^2J_{CPt}$ = 32.3 Hz, C_{Ar}P), 137.3 (d, $^1J_{CP}$ = 2.3 Hz, C_{Ar}P), 137.3 (dd, $^1J_{CP}$ = 2.3 Hz, $^2J_{CPt}$ = 28.9 Hz, C_{Ar}P), 137.7 (d, $^1J_{CP}$ = 2.3 Hz, C_{Ar}P), 137.7 (dd, $^1J_{CP}$ = 2.3 Hz, $^2J_{CPt}$ = 27.7 Hz, C_{Ar}P), 139.4 (C_{Ar}), 140.6 (bs, C_{Ar}P), 140.8 (bs, C_{Ar}P), 144.7 (d, $^3J_{CP}$ = 6.2 Hz, C_{Ar}CH), 144.7 (dd, $^3J_{CP}$ = 6.2 Hz, $^2J_{CPt}$ = 33.1 Hz, C_{Ar}CH), 150.8 (C_{Ar}O), 170.9 (dd, $^3J_{CP}$ = 2.3 Hz,

$^3J_{CP}$ = 4.6 Hz, CHCOO), 170.9 (ddd, $^3J_{CP}$ = 2.3 Hz, $^3J_{CP}$ = 4.6 Hz, $^2J_{CPt}$ = 46.9 Hz,
CHCOO), 196.1 (C$_{Ar}$$COC_{Ar}$).

^{31}P-NMR (161.83 MHz; C$_6$D$_6$): δ = 27.6 (d, $^2J_{PP}$ = 40.6 Hz), 27.6 (dd,
$^2J_{PP}$ = 40.6 Hz, $^1J_{PPt}$ = 4194.1 Hz), 28.5 (d, $^2J_{PP}$ = 40.6 Hz), 28.5 (dd, $^2J_{PP}$ = 40.6 Hz,
$^1J_{PPt}$ = 3630.4 Hz).

^{195}Pt-NMR (85.48 MHz; C$_6$D$_6$): δ = -5038.1 (dd, $^1J_{PPt}$ = 3630.4 Hz,
$^1J_{PPt}$ = 4195.8 Hz).

FAB-MS: m/z (%): 718.8 (100) [Pt(PPh$_3$)(PPh$_2$C$_6$H$_4$)$^+$].

ESI-MS: m/z (%): 758.9 [Pt(PPh$_3$)(PPh$_2$C$_6$H$_4$)CH$_3$CN$^+$].

elemental analysis for C$_{58}$H$_{46}$O$_3$P$_2$Pt (1048.01 g/mol):

calcd.: C 66.47, H 4.42, O 4.58, P 5.91, Pt 18.61

found: C 66.83, H 4.32

(1,4-Bis(diphenylphosphino)propane)-(η^2-2-benzoylphenyl cinnamate)platinum(0)

Pt(dppp)Ph-bp: C$_{49}$H$_{42}$O$_3$P$_2$Pt, M = 935.88 g/mol, off-white solid, NMR scale, **route:**
Pt-IV.

^{1}H-NMR (399.78 MHz; C$_6$D$_6$): δ = 1.17 - 1.52 (bm, 2H, CH$_2$CH_2CH$_2$), 1.85 - 2.14
(bm, 4H, CH_2CH$_2$CH_2), 3.60 (dpt, $^3J_{HH}$ = 9.2 Hz, $^3J_{HP}$ = 3.2 Hz, $^3J_{HP}$ = 9.2 Hz, 1H,
COCHCH), 3.60 (ddpt, $^3J_{HH}$ = 9.2 Hz, $^3J_{HP}$ = 3.2 Hz, $^3J_{HP}$ = 9.2 Hz, $^2J_{HPt}$ = 59.1 Hz,
1H, COCHCH), 4.09 (dpt, $^3J_{HH}$ = 9.2 Hz, $^3J_{HP}$ = 4.4 Hz, $^3J_{HP}$ = 9.2 Hz, 1H, COCHCH),
4.09 (ddpt, $^3J_{HH}$ = 9.2 Hz, $^3J_{HP}$ = 4.4 Hz, $^3J_{HP}$ = 9.2 Hz, $^2J_{HPt}$ = 54.0 Hz, 1H,

COCHCH), 6.60 (d, $^3J_{HH}$ = 7.1 Hz, 2H, H_{Ar}), 6.70 - 7.21 (m, 24H, H_{Ar}), 7.44 - 7.57 (m, 6H, H_{Ar}), 7.92 (d, $^3J_{HH}$ = 7.1 Hz, 2H, H_{Ar}).

^{31}P-NMR (161.83 MHz; C$_6$D$_6$): δ = 7.9 (d, $^2J_{PP}$ = 25.8 Hz), 7.9 (dd, $^2J_{PP}$ = 25.8 Hz, $^1J_{PPt}$ = 3220.2 Hz), 9.5 (d, $^2J_{PP}$ = 25.8 Hz), 9.5 (dd, $^2J_{PP}$ = 25.8 Hz, $^1J_{PPt}$ = 3829.6 Hz).

(1,3-Bis(di-*iso*-propylphosphino)propane)-(η^2-2-benzoylphenyl cinnamate)platinum(0)

Pt(dippp)Ph-bp: C$_{37}$H$_{50}$O$_3$P$_2$Pt, M = 799.82 *g/mol*, yellow solid, NMR scale, **route**: *Pt-IV*, crystallisation from pentane at -40 °C.

^{1}H-NMR (399.78 MHz; C$_6$D$_6$): δ = 0.48 (dd, $^3J_{HP}$ = 7.1 Hz, $^3J_{HH}$ = 16.0 Hz, 6H, CH_3), 0.76 (dd, $^3J_{HP}$ = 7.1 Hz, $^3J_{HH}$ = 11.4 Hz, 3H, CH_3), 2.11 - 2.21 (m, 9H, CH_3), 1.06 (dd, $^3J_{HP}$ = 7.1 Hz, $^3J_{HH}$ = 15.9 Hz, 8H, CH_3, PCH), 1.17 (dd, $^3J_{HP}$ = 7.0 Hz, $^3J_{HH}$ = 15.5 Hz, PCH_2CH$_2$, 4H), 1.30 - 1.69 (m, 3H, PCH, CH$_2$CH_2CH$_2$), 2.11 - 2.21 (m, 1H, CH$_2$CH_2CH$_2$), 3.12 (dpt, $^3J_{HH}$ = 2.4 Hz, $^3J_{HP}$ = 8.7 Hz, $^3J_{HP}$ = 8.7 Hz, 1H, CHCHCO), 3.12 (ddpt, $^3J_{HH}$ = 2.4 Hz, $^3J_{HP}$ = 8.7 Hz, $^3J_{HP}$ = 8.7 Hz, $^2J_{HPt}$ = 55.9 Hz, 1H, CHCHCO), 3.91 (dpt, $^3J_{HH}$ = 3.5 Hz, $^3J_{HP}$ = 8.5 Hz, $^3J_{HP}$ = 8.5 Hz, 1H, C$_{Ar}$CHCH), 3.91 (ddpt, $^3J_{HH}$ = 3.5 Hz, $^3J_{HP}$ = 8.5 Hz, $^3J_{HP}$ = 8.5 Hz, $^2J_{HPt}$ = 50.4 Hz, 1H, C$_{Ar}$CHCH), (m, 5H, H_{Ar}), 7.02 (t, $^3J_{HH}$ = 7.6 Hz, 2H, H_{Ar}), 7.08 (t, $^3J_{HH}$ = 7.5 Hz, 2H, H_{Ar}), 7.21 (dt, $^3J_{HH}$ = 1.6 Hz, $^3J_{HH}$ = 8.3 Hz, 1H, H_{Ar}), 7.50 (dd, $^3J_{HH}$ = 1.5 Hz, $^3J_{HH}$ = 7.6 Hz, 1H, H_{Ar}), 7.69 (d, $^3J_{HH}$ = 8.2 Hz, 1H, H_{Ar}), 7.97 (d, $^3J_{HH}$ = 7.2 Hz, 2H, H_{Ar}).

^{13}C-NMR (100.53 MHz; C$_6$D$_6$): δ = 16.2 (d, $^2J_{CP}$ = 4.3 Hz, CH_3), 16.2 (dd, $^2J_{CP}$ = 4.3 Hz, $^3J_{CPt}$ = 30.6 Hz, CH_3), 16.9 (d, $^2J_{CP}$ = 3.7 Hz, CH_3), 16.9 (dd,

$^2J_{CP}$ = 3.7 Hz, $^3J_{CPt}$ = 24.6 Hz, CH_3), 18.4 (d, $^2J_{CP}$ = 4.6 Hz, CH_3), 18.4 (m, CH_3), 18.8 (d, $^2J_{CP}$ = 5.5 Hz, CH_3), 18.8 (m, CH_3), 19.3 (d, $^2J_{CP}$ = 4.8 Hz, CH_3), 19.3 (m, CH_3), 19.7 (d, $^2J_{CP}$ = 4.6 Hz, CH_3), 19.7 (m, CH_3), 19.8 (d, $^2J_{CP}$ = 5.1 Hz, CH_3), 19.8 (m, CH_3), 20.1 (d, $^2J_{CP}$ = 5.8 Hz, CH_3), 20.1 (m, CH_3), 24.9 (t, $^2J_{CP}$ = 21.3 Hz, $CH_2CH_2CH_2$), 25.9(d, $^1J_{CP}$ = 3.1 Hz, PCH_2CH_2), 25.9 (dd, $^1J_{CP}$ = 3.1 Hz, $^2J_{CPt}$ = 33.6 Hz, PCH_2CH_2), 26.1 (d, $^1J_{CP}$ = 3.0 Hz, PCH_2CH_2), 26.1 (dd, $^1J_{CP}$ = 3.0 Hz, $^2J_{CPt}$ = 33.2 Hz, PCH_2CH_2), 27.5 (d, $^1J_{CP}$ = 2.8 Hz, PCH), 27.5 (dd, $^1J_{CP}$ = 2.8 Hz, $^2J_{CPt}$ = 32.0 Hz, PCH), 27.8 (d, $^1J_{CP}$ = 2.7 Hz, PCH), 27.8 (m, PCH), 28.0 (d, $^1J_{CP}$ = 3.5 Hz, PCH), 28.0 (m, PCH), 28.3 (d, $^1J_{CP}$ = 3.3 Hz, PCH), 28.3 (dd, $^1J_{CP}$ = 3.3 Hz, $^2J_{CPt}$ = 43.8 Hz, PCH), 40.0 (dd, $^2J_{CP}$ = 5.6 Hz, $^2J_{CP}$ = 27.7 Hz, $CHCHCO$), 40.0 (ddd, $^2J_{CP}$ = 5.6 Hz, $^2J_{CP}$ = 27.7 Hz, $^1J_{CPt}$ = 180.7 Hz, $CHCHCO$), 48.0 (dd, $^2J_{CP}$ = 1.8 Hz, $^2J_{CP}$ = 35.2 Hz, $C_{Ar}CHCH$), 48.0 (ddd, $^2J_{CP}$ = 1.8 Hz, $^2J_{CP}$ = 35.2 Hz, $^1J_{CPt}$ = 223.7 Hz, $C_{Ar}CHCH$), 123.0 (m, $C_{Ar}H$), 123.2 ($C_{Ar}H$), 124.2 ($C_{Ar}H$), 126.3 (d, $^5J_{CP}$ = 3.3 Hz, $C_{Ar}H$), 126.3 (dd, $^5J_{CP}$ = 3.3 Hz, $^4J_{CPt}$ = 16.2 Hz, $C_{Ar}H$), 127.5 (m, $C_{Ar}H$), 128.5 ($C_{Ar}H$), 129.9 ($C_{Ar}H$), 130.1 ($C_{Ar}H$), 131.0 ($C_{Ar}H$), 132.6 ($C_{Ar}H$), 133.1 (C_{Ar}), 138.8 (C_{Ar}), 148.2 (d, $^3J_{CP}$ = 1.1 Hz, $C_{Ar}CH$), 148.2 (dd, $^3J_{CP}$ = 1.1 Hz, $^2J_{CPt}$ = 33.1 Hz, $C_{Ar}CH$), 150.8 ($C_{Ar}O$), 171.9 (dd, $^3J_{CP}$ = 2.4 Hz, $^3J_{CP}$ = 4.9 Hz, $CHCOO$), 171.9 (ddd, $^3J_{CP}$ = 2.4 Hz, $^3J_{CP}$ = 4.9 Hz, $^2J_{CPt}$ = 43.8 Hz, $CHCOO$), 195.3 ($C_{Ar}COC_{Ar}$).

^{31}P-NMR (161.83 MHz; C_6D_6): δ = 27.5 (d, $^2J_{PP}$ = 8.5 Hz), 27.5 (dd, $^2J_{PP}$ = 8.5 Hz, $^1J_{PPt}$ = 3094.4 Hz), 27.7 (d, $^2J_{PP}$ = 8.5 Hz), 27.7 (dd, $^2J_{PP}$ = 8.5 Hz, $^1J_{PPt}$ = 3700.7 Hz).

^{195}Pt-NMR (85.48 MHz; C_6D_6): δ = -5215.7 (dd, $^1J_{PPt}$ = 3096.5 Hz, $^1J_{PPt}$ = 3703.3 Hz).

FAB-MS: m/z (%): 470.6 (100) [Pt($^iPr_2$P)(CH_2)$_3$($P^iPrCCH_3CH_2$)$^+$], 798.4 (10.07) [$C_{22}H_{16}O_3$Pt($^iPr_2$P)(CH_2)$_3$($P^iPrCCH_3CH_2$)$^+$].

ESI-MS: m/z (%): 800.3 (67) [$C_{22}H_{16}O_3$Pt($^iPr_2$P)(CH_2)$_3$(P^iPr_2)H$^+$], 822.2 (39) [$C_{22}H_{16}O_3$Pt($^iPr_2$P)(CH_2)$_3$(P^iPr_2)Na$^+$], 838.2 (13) [$C_{22}H_{16}O_3$Pt($^iPr_2$P)(CH_2)$_3$(P^iPr_2)K$^+$], 1621.2 (100) [($C_{22}H_{16}O_3$Pt($^iPr_2$P)(CH_2)$_3$(P^iPr_2))$_2$Na$^+$], 1637.1 (11) [($C_{22}H_{16}O_3$Pt($^iPr_2$P)(CH_2)$_3$(P^iPr_2))$_2$K$^+$].

(1,4-Bis(diphenylphosphino)butane)-(η^2-2-benzoylphenyl cinnamate)platinum(0)

Pt(dppb)Ph-bp: $C_{50}H_{44}O_3P_2Pt$, M = 949.91 g/mol, off-white solid, yield: >99%, **route**: *Pt-V*.

^{1}H-NMR (399.78 MHz; C_6D_6): δ = 1.35 (bs, 4H, CH_2CH_2P), 2.15 (bs, 4H, CH_2CH_2P), 3.28 (dpt, $^3J_{HH}$ = 9.46 Hz, $^3J_{HP}$ = 3.4 Hz, $^3J_{HP}$ = 9.46 Hz, 1H, CHCHCO), 3.28 (ddpt, $^3J_{HH}$ = 9.46 Hz, $^3J_{HP}$ = 3.4 Hz, $^3J_{HP}$ = 9.46 Hz, $^2J_{HPt}$ = 59.5 Hz, 1H, CHCHCO), 3.91 (dpt, $^3J_{HH}$ = 9.46 Hz, $^3J_{HP}$ = 4.3 Hz, $^3J_{HP}$ = 9.46 Hz, 1H, $C_{Ar}CH$CH), 3.91 (ddpt, $^3J_{HH}$ = 9.46 Hz, $^3J_{HP}$ =4.3 Hz, $^3J_{HP}$ = 9.46 Hz, $^2J_{HPt}$ = 54.2 Hz, 1H, $C_{Ar}CH$CH), 6.46 (d, $^3J_{HH}$ = 7.1 Hz, 2H, H_{Ar}), 6.69 - 6.95 (m, , H_{Ar}), 6.96 - 7.23 (m, 23H, H_{Ar}), 7.43 (bs, 2H, H_{Ar}), 7.49 - 7.72 (m, 5H, H_{Ar}), 7.86 (d, $^3J_{HH}$ = 7.2 Hz, 2H, H_{Ar}).

^{13}C-NMR (100.53 MHz; C_6D_6): δ = 23.6 (d, $^2J_{CP}$ = 41.5 Hz, CH_2CH_2P), 28.8 - 29.9 (m, CH_2CH_2P), 45.6 (dd, $^2J_{CP}$ = 6.9 Hz, $^2J_{CP}$ = 30.0 Hz, CHCHCO), 45.6 (ddd, $^2J_{CP}$ = 6.9 Hz, $^2J_{CP}$ = 30.0 Hz, $^1J_{CPt}$ = 193.0 Hz, CHCHCO), 56.0 (dd, $^2J_{CP}$ = 3.8 Hz, $^2J_{CP}$ = 27,7 Hz, $C_{Ar}CH$CH), 56.0 (ddd, $^2J_{CP}$ = 3.8 Hz, $^2J_{CP}$ = 27,7 Hz, $^1J_{CPt}$ = 218.3 Hz, $C_{Ar}CH$CH), 123.6 ($C_{Ar}H$), 124.1 ($C_{Ar}H$), 124.7 ($C_{Ar}H$), 126.7 ($C_{Ar}H$), 127.8 ($C_{Ar}H$), 128.8 ($C_{Ar}H$), 129.1 ($C_{Ar}H$), 129.5 ($C_{Ar}H$), 129.9 ($C_{Ar}H$), 130.1 ($C_{Ar}H$), 130.2 ($C_{Ar}H$), 130.3 ($C_{Ar}H$), 130.4 ($C_{Ar}H$), 131.4 ($C_{Ar}H$), 132.6 (d, $^3J_{CP}$ = 10.0 Hz, $C_{Ar}H$), 132.6 (dd, $^3J_{CP}$ = 10.0 Hz, $^2J_{CPt}$ = 19.3 Hz, $C_{Ar}H$), 132.8 (C_{Ar}), 133.2 (d, $^3J_{CP}$ = 9.2 Hz, $C_{Ar}H$), 133.2 (dd, $^3J_{CP}$ = 9.2 Hz, $^2J_{CPt}$ = 22.3 Hz, $C_{Ar}H$), 133.3 (d, $^3J_{CP}$ = 7.7 Hz, $C_{Ar}H$), 133.3 (dd, $^3J_{CP}$ = 7.7 Hz, $^2J_{CPt}$ = 13.8 Hz, $C_{Ar}H$), 133.9 (d, $^3J_{CP}$ = 9.2 Hz, $C_{Ar}H$), 133.9 (dd, $^3J_{CP}$ = 9.2 Hz, $^2J_{CPt}$ = 23.1 Hz, $C_{Ar}H$), 134.5 ($C_{Ar}H$), 134.6 ($C_{Ar}H$), 136.9 - 137.9 (m, $C_{Ar}P$), 138.7 - 139.5 (m, $C_{Ar}P$), 139.2 (C_{Ar}), 146.0 (d, $^3J_{CP}$ = 5.3 Hz, $C_{Ar}CH$), 146.0

(dd, $^3J_{CP}$ = 5.3 Hz, $^2J_{CPt}$ = 33.1 Hz, C_{Ar}CH), 150.7 (C_{Ar}O), 171.5 (d, $^3J_{CP}$ = 2.3 Hz, CHCOO), 171.5 (dd, $^3J_{CP}$ = 2.3 Hz, $^2J_{CPt}$ = 46.9 Hz, CHCOO), 195.5 ($C_{Ar}$$COC_{Ar}$).

^{31}P-NMR (161.83 MHz; C_6D_6): δ = 20.5 (d, $^2J_{PP}$ = 40.6 Hz), 20.5 (dd, $^2J_{PP}$ = 40.6 Hz, $^1J_{PPt}$ = 4012.9 Hz), 20.9 (d, $^2J_{PP}$ = 40.6 Hz), 20.9 (dd, $^2J_{PP}$ = 40.6 Hz, $^1J_{PPt}$ = 3356.9 Hz).

^{195}Pt-NMR (85.48 MHz; C_6D_6): δ = -5126.0 (dd, $^1J_{PPt}$ = 3358.1 Hz, $^1J_{PPt}$ = 4014.0 Hz).

FAB-MS: m/z (%): 718.3 (100) [Pt(PPh$_2$(CH$_2$)$_4$PPhC$_6$H$_4)^+$], (6.58) [Pt(PPh$_2$(CH$_2$)$_4$PPhC$_6$H$_4$)C$_{22}$H$_{16}$O$_3^+$].

ESI-MS: m/z (%): 661.1 (100) [Pt(PPh$_2$(CH$_2$)$_4$PPhC$_6$H$_4$)CH$_3$CN$^+$].

Bis(triphenylphosphine)-(η^2-2-benzoylphenyl 4-nitrocinnamate)platinum(0)

PtPhNO$_2$-bp: $C_{58}H_{45}NO_5P_2Pt$, M = 1093.01 g/mol, yellow solid, yield: 85%, **route**: *Pt-III*.

Single crystals suitable for X-ray analysis were obtained by slow diffusion of pentane into a toluene solution.

^{1}H-NMR (399.78 MHz; C_7D_8): δ = 3.33 (dpt, $^3J_{HH}$ = 8.9 Hz, $^3J_{HP}$ = 4.4 Hz, $^3J_{HP}$ = 8.9 Hz, 1H, CHCHCO), 3.33 (ddpt, $^3J_{HH}$ = 8.9 Hz, $^3J_{HP}$ = 4.4 Hz, $^3J_{HP}$ = 8.9 Hz, $^2J_{HPt}$ = 57.4 Hz, 1H, CHCHCO), 3.69 (dpt, $^3J_{HH}$ = 8.9 Hz, $^3J_{HP}$ = 4.8 Hz, $^3J_{HP}$ = 8.9 Hz, 1H $C_{Ar}C$HCH), 3.69 (ddpt, $^3J_{HH}$ = 8.9 Hz, $^3J_{HP}$ = 4.8 Hz, $^3J_{HP}$ = 8.9 Hz, $^2J_{HPt}$ = 50.2 Hz, 1H $C_{Ar}C$HCH), 5.96 (d, $^3J_{HP}$ = 7.3 Hz, 2H, H_{Ar}), 6.64 (d, $^3J_{HP}$ = 8.2 Hz, 1H, H_{Ar}), 6.77 - 7.08 (m, 31H, H_{Ar}), 7.36 - 7.45 (m, 5H, H_{Ar}), 7.55 (d, $^3J_{HP}$ = 8.7 Hz, 2H, H_{Ar}), 7.82 (d, $^3J_{HP}$ = 7.9 Hz, 2H, H_{Ar}).

^{31}P-NMR (161.83 MHz; C$_7$D$_8$): δ = 26.5 (d, $^2J_{PP}$ = 32.0 Hz), 26.5 (dd, $^2J_{PP}$ = 32.0 Hz, $^1J_{PPt}$ = 4071.8 Hz), 27.7 (d, $^2J_{PP}$ = 32.0 Hz), 27.7 (dd, $^2J_{PP}$ = 32.0 Hz, $^1J_{PPt}$ = 3703.3 Hz).

FAB-MS: m/z (%): 719.1 (100) [Pt(PPh$_3$)(PPh$_3$)$^+$], 1092.2 (3.74) [C$_{22}$H$_{15}$NO$_5$Pt(PPh$_3$)(PPh$_3$)$^+$].

elemental analysis for C$_{58}$H$_{45}$NO$_5$P$_2$Pt (1093.01 g/mol):

calcd.: C 63.73, H 4.15, N 1.28, O 7.32, P 5.67, Pt 17.85

found: C 65.55, H 4.19, N 1.43

Di(η^2-bicyclo[2.2.1]hept-2-ene)-(η^2-2-benzoylphenyl ethyl fumarate)platinum(0)

Pt(nbe)fu-bp: C$_{33}$H$_{36}$O$_5$Pt, M = 707.71 g/mol, off-white solid, yield: 91%, **route**: *Pt-II*.

^{1}H-NMR (399.78 MHz; C$_6$D$_6$): δ = -0.06 (d, $^2J_{HH}$ = 9.1 Hz, 2H, CHC*H$_2$*CH), 0.22 - 0.30 (m, 2H, CHC*H$_2$*CH), 0.94 (t, $^3J_{HH}$ = 7.1 Hz, 3H, C*H$_3$*), 1.02 (d, $^3J_{HH}$ = 9.3 Hz, 4H, CHC*H$_2$*C*H$_2$*CH), 1.34 - 1.46 (m, 4H, CHC*H$_2$*C*H$_2$*CH), 2.68 (s, 2H, C*H*CH$_2$CH$_2$C*H*), 2.92 (s, 2H, C*H*CH$_2$CH$_2$C*H*), 3.64 (s, 2H, CHC*H*C*H*CH), 3.64 (d, $^2J_{HPt}$ = 60.6 Hz, 2H, CHC*H*C*H*CH), 3.83 (s, 2H, CHC*H*C*H*CH), 3.83 (d, $^2J_{HPt}$ = 66.2 Hz, 2H, CHC*H*C*H*CH), 3.96 (q, $^3J_{HH}$ = 7.1 Hz, 2H, CH$_3$C*H$_2$*), 4.09 (d, $^3J_{HH}$ = 9.5 Hz, 1H, CHC*H*CO), 4.09 (dd, $^3J_{HH}$ = 9.5 Hz, $^2J_{HPt}$ = 60.4 Hz, 1H, CHC*H*CO), 4.16 (d, $^3J_{HH}$ = 9.5 Hz, 1H, C$_{Ar}$C*H*CH), 4.16 (dd, $^3J_{HH}$ = 9.5 Hz, $^2J_{HPt}$ = 61.4 Hz, 1H, C$_{Ar}$C*H*CH), 6.77 (t, $^3J_{HH}$ = 7.5 Hz, 1H, *H$_{Ar}$*), 6.98 - 7.08 (m, 3H, *H$_{Ar}$*), 7.09 - 7.14 (m, 2H, *H$_{Ar}$*), 7.22 (dd, $^3J_{HH}$ = 1.6 Hz, $^3J_{HH}$ = 7.6 Hz, 1H, *H$_{Ar}$*), 7.84 (d, $^3J_{HH}$ = 7.1 Hz, 2H, *H$_{Ar}$*).

^{13}C-NMR (100.53 MHz; C$_6$D$_6$): δ = 15.0 (CH$_3$), 27.7 (CHCH$_2$CH), 27.7 (d, $^2J_{CPt}$ = 42.3 Hz, CHCH$_2$CH), 40.7 (CHCH$_2$CH), 40.7 (d, $^3J_{CPt}$ = 36.9 Hz, CHCH$_2$CH), 43.0 (CHCH$_2$CH$_2$CH), 44.0 (CHCH$_2$CH$_2$CH), 49.1 (COCHCH), 49.1 (d, $^1J_{CPt}$ = 183.0 Hz, COCHCH), 50.5 (COCHCH), 50.5 (d, $^1J_{CPt}$ = 186.1 Hz, COCHCH), 60.0 (CH$_3$CH$_2$), 80.9 (bs, CHCHCHCH), 123.7 (C$_{Ar}$H), 125.5 (C$_{Ar}$H), 128.8 (C$_{Ar}$H), 130.2 (C$_{Ar}$H), 130.5 (C$_{Ar}$), 130.6 (C$_{Ar}$H), 131.7 (C$_{Ar}$H), 133.2 (C$_{Ar}$H), 138.4 (C$_{Ar}$), 149.7 (C$_{Ar}$O), 169.5 (CHCOO), 169.5 (d, $^2J_{CPt}$ = 43.0 Hz, CHCOO), 170.9 (CHCOO), 170.9 (d, $^2J_{CPt}$ = 40.7 Hz, CHCOO), 194.4 (C$_{Ar}$COC$_{Ar}$).

^{195}Pt-NMR (85.48 MHz; C$_6$D$_6$): δ = -6189.1.

FAB-MS: m/z (%): 518.7 (4.87) [C19H$_{15}$O$_5$Pt$^+$].

ESI-MS: m/z (%): 866.0 (80) [(C19H$_{16}$O$_5$)$_2$PtNa$^+$], 1060.8 (100) [(C19H$_{16}$O$_5$)$_2$Pt$_2$Na$^+$], 1154.8 (66) [(C19H$_{16}$O$_5$)$_2$Pt$_2$C$_7$H$_{10}$Na$^+$].

η^4-Divinyl tetramethyl siloxane-(η^2-2-benzoylphenyl ethyl fumarate)platinum(0)

Pt(dvtmsi)fu-bp: C$_{27}$H$_{34}$O$_6$PtSi$_2$, M = 705.80 g/mol.

10.0 mg (24.0 μmol) Pt(cod)$_2$ and 7.90 mg (24.0 μmol) 2-benzoylphenyl ethyl fumarate were dissolved in C$_6$D$_6$ and 4.50 mg (24.0 μmol) divinyltetramethylsiloxane was added after 15 min. After 1 h the solvent was removed in vacuo, the residue dissolved in pentane and crystallised at -40 °C to yield η^4-divinyltetramethylsiloxane-(η^2-2-benzoylphenyl ethyl fumarate)platinum(0) as white solid.

^{1}H-NMR (399.78 MHz; C$_6$D$_6$): δ = -0.39 (s, 3H, CH$_3$Si), -0.37 (s, 3H, CH$_3$Si), 0.30

(s, 3H, CH_3Si), 0.32 (s, 3H, CH_3Si), 0.91 (t, $^3J_{HH}$ = 7.1 Hz, 3H, CH_3CH$_2$), 3.25 - 3.55 (m, 4H, CH_2CH), 3.86 (dq, $^3J_{HH}$ = 7.1 Hz, $^2J_{HH}$ = 16.1 Hz, 2H, CH$_3$CH_2), 4.04 (d, $^3J_{HH}$ = 15.5 Hz, 1H, COCHCH), 4.04 (dd, $^3J_{HH}$ = 15.5 Hz, $^2J_{HPt}$ = 55.4 Hz, 1H, COCHCH), 4.28 (dd, $^4J_{HH}$ = 1.7 Hz, $^3J_{HH}$ = 15.5 Hz, 1H, COCHCH), 4.28 (ddd, $^4J_{HH}$ = 1.7 Hz, $^3J_{HH}$ = 15.5 Hz, $^2J_{HPt}$ = 54.2 Hz, 1H, COCHCH), 4.37 (d, $^3J_{HH}$ = 9.7 Hz, 1H, CH$_2$CH), 4.37 (dd, $^3J_{HH}$ = 9.7 Hz, $^2J_{HPt}$ = 37.8 Hz, 1H, CH$_2$CH), 4.53 (d, $^3J_{HH}$ = 9.7 Hz, 1H, CH$_2$CH), 4.53 (dd, $^3J_{HH}$ = 9.7 Hz, $^2J_{HPt}$ = 61.6 Hz, 1H, CH$_2$CH), 6.79 (t, $^3J_{HH}$ = 7.5 Hz, 1H, H_{Ar}), 6.95 - 7.14 (m, 6H, H_{Ar}), 7.22 (dd, $^3J_{HH}$ = 1.5 Hz, $^3J_{HH}$ = 7.6 Hz, 1H, H_{Ar}), 7.77 (d, $^3J_{HH}$ = 7.2 Hz, 1H, H_{Ar}).

^{13}C-NMR (100.53 MHz; C$_6$D$_6$): δ = -2.6 (CH_3Si), -2.2 (CH_3Si), 1.1 (CH_3Si), 1.2 (CH_3Si), 14.4 (CH_3CH$_2$), 49.5 (COCHCH), 49.5 (d, $^1J_{CPt}$ = 178.3 Hz, COCHCH), 50.8 (COCHCH), 50.8 (d, $^1J_{CPt}$ = 176.8 Hz, COCHCH), 60.3 (CH$_3$$CH_2$), 69.1 (CH$_2$$C$H), 69.1 (d, $^1J_{CPt}$ = 115.3 Hz, CH$_2$$C$H), 69.8 (CH$_2$$C$H), 69.8 (d, $^1J_{CPt}$ = 109.2 Hz, CH$_2$$C$H), 70.4 ($CH_2$CH), 70.4 (d, $^1J_{CPt}$ = 83.8 Hz, CH$_2$CH), 73.2 (CH$_2$CH), 73.2 (d, $^1J_{CPt}$ = 81.5 Hz, CH$_2$CH), 123.4 (C_{Ar}H), 125.5 (C_{Ar}H), 128.5 (C_{Ar}H), 130.0 (C_{Ar}H), 130.1 (C_{Ar}H), 131.4 (C_{Ar}H), 132.9 (C_{Ar}H), 133.0 (C_{Ar}), 138.0 (C_{Ar}), 149.0 (C_{Ar}O), 168.0 (CHCOO), 168.8 (CHCOO), 194.2 ($C_{Ar}$$COC_{Ar}$).

^{29}Si-NMR (79.43 MHz; C$_6$D$_6$): δ = 3.5 (s), 3.7 (s).

^{195}Pt-NMR (85.48 MHz; C$_6$D$_6$): δ = -6044.4.

Bis(triphenylphosphine)-(η^2-2-benzoylphenyl ethyl fumarate)nickel(0)

Nifu-bp: C$_{55}$H$_{46}$NiO$_5$P$_2$, M = 907.59 g/mol, orange solid, NMR scale, **route**: Ni.

^{1}H-NMR (399.78 MHz; C$_6$D$_6$): δ = 0.75 (t, $^3J_{HH}$ = 7.1 Hz, 3H, CH_3), 3.16 - 3.24 (m, 1H, COCHCH), 3.32 - 3.41 (m, 2H, COCHCH, CH$_3$CH_2), 3.84 (dq, $^3J_{HH}$ = 7.1 Hz, $^2J_{HH}$ = 11.0 Hz, 1H, CH$_3$CH_2), 6.76 (d, $^3J_{HH}$ = 8.2 Hz, 1H, H_{Ar}), 6.82 - 6.97 (m, 19H, H_{Ar}), 6.98 - 7.13 (m, 4H, H_{Ar}), 7.34 - 7.54 (m, 13H, H_{Ar}), 7.77 (d, $^3J_{HH}$ = 7.5 Hz, 2H, H_{Ar}).

^{13}C-NMR (100.53 MHz; C$_6$D$_6$): δ = 14.4 (CH_3), 49.8(dd, $^2J_{CP}$ = 1.3 Hz, $^2J_{CP}$ = 14.8 Hz, COCHCH), 50.9 (dd, $^2J_{CP}$ = 1.4 Hz, $^2J_{CP}$ = 15.9 Hz, COCHCH), 58.8 (CH$_3$$CH_2$), 123.9 ($C_{Ar}$H), 124.7 ($C_{Ar}$H), 128.3 ($C_{Ar}$H), 128.4 ($C_{Ar}$H), 128.5 ($C_{Ar}$H), 128.5 ($C_{Ar}$H), 129.5 (d, $^1J_{CP}$ = 1.8 Hz, C_{Ar}H), 129.5 (d, $^1J_{CP}$ = 1.8 Hz, C_{Ar}H), 129.7 (C_{Ar}H), 130.2 (C_{Ar}H), 131.0 (C_{Ar}H), 132.4 (C_{Ar}H), 132.9 (C_{Ar}), 134.2 (d, $^2J_{CP}$ = 4.2 Hz, C_{Ar}H), 134.3 (d, $^2J_{CP}$ = 4.4 Hz, C_{Ar}H), 134.8 (d, $^1J_{CP}$ = 3.0 Hz, C_{Ar}P), 135.0 (d, $^1J_{CP}$ = 3.1 Hz, C_{Ar}P), 135.2 (d, $^1J_{CP}$ = 3.0 Hz, C_{Ar}P), 135.3 (d, $^1J_{CP}$ = 3.1 Hz, C_{Ar}P), 138.7 (C_{Ar}), 150.0 (C_{Ar}O), 170.4 (dd, $^3J_{CP}$ = 1.8 Hz, $^3J_{CP}$ = 3.9 Hz, CHCOO), 173.0 (dd, $^3J_{CP}$ = 1.9 Hz, $^3J_{CP}$ = 3.8 Hz, CHCOO), 195.0 ($C_{Ar}$$COC_{Ar}$).

^{31}P-NMR (161.83 MHz; C$_6$D$_6$): δ = 33.4 (d, $^2J_{PP}$ = 22.3 Hz), 34.0 (d, $^2J_{PP}$ = 22.3 Hz).

FAB-MS: m/z (%): 324.8 (8.61) [C$_{19}$H$_{16}$O$_5$], 581.7 (7.77) [Ni(PPh$_3$)(PPh$_2$C$_6$H$_4$)$^+$], 725.6 (2.72) [C$_6$H$_6$O$_4$(PPh$_3$)(PPh$_2$C$_6$H$_4$)$^+$].

Bis(triphenylphosphine)-(η^2-2-benzoylphenyl ethyl fumarate)palladium(0)

Pdfu-bp: C$_{55}$H$_{46}$O$_5$P$_2$Pd, M = 955.32 g/mol, white solid, NMR scale, **route:** Pd-I.

^{1}H-NMR (399.78 MHz; C$_6$D$_6$): δ = 0.70 (t, $^3J_{HH}$ = 7.1 Hz, 3H, CH_3), 3.30 (dq, $^3J_{HH}$ = 7.1 Hz, $^2J_{HH}$ = 10.8 Hz, 1H, CH$_3$CH_2), 3.81 (dq, $^3J_{HH}$ = 7.1 Hz, $^2J_{HH}$ = 10.8 Hz, 1H, CH$_3$CH_2), 4.11 (ddd, $^3J_{HP}$ = 4.6 Hz, $^3J_{HP}$ = 8.1 Hz, $^3J_{HH}$ = 10.0 Hz, 1H, COCHCH),

4.33 (ddd, $^3J_{HP}$ = 4.2 Hz, $^3J_{HP}$ = 7.9 Hz, $^3J_{HH}$ = 10.0 Hz, 1H, COCHCH), 6.61 (d, $^3J_{HH}$ = 7.9 Hz, 1H, H_{Ar}), 6.80 - 7.01 (m, 21H, H_{Ar}), 7.03 - 7.13 (m, 2H, H_{Ar}), 7.31 - 7.45 (m, 12H, H_{Ar}), 7.47 (dd, $^3J_{HH}$ = 1.6 Hz, $^3J_{HH}$ = 7.5 Hz, 1H, H_{Ar}), 7.83 (d, $^3J_{HH}$ = 6.8 Hz, 2H, H_{Ar}).

^{13}C-NMR (100.53 MHz; C$_6$D$_6$): δ = 14.3 (CH_3), 58.1 (dd, $^2J_{CP}$ = 4.5 Hz, $^2J_{CP}$ = 22.1 Hz, COCHCH), 58.9 (CH_3CH_2), 60.3 (dd, $^2J_{CP}$ = 4.1 Hz, $^2J_{CP}$ =23.8 Hz, COCHCH), 123.9 (C_{Ar}H), 124.7 (C_{Ar}H), 128.4 (C_{Ar}H), 128.5 (C_{Ar}H), 128.5 (C_{Ar}H), 128.6 (C_{Ar}H), 128.6 (C_{Ar}H), 129.5 (d, $^1J_{CP}$ = 1.8 Hz, C_{Ar}P), 129.6 (d, $^1J_{CP}$ = 1.7 Hz, C_{Ar}P), 129.7 (C_{Ar}H), 130.0 (C_{Ar}H), 131.0 (C_{Ar}H), 132.5 (C_{Ar}), 132.8 (C_{Ar}H), 134.2 (d, $^2J_{CP}$ = 2.5 Hz, C_{Ar}P), 134.4 (d, $^2J_{CP}$ = 2.7 Hz, C_{Ar}P), 135.6 (d, $^1J_{CP}$ = 1.7 Hz, C_{Ar}P), 135.7 (d, $^1J_{CP}$ = 1.7 Hz, C_{Ar}P), 135.9 (d, $^1J_{CP}$ = 1.8 Hz, C_{Ar}P), 136.0 (d, $^1J_{CP}$ = 1.8 Hz, C_{Ar}P), 138.8 (C_{Ar}), 149.9 (C_{Ar}O), 169.0 (dd, $^3J_{CP}$ = 2.3 Hz, $^3J_{CP}$ = 5.0 Hz, CHCOO), 171.2 (dd, $^3J_{CP}$ = 2.1 Hz, $^3J_{CP}$ = 4.9 Hz, CHCOO), 195.1 ($C_{Ar}COC_{Ar}$).

^{31}P-NMR (161.83 MHz; C$_6$D$_6$): δ = 26.7 (s), 27.2 (s).

FAB-MS: m/z (%): 629.3 (100) [Pd(PPh$_3$)(PPh$_2$C$_6$H$_4$)$^+$].

elemental analysis for C$_{55}$H$_{46}$O$_5$P$_2$Pd (955.32 *g/mol*):

calcd.: C 69.15, H 4.85, O 8.37, P 6.48, Pd 11.14

found: C 70.19, H 4.66, P 10.2

Bis(triphenylphosphine)-(η^2-2-benzoylphenyl ethyl fumarate)platinum(0)

Ptfu-bp: C$_{55}$H$_{46}$O$_5$P$_2$Pt, M = 1043.98 *g/mol*, white solid, yield: 96%, **route:** *Pt-III*.

^{1}H-NMR (399.78 MHz; C$_6$D$_6$): δ = 0.72 (t, $^3J_{HH}$ = 7.1 Hz, 3H, CH_3), 3.38 (dq, $^3J_{HH}$ = 6.9 Hz, $^2J_{HH}$ = 10.7 Hz, 1H, CH$_3$CH_2), 3.45 (dpt, $^3J_{HH}$ = 3.3 Hz,

$^3J_{HP} = 9.1$ Hz, $^3J_{HP} = 9.1$ Hz, 1H, COCHCH), 3.45 (ddpt, $^3J_{HH} = 3.3$ Hz, $^3J_{HP} = 9.1$ Hz, $^3J_{HP} = 9.1$ Hz, $^2J_{HPt} = 56.2$ Hz, 1H, COCHCH), 3.68 (dpt, $^3J_{HH} = 3.3$ Hz, $^3J_{HP} = 8.8$ Hz, $^3J_{HP} = 8.8$ Hz, 1H, COCHCH), 3.68 (ddpt, $^3J_{HH} = 3.3$ Hz, $^3J_{HP} = 8.8$ Hz, $^3J_{HP} = 8.8$ Hz, $^2J_{HPt} = 56.7$ Hz, 1H, COCHCH), 3.85 (dq, $^3J_{HH} = 6.9$ Hz, $^2J_{HH} = 10.7$ Hz, 1H, CH$_3$CH_2), 6.65 (d, $^3J_{HH} = 9.2$ Hz, 1H, H_{Ar}), 6.79 - 7.05 (m, 23H, H_{Ar}), 7.06 - 7.14 (m, 5H, H_{Ar}), 7.33 - 7.45 (m, 12H, H_{Ar}), 7.51 (dd, $^3J_{HH} = 1.5$ Hz, $^3J_{HH} = 7.6$ Hz, 1H, H_{Ar}), 7.87 (d, $^3J_{HH} = 7.9$ Hz, 2H, H_{Ar}).

^{13}C-NMR (100.53 MHz; C$_6$D$_6$): $\delta = 14.8$ (CH$_3$), 47.8 (dd, $^2J_{CP} = 4.5$ Hz, $^2J_{CP} = 30.7$ Hz, COCHCH), 47.8 (ddd, $^2J_{CP} = 4.5$ Hz, $^2J_{CP} = 30.7$ Hz, $^1J_{CPt} = 209.1$ Hz, COCHCH), 49.9 (dd, $^2J_{CP} = 4.6$ Hz, $^2J_{CP} = 32.3$ Hz, COCHCH), 49.9 (ddd, $^2J_{CP} = 4.6$ Hz, $^2J_{CP} = 32.3$ Hz, $^1J_{CPt} = 213.7$ Hz, COCHCH), 59.1 (CH$_3$$CH_2$), 124.3 ($C_{Ar}$H), 124.9 ($C_{Ar}$H), 128.5 ($C_{Ar}$H), 128.9 ($C_{Ar}$H), 130.0 ($C_{Ar}$H), 130.1 ($C_{Ar}$H), 130.1 ($C_{Ar}$H), 130.4 ($C_{Ar}$H), 131.3 ($C_{Ar}$H), 132.9 ($C_{Ar}$H), 133.2 ($C_{Ar}$), 134.4 ($C_{Ar}$H), 134.5 ($C_{Ar}$H), 134.6 ($C_{Ar}$H), 134.7 ($C_{Ar}$H), 135.7 (d, $^1J_{CP} = 1.5$ Hz, C_{Ar}P), 135.7 (dd, $^1J_{CP} = 1.5$ Hz, $^2J_{CPt} = 30.3$ Hz, C_{Ar}P), 136.1 (d, $^1J_{CP} = 2.3$ Hz, C_{Ar}P), 136.1 (dd, $^1J_{CP} = 2.3$ Hz, $^2J_{CPt} = 30.0$ Hz, C_{Ar}P), 139.1 (C_{Ar}), 150.3 (C_{Ar}O), 170.9 (dd, $^3J_{CP} = 1.5$ Hz, $^3J_{CP} = 5.3$ Hz, CHCOO), 170.9 (ddd, $^3J_{CP} = 1.5$ Hz, $^3J_{CP} = 5.3$ Hz, $^2J_{CPt} = 40.8$ Hz, CHCOO), 172.9 (dd, $^3J_{CP} = 1.5$ Hz, $^3J_{CP} = 5.4$ Hz, CHCOO), 172.9 (ddd, $^3J_{CP} = 1.5$ Hz, $^3J_{CP} = 5.4$ Hz, $^2J_{CPt} = 36.1$ Hz, CHCOO), 195.5 (C$_{Ar}$$COC_{Ar}$).

^{31}P-NMR (161.83 MHz; C$_6$D$_6$): $\delta = 27.0$ (d, $^2J_{PP} = 26.8$ Hz), 27.0 (dd, $^2J_{PP} = 26.8$ Hz, $^1J_{PPt} = 3895.0$ Hz), 27.7 (d, $^2J_{PP} = 26.8$ Hz), 27.7 (dd, $^2J_{PP} = 26.8$ Hz, $^1J_{PPt} = 3826.6$ Hz).

^{195}Pt-NMR (85.48 MHz; C$_6$D$_6$): $\delta = -5107.4$ (dd, $^1J_{PPt} = 3826.1$ Hz, $^1J_{PPt} = 3895.7$ Hz).

FAB-MS: m/z (%): 718.7 (100) [Pt(PPh$_3$)(PPh$_2$C$_6$H$_4$)$^+$], 1044.1 (1.84) [C$_{19}$H$_{16}$O$_5$Pt(PPh$_3$)$_2$$^+$].

ESI-MS: m/z (%): 718.3 (19) [Pt(PPh$_3$)(PPh$_2$C$_6$H$_4$)$^+$], 759.9 (100) [Pt(PPh$_3$)(PPh$_2$C$_6$H$_4$)CH$_3$CN$^+$], 1043.8 (4) [C$_{19}$H$_{16}$O$_5$Pt(PPh$_3$)$_2$$^+$], 1065.9 (40) [C$_{19}H_{16}O_5$Pt(PPh$_3$)$_2Na^+$], 1081.8 (17) [C$_{19}H_{16}O_5$Pt(PPh$_3$)$_2K^+$], .

elemental analysis for $C_{55}H_{46}O_5P_2Pt$ (1043.98 g/mol):

calcd.: C 63.28, H 4.44, O 7.66, P 5.93, Pt 18.69

found: C 63.8, H 4.51, P 5.83

(1,2-Bis(diphenylphosphino)ethane)-(η^2-2-benzoylphenyl ethyl fumarate)platinum(0)

Pt(dppe)fu-bp: $C_{45}H_{40}O_5P_2Pt$, M = 917.82 g/mol, white solid, yield: >99%, **route**: *Pt-V*.

^{1}H-NMR (400.13 MHz; C_6D_6): δ = 0.81 (t, $^3J_{HH}$ = 7.1 Hz, 3H, CH_3), 1.24 - 1.48 (bm, 2H, PCH_2CH_2P), 1.48 - 1.70 (bm, 2H, PCH_2CH_2P), 3.72 - 4.25 (m, 4H, CH$_3$CH_2, COCHCH), 6.65 (dd, $^3J_{HH}$ = 0.9 Hz, $^3J_{HH}$ = 8.1 Hz, 1H, H_{Ar}), 6.75 - 7.23 (m, 21H, H_{Ar}), 7.33 - 7.43 (m, 6H, H_{Ar}), 7.94 (d, $^3J_{HH}$ = 6.9 Hz, 1H, H_{Ar}).

^{31}P-NMR (161.97 MHz; C_6D_6): δ = 48.3 (d, $^2J_{PP}$ = 4.2 Hz), 48.3 (dd, $^2J_{PP}$ = 4.2 Hz, $^1J_{PPt}$ = 3435.3 Hz), 48.3 (d, $^2J_{PP}$ = 4.2 Hz), 48.3 (dd, $^2J_{PP}$ = 4.2 Hz, $^1J_{PPt}$ = 3582.6 Hz).

^{195}Pt-NMR (85.48 MHz; C_6D_6): δ = -5265.4 (dd, $^1J_{PPt}$ = 3433.9 Hz, $^1J_{PPt}$ = 3585.5 Hz).

(1,3-Bis(diphenylphosphino)propane)-(η^2-2-benzoylphenyl ethyl fumarate)platinum(0)

Pt(dppp)fu-bp: $C_{46}H_{42}O_5P_2Pt$, $M = 931.85$ g/mol, off-white solid, NMR scale, **route**: *Pt-IV*.

^{1}H-NMR (399.78 MHz; C_6D_6): $\delta = 0.78$ (t, $^3J_{HH} = 7.1$ Hz, 3H, CH_3), 1.34 - 1.54 (bm, 2H, CH$_2$CH_2CH$_2$), 1.92 - 2.09 (bm, 4H, CH_2CH$_2$CH_2), 3.47 (dq, $^3J_{HH} = 7.4$ Hz, $^2J_{HH} = 11.3$ Hz, 1H, CH$_3$CH_2), 3.64 (dpt, $^3J_{HH} = 2.4$ Hz, $^3J_{HP} = 9.1$ Hz, $^3J_{HP} = 9.1$ Hz, 1H, COCHCH), 3.64 (ddpt, $^3J_{HH} = 2.4$ Hz, $^3J_{HP} = 9.1$ Hz, $^3J_{HP} = 9.1$ Hz, $^2J_{HPt} = 56.5$ Hz, 1H, COCHCH), 3.80 (dpt, $^3J_{HH} = 2.4$ Hz, $^3J_{HP} = 9.8$ Hz, $^3J_{HP} = 9.8$ Hz, 1H, COCHCH), 3.80 (ddpt, $^3J_{HH} = 2.4$ Hz, $^3J_{HP} = 9.8$ Hz, $^3J_{HP} = 9.8$ Hz, $^2J_{HPt} = 60.5$ Hz, 1H, COCHCH), 3.87 (dq, $^3J_{HH} = 7.4$ Hz, $^2J_{HH} = 13.9$ Hz, 1H, CH$_3$CH_2), 6.47 (d, $^3J_{HH} = 8.3$ Hz, 1H, H_{Ar}), 6.80 (t, $^3J_{HH} = 7.5$ Hz, 1H, H_{Ar}), 6.85 - 7.20 (m, 18H, H_{Ar}), 7.30 - 7.42 (m, 6H, H_{Ar}), 7.47 (d, $^3J_{HH} = 7.7$ Hz, 1H, H_{Ar}), 7.72 (dd, $^3J_{HH} = 11.0$ Hz, $^3J_{HH} = 19.1$ Hz, 1H, H_{Ar}), 7.89 (d, $^3J_{HH} = 7.1$ Hz, 2H, H_{Ar}).

^{13}C-NMR (100.53 MHz; C_6D_6): $\delta = 14.8$ (CH$_3$), 20.5 (CH$_2$$CH_2CH_2$), 28.3 (t, $^1J_{CP} = 29.9$ Hz, CH$_2$CH$_2$$CH_2$), 44.1 (dd, $^2J_{CP} = 4.6$ Hz, $^2J_{CP} = 30.8$ Hz, COCHCH), 44.1 (ddd, $^2J_{CP} = 4.6$ Hz, $^2J_{CP} = 30.8$ Hz, $^1J_{CPt} = 207.6$ Hz, COCHCH), 45.7 (dd, $^2J_{CP} = 3.9$ Hz, $^2J_{CP} = 32.3$ Hz, COCHCH), 45.7 (ddd, $^2J_{CP} = 3.9$ Hz, $^2J_{CP} = 32.3$ Hz, $^1J_{CPt} = 213.8$ Hz, COCHCH) 59.0 (CH$_3$$CH_2$), 124.0 ($C_{Ar}$H), 124.7 ($C_{Ar}$H), 128.9 ($C_{Ar}$H), 128.9 ($C_{Ar}$H), 129.0 ($C_{Ar}$H), 129.1 ($C_{Ar}$), 130.2 ($C_{Ar}$H), 130.3 ($C_{Ar}$H), 130.7 ($C_{Ar}$H), 131.2 ($C_{Ar}$H), 132.9 ($C_{Ar}$H), 133.1 ($C_{Ar}$), 133.7 (t, $^2J_{CP} = 13.1$ Hz, C_{Ar}H), 134.4 (m, C_{Ar}P), 134.6 (m, C_{Ar}P), 135.0 (m, C_{Ar}P), 135.5 (m, C_{Ar}P), 135.8 (m, C_{Ar}P), 136.0 (m, C_{Ar}P),

136.2 (m, $C_{Ar}P$), 136.4 (m, $C_{Ar}P$), 139.2 (C_{Ar}), 150.4 (C_{Ar}), 171.7 (d, $^3J_{CP}$ = 4.6 Hz, CHCOO), 171.7 (dd, $^3J_{CP}$ = 4.6 Hz, $^2J_{CPt}$ = 42.3 Hz, CHCOO), 173.5 (d, $^3J_{CP}$ = 3.8 Hz, CHCOO), 173.5 (dd, $^3J_{CP}$ = 3.8 Hz, $^2J_{CPt}$ = 35.4 Hz, CHCOO), 195.5 ($C_{Ar}COC_{Ar}$).

^{31}P-NMR (161.83 MHz; C$_6$D$_6$): δ = 7.7 (d, $^2J_{PP}$ = 11.4 Hz), 7.7 (dd, $^2J_{PP}$ = 11.4 Hz, $^1J_{PPt}$ = 3477.9 Hz), 8.1 (d, $^2J_{PP}$ = 11.4 Hz), 8.1 (dd, $^2J_{PP}$ = 11.4 Hz, $^1J_{PPt}$ = 3587.8 Hz).

^{195}Pt-NMR (85.48 MHz; C$_6$D$_6$): δ = -5196.5 (dd, $^1J_{PPt}$ = 3480.1 Hz, $^1J_{PPt}$ = 3578.7 Hz).

FAB-MS: m/z (%): 606.7 (100) [Pt(PPh$_2$)(CH$_2$)$_3$(PPhC$_6$H$_4$)$^+$], 930.7 (10.26) [C$_{19}$H$_{16}$O$_5$Pt(PPh$_2$)(CH$_2$)$_3$(PPhC$_6$H$_4$)$^+$].

ESI-MS: m/z (%): 933.2 (41) [Pt(PPh$_2$)(CH$_2$)$_3$(PPh$_2$)H$^+$], 955.0 (17) [Pt(PPh$_2$)(CH$_2$)$_3$(PPh$_2$)Na$^+$], 970.1 (12) [Pt(PPh$_2$)(CH$_2$)$_3$(PPh$_2$)K$^+$], 1885.9 (22) [(Pt(PPh$_2$)(CH$_2$)$_3$(PPh$_2$))$_2$Na$^+$], 1903.7 (2) [(Pt(PPh$_2$)(CH$_2$)$_3$(PPh$_2$))$_2$K$^+$].

elemental analysis for C$_{46}$H$_{42}$O$_5$P$_2$Pt (931.85 *g/mol*):

calcd.: C 59.29, H 4.54, O 8.58, P 6.65, Pt 20.93

found: C 59.81, H 4.32, P 7.25

(1,3-Bis(di-*iso*-propylphosphino)propane)-(η^2-2-benzoylphenyl ethyl fumarate)platinum(0)

Pt(dippp)fu-bp: C$_{34}$H$_{50}$O$_5$P$_2$Pt, M = 795.78 *g/mol*, white solid, NMR scale, **route:** *Pt-V*, crystallised from pentane at -40 °C.

^{31}P-NMR (161.83 MHz; C$_6$D$_6$): δ = 28.1 (s), 28.1 (d, $^1J_{PPt}$ = 3473.7 Hz), 28.1 (s), 28.1 (d, $^1J_{PPt}$ = 3360.2 Hz).

8.3.3. Fluoren-9-one-1-yl ester complexes

Bis(triphenylphosphine)-(η^2-9-fluorenone-1-yl cinnamate)platinum(0)

PtPh-fl: $C_{58}H_{44}O_3P_2Pt$, M $= 1045.99$ g/mol, light orange solid, NMR scale, **route:** Pt-III.

^{1}H-NMR (399.78 MHz; C_6D_6): $\delta = 4.49$ (dpt, $^3J_{HH} = 8.6$ Hz, $^3J_{HP} = 4.3$ Hz, $^3J_{HP} = 8.6$ Hz, 1H, COCHCH), 4.49 (ddpt, $^3J_{HH} = 8.6$ Hz, $^3J_{HP} = 4.3$ Hz, $^3J_{HP} = 8.6$ Hz, $^2J_{HPt} = 60.6$ Hz, 1H, COCHCH), 4.74 (dpt, $^3J_{HH} = 8.6$ Hz, $^3J_{HP} = 5.5$ Hz, $^3J_{HP} = 8.6$ Hz, 1H, C_{Ar}CHCH), 4.74 (ddpt, $^3J_{HH} = 8.6$ Hz, $^3J_{HP} = 5.5$ Hz, $^3J_{HP} = 8.6$ Hz, $^2J_{HPt} = 54.2$ Hz, 1H, C_{Ar}CHCH), 6.52 (d, $^3J_{HH} = 8.2$ Hz, 1H, H_{Ar}), 6.64 (d, $^3J_{HH} = 7.5$ Hz, 2H, H_{Ar}), 6.84 - 7.01 (m, 33H, H_{Ar}), 7.39 - 7.51 (m, 6H, H_{Ar}).

^{13}C-NMR (100.53 MHz; C_6D_6): $\delta = 49.4$ (dd, $^2J_{CP} = 5.8$ Hz, $^2J_{CP} = 27.7$ Hz, COCHCH), 49.4 (ddd, $^2J_{CP} = 5.8$ Hz, $^2J_{CP} = 27.7$ Hz, $^1J_{CPt} = 203.5$ Hz, COCHCH), 61.9 (dd, $^2J_{CP} = 3.3$ Hz, $^2J_{CP} = 32.4$ Hz, $C_{Ar}$$C$HCH), 61.9 (ddd, $^2J_{CP} = 3.3$ Hz, $^2J_{CP} = 32.4$ Hz, $^1J_{CPt} = 220.0$ Hz, $C_{Ar}$$C$HCH), 116.2 ($C_{Ar}$), 120.1 ($C_{Ar}$H), 124.2 ($C_{Ar}$H), 124.7 ($C_{Ar}$H), 126.5 (d, $^3J_{CP} = 3.1$ Hz, C_{Ar}H), 128.0 (C_{Ar}H), 128.2 (C_{Ar}H), 128.4 (C_{Ar}H), 128.6 (C_{Ar}H), 129.4 (d, $^3J_{CP} = 2.0$ Hz, C_{Ar}H), 129.6 (d, $^3J_{CP} = 1.9$ Hz, C_{Ar}H), 134.2 (d, $^2J_{CP} = 16.3$ Hz, C_{Ar}H), 134.5 (d, $^2J_{CP} = 13.2$ Hz, C_{Ar}H), 135.5 (dd, $^1J_{CP} = 43.0$ Hz, $^3J_{CP} = 3.3$ Hz, C_{Ar}), 137.2 (dd, $^1J_{CP} = 40.0$ Hz, $^3J_{CP} = 2.8$ Hz, C_{Ar}), 139.9 (C_{Ar}), 140.1 (C_{Ar}), 143.9 (C_{Ar}), 145.7 (C_{Ar}), 147.2 (C_{Ar}), 150.6 (C_{Ar}O), 144.6 (dd, $^3J_{CP} = 1.3$ Hz, $^3J_{CP} = 6.5$ Hz, C_{Ar}CH), 144.6 (ddd, $^3J_{CP} = 1.3$ Hz, $^3J_{CP} = 6.5$ Hz, $^2J_{CPt} = 36.8$ Hz, C_{Ar}CH), 170.7 (dd, $^3J_{CP} = 2.4$ Hz, $^3J_{CP} = 5.1$ Hz, CHCOO), 170.7 (ddd, $^3J_{CP} = 2.4$ Hz, $^3J_{CP} = 5.1$ Hz, $^2J_{CPt} = 41.4$ Hz, CHCOO), 190.1 ($C_{Ar}$$COC_{Ar}$).

^{31}P-NMR (161.83 MHz; C_6D_6): $\delta = 28.0$ (d, $^2J_{PP} = 41.6$ Hz), 28.0 (dd, $^2J_{PP} = 41.6$ Hz, $^1J_{PPt} = 4210.1$ Hz), 28.9 (d, $^2J_{PP} = 41.6$ Hz), 28.9 (dd, $^2J_{PP} = 41.6$ Hz, $^1J_{PPt} = 3630.4$ Hz).

[195]Pt-NMR (85.48 MHz; C_6D_6): $\delta = -5047.6$ (dd, $^1J_{PPt} = 3630.8$ Hz, $^1J_{PPt} = 4212.6$ Hz).

FAB-MS: m/z (%): 718.4 (100) $[Pt(PPh_3)(PPh_2C_6H_4)^+]$, 1044.1 (1.01) $[C_{22}H_{14}O_3Pt(PPh_3)(PPh_2C_6H_4)^+]$.

ESI-MS: m/z (%): 759.1 (100) $[Pt(PPh_3)(PPh_2C_6H_4)CH_3CN^+]$.

Bis(triphenylphosphine)-(η^2-9-fluorenone-1-yl ethyl fumarate)platinum(0)

Ptfu-fl: $C_{55}H_{44}O_5P_2Pt$, M = 1041.96 *g/mol*, light orange solid, NMR scale, **route**: *Pt-III*.

^{1}H-NMR (399.78 MHz; C_6D_6): $\delta = 0.71$ (t, $^3J_{HH} = 7.0$ Hz, 3H, CH_3), 3.46 (dq, $^3J_{HH} = 7.0$ Hz, $^2J_{HH} = 14.2$ Hz, 1H, CH_3CH_2), 3.91 (dq, $^3J_{HH} = 7.0$ Hz, $^2J_{HH} = 14.2$ Hz, 1H, CH_3CH_2), 4.43 (dpt, $^3J_{HH} = 3.3$ Hz, $^3J_{HP} = 9.1$ Hz, $^3J_{HP} = 9.1$ Hz, 1H, COCHCH), 4.43 (ddpt, $^3J_{HP} = 9.1$ Hz, $^3J_{HP} = 9.1$ Hz, $^2J_{HPt} = 55.2$ Hz, 1H, COCHCH), 4.58 (dpt, $^3J_{HH} = 3.2$ Hz, $^3J_{HP} = 9.2$ Hz, $^3J_{HP} = 9.2$ Hz, 1H, COCHCH), 4.58 (ddpt, $^3J_{HH} = 3.2$ Hz, $^3J_{HP} = 9.2$ Hz, $^3J_{HP} = 9.2$ Hz, $^2J_{HPt} = 57.2$ Hz, 1H, COCHCH), 6.43 (d, $^3J_{HH} = 8.2$ Hz, 1H, H_{Ar}), 6.71 - 6.79 (m, 2H, H_{Ar}), 6.81 - 6.87 (m, 1H, H_{Ar}), 6.87 - 6.99 (m, 19H, H_{Ar}), 7.01 - 7.06 (m, 1H, H_{Ar}), 7.43 (d, $^3J_{HH} = 7.3$ Hz, 1H, H_{Ar}), 7.49 - 7.63 (m, 12H, H_{Ar}).

^{13}C-NMR (100.53 MHz; C_6D_6): $\delta = 14.3$ (CH_3), 48.4 (dd, $^2J_{CP} = 5.0$ Hz, $^2J_{CP} = 30.7$ Hz, COCHCH), 48.4 (ddd, $^2J_{CP} = 5.0$ Hz, $^2J_{CP} = 30.7$ Hz, $^1J_{CPt} = 209.2$ Hz, COCHCH), 50.2 (dd, $^2J_{CP} = 4.7$ Hz, $^2J_{CP} = 32.4$ Hz, COCHCH), 50.2 (ddd,

$^2J_{CP}$ = 4.7 Hz, $^2J_{CP}$ = 32.4 Hz, $^1J_{CPt}$ = 213.8 Hz, COCHCH), 59.0 (CH$_3$$CH_2$), 116.3 ($C_{Ar}$H), 120.1 ($C_{Ar}$H), 124.2 ($C_{Ar}$H), 124.4 ($C_{Ar}$H), 125.1 ($C_{Ar}$), 128.2 ($C_{Ar}$H), 128.3 ($C_{Ar}$H), 128.4 ($C_{Ar}$H), 129.1 ($C_{Ar}$H), 129.5 - 130.0 (m, C_{Ar}H), 133.7 (C_{Ar}H), 134.2 - 134.7 (C_{Ar}H), 134.8 (C_{Ar}H), 134.9 (C_{Ar}), 135.5 (d, $^1J_{CP}$ = 2.6 Hz, C_{Ar}P), 135.5 (dd, $^1J_{CP}$ = 2.6 Hz, $^2J_{CPt}$ = 31.4 Hz, C_{Ar}P), 135.6 (d, $^1J_{CP}$ = 2.6 Hz, C_{Ar}P), 135.6 (dd, $^1J_{CP}$ = 2.6 Hz, $^2J_{CPt}$ = 29.6 Hz, C_{Ar}P), 136.0 (d, $^1J_{CP}$ = 2.7 Hz, C_{Ar}P), 136.0 (dd, $^1J_{CP}$ = 2.7 Hz, $^2J_{CPt}$ = 30.2 Hz, C_{Ar}P), 136.1 (d, $^1J_{CP}$ = 2.7 Hz, C_{Ar}P), 136.1 (dd, $^1J_{CP}$ = 2.7 Hz, $^2J_{CPt}$ = 29.2 Hz, C_{Ar}P), 143.9 (C_{Ar}), 145.7 (C_{Ar}), 150.2 (C_{Ar}O), 170.8 (dd, $^3J_{CP}$ = 2.1 Hz, $^3J_{CP}$ = 5.6Hz, CHCOO), 170.8 (ddd, $^3J_{CP}$ = 2.1 Hz, $^3J_{CP}$ = 5.6Hz, $^2J_{CPt}$ = 40.8 Hz, CHCOO), 173.3 (dd, $^3J_{CP}$ = 2.2 Hz, $^3J_{CP}$ = 5.1 Hz, CHCOO), 173.3 (ddd, $^3J_{CP}$ = 2.2 Hz, $^3J_{CP}$ = 5.1 Hz, $^2J_{CPt}$ = 36.1 Hz, CHCOO), 190.0 ($C_{Ar}$$COC_{Ar}$).

^{31}P-NMR (161.83 MHz; C$_6$D$_6$): δ = 27.3 (d, $^2J_{PP}$ = 27.2 Hz), 27.3 (dd, $^2J_{PP}$ = 27.2 Hz, $^1J_{PPt}$ = 3911.8 Hz), 28.0 (d, $^2J_{PP}$ = 27.2 Hz), 28.0 (dd, $^2J_{PP}$ = 27.2 Hz, $^1J_{PPt}$ = 3825.6 Hz).

^{195}Pt-NMR (85.48 MHz; C$_6$D$_6$): δ = -5111.1 (dd, $^1J_{PPt}$ = 3826.9 Hz, $^1J_{PPt}$ = 3914.2 Hz).

FAB-MS: m/z (%): 718.4 (100) [Pt(PPh$_3$)(PPh$_2$C$_6$H$_4$)$^+$].

ESI-MS: m/z (%): 760.1 (100) [Pt(PPh$_3$)(PPh$_3$)CH$_3$CN$^+$], 858.9 (50) [C$_{19}$H$_{14}$O$_5$Pt(PPh$_3$)CH$_3$CNK$^+$], 1042.0 (5) [C$_{19}$H$_{14}$O$_5$Pt(PPh$_3$)(PPh$_3$)CH$_3$H$^+$], 1064.0 (13) [C$_{19}$H$_{14}$O$_5$Pt(PPh$_3$)(PPh$_3$)CH$_3$Na$^+$], 1080.0 (5) [C$_{19}$H$_{14}$O$_5$Pt(PPh$_3$)(PPh$_3$)CH$_3$K$^+$].

8.3.4. 2-Benzoylphenyl amide complexes

Bis(triphenylphosphine)-(η^2-2-benzoylphenyl maleamide)platinum(0)

Ptma-bp: $C_{53}H_{43}NO_4P_2Pt$, M = 1014.91 g/mol, white solid, NMR scale, **route**: *Pt-III*.

^{1}H-NMR (399.78 MHz; C_6D_6): δ = 2.89 (dpt, $^3J_{HH}$ = 9.7 Hz, $^3J_{HP}$ = 2.0 Hz, $^3J_{HP}$ = 9.7 Hz, 1H, COCHCH), 2.89 (ddpt, $^3J_{HH}$ = 9.7 Hz, $^3J_{HP}$ = 2.9 Hz, $^3J_{HP}$ = 9.7 Hz, $^2J_{HPt}$ = 55.1 Hz, 1H, COCHCH), 3.59 (dpt, $^3J_{HH}$ = 9.7 Hz, $^3J_{HP}$ = 2.0 Hz, $^3J_{HP}$ = 9.7 Hz, 1H, COCHCH), 3.59 (ddpt, $^3J_{HH}$ = 9.7 Hz, $^3J_{HP}$ = 2.0 Hz, $^3J_{HP}$ = 9.7 Hz, $^2J_{HPt}$ = 52.5 Hz, 1H, COCHCH), 6.50 (t, $^3J_{HH}$ = 7.5 Hz, 1H, H_{Ar}), 6.69 - 6.87 (m, 11H, H_{Ar}), 6.88 - 7.10 (m, 12H, H_{Ar}), 7.27 - 7.39 (m, 6H, H_{Ar}), 7.46 (d, $^3J_{HH}$ = 7.2 Hz, 2H, H_{Ar}), 7.49 - 7.59 (m, 6H, H_{Ar}), 8.37 (d, $^3J_{HH}$ = 8.3 Hz, 1H, H_{Ar}), 10.53 (bs, 1H, NH), 13.82 (bs, 1H, OH).

^{13}C-NMR (100.53 MHz; C_6D_6): δ = 48.8 (dd, $^2J_{CP}$ = 5.6 Hz, $^2J_{CP}$ = 30.5 Hz, COCHCH), 48.8 (ddd, $^2J_{CP}$ = 5.6 Hz, $^2J_{CP}$ = 30.5 Hz, $^1J_{CPt}$ = 210.0 Hz, COCHCH), 51.3 (dd, $^2J_{CP}$ = 4.5 Hz, $^2J_{CP}$ = 33.4 Hz, COCHCH), 51.3 (ddd, $^2J_{CP}$ = 4.5 Hz, $^2J_{CP}$ = 33.4 Hz, $^1J_{CPt}$ = 220.1 Hz, COCHCH), 121.1 (C_{Ar}H), 122.1 (C_{Ar}H), 122.9 (C_{Ar}), 128.2 (C_{Ar}H), 128.2 (C_{Ar}H), 128.3 (C_{Ar}H), 129.6 (d, $^3J_{CP}$ = 2.0 Hz, C_{Ar}H), 129.9 (d, $^3J_{CP}$ = 2.2 Hz, C_{Ar}H), 130.0 (C_{Ar}H), 131.7 (dd, $^1J_{CP}$ = 38.4 Hz, $^3J_{CP}$ = 2.9 Hz, C_{Ar}), 131.9 (C_{Ar}H), 132.4 (C_{Ar}H), 132.5 (C_{Ar}H), 133.2 (C_{Ar}H), 133.6 (C_{Ar}H), 134.0 (d, $^2J_{CP}$ = 12.9 Hz, C_{Ar}H), 134.3 (d, $^2J_{CP}$ = 12.6 Hz, C_{Ar}H), 134.9 (dd, $^1J_{CP}$ = 46.1 Hz, $^3J_{CP}$ = 2.2 Hz, C_{Ar}), 135.4 (dd, $^1J_{CP}$ = 43.8 Hz, $^3J_{CP}$ = 1.9 Hz, C_{Ar}), 139.7 (C_{Ar}), 141.5 (C_{Ar}O), 171.2 (dd, $^3J_{CP}$ = 2.4 Hz, $^3J_{CP}$ = 5.3 Hz, CHCOO), 171.2 (ddd, $^3J_{CP}$ = 2.4 Hz, $^3J_{CP}$ = 5.3 Hz, $^2J_{CPt}$ = 40.1 Hz, CHCOO), 175.9 (dd, $^3J_{CP}$ = 2.6 Hz, $^3J_{CP}$ = 5.0 Hz, CHCOO), 175.9 (ddd, $^3J_{CP}$ = 2.6 Hz, $^3J_{CP}$ = 5.0 Hz, $^2J_{CPt}$ = 41.3 Hz, CHCOO), 198.6

$(C_{Ar}COC_{Ar})$.

^{31}P-NMR (161.83 MHz; C_6D_6): δ = 26.3 (d, $^2J_{PP}$ = 24.1 Hz), 26.3 (dd, $^2J_{PP}$ = 24.1 Hz, $^1J_{PPt}$ = 3706.9 Hz), 27.9 (d, $^2J_{PP}$ = 24.1 Hz), 27.9 (dd, $^2J_{PP}$ = 24.1 Hz, $^1J_{PPt}$ = 3883.1 Hz).

8.3.5. 2-Benzoylphenyl imide complexes

Bis(triphenylphosphine)-(η^2-N(2-benzoylphenyl) maleimide)platinum(0)

Ptmi-bp: $C_{53}H_{41}NO_3P_2Pt$, M = 996.92 g/mol, white solid, yield: >99%, **route**: *Pt-III*.

^{1}H-NMR (399.78 MHz; C_6D_6): δ = 3.47 (bs, 2H, COCHCHCO), 3.47 (bd, $^2J_{HPt}$ = 58.9 Hz, 2H, COCHCHCO), 6.79 - 7.11 (m, 24H, H_{Ar}), 7.28 - 7.37 (m, 1H, H_{Ar}), 7.40 - 7.65 (m, 12H, H_{Ar}), 6.61 (d, $^3J_{HH}$ = 7.8 Hz, 2H, H_{Ar}).

^{13}C-NMR (100.53 MHz; C_6D_6): δ = 48.6 (m, COCH), 48.6 (dm, $^1J_{CPt}$ = 223.7 Hz, COCH), 125.7 (bs, C_{Ar}H), 128.5 (C_{Ar}H), 128.5 (C_{Ar}H), 129.0 (C_{Ar}H), 129.8 (C_{Ar}H), 130.1 (C_{Ar}H), 130.9 (C_{Ar}H), 131.4 (C_{Ar}H), 132.1 (C_{Ar}H), 133.8 (C_{Ar}), 134.5 (C_{Ar}H), 134.6 (C_{Ar}H), 134.7 (C_{Ar}H), 134.8 (C_{Ar}H), 134.9 (C_{Ar}H), 135.3 (m, C_{Ar}P), 135.7 (m, C_{Ar}P), 136.7 (C_{Ar}P), 138.5 (C_{Ar}), 140.2 (bs, C_{Ar}N), 174.5 (CHCOO), 174.5 (d, $^1J_{CPt}$ = 36.9 Hz, CHCOO), 194.5 ($C_{Ar}COC_{Ar}$).

^{31}P-NMR (161.83 MHz; C_6D_6): δ = 27.8, 27.8 (d, $^1J_{PPt}$ = 3786.9 Hz).

^{195}Pt-NMR (85.48 MHz; C_6D_6): δ = -5036.37 (t, $^1J_{PPt}$ = 3770.1 Hz).

FAB-MS: m/z (%): 718.8 (100) [Pt(PPh$_3$)(PPh$_2$C$_6$H$_4$)$^+$], 997.2 (1.36) [$C_{17}H_{11}NO_3$Pt(PPh$_3$)$_2^+$].

ESI-MS: m/z (%): 759.0 (100) [Pt(PPh$_3$)(PPh$_2$C$_6$H$_4$)CH$_3$CN$^+$], 996.8 (3) [C$_{17}$H$_{11}$NO$_3$Pt(PPh$_3$)$_2$H$^+$], 1019.0 (8) [C$_{17}$H$_{11}$NO$_3$Pt(PPh$_3$)$_2$Na$^+$].

8.3.6. 2-(Diazo(phenyl)methyl)phenyl ester complexes

Bis(triphenylphosphine)-(η^2-2-(diazo(phenyl)methyl)phenyl cinnamate)platinum(0)

PtPh-bpN$_2$: C$_{58}$H$_{46}$N$_2$O$_2$P$_2$Pt, M = 1060.02 *g/mol*, light red solid, yield: 84%, **route:** *Pt-III.*

^{1}H-NMR (399.78 MHz; C$_6$D$_6$): δ = 3.91 (dpt, $^3J_{HH}$ = 8.6 Hz, $^3J_{HP}$ = 5.6 Hz, $^3J_{HP}$ = 8.6 Hz, 1H, CHC*H*CO), 3.91 (ddpt, $^3J_{HH}$ = 8.6 Hz, $^3J_{HP}$ = 5.6 Hz, $^3J_{HP}$ = 8.6 Hz, $^2J_{HPt}$ = 50.0 Hz, 1H, CHC*H*CO), 4.23 (dpt, $^3J_{HH}$ = 8.6 Hz, $^3J_{HP}$ = 5.4 Hz, $^3J_{HP}$ = 8.6 Hz, 1H, C$_{Ar}$C*H*CH), 4.23 (ddpt, $^3J_{HH}$ = 8.6 Hz, $^3J_{HP}$ = 5.4 Hz, $^3J_{HP}$ = 8.6 Hz, $^2J_{HPt}$ = 56.6 Hz, 1H, C$_{Ar}$C*H*CH), 6.22 (d, $^3J_{HH}$ = 7.0 Hz, 2H, H_{Ar}), 6.45 (d, $^3J_{HH}$ = 6.2 Hz, 2H, H_{Ar}), 6.67 - 7.24 (m, 32H, H_{Ar}), 7.33 - 7.55 (m, 6H, H_{Ar}).

^{31}P-NMR (161.83 MHz; C$_6$D$_6$): δ = 27.6 (d, $^2J_{PP}$ = 40.6 Hz), 27.6 (dd, $^2J_{PP}$ = 40.6 Hz, $^1J_{PPt}$ = 4185.3 Hz), 28.6 (d, $^2J_{PP}$ = 40.6 Hz), 28.6 (dd, $^2J_{PP}$ = 40.6 Hz, $^1J_{PPt}$ = 3634.4 Hz).

Bis(triphenylphosphine)-(η^2-2-(diazo(phenyl)methyl)phenyl 4-nitrocinnamate) palladium(0)

PdPhNO$_2$-bpN$_2$: C$_{58}$H$_{45}$N$_3$O$_4$P$_2$Pd, M = 1016.36 g/mol, red solid, yield: 98%, **route:** *Pd-I.*

^{1}H-NMR (400.13 MHz; C$_6$D$_6$): δ = 4.44 - 4.59 (m, 1H, CH*CH*CO), 4.91 - 5.04 (m, 1H, C$_{Ar}$*CH*CH), 6.20 (d, $^3J_{HP}$ = 8.5 Hz, 2H, H_{Ar}), 6.51 - 6.59 (m, 4H, H_{Ar}), 6.73 - 7.07 (m, 25H, H_{Ar}), 7.38 - 7.54 (m, 6H, H_{Ar}), 7.57 (d, $^3J_{HP}$ = 8.1 Hz, 2H, H_{Ar}).

^{31}P-NMR (161.97 MHz; C$_6$D$_6$): δ = 24.7 (d, $^2J_{PP}$ = 4.7 Hz), 26.6 (d, $^2J_{PP}$ = 4.7 Hz).

FT-IR (ATR, cm^{-1}): 2044 (m, N$_2$), 1712 (m, CO), 1584 (m), 1477 (w), 1433 (m), 1327 (s), 1168 (w), 1145 (w), 1108 (w), 1093 (m), 743 (s), 691 (s), 677 (m), 577 (m), 504 (s), 446 (w), 425 (w).

UV/vis (pentane, nm): 208 (s, Ar), 279 (m, N$_2$), 386 (w).

Bis(triphenylphosphine)-(η^2-2-(diazo(phenyl)methyl)phenyl 4-nitrocinnamate) platinum(0)

PtPhNO$_2$-bpN$_2$: C$_{58}$H$_{45}$N$_3$O$_4$P$_2$Pt, M = 1105.02 g/mol, orange solid, yield: 90%, **route:** *Pt-VI.*

^{1}H-NMR (400.13 MHz; C_6D_6): δ = 3.87 (dpt, $^3J_{HH}$ = 8.9 Hz, $^3J_{HP}$ = 3.9 Hz, $^3J_{HP}$ = 8.9 Hz, 1H, CHCHCO), 3.87 (ddpt, $^3J_{HH}$ = 8.9 Hz, $^3J_{HP}$ = 3.9 Hz, $^3J_{HP}$ = 8.9 Hz, $^2J_{HPt}$ = 51.0 Hz, 1H, CHCHCO), 4.23 (dpt, $^3J_{HH}$ = 8.9 Hz, $^3J_{HP}$ = 4.9 Hz, $^3J_{HP}$ = 8.9 Hz, 1H, C_{Ar}CHCH), 4.23 (ddpt, $^3J_{HH}$ = 8.9 Hz, $^3J_{HP}$ = 4.9 Hz, $^3J_{HP}$ = 8.9 Hz, $^2J_{HPt}$ = 50.2 Hz, 1H, C_{Ar}CHCH), 6.12 - 6.22 (m, 2H, H_{Ar}), 6.69 - 7.11 (m, 33H, H_{Ar}), 7.40 - 7.58 (m, 6H, H_{Ar}), 7.61 - 7.67 (m, 2H, H_{Ar}).

^{31}P-NMR (161.97 MHz; C_6D_6): δ = 25.8 (d, $^2J_{PP}$ = 32.4 Hz), 25.8 (dd, $^2J_{PP}$ = 32.4 Hz, $^1J_{PPt}$ = 4045.2 Hz), 27.1 (d, $^2J_{PP}$ = 32.4 Hz), 27.1 (dd, $^2J_{PP}$ = 32.4 Hz, $^1J_{PPt}$ = 3723.2 Hz).

FT-IR (ATR, cm^{-1}): 3060 (w), 2045 (m, N_2), 1702 (m, CO), 1586 (m), 1505 (m), 1479 (m), 1434 (m), 1330 (s), 1273 (w), 1180 (w), 1094 (s), 846 (w), 743 (m), 705 (s), 578 (w), 539 (w), 508 (s), 459 (w), 417 (w).

UV/vis (pentane, nm): 208 (s, Ar), 276 (m, N_2), 389 (m).

8.3.7. 9-Diazofluoren-1-yl ester complexes

Bis(triphenylphosphine)-(η^2-9-diazofluoren-1-yl 4-nitrocinnamate)platinum(0)

PtPhNO$_2$-flN$_2$: $C_{58}H_{43}N_3O_4P_2Pt$, M = 1103.01 g/mol, orange solid, yield: 91%, **route:** *Pt-VI.*

^{1}H-NMR (400.13 MHz; C_6D_6): δ = 3.94 (dpt, $^3J_{HH}$ = 9.1 Hz, $^3J_{HP}$ = 3.7 Hz, $^3J_{HP}$ = 9.1 Hz, 1H, CHCHCO), 3.94 (ddpt, $^3J_{HH}$ = 9.1 Hz, $^3J_{HP}$ = 3.7 Hz, $^3J_{HP}$ = 9.1 Hz, $^2J_{HPt}$ = 58.4 Hz, 1H, CHCHCO), 4.45 (dpt, $^3J_{HH}$ = 9.1 Hz, $^3J_{HP}$ = 4.1 Hz, $^3J_{HP}$ = 9.1 Hz, 1H, C_{Ar}CHCH), 4.45 (ddpt, $^3J_{HH}$ = 9.1 Hz, $^3J_{HP}$ = 4.1 Hz, $^3J_{HP}$ = 9.1 Hz,

$^2J_{HPt}$ = 51.8 Hz, 1H, $C_{Ar}CHCH$), 6.39 (d, $^3J_{HH}$ = 7.8 Hz, 2H, H_{Ar}), 6.68 - 7.03 (m, 24H, H_{Ar}), 7.04 - 7.17 (m, 7H, H_{Ar}), 7.48 - 7.59 (m, 6H, H_{Ar}), 7.69 (d, $^3J_{HH}$ = 8.7 Hz, 2H, H_{Ar}).

^{31}P-NMR (161.97 MHz; C_6D_6): δ = 26.1 (d, $^2J_{PP}$ = 30.7 Hz), 26.1 (dd, $^2J_{PP}$ = 30.7 Hz, $^1J_{PPt}$ = 4073.5 Hz), 26.7 (d, $^2J_{PP}$ = 30.7 Hz), 26.7 (dd, $^2J_{PP}$ = 30.7 Hz, $^1J_{PPt}$ = 3701.8 Hz).

FAB-MS: m/z (%): 719.0 (100) $[Pt(PPh_3)(PPh_3)^+]$, 1074.8 (1.01) $[C_{22}H_{13}NO_4Pt(PPh_3)(PPh_3)^+]$.

FT-IR (ATR, cm^{-1}): 3064 (w), 2965 (w), 2925 (w), 2859 (w), 2063 (s, N_2), 1713 (s, CO), 1586 (m), 1519 (m), 1434 (m), 1424 (m), 1329 (s), 1303 (w), 1263 (m), 1178 (w), 1138 (w), 1093 (s), 1018 (w), 844 (w), 802 (m), 748 (s), 693 (s), 538 (m), 518 (s), 508 (s).

8.3.8. Bis(triphenylphosphine) cinnamate complexes

Bis(triphenylphosphine)-(η^2-methyl 4-methoxycinnamate)nickel(0)

NiPhOMe-Me: $C_{47}H_{42}NiO_3P_2$, M = 775.48 g/mol, brick red solid, yield: 93%, **route:** *Ni.*

^{1}H-NMR (400.13 MHz; C_7D_8): δ = 3.20 (s, 3H, COOCH_3), 3.34 (s, 3H, OCH_3), 3.60 (dpt, $^3J_{HH}$ = 9.0 Hz, $^3J_{HP}$ = 3.7 Hz, $^3J_{HP}$ = 9.1 Hz, 1H, CHCHCO), 4.35 (dpt, $^3J_{HH}$ = 9.0 Hz, $^3J_{HP}$ = 5.1 Hz, $^3J_{HP}$ = 9.0 Hz, 1H, CC$_{Ar}$CHCH), 6.38 (d, $^3J_{HH}$ = 8.3 Hz, 2H, H_{Ar}), 6.50 (d, $^3J_{HH}$ = 8.6 Hz, 2H, H_{Ar}), 6.89 - 7.06 (m, 18H, H_{Ar}), 7.10 - 7.22 (m, 6H, H_{Ar}), 7.50 - 7.64 (m, 6H, H_{Ar}).

^{13}C-NMR (100.62 MHz; C_7D_8): δ = 49.7 (COOCH_3), 53.6 (dd, $^2J_{CP}$ = 2.0 Hz, $^2J_{CP}$ = 12.6 Hz, CHCHCO), 54.7 (OCH_3), 65.2 (d, $^2J_{CP}$ = 16.2 Hz, C_{Ar}CHCH), 114.4

(C_{Ar}H), 126.3 (d, $^4J_{CP}$ = 2.0 Hz, C_{Ar}H), 128.0 (C_{Ar}H), 128.1 (C_{Ar}H), 128.1 (C_{Ar}H), 128.2 (C_{Ar}H), 128.4 (C_{Ar}), 128.5 (C_{Ar}), 129.0 (d, $^3J_{CP}$ = 1.3 Hz, C_{Ar}H), 129.1 (d, $^3J_{CP}$ = 1.3 Hz, C_{Ar}H), 132.4 (d, $^1J_{CP}$ = 9.6 Hz, C_{Ar}), 134.2 (d, $^2J_{CP}$ = 7.6 Hz, C_{Ar}H), 134.4 (d, $^2J_{CP}$ = 8.3 Hz, C_{Ar}H), 135.1 (d, $^1J_{CP}$ = 3.0 Hz, C_{Ar}), 135.5 (d, $^1J_{CP}$ = 3.0 Hz, C_{Ar}), 136.1 (d, $^1J_{CP}$ = 5.5 Hz, C_{Ar}), 136.3 (d, $^1J_{CP}$ = 2.4 Hz, C_{Ar}), 136.6 (d, $^1J_{CP}$ = 2.3 Hz, C_{Ar}H), 157.4 (d, $^3J_{CP}$ = 1.5 Hz, C_{Ar}O), 172.5 (dd, $^3J_{CP}$ = 1.5 Hz, $^3J_{CP}$ = 3.3 Hz, CHCOO).

^{31}P-NMR (161.97 MHz; C$_7$D$_8$): δ = 31.8 (d, $^2J_{PP}$ = 36.8 Hz), 33.7 (d, $^2J_{PP}$ = 36.8 Hz).

FT-IR (ATR, cm^{-1}): 1668 (m, CO).

Bis(triphenylphosphine)-(η^2-methyl 4-methoxycinnamate)palladium(0)

PdPhOMe-Me: C$_{47}$H$_{42}$O$_3$P$_2$Pd, M = 823.20 *g/mol*, yellow solid, yield: 93%, **route**: *Pd-I*.

^{1}H-NMR (400.13 MHz; C$_7$D$_8$): δ = 3.22 (s, 3H, COOCH_3), 3.31 (s, 3H, OCH_3), 4.77 (bs, 1H, CHCHCO), 5.39 (bs, 1H, C$_{Ar}$CHCH), 6.40 - 6.49 (bm, 2H, H_{Ar}), 6.49 - 6.58 (bm, 2H, H_{Ar}), 6.86 - 7.05 (bm, 18H, H_{Ar}), 7.07 - 7.31 (bm, 6H, H_{Ar}), 7.37 - 7.60 (bm, 6H, H_{Ar}).

^{1}H-NMR (400.13 MHz; 246 K; C$_7$D$_8$): δ = 3.19 (s, 3H, COOCH_3), 3.27 (s, 3H, OCH_3), 4.86 (dpt, $^3J_{HH}$ = 10.1 Hz, $^3J_{HP}$ = 6.1 Hz, $^3J_{HP}$ = 6.1 Hz, 1H, CHCHCO), 5.40 (ddd, $^3J_{HH}$ = 10.1 Hz, $^3J_{HP}$ = 7.0 Hz, $^3J_{HP}$ = 6.1 Hz, 1H, C$_{Ar}$CHCH), 6.40 - 6.51 (m, 4H, H_{Ar}), 6.86 - 6.92 (m, 6H, H_{Ar}), 6.93 - 6.01 (m, 12H, H_{Ar}), 7.09 - 7.15 (m, 6H, H_{Ar}), 7.55 - 7.62 (m, 6H, H_{Ar}).

^{13}C-NMR (100.62 MHz; 246 K; C$_7$D$_8$): δ = 49.9 (OCH$_3$), 54.4 (OCH$_3$), 60.7 (dd, $^2J_{CP}$ = 5.4 Hz, $^2J_{CP}$ = 17.6 Hz, CHCHCO), 76.0 (dd, $^2J_{CP}$ = 1.7 Hz, $^2J_{CP}$ = 21.2 Hz, C$_{Ar}$$C$HCH), 113.6 ($C_{Ar}$H), 126.4 (bs, C_{Ar}CH), 128.1 (C_{Ar}), 128.2 (C_{Ar}), 128.4 (C_{Ar}),

128.9 (C_{Ar}), 129.2 (C_{Ar}), 134.1 (d, $^2J_{CP}$ = 14.2 Hz, C_{Ar}H), 134.3 (d, $^2J_{CP}$ = 15.1 Hz, C_{Ar}H), 134.6 (d, $^3J_{CP}$ = 6.2 Hz, C_{Ar}CH), 135.7 - 136.0 (m, C_{Ar}P), 137.6 (bs, C_{Ar}H), 157.5 (bs, C_{Ar}O), 170.9 - 171.0 (m, CHC(O)O).

^{31}P-NMR (161.97 MHz; C_7D_8): δ = 24.8 (bs), 25.9 (bs).

^{31}P-NMR (161.97 MHz; 246 K; C_7D_8): δ = 24.1 (d, $^2J_{PP}$ = 21.7 Hz), 25.8 (d, $^2J_{PP}$ = 21.7 Hz).

FT-IR (ATR, cm^{-1}): 3057 (m), 2935 (m), 1678 (s, CO), 1605 (w), 1515 (m), 1482 (m), 1432 (s), 1275 (m), 1244 (w), 1153 (s), 1091 (w), 1030 (w), 816 (m), 742 (s), 696 (s), 528 (w), 512 (m), 500 (s), 407 (m).

elemental analysis for $C_{47}H_{42}O_3P_2Pd$ (823.20 g/mol):

calcd.: C 68.57, H 5.24, O 5.83, P 7.53, Pd 12.93

found: C 68.60, H 5.01

Bis(triphenylphosphine)-(η^2-methyl 4-methoxycinnamate)platinum(0)

PtPhOMe-Me: $C_{47}H_{42}O_3P_2Pt$, M = 911.86 g/mol, white solid, yield: 94%, **route:** *Pt-VI.*

^{1}H-NMR (400.13 MHz; C_6D_6): δ = 3.24 (s, 3H, COOCH_3), 3.31 (s, 3H, OCH_3), 4.06 (pdd, $^3J_{HH}$ = 8.8 Hz, $^3J_{HP}$ = 4.8 Hz, $^3J_{HP}$ = 8.8 Hz, 1H, CHCHCO), 4.06 (ddpt, $^3J_{HH}$ = 8.8 Hz, $^3J_{HP}$ = 4.8 Hz, $^3J_{HP}$ = 8.8 Hz, $^2J_{HPt}$ = 61.7 Hz, 1H, CHCHCO), 4.55 (pdd, $^3J_{HH}$ = 8.8 Hz, $^3J_{HP}$ = 6.1 Hz, $^3J_{HP}$ = 8.8 Hz, 1H, C_{Ar}CHCH), 4.55 (ddpt, $^3J_{HH}$ = 8.8 Hz, $^3J_{HP}$ = 6.1 Hz, $^3J_{HP}$ = 8.8 Hz, $^2J_{HPt}$ = 53.5 Hz, 1H, CC$_{Ar}$CHCH), 6.46 - 6.60 (m, 4H, H_{Ar}), 6.82 - 7.02 (m, 18H, H_{Ar}), 7.17 - 7.28 (m, 6H, H_{Ar}), 7.52 - 7.65 (m, 6H, H_{Ar}).

^{13}C-NMR (100.62 MHz; C$_6$D$_6$): δ = 50.1 (COOCH$_3$), 50.2 (dd, $^2J_{CP}$ = 6.6 Hz, $^2J_{CP}$ = 28.6 Hz, CHCHCO), 50.2 (ddd, $^2J_{CP}$ = 6.6 Hz, $^2J_{CP}$ = 28.6 Hz, $^1J_{CPt}$ = 192.2 Hz, CHCHCO), 54.8 (OCH$_3$), 60.9 (dd, $^2J_{CP}$ = 5.2 Hz, $^2J_{CP}$ = 33.0 Hz, $C_{Ar}$$C$HCH), 60.9 (ddd, $^2J_{CP}$ = 5.2 Hz, $^2J_{CP}$ = 33.0 Hz, $^1J_{CPt}$ = 197.4 Hz, $C_{Ar}$$C$HCH), 113.5 ($C_{Ar}$H), 127.1 (bs, C_{Ar}H), 127.1 (dd, $^5J_{CP}$ = 3.3 Hz, $^4J_{CPt}$ = 14.3 Hz, C_{Ar}H), 128.0 (C_{Ar}H), 128.0 (C_{Ar}H), 128.2 (C_{Ar}H), 129.3 (C_{Ar}), 129.5 (C_{Ar}), 134.0 - 134.6 (m, C_{Ar}H), 135.4 - 136.6 (m, C_{Ar}H), 137.0 - 137.8 (m, C_{Ar}H), 157.4 (bs, C_{Ar}O, C_{Ar}CH), 157.4 (bd, $^2J_{CPt}$ = 9.0 Hz, C_{Ar}O, C_{Ar}CH), 173.5 (bs, CHCOO), 173.5 (dm, $^2J_{CPt}$ = 47.8 Hz, CHCOO).

^{31}P-NMR (161.97 MHz; C$_6$D$_6$): δ = 28.4 (dd, $^2J_{PP}$ = 45.6 Hz, $^1J_{PPt}$ = 3613.6 Hz), 28.5 (s), 28.6 (dd, $^2J_{PP}$ = 45.6 Hz, $^1J_{PPt}$ = 4198.5 Hz).

^{195}Pt-NMR (85.48 MHz; C$_6$D$_6$): δ = -5058.1 (dd, $^1J_{PPt}$ = 3613.5 Hz, $^1J_{PPt}$ = 4199.4 Hz).

FAB-MS: m/z (%): 719.3 (100) [Pt(PPh$_3$)$_2$$^+$], 911.4 (1.01) [(C$_{11}H_{12}O_3$)Pt(PPh$_3$)$_2$$^+$].

FT-IR (ATR, cm^{-1}): 3061 (w), 2939 (w), 1682 (s, CO), 1591 (w), 1509 (m), 1433 (s), 1268 (m), 1243 (m), 1150 (s), 1092 (m), 1015 (m), 924 (w), 822 (m), 742 (s), 692 (s), 540 (m), 516 (s), 500 (s), 417 (m).

elemental analysis for C$_{47}$H$_{42}$O$_3$P$_2$Pt (911.86 *g/mol*):

calcd.: C 61.91, H 4.64, O 5.26, P 6.79, Pt 21.39

found: C 61.17, H 4.48

Bis(triphenylphosphine)-(η^2-ethyl 4-methoxycinnamate)nickel(0)

NiPhOMe-Et: C$_{48}$H$_{44}$NiO$_3$P$_2$, M = 789.50 *g/mol*, orange solid, yield: 84%, **route**: *Ni*.

^{1}H-NMR (400.13 MHz; C$_7$D$_8$): δ = 0.88 (t, $^3J_{HH}$ = 7.1 Hz, 3H, CH$_2$CH_3), 3.34 (s, 3H, OCH_3), 3.50 - 3.69 (m, 2H, CHCHCO, CH_2CH$_3$), 4.00 (dq, $^2J_{HH}$ = 10.7 Hz,

$^3J_{HH}$ = 7.1 Hz, 1H, CH$_2$CH$_3$), 4.36 (dpt, $^3J_{HH}$ = 9.0 Hz, $^3J_{HP}$ = 5.0 Hz, $^3J_{HP}$ = 9.0 Hz, 1H, C$_{Ar}$CHCH), 6.38 (d, $^3J_{HH}$ = 8.3 Hz, 2H, H_{Ar}), 6.50 (d, $^3J_{HH}$ = 8.6 Hz, 2H, H_{Ar}), 6.89 - 7.08 (m, 18H, H_{Ar}), 7.08 - 7.24 (m, 6H, H_{Ar}), 7.51 - 7.63 (m, 6H, H_{Ar}).

^{13}C-NMR (100.62 MHz; C$_7$D$_8$): δ = 14.4 (CH$_2$$CH_3$), 53.8 (dd, $^2J_{CP}$ = 2.0 Hz, $^2J_{CP}$ = 12.5 Hz, CHCHCO), 54.7 (OCH$_3$), 58.6 (CH$_2$CH$_3$), 65.4 (d, $^2J_{CP}$ = 16.0 Hz, C$_{Ar}$$C$HCH), 114.4 ($C_{Ar}$H), 126.3 (d, $^4J_{CP}$ = 2.1 Hz, C_{Ar}H), 128.0 (C_{Ar}H), 128.1 (C_{Ar}H), 128.4 (C_{Ar}), 128.5 (C_{Ar}), 128.9 (d, $^3J_{CP}$ = 1.3 Hz, C_{Ar}H), 129.1 (d, $^3J_{CP}$ = 1.2 Hz, C_{Ar}H), 132.4 (d, $^1J_{CP}$ = 9.5 Hz, C_{Ar}), 134.2 (d, $^2J_{CP}$ = 12.9 Hz, C_{Ar}H), 134.4 (d, $^2J_{CP}$ = 13.6 Hz, C_{Ar}H), 135.2 (d, $^1J_{CP}$ = 3.0 Hz, C_{Ar}), 135.5 (d, $^1J_{CP}$ = 3.1 Hz, C_{Ar}), 136.1 (d, $^1J_{CP}$ = 5.5 Hz, C_{Ar}), 136.4 (d, $^1J_{CP}$ = 2.4 Hz, C_{Ar}), 136.7 (d, $^1J_{CP}$ = 2.4 Hz, C_{Ar}H), 157.4 (d, $^3J_{CP}$ = 1.5 Hz, C_{Ar}O), 172.3 (dd, $^3J_{CP}$ = 1.4 Hz, $^3J_{CP}$ = 3.3 Hz, CHCOO).

^{31}P-NMR (161.97 MHz; C$_7$D$_8$): δ = 31.7 (d, $^2J_{PP}$ = 37.5 Hz), 33.6 (d, $^2J_{PP}$ = 37.5 Hz).

FT-IR (ATR, cm^{-1}): 1666 (m, CO).

Bis(triphenylphosphine)-(η^2-ethyl 4-methoxycinnamate)palladium(0)

PdPhOMe-Et: C$_{48}$H$_{44}$O$_3$P$_2$Pd, M = 837.23 g/mol, yellow solid, yield: 78%, **route:** Pd-I.

^{1}H-NMR (400.13 MHz; C$_7$D$_8$): δ = 0.88 (t, $^3J_{HH}$ = 6.9 Hz, 3H, CH$_2$CH$_3$), 3.30 (s, 3H, OCH$_3$), 3.66 (bs, 1H, CH$_2$CH$_3$), 3.99 (bs, 1H, CH$_2$CH$_3$), 4.78 (bs, 1H, CHCHCO), 5.37 (bs, 1H, C$_{Ar}$CHCH), 6.38 - 6.47 (bm, 2H, H_{Ar}), 6.47 - 6.61 (bm, 2H, H_{Ar}), 6.84 - 7.05 (bm, 18H, H_{Ar}), 7.05 - 7.27 (bm, 6H, H_{Ar}), 7.33 - 7.63 (bm, 6H, H_{Ar}).

^{1}H-NMR (400.13 MHz; 246 K; C$_7$D$_8$): δ = 0.83 (t, $^3J_{HH}$ = 7.1 Hz, 3H, CH$_2$CH$_3$), 3.26 (s, 3H, OCH$_3$), 3.58 (dq, $^2J_{HH}$ = 7.1 Hz, $^3J_{HH}$ = 10.7 Hz, 1H, CH$_2$CH$_3$), 4.00 (dq,

$^2J_{HH}$ = 7.1 Hz, $^3J_{HH}$ = 10.7 Hz, 1H, CH_2CH$_3$), 4.90 (dpt, $^3J_{HH}$ = 10.4 Hz, $^3J_{HP}$ = 6.1 Hz, $^3J_{HP}$ = 6.1 Hz, 1H, CHCHCO), 5.40 (dpt, $^3J_{HH}$ = 10.4 Hz, $^3J_{HP}$ = 7.0 Hz, $^3J_{HP}$ = 7.0 Hz, 1H, C$_{Ar}$CHCH), 6.36 - 6.52 (m, 4H, H_{Ar}), 6.87 - 6.93 (m, 6H, H_{Ar}), 6.93 - 7.00 (m, 12H, H_{Ar}), 7.07 - 7.14 (m, 6H, H_{Ar}), 7.57 - 7.63 (m, 6H, H_{Ar}).

^{13}C-NMR (100.62 MHz; C$_7$D$_8$): δ = 14.4 (CH$_2$CH_3), 54.7 (OCH_3), 59.1 (bs, CH_2CH$_3$), 114.0 (C_{Ar}H), 128.2 (C_{Ar}H), 128.2 (C_{Ar}H), 128.2 (C_{Ar}H), 128.3 (C_{Ar}H), 129.2 (C_{Ar}H), 134.2 (bs, C_{Ar}H), 134.4 (bs, C_{Ar}H), 136.1 - 137.0 (m, C_{Ar}H).

^{13}C-NMR (100.62 MHz; 246 K; C$_7$D$_8$): δ = 14.3 (CH$_2$CH_3), 54.4 (OCH_3), 58.8 (CH_2CH$_3$), 60.9 (dd, $^2J_{CP}$ = 5.2 Hz, $^2J_{CP}$ = 16.6 Hz, CHCHCO), 76.3 (dd, $^2J_{CP}$ = 2.1 Hz, $^2J_{CP}$ = 21.8 Hz, C$_{Ar}$CHCH), 113.6 (C_{Ar}H), 126.4 (d, $^3J_{CP}$ = 2.2 Hz, C_{Ar}CH), 128.1 (C_{Ar}), 128.3 (C_{Ar}), 128.4 (C_{Ar}), 128.9 (C_{Ar}), 129.1 (C_{Ar}), 129.2 (C_{Ar}), 129.8 (C_{Ar}H), 134.1 ($^2J_{CP}$ = 14.5 Hz, C_{Ar}H), 134.3 ($^2J_{CP}$ = 15.5 Hz, C_{Ar}H), 134.7 (d, $^3J_{CP}$ = 6.3 Hz, C_{Ar}CH), 135.9 (dd, $^1J_{CP}$ = 1.8 Hz, $^3J_{CP}$ = 27.5 Hz, C_{Ar}P), 137.6 (dd, $^1J_{CP}$ = 1.6 Hz, $^3J_{CP}$ = 26.9 Hz, C_{Ar}P), 157.5 - 157.6 (m, C_{Ar}O), 170.7 (dd, $^3J_{CP}$ = 2.2 Hz, $^3J_{CP}$ = 4.0 Hz, CHCOO).

^{31}P-NMR (161.97 MHz; C$_7$D$_8$): δ = 24.5 (bs), 25.7 (bs).

^{31}P-NMR (161.97 MHz; 246 K; C$_7$D$_8$): δ = 23.9 (d, $^2J_{PP}$ = 22.6 Hz), 25.6 (d, $^2J_{PP}$ = 22.6 Hz).

FT-IR (ATR, cm^{-1}): 3037 (w), 1673 (s, CO), 1604 (w), 1512 (m), 1476 (m), 1432 (m), 1237 (s), 1219 (m), 1168 (w), 1092 (w), 1036 (m), 816 (m), 740 (s), 692 (s), 500 (s), 503 (m), 494 (m), 428 (m).

elemental analysis for C$_{48}$H$_{44}$O$_3$P$_2$Pd (837.23 *g/mol*):

calcd.: C 68.86, H 5.30, O 5.73, P 7.40, Pd 12.71

found: C 68.83, H 5.13

Bis(triphenylphosphine)-(η^2-ethyl 4-methoxycinnamate)platinum(0)

PtPhOMe-Et: $C_{48}H_{44}O_3P_2Pt$, M = 925.89 g/mol, white solid, yield: 97%, **route**: Pt-VI.

^{1}H-NMR (400.13 MHz; C_6D_6): δ = 0.84 (t, $^3J_{HH}$ = 7.1 Hz, 3H, CH_2CH_3), 3.31 (s, 3H, OCH_3), 3.68 (dq, $^2J_{HH}$ = 10.7 Hz, $^3J_{HH}$ = 7.1 Hz, 1H, CH_2CH_3), 4.05 (dq, $^2J_{HH}$ = 10.7 Hz, $^3J_{HH}$ = 7.1 Hz, 1H, CH_2CH_3), 4.10 (dpt, $^3J_{HH}$ = 8.8 Hz, $^3J_{HP}$ = 4.8 Hz, $^3J_{HP}$ = 8.8 Hz, 1H, CHCHCO), 4.10 (ddpt, $^3J_{HH}$ = 8.8 Hz, $^3J_{HP}$ = 4.8 Hz, $^3J_{HP}$ = 8.8 Hz, $^2J_{HPt}$ = 62.0 Hz, 1H, CHCHCO), 4.54 (dpt, $^3J_{HH}$ = 8.8 Hz, $^3J_{HP}$ = 6.1 Hz, $^3J_{HP}$ = 8.8 Hz, 1H, $C_{Ar}CH$CH), 4.54 (ddpt, $^3J_{HH}$ = 8.8 Hz, $^3J_{HP}$ = 6.1 Hz, $^3J_{HP}$ = 8.8 Hz, $^2J_{HPt}$ = 52.7 Hz, 1H, $C_{Ar}CH$CH), 6.45 - 6.57 (m, 4H, H_{Ar}), 6.83 - 7.04 (m, 18H, H_{Ar}), 7.17 - 7.28 (m, 6H, H_{Ar}), 7.54 - 7.68 (m, 6H, H_{Ar}).

^{13}C-NMR (100.62 MHz; C_6D_6): δ = 14.5 (CH_2CH_3), 50.4 (bd, $^2J_{CP}$ = 28.1 Hz, CHCHCO), 50.4 (dbd, $^2J_{CP}$ = 28.1 Hz, $^1J_{CPt}$ = 194.7 Hz, CHCHCO), 54.8 (OCH_3), 58.8 (CH_2CH_3), 60.8 (bd, $^2J_{CP}$ = 64.4 Hz, $C_{Ar}CH$CH), 60.8 (dbd, $^2J_{CP}$ = 64.4 Hz, $^1J_{CPt}$ = 194.8 Hz, $C_{Ar}CH$CH), 113.5 ($C_{Ar}H$), 127.1 (d, $^5J_{CP}$ = 2.2 Hz, $C_{Ar}H$), 127.1 (dd, $^5J_{CP}$ = 2.2 Hz, $^4J_{CPt}$ = 12.4 Hz, $C_{Ar}H$), 127.9 ($C_{Ar}H$), 128.2 ($C_{Ar}H$), 129.3 (C_{Ar}), 129.5 (C_{Ar}), 134.0 - 134.6 (m, $C_{Ar}H$), 135.5 - 136.5 (m, $C_{Ar}H$), 136.8 - 138.1 (m, $C_{Ar}H$), 157.4 (d, $^3J_{CP}$ = 1.6 Hz, $C_{Ar}CH$), 157.4 (bm, $C_{Ar}O$), 173.2 (bd, $^3J_{CP}$ 3.0 Hz, CHCOO), 173.2 (bd, $^2J_{CPt}$ = 47.2 Hz, CHCOO).

^{31}P-NMR (161.97 MHz; C_6D_6): δ = 28.1 (d, $^2J_{PP}$ = 47.6 Hz), 28.1 (dd, $^2J_{PP}$ = 47.6 Hz, $^1J_{PPt}$ = 4171.6 Hz), 28.7 (d, $^2J_{PP}$ = 47.6 Hz), 28.7 (dd, $^2J_{PP}$ = 47.6 Hz, $^1J_{PPt}$ = 3654.2 Hz).

^{195}Pt-NMR (85.48 MHz; C_6D_6): δ = -5053.8 (dd, $^1J_{PPt}$ = 3631.5 Hz, $^1J_{PPt}$ = 4188.7 Hz).

FAB-MS: m/z (%): 718.9 (100) [Pt(PPh$_3$)$_2^+$], 925.0 (1.32) [(C$_{12}$H$_{14}$O$_3$)Pt(PPh$_3$)$_2^+$].

FT-IR (ATR, cm^{-1}): 3068 (w), 2977 (w), 2894 (w), 1674 (s, CO), 1509 (m), 1484 (w), 1433 (m), 1246 (m), 1232 (m), 1158 (w), 1094 (m), 1035 (m), 832 (w), 741 (m), 692 (s), 538 (w), 529 (s), 506 (s), 423 (m).

elemental analysis for C$_{48}$H$_{44}$O$_3$P$_2$Pt (925.89 *g/mol*):

calcd.: C 62.27, H 4.79, O 5.18, P 6.69, Pt 21.07

found: C 62.15, H 4.78

Bis(triphenylphosphine)-(η^2-*iso*-propyl 4-methoxycinnamate)nickel(0)

NiPhOMe-^{i}Pr: C$_{49}$H$_{46}$NiO$_3$P$_2$, M = 803.53 *g/mol*, orange red solid, yield: 94%, **route:** *Ni*.

^{1}H-NMR (400.13 MHz; C$_7$D$_8$): δ = 0.73 (d, $^3J_{HH}$ = 6.2 Hz, 3H, CHCH_3), 1.21 (d, $^3J_{HH}$ = 6.2 Hz, 3H, CHCH_3), 3.34 (s, 3H, OCH_3), 3.55 (dpt, $^3J_{HH}$ = 9.0 Hz, $^3J_{HP}$ = 3.5 Hz, $^3J_{HP}$ = 9.0 Hz, 1H, CHCHCO), 4.37 (dpt, $^3J_{HH}$ = 9.0 Hz, $^3J_{HP}$ = 5.0 Hz, $^3J_{HP}$ = 9.0 Hz, 1H, C$_{Ar}$CHCH), 4.85 (hp, $^3J_{HH}$ = 6.2 Hz, 1H, CH(CH$_3$)$_2$), 6.39 (d, $^3J_{HH}$ = 8.3 Hz, 2H, H_{Ar}), 6.50 (d, $^3J_{HH}$ = 8.6 Hz, 2H, H_{Ar}), 6.89 - 7.03 (m, 18H, H_{Ar}), 7.06 - 7.22 (m, 6H, H_{Ar}), 7.46 - 7.71 (m, 6H, H_{Ar}).

^{13}C-NMR (100.62 MHz; C$_7$D$_8$): δ = 21.7 (CHCH_3), 22.6 (CHCH_3), 54.0 (dd, $^2J_{CP}$ = 2.0 Hz, $^2J_{CP}$ = 12.8 Hz, CHCHCO), 54.7 (OCH$_3$), 65.7 (OCH(CH$_3$)$_2$), 65.8 (d, $^2J_{CP}$ = 16.1 Hz, C$_{Ar}$$C$HCH), 114.4 ($C_{Ar}$H), 126.2 (d, $^4J_{CP}$ = 1.8 Hz, C_{Ar}H), 128.0 (C_{Ar}H), 128.1 (C_{Ar}H), 128.4 (C_{Ar}), 128.5 (C_{Ar}), 128.9 (d, $^3J_{CP}$ = 1.2 Hz, C_{Ar}H), 129.2 (d, $^3J_{CP}$ = 0.9 Hz, C_{Ar}H), 132.4 (d, $^1J_{CP}$ = 9.5 Hz, C_{Ar}), 134.2 (d, $^2J_{CP}$ = 12.6 Hz, C_{Ar}H), 134.5 (d, $^2J_{CP}$ = 13.4 Hz, C_{Ar}H), 135.2 (d, $^1J_{CP}$ = 3.0 Hz, C_{Ar}), 135.5 (d, $^1J_{CP}$ = 3.1 Hz, C_{Ar}), 136.1 (d, $^1J_{CP}$ = 5.5 Hz, C_{Ar}), 136.5 (d, $^1J_{CP}$ = 2.5 Hz, C_{Ar}), 136.8

(d, $^1J_{CP}$ = 2.5 Hz, C_{Ar}H), 157.4 (d, $^3J_{CP}$ = 1.3 Hz, C_{Ar}O), 172.2 (dd, $^3J_{CP}$ = 1.1 Hz, $^3J_{CP}$ = 3.2 Hz, CHCOO).

^{31}P-NMR (161.97 MHz; C$_7$D$_8$): δ = 31.5 (d, $^2J_{PP}$ = 37.1 Hz), 33.5 (d, $^2J_{PP}$ = 37.1 Hz).

FT-IR (ATR, cm^{-1}): 1664 (m, CO).

elemental analysis for C$_{49}$H$_{46}$NiO$_3$P$_2$ (803.53 g/mol):

calcd.: C 73.24, H 5.77, Ni 7.30 , O 5.97, P 7.71

found: C 70.64, H 5.86

Bis(triphenylphosphine)-(η^2-iso-propyl 4-methoxycinnamate)palladium(0)

PdPhOMe-^{i}Pr: C$_{49}$H$_{46}$O$_3$P$_2$Pd, M = 851.26 g/mol, light yellow solid, yield: 86%, route: Pd-I.

^{1}H-NMR (400.13 MHz; C$_7$D$_8$): δ = 0.79 (bs, 3H, CHCH_3), 1.15 (bs, 3H, CHCH_3), 3.30 (s, 3H, OCH_3), 4.79 (bs, 1H, CHCHCO), 4.92 (bs, 1H, CH(CH$_3$)$_2$), 5.35 (bs, 1H, C$_{Ar}$CHCH), 6.35 - 6.47 (bm, 2H, H_{Ar}), 6.47 - 6.62 (bm, 2H, H_{Ar}), 6.84 - 7.05 (bm, 18H, H_{Ar}), 7.05 - 7.27 (bm, 6H, H_{Ar}), 7.36 - 7.67 (bm, 6H, H_{Ar}).

^{1}H-NMR (400.13 MHz; 246 K; C$_7$D$_8$): δ = 0.70 (d, $^3J_{HH}$ = 6.1 Hz, 3H, CHCH_3), 1.20 (d, $^3J_{HH}$ = 6.1 Hz, 3H, CHCH_3), 3.26 (s, 3H, OCH_3), 4.85 - 4.97 (m, 2H, CH(CH$_3$)$_2$, CHCHCO), 5.39 (dpt, $^3J_{HH}$ = 10.1 Hz, $^3J_{HP}$ = 7.0 Hz, $^3J_{HP}$ = 7.0 Hz, 1H, C$_{Ar}$CHCH), 6.35 - 6.49 (m, 4H, H_{Ar}), 6.85 - 7.01 (m, 18H, H_{Ar}), 7.05 - 7.12 (m, 6H, H_{Ar}), 7.58 - 7.64 (m, 6H, H_{Ar}).

^{13}C-NMR (100.62 MHz; C$_7$D$_8$): δ = 21.5 - 22.6 (bm, CHCH$_3$), 54.7 (OCH$_3$), 61.7 (bs, CHCHCO), 66.1 (bs, OCH(CH$_3$)$_2$), 77.1 (bs, C$_{Ar}$$C$HCH), 114.0 ($C_{Ar}$H), 128.2 ($C_{Ar}$H), 128.2 ($C_{Ar}$H), 128.2 ($C_{Ar}$H), 128.3 - 138.4 (m, C_{Ar}H), 129.2 (C_{Ar}H), 134.0 - 134.7 (m, C_{Ar}H), 139.7 (bs, C_{Ar}CH).

^{13}C-NMR (100.62 MHz; 246 K; C_7D_8): δ = 21.4 (CHCH_3), 22.5 (CHCH_3), 54.4 (OCH_3), 61.1 (dd, $^2J_{CP}$ = 5.9 Hz, $^2J_{CP}$ = 17.3 Hz, CHCHCO), 65.8 (OCH(CH$_3$)$_2$), 76.9 (dd, $^2J_{CP}$ = 2.7 Hz, $^2J_{CP}$ = 21.6 Hz, $C_{Ar}CH$CH), 113.7 (C_{Ar}H), 126.4 (d, $^3J_{CP}$ = 1.9 Hz, C_{Ar}H), 128.1 (C_{Ar}), 128.3 (C_{Ar}), 128.3 (C_{Ar}), 128.9 (C_{Ar}), 129.2 (C_{Ar}), 129.8 (C_{Ar}H), 134.1 (d, $^2J_{CP}$ = 14.4 Hz, C_{Ar}H), 134.5 (d, $^2J_{CP}$ = 15.6 Hz, C_{Ar}H), 134.8 (d, $^3J_{CP}$ = 6.7 Hz, C_{Ar}CH), 135.9 (dd, $^1J_{CP}$ = 27.3 Hz, $^3J_{CP}$ = 1.9 Hz, C_{Ar}P), 137.7 (dd, $^1J_{CP}$ = 26.2 Hz, $^3J_{CP}$ = 2.0 Hz, C_{Ar}P), 157.5 - 157.6 (m, C_{Ar}O), 170.4 (dd, $^3J_{CP}$ = 2.3 Hz, $^3J_{CP}$ = 3.8 Hz, HCOO).

^{31}P-NMR (161.97 MHz; C_7D_8): δ = 23.4 (bs), 24.7 (bs).

^{31}P-NMR (161.97 MHz; 246 K; C_7D_8): δ = 23.5 (d, $^2J_{PP}$ = 23.7 Hz), 25.3 (d, $^2J_{PP}$ = 23.7 Hz).

FT-IR (ATR, cm^{-1}): 3063 (m), 2972 (m), 2926 (m), 2835 (w), 1669 (s, CO), 1599 (m), 1512 (m), 1477 (m), 1433 (s), 1271 (m), 1244 (m), 1154 (s), 1107 (m), 1091 (m), 1035 (w), 827 (m), 739 (s), 690 (s), 524 (m), 512 (m), 498 (s), 408 (s).

elemental analysis for $C_{49}H_{46}O_3P_2Pd$ (851.26 g/mol):

calcd.: C 69.14, H 5.45, O 5.64, P 7.28, Pd 12.50

found: C 69.13, H 5.33

Bis(triphenylphosphine)-(η^2-*iso*-propyl 4-methoxycinnamate)platinum(0)

PtPhOMe-^{i}Pr: $C_{49}H_{46}O_3P_2Pt$, M = 939.91 g/mol, white solid, yield: 99%, **route:** *Pt-VI.*

^{1}H-NMR (400.13 MHz; C_6D_6): δ = 0.75 (d, $^3J_{HH}$ = 6.3 Hz, 3H, CHCH_3), 1.13 (d, $^3J_{HH}$ = 6.2 Hz, 3H, CHCH_3), 3.30 (s, 3H, OCH_3), 4.14 (dpt, $^3J_{HH}$ = 8.3 Hz, $^3J_{HP}$ = 4.3 Hz, $^3J_{HP}$ = 8.3 Hz, 1H, CHCHCO), 4.14 (ddpt, $^3J_{HH}$ = 8.3 Hz, $^3J_{HP}$ = 4.3 Hz,

$^3J_{HP} = 8.3$ Hz, $^2J_{HPt} = 62.9$ Hz, 1H, CHCHCO), 4.52 (dpt, $^3J_{HH} = 8.3$ Hz, $^3J_{HP} = 5.4$ Hz,

$^3J_{HP} = 8.3$ Hz, 1H, C$_{Ar}$CHCH), 4.52 (ddpt, $^3J_{HH} = 8.3$ Hz, $^3J_{HP} = 5.4$ Hz, $^3J_{HP} = 8.3$ Hz,

$^2J_{HPt} = 51.9$ Hz, 1H, C$_{Ar}$CHCH), 4.99 (hp, $^3J_{HH} = 6.3$ Hz, 1H, CH(CH$_3$)$_2$), 6.38 - 6.63

(m, 4H, H_{Ar}), 6.80 - 7.07 (m, 18H, H_{Ar}), 7.16 - 7.23 (m, 6H, H_{Ar}), 7.52 - 7.72 (m, 6H,

H_{Ar}).

^{13}C-NMR (100.62 MHz; C$_6$D$_6$): δ = 21.8 (CHCH_3), 22.6 (CHCH_3), 50.7 (dd,

$^2J_{CP} = 5.0$ Hz, $^2J_{CP} = 27.1$ Hz, CHCHCO), 50.7 (ddd, $^2J_{CP} = 5.0$ Hz, $^2J_{CP} = 27.1$ Hz,

$^1J_{CPt} = 200.9$ Hz, CHCHCO), 54.8 (OCH$_3$), 61.8 (dd, $^2J_{CP} = 2.7$ Hz, $^2J_{CP} = 30.9$ Hz,

C$_{Ar}$$C$HCH), 61.8 (ddd, $^2J_{CP} = 2.7$ Hz, $^2J_{CP} = 30.9$ Hz, $^1J_{CPt} = 194.5$ Hz, C$_{Ar}$$C$HCH),

65.7 (OCH(CH$_3$)$_2$), 113.5 (d, $^5J_{CP} = 1.7$ Hz, C_{Ar}H), 127.0 (d, $^5J_{CP} = 3.0$ Hz, C_{Ar}H), 127.0

(dd, $^5J_{CP} = 3.0$ Hz, $^4J_{CPt} = 13.6$ Hz, C_{Ar}H), 127.9 (C_{Ar}H), 128.0 (C_{Ar}H), 128.2 (C_{Ar}H),

128.2 (C_{Ar}H), 129.2 (d, $^1J_{CP} = 1.6$ Hz, C_{Ar}), 129.5 (d, $^1J_{CP} = 1.4$ Hz, C_{Ar}), 134.2 (C_{Ar}H),

134.2 (d, $^3J_{CPt} = 21.8$ Hz, C_{Ar}H), 134.3 (C_{Ar}H), 134.3 (d, $^3J_{CPt} = 24.0$ Hz, C_{Ar}H), 134.5

(C_{Ar}H), 134.5 (d, $^3J_{CPt} = 18.6$ Hz, C_{Ar}H), 134.6 (C_{Ar}H), 134.6 (d, $^3J_{CPt} = 19.0$ Hz,

C_{Ar}H), 135.8 (d, $^1J_{CP} = 4.5$ Hz, C_{Ar}P), 135.8 (dd, $^1J_{CP} = 4.5$ Hz, $^2J_{CPt} = 33.4$ Hz,

C_{Ar}P), 136.2 (d, $^1J_{CP} = 4.5$ Hz, C_{Ar}P), 136.2 (dd, $^1J_{CP} = 4.5$ Hz, $^2J_{CPt} = 32.6$ Hz,

C_{Ar}P), 137.2 (dd, $^1J_{CP} = 6.4$ Hz, $^3J_{CP} = 1.0$ Hz, C_{Ar}P), 137.2 (ddd, $^1J_{CP} = 6.4$ Hz,

$^3J_{CP} = 1.0$ Hz, $^2J_{CPt} = 32.6$ Hz, C_{Ar}P), 137.5 (d, $^1J_{CP} = 4.3$ Hz, C_{Ar}P), 137.5 (dd,

$^1J_{CP} = 4.3$ Hz, $^2J_{CPt} = 26.1$ Hz, C_{Ar}P), 137.9 (d, $^1J_{CP} = 4.3$ Hz, C_{Ar}P), 137.9 (dd,

$^1J_{CP} = 4.3$ Hz, $^2J_{CPt} = 28.5$ Hz, C_{Ar}P), 157.4 (d, $^3J_{CP} = 2.3$ Hz, C_{Ar}O), 157.4 (bm,

C_{Ar}CH), 172.9 (dd, $^3J_{CP}$ 2.1 Hz, $^3J_{CP} = 4.5$ Hz, CHCOO), 172.9 (ddd, $^3J_{CP} = 2.1$ Hz,

$^3J_{CP} = 4.5$ Hz, $^2J_{CPt} = 46.7$ Hz, CHCOO).

^{31}P-NMR (161.97 MHz; C$_6$D$_6$): δ = 28.3 (d, $^2J_{PP} = 46.6$ Hz), 28.3 (dd, $^2J_{PP} = 46.6$ Hz,

$^1J_{PPt} = 4188.5$ Hz), 28.7 (d, $^2J_{PP} = 46.6$ Hz), 28.7 (dd, $^2J_{PP} = 3631.4$ Hz,

$^1J_{PPt} = 46.6$ Hz).

^{195}Pt-NMR (85.48 MHz; C$_6$D$_6$): δ = -5048.7 (dd, $^1J_{PPt} = 3654.1$ Hz,

$^1J_{PPt} = 4171.7$ Hz).

FAB-MS: m/z (%): 719.3 (100) [Pt(PPh$_3$)$_2$$^+$], 939.3 (1.39) [(C$_{13}H_{16}O_3$)Pt(PPh$_3$)$_2$$^+$].

FT-IR (ATR, cm^{-1}): 3054 (w), 2979 (w), 1676 (s, CO), 1510 (m), 1434 (m), 1301 (w),

1263 (m), 1240 (w), 1153 (s), 1114 (m), 1093 (s), 1036 (w), 829 (m), 740 (m), 687 (s), 535 (m), 516 (s), 498 (s), 422 (m).

elemental analysis for $C_{49}H_{46}O_3P_2Pt$ (939.91 g/mol):

calcd.: C 62.61, H 4.93, O 5.11, P 6.59, Pt 20.75

found: C 62.40, H 4.63

Bis(triphenylphosphine)-(η^2-*tert*-butyl 4-methoxycinnamate)nickel(0)

NiPhOMe-^{t}Bu: $C_{50}H_{48}NiO_3P_2$, M = 817.56 g/mol, brick red solid, yield: 70%, **route**: *Ni*.

^{1}H-NMR (400.13 MHz; C_7D_8): $\delta = 1.32$ (s, 9H, C(CH_3)$_3$), 3.34 (s, 3H, OCH_3), 3.52 (dpt, $^3J_{HH} = 9.2$ Hz, $^3J_{HP} = 3.4$ Hz, $^3J_{HP} = 9.2$ Hz, 1H, CHCHCO), 4.35 (dpt, $^3J_{HH} = 9.2$ Hz, $^3J_{HP} = 5.0$ Hz, $^3J_{HP} = 9.2$ Hz, 1H, C_{Ar}CHCH), 6.44 (d, $^3J_{HH} = 8.5$ Hz, 2H, H_{Ar}), 6.48 (d, $^3J_{HH} = 8.7$ Hz, 2H, H_{Ar}), 6.91 - 7.05 (m, 18H, H_{Ar}), 7.05 - 7.18 (m, 6H, H_{Ar}), 7.49 - 7.62 (m, 6H, H_{Ar}).

^{13}C-NMR (100.62 MHz; C_7D_8): $\delta = 28.5$ (CCH$_3$), 54.7 (OCH$_3$), 55.0 (dd, $^2J_{CP} = 2.0$ Hz, $^2J_{CP} = 13.2$ Hz, CHCHCO), 66.2 (d, $^2J_{CP} = 16.1$ Hz, $C_{Ar}C$HCH), 77.4 (OC(CH$_3$)$_3$), 114.5 (C_{Ar}H), 126.1 (d, $^4J_{CP} = 1.9$ Hz, C_{Ar}H), 128.1 (C_{Ar}H), 128.4 (C_{Ar}), 128.5 (C_{Ar}), 132.4 (d, $^1J_{CP} = 9.5$ Hz, C_{Ar}), 134.3 (d, $^2J_{CP} = 12.5$ Hz, C_{Ar}H), 134.5 (d, $^2J_{CP} = 13.5$ Hz, C_{Ar}H), 135.2 (d, $^1J_{CP} = 3.1$ Hz, C_{Ar}), 135.5 (d, $^1J_{CP} = 3.1$ Hz, C_{Ar}), 136.4 (d, $^1J_{CP} = 5.6$ Hz, C_{Ar}), 136.4 (d, $^1J_{CP} = 2.6$ Hz, C_{Ar}), 136.8 (d, $^1J_{CP} = 2.7$ Hz, C_{Ar}H), 157.4 (d, $^3J_{CP} = 1.3$ Hz, C_{Ar}O), 172.7 (dd, $^3J_{CP} = 1.3$ Hz, $^3J_{CP} = 3.2$ Hz, CHCOO).

^{31}P-NMR (161.97 MHz; C_7D_8): $\delta = 30.8$ (d, $^2J_{PP} = 39.4$ Hz), 32.8 (d, $^2J_{PP} = 39.4$ Hz).

FT-IR (ATR, cm^{-1}): 1668 (m, CO).

elemental analysis for $C_{50}H_{48}NiO_3P_2$ (817.56 g/mol):

calcd.: C 73.45, H 5.92, Ni 7.18 , O 5.87, P 7.58

found: C 71.93, H 5.88

Bis(triphenylphosphine)-(η^2-*tert*-butyl 4-methoxycinnamate)palladium(0)

PdPhOMe-^{t}Bu: $C_{50}H_{48}O_3P_2Pd$, M = 865.28 g/mol, yellow solid, yield: 79%, **route**: *Pd-I*.

^{1}H-NMR (400.13 MHz; C_7D_8): δ = 1.33 (s, 9H, C(CH_3)$_3$), 3.29 (s, 3H, OCH_3), 4.75 (bs, 1H, CHCHCO), 5.29 (bs, 1H, C_{Ar}CHCH), 6.34 - 6.45 (bm, 2H, H_{Ar}), 6.45 - 6.62 (bm, 2H, H_{Ar}), 6.83 - 7.04 (bm, 18H, H_{Ar}), 7.04 - 7.20 (bm, 6H, H_{Ar}), 7.41 - 7.66 (bm, 6H, H_{Ar}).

^{1}H-NMR (400.13 MHz; 246 K; C_7D_8): δ = 1.32 (s, 9H, C(CH_3)$_3$), 3.25 (s, 3H, OCH_3), 4.83 (dpt, $^3J_{HH}$ = 10.2 Hz, $^3J_{HP}$ = 6.1 Hz, $^3J_{HP}$ = 6.1 Hz, 1H, CHCHCO), 5.33 (dpt, $^3J_{HH}$ = 10.2 Hz, $^3J_{HP}$ = 7.0 Hz, $^3J_{HP}$ = 7.0 Hz, 1H, C_{Ar}CHCH), 6.33 - 6.48 (m, 4H, H_{Ar}), 6.85 - 7.01 (m, 18H, H_{Ar}), 7.03 - 7.11 (m, 6H, H_{Ar}), 7.54 - 7.64 (m, 6H, H_{Ar}).

^{13}C-NMR (100.62 MHz; C_7D_8): δ = 28.4 (CC_{H3}), 54.6 (OCH$_3$), 63.2 (bs, CHCHCO), 76.7 - 78.7 (bm, $C_{Ar}$$CHCH, OC$(CH$_3$)$_3$), 114.1 ($C_{Ar}$H), 128.2 ($C_{Ar}$H), 128.2 ($C_{Ar}$H), 128.2 ($C_{Ar}$H), 129.2 ($C_{Ar}$H), 133.9 - 135.0 (m, C_{Ar}H), 139.8 (bs, C_{Ar}CH).

^{13}C-NMR (100.62 MHz; 246 K; C_7D_8): δ = 28.1 (CC_{H3}), 54.4 (OCH$_3$), 62.2 (dd, $^2J_{CP}$ = 5.4 Hz, $^2J_{CP}$ = 16.1 Hz, CHCHCO), 77.3 (dd, $^2J_{CP}$ = 3.6 Hz, $^2J_{CP}$ = 8.9 Hz, $C_{Ar}$$C$HCH), 77.6 (O$C$(CH$_3$)$_3$), 113.7 ($C_{Ar}$H), 126.2 (d, $^3J_{CP}$ = 1.6 Hz, C_{Ar}H), 128.2 (C_{Ar}), 128.3 (C_{Ar}), 129.2 (C_{Ar}), 129.2 (C_{Ar}), 129.8 (C_{Ar}), 134.1 (d, $^2J_{CP}$ = 14.0 Hz, C_{Ar}H), 134.5 (d, $^2J_{CP}$ = 15.7 Hz, C_{Ar}H), 142.5 - 142.6 (m, C_{Ar}CH), 135.9 (dd,

$^1J_{CP} = 26.8$ Hz, $^3J_{CP} = 1.5$ Hz, C_{Ar}P), 137.6 (dd, $^1J_{CP} = 26.6$ Hz, $^3J_{CP} = 1.7$ Hz, C_{Ar}P), 157.5 - 157.6 (m, C_{Ar}O), 170.7 - 170.8 (m, CHCOO).

^{31}P-NMR (161.97 MHz; C$_7$D$_8$): $\delta = 22.9$ (bs), 24.3 (bs).

^{31}P-NMR (161.97 MHz; 246 K; C$_7$D$_8$): $\delta = 23.1$ (d, $^2J_{PP} = 25.0$ Hz), 24.9 (d, $^2J_{PP} = 25.0$ Hz).

FT-IR (ATR, cm^{-1}): 3049 (m), 2973 (m), 1670 (m, CO), 1595 (w), 1514 (w), 1478 (m), 1433 (m), 1274 (w), 1246 (w), 1134 (s), 1091 (m), 1027 (w), 827 (m), 741 (s), 691 (s), 501 (s) 405 (m).

elemental analysis for C$_{50}$H$_{48}$O$_3$P$_2$Pd (865.28 g/mol):

calcd.: C 69.40, H 5.59, O 5.55, P 7.16, Pd 12.30

found: C 69.68, 5.59

Bis(triphenylphosphine)-(η^2-*tert*-butyl 4-methoxycinnamate)platinum(0)

PtPhOMe-^{t}Bu: C$_{50}$H$_{48}$O$_3$P$_2$Pt, M = 953.94 g/mol, white solid, yield: 99%, **route**: *Pt-VI.*

^{1}H-NMR (400.13 MHz; C$_6$D$_6$): $\delta = 1.32$ (s, 9H, C(CH_3)$_3$), 3.29 (s, 3H, OCH_3), 4.10 (dpt, $^3J_{HH} = 8.6$ Hz, $^3J_{HP} = 4.9$ Hz, $^3J_{HP} = 8.6$ Hz, 1H, CHCHCO), 4.10 (ddpt, $^3J_{HH} = 8.6$ Hz, $^3J_{HP} = 4.9$ Hz, $^3J_{HP} = 8.6$ Hz, $^2J_{HPt} = 63.4$ Hz, 1H, CHCHCO), 4.47 (dpt, $^3J_{HH} = 8.6$ Hz, $^3J_{HP} = 6.1$ Hz, $^3J_{HP} = 8.6$ Hz, 1H, C$_{Ar}$CHCH), 4.47 (ddpt, $^3J_{HH} = 8.6$ Hz, $^3J_{HP} = 6.1$ Hz, $^3J_{HP} = 8.6$ Hz, $^2J_{HPt} = 51.8$ Hz, 1H, C$_{Ar}$CHCH), 6.47 (s, 4H, H_{Ar}), 6.79 - 7.02 (m, 18H, H_{Ar}), 7.06 - 7.26 (m, 6H, H_{Ar}), 7.53 - 7.69 (m, 6H, H_{Ar}).

^{13}C-NMR (100.62 MHz; C$_6$D$_6$): $\delta = 28.5$ (CCH$_3$), 51.6 (dd, $^2J_{CP} = 6.2$ Hz, $^2J_{CP} = 28.6$ Hz, CHCHCO), 51.6 (ddd, $^2J_{CP} = 6.2$ Hz, $^2J_{CP} = 28.6$ Hz, $^1J_{CPt} = 205.9$ Hz,

CHCHCO), 54.8 (OCH_3), 62.6 (dd, $^2J_{CP}$ = 4.3 Hz, $^2J_{CP}$ = 32.2 Hz, $C_{Ar}CH$CH), 62.6 (ddd, $^2J_{CP}$ = 4.3 Hz, $^2J_{CP}$ = 32.2 Hz, $^1J_{CPt}$ = 196.0 Hz, $C_{Ar}CH$CH), 77.6 (OC(CH$_3$)$_3$), 113.6 (d, $^5J_{CP}$ = 1.8 Hz, C_{Ar}H), 126.9 (d, $^5J_{CP}$ = 3.0 Hz, C_{Ar}H), 126.9 (dd, $^5J_{CP}$ = 3.0 Hz, $^4J_{CPt}$ = 13.2 Hz, C_{Ar}H), 127.9 (C_{Ar}H), 128.0 (C_{Ar}H), 128.2 (C_{Ar}H), 128.2 (C_{Ar}H), 129.2 (d, $^1J_{CP}$ = 1.7 Hz, C_{Ar}), 129.5 (d, $^1J_{CP}$ = 1.5 Hz, C_{Ar}), 134.2 (C_{Ar}H), 134.2 (d, $^3J_{CPt}$ = 20.6 Hz, C_{Ar}H), 134.3 (C_{Ar}H), 134.3 (d, $^3J_{CPt}$ = 20.4 Hz, C_{Ar}H), 134.6 (C_{Ar}H), 134.6 (d, $^3J_{CPt}$ = 19.0 Hz, C_{Ar}H), 134.7 (C_{Ar}H), 134.7 (d, $^3J_{CPt}$ = 19.2 Hz, C_{Ar}H), 135.8 (d, $^1J_{CP}$ = 3.0 Hz, C_{Ar}P), 135.8 (dd, $^1J_{CP}$ = 3.0 Hz, $^2J_{CPt}$ = 30.2 Hz, C_{Ar}P), 136.2 (d, $^1J_{CP}$ = 3.1 Hz, C_{Ar}P), 136.2 (dd, $^1J_{CP}$ = 3.1 Hz, $^2J_{CPt}$ = 27.4 Hz, C_{Ar}P), 137.2 (dd, $^1J_{CP}$ = 6.7 Hz, $^3J_{CP}$ = 1.5 Hz, C_{Ar}P), 137.2 (ddd, $^1J_{CP}$ = 6.7 Hz, $^3J_{CP}$ = 1.5 Hz, $^2J_{CPt}$ = 28.8 Hz, C_{Ar}P), 137.5 (d, $^1J_{CP}$ = 2.9 Hz, C_{Ar}P), 137.5 (dd, $^1J_{CP}$ = 2.9 Hz, $^2J_{CPt}$ = 30.1 Hz, C_{Ar}P), 137.9 (d, $^1J_{CP}$ = 3.0 Hz, C_{Ar}P), 137.9 (dd, $^1J_{CP}$ = 3.0 Hz, $^2J_{CPt}$ = 28.4 Hz, C_{Ar}P), 157.4 (d, $^3J_{CP}$ = 2.4 Hz, C_{Ar}O), 157.4 (bd, C_{Ar}CH), 173.1 (dd, $^3J_{CP}$ 2.3 Hz, $^3J_{CP}$ = 4.7 Hz, CHCOO), 173.1 (dm, $^2J_{CPt}$ = 46.5 Hz, CHCOO).

^{31}P-NMR (161.97 MHz; C$_6$D$_6$): δ = 27.5 (d, $^2J_{PP}$ = 48.7 Hz), 27.5 (dd, $^2J_{PP}$ = 48.7 Hz, $^1J_{PPt}$ = 4127.4 Hz), 28.7 (d, $^2J_{PP}$ = 48.7 Hz), 27.5 (dd, $^2J_{PP}$ = 48.7 Hz, $^1J_{PPt}$ = 3699.8 Hz).

^{195}Pt-NMR (85.48 MHz; C$_6$D$_6$): δ = -5049.2 (dd, $^1J_{PPt}$ = 3682.4 Hz, $^1J_{PPt}$ = 4144.9 Hz).

FAB-MS: m/z (%): 719.3 (100) [Pt(PPh$_3$)$_2$$^+$], 954.4 (1.13) [(C$_{14}H_{18}O_3$)Pt(PPh$_3$)$_2$$^+$].

FT-IR (ATR, cm^{-1}): 3056 (w), 2965 (w), 1677 (m, CO), 1510 (m), 1434 (m), 1273 (w), 1243 (w), 1132 (s), 1092 (m), 1023 (w), 827 (m), 743 (m), 695 (s), 546 (m), 516 (s), 506 (s), 417 (m).

elemental analysis for C$_{50}$H$_{48}$O$_3$P$_2$Pt (953.94 *g/mol*):

calcd.: C 62.95, H 5.07, O 5.03, P 6.49, Pt 20.45

found: C 62.92, H 5.15

Bis(triphenylphosphine)-(η^2-methyl 4-methylcinnamate)nickel(0)

NiPhMe-Me: $C_{47}H_{42}NiO_2P_2$, M $= 759.48$ *g/mol*, brick red solid, yield: 95%, **route:** *Ni*.

Single crystals suitable for X-ray analysis were obtained by slow diffusion of pentane into a toluene solution.

^{1}H-NMR (400.13 MHz; C_7D_8): $\delta = 2.07$ (s, 3H, CCH_3), 3.20 (s, 3H, COOCH_3), 3.58 (dpt, $^3J_{HH} = 8.9$ Hz, $^3J_{HP} = 3.8$ Hz, $^3J_{HP} = 8.9$ Hz, 1H, CHCHCO), 4.33 (dpt, $^3J_{HH} = 8.9$ Hz, $^3J_{HP} = 5.2$ Hz, $^3J_{HP} = 8.9$ Hz, 1H, CC$_{Ar}$CHCH), 6.35 (d, $^3J_{HH} = 7.7$ Hz, 2H, H_{Ar}), 6.68 (d, $^3J_{HH} = 7.8$ Hz, 2H, H_{Ar}), 6.86 - 7.05 (m, 18H, H_{Ar}), 7.05 - 7.23 (m, 6H, H_{Ar}), 7.46 - 7.70 (m, 6H, H_{Ar}).

^{13}C-NMR (100.62 MHz; C_7D_8): $\delta = 21.2$ (CCH$_3$), 49.7 (COOCH$_3$), 53.4 (dd, $^2J_{CP} = 2.0$ Hz, $^2J_{CP} = 12.7$ Hz, CHCHCO), 65.5 (d, $^2J_{CP} = 16.0$ Hz, C$_{Ar}$$C$HCH), 125.3 ($C_{Ar}$H), 128.0 ($C_{Ar}$H), 128.1 ($C_{Ar}$H), 128.1 ($C_{Ar}$H), 128.2 ($C_{Ar}$H), 128.9 (d, $^3J_{CP} = 1.3$ Hz, C_{Ar}H), 129.2 (d, $^3J_{CP} = 1.2$ Hz, C_{Ar}H), 129.4 (d, $^2J_{CP} = 0.6$ Hz, C_{Ar}H), 132.8 (d, $^1J_{CP} = 1.5$ Hz, C_{Ar}), 134.2 (d, $^2J_{CP} = 7.8$ Hz, C_{Ar}H), 134.4 (d, $^2J_{CP} = 8.5$ Hz, C_{Ar}H), 135.0 (d, $^1J_{CP} = 2.9$ Hz, C_{Ar}), 135.4 (d, $^1J_{CP} = 2.9$ Hz, C_{Ar}), 136.3 (d, $^1J_{CP} = 2.3$ Hz, C_{Ar}), 136.6 (d, $^1J_{CP} = 2.3$ Hz, C_{Ar}), 140.5 (d, $^3J_{CP} = 5.5$ Hz, C_{Ar}), 172.6 (dd, $^3J_{CP} = 1.4$ Hz, $^3J_{CP} = 3.3$ Hz, CHCOO).

^{31}P-NMR (161.97 MHz; C_7D_8): $\delta = 31.6$ (d, $^2J_{PP} = 35.8$ Hz), 34.2 (d, $^2J_{PP} = 35.8$ Hz).

FT-IR (ATR, cm^{-1}): 1664 (m, CO).

Bis(triphenylphosphine)-(η^2-methyl 4-methylcinnamate)palladium(0)

PdPhMe-Me: $C_{47}H_{42}O_2P_2Pd$, M = 807.20 g/mol, light yellow solid, yield: 92%, **route**: *Pd-I*.

^{1}H-NMR (400.13 MHz; C_7D_8): δ = 2.08 (s, 3H, CH_3), 3.21 (s, 3H, OCH_3), 4.75 (bs, 1H, CHCHCO), 5.33 (bs, 1H, $C_{Ar}CH$CH), 6.46 - 6.56 (bm, 2H, H_{Ar}), 6.63 - 6.69 (bm, 2H, H_{Ar}), 6.83 - 7.04 (bm, 18H, H_{Ar}), 7.05 - 7.21 (bm, 6H, H_{Ar}), 7.40 - 7.60 (bm, 6H, H_{Ar}).

^{1}H-NMR (400.13 MHz; 246 K; C_6D_6): δ = 2.09 (s, 3H, CH_3), 3.19 (s, 3H, OCH_3), 4.86 (dpt, $^3J_{HH}$ = 10.4 Hz, $^3J_{HP}$ = 6.2 Hz, $^3J_{HP}$ = 6.2 Hz, 1H, CHCHCO), 5.39 (dpt, $^3J_{HH}$ = 10.4 Hz, $^3J_{HP}$ = 7.0 Hz, $^3J_{HP}$ = 7.0 Hz, 1H, $C_{Ar}CH$CH), 6.46 - 6.50 (m, 2H, H_{Ar}), 6.64 - 6.69 (m, 2H, H_{Ar}), 6.85 - 6.91 (m, 6H, H_{Ar}), 6.93 - 6.99 (m, 12H, H_{Ar}), 7.07 - 7.14 (m, 6H, H_{Ar}), 7.53 - 7.59 (m, 6H, H_{Ar}).

^{13}C-NMR (100.62 MHz; 246 K; C_6D_6): δ = 20.6 (CH_3), 50.0 (OOCH_3), 60.7 (dd, $^2J_{CP}$ = 5.4 Hz, $^2J_{CP}$ = 15.9 Hz, CHCHCO), 76.0 (dd, $^2J_{CP}$ = 3.2 Hz, $^2J_{CP}$ = 25.9 Hz, $C_{Ar}CH$CH), 128.2 (C_{Ar}), 128.3 (C_{Ar}), 128.4 (C_{Ar}), 128.9 (C_{Ar}), 133.6 (d, $^3J_{CP}$ = 2.2 Hz, C_{Ar}P), 134.1 (d, $^2J_{CP}$ = 11.7 Hz, C_{Ar}H), 134.3 (d, $^2J_{CP}$ = 14.8 Hz, C_{Ar}H), 135.6 - 135.9 (m, C_{Ar}P), 137.2 (d, $^3J_{CP}$ = 1.3 Hz, C_{Ar}P), 139.3 (d, $^3J_{CP}$ = 7.0 Hz, C_{Ar}CH), 171.1 (dd, $^3J_{CP}$ = 3.3 Hz, $^3J_{CP}$ = 1.7 Hz, CHCOO).

^{31}P-NMR (161.97 MHz; C_7D_8): δ = 23.8 (bs), 25.4 (bs).

^{31}P-NMR (161.97 MHz; 246 K; C_6D_6): δ = 24.1 (d, $^2J_{PP}$ = 19.7 Hz), 26.2 (d, $^2J_{PP}$ = 19.7 Hz).

FT-IR (ATR, cm^{-1}): 3050 (m), 1686 (m, CO), 1591 (w), 1477 (m), 1432 (s), 1303 (w), 1257 (w), 1149 (s), 1091 (m), 1015 (w), 833 (w), 742 (s), 682 (s), 523 (m), 511 (m), 501 (s), 406 (m).

elemental analysis for $C_{47}H_{42}O_2P_2Pd$ (807.20 g/mol):

calcd.: C 69.93, H 5.24, O 3.96, P 7.67, Pd 13.18

found: C 69.54, H 5.51

Bis(triphenylphosphine)-(η^2-methyl 4-methylcinnamate)platinum(0)

PtPhMe-Me: $C_{47}H_{42}O_2P_2Pt$, M $= 895.86$ g/mol, white solid, yield: 96%, **route**: *Pt-VI*.

^{1}H-NMR (400.13 MHz; C_6D_6): $\delta = 2.11$ (s, 3H, CCH_3), 3.23 (s, 3H, OCH_3), 4.09 (dpt, $^3J_{HH} = 8.9$ Hz, $^3J_{HP} = 4.8$ Hz, $^3J_{HP} = 8.9$ Hz, 1H, CHCHCO), 4.09 (ddpt, $^3J_{HH} = 8.9$ Hz, $^3J_{HP} = 4.8$ Hz, $^3J_{HP} = 8.9$ Hz, $^2J_{HPt} = 61.0$ Hz, 1H, CHCHCO), 4.56 (dpt, $^3J_{HH} = 8.9$ Hz, $^3J_{HP} = 5.8$ Hz, $^3J_{HP} = 8.9$ Hz, 1H, C_{Ar}CHCH), 4.56 (ddpt, $^3J_{HH} = 8.9$ Hz, $^3J_{HP} = 5.8$ Hz, $^3J_{HP} = 8.9$ Hz, $^2J_{HPt} = 53.7$ Hz, 1H, C_{Ar}CHCH), 6.55 (d, $^3J_{HH} = 7.4$ Hz, 2H, H_{Ar}), 6.73 (d, $^3J_{HH} = 7.9$ Hz, 2H, H_{Ar}), 6.83 - 7.01 (m, 18H, H_{Ar}), 7.22 (t, $^3J_{HH} = 8.7$ Hz, 6H, H_{Ar}), 7.54 - 7.61 (m, 6H, H_{Ar}).

^{13}C-NMR (100.62 MHz; C_6D_6): $\delta = 21.1$ (CCH$_3$), 50.1 (COOCH$_3$), 50.2 (dd, $^2J_{CP} = 6.6$ Hz, $^2J_{CP} = 28.6$ Hz, CHCHCO), 50.2 (ddd, $^2J_{CP} = 6.6$ Hz, $^2J_{CP} = 28.6$ Hz, $^1J_{CPt} = 195.1$ Hz, CHCHCO), 61.1 (dd, $^2J_{CP} = 2.5$ Hz, $^2J_{CP} = 25.3$ Hz, $C_{Ar}C$HCH), 61.1 (ddd, $^2J_{CP} = 2.5$ Hz, $^2J_{CP} = 25.3$ Hz, $^1J_{CPt} = 197.3$ Hz, $C_{Ar}C$HCH), 126.2 (d, $^5J_{CP} = 2.4$ Hz, C_{Ar}H), 126.2 (dd, $^5J_{CP} = 2.4$ Hz, $^4J_{CPt} = 14.2$ Hz, C_{Ar}H), 127.9 (C_{Ar}H), 128.0 (C_{Ar}H), 128.2 (C_{Ar}H), 128.6 (d, $^5J_{CP} = 1.2$ Hz, C_{Ar}H), 129.3 (d, $^5J_{CP} = 1.0$ Hz, C_{Ar}H), 129.5 (d, $^5J_{CP} = 0.8$ Hz, C_{Ar}H), 133.0 (d, $^5J_{CP} = 1.9$ Hz, C_{Ar}H), 133.0 (dd, $^5J_{CP} = 1.9$ Hz, $^4J_{CPt} = 8.4$ Hz, C_{Ar}H), 134.0 - 134.6 (m, C_{Ar}H), 135.5 - 136.4 (m, C_{Ar}H), 136.9 - 137.8 (m, C_{Ar}H), 141.8 (d, $^3J_{CP} = 5.1$ Hz, C_{Ar}CH), 141.8 (dd, $^3J_{CP} = 5.1$ Hz, $^2J_{CPt} = 35.2$ Hz, C_{Ar}CH), 173.5 (dd, $^3J_{CP}$ 2.6 Hz, $^3J_{CP} = 4.8$ Hz, CHCOO), 173.5 (ddd, $^3J_{CP} = 2.6$ Hz, $^3J_{CP} = 4.8$ Hz, $^2J_{CPt} = 46.4$ Hz, CHCOO).

^{31}P-NMR (161.97 MHz; C_6D_6): $\delta = 28.2$ (d, $^2J_{PP} = 44.4$ Hz), 28.2 (dd, $^2J_{PP} = 44.4$ Hz, $^1J_{PPt} = 4177.0$ Hz), 28.4 (d, $^2J_{PP} = 44.4$ Hz), 28.4 (dd, $^2J_{PP} = 44.4$ Hz, $^1J_{PPt} = 3632.4$ Hz).

^{195}Pt-NMR (85.48 MHz; C_6D_6): $\delta = $ -5061.8 (dd, $^1J_{PPt} = 3632.2$ Hz, $^1J_{PPt} = 4176.9$ Hz).

FAB-MS: m/z (%): 719.0 (100) [Pt(PPh$_3$)$_2^+$], 895.0 (2.18) [(C$_{11}$H$_{12}$O$_2$)Pt(PPh$_3$)$_2^+$].

FT-IR (ATR, cm^{-1}): 3056 (w), 1692 (s, CO), 1432 (m), 1319 (w), 1258 (w), 1147 (s), 1093 (m), 1031 (w), 894 (w), 849 (w), 803 (w), 743 (s), 690 (s), 576 (w), 532 (m), 515 (m), 500 (s), 425 (m).

elemental analysis for C$_{47}$H$_{42}$O$_2$P$_2$Pt (895.86 g/mol):

calcd.: C 63.01, H 4.73, O 3.57, P 6.91, Pt 21.78

found: C 62.78, H 4.69

Bis(triphenylphosphine)-(η^2-ethyl 4-methylcinnamate)nickel(0)

NiPhMe-Et: C$_{48}$H$_{44}$NiO$_2$P$_2$, M = 773.50 g/mol, brick red solid, yield: 72%, **route:** *Ni.*

^{1}H-NMR (400.13 MHz; C$_7$D$_8$): $\delta = 0.87$ (t, $^3J_{HH} = 7.1$ Hz, 3H, CH$_2$CH_3), 2.07 (s, 3H, CCH_3), 3.48 - 3.67 (m, 2H, CHCHCO, CH_2CH$_3$), 3.99 (dq, $^2J_{HH} = 10.7$ Hz, $^3J_{HH} = 7.1$ Hz, 1H, CH_2CH$_3$), 4.35 (dpt, $^3J_{HH} = 9.2$ Hz, $^3J_{HP} = 5.1$ Hz, $^3J_{HP} = 9.2$ Hz, 1H, C$_{Ar}$CHCH), 6.35 (d, $^3J_{HH} = 7.6$ Hz, 2H, H_{Ar}), 6.69 (d, $^3J_{HH} = 7.9$ Hz, 2H, H_{Ar}), 6.90 - 7.05 (m, 18H, H_{Ar}), 7.05 - 7.17 (m, 6H, H_{Ar}), 7.50 - 7.63 (m, 6H, H_{Ar}).

^{13}C-NMR (100.62 MHz; C$_7$D$_8$): $\delta = 14.4$ (CH$_2$$CH_3$), 21.2 (C$CH_3$), 53.7 (dd, $^2J_{CP} = 2.0$ Hz, $^2J_{CP} = 12.7$ Hz, CHCHCO), 58.7 (CH$_2$CH$_3$), 65.7 (d, $^2J_{CP} = 16.0$ Hz, $C_{Ar}$$C$HCH), 125.3 ($C_{Ar}$H), 128.1 ($C_{Ar}$H), 128.1 ($C_{Ar}$H), 128.4 ($C_{Ar}$), 128.5 ($C_{Ar}$), 128.9

(d, $^3J_{CP}$ = 1.4 Hz, C_{Ar}H), 129.2 (d, $^3J_{CP}$ = 1.3 Hz, C_{Ar}H), 129.4 (d, $^2J_{CP}$ = 0.9 Hz, C_{Ar}H), 132.7 (d, $^1J_{CP}$ = 1.6 Hz, C_{Ar}), 134.2 (d, $^2J_{CP}$ = 12.7 Hz, C_{Ar}H), 134.4 (d, $^2J_{CP}$ = 13.5 Hz, C_{Ar}H), 135.1 (d, $^1J_{CP}$ = 2.9 Hz, C_{Ar}), 135.4 (d, $^1J_{CP}$ = 3.0 Hz, C_{Ar}), 136.3 (d, $^1J_{CP}$ = 2.4 Hz, C_{Ar}), 136.7 (d, $^1J_{CP}$ = 2.3 Hz, C_{Ar}), 140.5 (d, $^3J_{CP}$ = 5.5 Hz, C_{Ar}), 172.4 (dd, $^3J_{CP}$ = 1.4 Hz, $^3J_{CP}$ = 3.4 Hz, CHCOO).

^{31}P-NMR (161.97 MHz; C$_7$D$_8$): δ = 31.5 (d, $^2J_{PP}$ = 36.5 Hz), 34.1 (d, $^2J_{PP}$ = 36.5 Hz).

FT-IR (ATR, cm^{-1}): 1669 (m, CO).

elemental analysis for C$_{48}$H$_{44}$NiO$_2$P$_2$ (773.50 g/mol):

calcd.: C 74.53, H 5.73, Ni 7.59, O 4.14, P 8.01

found: C 71.76, H 5.70

Bis(triphenylphosphine)-(η^2-ethyl 4-methylcinnamate)palladium(0)

PdPhMe-Et: C$_{48}$H$_{44}$O$_2$P$_2$Pd, M = 821.23 g/mol, yellow solid, yield: 74%, **route**: *Pd-I.*

^{1}H-NMR (400.13 MHz; C$_7$D$_8$): δ = 0.87 (t, $^3J_{HH}$ = 7.0 Hz, 3H, CH$_2$CH_3), 2.08 (s, 3H, CH_3), 3.62 (bs, 1H, CH_2CH$_3$), 3.99 (bs, 1H, CH_2CH$_3$), 4.78 (bs, 1H, CHCHCO), 5.33 (bs, 1H, C$_{Ar}$CHCH), 6.43 - 6.58 (bm, 2H, H_{Ar}), 6.60 - 6.70 (bm, 2H, H_{Ar}), 6.83 - 7.04 (bm, 18H, H_{Ar}), 7.04 - 7.19 (bm, 6H, H_{Ar}), 7.44 - 7.64 (bm, 6H, H_{Ar}).

^{1}H-NMR (400.13 MHz; 246 K; C$_7$D$_8$): δ = 0.82 (t, $^3J_{HH}$ = 7.1 Hz, 3H, CH$_2$CH_3), 2.09 (s, 3H, CH_3), 3.58 (dq, $^2J_{HH}$ = 10.7 Hz, $^3J_{HH}$ = 7.1 Hz, 1H, CH_2CH$_3$), 4.00 (dq, $^2J_{HH}$ = 10.7 Hz, $^3J_{HH}$ = 7.1 Hz, 1H, CH_2CH$_3$), 4.90 (dpt, $^3J_{HH}$ = 10.4 Hz, $^3J_{HP}$ = 6.2 Hz, $^3J_{HP}$ = 6.2 Hz, 1H, CHCHCO), 5.39 (dpt, $^3J_{HH}$ = 10.4 Hz, $^3J_{HP}$ = 7.0 Hz, $^3J_{HP}$ = 7.0 Hz, 1H, C$_{Ar}$CHCH), 6.45 - 6.49 (m, 2H, H_{Ar}), 6.63 - 6.67 (m, 2H, H_{Ar}), 6.86 - 6.92 (m, 6H, H_{Ar}), 6.94 - 7.00 (m, 12H, H_{Ar}), 7.06 - 7.12 (m, 6H, H_{Ar}), 7.55 - 7.61 (m, 6H, H_{Ar}).

^{13}C-NMR (100.62 MHz; C$_7$D$_8$): δ = 14.4 (CH$_2$CH$_3$), 21.1 (CCH$_3$), 58.9 (bs, CH$_2$CH$_3$), 61.5 (bs, CHCHCO), 76.7 (bs, C$_{Ar}$CHCH), 128.2 (C$_{Ar}$H), 128.2 (C$_{Ar}$H), 128.2 (C$_{Ar}$H), 128.3 - 128.4 (m, C$_{Ar}$H), 134.1 (bs, C$_{Ar}$H), 134.4 (bs, C$_{Ar}$H), 136.0 - 136.6 (m, C$_{Ar}$H), 139.9 (bs, C$_{Ar}$CH).

^{13}C-NMR (100.62 MHz; 246 K; C$_7$D$_8$): δ = 14.2 (CH$_2$CH$_3$), 21.1 (CCH$_3$), 58.9 (CH$_2$CH$_3$), 60.9 (dd, $^2J_{CP}$ = 6.1 Hz, $^2J_{CP}$ = 16.8 Hz, CHCHCO), 76.3 (dd, $^2J_{CP}$ = 3.2 Hz, $^2J_{CP}$ = 18.4 Hz, C$_{Ar}$CHCH), 125.2 (bs, C$_{Ar}$H), 125.5 (bs, C$_{Ar}$H), 128.0 (C$_{Ar}$), 128.2 (C$_{Ar}$), 128.3 (C$_{Ar}$), 128.4 (C$_{Ar}$), 128.9 (C$_{Ar}$), 129.2 (C$_{Ar}$), 133.6 (d, $^3J_{CP}$ = 2.1 Hz, C$_{Ar}$P), 134.1 (d, $^2J_{CP}$ = 14.4 Hz, C$_{Ar}$H), 134.3 (d, $^2J_{CP}$ = 15.8 Hz, C$_{Ar}$H), 135.8 (dd, $^1J_{CP}$ = 27.8 Hz, $^3J_{CP}$ = 1.7 Hz, C$_{Ar}$P), 137.6 (d, $^3J_{CP}$ = 1.8 Hz, C$_{Ar}$P), 139.4 (d, $^3J_{CP}$ = 7.2 Hz, C$_{Ar}$CH), 170.8 (dd, $^3J_{CP}$ = 3.9 Hz, $^3J_{CP}$ = 2.3 Hz, CHCOO).

^{31}P-NMR (161.97 MHz; C$_7$D$_8$): δ = 24.3 (bd, $^2J_{PP}$ = 19.3 Hz), 26.1 (bd, $^2J_{PP}$ = 19.3 Hz).

^{31}P-NMR (161.97 MHz; 246 K; C$_7$D$_8$): δ = 23.8 (d, $^2J_{PP}$ = 20.8 Hz), 25.9 (d, $^2J_{PP}$ = 20.8 Hz).

FT-IR (ATR, cm^{-1}): 3054 (w), 2979 (w), 1676 (s, CO), 1476 (m), 1432 (m), 1239 (w), 1218 (m), 1092 (m), 1048 (m), 810 (m), 739 (s), 692 (s), 525 (w), 494 (s), 403 (m).

elemental analysis for C$_{48}$H$_{44}$O$_2$P$_2$Pd (821.23 g/mol):

calcd.: C 70.20, H 5.40, O 3.90, P 7.54, Pd 12.96

found: C 70.16, H 5.44

Bis(triphenylphosphine)-(η^2-ethyl 4-methylcinnamate)platinum(0)

PtPhMe-Et: C$_{48}$H$_{44}$O$_2$P$_2$Pt, M = 909.89 g/mol, white solid, yield: 98%, **route**: *Pt-VI*.

^{1}H-NMR (400.13 MHz; C_6D_6): $\delta = 0.83$ (t, $^3J_{HH} = 7.1$ Hz, 3H, CH_2CH_3), 2.11 (s, 3H, CCH_3), 3.67 (dq, $^2J_{HH} = 10.8$ Hz, $^3J_{HH} = 7.1$ Hz, 1H, CH_2CH_3), 4.05 (dq, $^2J_{HH} = 10.8$ Hz, $^3J_{HH} = 7.1$ Hz, 1H, CH_2CH_3), 4.13 (dpt, $^3J_{HH} = 8.9$ Hz, $^3J_{HP} = 4.2$ Hz, $^3J_{HP} = 8.9$ Hz, 1H, CHCHCO), 4.13 (ddpt, $^3J_{HH} = 8.9$ Hz, $^3J_{HP} = 4.2$ Hz, $^3J_{HP} = 8.9$ Hz, $^2J_{HPt} = 62.4$ Hz, 1H, CHCHCO), 4.55 (dpt, $^3J_{HH} = 8.9$ Hz, $^3J_{HP} = 6.0$ Hz, $^3J_{HP} = 8.9$ Hz, 1H, C_{Ar}CHCH), 4.55 (ddpt, $^3J_{HH} = 8.9$ Hz, $^3J_{HP} = 6.0$ Hz, $^3J_{HP} = 8.9$ Hz, $^2J_{HPt} = 53.1$ Hz, 1H, C_{Ar}CHCH), 6.54 (d, $^3J_{HH} = 7.5$ Hz, 2H, H_{Ar}), 6.72 (d, $^3J_{HH} = 7.8$ Hz, 2H, H_{Ar}), 6.83 - 6.99 (m, 18H, H_{Ar}), 7.21(t, $^3J_{HH} = 8.6$ Hz, 6H, H_{Ar}), 7.55 - 7.64 (m, 6H, H_{Ar}).

^{13}C-NMR (100.62 MHz; C_6D_6): $\delta = 14.5$ (CH_2CH_3), 21.1 (CCH_3), 50.4 (dd, $^2J_{CP} = 3.6$ Hz, $^2J_{CP} = 25.6$ Hz, CHCHCO), 50.4 (ddd, $^2J_{CP} = 3.6$ Hz, $^2J_{CP} = 25.6$ Hz, $^1J_{CPt} = 202.6$ Hz, CHCHCO), 58.8 (CH_2CH_3), 61.5 (dd, $^2J_{CP} = 1.5$ Hz, $^2J_{CP} = 29.3$ Hz, $C_{Ar}$$C$HCH), 61.5 (ddd, $^2J_{CP} = 1.5$ Hz, $^2J_{CP} = 29.3$ Hz, $^1J_{CPt} = 195.2$ Hz, $C_{Ar}$$C$HCH), 126.2 (d, $^5J_{CP} = 2.9$ Hz, C_{Ar}H), 126.2 (dd, $^5J_{CP} = 2.9$ Hz, $^4J_{CPt} = 13.9$ Hz, C_{Ar}H), 127.9 (C_{Ar}H), 128.0 (C_{Ar}H), 128.2 (C_{Ar}H), 128.6 (d, $^5J_{CP} = 1.8$ Hz, C_{Ar}H), 129.2 (d, $^5J_{CP} = 1.5$ Hz, C_{Ar}H), 129.5 (d, $^5J_{CP} = 1.4$ Hz, C_{Ar}H), 132.9 (d, $^5J_{CP} = 2.5$ Hz, C_{Ar}H), 132.9 (dd, $^5J_{CP} = 2.5$ Hz, $^4J_{CPt} = 9.9$ Hz, C_{Ar}H), 134.2 (d, $^2J_{CP} = 1.0$ Hz, C_{Ar}P), 134.2 (dd, $^2J_{CP} = 1.0$ Hz, $^3J_{CPt} = 23.4$ Hz, C_{Ar}P), 134.3 (d, $^2J_{CP} = 1.0$ Hz, C_{Ar}P), 134.3 (dd, $^2J_{CP} = 1.0$ Hz, $^3J_{CPt} = 22.7$ Hz, C_{Ar}P), 134.4 (d, $^2J_{CP} = 1.0$ Hz, C_{Ar}P), 134.4 (dd, $^2J_{CP} = 1.0$ Hz, $^3J_{CPt} = 17.1$ Hz, C_{Ar}P), 134.5 (d, $^2J_{CP} = 1.0$ Hz, C_{Ar}P), 134.5 (dd, $^2J_{CP} = 1.0$ Hz, $^3J_{CPt} = 18.5$ Hz, C_{Ar}P), 135.8 (d, $^1J_{CP} = 5.8$ Hz, C_{Ar}P), 135.8 (dd, $^1J_{CP} = 5.8$ Hz, $^2J_{CPt} = 32.8$ Hz, C_{Ar}P), 136.1 (d, $^1J_{CP} = 5.8$ Hz, C_{Ar}P), 136.1 (dd, $^1J_{CP} = 5.8$ Hz, $^2J_{CPt} = 32.2$ Hz, C_{Ar}P), 137.3 (d, $^1J_{CP} = 5.6$ Hz, C_{Ar}P), 137.3 (dd, $^1J_{CP} = 5.6$ Hz, $^2J_{CPt} = 28.7$ Hz, C_{Ar}P), 137.7 (d, $^1J_{CP} = 5.6$ Hz, C_{Ar}P), 137.7 (dd, $^1J_{CP} = 5.6$ Hz, $^2J_{CPt} = 28.5$ Hz, C_{Ar}P), 141.9 (d, $^3J_{CP} = 5.4$ Hz, C_{Ar}CH), 141.9 (dd, $^3J_{CP} = 5.4$ Hz, $^2J_{CPt} = 36.3$ Hz, C_{Ar}CH), 173.3 (dd, $^3J_{CP}$ 1.9 Hz, $^3J_{CP} = 4.4$ Hz, CHCOO), 173.3 (ddd, $^3J_{CP} = 1.9$ Hz, $^3J_{CP} = 4.4$ Hz, $^2J_{CPt} = 46.3$ Hz, CHCOO).

^{31}P-NMR (161.97 MHz; C_6D_6): $\delta = 28.1$ (d, $^2J_{PP} = 45.4$ Hz), 28.1 (dd, $^2J_{PP} = 45.4$ Hz, $^1J_{PPt} = 4167.4$ Hz), 28.5 (d, $^2J_{PP} = 45.4$ Hz), 28.5 (dd, $^2J_{PP} = 45.4$ Hz,

$^1J_{PPt} = 3650.0$ Hz).

^{195}Pt-NMR (85.48 MHz; C$_6$D$_6$): $\delta = $ -5057.6 (dd, $^1J_{PPt} = $ 3649.7 Hz, $^1J_{PPt} = 4167.1$ Hz).

FAB-MS: m/z (%): 719.2 (100) [Pt(PPh$_3$)$_2$$^+$], 909.3 (1.55) [(C$_{12}H_{14}O_2$)Pt(PPh$_3$)$_2$$^+$].

FT-IR (ATR, cm^{-1}): 3060 (w), 2984 (w), 1681 (m, CO), 1514 (w), 1478 (w), 1433 (m), 1271 (w), 1141 (m), 1092 (m), 1029 (w), 817 (w), 741 (m), 691 (s), 535 (m), 514 (s), 423 (m).

elemental analysis for C$_{48}$H$_{44}$O$_2$P$_2$Pt (909.89 g/mol):

calcd.: C 63.36, H 4.87, O 3.52, P 6.81, Pt 21.44

found: C 63.34, H 5.19

Bis(triphenylphosphine)-(η^2-*iso*-propyl 4-methylcinnamate)nickel(0)

NiPhMe-^{i}Pr: C$_{49}$H$_{46}$NiO$_2$P$_2$, M = 787.53 g/mol, brick red solid, yield: 76%, **route:** *Ni.*

^{1}H-NMR (400.13 MHz; C$_7$D$_8$): $\delta = $ 0.72 (d, $^3J_{HH} = 6.2$ Hz, 3H, CHCH_3), 1.20 (d, $^3J_{HH} = 6.2$ Hz, 3H, CHCH_3), 2.07 (s, 3H, CCH_3), 3.54 (dpt, $^3J_{HH} = 9.0$ Hz, $^3J_{HP} = 3.5$ Hz, $^3J_{HP} = 9.0$ Hz, 1H, CHCHCO), 4.35 (dpt, $^3J_{HH} = 9.0$ Hz, $^3J_{HP} = 5.2$ Hz, $^3J_{HP} = 9.0$ Hz, 1H, C$_{Ar}$CHCH), 4.85 (hp, $^3J_{HH} = 6.2$ Hz, 1H, CH(CH$_3$)$_2$), 6.36 (d, $^3J_{HH} = 7.6$ Hz, 2H, H_{Ar}), 6.69 (d, $^3J_{HH} = 7.8$ Hz, 2H, H_{Ar}), 6.90 - 7.04 (m, 18H, H_{Ar}), 7.05 - 7.15 (m, 6H, H_{Ar}), 7.50 - 7.65 (m, 6H, H_{Ar}).

^{13}C-NMR (100.62 MHz; C$_7$D$_8$): $\delta = $ 21.2 (CCH$_3$), 21.7 (CHCH$_3$), 22.6 (CHCH$_3$), 54.0 (dd, $^2J_{CP} = 1.9$ Hz, $^2J_{CP} = 12.8$ Hz, CHCHCO), 65.7 (OCH(CH$_3$)$_2$), 66.1 (d, $^2J_{CP} = 15.8$ Hz, C$_{Ar}$$C$HCH), 125.2 (d, $^4J_{CP} = 1.8$ Hz, C$_{Ar}$H), 128.0 (C$_{Ar}$H), 128.0

(C_{Ar}H), 128.1 (C_{Ar}H), 128.9 (d, $^3J_{CP} = 1.3$ Hz, C_{Ar}H), 129.2 (d, $^3J_{CP} = 1.2$ Hz, C_{Ar}H), 129.4 (d, $^3J_{CP} = 0.4$ Hz, C_{Ar}H), 132.7 (d, $^1J_{CP} = 1.6$ Hz, C_{Ar}), 134.2 (d, $^2J_{CP} = 12.6$ Hz, C_{Ar}H), 134.5 (d, $^2J_{CP} = 13.4$ Hz, C_{Ar}H), 135.1 (d, $^1J_{CP} = 2.9$ Hz, C_{Ar}), 135.4 (d, $^1J_{CP} = 3.0$ Hz, C_{Ar}), 136.4 (d, $^1J_{CP} = 2.5$ Hz, C_{Ar}), 136.7 (d, $^1J_{CP} = 2.4$ Hz, C_{Ar}), 140.5 (d, $^3J_{CP} = 5.5$ Hz, C_{Ar}), 172.3 (dd, $^3J_{CP} = 1.2$ Hz, $^3J_{CP} = 3.3$ Hz, CHCOO).

^{31}P-NMR (161.97 MHz; C$_7$D$_8$): $\delta = 31.3$ (d, $^2J_{PP} = 37.1$ Hz), 34.0 (d, $^2J_{PP} = 37.1$ Hz).

FT-IR (ATR, cm^{-1}): 1665 (m, CO).

Bis(triphenylphosphine)-(η^2-*iso*-propyl 4-methylcinnamate)palladium(0)

PdPhMe-^{i}Pr: C$_{49}$H$_{46}$O$_2$P$_2$Pd, M $= 835.26$ *g/mol*, deep yellow solid, yield: $>99\%$, **route**: *Pd-I.*

^{1}H-NMR (400.13 MHz; C$_7$D$_8$): $\delta = 0.76$ (bs, 3H, CHCH_3), 1.15 (bs, 3H, CHCH_3), 2.07 (s, 3H, CH_3), 4.77 (bs, 1H, CHCHCO), 4.91 (bs, 1H, CH(CH$_3$)$_2$), 5.30 (bs, 1H, C_{Ar}CHCH), 6.43 - 6.57 (bm, 2H, H_{Ar}), 6.61 - 6.67 (bm, 2H, H_{Ar}), 6.83 - 7.04 (bm, 18H, H_{Ar}), 7.04 - 7.22 (bm, 6H, H_{Ar}), 7.36 - 7.65 (bm, 6H, H_{Ar}).

^{1}H-NMR (400.13 MHz; 246 K; C$_7$D$_8$): $\delta = 0.70$ (d, $^3J_{HH} = 6.2$ Hz, 3H, CHCH_3), 1.19 (d, $^3J_{HH} = 6.2$ Hz, 3H, CHCH_3), 2.08 (s, 3H, CH_3), 4.87 - 4.94 (m, 2H, CH(CH$_3$)$_2$, CHCHCO), 5.37 (dpt, $^3J_{HH} = 10.1$ Hz, $^3J_{HP} = 7.0$ Hz, $^3J_{HP} = 7.0$ Hz, 1H, C_{Ar}CHCH), 6.44 - 6.49 (m, 2H, H_{Ar}), 6.60 - 6.64 (m, 2H, H_{Ar}), 6.87 - 6.93 (m, 6H, H_{Ar}), 6.93 - 7.00 (m, 12H, H_{Ar}), 7.03 - 7.10 (m, 6H, H_{Ar}), 7.56 - 7.63 (m, 6H, H_{Ar}).

^{13}C-NMR (100.62 MHz; C$_7$D$_8$): $\delta = 21.1$ (CCH$_3$), 21.7 (bs, CHCH$_3$), 22.5 (bs, CHCH$_3$), 61.7 (bs, CHCHCO), 66.0 (bs, OCH(CH$_3$)$_2$), 77.0 (bs, $C_{Ar}$$C$HCH), 128.2 ($C_{Ar}$H), 128.2 ($C_{Ar}$H), 128.2 ($C_{Ar}$H), 128.3($C_{Ar}$H), 129.1 ($C_{Ar}$H), 134.0 - 134.7 (m, C_{Ar}H), 139.8 (bs, C_{Ar}CH).

^{13}C-NMR (100.62 MHz; 246 K; C$_7$D$_8$): δ = 21.1 (CCH_3), 21.4 (CHCH_3), 22.4 (CHCH_3), 61.1 (dd, $^2J_{CP}$ = 7.2 Hz, $^2J_{CP}$ = 16.8 Hz, CHCHCO), 65.9 (OCH(CH$_3$)$_2$), 76.9 (dd, $^2J_{CP}$ = 2.3 Hz, $^2J_{CP}$ = 21.1 Hz, C$_{Ar}$$CH$CH), 128.0 ($C_{Ar}$), 128.2 ($C_{Ar}$), 128.3 ($C_{Ar}$), 128.9 ($C_{Ar}$), 129.2 ($C_{Ar}$), 133.6 (bs, C_{Ar}P), 134.1 (d, $^2J_{CP}$ = 14.5 Hz, C_{Ar}H), 134.5 (d, $^2J_{CP}$ = 15.5 Hz, C_{Ar}H), 135.8 (dd, $^1J_{CP}$ = 27.6 Hz, $^3J_{CP}$ = 1.8 Hz, C_{Ar}P), 137.7 (d, $^3J_{CP}$ = 1.8 Hz, C_{Ar}P), 139.4 (d, $^3J_{CP}$ = 6.9 Hz, C_{Ar}CH), 170.6 (dd, $^3J_{CP}$ = 3.9 Hz, $^3J_{CP}$ = 2.4 Hz, CHCOO).

^{31}P-NMR (161.97 MHz; C$_7$D$_8$): δ = 24.0 (bd, $^2J_{PP}$ = 18.3 Hz), 25.8 (bd, $^2J_{PP}$ = 18.3 Hz).

^{31}P-NMR (161.97 MHz; 246 K; C$_7$D$_8$): δ = 23.5 (d, $^2J_{PP}$ = 21.6 Hz), 25.6 (d, $^2J_{PP}$ = 21.6 Hz).

FT-IR (ATR, cm^{-1}): 3053 (m), 1674 (s, CO), 1477 (m), 1432 (s), 1377 (w), 1271 (m), 1151 (s), 1104 (s), 1091(m), 998 (w), 801 (m), 740 (s), 695 (s), 524 (m), 512 (m), 498 (s), 422 (s).

elemental analysis for C$_{49}$H$_{46}$O$_2$P$_2$Pd (835.26 g/mol):

calcd.: C 70.46, H 5.55, O 3.83, P 7.42, Pd 12.74

found: C 70.23, H 5.60

Bis(triphenylphosphine)-(η^2-*iso*-propyl 4-methylcinnamate)platinum(0)

PtPhMe-^{i}Pr: C$_{49}$H$_{46}$O$_2$P$_2$Pt, M = 923.91 g/mol, white solid, yield: 74%, **route**: *Pt-VI*.

^{1}H-NMR (400.13 MHz; C$_6$D$_6$): δ = 0.75 (d, $^3J_{HH}$ = 6.2 Hz, 3H, CHCH_3), 1.12 (d, $^3J_{HH}$ = 6.2 Hz, 3H, CHCH_3), 2.10 (s, 3H, CCH_3), 4.16 (dpt, $^3J_{HH}$ = 8.6 Hz, $^3J_{HP}$ = 4.6 Hz, $^3J_{HP}$ = 8.6 Hz, 1H, CHCHCO), 4.16 (ddpt, $^3J_{HH}$ = 8.6 Hz, $^3J_{HP}$ = 4.6 Hz, $^3J_{HP}$ = 8.6 Hz, $^2J_{HPt}$ = 62.7 Hz, 1H, CHCHCO), 4.53 (dpt, $^3J_{HH}$ = 8.6 Hz, $^3J_{HP}$ = 5.7 Hz,

$^3J_{HP}$ = 8.6 Hz, 1H, C$_{Ar}$CHCH), 4.53 (ddpt, $^3J_{HH}$ = 8.6 Hz, $^3J_{HP}$ = 5.7 Hz, $^3J_{HP}$ = 8.6 Hz,
$^2J_{HPt}$ = 52.4 Hz, 1H, C$_{Ar}$CHCH), 4.98 (hp, $^3J_{HH}$ = 6.2 Hz, 1H, CH(CH$_3$)$_2$), 6.51 (d,
$^3J_{HH}$ = 7.1 Hz, 2H, H_{Ar}), 6.69 (d, $^3J_{HH}$ = 7.8 Hz, 2H, H_{Ar}), 6.82 - 7.02 (m, 18H, H_{Ar}),
7.13 - 7.24 (m, 6H, H_{Ar}), 7.56 - 7.67 (m, 6H, H_{Ar}).

^{13}C-NMR (100.62 MHz; C$_6$D$_6$): δ = 21.1 (CCH$_3$), 21.8 (CHCH$_3$), 22.6 (CHCH$_3$),
50.6 (dd, $^2J_{CP}$ = 5.9 Hz, $^2J_{CP}$ = 28.6 Hz, CHCHCO), 50.6 (ddd, $^2J_{CP}$ = 5.9 Hz,
$^2J_{CP}$ = 28.6 Hz, $^1J_{CPt}$ = 203.9 Hz, CHCHCO), 62.0 (dd, $^2J_{CP}$ = 3.7 Hz, $^2J_{CP}$ = 31.7 Hz,
C$_{Ar}$$C$HCH), 62.0 (ddd, $^2J_{CP}$ = 3.7 Hz, $^2J_{CP}$ = 31.7 Hz, $^1J_{CPt}$ = 195.2 Hz, C$_{Ar}$$C$HCH),
65.8 (OCH(CH$_3$)$_2$), 126.1 (d, $^5J_{CP}$ = 3.0 Hz, C_{Ar}H), 126.1 (dd, $^5J_{CP}$ = 3.0 Hz,
$^4J_{CPt}$ = 13.7 Hz, C_{Ar}H), 127.9 (C_{Ar}H), 128.0 (C_{Ar}H), 128.1 (C_{Ar}H), 128.2 (C_{Ar}H), 128.6
(d, $^5J_{CP}$ = 2.0 Hz, C_{Ar}H), 129.2 (d, $^5J_{CP}$ = 1.6 Hz, C_{Ar}H), 129.5 (d, $^5J_{CP}$ = 1.5 Hz, C_{Ar}H),
133.0 (d, $^5J_{CP}$ = 2.6 Hz, C_{Ar}H), 133.0 (dd, $^5J_{CP}$ = 2.6 Hz, $^4J_{CPt}$ = 6.2 Hz, C_{Ar}H), 134.2
(C_{Ar}P), 134.2 (d, $^3J_{CPt}$ = 21.2 Hz, C_{Ar}P), 134.3 (C_{Ar}P), 134.3 (d, $^3J_{CPt}$ = 21.3 Hz, C_{Ar}P),
134.5 (C_{Ar}P), 134.5 (d, $^3J_{CPt}$ = 19.7 Hz, C_{Ar}P), 134.6 (C_{Ar}P), 134.6 (d, $^3J_{CPt}$ = 18.4 Hz,
C_{Ar}P), 135.7 (d, $^1J_{CP}$ = 3.5 Hz, C_{Ar}P), 135.7 (dd, $^1J_{CP}$ = 3.5 Hz, $^2J_{CPt}$ = 33.3 Hz, C_{Ar}P),
136.1 (d, $^1J_{CP}$ = 3.4 Hz, C_{Ar}P), 136.1 (dd, $^1J_{CP}$ = 3.4 Hz, $^2J_{CPt}$ = 31.8 Hz, C_{Ar}P), 137.4
(d, $^1J_{CP}$ = 3.3 Hz, C_{Ar}P), 137.4 (dd, $^1J_{CP}$ = 3.3 Hz, $^2J_{CPt}$ = 29.4 Hz, C_{Ar}P), 137.8 (d,
$^1J_{CP}$ = 3.3 Hz, C_{Ar}P), 137.8 (dd, $^1J_{CP}$ = 3.3 Hz, $^2J_{CPt}$ = 27.7 Hz, C_{Ar}P), 141.9 (dd,
$^3J_{CP}$ = 1.3 Hz, $^3J_{CP}$ = 6.6 Hz, C_{Ar}CH), 141.9 (ddd, $^3J_{CP}$ = 1.3 Hz, $^3J_{CP}$ = 6.6 Hz,
$^2J_{CPt}$ = 34.1 Hz, C_{Ar}CH), 173.0 (dd, $^3J_{CP}$ 2.2 Hz, $^3J_{CP}$ = 4.8 Hz, CHCOO), 173.0 (ddd,
$^3J_{CP}$ = 2.2 Hz, $^3J_{CP}$ = 4.8 Hz, $^2J_{CPt}$ = 45.9 Hz, CHCOO).

^{31}P-NMR (161.97 MHz; C$_6$D$_6$): δ = 27.8 (d, $^2J_{PP}$ = 46.4 Hz), 27.8 (dd,
$^2J_{PP}$ = 46.4 Hz, $^1J_{PPt}$ = 4151.5 Hz), 28.6 (d, $^2J_{PP}$ = 46.4 Hz), 28.6 (dd, $^2J_{PP}$ = 46.4 Hz,
$^1J_{PPt}$ = 3670.7 Hz).

^{195}Pt-NMR (85.48 MHz; C$_6$D$_6$): δ = -5051.9 (dd, $^1J_{PPt}$ = 3670.8 Hz,
$^1J_{PPt}$ = 4151.7 Hz).

FAB-MS: m/z (%): 719.2 (100) [Pt(PPh$_3$)$_2$$^+$], 923.3 (1.55) [(C$_{13}H_{16}O_2$)Pt(PPh$_3$)$_2$$^+$].

FT-IR (ATR, cm^{-1}): 3042 (w), 2966 (w), 1678 (s, CO), 1514 (w), 1476 (w), 1432 (m),
1268 (m), 1148 (s), 1091 (m), 999 (w), 916 (w), 814 (m), 742 (s), 692 (s), 533 (m), 499

(s), 423 (m).

elemental analysis for $C_{49}H_{46}O_2P_2Pt$ (923.91 g/mol):

calcd.: C 63.70, H 5.02, O 3.46, P 6.70, Pt 21.11

found: C 63.58, H 5.04

Bis(triphenylphosphine)-(η^2-*tert*-butyl 4-methylcinnamate)nickel(0)

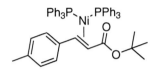

NiPhMe-^{t}Bu: $C_{50}H_{48}NiO_2P_2$, M = 801.56 g/mol, orange solid, yield: 82%, **route**: *Ni*.

^{1}H-NMR (400.13 MHz; C_7D_8): $\delta = 1.31$ (s, 9H, C(CH_3)$_3$), 2.07 (s, 3H, CCH_3), 3.51 (dpt, $^3J_{HH} = 9.1$ Hz, $^3J_{HP} = 3.1$ Hz, $^3J_{HP} = 9.1$ Hz, 1H, CHCHCO), 4.33 (dpt, $^3J_{HH} = 9.1$ Hz, $^3J_{HP} = 4.9$ Hz, $^3J_{HP} = 9.1$ Hz, 1H, C_{Ar}CHCH), 6.41 (d, $^3J_{HH} = 7.6$ Hz, 2H, H_{Ar}), 6.67 (d, $^3J_{HH} = 7.6$ Hz, 2H, H_{Ar}), 6.87 - 7.02 (m, 18H, H_{Ar}), 7.02 - 7.16 (m, 6H, H_{Ar}), 7.46 - 7.68 (m, 6H, H_{Ar}).

^{13}C-NMR (100.62 MHz; C_7D_8): $\delta = 21.2$ (CCH$_3$), 28.5 (CCH$_3$), 54.9 (dd, $^2J_{CP} = 1.8$ Hz, $^2J_{CP} = 13.4$ Hz, CHCHCO), 66.5 (d, $^2J_{CP} = 15.9$ Hz, $C_{Ar}$$C$HCH), 77.5 (O$C$(CH$_3$)$_3$), 128.0 ($C_{Ar}$H), 128.0 ($C_{Ar}$H), 128.1 ($C_{Ar}$H),129.1 (bs, C_{Ar}H), 129.5 (bs, C_{Ar}H), 132.7 (d, $^2J_{CP} = 1.5$ Hz, C_{Ar}), 134.2 (d, $^2J_{CP} = 12.5$ Hz, C_{Ar}H), 134.5 (d, $^2J_{CP} = 13.5$ Hz, C_{Ar}H), 135.1 (d, $^1J_{CP} = 3.0$ Hz, C_{Ar}), 135.4 (d, $^1J_{CP} = 3.0$ Hz, C_{Ar}), 136.3 (d, $^1J_{CP} = 2.5$ Hz, C_{Ar}), 136.7 (d, $^1J_{CP} = 2.5$ Hz, C_{Ar}), 140.8 (d, $^3J_{CP} = 5.3$ Hz, C_{Ar}), 172.8 (dd, $^3J_{CP} = 1.0$ Hz, $^3J_{CP} = 3.2$ Hz, CHCOO).

^{31}P-NMR (161.97 MHz; C_7D_8): $\delta = 30.6$ (d, $^2J_{PP} = 38.3$ Hz), 33.3 (d, $^2J_{PP} = 38.3$ Hz).

FT-IR (ATR, cm^{-1}): 1668 (m, CO).

elemental analysis for $C_{50}H_{48}NiO_2P_2$ (801.56 g/mol):

calcd.: C 74.92, H 6.04, Ni 7.32, O 3.99, P 7.73

found: C 73.77, H 6.04

Bis(triphenylphosphine)-(η^2-*tert*-butyl 4-methylcinnamate)palladium(0)

PdPhMe-^{t}Bu: $C_{50}H_{48}O_2P_2Pd$, M $= 849.28$ g/mol, yellow solid, yield: $>99\%$, **route:** *Pd-I*.

^{1}H-NMR (400.13 MHz; C_7D_8): $\delta = 1.32$ (s, 9H, $C(CH_3)_3$), 2.07 (s, 3H, CH_3), 4.73 (bs, 1H, CHCHCO), 5.26 (bs, 1H, $C_{Ar}CH$CH), 6.42 - 6.57 (bm, 2H, H_{Ar}), 6.58 - 6.68 (bm, 2H, H_{Ar}), 6.83 - 7.03 (bm, 18H, H_{Ar}), 7.03 - 7.17 (bm, 6H, H_{Ar}), 7.44 - 7.62 (bm, 6H, H_{Ar}).

^{1}H-NMR (400.13 MHz; 246 K; C_7D_8): $\delta = 1.32$ (s, 9H, $C(CH_3)_3$), 2.07 (s, 3H, CH_3), 4.85 (dpt, $^3J_{HH} = 10.4$ Hz, $^3J_{HP} = 6.1$ Hz, $^3J_{HP} = 6.1$ Hz, 1H, CHCHCO), 5.32 (dpt, $^3J_{HH} = 10.4$ Hz, $^3J_{HP} = 7.0$ Hz, $^3J_{HP} = 7.0$ Hz, 1H, $C_{Ar}CH$CH), 6.43 - 6.49 (m, 2H, H_{Ar}), 6.58 - 6.63 (m, 2H, H_{Ar}), 6.86 - 7.01 (m, 18H, H_{Ar}), 7.02 - 7.08 (m, 6H, H_{Ar}), 7.54 - 7.60 (m, 6H, H_{Ar}).

^{13}C-NMR (100.62 MHz; C_7D_8): $\delta = 21.1$ (CCH_3), 28.3 (CCH_3), 63.0 (bs, CHCHCO), 77.1 - 78.3 (bm, $C_{Ar}CH$CH, OC(CH$_3$)$_3$), 128.2 (C_{Ar}H), 128.2 (C_{Ar}H), 128.2 (C_{Ar}H), 128.3 (C_{Ar}H), 129.2 (C_{Ar}H), 133.9 - 135.0 (m, C_{Ar}H), 136.0 - 136.7 (m, C_{Ar}H), 139.8 (bs, C_{Ar}CH).

^{13}C-NMR (100.62 MHz; 246 K; C_7D_8): $\delta = 20.8$ (CCH_3), 28.1 ((CCH_3)), 62.2 (dd, $^2J_{CP} = 5.3$ Hz, $^2J_{CP} = 16.7$ Hz, CHCHCO), 77.3 - 77.6 (m, $C_{Ar}CH$CH), 77.6 (OC(CH$_3$)$_3$), 125.2 (d, $^3J_{CP} = 1.3$ Hz, C_{Ar}H), 125.5 (d, $^3J_{CP} = 1.5$ Hz, C_{Ar}H), 128.0 (C_{Ar}), 128.2 (C_{Ar}), 128.3 (C_{Ar}), 129.2 (C_{Ar}), 133.5 (d, $^3J_{CP} = 1.9$ Hz, C_{Ar}P), 134.2 (d, $^2J_{CP} = 14.2$ Hz, C_{Ar}H), 134.5 (d, $^2J_{CP} = 15.7$ Hz, C_{Ar}H), 135.8 (dd, $^1J_{CP} = 26.9$ Hz, $^3J_{CP} = 1.5$ Hz, C_{Ar}P), 137.7 (d, $^3J_{CP} = 2.1$ Hz, C_{Ar}P), 139.6 (d, $^3J_{CP} = 6.4$ Hz, C_{Ar}CH), 170.9 (dd, $^3J_{CP} = 3.5$ Hz, $^3J_{CP} = 2.1$ Hz, CHCOO).

^{31}P-NMR (161.97 MHz; C$_7$D$_8$): δ = 22.8 (bd, $^2J_{PP}$ = 20.1 Hz), 24.6 (bd, $^2J_{PP}$ = 20.1 Hz).

^{31}P-NMR (161.97 MHz; 246 K; C$_7$D$_8$): δ = 23.0 (d, $^2J_{PP}$ = 23.0 Hz), 25.2 (d, $^2J_{PP}$ = 23.0 Hz).

FAB-MS: m/z (%): 368.0 (14.78) [Pd(PPh$_3$)$^+$], 630.1 (18.83) [Pd(PPh$_3$)$_2$$^+$].

FT-IR (ATR, cm^{-1}): 3053 (m), 2977 (m), 1674 (s, CO), 1477 (m), 1432 (s), 1316 (w), 1279 (m), 1133 (s), 1091 (s), 998 (w), 810 (m), 741 (s), 692 (s), 525 (m), 512 (m), 498 (s), 407 (m).

elemental analysis for C$_{50}$H$_{48}$O$_2$P$_2$Pd (849.28 g/mol):

calcd.: C 70.71, H 5.70, O 3.77, P 7.29, Pd 12.53

found: C 70.54, H 5.47

Bis(triphenylphosphine)-(η^2-*tert*-butyl 4-methylcinnamate)platinum(0)

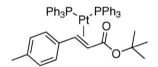

PtPhMe-^{t}Bu: C$_{50}$H$_{48}$O$_2$P$_2$Pt, M = 937.94 g/mol, white solid, yield: 95%, **route**: *Pt-VI*.

^{1}H-NMR (400.13 MHz; C$_6$D$_6$): δ = 1.31 (s, 9H, C(CH$_3$)$_3$), 2.10 (s, 3H, CCH$_3$), 4.13 (dpt, $^3J_{HH}$ = 8.7 Hz, $^3J_{HP}$ = 5.0 Hz, $^3J_{HP}$ = 8.7 Hz, 1H, CHCHCO), 4.13 (ddpt, $^3J_{HH}$ = 8.7 Hz, $^3J_{HP}$ = 5.0 Hz, $^3J_{HP}$ = 8.7 Hz, $^2J_{HPt}$ = 62.8 Hz, 1H, CHCHCO), 4.48 (dpt, $^3J_{HH}$ = 8.7 Hz, $^3J_{HP}$ = 6.0 Hz, $^3J_{HP}$ = 8.7 Hz, 1H, C$_{Ar}$CHCH), 4.48 (ddpt, $^3J_{HH}$ = 8.7 Hz, $^3J_{HP}$ = 6.0 Hz, $^3J_{HP}$ = 8.7 Hz, $^2J_{HPt}$ = 52.0 Hz, 1H, C$_{Ar}$CHCH), 6.48 (d, $^3J_{HH}$ = 6.7 Hz, 2H, H$_{Ar}$), 6.68 (d, $^3J_{HH}$ = 7.9 Hz, 2H, H$_{Ar}$), 6.84 - 7.04 (m, 18H, H$_{Ar}$), 7.09 - 7.26 (m, 6H, H$_{Ar}$), 7.49 - 7.73 (m, 6H, H$_{Ar}$).

^{13}C-NMR (100.62 MHz; C$_6$D$_6$): δ = 21.1 (CCH$_3$), 28.5 (CCH$_3$), 51.6 (dd, $^2J_{CP}$ = 6.3 Hz, $^2J_{CP}$ = 28.9 Hz, CHCHCO), 51.6 (ddd, $^2J_{CP}$ = 6.3 Hz, $^2J_{CP}$ = 28.9 Hz,

$^1J_{CPt} = 218.6$ Hz, CH*CH*CO), 62.8 (dd, $^2J_{CP} = 4.5$ Hz, $^2J_{CP} = 32.2$ Hz, C_{Ar}*CH*CH), 62.8 (ddd, $^2J_{CP} = 4.5$ Hz, $^2J_{CP} = 32.2$ Hz, $^1J_{CPt} = 192.2$ Hz, C_{Ar}*CH*CH), 77.6 (O*C*(CH$_3$)$_3$), 126.0 (d, $^5J_{CP} = 3.0$ Hz, C_{Ar}H), 126.0 (dd, $^5J_{CP} = 3.0$ Hz, $^4J_{CPt} = 13.4$ Hz, C_{Ar}H), 128.0 (C_{Ar}H), 128.1 (C_{Ar}H), 128.2 (C_{Ar}H), 128.6 (d, $^5J_{CP} = 1.9$ Hz, C_{Ar}H), 129.2 (d, $^5J_{CP} = 1.7$ Hz, C_{Ar}H), 129.5 (d, $^5J_{CP} = 1.4$ Hz, C_{Ar}H), 132.9 (d, $^5J_{CP} = 2.6$ Hz, C_{Ar}H), 132.9 (bm, C_{Ar}H), 134.2 (C_{Ar}P), 134.2 (d, $^3J_{CPt} = 20.3$ Hz, C_{Ar}P), 134.3 (C_{Ar}P), 134.3 (d, $^3J_{CPt} = 20.7$ Hz, C_{Ar}P), 134.6 (C_{Ar}P), 134.6 (d, $^3J_{CPt} = 19.1$ Hz, C_{Ar}P), 134.7 (C_{Ar}P), 134.7 (d, $^3J_{CPt} = 19.2$ Hz, C_{Ar}P), 135.7 (d, $^1J_{CP} = 2.7$ Hz, C_{Ar}P), 135.7 (dd, $^1J_{CP} = 2.7$ Hz, $^2J_{CPt} = 31.9$ Hz, C_{Ar}P), 136.1 (d, $^1J_{CP} = 2.8$ Hz, C_{Ar}P), 136.1 (dd, $^1J_{CP} = 2.8$ Hz, $^2J_{CPt} = 31.8$ Hz, C_{Ar}P), 137.5 (d, $^1J_{CP} = 2.7$ Hz, C_{Ar}P), 137.5 (dd, $^1J_{CP} = 2.7$ Hz, $^2J_{CPt} = 27.1$ Hz, C_{Ar}P), 137.8 (d, $^1J_{CP} = 2.7$ Hz, C_{Ar}P), 137.8 (dd, $^1J_{CP} = 2.7$ Hz, $^2J_{CPt} = 29.0$ Hz, C_{Ar}P), 141.9 (dd, $^3J_{CP} = 1.4$ Hz, $^3J_{CP} = 6.7$ Hz, C_{Ar}CH), 141.9 (ddd, $^3J_{CP} = 1.4$ Hz, $^3J_{CP} = 6.7$ Hz, $^2J_{CPt} = 35.3$ Hz, C_{Ar}CH), 173.2 (dd, $^3J_{CP}$ 2.3 Hz, $^3J_{CP} = 4.8$ Hz, CH*C*OO), 173.2 (ddd, $^3J_{CP} = 2.3$ Hz, $^3J_{CP} = 4.8$ Hz, $^2J_{CPt} = 46.1$ Hz, CH*C*OO).

^{31}P-NMR (161.97 MHz; C$_6$D$_6$): $\delta = 27.2$ (d, $^2J_{PP} = 47.2$ Hz), 27.2 (dd, $^2J_{PP} = 47.2$ Hz, $^1J_{PPt} = 4106.8$ Hz), 28.6 (d, $^2J_{PP} = 47.2$ Hz), 28.6 (dd, $^2J_{PP} = 47.2$ Hz, $^1J_{PPt} = 3716.6$ Hz).

^{195}Pt-NMR (85.48 MHz; C$_6$D$_6$): $\delta = -5052.7$ (dd, $^1J_{PPt} = 3728.6$ Hz, $^1J_{PPt} = 4095.2$ Hz).

FAB-MS: m/z (%): 718.9 (100) [Pt(PPh$_3$)$_2^+$], 937.1 (2.17) [(C$_{14}$H$_{18}$O$_2$)Pt(PPh$_3$)$_2^+$].

FT-IR (ATR, cm^{-1}): 3042 (w), 2966 (w), 1678 (s, CO), 1500 (w), 1469 (w), 1433 (m), 1303 (w), 1257 (w), 1132 (s), 1091 (m), 969 (w), 848 (w), 803 (m), 742 (s), 694 (s), 575 (w), 538 (w), 517 (m), 500 (s), 439 (w), 416 (m).

elemental analysis for C$_{50}$H$_{48}$O$_2$P$_2$Pt (937.94 *g/mol*):

calcd.: C 64.03, H 5.16, O 3.41, P 6.60, Pt 20.80

found: C 63.91, H 5.34

Bis(triphenylphosphine)-(η^2-methyl cinnamate)nickel(0)

NiPh-Me: $C_{46}H_{40}NiO_2P_2$, M = 745.45 g/mol, brick red solid, yield: 90%, **route:** Ni.

^{1}H-NMR (399.78 MHz; C_7D_8): δ = 3.19 (s, 3H, COOCH_3), 3.58 (dpt, $^3J_{HH}$ = 8.2 Hz, $^3J_{HP}$ = 3.9 Hz, $^3J_{HP}$ = 9.7 Hz, 1H, CHCHCO), 4.30 (ddd, $^3J_{HH}$ = 8.2 Hz, $^3J_{HP}$ = 5.2 Hz, $^3J_{HP}$ = 9.5 Hz, 1H, CC$_{Ar}$$CH$CH), 6.44 (d, $^3J_{HH}$ = 7.7 Hz, 2H, H_{Ar}), 6.83 - 6.90 (m, 2H, H_{Ar}), 6.90 - 7.05 (m, 19H, H_{Ar}), 7.05 - 7.20 (m, 6H, H_{Ar}), 7.47 - 7.64 (m, 6H, H_{Ar}).

^{13}C-NMR (100.53 MHz; C_7D_8): δ = 49.7 (COOCH$_3$), 53.4 (dd, $^2J_{CP}$ = 2.1 Hz, $^2J_{CP}$ = 13.0 Hz, CHCHCO), 65.2 (d, $^2J_{CP}$ = 16.1 Hz, C$_{Ar}$$C$HCH), 123.6 (bs, C_{Ar}H), 125.5 (d, $^4J_{CP}$ = 2.0 Hz, C_{Ar}H), 128.0 (C_{Ar}H), 128.1 (C_{Ar}H), 128.2 (C_{Ar}H), 128.4 (C_{Ar}), 128.5 (C_{Ar}), 129.0 (d, $^3J_{CP}$ = 1.3 Hz, C_{Ar}H), 129.2 (d, $^3J_{CP}$ = 1.3 Hz, C_{Ar}H), 132.4 (d, $^1J_{CP}$ = 9.5 Hz, C_{Ar}), 134.2 (d, $^2J_{CP}$ = 9.4 Hz, C_{Ar}H), 134.4 (d, $^2J_{CP}$ = 10.1 Hz, C_{Ar}H), 134.9 (d, $^1J_{CP}$ = 2.9 Hz, C_{Ar}), 135.3 (d, $^1J_{CP}$ = 2.9 Hz, C_{Ar}), 136.2 (d, $^1J_{CP}$ = 2.3 Hz, C_{Ar}), 136.5 (d, $^1J_{CP}$ = 2.3 Hz, C_{Ar}), 143.7 (d, $^3J_{CP}$ = 5.5 Hz, C_{Ar}), 172.7 (dd, $^3J_{CP}$ = 1.5 Hz, $^3J_{CP}$ = 3.4 Hz, CHCOO).

^{31}P-NMR (161.83 MHz; C_7D_8): δ = 32.1 (d, $^2J_{PP}$ = 34.6 Hz), 34.2 (d, $^2J_{PP}$ = 34.6 Hz).

FT-IR (ATR, cm^{-1}): 1669 (m, CO).

Bis(triphenylphosphine)-(η^2-methyl cinnamate)palladium(0)

PdPh-Me: $C_{46}H_{40}O_2P_2Pd$, M $= 793.18$ g/mol, light yellow solid, yield: 94%, **route**: Pd-I.

^{1}H-NMR (400.13 MHz; C_7D_8): $\delta = 3.20$ (s, 3H, CH_3), 4.72 (bs, 1H, CHCHCO), 5.26 (bs, 1H, $C_{Ar}CH$CH), 6.53 - 6.62 (bm, 2H, H_{Ar}), 6.81 - 7.02 (bm, 21H, H_{Ar}), 7.05 - 7.21 (bm, 6H, H_{Ar}), 7.39 - 7.60 (bm, 6H, H_{Ar}).

^{1}H-NMR (399.78 MHz; 246 K; C_7D_8): $\delta = 3.18$ (s, 3H, CH_3), 4.84 (dpt, $^3J_{HH} = 10.4$ Hz, $^3J_{HP} = 6.2$ Hz, $^3J_{HP} = 6.2$ Hz, 1H, CHCHCO), 5.33 (dpt, $^3J_{HH} = 10.4$ Hz, $^3J_{HP} = 7.0$ Hz, $^3J_{HP} = 7.0$ Hz, 1H, $C_{Ar}CH$CH), 6.53 - 6.58 (m, 2H, H_{Ar}), 6.84 - 6.90 (m, 9H, H_{Ar}), 6.92 - 6.98 (m, 12H, H_{Ar}), 7.06 - 7.12 (m, 6H, H_{Ar}), 7.51 - 7.59 (m, 6H, H_{Ar}).

^{13}C-NMR (100.62 MHz; C_7D_8): $\delta =50.0$ (OCH_3), 61.2 (bs, CHCHCO), 75.8 (bs, $C_{Ar}CH$CH), 124.5 (bs, $C_{Ar}H$), 125.7 (bs, $C_{Ar}H$), 128.2 ($C_{Ar}H$), 128.2 ($C_{Ar}H$), 128.2 ($C_{Ar}H$), 128.3 ($C_{Ar}H$), 129.2 ($C_{Ar}H$), 134.2 (bs, $C_{Ar}H$), 134.4 (bs, $C_{Ar}H$), 136.0 (bs, $C_{Ar}H$), 136.2 (bs, $C_{Ar}H$), 142.7 (bs, $C_{Ar}CH$), 171.1 (bs, CHCOO).

^{13}C-NMR (100.53 MHz; 246 K; C_7D_8): $\delta = 50.0$ (OCH_3), 60.7 (dd, $^2J_{CP} = 5.9$ Hz, $^2J_{CP} = 18.2$ Hz, CHCHCO), 75.4 (dd, $^2J_{CP} = 3.7$ Hz, $^2J_{CP} = 18.3$ Hz, $C_{Ar}CH$CH), 124.4 (bs, $C_{Ar}H$), 125.4 (bs, $C_{Ar}H$), 128.3 (bs, C_{Ar}), 128.4 (C_{Ar}), 128.4 (C_{Ar}), 129.2 (C_{Ar}), 134.1 (d, $^2J_{CP} = 14.5$ Hz, $C_{Ar}H$), 134.2 (d, $^2J_{CP} = 15.2$ Hz, $C_{Ar}H$), 135.4 - 135.8 (m, $C_{Ar}P$), 137.1 (d, 1.3 Hz,), 142.4 (d, $^3J_{CP} = 6.7$ Hz, $C_{Ar}CH$), 171.1 - 171.2 (m, CHCOO).

^{31}P-NMR (161.97 MHz; C_7D_8): $\delta = 24.9$ (bd, $^2J_{PP} = 17.5$ Hz), 26.3 (bd, $^2J_{PP} = 17.5$ Hz).

^{31}P-NMR (161.83 MHz; 246 K; C_7D_8): $\delta = 24.4$ (d, $^2J_{PP} = 17.6$ Hz), 26.2 (d, $^2J_{PP} = 17.6$ Hz).

FT-IR (ATR, cm^{-1}): 3066 (m), 1687 (s, CO), 1591 (w), 1478 (m), 1432 (s), 1319 (w), 1258 (w), 1150 (s), 1092 (m), 1031 (w), 743 (s), 682 (s), 523 (m), 512 (m), 494 (s), 410 (m).

elemental analysis for $C_{46}H_{40}O_2P_2Pd$ (793.18 g/mol):

calcd.: C 69.66, H 5.08, O 4.03, P 7.81, Pd 13.42

found: C 69.43, H 5.15

Bis(triphenylphosphine)-(η^2-methyl cinnamate)platinum(0)

PtPh-Me: $C_{46}H_{40}O_2P_2Pt$, M = 881.83 g/mol, white solid, yield: 89%, **route**: *Pt-VI*.

^{1}H-NMR (399.78 MHz; C_6D_6): $\delta = 3.23$ (s, 3H, CH_3), 4.07 (dpt, $^3J_{HH} = 8.9$ Hz, $^3J_{HP} = 4.3$ Hz, $^3J_{HP} = 8.9$ Hz, 1H, CHCHCO), 4.07 (ddpt, $^3J_{HH} = 8.9$ Hz, $^3J_{HP} = 4.3$ Hz, $^3J_{HP} = 8.9$ Hz, $^2J_{HPt} = 60.8$ Hz, 1H, CHCHCO), 4.51 (dpt, $^3J_{HH} = 8.9$ Hz, $^3J_{HP} = 3.6$ Hz, $^3J_{HP} = 8.9$ Hz, 1H, C_{Ar}CHCH), 4.51 (ddpt, $^3J_{HH} = 8.9$ Hz, $^3J_{HP} = 3.6$ Hz, $^3J_{HP} = 8.9$ Hz, $^2J_{HPt} = 54.0$ Hz, 1H, C_{Ar}CHCH), 6.62 (d, $^3J_{HH} = 7.9$ Hz, 2H, H_{Ar}), 6.82 - 7.02 (m, 22H, H_{Ar}), 7.17 - 7.25 (m, 3H, H_{Ar}), 7.48 (d, $^3J_{HH} = 7.0$ Hz, 4H, H_{Ar}), 7.53 - 7.61 (m, 4H, H_{Ar}).

^{13}C-NMR (100.53 MHz; C_6D_6): $\delta = 50.5$ (COOCH_3), 50.7 (dd, $^2J_{CP} = 3.1$ Hz, $^2J_{CP} = 19.2$ Hz, CHCHCO), 50.7 (ddd, $^2J_{CP} = 3.1$ Hz, $^2J_{CP} = 19.2$ Hz, $^1J_{CPt} = 196.8$ Hz, CHCHCO), 61.3 (dd, $^2J_{CP} = 5.3$ Hz, $^2J_{CP} = 23.0$ Hz, $C_{Ar}C$HCH), 61.3 (ddd, $^2J_{CP} = 5.3$ Hz, $^2J_{CP} = 23.0$ Hz, $^1J_{CPt} = 199.1$ Hz, $C_{Ar}C$HCH), 124.3 (C_{Ar}H), 126.6 (C_{Ar}H), 128.3 (C_{Ar}H), 128.5 (C_{Ar}H), 128.8 (C_{Ar}H), 128.9 (C_{Ar}H), 129.6 (C_{Ar}H), 129.8 (C_{Ar}H), 131.9 (C_{Ar}H), 131.9 (C_{Ar}H), 132.7 (C_{Ar}H), 132.8 (C_{Ar}H), 134.5 (C_{Ar}H), 134.6

(C_{Ar}H), 134.6 (C_{Ar}H),134.6 (C_{Ar}H), 134.7 (C_{Ar}H), 134.7 (C_{Ar}H), 136.0 (C_{Ar}P), 136.0 (d, $^2J_{CPt}$ = 27.8 Hz, C_{Ar}P), 136.3 (C_{Ar}P), 136.3 (d, $^2J_{CPt}$ = 30.8 Hz, C_{Ar}P), 137.4 (d, C_{Ar}P), 137.4 (d, $^2J_{CPt}$ = 25.4 Hz, C_{Ar}P), 137.8 (C_{Ar}P), 137.8 (d, $^2J_{CPt}$ = 26.9 Hz, C_{Ar}P), 140.9 - 141.5 (bm, C_{Ar}P), 145.3 (d, $^3J_{CP}$ = 4.7 Hz, C_{Ar}CH), 145.3 (dd, $^3J_{CP}$ = 4.7 Hz, $^2J_{CPt}$ = 35.4 Hz, C_{Ar}CH), 173.9 (d, $^3J_{CP}$ = 2.3 Hz, CHCOO), 173.9 (dd, $^3J_{CP}$ = 2.3 Hz, $^2J_{CPt}$ = 46.1 Hz, CHCOO).

^{31}P-NMR (161.83 MHz; C$_6$D$_6$): δ = 28.7 (d, $^2J_{PP}$ = 42.6 Hz), 28.7 (dd, $^2J_{PP}$ = 42.6 Hz, $^1J_{PPt}$ = 4151.5 Hz), 29.1 (d, $^2J_{PP}$ = 42.6 Hz), 29.1 (dd, $^2J_{PP}$ = 42.6 Hz, $^1J_{PPt}$ = 3643.3 Hz).

^{195}Pt-NMR (85.48 MHz; C$_6$D$_6$): δ = -5067.7 (dd, $^1J_{PPt}$ = 3637.8 Hz, $^1J_{PPt}$ = 4149.9 Hz).

FAB-MS: m/z (%): 719.1 (100) [Pt(PPh$_3$)$_2$$^+$], 881.2 (1.42) [(C$_{10}H_{10}O_2$)Pt(PPh$_3$)$_2$$^+$].

FT-IR (ATR, cm^{-1}): 3056 (w), 2935 (w), 1693 (s, CO), 1576 (w), 1470 (w), 1432 (m), 1318 (w), 1258 (w), 1147 (s), 1092 (m), 1023 (w), 894 (w), 743 (s), 689 (s), 576 (w), 532 (m), 515 (m), 492 (s), 453 (w), 424 (m).

elemental analysis for C$_{46}$H$_{40}$O$_2$P$_2$Pt (881.83 *g/mol*):
calcd.: C 62.65, H 4.57, O 3.63, P 7.02, Pt 22.12
found: C 62.51, H 4.71

Bis(triphenylphosphine)-(η^2-ethyl cinnamate)nickel(0)

NiPh-Et: C$_{47}$H$_{42}$NiO$_2$P$_2$, M = 759.48 *g/mol*, brick red solid, yield: 78%, **route:** *Ni*.

^{1}H-NMR (400.13 MHz; C$_7$D$_8$): δ = 0.87 (t, $^3J_{HH}$ = 7.1 Hz, 3H, CH$_2$CH_3), 3.43 - 3.69 (m, 2H, CHCHCO, CH_2CH$_3$), 3.99 (dq, $^2J_{HH}$ = 10.7 Hz, $^3J_{HH}$ = 7.1 Hz, 1H, CH_2CH$_3$),

4.32 (dpt, $^3J_{HH}$ = 8.8 Hz, $^3J_{HP}$ = 5.2 Hz, $^3J_{HP}$ = 8.8 Hz, 1H, $C_{Ar}CHCH$), 6.44 (d, $^3J_{HH}$ = 7.6 Hz, 2H, H_{Ar}), 6.84 - 6.91 (m, 2H, H_{Ar}), 6.91 - 7.04 (m, 19H, H_{Ar}), 7.06 - 7.15 (m, 6H, H_{Ar}), 7.51 - 7.62 (m, 6H, H_{Ar}).

^{13}C-NMR (100.62 MHz; C_7D_8): δ = 14.4 (CH_2CH_3), 53.6 (dd, $^2J_{CP}$ = 1.9 Hz, $^2J_{CP}$ = 12.9 Hz, $CHCHCO$), 58.7 (CH_2CH_3), 65.4 (d, $^2J_{CP}$ = 16.0 Hz, $C_{Ar}CHCH$), 123.6 (d, $^4J_{CP}$ = 1.1 Hz, $C_{Ar}H$), 125.3 ($C_{Ar}H$), 128.0 ($C_{Ar}H$), 128.1 ($C_{Ar}H$), 128.1 ($C_{Ar}H$), 128.6 ($C_{Ar}H$), 129.0 (d, $^3J_{CP}$ = 1.3 Hz, $C_{Ar}H$), 129.2 (d, $^3J_{CP}$ = 1.3 Hz, $C_{Ar}H$), 134.2 (d, $^2J_{CP}$ = 12.6 Hz, $C_{Ar}H$), 134.4 (d, $^2J_{CP}$ = 13.4 Hz, $C_{Ar}H$), 134.9 (d, $^1J_{CP}$ = 2.9 Hz, C_{Ar}), 135.2 (d, $^1J_{CP}$ = 2.9 Hz, C_{Ar}), 136.2 (d, $^1J_{CP}$ = 2.3 Hz, C_{Ar}), 136.5 (d, $^1J_{CP}$ = 2.4 Hz, C_{Ar}), 143.6 (d, $^3J_{CP}$ = 5.4 Hz, C_{Ar}), 172.4 (dd, $^3J_{CP}$ = 1.4 Hz, $^3J_{CP}$ = 3.4 Hz, $CHCOO$).

^{31}P-NMR (161.97 MHz; C_7D_8): δ = 31.9 (d, $^2J_{PP}$ = 35.3 Hz), 34.1 (d, $^2J_{PP}$ = 35.3 Hz).

FT-IR (ATR, cm^{-1}): 1668 (m, CO).

elemental analysis for $C_{47}H_{42}NiO_2P_2$ (759.48 g/mol):

calcd.: C 74.33, H 5.57, Ni 7.73, O 4.21, P 8.16

found: C 76.42, H 5.60

Bis(triphenylphosphine)-(η^2-ethyl cinnamate)palladium(0)

PdPh-Et: $C_{47}H_{42}O_2P_2Pd$, M = 807.20 g/mol, light yellow solid, yield: 87%, **route**: Pd-I.

^{1}H-NMR (400.13 MHz; C_7D_8): δ = 0.86 (t, $^3J_{HH}$ = 7.1 Hz, 3H, CH_2CH_3), 3.59 (bs, 1H, CH_2CH_3), 3.99 (bs, 1H, CH_2CH_3), 4.75 (bs, 1H, $CHCHCO$), 5.25 (bs, 1H, $C_{Ar}CHCH$), 6.52 - 6.63 (bm, 2H, H_{Ar}), 6.78 - 7.02 (bm, 21H, H_{Ar}), 7.03 - 7.21 (bm, 6H, H_{Ar}), 7.44 - 7.62 (bm, 6H, H_{Ar}).

^{1}H-NMR (400.13 MHz; 246 K; C_7D_8): δ = 0.82 (t, $^3J_{HH}$ = 7.1 Hz, 3H, CH_2CH_3), 3.55 (dq, $^2J_{HH}$ = 10.7 Hz, $^3J_{HH}$ = 7.1 Hz, 1H, CH_2CH_3), 3.99 (dq, $^2J_{HH}$ = 10.7 Hz, $^3J_{HH}$ = 7.1 Hz, 1H, CH_2CH_3), 4.89 (dpt, $^3J_{HH}$ = 10.4 Hz, $^3J_{HP}$ = 6.2 Hz, $^3J_{HP}$ = 6.2 Hz, 1H, CHCHCO), 5.32 (dpt, $^3J_{HH}$ = 10.4 Hz, $^3J_{HP}$ = 7.0 Hz, $^3J_{HP}$ = 7.0 Hz, 1H, C_{Ar}CHCH), 6.52 - 6.57 (m, 2H, H_{Ar}), 6.84 - 6.91 (m, 9H, H_{Ar}), 6.94 - 6.98 (m, 12H, H_{Ar}), 7.04 - 7.11 (m, 6H, H_{Ar}), 7.53 - 7.61 (m, 6H, H_{Ar}).

^{13}C-NMR (100.62 MHz; C_7D_8): δ = 14.4 (CH_2CH$_3$), 58.9 (bs, CH$_2$CH$_3$), 61.2 (bs, CHCHCO), 76.2 (bs, C_{Ar}CHCH), 124.4 (bs, C_{Ar}H), 125.7 (bs, C_{Ar}H), 128.2 (C_{Ar}H), 128.3 (C_{Ar}H), 129.0 (C_{Ar}H), 129.1 (C_{Ar}H), 133.9 - 134.7 (m, C_{Ar}H), 135.9 (bs, C_{Ar}H), 136.1 (bs, C_{Ar}H), 142.7 (bs, C_{Ar}CH), 170.8 (bs, CHCOO).

^{13}C-NMR (100.62 MHz; 246 K; C_7D_8): δ = 14.2 (CH_2CH$_3$), 58.9 (CH$_2$CH$_3$), 60.8 (dd, $^2J_{CP}$ = 5.5 Hz, $^2J_{CP}$ = 17.4 Hz, CHCHCO), 75.8 (dd, $^2J_{CP}$ = 2.0 Hz, $^2J_{CP}$ = 22.0 Hz, C_{Ar}CHCH), 124.4 (bs, C_{Ar}H), 125.4 (d, $^3J_{CP}$ = 2.0 Hz, C_{Ar}H), 128.1 (C_{Ar}), 128.3 (C_{Ar}), 128.4 (C_{Ar}), 129.0 (C_{Ar}), 129.2 (C_{Ar}), 134.1 (d, $^2J_{CP}$ = 14.6 Hz, C_{Ar}H), 134.3 (d, $^2J_{CP}$ = 15.4 Hz, C_{Ar}H), 135.6 (dd, $^1J_{CP}$ = 28.1 Hz, $^3J_{CP}$ = 1.6 Hz, C_{Ar}P), 137.3 (dd, $^1J_{CP}$ = 26.9 Hz, $^3J_{CP}$ = 1.6 Hz, C_{Ar}P), 142.4 (d, $^3J_{CP}$ = 6.3 Hz, C_{Ar}CH), 170.8 (dd, $^3J_{CP}$ = 3.8 Hz, $^3J_{CP}$ = 2.4 Hz, CHCOO).

^{31}P-NMR (161.97 MHz; C_7D_8): δ = 24.7 (bd, $^2J_{PP}$ = 19.5 Hz), 26.1 (bd, $^2J_{PP}$ = 19.5 Hz).

^{31}P-NMR (161.97 MHz; 246 K; C_7D_8): δ = 24.1 (d, $^2J_{PP}$ = 18.5 Hz), 26.0 (d, $^2J_{PP}$ = 18.5 Hz).

FT-IR (ATR, cm^{-1}): 3040 (m), 1676 (m, CO), 1586 (w), 1477 (m), 1432 (s), 1313 (w), 1268 (w), 1146 (s), 1092 (m), 1025 (w), 813 (w), 737 (s), 692 (s), 513 (s), 419 (m).

elemental analysis for $C_{47}H_{42}O_2P_2Pd$ (807.20 *g/mol*):

calcd.: C 69.93, H 5.24, O 3.96, P 7.67, Pd 13.18

found: C 69.46, H 5.31

Bis(triphenylphosphine)-(η^2-ethyl cinnamate)platinum(0)

PtPh-Et: $C_{47}H_{42}O_2P_2Pt$, M = 895.86 g/mol, white solid, yield: 58%, **route**: *Pt-VI*.

^{1}H-NMR (400.13 MHz; C_6D_6): $\delta = 0.83$ (t, $^3J_{HH} = 7.1$ Hz, 3H, CH$_2$CH_3), 3.67 (dq, $^2J_{HH} = 10.7$ Hz, $^3J_{HH} = 7.1$ Hz, 1H, CH_2CH$_3$), 4.05 (dq, $^2J_{HH} = 10.7$ Hz, $^3J_{HH} = 7.1$ Hz, 1H, CH_2CH$_3$), 4.12 (dpt, $^3J_{HH} = 8.9$ Hz, $^3J_{HP} = 4.3$ Hz, $^3J_{HP} = 8.9$ Hz, 1H, CHCHCO), 4.12 (ddpt, $^3J_{HH} = 8.9$ Hz, $^3J_{HP} = 4.3$ Hz, $^3J_{HP} = 8.9$ Hz, $^2J_{HPt} = 61.8$ Hz, 1H, CHCHCO), 4.51 (dpt, $^3J_{HH} = 8.9$ Hz, $^3J_{HP} = 5.7$ Hz, $^3J_{HP} = 8.9$ Hz, 1H, C$_{Ar}$CHCH), 4.51 (ddpt, $^3J_{HH} = 8.9$ Hz, $^3J_{HP} = 5.7$ Hz, $^3J_{HP} = 8.9$ Hz, $^2J_{HPt} = 53.1$ Hz, 1H, C$_{Ar}$CHCH), 6.61 (d, $^3J_{HH} = 7.9$ Hz, 2H, H_{Ar}), 6.79 - 7.00 (m, 21H, H_{Ar}), 7.20 (t, $^3J_{HH} = 8.6$ Hz, 6H, H_{Ar}), 7.52 - 7.68 (m, 6H, H_{Ar}).

^{13}C-NMR (100.62 MHz; C_6D_6): $\delta = 14.5$ (CH$_2$CH$_3$), 50.6 (dd, $^2J_{CP} = 5.1$ Hz, $^2J_{CP} = 25.7$ Hz, CHCHCO), 50.6 (ddd, $^2J_{CP} = 5.1$ Hz, $^2J_{CP} = 25.7$ Hz, $^1J_{CPt} = 208.4$ Hz, CHCHCO), 58.9 (CH$_2$CH$_3$), 61.3 (dd, $^2J_{CP} = 2.2$ Hz, $^2J_{CP} = 28.6$ Hz, C$_{Ar}$CHCH), 61.3 (ddd, $^2J_{CP} = 2.2$ Hz, $^2J_{CP} = 28.6$ Hz, $^1J_{CPt} = 205.4$ Hz, C$_{Ar}$CHCH), 124.0 (d, $^5J_{CP} = 2.1$ Hz, C$_{Ar}$H), 124.0 (bm, C$_{Ar}$H), 126.2 (d, $^5J_{CP} = 2.8$ Hz, C$_{Ar}$H), 126.2 (dd, $^5J_{CP} = 2.8$ Hz, $^4J_{CPt} = 15.3$ Hz, C$_{Ar}$H), 127.9 (C$_{Ar}$H), 128.2 (C$_{Ar}$H), 129.3 (d, $^5J_{CP} = 1.5$ Hz, C$_{Ar}$H), 129.5 (d, $^5J_{CP} = 1.4$ Hz, C$_{Ar}$H), 134.1 (d, $^2J_{CP} = 1.4$ Hz, C$_{Ar}$P), 134.1 (dd, $^2J_{CP} = 1.4$ Hz, $^3J_{CPt} = 20.9$ Hz, C$_{Ar}$P), 134.2 (d, $^2J_{CP} = 1.1$ Hz, C$_{Ar}$P), 134.2 (dd, $^2J_{CP} = 1.1$ Hz, $^3J_{CPt} = 20.5$ Hz, C$_{Ar}$P), 134.3 (d, $^2J_{CP} = 1.4$ Hz, C$_{Ar}$P), 134.3 (dd, $^2J_{CP} = 1.4$ Hz, $^3J_{CPt} = 21.3$ Hz, C$_{Ar}$P), 134.5 (d, $^2J_{CP} = 1.4$ Hz, C$_{Ar}$P), 134.5 (dd, $^2J_{CP} = 1.4$ Hz, $^3J_{CPt} = 21.2$ Hz, C$_{Ar}$P), 135.6 (d, $^1J_{CP} = 6.4$ Hz, C$_{Ar}$P), 135.6 (dd, $^1J_{CP} = 6.4$ Hz, $^2J_{CPt} = 32.5$ Hz, C$_{Ar}$P), 136.0 (d, $^1J_{CP} = 6.4$ Hz, C$_{Ar}$P), 136.0 (dd, $^1J_{CP} = 6.4$ Hz, $^2J_{CPt} = 31.9$ Hz, C$_{Ar}$P), 137.2 (d, $^1J_{CP} = 6.4$ Hz, C$_{Ar}$P), 137.2 (dd, $^1J_{CP} = 6.4$ Hz, $^2J_{CPt} = 29.2$ Hz, C$_{Ar}$P), 137.6 (d, $^1J_{CP} = 6.4$ Hz, C$_{Ar}$P), 137.6

(dd, $^1J_{CP}$ = 6.4 Hz, $^2J_{CPt}$ = 28.6 Hz, C_{Ar}P), 145.0 (d, $^3J_{CP}$ = 5.8 Hz, C_{Ar}CH), 145.0 (dd, $^3J_{CP}$ = 5.8 Hz, $^2J_{CPt}$ = 36.8 Hz, C_{Ar}CH), 173.3 (dd, $^3J_{CP}$ 2.2 Hz, $^3J_{CP}$ = 5.8 Hz, CHCOO), 173.3 (ddd, $^3J_{CP}$ = 2.2 Hz, $^3J_{CP}$ = 5.8 Hz, $^2J_{CPt}$ = 48.4 Hz, CHCOO).

^{31}P-NMR (161.97 MHz; C$_6$D$_6$): δ = 28.1 (d, $^2J_{PP}$ = 43.8 Hz), 28.1 (dd, $^2J_{PP}$ = 43.8 Hz, $^1J_{PPt}$ = 4139.6 Hz), 28.4 (d, $^2J_{PP}$ = 43.8 Hz), 28.4 (dd, $^2J_{PP}$ = 43.8 Hz, $^1J_{PPt}$ = 3654.4 Hz).

^{195}Pt-NMR (85.48 MHz; C$_6$D$_6$): δ = -5063.6 (dd, $^1J_{PPt}$ = 3654.3 Hz, $^1J_{PPt}$ = 4139.6 Hz).

FAB-MS: m/z (%): 719.0 (100) [Pt(PPh$_3$)$_2$$^+$], 895.0 (1.92) [(C$_{11}H_{12}O_2$)Pt(PPh$_3$)$_2$$^+$].

FT-IR (ATR, cm^{-1}): 3066 (w), 2975 (w), 1688 (s, CO), 1591 (w), 1484 (w), 1433 (m), 1223 (m), 1091 (w), 1038 (w), 742 (m), 691 (s), 545 (w), 507 (s), 499 (m), 424 (m).

elemental analysis for C$_{47}$H$_{42}$O$_2$P$_2$Pt (895.86 g/mol):

calcd.: C 63.01, H 4.73, O 3.57, P 6.91, Pt 21.78

found: C 62.69, H 4.65

Bis(triphenylphosphine)-(η^2-*iso*-propyl cinnamate)nickel(0)

NiPh-^{i}Pr: C$_{48}$H$_{44}$NiO$_2$P$_2$, M = 773.50 g/mol, brick red solid, yield: 89%, **route:** *Ni*.

^{1}H-NMR (400.13 MHz; C$_7$D$_8$): δ = 0.73 (d, $^3J_{HH}$ = 6.. Hz, 3H, CHCH_3), 1.20 (d, $^3J_{HH}$ = 6.1 Hz, 3H, CHCH_3), 3.53 (dpt, $^3J_{HH}$ = 8.9 Hz, $^3J_{HP}$ = 3.7 Hz, $^3J_{HP}$ = 8.9 Hz, 1H, CHCHCO), 4.32 (dpt, $^3J_{HH}$ = 8.9 Hz, $^3J_{HP}$ = 5.1 Hz, $^3J_{HP}$ = 8.9 Hz, 1H, C$_{Ar}$CHCH), 4.85 (hp, $^3J_{HH}$ = 6.1Hz, 1H, CH(CH$_3$)$_2$), 6.44 (d, $^3J_{HH}$ = 7.5 Hz, 2H, H_{Ar}), 6.83 - 6.90 (m, 2H, H_{Ar}), 6.90 - 7.04 (m, 19H, H_{Ar}), 7.04 - 7.17 (m, 6H, H_{Ar}), 7.50 - 7.64 (m, 6H, H_{Ar}).

^{13}C-NMR (100.62 MHz; C$_7$D$_8$): δ = 21.7 (CHCH_3), 22.6 (CHCH_3), 53.9 (dd, $^2J_{CP}$ = 1.9 Hz, $^2J_{CP}$ = 13.0 Hz, CHCHCO), 65.8 (OCH(CH$_3$)$_2$), 65.8 (d, $^2J_{CP}$ = 15.8 Hz, $C_{Ar}CH$CH), 123.6 (d, $^4J_{CP}$ = 1.0 Hz, C_{Ar}H), 125.3 (C_{Ar}H), 128.0 (C_{Ar}H), 128.1 (C_{Ar}H), 128.1 (C_{Ar}H), 128.7 (bs, C_{Ar}H), 128.9 (d, $^3J_{CP}$ = 1.3 Hz, C_{Ar}H), 129.2 (d, $^3J_{CP}$ = 1.3 Hz, C_{Ar}H), 134.2 (d, $^2J_{CP}$ = 12.6 Hz, C_{Ar}H), 134.5 (d, $^2J_{CP}$ = 13.4 Hz, C_{Ar}H), 134.9 (d, $^1J_{CP}$ = 2.9 Hz, C_{Ar}), 135.3 (d, $^1J_{CP}$ = 2.9 Hz, C_{Ar}), 136.3 (d, $^1J_{CP}$ = 2.4 Hz, C_{Ar}), 136.6 (d, $^1J_{CP}$ = 2.4 Hz, C_{Ar}), 143.6 (d, $^3J_{CP}$ = 5.4 Hz, C_{Ar}), 172.3 (dd, $^3J_{CP}$ = 1.2 Hz, $^3J_{CP}$ = 3.4 Hz, CHCOO).

^{31}P-NMR (161.97 MHz; C$_7$D$_8$): δ = 31.5 (d, $^2J_{PP}$ = 36.0 Hz), 33.7 (d, $^2J_{PP}$ = 36.0 Hz).

FT-IR (ATR, cm^{-1}): 1671 (m, CO).

Bis(triphenylphosphine)-(η^2-*iso*-propyl cinnamate)palladium(0)

PdPh-^{i}Pr: C$_{48}$H$_{44}$O$_2$P$_2$Pd, M = 821.23 g/mol, light yellow solid, yield: 93%, **route:** Pd-I.

^{1}H-NMR (400.13 MHz; C$_7$D$_8$): δ = 0.75 (bs, 3H, CHCH_3), 1.15 (bs, 3H, CHCH_3), 4.76 (bs, 1H, CHCHCO), 4.88 (hp, $^3J_{HH}$ = 6.2 Hz, 1H, CH(CH$_3$)$_2$), 5.25 (bs, 1H, C$_{Ar}CH$CH), 6.50 - 6.63 (bm, 2H, H_{Ar}), 6.78 - 6.89 (bm, 3H, H_{Ar}), 6.89 - 7.03 (bm, 18H, H_{Ar}), 7.03 - 7.16 (bm, 6H, H_{Ar}), 7.48 - 7.62 (bm, 6H, H_{Ar}).

^{1}H-NMR (400.13 MHz; 246 K; C$_7$D$_8$): δ = 0.70 (d, $^3J_{HH}$ = 6.2 Hz, 3H, CHCH_3), 1.18 (d, $^3J_{HH}$ = 6.2 Hz, 3H, CHCH_3), 4.85 - 4.93 (m, 2H, CH(CH$_3$)$_2$, CHCHCO), 5.31 (dpt, $^3J_{HH}$ = 10.2 Hz, $^3J_{HP}$ = 7.0 Hz, $^3J_{HP}$ = 7.0 Hz, 1H, C$_{Ar}CH$CH), 6.51 - 6.57 (m, 2H, H_{Ar}), 6.85 - 6.93 (m, 9H, H_{Ar}), 6.94 - 6.99 (m, 12H, H_{Ar}), 7.03 - 7.09 (m, 6H, H_{Ar}), 7.55 - 7.62 (m, 6H, H_{Ar}).

^{13}C-NMR (100.62 MHz; C$_7$D$_8$): δ = 21.7 (bs, CHCH_3), 22.5 (bs, CHCH_3), 61.8 (bs, CHCHCO), 65.9 (bs, OCH(CH$_3$)$_2$), 76.8 (bs, C$_{Ar}$$CH$CH), 124.5 (bs, C_{Ar}H), 125.6 (bs, C_{Ar}H), 128.2 (C_{Ar}H), 128.2 (C_{Ar}H), 128.2 (C_{Ar}H), 128.4 (C_{Ar}H), 129.0 (C_{Ar}H), 129.1 (C_{Ar}H), 129.2 (C_{Ar}H), 133.9 - 134.8 (m, C_{Ar}H), 135.9 (bs, C_{Ar}H), 136.2 (bs, C_{Ar}H), 142.7 (bs, C_{Ar}CH), 170.6 (bs, CHCOO).

^{13}C-NMR (100.62 MHz; 246 K; C$_7$D$_8$): δ = 21.4 (CHCH_3), 22.4 (CHCH_3), 61.0 (dd, $^2J_{CP}$ = 5.0 Hz, $^2J_{CP}$ = 18.2 Hz, CHCHCO), 65.9 (OCH(CH$_3$)$_2$), 76.3 (dd, $^2J_{CP}$ = 2.6 Hz, $^2J_{CP}$ = 22.0 Hz, C$_{Ar}$$CH$CH), 124.4 ($C_{Ar}$), 128.1 ($C_{Ar}$), 128.3 ($C_{Ar}$), 128.3 ($C_{Ar}$), 128.9 ($C_{Ar}$), 129.3 ($C_{Ar}$), 134.1 (d, $^2J_{CP}$ = 14.5 Hz, C_{Ar}H), 134.4 (d, $^2J_{CP}$ = 15.5 Hz, C_{Ar}H), 135.6 (dd, $^1J_{CP}$ = 28.0 Hz, $^3J_{CP}$ = 1.6 Hz, C_{Ar}P), 137.5 (d, $^3J_{CP}$ = 1.8 Hz, C_{Ar}P), 142.4 (d, $^3J_{CP}$ = 6.9 Hz, C_{Ar}CH), 170.6 (dd, $^3J_{CP}$ = 2.4 Hz, $^3J_{CP}$ = 4.0 Hz, CHCOO).

^{31}P-NMR (161.97 MHz; C$_7$D$_8$): δ = 24.4 (bd, $^2J_{PP}$ = 20.9 Hz), 25.9 (bd, $^2J_{PP}$ = 20.9 Hz).

^{31}P-NMR (161.97 MHz; 246 K; C$_7$D$_8$): δ = 23.8 (d, $^2J_{PP}$ = 19.5 Hz), 25.9 (d, $^2J_{PP}$ = 19.5 Hz).

FT-IR (ATR, cm^{-1}): 3049 (w), 1671 (s, CO), 1591 (w), 1478 (s), 1432 (s), 1273 (m), 1153 (s), 1110 (s), 1092 (s), 1030 (w), 1000 (w), 818 (w), 742 (s), 692 (s), 522 (m), 501 (s), 424 (m).

elemental analysis for C$_{48}$H$_{44}$O$_2$P$_2$Pd (821.23 *g/mol*):

calcd.: C 70.20, H 5.40, O 3.90, P 7.54, Pd 12.96

found: C 69.93, H 5.35

Bis(triphenylphosphine)-(η^2-*iso*-propyl cinnamate)platinum(0)

PtPh-^{i}Pr: C$_{48}$H$_{44}$O$_2$P$_2$Pt, M = 909.89 *g/mol*, white solid, yield: 86%, **route**: *Pt-VI*.

^{1}H-NMR (400.13 MHz; C_6D_6): $\delta = 0.75$ (d, $^{3}J_{HH} = 6.3$ Hz, 3H, CHCH_3), 1.12 (d, $^{3}J_{HH} = 6.3$ Hz, 3H, CHCH_3), 4.15 (dpt, $^{3}J_{HH} = 8.6$ Hz, $^{3}J_{HP} = 4.5$ Hz, $^{3}J_{HP} = 8.6$ Hz, 1H, CHCHCO), 4.15 (ddpt, $^{3}J_{HH} = 8.6$ Hz, $^{3}J_{HP} = 4.5$ Hz, $^{3}J_{HP} = 8.6$ Hz, $^{2}J_{HPt} = 61.1$ Hz, 1H, CHCHCO), 4.49 (dpt, $^{3}J_{HH} = 8.6$ Hz, $^{3}J_{HP} = 5.4$ Hz, $^{3}J_{HP} = 8.6$ Hz, 1H, $C_{Ar}CH$CH), 4.49 (ddpt, $^{3}J_{HH} = 8.6$ Hz, $^{3}J_{HP} = 5.4$ Hz, $^{3}J_{HP} = 8.6$ Hz, $^{2}J_{HPt} = 52.7$ Hz, 1H, $C_{Ar}CH$CH), 4.98 (hp, $^{3}J_{HH} = 6.3$ Hz, 1H, CH(CH$_3$)$_2$), 6.59 (d, $^{3}J_{HH} = 7.9$ Hz, 2H, H_{Ar}), 6.82 - 7.00 (m, 21H, H_{Ar}), 7.12 - 7.22 (m, 6H, H_{Ar}), 7.54 - 7.66 (m, 6H, H_{Ar}).

^{13}C-NMR (100.62 MHz; C_6D_6): $\delta = 21.8$ (CHCH_3), 22.6 (CHCH_3), 50.8 (dd, $^{2}J_{CP} = 5.1$ Hz, $^{2}J_{CP} = 27.9$ Hz, CHCHCO), 50.8 (ddd, $^{2}J_{CP} = 5.1$ Hz, $^{2}J_{CP} = 27.9$ Hz, $^{1}J_{CPt} = 212.8$ Hz, CHCHCO), 61.8 (dd, $^{2}J_{CP} = 3.6$ Hz, $^{2}J_{CP} = 31.5$ Hz, $C_{Ar}C$HCH), 61.8 (ddd, $^{2}J_{CP} = 3.6$ Hz, $^{2}J_{CP} = 31.5$ Hz, $^{1}J_{CPt} = 196.6$ Hz, $C_{Ar}C$HCH), 65.8 (OCH(CH$_3$)$_2$), 124.0 (d, $^{5}J_{CP} = 2.4$ Hz, C_{Ar}H), 124.0 (bm, C_{Ar}H), 126.2 (d, $^{5}J_{CP} = 3.0$ Hz, C_{Ar}H), 126.2 (dd, $^{5}J_{CP} = 3.0$ Hz, $^{4}J_{CPt} = 15.0$ Hz, C_{Ar}H), 127.9 (C_{Ar}H), 128.2 (C_{Ar}H), 128.3 (C_{Ar}H), 129.3 (d, $^{5}J_{CP} = 1.8$ Hz, C_{Ar}H), 129.6 (d, $^{5}J_{CP} = 1.6$ Hz, C_{Ar}H), 134.1 (C_{Ar}P), 134.1 (d, $^{3}J_{CPt} = 21.2$ Hz, C_{Ar}P), 134.3 (C_{Ar}P), 134.3 (d, $^{3}J_{CPt} = 20.9$ Hz, C_{Ar}P), 134.5 (C_{Ar}P), 134.5 (d, $^{3}J_{CPt} = 19.3$ Hz, C_{Ar}P), 134.6 (C_{Ar}P), 134.6 (d, $^{3}J_{CPt} = 18.7$ Hz, C_{Ar}P), 135.6 (d, $^{1}J_{CP} = 3.5$ Hz, C_{Ar}P), 135.6 (dd, $^{1}J_{CP} = 3.5$ Hz, $^{2}J_{CPt} = 33.4$ Hz, C_{Ar}P), 136.0 (d, $^{1}J_{CP} = 3.5$ Hz, C_{Ar}P), 136.0 (dd, $^{1}J_{CP} = 3.5$ Hz, $^{2}J_{CPt} = 31.8$ Hz, C_{Ar}P), 137.3 (d, $^{1}J_{CP} = 3.4$ Hz, C_{Ar}P), 137.3 (dd, $^{1}J_{CP} = 3.4$ Hz, $^{2}J_{CPt} = 29.8$ Hz, C_{Ar}P), 137.7 (d, $^{1}J_{CP} = 3.5$ Hz, C_{Ar}P), 137.7 (dd, $^{1}J_{CP} = 3.5$ Hz, $^{2}J_{CPt} = 28.5$ Hz, C_{Ar}P), 145.0 (dd, $^{3}J_{CP} = 1.5$ Hz, $^{3}J_{CP} = 5.9$ Hz, C_{Ar}CH), 145.0 (ddd, $^{3}J_{CP} = 1.5$ Hz, $^{3}J_{CP} = 5.9$ Hz, $^{2}J_{CPt} = 35.2$ Hz, C_{Ar}CH), 173.0 (dd, $^{3}J_{CP}$ 2.2 Hz, $^{3}J_{CP} = 5.1$ Hz, CHCOO), 173.0 (ddd, $^{3}J_{CP} = 2.2$ Hz, $^{3}J_{CP} = 5.1$ Hz, $^{2}J_{CPt} = 48.4$ Hz, CHCOO).

^{31}P-NMR (161.97 MHz; C_6D_6): $\delta = 27.8$ (d, $^{2}J_{PP} = 44.7$ Hz), 27.8 (dd, $^{2}J_{PP} = 44.7$ Hz, $^{1}J_{PPt} = 4123.2$ Hz), 28.5 (d, $^{2}J_{PP} = 44.7$ Hz), 28.5 (dd, $^{2}J_{PP} = 44.7$ Hz, $^{1}J_{PPt} = 3674.0$ Hz).

^{195}Pt-NMR (85.48 MHz; C_6D_6): $\delta = -5059.1$ (dd, $^{1}J_{PPt} = 3673.8$ Hz, $^{1}J_{PPt} = 4122.8$ Hz).

FAB-MS: m/z (%): 719.0 (100) [Pt(PPh$_3$)$_2^{+}$], 909.1 (1.30) [(C$_{12}$H$_{14}$O$_2$)Pt(PPh$_3$)$_2^{+}$].

FT-IR (ATR, cm^{-1}): 3046 (w), 2955 (w), 2848 (w), 1682 (s, CO), 1591 (w), 1432 (m), 1265 (m), 1155 (s), 1106 (s), 1093 (s), 985 (m), 742 (m), 692 (s), 560 (w), 509 (s), 501 (m), 424 (m).

elemental analysis for C$_{48}$H$_{44}$O$_2$P$_2$Pt (909.89 *g/mol*):

calcd.: C 63.36, H 4.87, O 3.52, P 6.81, Pt 21.44

found: C 63.65, H 5.31

Bis(triphenylphosphine)-(η^2-*tert*-butyl cinnamate)nickel(0)

NiPh-^{t}Bu: C$_{49}$H$_{46}$NiO$_2$P$_2$, M $= 787.53$ *g/mol*, brick red solid, yield: 90%, **route**: *Ni*.

^{1}H-NMR (400.13 MHz; C$_7$D$_8$): $\delta = 1.31$ (s, 9H, C(CH$_3$)$_3$), 3.50 (dpt, $^3J_{HH} = 8.9$ Hz, $^3J_{HP} = 3.6$ Hz, $^3J_{HP} = 8.9$ Hz, 1H, CHC*H*CO), 4.30 (dpt, $^3J_{HH} = 8.9$ Hz, $^3J_{HP} = 5.3$ Hz, $^3J_{HP} = 8.9$ Hz, 1H, C$_{Ar}$C*H*CH), 6.50 (d, $^3J_{HH} = 7.5$ Hz, 2H, H_{Ar}), 6.82 - 6.90 (m, 2H, H_{Ar}), 6.91 - 7.04 (m, 19H, H_{Ar}), 7.04 - 7.13 (m, 6H, H_{Ar}), 7.49 - 7.60 (m, 6H, H_{Ar}).

^{13}C-NMR (100.62 MHz; C$_7$D$_8$): $\delta = 28.5$ (C*C*H$_3$), 54.8 (dd, $^2J_{CP} = 1.8$ Hz, $^2J_{CP} = 13.5$ Hz, CH*C*HCO), 66.2 (d, $^2J_{CP} = 16.0$ Hz, C$_{Ar}$*C*HCH), 77.5 (O*C*(CH$_3$)$_3$), 123.6 (d, $^4J_{CP} = 1.2$ Hz, C_{Ar}H), 125.3 (d, $^4J_{CP} = 1.9$ Hz, C_{Ar}H), 128.0 (C_{Ar}H), 128.1 (C_{Ar}H), 128.1 (C_{Ar}H), 128.7 (bs, C_{Ar}H), 128.9 (d, $^3J_{CP} = 1.3$ Hz, C_{Ar}H), 129.2 (d, $^3J_{CP} = 1.2$ Hz, C_{Ar}H), 134.2 (d, $^2J_{CP} = 12.5$ Hz, C_{Ar}H), 134.5 (d, $^2J_{CP} = 13.5$ Hz, C_{Ar}H), 135.0 (d, $^1J_{CP} = 3.0$ Hz, C_{Ar}), 135.3 (d, $^1J_{CP} = 3.0$ Hz, C_{Ar}), 136.3 (d, $^1J_{CP} = 2.6$ Hz, C_{Ar}), 136.6 (d, $^1J_{CP} = 2.6$ Hz, C_{Ar}), 143.9 (d, $^3J_{CP} = 5.4$ Hz, C_{Ar}), 172.8 (dd, $^3J_{CP} = 1.2$ Hz, $^3J_{CP} = 3.4$ Hz, CH*C*OO).

^{31}P-NMR (161.97 MHz; C$_7$D$_8$): $\delta = 31.1$ (d, $^2J_{PP} = 37.0$ Hz), 33.3 (d, $^2J_{PP} = 37.0$ Hz).

FT-IR (ATR, cm^{-1}): 1668 (m, CO).

Bis(triphenylphosphine)-(η^2-*tert*-butyl cinnamate)palladium(0)

PdPh-^{t}Bu: $C_{49}H_{46}O_2P_2Pd$, M = 835.26 g/mol, light yellow solid, yield: 95%, **route:** Pd-I.

^{1}H-NMR (400.13 MHz; C_7D_8): δ = 1.30 (s, 9H, C(CH_3)$_3$), 4.71 (bs, 1H, CHCHCO), 5.20 (bs, 1H, C$_{Ar}$CHCH), 6.50 - 6.58 (bm, 2H, H_{Ar}), 6.77 - 6.89 (bm, 3H, H_{Ar}), 6.89 - 7.02 (bm, 18H, H_{Ar}), 7.02 - 7.15 (bm, 6H, H_{Ar}), 7.42 - 7.61 (bm, 6H, H_{Ar}).

^{1}H-NMR (400.13 MHz; 246 K; C_7D_8): δ = 1.31 (s, 9H, C(CH_3)$_3$), 4.83 (dpt, $^3J_{HH}$ = 10.1 Hz, $^3J_{HP}$ = 6.1 Hz, $^3J_{HP}$ = 6.1 Hz, 1H, CHCHCO), 5.25 (dpt, $^3J_{HH}$ = 10.1 Hz, $^3J_{HP}$ = 7.0 Hz, $^3J_{HP}$ = 7.0 Hz, 1H, C$_{Ar}$CHCH), 6.50 - 6.56 (m, 2H, H_{Ar}), 6.87 - 6.93 (m, 9H, H_{Ar}), 6.93 - 6.98 (m, 12H, H_{Ar}), 7.01 - 7.07 (m, 6H, H_{Ar}), 7.52 - 7.60 (m, 6H, H_{Ar}).

^{13}C-NMR (100.62 MHz; C_7D_8): δ = 28.3 (CCH$_3$), 63.0 (bs, CHCHCO), 77.1 (bs, C$_{Ar}$$C$HCH), 77.8 (bs, O$C$(CH$_3$)$_3$), 124.5 (bs, C_{Ar}H), 125.6 (bs, C_{Ar}H), 128.2 (C_{Ar}H), 128.2 (C_{Ar}H), 128.2 (C_{Ar}H), 128.4 (C_{Ar}H), 128.9 (C_{Ar}H), 129.2 (C_{Ar}H), 133.8 - 134.9 (m, C_{Ar}H), 135.8 (bs, C_{Ar}H), 136.2 (bs, C_{Ar}H), 142.9 (bs, C_{Ar}CH), 170.9 (bs, CHCOO).

^{13}C-NMR (100.62 MHz; 246 K; C_7D_8): δ = 28.1 (CCH$_3$), 62.1 (dd, $^2J_{CP}$ = 5.7 Hz, $^2J_{CP}$ = 17.5 Hz, CHCHCO), 76.8 (dd, $^2J_{CP}$ = 2.8 Hz, $^2J_{CP}$ = 20.8 Hz, C$_{Ar}$$C$HCH), 77.7 (O$C$(CH$_3$)$_3$), 124.4 (bs, C_{Ar}H), 128.2 (C_{Ar}), 128.3 (C_{Ar}), 128.9 (C_{Ar}), 129.2 (C_{Ar}), 134.1 (d, $^2J_{CP}$ = 14.4 Hz, C_{Ar}H), 134.5 (d, $^2J_{CP}$ = 15.7 Hz, C_{Ar}H), 135.6 (dd, $^1J_{CP}$ = 27.5 Hz, $^3J_{CP}$ = 1.8 Hz, C_{Ar}P), 137.4 (dd, $^1J_{CP}$ = 26.7 Hz, $^3J_{CP}$ = 2.1 Hz, C_{Ar}P), 142.6 (d, $^3J_{CP}$ = 5.9 Hz, C_{Ar}CH), 170.9 (dd, $^3J_{CP}$ = 2.2 Hz, $^3J_{CP}$ = 3.9 Hz, CHCOO).

^{31}P-NMR (161.97 MHz; C_7D_8): δ = 24.0 (bd, $^2J_{PP}$ = 21.6 Hz), 25.5 (bd, $^2J_{PP}$ = 21.6 Hz).

^{31}P-NMR (161.97 MHz; 246 K; C$_7$D$_8$): δ = 23.5 (d, $^2J_{PP}$ = 20.6 Hz), 25.2 (d, $^2J_{PP}$ = 20.6 Hz).

FT-IR (ATR, cm^{-1}): 3047 (w), 1676 (s, CO), 1575 (w), 1478 (m), 1432 (s), 1287 (w), 1257 (w), 1130 (s), 1091 (m), 938 (w), 817 (w), 741 (s), 681 (s), 512 (m), 501 (s), 408 (s).

elemental analysis for C$_{49}$H$_{46}$O$_2$P$_2$Pd (835.26 g/mol):

calcd.: C 70.46, H 5.55, O 3.83, P 7.42, Pd 12.74

found: C 70.46, H 5.56

Bis(triphenylphosphine)-(η^2-*tert*-butyl cinnamate)platinum(0)

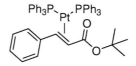

PtPh-^{t}Bu: C$_{49}$H$_{46}$O$_2$P$_2$Pt, M = 923.91 g/mol, white solid, yield: 89%, **route**: *Pt-VI*.

^{1}H-NMR (400.13 MHz; C$_6$D$_6$): δ = 1.31 (s, 9H, C(CH$_3$)$_3$), 4.12 (dpt, $^3J_{HH}$ = 8.7 Hz, $^3J_{HP}$ = 4.8 Hz, $^3J_{HP}$ = 8.7 Hz, 1H, CHCHCO), 4.12 (ddpt, $^3J_{HH}$ = 8.7 Hz, $^3J_{HP}$ = 4.8 Hz, $^3J_{HP}$ = 8.7 Hz, $^2J_{HPt}$ = 62.2 Hz, 1H, CHCHCO), 4.43 (dpt, $^3J_{HH}$ = 8.7 Hz, $^3J_{HP}$ = 5.8 Hz, $^3J_{HP}$ = 8.7 Hz, 1H, C$_{Ar}$CHCH), 4.43 (ddpt, $^3J_{HH}$ = 8.7 Hz, $^3J_{HP}$ = 5.8 Hz, $^3J_{HP}$ = 8.7 Hz, $^2J_{HPt}$ = 52.2 Hz, 1H, C$_{Ar}$CHCH), 6.55 (d, $^3J_{HH}$ = 7.9 Hz, 2H, H_{Ar}), 6.75 - 7.02 (m, 21H, H_{Ar}), 7.08 - 7.22 (m, 6H, H_{Ar}), 7.58 - 7.70 (m, 6H, H_{Ar}).

^{13}C-NMR (100.62 MHz; C$_6$D$_6$): δ = 28.5 (CCH$_3$), 51.7 (dd, $^2J_{CP}$ = 6.3 Hz, $^2J_{CP}$ = 29.3 Hz, CHCHCO), 51.7 (ddd, $^2J_{CP}$ = 6.3 Hz, $^2J_{CP}$ = 29.3 Hz, $^1J_{CPt}$ = 218.6 Hz, CHCHCO), 62.7 (dd, $^2J_{CP}$ = 4.5 Hz, $^2J_{CP}$ = 32.3 Hz, C$_{Ar}$$C$HCH), 62.7 (ddd, $^2J_{CP}$ = 4.5 Hz, $^2J_{CP}$ = 32.3 Hz, $^1J_{CPt}$ = 195.2 Hz, C$_{Ar}$$C$HCH), 77.6 (O$C$(CH$_3$)$_3$), 123.9 (d, $^5J_{CP}$ = 2.3 Hz, C$_{Ar}$H), 123.9 (bm, C$_{Ar}$H), 126.1 (d, $^5J_{CP}$ = 3.0 Hz, C$_{Ar}$H), 126.1 (dd, $^5J_{CP}$ = 3.0 Hz, $^4J_{CPt}$ = 13.7 Hz, C$_{Ar}$H), 127.9 (C$_{Ar}$H), 128.0 (C$_{Ar}$H), 128.2 (C$_{Ar}$H), 128.3 (C$_{Ar}$H), 129.3 (d, $^5J_{CP}$ = 1.7 Hz, C$_{Ar}$H), 129.5 (d, $^5J_{CP}$ = 1.5 Hz, C$_{Ar}$H), 134.2 (C$_{Ar}$P), 134.2 (d, $^3J_{CPt}$ = 20.2 Hz, C$_{Ar}$P), 134.3 (C$_{Ar}$P), 134.3 (d, $^3J_{CPt}$ = 20.4 Hz, C$_{Ar}$P), 134.6

(C_{Ar}P), 134.6 (d, $^3J_{CPt} = 19.1$ Hz, C_{Ar}P), 134.7 (C_{Ar}P), 134.7 (d, $^3J_{CPt} = 19.3$ Hz, C_{Ar}P), 135.6 (d, $^1J_{CP} = 2.8$ Hz, C_{Ar}P), 135.6 (dd, $^1J_{CP} = 2.8$ Hz, $^2J_{CPt} = 33.7$ Hz, C_{Ar}P), 136.0 (d, $^1J_{CP} = 2.8$ Hz, C_{Ar}P), 136.0 (dd, $^1J_{CP} = 2.8$ Hz, $^2J_{CPt} = 37.6$ Hz, C_{Ar}P), 137.3 (d, $^1J_{CP} = 2.8$ Hz, C_{Ar}P), 137.3 (dd, $^1J_{CP} = 2.8$ Hz, $^2J_{CPt} = 32.2$ Hz, C_{Ar}P), 137.7 (d, $^1J_{CP} = 2.7$ Hz, C_{Ar}P), 137.7 (dd, $^1J_{CP} = 2.7$ Hz, $^2J_{CPt} = 32.8$ Hz, C_{Ar}P), 145.0 (dd, $^3J_{CP} = 1.5$ Hz, $^3J_{CP} = 6.6$ Hz, C_{Ar}CH), 145.0 (ddd, $^3J_{CP} = 1.5$ Hz, $^3J_{CP} = 6.6$ Hz, $^2J_{CPt} = 34.6$ Hz, C_{Ar}CH), 173.2 (dd, $^3J_{CP}$ 2.3 Hz, $^3J_{CP} = 4.8$ Hz, CHCOO), 173.2 (ddd, $^3J_{CP} = 2.3$ Hz, $^3J_{CP} = 4.8$ Hz, $^2J_{CPt} = 45.9$ Hz, CHCOO).

^{31}P-NMR (161.97 MHz; C$_6$D$_6$): $\delta = 27.2$ (d, $^2J_{PP} = 45.5$ Hz), 27.2 (dd, $^2J_{PP} = 45.5$ Hz, $^1J_{PPt} = 4077.4$ Hz), 28.5 (d, $^2J_{PP} = 45.5$ Hz), 28.5 (dd, $^2J_{PP} = 45.5$ Hz, $^1J_{PPt} = 3720.0$ Hz).

^{195}Pt-NMR (85.48 MHz; C$_6$D$_6$): $\delta = -5060.2$ (dd, $^1J_{PPt} = 3730.3$ Hz, $^1J_{PPt} = 4067.2$ Hz).

FAB-MS: m/z (%): 719.0 (100) [Pt(PPh$_3$)$_2$$^+$], 923.1 (2.08) [(C$_{13}H_{16}O_2$)Pt(PPh$_3$)$_2$$^+$].

FT-IR (ATR, cm^{-1}): 3044 (w), 2968 (w), 1681 (s, CO), 1591 (w), 1485 (w), 1425 (m), 1258 (m), 1130 (s), 1084 (m), 955 (w), 834 (m), 743 (s), 682 (s), 561 (w), 539 (m), 516 (m), 500 (s), 410 (m).

elemental analysis for C$_{49}$H$_{46}$O$_2$P$_2$Pt (923.91 g/mol):

calcd.: C 63.70, H 5.02, O 3.46, P 6.70, Pt 21.11

found: C 63.44, H 5.14

Bis(triphenylphosphine)-(η^2-methyl 4-chlorocinnamate)palladium(0)

PdPhCl-Me: C$_{46}$H$_{39}$ClO$_2$P$_2$Pd, M = 827.62 g/mol, light yellow solid, yield: >99%, **route:** *Pd-I.*

^{1}H-NMR (400.13 MHz; C$_7$D$_8$): $\delta = 3.19$ (s, 3H, CH_3), 4.55 (bs, 1H, CHCHCO), 5.13 (bs, 1H, C$_{Ar}$CHCH), 6.28 - 6.40 (bm, 2H, H_{Ar}), 6.70 - 6.80 (bm, 2H, H_{Ar}), 6.85 - 7.03 (bm, 18H, H_{Ar}), 7.03 - 7.12 (bm, 6H, H_{Ar}), 7.40 - 7.56 (bm, 6H, H_{Ar}).

^{1}H-NMR (400.13 MHz; 246 K; C$_7$D$_8$): $\delta = 3.16$ (s, 3H, CH_3), 4.69 (dpt, $^3J_{HH} = 10.3$ Hz, $^3J_{HP} = 6.2$ Hz, $^3J_{HP} = 6.2$ Hz, 1H, CHCHCO), 5.19 (dpt, $^3J_{HH} = 10.3$ Hz, $^3J_{HP} = 7.3$ Hz, $^3J_{HP} = 7.3$ Hz, 1H, C$_{Ar}$CHCH), 6.27 (d, $^3J_{HH} = 7.4$ Hz, 2H, H_{Ar}), 6.74 (d, $^3J_{HH} = 8.5$ Hz, 2H, H_{Ar}), 6.82 - 6.99 (m, 18H, H_{Ar}), 7.02 - 7.10 (m, 6H, H_{Ar}), 7.48 - 7.58 (m, 6H, H_{Ar}).

^{13}C-NMR (100.62 MHz; C$_7$D$_8$): $\delta = 50.1$ (OCH$_3$), 61.6 (bs, CHCHCO), 74.0 (bs, C$_{Ar}$$C$HCH), 126.6 (bs, C_{Ar}H), 128.3 - 124.4 (m, C_{Ar}H), 128.5 (C_{Ar}H), 129.1 (C_{Ar}H), 134.2 (bs, C_{Ar}H), 134.3 (bs, C_{Ar}H), 135.7 (bs, C_{Ar}H), 136.0 (bs, C_{Ar}H), 141.4 (bs, C_{Ar}CH), 171.2 (bs, CHCOO).

^{13}C-NMR (100.62 MHz; 246 K; C$_7$D$_8$): $\delta = 50.1$ (s, OCH$_3$), 59.9 (dd, $^2J_{CP} = 3.6$ Hz, $^2J_{CP} = 18.2$ Hz, CHCHCO), 73.5 (dd, $^2J_{CP} = 2.1$ Hz, $^2J_{CP} = 22.9$ Hz, C$_{Ar}$$C$HCH), 126.4 (bs, C_{Ar}H), 128.2 (C_{Ar}H), 128.4 (C_{Ar}H), 128.5 (C_{Ar}H), 129.2 (d, $^3J_{CP} = 20.0$ Hz, C_{Ar}H), 134.0 (d, $^2J_{CP} = 14.6$ Hz, C_{Ar}H), 134.2 (d, $^2J_{CP} = 15.3$ Hz, C_{Ar}H), 135.3 (dd, $^1J_{CP} = 28.7$ Hz, $^3J_{CP} = 1.6$ Hz, C_{Ar}P), 136.9 (dd, $^1J_{CP} = 27.6$ Hz, $^3J_{CP} = 1.6$ Hz, C_{Ar}P), 141.0 (d, $^3J_{CP} = 6.4$ Hz, C_{Ar}CH), 171.1 (dd, $^3J_{CP} = 2.3$ Hz, $^3J_{CP} = 4.0$ Hz, CHCOO).

^{31}P-NMR (161.97 MHz; C$_7$D$_8$): $\delta = 24.8$ (d, $^2J_{PP} = 15.2$ Hz), 26.3 (d, $^2J_{PP} = 15.2$ Hz).

^{31}P-NMR (161.97 MHz; 246 K; C$_7$D$_8$): $\delta = 24.2$ (d, $^2J_{PP} = 15.4$ Hz), 26.1 (d, $^2J_{PP} = 15.4$ Hz).

FT-IR (ATR, cm^{-1}): 3075 (w), 1681 (m, CO), 1589 (w), 1476 (m), 1432 (s), 1309 (m), 1255 (w), 1145 (s), 1090 (s), 1028 (w), 808 (w), 740 (s), 692 (s), 502 (s), 402 (m).

Bis(triphenylphosphine)-(η^2-methyl 4-chlorocinnamate)platinum(0)

PtPhCl-Me: $C_{46}H_{39}ClO_2P_2Pt$, M $= 916.28$ g/mol, white solid, yield: $>99\%$, **route**: *Pt-VI*.

^{1}H-NMR (400.13 MHz; C_6D_6): $\delta = 3.22$ (s, 3H, OCH_3), 3.91 (dpt, $^3J_{HH} = 8.4$ Hz, $^3J_{HP} = 4.6$ Hz, $^3J_{HP} = 8.4$ Hz, 1H, CHCHCO), 3.91 (ddpt, $^3J_{HH} = 8.4$ Hz, $^3J_{HP} = 4.6$ Hz, $^3J_{HP} = 8.4$ Hz, $^2J_{HPt} = 60.7$ Hz, 1H, CHCHCO), 4.40 (dpt, $^3J_{HH} = 8.4$ Hz, $^3J_{HP} = 5.6$ Hz, $^3J_{HP} = 8.4$ Hz, 1H, $C_{Ar}CH$CH), 4.40 (ddpt, $^3J_{HH} = 8.4$ Hz, $^3J_{HP} = 5.6$ Hz, $^3J_{HP} = 8.4$ Hz, $^2J_{HPt} = 54.3$ Hz, 1H, $C_{Ar}CH$CH), 6.36 (d, $^3J_{HH} = 8.0$ Hz, 2H, H_{Ar}), 6.80 - 6.93 (m, 10H, H_{Ar}), 6.93 - 7.00 (m, 9H, H_{Ar}), 7.12 - 7.22 (m, 7H, H_{Ar}), 7.49 - 7.59 (m, 6H, H_{Ar}).

^{13}C-NMR (100.62 MHz; C_6D_6): $\delta = 49.7$ (d, $^2J_{CP} = 23.9$ Hz, CHCHCO), 49.7 (dd, $^2J_{CP} = 23.9$ Hz, $^1J_{CPt} = 199.3$ Hz, CHCHCO), 50.2 (COOCH$_3$), 59.4 (d, $^2J_{CP} = 28.1$ Hz, $C_{Ar}C$HCH), 59.4 (dd, $^2J_{CP} = 28.1$ Hz, $^1J_{CPt} = 197.2$ Hz, $C_{Ar}C$HCH), 127.2 (d, $^5J_{CP} = 2.7$ Hz, C_{Ar}H), 127.2 (dd, $^5J_{CP} = 2.7$ Hz, $^4J_{CPt} = 14.7$ Hz, C_{Ar}H), 127.9 (C_{Ar}H), 128.2 (C_{Ar}H), 128.6 (C_{Ar}), 129.4 (d, $^5J_{CP} = 1.0$ Hz, C_{Ar}H), 129.6 (d, $^5J_{CP} = 0.9$ Hz, C_{Ar}H), 134.1 (d, $^2J_{CP} = 1.8$ Hz, C_{Ar}P), 134.1 (dd, $^2J_{CP} = 1.8$ Hz, $^3J_{CPt} = 21.2$ Hz, C_{Ar}P), 134.2 (d, $^2J_{CP} = 1.8$ Hz, C_{Ar}P), 134.2 (dd, $^2J_{CP} = 1.8$ Hz, $^3J_{CPt} = 18.6$ Hz, C_{Ar}P), 134.2 (d, $^2J_{CP} = 1.9$ Hz, C_{Ar}P), 134.2 (dd, $^2J_{CP} = 1.9$ Hz, $^3J_{CPt} = 19.2$ Hz, C_{Ar}P), 134.3 (d, $^2J_{CP} = 1.9$ Hz, C_{Ar}P), 134.3 (dd, $^2J_{CP} = 1.9$ Hz, $^3J_{CPt} = 18.4$ Hz, C_{Ar}P), 135.4 (d, $^2J_{CP} = 7.8$ Hz, C_{Ar}P), 135.4 (dd, $^2J_{CP} = 7.8$ Hz, $^3J_{CPt} = 32.6$ Hz, C_{Ar}P), 135.8 (d, $^2J_{CP} = 7.7$ Hz, C_{Ar}P), 135.8 (dd, $^2J_{CP} = 7.7$ Hz, $^3J_{CPt} = 29.7$ Hz, C_{Ar}P), 136.8 (d, $^2J_{CP} = 7.9$ Hz, C_{Ar}P), 136.8 (dd, $^2J_{CP} = 7.9$ Hz, $^3J_{CPt} = 29.4$ Hz, C_{Ar}P), 137.2 (d, $^2J_{CP} = 7.9$ Hz, C_{Ar}P), 137.2 (dd, $^2J_{CP} = 7.9$ Hz, $^3J_{CPt} = 28.4$ Hz, C_{Ar}P), 143.6 (d, $^3J_{CP} = 5.5$ Hz, C_{Ar}CH), 143.6 (dd, $^3J_{CP} = 5.5$ Hz, $^2J_{CPt} = 35.7$ Hz, C_{Ar}CH), 173.5 (dd, $^3J_{CP}$ 1.3 Hz, $^3J_{CP} = 4.0$ Hz, CHCOO), 173.5 (ddd, $^3J_{CP} = 1.3$ Hz, $^3J_{CP} = 4.0$ Hz, $^2J_{CPt} = 43.2$ Hz, CHCOO).

^{31}P-NMR (161.97 MHz; C$_6$D$_6$): δ = 27.6 (d, $^2J_{PP}$ = 41.0 Hz), 27.6 (dd, $^2J_{PP}$ = 41.0 Hz, $^1J_{PPt}$ = 4121.1 Hz), 28.0 (d, $^2J_{PP}$ = 41.0 Hz), 28.0 (dd, $^2J_{PP}$ = 41.0 Hz, $^1J_{PPt}$ = 3645.8 Hz).

^{195}Pt-NMR (85.48 MHz; C$_6$D$_6$): δ = -5067.0 (dd, $^1J_{PPt}$ = 3645.0 Hz, $^1J_{PPt}$ = 4119.5 Hz).

FAB-MS: m/z (%): 719.2 (100) [Pt(PPh$_3$)(PPh$_3$)$^+$].

FT-IR (ATR, cm^{-1}): 3051 (w), 1687 (m, CO), 1469 (w), 1433 (s), 1378 (w), 1287 (w), 1242 (w), 1148 (s), 1093 (s), 1015 (w), 818 (w), 742 (m), 693 (s), 530 (w), 508 (s), 409 (m).

Bis(triphenylphosphine)-(η^2-ethyl 4-chlorocinnamate)palladium(0)

PdPhCl-Et: C$_{47}$H$_{41}$ClO$_2$P$_2$Pd, M = 841.65 *g/mol*, light yellow solid, yield: 97%, **route**: *Pd-I*.

^{1}H-NMR (400.13 MHz; C$_7$D$_8$): δ = 0.87 (bs, 3H, CH$_2$CH_3), 3.59 (bs, 1H, CH_2CH$_3$), 3.98 (bs, 1H, CH_2CH$_3$), 4.58 (bs, 1H, CHCHCO), 5.13 (bs, 1H, C$_{Ar}$CHCH), 6.27 - 6.40 (bm, 2H, H_{Ar}), 6.68 - 6.79 (bm, 2H, H_{Ar}), 6.86 - 7.02 (bm, 18H, H_{Ar}), 7.02 - 7.15 (bm, 6H, H_{Ar}), 7.41 - 7.61 (bm, 6H, H_{Ar}).

^{1}H-NMR (400.13 MHz; 246 K; C$_7$D$_8$): δ = 0.82 (t, $^3J_{HH}$ = 7.1 Hz, 3H, CH$_2$CH_3), 3.54 (dq, $^2J_{HH}$ = 10.7 Hz, $^3J_{HH}$ = 7.1 Hz, 1H, CH_2CH$_3$), 3.99 (dq, $^2J_{HH}$ = 10.7 Hz, $^3J_{HH}$ = 7.1 Hz, 1H, CH_2CH$_3$), 4.74 (dpt, $^3J_{HH}$ = 10.3 Hz, $^3J_{HP}$ = 6.2 Hz, $^3J_{HP}$ = 6.2 Hz, 1H, CHCHCO), 5.19 (dpt, $^3J_{HH}$ = 10.3 Hz, $^3J_{HP}$ = 7.1 Hz, $^3J_{HP}$ = 7.1 Hz, 1H, C$_{Ar}$CHCH), 6.26 (d, $^3J_{HH}$ = 7.6 Hz, 2H, H_{Ar}), 6.72 (d, $^3J_{HH}$ = 8.6 Hz, 2H, H_{Ar}), 6.82 - 7.00 (m, 18H, H_{Ar}), 7.00 - 7.10 (m, 6H, H_{Ar}), 7.49 - 7.60 (m, 6H, H_{Ar}).

^{13}C-NMR (100.62 MHz; C$_7$D$_8$): δ = 14.4 (CH$_2$$CH_3$), 59.0 (bs, CH$_2CH_3$), 60.5 (bs, CHCHCO), 74.4 (bs, C$_{Ar}$$C$HCH), 126.5 (bs, C_{Ar}H), 128.3 (C_{Ar}H), 128.4 (C_{Ar}H), 128.5 (C_{Ar}H), 128.5 (C_{Ar}H), 133.8 - 134.7 (m, C_{Ar}H), 135.5 (bs, C_{Ar}H), 136.0 (bs, C_{Ar}H), 141.4 (bs, C_{Ar}CH), 170.7 (bs, CHCOO).

^{13}C-NMR (100.62 MHz; 246 K; C$_7$D$_8$): δ = 14.2 (CH$_2$$CH_3$), 59.0 ($CH_2CH_3$), 60.1 (dd, $^2J_{CP}$ = 5.9 Hz, $^2J_{CP}$ = 17.7 Hz, CHCHCO), 73.9 (dd, $^2J_{CP}$ = 2.4 Hz, $^2J_{CP}$ = 22.0 Hz, C$_{Ar}$$C$HCH), 126.4 (bs, C_{Ar}H), 128.2 (C_{Ar}H), 128.3 (C_{Ar}H), 128.3 (C_{Ar}H), 128.4 (C_{Ar}H), 128.5 (C_{Ar}H), 129.2 (d, $^3J_{CP}$ = 22.9 Hz, C_{Ar}H), 134.0 (d, $^2J_{CP}$ = 14.6 Hz, C_{Ar}H), 134.3 (d, $^2J_{CP}$ = 15.3 Hz, C_{Ar}H), 135.3 (dd, $^1J_{CP}$ = 28.4 Hz, $^3J_{CP}$ = 1.6 Hz, C_{Ar}P), 137.1 (dd, $^1J_{CP}$ = 27.5 Hz, $^3J_{CP}$ = 1.7 Hz, C_{Ar}P), 141.0 (d, $^3J_{CP}$ = 5.7 Hz, C_{Ar}CH), 170.8 (dd, $^3J_{CP}$ = 2.2 Hz, $^3J_{CP}$ = 4.2 Hz, CHCOO).

^{31}P-NMR (161.97 MHz; C$_7$D$_8$): δ = 23.9 (d, $^2J_{PP}$ = 17.2 Hz), 25.5 (d, $^2J_{PP}$ = 17.2 Hz).

^{31}P-NMR (161.97 MHz; 246 K; C$_7$D$_8$): δ = 23.9 (d, $^2J_{PP}$ = 16.5 Hz), 26.0 (d, $^2J_{PP}$ = 16.5 Hz).

FT-IR (ATR, cm^{-1}): 3060 (w), 1681 (m, CO), 1589 (w), 1476 (m), 1433 (s), 1307 (m), 1211 (w), 1147 (m), 1091 (s), 1036 (w), 817 (w), 741 (s), 692 (s), 503 (s), 410 (m).

Bis(triphenylphosphine)-(η^2-ethyl 4-chlorocinnamate)platinum(0)

PtPhCl-Et: C$_{47}$H$_{41}$ClO$_2$P$_2$Pt, M = 930.31 g/mol, white solid, yield: >99%, **route**: Pt-VI.

^{1}H-NMR (400.13 MHz; C$_6$D$_6$): δ = 0.84 (t, $^3J_{HH}$ = 7.1 Hz, 3H, CH$_2$CH_3), 3.66 (dq, $^2J_{HH}$ = 10.5 Hz, $^3J_{HH}$ = 7.1 Hz, 1H, CH_2CH$_3$), 3.95 (dpt, $^3J_{HH}$ = 8.7 Hz, $^3J_{HP}$ = 3.7 Hz, $^3J_{HP}$ = 8.7 Hz, 1H, CHCHCO), 3.95 (ddpt, $^3J_{HH}$ = 8.7 Hz, $^3J_{HP}$ = 3.7 Hz, $^3J_{HP}$ = 8.7 Hz, $^2J_{HPt}$ = 61.2 Hz, 1H, CHCHCO), 4.05 (dq, $^2J_{HH}$ = 10.5 Hz, $^3J_{HH}$ = 7.1 Hz, 1H, CH_2CH$_3$),

4.39 (dpt, $^3J_{HH}$ = 8.7 Hz, $^3J_{HP}$ = 5.5 Hz, $^3J_{HP}$ = 8.7 Hz, 1H, C$_{Ar}$CHCH), 4.39 (ddpt, $^3J_{HH}$ = 8.7 Hz, $^3J_{HP}$ = 5.5 Hz, $^3J_{HP}$ = 8.7 Hz, $^2J_{HPt}$ = 53.5 Hz, 1H, C$_{Ar}$CHCH), 6.34 (d, $^3J_{HH}$ = 8.1 Hz, 2H, H_{Ar}), 6.78 - 6.93 (m, 10H, H_{Ar}), 6.93 - 7.01 (m, 9H, H_{Ar}), 7.11 - 7.21 (m, 7H, H_{Ar}), 7.52 - 7.60 (m, 6H, H_{Ar}).

^{13}C-NMR (100.62 MHz; C$_6$D$_6$): δ = 14.5 (CH$_2$CH$_3$), 49.9 (dd, $^2J_{CP}$ = 4.4 Hz, $^2J_{CP}$ = 27.6 Hz, CHCHCO), 49.9 (ddd, $^2J_{CP}$ = 4.4 Hz, $^2J_{CP}$ = 27.6 Hz, $^1J_{CPt}$ = 205.0 Hz, CHCHCO), 59.0 (CH$_2$CH$_3$), 59.8 (dd, $^2J_{CP}$ = 2.9 Hz, $^2J_{CP}$ = 31.1 Hz, C$_{Ar}$CHCH), 59.8 (ddd, $^2J_{CP}$ = 2.9 Hz, $^2J_{CP}$ = 31.1 Hz, $^1J_{CPt}$ = 196.2 Hz, C$_{Ar}$CHCH), 127.2 (d, $^5J_{CP}$ = 3.0 Hz, C_{Ar}H), 127.2 (dd, $^5J_{CP}$ = 3.0 Hz, $^4J_{CPt}$ = 15.4 Hz, C_{Ar}H), 127.9 (C_{Ar}H), 128.0 (C_{Ar}H), 128.2 (C_{Ar}H), 128.6 (C_{Ar}), 129.4 (d, $^5J_{CP}$ = 1.3 Hz, C_{Ar}H), 129.6 (d, $^5J_{CP}$ = 1.4 Hz, C_{Ar}H), 134.1(C_{Ar}P), 134.1 (d, $^3J_{CPt}$ = 21.8 Hz, C_{Ar}P), 134.2 (C_{Ar}P), 134.2 (d, $^3J_{CPt}$ = 24.2 Hz, C_{Ar}P), 134.3 (C_{Ar}P), 134.3 (d, $^3J_{CPt}$ = 19.0 Hz, C_{Ar}P), 134.4 (C_{Ar}P), 134.4 (d, $^3J_{CPt}$ = 19.4 Hz, C_{Ar}P), 135.4 (d, $^2J_{CP}$ = 4.3 Hz, C_{Ar}P), 135.4 (dd, $^2J_{CP}$ = 4.3 Hz, $^3J_{CPt}$ = 29.8 Hz, C_{Ar}P), 135.8 (d, $^2J_{CP}$ = 4.3 Hz, C_{Ar}P), 135.8 (dd, $^2J_{CP}$ = 4.3 Hz, $^3J_{CPt}$ = 33.1 Hz, C_{Ar}P), 136.9 (d, $^2J_{CP}$ = 4.4 Hz, C_{Ar}P), 136.9 (dd, $^2J_{CP}$ = 4.4 Hz, $^3J_{CPt}$ = 29.2 Hz, C_{Ar}P), 137.3 (d, $^2J_{CP}$ = 4.2 Hz, C_{Ar}P), 137.3 (dd, $^2J_{CP}$ = 4.2 Hz, $^3J_{CPt}$ = 28.5 Hz, C_{Ar}P), 143.6 (dd, $^3J_{CP}$ = 1.0 Hz, $^3J_{CP}$ = 5.8 Hz, C_{Ar}CH), 143.6 (ddd, $^3J_{CP}$ = 1.0 Hz, $^3J_{CP}$ = 5.8 Hz, $^2J_{CPt}$ = 33.6 Hz, C_{Ar}CH), 173.2 (dd, $^3J_{CP}$ 2.3 Hz, $^3J_{CP}$ = 4.7 Hz, CHCOO), 173.2 (ddd, $^3J_{CP}$ = 2.3 Hz, $^3J_{CP}$ = 4.7 Hz, $^2J_{CPt}$ = 42.2 Hz, CHCOO).

^{31}P-NMR (161.97 MHz; C$_6$D$_6$): δ = 27.6 (d, $^2J_{PP}$ = 41.8 Hz), 27.6 (dd, $^2J_{PP}$ = 41.8 Hz, $^1J_{PPt}$ = 4111.3 Hz), 28.1 (d, $^2J_{PP}$ = 41.8 Hz), 28.1 (dd, $^2J_{PP}$ = 41.8 Hz, $^1J_{PPt}$ = 3661.9 Hz).

^{195}Pt-NMR (85.48 MHz; C$_6$D$_6$): δ = -5063.5 (dd, $^1J_{PPt}$ = 3661.0 Hz, $^1J_{PPt}$ = 4110.5 Hz).

FAB-MS: m/z (%): 719.1 (100) [Pt(PPh$_3$)(PPh$_2$C$_6$H$_4$)$^+$], 929.9 (1.09) [(C$_{11}$H11ClO$_2$)Pt(PPh$_3$)(PPh$_2$C$_6$H$_4$)$^+$].

FT-IR (ATR, cm^{-1}): 3045 (w), 2969 (w), 1681 (s, CO), 1575 (w), 1478 (m), 1433 (s), 1378 (w), 1294 (w), 1241 (w), 1144 (s), 1091 (s), 1022 (w), 802 (m), 741 (s), 692 (s), 484 (s), 450 (w), 408 (m).

Bis(triphenylphosphine)-(η^2-*iso*-propyl 4-chlorocinnamate)palladium(0)

PdPhCl-^{i}Pr: $C_{48}H_{43}ClO_2P_2Pd$, M $=$ 855.67 g/mol, light yellow solid, yield: $>99\%$, **route**: *Pd-I*.

^{1}H-NMR (400.13 MHz; C_7D_8): $\delta = 0.75$ (bs, 3H, CHCH_3), 1.14 (bs, 3H, CHCH_3), 4.58 (bs, 1H, CHCHCO), 4.88 (bs, 1H, CH(CH$_3$)$_2$), 5.11 (bs, 1H, C$_{Ar}$CHCH), 6.27 - 6.37 (bm, 2H, H_{Ar}), 6.69 - 6.80 (bm, 2H, H_{Ar}), 6.87 - 7.12 (bm, 24H, H_{Ar}), 7.43 - 7.62 (bm, 6H, H_{Ar}).

^{1}H-NMR (400.13 MHz; 246 K; C_7D_8): $\delta = 0.69$ (d, $^3J_{HH} = 6.2$ Hz, 3H, CHCH_3), 1.18 (d, $^3J_{HH} = 6.2$ Hz, 3H, CHCH_3), 4.75 (dpt, $^3J_{HH} = 10.2$ Hz, $^3J_{HP} = 6.2$ Hz, $^3J_{HP} = 6.2$ Hz, 1H, CHCHCO), 4.88 (hp, $^3J_{HH} = 6.2$ Hz, 1H, CH(CH$_3$)$_2$), 5.17 (dpt, $^3J_{HH} = 10.2$ Hz, $^3J_{HP} = 6.9$ Hz, $^3J_{HP} = 6.9$ Hz, 1H, C$_{Ar}$CHCH), 6.25 (d, $^3J_{HH} = 7.7$ Hz, 2H, H_{Ar}), 6.69 (d, $^3J_{HH} = 8.6$ Hz, 2H, H_{Ar}), 6.84 - 7.07 (m, 24H, H_{Ar}), 7.50 - 7.62 (m, 6H, H_{Ar}).

^{13}C-NMR (100.62 MHz; C_7D_8): $\delta = 21.7$ (bs, CHCH_3), 22.4 (bs, CHCH_3), 61.4 (bs, CHCHCO), 66.2 (bs, OCH(CH$_3$)$_2$), 74.8 (bs, C$_{Ar}$CHCH), 126.5 - 126.6 (m, C_{Ar}H), 128.3 (C_{Ar}H), 128.5 (C_{Ar}H), 133.9 - 134.8 (m, C_{Ar}H), 135.5 - 136.2 (m, C_{Ar}H), 141.4 (bs, C_{Ar}CH), 170.6 (bs, CHCOO).

^{13}C-NMR (100.62 MHz; 246 K; C_7D_8): $\delta = 21.3$ (CHCH_3), 22.4 (CHCH_3), 60.3 (dd, $^2J_{CP} = 6.1$ Hz, $^2J_{CP} = 18.3$ Hz, CHCHCO), 66.1 (OCH(CH$_3$)$_2$), 74.5 (dd, $^2J_{CP} = 2.6$ Hz, $^2J_{CP} = 20.5$ Hz, C$_{Ar}$$C$HCH), 126.3 (bs, C_{Ar}H), 128.3 (C_{Ar}H), 128.4 (C_{Ar}H), 128.5 (C_{Ar}H), 129.2 (d, $^3J_{CP} = 26.4$ Hz, C_{Ar}H), 134.0 (d, $^2J_{CP} = 14.5$ Hz, C_{Ar}H), 134.4 (d, $^2J_{CP} = 15.4$ Hz, C_{Ar}H), 135.3 (dd, $^1J_{CP} = 28.3$ Hz, $^3J_{CP} = 1.5$ Hz, C_{Ar}P), 137.2 (dd,

$^1J_{CP} = 31.0$ Hz, $^3J_{CP} = 1.8$ Hz, $C_{Ar}P$), 141.1 (d, $^3J_{CP} = 6.5$ Hz, $C_{Ar}CH$), 170.6 (dd, $^3J_{CP} = 1.9$ Hz, $^3J_{CP} = 3.8$ Hz, $CHCOO$).

^{31}P-NMR (161.97 MHz; C$_7$D$_8$): $\delta = 23.4$ (d, $^2J_{PP} = 16.5$ Hz), 25.3 (d, $^2J_{PP} = 16.5$ Hz).

^{31}P-NMR (161.97 MHz; 246 K; C$_7$D$_8$): $\delta = 23.5$ (d, $^2J_{PP} = 17.4$ Hz), 25.7 (d, $^2J_{PP} = 17.4$ Hz).

FT-IR (ATR, cm^{-1}): 3056 (w), 2980 (w), 1677 (s, CO), 1589 (w), 1476 (m), 1433 (s), 1302 (w), 1256 (w), 1153 (m), 1107 (w), 1091 (s), 983 (w), 908 (w), 817 (m), 741 (s), 688 (s), 514 (s), 415 (m).

Bis(triphenylphosphine)-(η^2-*iso*-propyl 4-chlorocinnamate)platinum(0)

PtPhCl-^{i}Pr: C$_{48}$H$_{43}$ClO$_2$P$_2$Pt, M $= 944.33$ g/mol, white solid, yield: >99%, **route**: *Pt-VI*.

^{1}H-NMR (400.13 MHz; C$_6$D$_6$): $\delta = 0.76$ (d, $^3J_{HH} = 6.2$ Hz, 3H, CHCH_3), 1.12 (d, $^3J_{HH} = 6.2$ Hz, 3H, CHCH_3), 3.99 (dpt, $^3J_{HH} = 8.5$ Hz, $^3J_{HP} = 4.6$ Hz, $^3J_{HP} = 8.5$ Hz, 1H, CHCHCO), 3.99 (ddpt, $^3J_{HH} = 8.5$ Hz, $^3J_{HP} = 4.6$ Hz, $^3J_{HP} = 8.5$ Hz, $^2J_{HPt} = 60.9$ Hz, 1H, CHCHCO), 4.37 (dpt, $^3J_{HH} = 8.5$ Hz, $^3J_{HP} = 5.8$ Hz, $^3J_{HP} = 8.5$ Hz, 1H, C$_{Ar}$CHCH), 4.37 (ddpt, $^3J_{HH} = 8.5$ Hz, $^3J_{HP} = 5.8$ Hz, $^3J_{HP} = 8.5$ Hz, $^2J_{HPt} = 52.4$ Hz, 1H, C$_{Ar}$CHCH), 4.97 (hp, $^3J_{HH} = 6.2$ Hz, 1H, CH(CH$_3$)$_2$), 6.32 (d, $^3J_{HH} = 8.1$ Hz, 2H, H_{Ar}), 6.79 (d, $^3J_{HH} = 7.8$ Hz, 2H, H_{Ar}), 6.82 - 6.93 (m, 9H, H_{Ar}), 6.93 - 7.01 (m, 8H, H_{Ar}), 7.10 - 7.21 (m, 7H, H_{Ar}), 7.54 - 7.62 (m, 6H, H_{Ar}).

^{13}C-NMR (100.62 MHz; C$_6$D$_6$): $\delta = 21.8$ (CHCH_3), 22.6 (CHCH_3), 50.2 (dd, $^2J_{CP} = 6.1$ Hz, $^2J_{CP} = 29.3$ Hz, CHCHCO), 50.2 (ddd, $^2J_{CP} = 6.1$ Hz, $^2J_{CP} = 29.3$ Hz, $^1J_{CPt} = 198.2$ Hz, CHCHCO), 60.3 (dd, $^2J_{CP} = 4.3$ Hz, $^2J_{CP} = 32.6$ Hz, C$_{Ar}$$C$HCH), 60.3 (ddd, $^2J_{CP} = 4.3$ Hz, $^2J_{CP} = 32.6$ Hz, $^1J_{CPt} = 186.4$ Hz, C$_{Ar}$$C$HCH), 66.0 (O$C$H(CH$_3$)$_2$),

127.2 (d, $^5J_{CP}$ = 3.1 Hz, C_{Ar}H), 127.2 (dd, $^5J_{CP}$ = 3.1 Hz, $^4J_{CPt}$ = 14.6 Hz, C_{Ar}H),

128.0 (C_{Ar}H), 128.2 (C_{Ar}H), 128.6 (C_{Ar}), 129.4 (d, $^5J_{CP}$ = 1.8 Hz, C_{Ar}H), 129.7 (d,

$^5J_{CP}$ = 1.6 Hz, C_{Ar}H), 134.0 (C_{Ar}P), 134.0 (d, $^3J_{CPt}$ = 21.0 Hz, C_{Ar}P), 134.2 (C_{Ar}P),

134.2 (d, $^3J_{CPt}$ = 20.8 Hz, C_{Ar}P), 134.4 (C_{Ar}P), 134.4 (d, $^3J_{CPt}$ = 19.2 Hz, C_{Ar}P), 134.6

(C_{Ar}P), 134.6 (d, $^3J_{CPt}$ = 19.0 Hz, C_{Ar}P), 135.3 (d, $^2J_{CP}$ = 2.8 Hz, C_{Ar}P), 135.3 (dd,

$^2J_{CP}$ = 2.8 Hz, $^3J_{CPt}$ = 31.5 Hz, C_{Ar}P), 135.8 (d, $^2J_{CP}$ = 2.8 Hz, C_{Ar}P), 135.8 (dd,

$^2J_{CP}$ = 2.8 Hz, $^3J_{CPt}$ = 32.5 Hz, C_{Ar}P), 137.0 (d, $^2J_{CP}$ = 2.9 Hz, C_{Ar}P), 137.0 (dd,

$^2J_{CP}$ = 2.9 Hz, $^3J_{CPt}$ = 30.0 Hz, C_{Ar}P), 137.4 (d, $^2J_{CP}$ = 2.9 Hz, C_{Ar}P), 137.4 (dd,

$^2J_{CP}$ = 2.9 Hz, $^3J_{CPt}$ = 37.4 Hz, C_{Ar}P), 143.6 (dd, $^3J_{CP}$ = 1.6 Hz, $^3J_{CP}$ = 6.7 Hz,

C_{Ar}CH), 143.6 (ddd, $^3J_{CP}$ = 1.6 Hz, $^3J_{CP}$ = 6.7 Hz, $^2J_{CPt}$ = 21.0 Hz, C_{Ar}CH), 172.9

(dd, $^3J_{CP}$ 2.1 Hz, $^3J_{CP}$ = 4.9 Hz, CHCOO), 172.9 (ddd, $^3J_{CP}$ = 2.1 Hz, $^3J_{CP}$ = 4.9 Hz,

$^2J_{CPt}$ = 41.6 Hz, CHCOO).

^{31}P-NMR (161.97 MHz; C_6D_6): δ = 27.3 (d, $^2J_{PP}$ = 42.8 Hz), 27.3 (dd,

$^2J_{PP}$ = 42.8 Hz, $^1J_{PPt}$ = 4098.7 Hz), 28.3 (d, $^2J_{PP}$ = 42.8 Hz), 28.3 (dd, $^2J_{PP}$ = 42.8 Hz,

$^1J_{PPt}$ = 3691.2 Hz).

^{195}Pt-NMR (85.48 MHz; C_6D_6): δ = -5058.8 (dd, $^1J_{PPt}$ = 3676.7 Hz,

$^1J_{PPt}$ = 4098.4 Hz).

FAB-MS: m/z (%): 719.1 (100) [Pt(PPh$_3$)(PPh$_2$C$_6$H$_4$)$^+$].

FT-IR (ATR, cm^{-1}): 3058 (w), 2967 (w), 1679 (m, CO), 1575 (w), 1484 (m), 1433 (s), 1393 (w), 1302 (w), 1257 (w), 1151 (m), 1092 (s), 999 (m), 817 (m), 741 (s), 692 (s), 514 (s), 423 (m).

Bis(triphenylphosphine)-(η^2-*tert*-butyl 4-chlorocinnamate)palladium(0)

PdPhCl-^{t}Bu: C$_{49}$H$_{45}$ClO$_2$P$_2$Pd, M = 869.70 g/mol, light yellow solid, yield: >99%, **route**: *Pd-I*.

^{1}H-NMR (400.13 MHz; C_7D_8): δ = 1.30 (s, 9H, $C(CH_3)_3$), 4.56 (bs, 1H, CHCHCO), 5.07 (bs, 1H, $C_{Ar}$$C$HCH), 6.23 - 6.40 (bm, 2H, H_{Ar}), 6.66 - 6.79 (bm, 2H, H_{Ar}), 6.83 - 7.14 (bm, 24H, H_{Ar}), 7.37 - 7.61 (bm, 6H, H_{Ar}).

^{1}H-NMR (400.13 MHz; 246 K; C_7D_8): δ = 1.30 (s, 9H, $C(CH_3)_3$), 4.69 (dpt, $^3J_{HH}$ = 10.3 Hz, $^3J_{HP}$ = 6.1 Hz, $^3J_{HP}$ = 6.1 Hz, 1H, CHCHCO), 5.11 (dpt, $^3J_{HH}$ = 10.3 Hz, $^3J_{HP}$ = 7.0 Hz, $^3J_{HP}$ = 7.0 Hz, 1H, $C_{Ar}$$C$HCH), 6.24 (d, $^3J_{HH}$ = 8.5 Hz, 2H, H_{Ar}), 6.68 (d, $^3J_{HH}$ = 8.6 Hz, 2H, H_{Ar}), 6.83 - 7.07 (m, 24H, H_{Ar}), 7.49 - 7.58 (m, 6H, H_{Ar}).

^{13}C-NMR (100.62 MHz; C_7D_8): δ = 28.3 (CCH_3), 62.4 (bs, CHCHCO), 75.3 (bs, $C_{Ar}$$C$HCH), 78.0 (bs, O$C(CH_3)_3$), 126.2 - 126.8 (m, C_{Ar}H), 128.5 (C_{Ar}H), 133.8 - 134.9 (m, C_{Ar}H), 135.6 - 136.2 (m, C_{Ar}H), 141.7 (bs, C_{Ar}CH), 170.8 (bs, CHCOO).

^{13}C-NMR (100.62 MHz; 246 K; C_7D_8): δ = 28.1 (CH_3), 61.4 (dd, $^2J_{CP}$ = 6.2 Hz, $^2J_{CP}$ = 17.9 Hz, CHCHCO), 75.2 (dd, $^2J_{CP}$ = 3.5 Hz, $^2J_{CP}$ = 20.9 Hz, $C_{Ar}$$C$HCH), 77.8 (O$C(CH_3)_3$) 126.2 (bs, C_{Ar}H), 128.3 (C_{Ar}H), 128.4 (C_{Ar}H), 128.5 (C_{Ar}H), 129.2 (d, $^3J_{CP}$ = 31.0 Hz, C_{Ar}H), 134.0 (d, $^2J_{CP}$ = 14.4 Hz, C_{Ar}H), 134.2 (C_{Ar}), 134.5 (d, $^2J_{CP}$ = 15.5 Hz, C_{Ar}H), 135.3 (dd, $^1J_{CP}$ = 27.7 Hz, $^3J_{CP}$ = 1.6 Hz, C_{Ar}P), 137.2 (dd, $^1J_{CP}$ = 30.2 Hz, $^3J_{CP}$ = 1.8 Hz, C_{Ar}P), 141.2 (d, $^3J_{CP}$ = 6.2 Hz, C_{Ar}CH), 170.8 (dd, $^3J_{CP}$ = 1.9 Hz, $^3J_{CP}$ = 4.2 Hz, CHCOO).

^{31}P-NMR (161.97 MHz; C_7D_8): δ = 23.0 (d, $^2J_{PP}$ = 18.1 Hz), 24.9 (d, $^2J_{PP}$ = 18.1 Hz).

^{31}P-NMR (161.97 MHz; 246 K; C_7D_8): δ = 23.1 (d, $^2J_{PP}$ = 18.5 Hz), 25.3 (d, $^2J_{PP}$ = 18.5 Hz).

FT-IR (ATR, cm^{-1}): 3045 (w), 1681 (m, CO), 1574 (w), 1476 (m), 1433 (s), 1292 (w), 1309 (m), 1256 (w), 1135 (s), 1090 (s), 1229 (w), 968 (w), 817 (m), 741 (s), 692 (s), 503 (s), 415 (m).

Bis(triphenylphosphine)-(η^2-*tert*-butyl 4-chlorocinnamate)platinum(0)

PtPhCl-^{t}Bu: $C_{49}H_{45}ClO_2P_2Pt$, $M = 958.36$ g/mol, white solid, yield: >99%, **route:** *Pt-VI*.

^{1}H-NMR (400.13 MHz; C_6D_6): $\delta = 1.31$ (s, 9H, C(CH_3)$_3$), 3.96 (dpt, $^3J_{HH} = 8.5$ Hz, $^3J_{HP} = 4.9$ Hz, $^3J_{HP} = 8.5$ Hz, 1H, CHCHCO), 3.96 (ddpt, $^3J_{HH} = 8.5$ Hz, $^3J_{HP} = 4.9$ Hz, $^3J_{HP} = 8.5$ Hz, $^2J_{HPt} = 62.4$ Hz, 1H, CHCHCO), 4.32 (dpt, $^3J_{HH} = 8.5$ Hz, $^3J_{HP} = 6.1$ Hz, $^3J_{HP} = 8.5$ Hz, 1H, C$_{Ar}$CHCH), 4.32 (ddpt, $^3J_{HH} = 8.5$ Hz, $^3J_{HP} = 6.1$ Hz, $^3J_{HP} = 8.5$ Hz, $^2J_{HPt} = 52.8$ Hz, 1H, C$_{Ar}$CHCH), 6.28 (d, $^3J_{HH} = 8.4$ Hz, 2H, H_{Ar}), 6.78 (d, $^3J_{HH} = 7.8$ Hz, 2H, H_{Ar}), 6.82 - 7.03 (m, 17H, H_{Ar}), 7.07 - 7.19 (m, 7H, H_{Ar}), 7.50 - 7.62 (m, 6H, H_{Ar}).

^{13}C-NMR (100.62 MHz; C_6D_6): $\delta = 28.5$ (CH_3), 51.1 (dd, $^2J_{CP} = 6.0$ Hz, $^2J_{CP} = 30.2$ Hz, CHCHCO), 51.1 (ddd, $^2J_{CP} = 6.0$ Hz, $^2J_{CP} = 30.2$ Hz, $^1J_{CPt} = 200.6$ Hz, CHCHCO), 61.1 (dd, $^2J_{CP} = 4.8$ Hz, $^2J_{CP} = 32.9$ Hz, C$_{Ar}$$C$HCH), 61.1 (ddd, $^2J_{CP} = 4.8$ Hz, $^2J_{CP} = 32.9$ Hz, $^1J_{CPt} = 190.2$ Hz, C$_{Ar}$$C$HCH), 77.8 (O$C$(CH$_3$)$_3$), 127.1 (d, $^5J_{CP} = 3.1$ Hz, C$_{Ar}$H), 127.1 (dd, $^5J_{CP} = 3.1$ Hz, $^4J_{CPt} = 14.6$ Hz, C$_{Ar}$H), 128.0 (C_{Ar}H), 128.2 (C_{Ar}H), 128.6 (C_{Ar}), 129.4 (d, $^5J_{CP} = 1.6$ Hz, C$_{Ar}$H), 129.6 (d, $^5J_{CP} = 1.3$ Hz, C$_{Ar}$H), 134.1 (C_{Ar}P), 134.1 (d, $^3J_{CPt} = 20.2$ Hz, C$_{Ar}$P), 134.2 (C_{Ar}P), 134.2 (d, $^3J_{CPt} = 20.0$ Hz, C$_{Ar}$P), 134.5 (C_{Ar}P), 134.5 (d, $^3J_{CPt} = 19.2$ Hz, C$_{Ar}$P), 134.6 (C_{Ar}P), 134.6 (d, $^3J_{CPt} = 18.4$ Hz, C$_{Ar}$P), 135.4 (d, $^2J_{CP} = 2.5$ Hz, C$_{Ar}$P), 135.4 (dd, $^2J_{CP} = 2.5$ Hz, $^3J_{CPt} = 32.5$ Hz, C$_{Ar}$P), 135.8 (d, $^2J_{CP} = 2.5$ Hz, C$_{Ar}$P), 135.8 (dd, $^2J_{CP} = 2.5$ Hz, $^3J_{CPt} = 33.2$ Hz, C$_{Ar}$P), 137.1 (d, $^2J_{CP} = 2.5$ Hz, C$_{Ar}$P), 137.1 (dd, $^2J_{CP} = 2.5$ Hz, $^3J_{CPt} = 30.5$ Hz, C$_{Ar}$P), 137.5 (d, $^2J_{CP} = 2.5$ Hz, C$_{Ar}$P), 137.5 (dd, $^2J_{CP} = 2.5$ Hz, $^3J_{CPt} = 31.0$ Hz, C$_{Ar}$P), 143.7 (dd, $^3J_{CP} = 1.7$ Hz, $^3J_{CP} = 4.1$ Hz, C$_{Ar}$CH), 143.7 (ddd, $^3J_{CP} = 1.7$ Hz, $^3J_{CP} = 4.1$ Hz, $^2J_{CPt} = 33.8$ Hz, C$_{Ar}$CH), 173.1

(dd, $^3J_{CP}$ 2.1 Hz, $^3J_{CP}$ = 4.5 Hz, CHCOO), 173.1 (ddd, $^3J_{CP}$ = 2.1 Hz, $^3J_{CP}$ = 4.5 Hz, $^2J_{CPt}$ = 43.2 Hz, CHCOO).

^{31}P-NMR (161.97 MHz; C$_6$D$_6$): δ = 26.8 (d, $^2J_{PP}$ = 43.8 Hz), 26.8 (dd, $^2J_{PP}$ = 43.8 Hz, $^1J_{PPt}$ = 4141.4 Hz), 28.3 (d, $^2J_{PP}$ = 43.8 Hz), 28.3 (dd, $^2J_{PP}$ = 43.8 Hz, $^1J_{PPt}$ = 3636.3 Hz).

^{195}Pt-NMR (85.48 MHz; C$_6$D$_6$): δ = -5059.8 (dd, $^1J_{PPt}$ = 3729.5 Hz, $^1J_{PPt}$ = 4047.7 Hz).

FAB-MS: m/z (%): 719.0 (100) [Pt(PPh$_3$)(PPh$_2$C$_6$H$_4$)$^+$].

FT-IR (ATR, cm^{-1}): 3060 (w), 2969 (w), 1681 (m, CO), 1478 (m), 1433 (s), 1393 (w), 1355 (w), 1302 (w), 1257 (w), 1133 (s), 1091 (s), 984 (w), 817 (m), 741 (s), 692 (s), 590 (w), 529 (m), 507 (s), 423 (m).

Bis(triphenylphosphine)-(η^2-methyl 4-trifluoromethylcinnamate)palladium(0)

PdPhCF$_3$-Me: C$_{47}$H$_{39}$F$_3$O$_2$P$_2$Pd, M = 861.17 *g/mol*, deep yellow solid, yield: >99%, **route**: *Pd-I*.

^{1}H-NMR (400.13 MHz; C$_6$D$_6$): δ = 3.20 (s, 3H, CH_3), 4.58 - 4.70 (bm, 1H, CHCHCO), 5.17 - 5.29 (bm, 1H, C$_{Ar}$CHCH), 6.44 (bd, $^3J_{HH}$ = 7.8 Hz, 2H, H_{Ar}), 6.83 - 7.03 (bm, 20H, H_{Ar}), 7.04 - 7.13 (bm, 6H, H_{Ar}), 7.47 - 7.59 (bm, 6H, H_{Ar}).

^{1}H-NMR (400.13 MHz; 278 K; C$_6$D$_6$): δ = 3.19 (s, 3H, CH_3), 4.69 (dpt, $^3J_{HH}$ = 10.2 Hz, $^3J_{HP}$ = 6.3 Hz, $^3J_{HP}$ = 6.3 Hz, 1H, CHCHCO), 5.26 (dpt, $^3J_{HH}$ = 10.2 Hz, $^3J_{HP}$ = 6.4 Hz, $^3J_{HP}$ = 6.4 Hz, 1H, C$_{Ar}$CHCH), 6.41 (d, $^3J_{HH}$ = 7.9 Hz, 2H, H_{Ar}), 6.81 - 7.02 (m, 20H, H_{Ar}), 7.03 - 7.12 (m, 6H, H_{Ar}), 7.49 - 7.60 (m, 6H, H_{Ar}).

^{13}C-NMR (100.62 MHz; C$_6$D$_6$): δ = 50.3 (s, OCH$_3$), 60.3 (dd, $^2J_{CP}$ = 6.5 Hz, $^2J_{CP}$ = 18.5 Hz, CHCHCO), 73.1 (dd, $^2J_{CP}$ = 2.2 Hz, $^2J_{CP}$ = 21.9 Hz, C$_{Ar}$$C$HCH),

125.3 (bs, C_{Ar}H), 127.9 (C_{Ar}H), 128.2 (C_{Ar}H), 128.5 (C_{Ar}H), 128.6 (C_{Ar}H), 129.4 (d, $^3J_{CP} = 15.5$ Hz, C_{Ar}H), 134.1 (d, $^2J_{CP} = 14.9$ Hz, C_{Ar}H), 134.3 (d, $^2J_{CP} = 15.2$ Hz, C_{Ar}H), 135.5 (d, $^1J_{CP} = 27.4$ Hz, C_{Ar}P), 137.0 (d, $^1J_{CP} = 27.9$ Hz, C_{Ar}P), 146.7 - 146.8 (bm, C_{Ar}CH), 171.1 - 171.6 (bm, CHCOO).

^{19}F-NMR (376.50 MHz; C_6D_6): $\delta = $ -61.4.

^{31}P-NMR (161.97 MHz; C_6D_6): $\delta = 25.1$ (d, $^2J_{PP} = 12.6$ Hz), 26.7 (d, $^2J_{PP} = 12.6$ Hz).

^{31}P-NMR (161.97 MHz; 278 K; C_6D_6): $\delta = 24.9$ (d, $^2J_{PP} = 12.5$ Hz), 26.6 (dd, $^1J_{PF} = 2.9$ Hz, $^2J_{PP} = 12.5$ Hz).

FT-IR (ATR, cm^{-1}): 3041 (w), 1693 (m, CO), 1609 (m), 1479 (m), 1433 (m), 1318 (s), 1276 (m), 1152 (m), 1115 (m), 1103 (w), 1064 (s), 1026 (w), 1012 (w), 830 (m), 741 (m), 693 (s), 595 (w), 527 (w), 512 (m), 505 (s), 410 (m).

Bis(triphenylphosphine)-(η^2-methyl 4-trifluoromethylcinnamate)platinum(0)

PtPhCF$_3$-Me: $C_{47}H_{39}F_3O_2P_2Pt$, M = 949.83 g/mol, white solid, yield: >99%, **route**: Pt-VI.

^{1}H-NMR (400.13 MHz; C_6D_6): $\delta = 3.23$ (s, 3H, OCH_3), 3.91 (dpt, $^3J_{HH} = 8.7$ Hz, $^3J_{HP} = 4.3$ Hz, $^3J_{HP} = 8.7$ Hz, 1H, CHCHCO), 3.91 (ddpt, $^3J_{HH} = 8.7$ Hz, $^3J_{HP} = 4.3$ Hz, $^3J_{HP} = 8.7$ Hz, $^2J_{HPt} = 59.6$ Hz, 1H, CHCHCO), 4.43 (dpt, $^3J_{HH} = 8.7$ Hz, $^3J_{HP} = 5.4$ Hz, $^3J_{HP} = 8.7$ Hz, 1H, C_{Ar}CHCH), 4.43 (ddpt, $^3J_{HH} = 8.7$ Hz, $^3J_{HP} = 5.4$ Hz, $^3J_{HP} = 8.7$ Hz, $^2J_{HPt} = 54.0$ Hz, 1H, C_{Ar}CHCH), 6.42 (d, $^3J_{HH} = 7.8$ Hz, 2H, H_{Ar}), 6.80 - 6.93 (m, 9H, H_{Ar}), 6.93 - 7.00 (m, 8H, H_{Ar}), 7.03 (d, $^3J_{HH} = 7.9$ Hz, 2H, H_{Ar}), 7.10 - 7.19 (m, 7H, H_{Ar}), 7.47 - 7.59 (m, 6H, H_{Ar}).

^{13}C-NMR (100.62 MHz; C_6D_6): $\delta = 49.6$ (d, $^2J_{CP} = 23.9$ Hz, CHCHCO), 49.6 (dd, $^2J_{CP} = 23.9$ Hz, $^1J_{CPt} = 200.1$ Hz, CHCHCO), 50.3 (COOCH$_3$), 59.2 (d, $^2J_{CP} = 26.9$ Hz,

C_{Ar}CHCH), 59.2 (dd, $^2J_{CP}$ = 26.9 Hz, $^1J_{CPt}$ = 197.1 Hz, C_{Ar}CHCH), 124.7 (bm, C_{Ar}H),
125.7 (d, $^5J_{CP}$ = 2.4 Hz, C_{Ar}H), 125.7 (dd, $^5J_{CP}$ = 2.4 Hz, $^4J_{CPt}$ = 14.6 Hz, C_{Ar}H),
127.9 (C_{Ar}H), 128.2 (C_{Ar}H), 128.6 (C_{Ar}), 129.5 (d, $^5J_{CP}$ = 1.2 Hz, C_{Ar}H), 129.7 (d,
$^5J_{CP}$ = 1.0 Hz, C_{Ar}H), 134.0 (d, $^2J_{CP}$ = 2.1 Hz, C_{Ar}P), 134.0 (dd, $^2J_{CP}$ = 2.1 Hz,
$^3J_{CPt}$ = 20.6 Hz, C_{Ar}P), 134.1 (d, $^2J_{CP}$ = 2.1 Hz, C_{Ar}P), 134.1 (dd, $^2J_{CP}$ = 2.1 Hz,
$^3J_{CPt}$ = 21.8 Hz, C_{Ar}P), 134.2 (d, $^2J_{CP}$ = 2.0 Hz, C_{Ar}P), 134.2 (dd, $^2J_{CP}$ = 2.0 Hz,
$^3J_{CPt}$ = 23.2 Hz, C_{Ar}P), 134.3 (d, $^2J_{CP}$ = 2.3 Hz, C_{Ar}P), 134.3 (dd, $^2J_{CP}$ = 2.3 Hz,
$^3J_{CPt}$ = 18.8 Hz, C_{Ar}P), 135.2 (d, $^2J_{CP}$ = 8.7 Hz, C_{Ar}P), 135.2 (dd, $^2J_{CP}$ = 8.7 Hz,
$^3J_{CPt}$ = 43.5 Hz, C_{Ar}P), 135.5 (d, $^2J_{CP}$ = 8.6 Hz, C_{Ar}P), 135.5 (dd, $^2J_{CP}$ = 8.6 Hz,
$^3J_{CPt}$ = 42.9 Hz, C_{Ar}P), 136.6 (d, $^2J_{CP}$ = 9.0 Hz, C_{Ar}P), 136.6 (dd, $^2J_{CP}$ = 9.0 Hz,
$^3J_{CPt}$ = 44.3 Hz, C_{Ar}P), 136.9 (d, $^2J_{CP}$ = 8.9 Hz, C_{Ar}P), 136.9 (dd, $^2J_{CP}$ = 8.9 Hz,
$^3J_{CPt}$ = 41.1 Hz, C_{Ar}P), 149.2 (dd, $^3J_{CP}$ = 1.0 Hz, $^3J_{CP}$ = 5.6 Hz, C_{Ar}CH), 149.2 (ddd,
$^3J_{CP}$ = 1.0 Hz, $^3J_{CP}$ = 5.6 Hz, $^2J_{CPt}$ = 39.2 Hz, C_{Ar}CH), 173.4 (dd, $^3J_{CP}$ 0.6 Hz,
$^3J_{CP}$ = 3.9 Hz, CHCOO), 173.4 (ddd, $^3J_{CP}$ = 0.6 Hz, $^3J_{CP}$ = 3.9 Hz, $^2J_{CPt}$ = 43.8 Hz,
CHCOO).

^{19}F-NMR (376.50 MHz; C_6D_6): δ = -61.26.

^{31}P-NMR (161.97 MHz; C_6D_6): δ = 27.5 (d, $^2J_{PP}$ = 38.2 Hz), 27.5 (dd, $^2J_{PP}$ = 38.2 Hz,
$^1J_{PPt}$ = 4077.6 Hz), 27.8 (dd, $^1J_{PF}$ = 2.2 Hz, $^2J_{PP}$ = 38.2 Hz), 27.8 (ddd, $^1J_{PF}$ = 2.2 Hz,
$^2J_{PP}$ = 38.2 Hz, $^1J_{PPt}$ = 3684.8 Hz).

^{195}Pt-NMR (85.48 MHz; C_6D_6): δ = -5070.7 (dd, $^1J_{PPt}$ = 3684.3 Hz,
$^1J_{PPt}$ = 4077.2 Hz).

FAB-MS: m/z (%): 719.1 (100) [Pt(PPh$_3$)(PPh$_2$C$_6$H$_4$)$^+$].

FT-IR (ATR, cm^{-1}): 3052 (w), 1688 (m, CO), 1609 (w), 1434 (m), 1322 (s), 1270 (m),
1151 (m), 1113 (w), 1094 (m), 1064 (m), 1013 (w), 835 (m), 742 (m), 693 (s), 606 (w),
508 (s), 424 (m).

Bis(triphenylphosphine)-(η^2-ethyl 4-trifluoromethylcinnamate)palladium(0)

PdPhCF$_3$-Et: C$_{48}$H$_{41}$F$_3$O$_2$P$_2$Pd, M = 875.20 g/mol, deep yellow solid, yield: 83%, **route**: *Pd-I*.

^{1}H-NMR (400.13 MHz; C$_6$D$_6$): $\delta = 0.84$ (bt, $^3J_{HH} = 7.0$ Hz, 3H, CH$_2$CH_3), 3.54 - 3.70 (bm, 1H, CH_2CH$_3$), 3.95 - 4.10 (bm, 1H, CH_2CH$_3$), 4.60 - 4.75 (bm, 1H, CHCHCO), 5.15 - 5.32 (bm, 1H, C$_{Ar}$CHCH), 6.43 (bd, $^3J_{HH} = 7.8$ Hz, 2H, H_{Ar}), 6.84 - 7.02 (bm, 20H, H_{Ar}), 7.02 - 7.13 (bm, 6H, H_{Ar}), 7.50 - 7.60 (bm, 6H, H_{Ar}).

^{1}H-NMR (400.13 MHz; 278 K; C$_6$D$_6$): $\delta = 0.82$ (t, $^3J_{HH} = 7.1$ Hz, 3H, CH$_2$CH_3), 3.61 (dq, $^2J_{HH} = 10.8$ Hz, $^3J_{HH} = 7.1$ Hz, 1H, CH_2CH$_3$), 4.03 (dq, $^2J_{HH} = 10.8$ Hz, $^3J_{HH} = 7.1$ Hz, 1H, CH_2CH$_3$), 4.74 (dpt, $^3J_{HH} = 10.2$ Hz, $^3J_{HP} = 6.3$ Hz, $^3J_{HP} = 6.3$ Hz, 1H, CHCHCO), 5.26 (dpt, $^3J_{HH} = 10.2$ Hz, $^3J_{HP} = 7.5$ Hz, $^3J_{HP} = 7.5$ Hz, 1H, C$_{Ar}$CHCH), 6.41 (d, $^3J_{HH} = 7.9$ Hz, 2H, H_{Ar}), 6.82 - 7.01 (m, 20H, H_{Ar}), 7.01 - 7.10 (m, 6H, H_{Ar}), 7.53 - 7.61 (m, 6H, H_{Ar}).

^{13}C-NMR (100.62 MHz; C$_6$D$_6$): $\delta = 14.4$ (CH$_2$$CH_3$), 59.2 ($CH_2CH_3$), 60.6 (dd, $^2J_{CP} = 4.5$ Hz, $^2J_{CP} = 17.8$ Hz, CHCHCO), 73.5 (dd, $^2J_{CP} = 2.6$ Hz, $^2J_{CP} = 21.9$ Hz, C$_{Ar}$$C$HCH), 125.2 (bs, C_{Ar}H), 127.9 (C_{Ar}H), 128.2 (C_{Ar}H), 128.5 (C_{Ar}H), 128.6 (C_{Ar}H), 129.4 (d, $^3J_{CP} = 18.4$ Hz, C_{Ar}H), 134.1 (d, $^2J_{CP} = 14.3$ Hz, C_{Ar}H), 134.4 (d, $^2J_{CP} = 15.1$ Hz, C_{Ar}H), 135.6 (d, $^1J_{CP} = 28.0$ Hz, C_{Ar}P), 137.1 (d, $^1J_{CP} = 27.3$ Hz, C_{Ar}P), 146.6 - 146.9 (bm, C_{Ar}CH), 171.1 - 171.2 (bm, CHCOO).

^{19}F-NMR (376.50 MHz; C$_6$D$_6$): $\delta = -61.4$.

^{31}P-NMR (161.97 MHz; C$_6$D$_6$): $\delta = 24.9$ (d, $^2J_{PP} = 13.5$ Hz), 26.6 (d, $^2J_{PP} = 13.5$ Hz).

^{31}P-NMR (161.97 MHz; 278 K; C$_6$D$_6$): $\delta = 24.6$ (d, $^2J_{PP} = 13.4$ Hz), 26.5 (dd, $^1J_{PF} = 2.9$ Hz, $^2J_{PP} = 13.4$ Hz).

FT-IR (ATR, cm^{-1}): 3045 (w), 1682 (m, CO), 1608 (m), 1478 (m), 1434 (m), 1320 (s), 1274 (m), 1224 (w), 1148 (m), 1115 (w), 1103 (w), 1092 (m), 1064 (s), 1012 (w), 830 (m), 741 (s), 692 (s), 595 (w), 504 (s), 415 (m).

Bis(triphenylphosphine)-(η^2-ethyl 4-trifluoromethylcinnamate)platinum(0)

PtPhCF$_3$-Et: C$_{48}$H$_{41}$F$_3$O$_2$P$_2$Pt, M = 963.86 *g/mol*, white solid, yield: >99%, **route**: *Pt-VI*.

^{1}H-NMR (400.13 MHz; C$_6$D$_6$): δ = 0.85 (t, $^3J_{HH}$ = 7.1 Hz, 3H, CH$_2$C*H*$_3$), 3.66 (dq, $^2J_{HH}$ = 9.9 Hz, $^3J_{HH}$ = 7.1 Hz, 1H, C*H*$_2$CH$_3$), 3.95 (dpt, $^3J_{HH}$ = 8.2 Hz, $^3J_{HP}$ = 3.3 Hz, $^3J_{HP}$ = 8.2 Hz, 1H, CHC*H*CO), 3.95 (ddpt, $^3J_{HH}$ = 8.2 Hz, $^3J_{HP}$ = 3.3 Hz, $^3J_{HP}$ = 8.2 Hz, $^2J_{HPt}$ = 60.4 Hz, 1H, CHC*H*CO), 4.05 (dq, $^2J_{HH}$ = 9.9 Hz, $^3J_{HH}$ = 7.1 Hz, 1H, C*H*$_2$CH$_3$), 4.42 (dpt, $^3J_{HH}$ = 8.2 Hz, $^3J_{HP}$ = 5.0 Hz, $^3J_{HP}$ = 8.2 Hz, 1H, C$_{Ar}$C*H*CH), 4.42 (ddpt, $^3J_{HH}$ = 8.2 Hz, $^3J_{HP}$ = 5.0 Hz, $^3J_{HP}$ = 8.2 Hz, $^2J_{HPt}$ = 54.0 Hz, 1H, C$_{Ar}$C*H*CH), 6.41 (d, $^3J_{HH}$ = 7.7 Hz, 2H, H_{Ar}), 6.81 - 6.93 (m, 9H, H_{Ar}), 6.93 - 7.00 (m, 8H, H_{Ar}), 7.02 (d, $^3J_{HH}$ = 7.9 Hz, 2H, H_{Ar}), 7.08 - 7.18 (m, 7H, H_{Ar}), 7.50 - 7.61 (m, 6H, H_{Ar}).

^{13}C-NMR (100.62 MHz; C$_6$D$_6$): δ = 14.5 (CH$_2$*C*H$_3$), 49.9 (dd, $^2J_{CP}$ = 4.5 Hz, $^2J_{CP}$ = 27.9 Hz, CH*C*HCO), 49.9 (ddd, $^2J_{CP}$ = 4.5 Hz, $^2J_{CP}$ = 27.9 Hz, $^1J_{CPt}$ = 204.3 Hz, CH*C*HCO), 59.1 (*C*H$_2$CH$_3$), 59.5 (dd, $^2J_{CP}$ = 2.6 Hz, $^2J_{CP}$ = 30.7 Hz, C$_{Ar}$*C*HCH), 59.5 (ddd, $^2J_{CP}$ = 2.6 Hz, $^2J_{CP}$ = 30.7 Hz, $^1J_{CPt}$ = 199.9 Hz, C$_{Ar}$*C*HCH), 124.7 (bm, *C*$_{Ar}$H), 125.7 (d, $^5J_{CP}$ = 2.9 Hz, *C*$_{Ar}$H), 125.7 (dd, $^5J_{CP}$ = 2.9 Hz, $^4J_{CPt}$ = 14.1 Hz, *C*$_{Ar}$H), 127.9 (*C*$_{Ar}$H), 128.0 (*C*$_{Ar}$H), 128.2 (*C*$_{Ar}$H), 128.2 (*C*$_{Ar}$H), 128.6 (*C*$_{Ar}$), 129.5 (d, $^5J_{CP}$ = 1.5 Hz, *C*$_{Ar}$H), 129.7 (d, $^5J_{CP}$ = 1.4 Hz, *C*$_{Ar}$H), 134.0 (*C*$_{Ar}$P), 134.0 (d, $^3J_{CPt}$ = 21.4 Hz, *C*$_{Ar}$P), 134.1 (*C*$_{Ar}$P), 134.1 (d, $^3J_{CPt}$ = 16.6 Hz, *C*$_{Ar}$P), 134.3 (*C*$_{Ar}$P), 134.3 (d, $^3J_{CPt}$ = 15.4 Hz, *C*$_{Ar}$P), 134.4 (*C*$_{Ar}$P), 134.4 (d, $^3J_{CPt}$ = 19.4 Hz, *C*$_{Ar}$P), 135.1 (d, $^2J_{CP}$ = 4.2 Hz, *C*$_{Ar}$P), 135.1 (dd, $^2J_{CP}$ = 4.2 Hz, $^3J_{CPt}$ = 31.6 Hz, *C*$_{Ar}$P), 135.6 (d, $^2J_{CP}$ = 4.2 Hz, *C*$_{Ar}$P),

135.6 (dd, $^2J_{CP}$ = 4.2 Hz, $^3J_{CPt}$ = 31.4 Hz, $C_{Ar}P$), 136.7 (d, $^2J_{CP}$ = 4.3 Hz, $C_{Ar}P$), 136.7 (dd, $^2J_{CP}$ = 4.3 Hz, $^3J_{CPt}$ = 29.2 Hz, $C_{Ar}P$), 137.1 (d, $^2J_{CP}$ = 4.2 Hz, $C_{Ar}P$), 137.1 (dd, $^2J_{CP}$ = 4.2 Hz, $^3J_{CPt}$ = 27.7 Hz, $C_{Ar}P$), 149.2 (d, $^3J_{CP}$ = 6.2 Hz, $C_{Ar}CH$), 149.2 (dd, $^3J_{CP}$ = 6.2 Hz, $^2J_{CPt}$ = 38.6 Hz, $C_{Ar}CH$), 173.1 (dd, $^3J_{CP}$ 1.9 Hz, $^3J_{CP}$ = 4.7 Hz, CHCOO), 173.1 (ddd, $^3J_{CP}$ = 1.9 Hz, $^3J_{CP}$ = 4.7 Hz, $^2J_{CPt}$ = 42.6 Hz, CHCOO).

^{19}F-NMR (376.50 MHz; C_6D_6): δ = -61.24.

^{31}P-NMR (161.97 MHz; C_6D_6): δ = 27.4 (d, $^2J_{PP}$ = 39.0 Hz), 27.4 (dd, $^2J_{PP}$ = 39.0 Hz, $^1J_{PPt}$ = 4068.2 Hz), 28.0 (dd, $^1J_{PF}$ = 2.7 Hz, $^2J_{PP}$ = 39.0 Hz), 28.0 (ddd, $^1J_{PF}$ = 2.7 Hz, $^2J_{PP}$ = 39.0 Hz, $^1J_{PPt}$ = 3701.2 Hz).

^{195}Pt-NMR (85.48 MHz; C_6D_6): δ = -5066.6 (dd, $^1J_{PPt}$ = 3699.3 Hz, $^1J_{PPt}$ = 4067.5 Hz).

FAB-MS: m/z (%): 719.2 (100) [Pt(PPh$_3$)(PPh$_2$C$_6$H$_4$)$^+$].

FT-IR (ATR, cm^{-1}): 3046 (w), 1682 (m, CO), 1608 (w), 1433 (m), 1322 (s), 1268 (m), 1148 (m), 1112 (m), 1094 (s), 1064 (s), 1027 (w), 1013 (w), 835 (m), 742 (m), 692 (s), 590 (w), 507 (s), 423 (m).

Bis(triphenylphosphine)-(η^2-*iso*-propyl 4-trifluoromethylcinnamate)palladium(0)

PdPhCF$_3$-^{i}Pr: C$_{49}$H$_{43}$F$_3$O$_2$P$_2$Pd, M = 889.23 *g/mol*, deep yellow solid, yield: >99%, route: *Pd-I*.

^{1}H-NMR (400.13 MHz; C_6D_6): δ = 0.75 (bd, $^3J_{HH}$ = 5.0 Hz, 3H, CHCH_3), 1.14 (bd, $^3J_{HH}$ = 5.0 Hz, 3H, CHCH_3), 4.63 - 4.77 (bm, 1H, CHCHCO), 4.86 - 5.01 (bm, 1H, CH(CH$_3$)$_2$), 5.10 - 5.27 (bm, 1H, C$_{Ar}$CHCH), 6.42 (bd, $^3J_{HH}$ = 7.1 Hz, 2H, H_{Ar}), 6.85 - 7.00 (bm, 20H, H_{Ar}), 7.00 - 7.11 (bm, 6H, H_{Ar}), 7.51 - 7.62 (bm, 6H, H_{Ar}).

^{1}H-NMR (400.13 MHz; 278 K; C_6D_6): δ = 0.72 (d, $^3J_{HH}$ = 6.2 Hz, 3H, CHCH_3), 1.14 (d, $^3J_{HH}$ = 6.2 Hz, 3H, CHCH_3), 4.75 (dpt, $^3J_{HH}$ = 10.2 Hz, $^3J_{HP}$ = 6.3 Hz, $^3J_{HP}$ = 6.3 Hz, 1H, CHCHCO), 4.94 (hp, $^3J_{HH}$ = 6.2 Hz, 1H, CH(CH$_3$)$_2$), 5.23 (dpt, $^3J_{HH}$ = 10.2 Hz, $^3J_{HP}$ = 7.2 Hz, $^3J_{HP}$ = 7.2 Hz, 1H, C$_{Ar}$CHCH), 6.39 (d, $^3J_{HH}$ = 7.9 Hz, 2H, H_{Ar}), 6.84 - 7.00 (m, 20H, H_{Ar}), 7.00 - 7.10 (m, 6H, H_{Ar}), 7.53 - 7.63 (m, 6H, H_{Ar}).

^{13}C-NMR (100.62 MHz; C_6D_6): δ = 21.7 (CHCH$_3$), 22.5 (CHCH$_3$), 60.9 (dd, $^2J_{CP}$ = 3.0 Hz, $^2J_{CP}$ = 18.4 Hz, CHCHCO), 66.4 (OCH(CH$_3$)$_2$), 74.0 (dd, $^2J_{CP}$ = 3.4 Hz, $^2J_{CP}$ = 22.1 Hz, C$_{Ar}$$C$HCH), 125.2 ($C_{Ar}$H), 125.3 (bs, C_{Ar}H), 127.9 (C_{Ar}H), 128.2 (C_{Ar}H), 128.4 (C_{Ar}H), 128.5 (C_{Ar}H), 128.6 (C_{Ar}H), 128.6 (C_{Ar}H), 129.4 (d, $^3J_{CP}$ = 22.3 Hz, C_{Ar}H), 129.6 (C_{Ar}H), 134.1 (d, $^2J_{CP}$ = 14.5 Hz, C_{Ar}H), 134.3 (C_{Ar}), 134.5 (d, $^2J_{CP}$ = 15.4 Hz, C_{Ar}H), 135.5 (d, $^1J_{CP}$ = 28.1 Hz, C_{Ar}P), 137.2 (d, $^1J_{CP}$ = 28.4 Hz, C_{Ar}P), 146.7 - 146.9 (bm, C_{Ar}CH), 170.8 - 171.0 (bm, CHCOO).

^{19}F-NMR (376.50 MHz; C_6D_6): δ = -61.4.

^{31}P-NMR (161.97 MHz; C_6D_6): δ = 24.6 (d, $^2J_{PP}$ = 14.8 Hz), 26.5 (d, $^2J_{PP}$ = 14.8 Hz).

^{31}P-NMR (161.97 MHz; 278 K; C_6D_6): δ = 24.3 (d, $^2J_{PP}$ = 14.3 Hz), 26.3 (dd, $^1J_{PF}$ = 2.8 Hz, $^2J_{PP}$ = 14.3 Hz).

FT-IR (ATR, cm^{-1}):3060 (w), 2969 (w), 1681 (m, CO), 1608 (m), 1478 (m), 1434 (m), 1321 (s), 1274 (m), 1155 (m), 1102 (s), 1064 (s), 1013 (w), 997 (w), 830 (m), 741 (s), 692 (s), 595 (w), 504 (s), 417 (m).

Bis(triphenylphosphine)-(η^2-*iso*-propyl 4-trifluoromethylcinnamate)platinum(0)

PtPhCF$_3$-^{i}Pr: $C_{49}H_{43}F_3O_2P_2Pt$, M = 977.89 g/mol, white solid, yield: >99%, **route**: *Pt-VI*.

^{1}H-NMR (400.13 MHz; C$_6$D$_6$): δ = 0.76 (d, $^3J_{HH}$ = 6.1 Hz, 3H, CHCH_3), 1.13 (d, $^3J_{HH}$ = 6.1 Hz, 3H, CHCH_3), 3.99 (dpt, $^3J_{HH}$ = 8.4 Hz, $^3J_{HP}$ = 4.6 Hz, $^3J_{HP}$ = 8.4 Hz, 1H, CHCHCO), 3.99 (ddpt, $^3J_{HH}$ = 8.4 Hz, $^3J_{HP}$ = 4.6 Hz, $^3J_{HP}$ = 8.4 Hz, $^2J_{HPt}$ = 59.8 Hz, 1H, CHCHCO), 4.39 (dpt, $^3J_{HH}$ = 8.4 Hz, $^3J_{HP}$ = 5.7 Hz, $^3J_{HP}$ = 8.4 Hz, 1H, C$_{Ar}$CHCH), 4.39 (ddpt, $^3J_{HH}$ = 8.4 Hz, $^3J_{HP}$ = 5.7 Hz, $^3J_{HP}$ = 8.4 Hz, $^2J_{HPt}$ = 52.2 Hz, 1H, C$_{Ar}$CHCH), 4.97 (hp, $^3J_{HH}$ = 6.1 Hz, 1H, CH(CH$_3$)$_2$), 6.38 (d, $^3J_{HH}$ = 7.6 Hz, 2H, H_{Ar}), 6.83 - 6.93 (m, 9H, H_{Ar}), 6.94 - 7.04 (m, 10H, H_{Ar}), 7.06 - 7.18 (m, 7H, H_{Ar}), 7.51 - 7.63 (m, 6H, H_{Ar}).

^{13}C-NMR (100.62 MHz; C$_6$D$_6$): δ = 21.8 (CHCH$_3$), 22.5 (CHCH$_3$), 50.1 (dd, $^2J_{CP}$ = 5.9 Hz, $^2J_{CP}$ = 30.6 Hz, CHCHCO), 50.1 (ddd, $^2J_{CP}$ = 5.9 Hz, $^2J_{CP}$ = 30.6 Hz, $^1J_{CPt}$ = 209.1 Hz, CHCHCO), 60.0 (dd, $^2J_{CP}$ = 4.4 Hz, $^2J_{CP}$ = 32.7 Hz, C$_{Ar}$$C$HCH), 60 0 (ddd, $^2J_{CP}$ = 4.4 Hz, $^2J_{CP}$ = 32.7 Hz, $^1J_{CPt}$ = 186.9 Hz, C$_{Ar}$$C$HCH), 66.1 (O$C$H(CH$_3$)$_2$), 124.7 (bm, C_{Ar}H), 125.7 (d, $^5J_{CP}$ = 2.9 Hz, C_{Ar}H), 125.7 (dd, $^5J_{CP}$ = 2.9 Hz, $^4J_{CPt}$ = 14.2 Hz, C_{Ar}H), 127.9 (C_{Ar}H), 128.0 (C_{Ar}H), 128.2 (C_{Ar}H), 128.6 (C_{Ar}), 129.5 (d, $^5J_{CP}$ = 1.8 Hz, C_{Ar}H), 129.8 (d, $^5J_{CP}$ = 1.6 Hz, C_{Ar}H), 134.0 (C_{Ar}P), 134.0 (d, $^3J_{CPt}$ = 20.6 Hz, C_{Ar}P), 134.1 (C_{Ar}P), 134.1 (d, $^3J_{CPt}$ = 25.2 Hz, C_{Ar}P), 134.4 (C_{Ar}P), 134.4 (d, $^3J_{CPt}$ = 19.4 Hz, C_{Ar}P), 134.5 (C_{Ar}P), 134.5 (d, $^3J_{CPt}$ = 19.4 Hz, C_{Ar}P), 135.1 (d, $^2J_{CP}$ = 2.7 Hz, C_{Ar}P), 135.1 (dd, $^2J_{CP}$ = 2.7 Hz, $^3J_{CPt}$ = 32.4 Hz, C_{Ar}P), 135.5 (d, $^2J_{CP}$ = 2.7 Hz, C_{Ar}P), 135.5 (dd, $^2J_{CP}$ = 2.7 Hz, $^3J_{CPt}$ = 33.3 Hz, C_{Ar}P), 136.8 (d, $^2J_{CP}$ = 2.8 Hz, C_{Ar}P), 136.8 (dd, $^2J_{CP}$ = 2.8 Hz, $^3J_{CPt}$ = 30.8 Hz, C_{Ar}P), 137.2 (d, $^2J_{CP}$ = 2.8 Hz, C_{Ar}P), 137.2 (dd, $^2J_{CP}$ = 2.8 Hz, $^3J_{CPt}$ = 28.6 Hz, C_{Ar}P), 149.3 (d, $^3J_{CP}$ = 6.6 Hz, C_{Ar}CH), 149.3 (dd, $^3J_{CP}$ = 6.6 Hz, $^2J_{CPt}$ = 39.4 Hz, C_{Ar}CH), 172.8 (dd, $^3J_{CP}$ 2.0 Hz, $^3J_{CP}$ = 5.0 Hz, CHCOO), 172.8 (ddd, $^3J_{CP}$ = 2.0 Hz, $^3J_{CP}$ = 5.0 Hz, $^2J_{CPt}$ = 42.0 Hz, CHCOO).

^{19}F-NMR (376.50 MHz; C$_6$D$_6$): δ = -61.24.

^{31}P-NMR (161.97 MHz; C$_6$D$_6$): δ = 27.2 (d, $^2J_{PP}$ = 40.0 Hz), 27.2 (dd, $^2J_{PP}$ = 40.0 Hz, $^1J_{PPt}$ = 4041.8 Hz), 28.2 (dd, $^1J_{PF}$ = 2.8 Hz, $^2J_{PP}$ = 40.0 Hz), 28.2 (ddd, $^1J_{PF}$ = 2.8 Hz, $^2J_{PP}$ = 40.0 Hz, $^1J_{PPt}$ = 3727.6 Hz).

^{195}Pt-NMR (85.48 MHz; C$_6$D$_6$): δ = -5061.3 (dd, $^1J_{PPt}$ = 3737.7 Hz,

$^1J_{PPt}$ = 4031.3 Hz).

FAB-MS: m/z (%): 719.0 (100) $[Pt(PPh_3)(PPh_2C_6H_4)^+]$.

FT-IR (ATR, cm^{-1}): 3045 (w), 2969 (w), 1681 (m, CO), 1608 (w), 1433 (m), 1321 (s), 1267 (m), 1153 (m), 1095 (s), 1064 (s), 997 (w), 835 (m), 741 (m), 692 (s), 590 (w), 544 (m), 506 (s), 423 (m).

Bis(triphenylphosphine)-(η^2-*tert*-butyl 4-trifluoromethylcinnamate)palladium(0)

PdPhCF$_3$-^{t}Bu: $C_{50}H_{45}F_3lO_2P_2Pd$, M = 903.25 *g/mol*, deep yellow solid, yield: 95%, **route:** *Pd-I*.

^{1}H-NMR (400.13 MHz; C$_6$D$_6$): δ = 1.31 (s, 9H, C(CH_3)$_3$), 4.51 - 4.75 (bm, 1H, CHCHCO), 5.06 - 5.28 (bm, 1H, C$_{Ar}$CHCH), 6.41 (bd, $^3J_{HH}$ = 7.1 Hz, 2H, H_{Ar}), 6.82 - 7.13 (bm, 26H, H_{Ar}), 7.49 - 7.60 (bm, 6H, H_{Ar}).

^{1}H-NMR (400.13 MHz; 278 K; C$_6$D$_6$): δ = 1.31 (s, 9H, C(CH_3)$_3$), 4.70 (dpt, $^3J_{HH}$ = 10.2 Hz, $^3J_{HP}$ = 6.2 Hz, $^3J_{HP}$ = 6.2 Hz, 1H, CHCHCO), 5.18 (dpt, $^3J_{HH}$ = 10.2 Hz, $^3J_{HP}$ = 7.2 Hz, $^3J_{HP}$ = 7.2 Hz, 1H, C$_{Ar}$CHCH), 6.38 (d, $^3J_{HH}$ = 8.0 Hz, 2H, H_{Ar}), 6.82 - 7.14 (m, 26H, H_{Ar}), 7.47 - 7.62 (m, 6H, H_{Ar}).

^{13}C-NMR (100.62 MHz; C$_6$D$_6$): δ = 28.4 (CH_3), 62.0 (dd, $^2J_{CP}$ = 6.0 Hz, $^2J_{CP}$ = 23.7 Hz, CHCHCO), 74.6 (dd, $^2J_{CP}$ = 3.7 Hz, $^2J_{CP}$ = 22.3 Hz, C$_{Ar}$CHCH), 78.2 (OC(CH$_3$)$_3$) 125.1 (bs, C$_{Ar}$H), 125.4 (bs, C$_{Ar}$H), 127.9 (C$_{Ar}$H), 128.2 (C$_{Ar}$H), 128.5 (C$_{Ar}$H), 128.6 (C$_{Ar}$H), 129.4 (d, $^3J_{CP}$ = 21.1 Hz, C$_{Ar}$H), 134.1 (d, $^2J_{CP}$ = 14.5 Hz, C$_{Ar}$H), 134.4 (C$_{Ar}$), 134.6 (d, $^2J_{CP}$ = 15.0 Hz, C$_{Ar}$H), 135.6 (d, $^1J_{CP}$ = 29.2 Hz, C$_{Ar}$P), 137.2 (d, $^1J_{CP}$ = 27.6 Hz, C$_{Ar}$P), 146.8 - 147.1 (bm, C$_{Ar}$CH), 171.0 - 171.3 (bm, CHCOO).

^{19}F-NMR (376.50 MHz; C$_6$D$_6$): δ = -61.4.

^{31}P-NMR (161.97 MHz; C$_6$D$_6$): $\delta = 24.1$ (d, $^2J_{PP} = 15.7$ Hz), 26.1 (d, $^2J_{PP} = 15.7$ Hz).

^{31}P-NMR (161.97 MHz; 278 K; C$_6$D$_6$): $\delta = 23.9$ (d, $^2J_{PP} = 15.4$ Hz), 26.0 (dd, $^1J_{PF} = 3.0$ Hz, $^2J_{PP} = 15.4$ Hz).

FT-IR (ATR, cm^{-1}): 3045 (w), 2666 (w), 1681 (m, CO), 1608 (m), 1478 (m), 1434 (m), 1319 (s), 1277 (m), 1155 (m), 1136 (m), 1115 (m), 1102 (m), 1064 (s), 1013 (w), 967 (w), 831 (m), 741 (s), 692 (s), 595 (w), 526 (w), 503 (s), 415 (m).

Bis(triphenylphosphine)-(η^2-*tert*-butyl 4-trifluoromethylcinnamate)platinum(0)

PtPhCF$_3$-^{t}Bu: C$_{50}$H$_{45}$ClO$_2$P$_2$Pt, M $= 991.91$ *g/mol*, white solid, yield: 98%, **route**: *Pt-VI*.

^{1}H-NMR (400.13 MHz; C$_6$D$_6$): $\delta = 1.32$ (s, 9H, C(CH$_3$)$_3$), 3.96 (dpt, $^3J_{HH} = 8.4$ Hz, $^3J_{HP} = 5.0$ Hz, $^3J_{HP} = 8.4$ Hz, 1H, CHCHCO), 3.96 (ddpt, $^3J_{HH} = 8.4$ Hz, $^3J_{HP} = 5.0$ Hz, $^3J_{HP} = 8.4$ Hz, $^2J_{HPt} = 60.4$ Hz, 1H, CHCHCO), 4.34 (dpt, $^3J_{HH} = 8.4$ Hz, $^3J_{HP} = 6.4$ Hz, $^3J_{HP} = 8.4$ Hz, 1H, C$_{Ar}$CHCH), 4.34 (ddpt, $^3J_{HH} = 8.4$ Hz, $^3J_{HP} = 6.4$ Hz, $^3J_{HP} = 8.4$ Hz, $^2J_{HPt} = 51.6$ Hz, 1H, C$_{Ar}$CHCH), 6.36 (d, $^3J_{HH} = 7.4$ Hz, 2H, H_{Ar}), 6.80 - 7.03 (m, 19H, H_{Ar}), 7.04 - 7.20 (m, 7H, H_{Ar}), 7.50 - 7.62 (m, 6H, H_{Ar}).

^{13}C-NMR (100.62 MHz; C$_6$D$_6$): $\delta = 28.5$ (CH$_3$), 51.1 (dd, $^2J_{CP} = 6.5$ Hz, $^2J_{CP} = 30.4$ Hz, CHCHCO), 51.1 (ddd, $^2J_{CP} = 6.5$ Hz, $^2J_{CP} = 30.4$ Hz, $^1J_{CPt} = 218.4$ Hz, CHCHCO), 60.8 (dd, $^2J_{CP} = 5.5$ Hz, $^2J_{CP} = 31.7$ Hz, C$_{Ar}$$C$HCH), 60.8 (ddd, $^2J_{CP} = 5.5$ Hz, $^2J_{CP} = 31.7$ Hz, $^1J_{CPt} = 199.9$ Hz, C$_{Ar}$$C$HCH), 78.0 (O$C$(CH$_3$)$_3$), 124.8 (bm, C_{Ar}H), 125.7 (d, $^5J_{CP} = 3.0$ Hz, C_{Ar}H), 125.7 (dd, $^5J_{CP} = 3.0$ Hz, $^4J_{CPt} = 10.7$ Hz, C_{Ar}H), 127.9 (C_{Ar}H), 128.0 (C_{Ar}H), 128.2 (C_{Ar}H), 128.2 (C_{Ar}H), 128.6 (C_{Ar}), 129.5 (d, $^5J_{CP} = 1.6$ Hz, C_{Ar}H), 129.7 (d, $^5J_{CP} = 1.4$ Hz, C_{Ar}H), 134.0 (C_{Ar}P), 134.0 (d, $^3J_{CPt} = 20.0$ Hz, C_{Ar}P), 134.1 (C_{Ar}P), 134.1 (d, $^3J_{CPt} = 19.8$ Hz, C_{Ar}P), 134.5 (C_{Ar}P),

134.5 (d, $^3J_{CPt}$ = 19.4 Hz, C_{Ar}P), 134.6 (C_{Ar}P), 134.6 (d, $^3J_{CPt}$ = 19.0 Hz, C_{Ar}P), 135.1 (d, $^2J_{CP}$ = 2.4 Hz, C_{Ar}P), 135.1 (dd, $^2J_{CP}$ = 2.4 Hz, $^3J_{CPt}$ = 30.0 Hz, C_{Ar}P), 135.5 (d, $^2J_{CP}$ = 2.3 Hz, C_{Ar}P), 135.5 (dd, $^2J_{CP}$ = 2.3 Hz, $^3J_{CPt}$ = 34.8 Hz, C_{Ar}P), 136.9 (d, $^2J_{CP}$ = 2.4 Hz, C_{Ar}P), 136.9 (dd, $^2J_{CP}$ = 2.4 Hz, $^3J_{CPt}$ = 31.5 Hz, C_{Ar}P), 137.3 (d, $^2J_{CP}$ = 2.4 Hz, C_{Ar}P), 137.3 (dd, $^2J_{CP}$ = 2.4 Hz, $^3J_{CPt}$ = 29.4 Hz, C_{Ar}P), 149.3 (dd, $^3J_{CP}$ = 1.5 Hz, $^3J_{CP}$ = 7.5 Hz, C_{Ar}CH), 149.3 (ddd, $^3J_{CP}$ = 1.5 Hz, $^3J_{CP}$ = 7.5 Hz, $^2J_{CPt}$ = 39.7 Hz, C_{Ar}CH), 173.1 (dd, $^3J_{CP}$ 2.1 Hz, $^3J_{CP}$ = 4.6 Hz, CHCOO), 173.1 (ddd, $^3J_{CP}$ = 2.1 Hz, $^3J_{CP}$ = 4.6 Hz, $^2J_{CPt}$ = 41.6 Hz, CHCOO).

^{19}F-NMR (376.50 MHz; C_6D_6): δ = -61.23.

^{31}P-NMR (161.97 MHz; C_6D_6): δ = 26.7 (d, $^2J_{PP}$ = 40.7 Hz), 26.7 (dd, $^2J_{PP}$ = 40.7 Hz, $^1J_{PPt}$ = 4005.8 Hz), 28.2 (dd, $^1J_{PF}$ = 2.8 Hz, $^2J_{PP}$ = 40.7 Hz), 28.2 (ddd, $^1J_{PF}$ = 2.8 Hz, $^2J_{PP}$ = 40.7 Hz, $^1J_{PPt}$ = 3764.6 Hz).

^{195}Pt-NMR (85.48 MHz; C_6D_6): δ = -5062.3 (dd, $^1J_{PPt}$ = 3761.4 Hz, $^1J_{PPt}$ = 4003.3 Hz).

FAB-MS: m/z (%): 719.0 (100) [Pt(PPh$_3$)(PPh$_2$C$_6$H$_4$)$^+$].

FT-IR (ATR, cm^{-1}): 3044 (w), 2954 (w), 1681 (s, CO), 1609 (m), 1478 (m), 1433 (s), 1392 (w), 1321 (s), 1272 (s), 1153 (w), 1135 (m), 1113 (m), 1093 (s), 1063 (s), 997 (m), 834 (s), 740 (s), 691 (s), 604 (w), 536 (m), 506 (s), 422 (m).

Bis(triphenylphosphine)-(η^2-methyl 4-nitrocinnamate)palladium(0)

PdPhNO$_2$-Me: C$_{46}$H$_{39}$NO$_4$P$_2$Pd, M = 838.17 *g/mol*, orange red solid, yield: 93%, **route**: *Pd-I*.

^{1}H-NMR (400.13 MHz; C_6D_6): δ = 3.18 (s, 3H, OCH_3), 4.51 (dpt, $^3J_{HH}$ = 10.1 Hz, $^3J_{HP}$ = 6.3 Hz, $^3J_{HP}$ = 6.3 Hz, 1H, CHCHCO), 5.19 (ddd, $^3J_{HH}$ = 10.1 Hz,

$^3J_{HP} = 8.0$ Hz, $^3J_{HP} = 5.5$ Hz, 1H, $C_{Ar}CHCH$), 6.25 - 6.30 (m, 2H, H_{Ar}), 6.77 - 7.00 (m, 18H, H_{Ar}), 7.01 - 7.09 (m, 6H, H_{Ar}), 7.45 - 7.53 (m, 6H, H_{Ar}), 7.59 - 7.64 (m, 2H, H_{Ar}).

^{13}C-NMR (100.62 MHz; C_6D_6): $\delta = 50.4$ (s, $CHCHCO$), 59.5 (dd, $^2J_{CP} = 5.3$ Hz, $^2J_{CP} = 19.6$ Hz, $CHCHCO$), 72.3 (dd, $^2J_{CP} = 2.9$ Hz, $^2J_{CP} = 22.3$ Hz, $C_{Ar}CHCH$), 123.9 (d, $^3J_{CP} = 1.3$ Hz, $C_{Ar}H$), 124.8 (d, $^3J_{CP} = 2.6$ Hz, $C_{Ar}H$), 128.4 (C_{Ar}), 128.4 (d, $^3J_{CP} = 1.1$ Hz, C_{Ar}), 28.5 ($C_{Ar}H$), 128.6 ($C_{Ar}H$), 129.6 (dd, $^1J_{CP} = 12.2$ Hz, $^3J_{CP} = 1.5$ Hz, $C_{Ar}P$), 134.0 (d, $^2J_{CP} = 14.5$ Hz, $C_{Ar}H$), 134.3 (d, $^2J_{CP} = 15.1$ Hz, $C_{Ar}H$), 135.1 (dd, $^1J_{CP} = 29.3$ Hz, $^3J_{CP} = 1.5$ Hz, $C_{Ar}P$), 136.54 (dd, $^1J_{CP} = 28.6$ Hz, $^3J_{CP} = 1.6$ Hz, $C_{Ar}P$), 144.3 - 144.4 (m, $C_{Ar}CH$), 150.0 (d, $^3J_{CP} = 5.3$ Hz, $C_{Ar}N$), 171.5 (dd, $^3J_{CP} = 2.2$ Hz, $^3J_{CP} = 4.4$ Hz, $CHCOO$).

^{31}P-NMR (161.97 MHz; C_6D_6): $\delta = 25.0$ (d, $^2J_{PP} = 7.0$ Hz), 27.0 (d, $^2J_{PP} = 7.0$ Hz).

FAB-MS: m/z (%): 368.0 (6.65) $[Pd(PPh_3)^+]$, 630.1 (29.88) $[Pd(PPh_3)_2{}^+]$.

FT-IR (ATR, cm^{-1}): 3051 (w), 1687 (m, CO), 1584 (m), 1504 (m), 1476 (m), 1431 (s), 1327 (s), 1305 (w), 1278 (m), 1145 (m), 1109 (w), 1091 (m), 1032 (w), 850 (m), 744 (m), 690 (s), 500 (s), 401 (m).

elemental analysis for $C_{46}H_{39}NO_4P_2Pd$ (838.17 g/mol):

calcd.: C 65.92, H 4.69, N 1.67, O 7.64, P 7.39, Pd 12.70

found: C 66.04, H 4.80, N 1.68

Bis(triphenylphosphine)-(η^2-methyl 4-nitrocinnamate)platinum(0)

PtPhNO$_2$-Me: $C_{46}H_{39}NO_4P_2Pt$, M = 926.83 g/mol, yellow solid, yield: 83%, **route:** *Pt-III.*

^{1}H-NMR (400.13 MHz; C$_6$D$_6$): δ = 3.20 (s, 3H, OCH_3), 3.81 (dpt, $^3J_{HH}$ = 8.8 Hz, $^3J_{HP}$ = 3.7 Hz, $^3J_{HP}$ = 8.8 Hz, 1H, CHCHCO), 3.81 (ddpt, $^3J_{HH}$ = 8.8 Hz, $^3J_{HP}$ = 3.7 Hz, $^3J_{HP}$ = 8.8 Hz, $^2J_{HPt}$ = 58.7 Hz, 1H, CHCHCO), 4.42 (dpt, $^3J_{HH}$ = 8.8 Hz, $^3J_{HP}$ = 3.9 Hz, $^3J_{HP}$ = 8.8 Hz, 1H, C$_{Ar}$CHCH), 4.42 (ddpt, $^3J_{HH}$ = 8.8 Hz, $^3J_{HP}$ = 3.9 Hz, $^3J_{HP}$ = 8.8 Hz, $^2J_{HPt}$ = 54.5 Hz, 1H, C$_{Ar}$CHCH), 6.25 (d, $^3J_{HH}$ = 7.4 Hz, 2H, H_{Ar}), 6.72 - 7.14 (m, 24H, H_{Ar}), 7.44 - 7.54 (m, 6H, H_{Ar}), 7.64 (d, $^3J_{HH}$ = 8.6 Hz, 2H, H_{Ar}).

^{13}C-NMR (100.62 MHz; C$_6$D$_6$): δ = 49.3 (dd, $^2J_{CP}$ = 4.4 Hz, $^2J_{CP}$ = 28.6 Hz, CHCHCO), 49.3 (ddd, $^2J_{CP}$ = 4.4 Hz, $^2J_{CP}$ = 28.6 Hz, $^1J_{CPt}$ = 208.0 Hz, CHCHCO), 50.4 (COOCH$_3$), 58.9 (dd, $^2J_{CP}$ = 2.9 Hz, $^2J_{CP}$ = 30.0 Hz, C$_{Ar}$$C$HCH), 58.9 (ddd, $^2J_{CP}$ = 2.9 Hz, $^2J_{CP}$ = 30.0 Hz, $^1J_{CPt}$ = 191.6 Hz, C$_{Ar}$$C$HCH), 123.4 (d, $^5J_{CP}$ = 1.3 Hz, C_{Ar}H), 123.4 (bm, C_{Ar}H), 125.3 (d, $^5J_{CP}$ = 2.9 Hz, C_{Ar}H), 125.3 (dd, $^5J_{CP}$ = 2.9 Hz, $^4J_{CPt}$ = 13.2 Hz, C_{Ar}H), 128.2 (C_{Ar}H), 128.4 (C_{Ar}H), 129.7 (d, $^5J_{CP}$ = 1.9 Hz, C_{Ar}H), 129.8 (d, $^5J_{CP}$ = 2.0 Hz, C_{Ar}H), 133.9 (C_{Ar}P), 133.9 (d, $^3J_{CPt}$ = 21.3 Hz, C_{Ar}P), 134.0 (d, $^2J_{CP}$ = 0.5 Hz, C_{Ar}P), 134.0 (dd, $^2J_{CP}$ = 0.5 Hz, $^3J_{CPt}$ = 20.6 Hz, C_{Ar}P), 134.1 (C_{Ar}P), 134.1 (d, $^3J_{CPt}$ = 20.7 Hz, C_{Ar}P), 134.2 (d, $^2J_{CP}$ = 0.6 Hz, C_{Ar}P), 134.2 (dd, $^2J_{CP}$ = 0.6 Hz, $^3J_{CPt}$ = 20.1 Hz, C_{Ar}P), 134.7 (d, $^1J_{CP}$ = 3.8 Hz, C_{Ar}P), 134.7 (dd, $^1J_{CP}$ = 3.8 Hz, $^2J_{CPt}$ = 31.9 Hz, C_{Ar}P), 135.2 (d, $^1J_{CP}$ = 3.8 Hz, C_{Ar}P), 135.2 (dd, $^1J_{CP}$ = 3.8 Hz, $^2J_{CPt}$ = 30.5 Hz, C_{Ar}P), 136.1 (d, $^1J_{CP}$ = 4.0 Hz, C_{Ar}P), 136.1 (dd, $^1J_{CP}$ = 4.0 Hz, $^2J_{CPt}$ = 30.1 Hz, C_{Ar}P), 136.5 (d, $^1J_{CP}$ = 4.0 Hz, C_{Ar}P), 136.5 (dd, $^1J_{CP}$ = 4.0 Hz, $^2J_{CPt}$ = 28.2 Hz, C_{Ar}P), 144.3 (d, $^3J_{CP}$ = 3.1 Hz, C_{Ar}N), 153.0 (dd, $^3J_{CP}$ = 1.5 Hz, $^3J_{CP}$ = 6.4 Hz, C_{Ar}CH), 153.0 (ddd, $^3J_{CP}$ = 1.5 Hz, $^3J_{CP}$ = 6.4 Hz, $^2J_{CPt}$ = 40.1 Hz, C_{Ar}CH), 173.4 (dd, $^3J_{CP}$ 2.0 Hz, $^3J_{CP}$ = 4.9 Hz, CHCOO), 173.4 (ddd, $^3J_{CP}$ = 2.0 Hz, $^3J_{CP}$ = 4.9 Hz, $^2J_{CPt}$ = 42.0 Hz, CHCOO).

^{31}P-NMR (161.97 MHz; C$_6$D$_6$): δ = 26.8 (d, $^2J_{PP}$ = 33.8 Hz), 26.8 (dd, $^2J_{PP}$ = 33.8 Hz, $^1J_{PPt}$ = 4026.8 Hz), 27.2 (d, $^2J_{PP}$ = 33.8 Hz), 27.2 (dd, $^2J_{PP}$ = 33.8 Hz, $^1J_{PPt}$ = 3730.3 Hz).

^{195}Pt-NMR (85.48 MHz; C$_6$D$_6$): δ = -5065.1 (dd, $^1J_{PPt}$ = 3730.2 Hz, $^1J_{PPt}$ = 4027.3 Hz).

FAB-MS: m/z (%): 719.0 (100) [Pt(PPh$_3$)$_2$$^+$], 926.1 (2.33) [(C$_{10}H_9NO_4$)Pt(PPh$_3$)$_2$$^+$].

FT-IR (ATR, cm^{-1}): 3054 (m), 1720 (m, CO), 1585 (m), 1509 (m), 1476 (m), 1431 (S), 1337 (s), 1152 (m), 1081 (m), 1026 (w), 995 (w), 847 (m), 741 (s), 691 (s), 681 (s), 436 (w), 408 (m).

elemental analysis for C$_{46}$H$_{39}$NO$_4$P$_2$Pt (926.83 g/mol):

calcd.: C 59.61, H 4.24, N 1.51, O 6.90, P 6.68, Pt 21.05

found: C 57.03, H 4.11, N 1.45

Bis(triphenylphosphine)-(η^2-ethyl 4-nitrocinnamate)palladium(0)

PdPhNO$_2$-Et: C$_{47}$H$_{41}$NO$_4$P$_2$Pd, M $=$ 852.20 g/mol, orange solid, yield: 91%, **route:** Pd-I.

^{1}H-NMR (400.13 MHz; C$_6$D$_6$): $\delta = 0.85$ (t, $^3J_{HH} = 7.1$ Hz, 3H, CH$_2$CH_3), 3.58 (dq, $^2J_{HH} = 10.8$ Hz, $^3J_{HH} = 7.1$ Hz, 1H, CH_2CH$_3$), 4.02 (dq, $^2J_{HH} = 10.8$ Hz, $^3J_{HH} = 7.1$ Hz, 1H, CH_2CH$_3$), 4.53 (dpt, $^3J_{HH} = 10.1$ Hz, $^3J_{HP} = 6.4$ Hz, $^3J_{HP} = 6.4$ Hz, 1H, CHCHCO), 5.18 (ddd, $^3J_{HH} = 10.1$ Hz, $^3J_{HP} = 7.9$ Hz, $^3J_{HP} = 5.5$ Hz, 1H, C$_{Ar}$CHCH), 6.24 - 6.32 (m, 2H, H_{Ar}), 6.78 - 6.86 (m, 6H, H_{Ar}), 6.78 - 6.86 (m, 12H, H_{Ar}), 6.87 - 7.01 (m, 6H, H_{Ar}), 7.01 - 7.11 (m, 6H, H_{Ar}), 7.46 - 7.56 (m, 6H, H_{Ar}), 7.57 - 7.64 (m, 2H, H_{Ar}).

^{13}C-NMR (100.62 MHz; C$_6$D$_6$): $\delta = 14.4$ (CH$_2$CH_3), 59.3 (CH_2CH$_3$), 59.9 (dd, $^2J_{CP} = 4.9$ Hz, $^2J_{CP} = 19.8$ Hz, CHCHCO), 72.6 (dd, $^2J_{CP} = 2.6$ Hz, $^2J_{CP} = 21.9$ Hz, C$_{Ar}$$C$HCH), 123.9 (bs, C_{Ar}H), 124.8 (bs, C_{Ar}H), 128.4 (C_{Ar}), 128.5 (C_{Ar}), 128.6 (C_{Ar}H), 129.4 - 129.7 (m, C_{Ar}P), 134.1 (d, $^2J_{CP} = 14.6$ Hz, C_{Ar}H), 134.3 (d, $^2J_{CP} = 15.1$ Hz, C_{Ar}H), 135.0 - 135.3 (m, C_{Ar}P), 136.7 (dd, $^1J_{CP} = 28.5$ Hz, $^3J_{CP} = 1.1$ Hz, C_{Ar}P), 144.4 (bs, C_{Ar}CH), 150.0 (d, $^3J_{CP} = 5.3$ Hz, C_{Ar}N), 171.2 - 171.3 (m, CHCOO).

^{31}P-NMR (161.97 MHz; C$_6$D$_6$): $\delta = 24.9$ (d, $^2J_{PP} = 8.1$ Hz), 27.1 (d, $^2J_{PP} = 8.1$ Hz).

FAB-MS: m/z (%): 368.0 (15.11) [Pd(PPh$_3$)$^+$], 630.0 (23.47) [Pd(PPh$_3$)$_2$$^+$].

FT-IR (ATR, cm^{-1}): 3051 (w), 2960 (w), 1687 (s, CO), 1583 (m), 1495 (m), 1477 (w), 1432 (m), 1291 (s), 1234 (s), 1212 (W), 1107 (m), 1096 (m), 1044 (m), 927 (w), 851 (m), 746 (m), 739 (s), 514 (s), 527 (m), 514 (s), 493 (s), 449 (w), 418 (m),

elemental analysis for C$_{47}$H$_{41}$NO$_4$P$_2$Pd (852.20 *g/mol*):

calcd.: C 66.24, H 4.85, N 1.64, O 7.51, P 7.27, Pd 12.49

found: C 66.15, H 5.35, N 1.58

Bis(triphenylphosphine)-(η^2-ethyl 4-nitrocinnamate)platinum(0)

PtPhNO$_2$-Et: C$_{47}$H$_{41}$NO$_4$P$_2$Pt, M = 940.86 *g/mol*, orange solid, yield: 81%, **route**: *Pt-III.*

^{1}H-NMR (400.13 MHz; C$_6$D$_6$): δ = 0.84 (t, $^3J_{HH}$ = 7.1 Hz, 3H, CH$_2$C*H*$_3$), 3.63 (dq, $^2J_{HH}$ = 10.8 Hz, $^3J_{HH}$ = 7.2 Hz, 1H, C*H*$_2$CH$_3$), 3.86 (dpt, $^3J_{HH}$ = 8.9 Hz, $^3J_{HP}$ = 4.3 Hz, $^3J_{HP}$ = 8.9 Hz, 1H, CHC*H*CO), 3.86 (ddpt, $^3J_{HH}$ = 8.9 Hz, $^3J_{HP}$ = 4.3 Hz, $^3J_{HP}$ = 8.9 Hz, $^2J_{HPt}$ = 58.8 Hz, 1H, CHC*H*CO), 4.04 (dq, $^2J_{HH}$ = 10.8 Hz, $^3J_{HH}$ = 7.2 Hz, 1H, C*H*$_2$CH$_3$), 4.41 (dpt, $^3J_{HH}$ = 8.9 Hz, $^3J_{HP}$ = 4.5 Hz, $^3J_{HP}$ = 8.9 Hz, 1H, C$_{Ar}$C*H*CH), 4.41 (ddpt, $^3J_{HH}$ = 8.9 Hz, $^3J_{HP}$ = 4.5 Hz, $^3J_{HP}$ = 8.9 Hz, $^2J_{HPt}$ = 55.4 Hz, 1H, C$_{Ar}$C*H*CH), 6.24 (d, $^3J_{HH}$ = 7.4 Hz, 2H, *H*$_{Ar}$), 6.65 - 7.22 (m, 24H, *H*$_{Ar}$), 7.32 - 7.60 (m, 6H, *H*$_{Ar}$), 7.63 (d, $^3J_{HH}$ = 8.7 Hz, 2H, *H*$_{Ar}$).

^{13}C-NMR (100.62 MHz; C$_6$D$_6$): δ = 14.44 (CH$_2$*C*H$_3$), 49.6 (dd, $^2J_{CP}$ = 5.4 Hz, $^2J_{CP}$ = 29.9 Hz, CH*C*HCO), 49.6 (ddd, $^2J_{CP}$ = 5.4 Hz, $^2J_{CP}$ = 29.9 Hz, $^1J_{CPt}$ = 193.8 Hz, CH*C*HCO), 59.2 (*C*H$_2$CH$_3$), 59.2 (dd, $^2J_{CP}$ = 4.3 Hz, $^2J_{CP}$ = 31.1 Hz, C$_{Ar}$*C*HCH), 59.2 (ddd, $^2J_{CP}$ = 4.3 Hz, $^2J_{CP}$ = 31.1 Hz, $^1J_{CPt}$ = 190.7 Hz, C$_{Ar}$*C*HCH), 123.4 (d, $^5J_{CP}$ = 2.0 Hz, *C*$_{Ar}$H), 123.4 (bm, *C*$_{Ar}$H), 125.3 (d, $^5J_{CP}$ = 2.9 Hz, *C*$_{Ar}$H), 125.3 (dd,

$^5J_{CP}$ = 2.9 Hz, $^4J_{CPt}$ = 12.4 Hz, C_{Ar}H), 127.9 (C_{Ar}H), 128.1 (C_{Ar}H), 128.4 (C_{Ar}H),

129.7 (d, $^5J_{CP}$ = 1.8 Hz, C_{Ar}H), 129.8 (d, $^5J_{CP}$ = 1.8 Hz, C_{Ar}H), 133.9 (C_{Ar}P), 133.9 (d,

$^3J_{CPt}$ = 20.8 Hz, C_{Ar}P), 134.0 (C_{Ar}P), 134.0 (d, $^3J_{CPt}$ = 21.4 Hz, C_{Ar}P), 134.2 (C_{Ar}P),

134.2 (d, $^3J_{CPt}$ = 19.8 Hz, C_{Ar}P), 134.3 (C_{Ar}P), 134.3 (d, $^3J_{CPt}$ = 19.6 Hz, C_{Ar}P), 134.7

(d, $^1J_{CP}$ = 2.7 Hz, C_{Ar}P), 137.7 (dd, $^1J_{CP}$ = 2.7 Hz, $^2J_{CPt}$ = 29.7 Hz, C_{Ar}P), 135.2

(d, $^1J_{CP}$ = 2.7 Hz, C_{Ar}P), 135.2 (dd, $^1J_{CP}$ = 2.7 Hz, $^2J_{CPt}$ = 30.6 Hz, C_{Ar}P), 136.2

(d, $^1J_{CP}$ = 2.8 Hz, C_{Ar}P), 136.2 (dd, $^1J_{CP}$ = 2.8 Hz, $^2J_{CPt}$ = 29.8 Hz, C_{Ar}P), 136.7

(d, $^1J_{CP}$ = 2.8 Hz, C_{Ar}P), 136.7 (dd, $^1J_{CP}$ = 2.8 Hz, $^2J_{CPt}$ = 29.8 Hz, C_{Ar}P), 144.3

(d, $^3J_{CP}$ = 3.2 Hz, C_{Ar}N), 153.0 (dd, $^3J_{CP}$ = 1.4 Hz, $^3J_{CP}$ = 6.6 Hz, C_{Ar}CH), 153.0

(ddd, $^3J_{CP}$ = 1.4 Hz, $^3J_{CP}$ = 6.6 Hz, $^2J_{CPt}$ = 38.9 Hz, C_{Ar}CH), 173.1 (dd, $^3J_{CP}$ 2.1 Hz,

$^3J_{CP}$ = 5.0 Hz, CHCOO), 173.1 (ddd, $^3J_{CP}$ = 2.1 Hz, $^3J_{CP}$ = 5.0 Hz, $^2J_{CPt}$ = 42.1 Hz,

CHCOO).

^{31}P-NMR (161.97 MHz; C$_6$D$_6$): δ = 26.7 (d, $^2J_{PP}$ = 34.9 Hz), 26.7 (dd, $^2J_{PP}$ = 34.9 Hz, $^1J_{PPt}$ = 4008.8 Hz), 27.5 (d, $^2J_{PP}$ = 34.9 Hz), 27.5 (dd, $^2J_{PP}$ = 34.9 Hz, $^1J_{PPt}$ = 3753.5 Hz).

^{195}Pt-NMR (85.48 MHz; C$_6$D$_6$): δ = -5057.7 (dd, $^1J_{PPt}$ = 3735.7 Hz, $^1J_{PPt}$ = 4026.5 Hz).

FAB-MS: m/z (%): 719.1 (100) [Pt(PPh$_3$)$_2$$^+$], 941.1 (2.14) [(C$_{11}H_{11}NO_4$)Pt(PPh$_3$)$_2$$^+$].

FT-IR (ATR, cm^{-1}): 3068 (w), 2977 (w), 1689 (s), 1583 (m), 1493 (m), 1433 (m), 1301 (s), 1231 (m), 1180 (w), 1108 (m), 1098 (m), 1090 (w), 1028 (m), 952 (w), 877 (m), 831 (m), 740 (m), 680 (s), 539 (m), 519 (s), 498 (s), 421 (m).

elemental analysis for C$_{47}$H$_{41}$NO$_4$P$_2$Pt (940.86 *g/mol*):

calcd.: C 60.00, H 4.39, N 1.49, O 6.80, P 6.58, Pt 20.73

found: C 56.66, H 4.13, N 1.23

Bis(triphenylphosphine)-(η^2-*iso*-propyl 4-nitrocinnamate)palladium(0)

PdPhNO$_2$-^{i}Pr: C$_{48}$H$_{43}$NO$_4$P$_2$Pd, M = 866.23 *g/mol*, orange solid, yield: 89%, **route**: *Pd-I*.

^{1}H-NMR (400.13 MHz; C$_6$D$_6$): δ = 0.75 (d, $^3J_{HH}$ = 6.2 Hz, 3H, CHCH$_3$), 1.14 (d, $^3J_{HH}$ = 6.2 Hz, 3H, CHCH$_3$), 4.56 (dpt, $^3J_{HH}$ = 10.0 Hz, $^3J_{HP}$ = 6.3 Hz, $^3J_{HP}$ = 6.3 Hz, 1H, CHCHCO), 4.9 (hp, $^3J_{HH}$ = 6.2 Hz, 1H, CH(CH$_3$)$_2$), 5.15 (ddd, $^3J_{HH}$ = 10.0 Hz, $^3J_{HP}$ = 7.8 Hz, $^3J_{HP}$ = 5.8 Hz, 1H, C$_{Ar}$CHCH), 6.23 - 6.29 (m, 2H, H_{Ar}), 6.79 - 6.87 (m, 6H, H_{Ar}), 6.87 - 7.06 (m, 18H, H_{Ar}), 7.47 - 7.57 (m, 6H, H_{Ar}), 7.57 - 7.63 (m, 2H, H_{Ar}).

^{13}C-NMR (100.62 MHz; C$_6$D$_6$): δ = 21.6 (CHCH$_3$), 22.5 (CHCH$_3$), 60.1 (dd, $^2J_{CP}$ = 4.9 Hz, $^2J_{CP}$ = 20.1 Hz, CHCHCO), 66.5 (OCH(CH$_3$)$_2$), 73.2 (dd, $^2J_{CP}$ = 3.1 Hz, $^2J_{CP}$ = 22.0 Hz, C$_{Ar}$CHCH), 123.9 (bs, C$_{Ar}$H), 124.8 (d, $^3J_{CP}$ = 1.8 Hz, C$_{Ar}$H), 128.3 (C$_{Ar}$), 128.4 (C$_{Ar}$), 128.5 (C$_{Ar}$H), 128.6 (C$_{Ar}$H), 129.6 (dd, $^1J_{CP}$ = 18.1 Hz, $^3J_{CP}$ = 1.3 Hz, C$_{Ar}$P), 134.0 (d, $^2J_{CP}$ = 14.5 Hz, C$_{Ar}$H), 134.4 (d, $^2J_{CP}$ = 15.2 Hz, C$_{Ar}$H), 135.1 (dd, $^1J_{CP}$ = 28.9 Hz, $^3J_{CP}$ = 1.6 Hz, C$_{Ar}$P), 136.8 (dd, $^1J_{CP}$ = 28.5 Hz, $^3J_{CP}$ = 1.8 Hz, C$_{Ar}$P), 144.4 - 144.5 (m, C$_{Ar}$CH), 150.0 (d, $^3J_{CP}$ = 5.4 Hz, C$_{Ar}$NO$_2$), 170.9 (dd, $^3J_{CP}$ = 2.0 Hz, $^3J_{CP}$ = 4.4 Hz, CHCOO).

^{31}P-NMR (161.97 MHz; C$_6$D$_6$): δ = 24.5 (d, $^2J_{PP}$ = 8.8 Hz), 27.0 (d, $^2J_{PP}$ = 8.8 Hz).

FAB-MS: m/z (%): 368.0 (6.17) [Pd(PPh$_3$)$^+$], 630.1 (25.35) [Pd(PPh$_3$)$_2$$^+$].

FT-IR (ATR, cm^{-1}): 3056 (m), 2980 (m), 1677 (s, CO), 1585 (m), 1504 (m), 1475 (m), 1433 (m), 1326 (s), 1307 (w), 1278 (m), 1174 (s), 1111 (w), 1097 (s), 1000 (w), 849 (m), 743 (s), 692 (s), 526 (w), 512 (m), 502 (s), 486 (m), 433 (m).

elemental analysis for C$_{48}$H$_{43}$NO$_4$P$_2$Pd (866.23 *g/mol*):

calcd.: C 66.55, H 5.00, N 1.62, O 7.39, P 7.15, Pd 12.29

found: C 66.33, H 5.09, 1.56

Bis(triphenylphosphine)-(η^2-*iso*-propyl 4-nitrocinnamate)platinum(0)

PtPhNO$_2$-^{i}Pr: C$_{48}$H$_{43}$NO$_4$P$_2$Pt, M = 954.88 *g/mol*, yellow solid, yield: 79%, **route:** *Pt-III.*

Single crystals suitable for X-ray analysis were obtained by slow diffusion of pentane into a toluene solution.

^{1}H-NMR (400.13 MHz; C$_6$D$_6$): δ = 0.76 (d, $^3J_{HH}$ = 6.3 Hz, 3H, CHCH_3), 1.13 (d, $^3J_{HH}$ = 6.3 Hz, 3H, CHCH_3), 3.90 (dpt, $^3J_{HH}$ = 8.8 Hz, $^3J_{HP}$ = 4.6 Hz, $^3J_{HP}$ = 8.8 Hz, 1H, CHCHCO), 3.90 (ddpt, $^3J_{HH}$ = 8.8 Hz, $^3J_{HP}$ = 4.6 Hz, $^3J_{HP}$ = 8.8 Hz, $^2J_{HPt}$ = 59.0 Hz, 1H, CHCHCO), 4.39 (dpt, $^3J_{HH}$ = 9.0 Hz, $^3J_{HP}$ = 4.9 Hz, $^3J_{HP}$ = 9.0 Hz, 1H, C$_{Ar}$CHCH), 4.39 (ddpt, $^3J_{HH}$ = 9.0 Hz, $^3J_{HP}$ = 4.9 Hz, $^3J_{HP}$ = 9.0 Hz, $^2J_{HPt}$ = 51.9 Hz, 1H, C$_{Ar}$CHCH), 4.96 (hp, $^3J_{HH}$ = 6.3 Hz, 1H, CH(CH$_3$)$_2$), 6.20 - 6.26 (m, 2H, H_{Ar}), 6.78 - 6.91 (m, 9H, H_{Ar}), 6.91 - 7.00 (m, 9H, H_{Ar}), 7.00 - 7.11 (m, 6H, H_{Ar}), 7.49 - 7.57 (m, 6H, H_{Ar}), 7.62 (d, $^3J_{HH}$ = 8.7 Hz, 2H, H_{Ar}).

^{13}C-NMR (100.62 MHz; C$_6$D$_6$): δ = 21.7 (CHCH_3), 22.5 (CHCH_3), 49.9 (dd, $^2J_{CP}$ = 5.7 Hz, $^2J_{CP}$ = 30.6 Hz, CHCHCO), 49.9 (ddd, $^2J_{CP}$ = 5.7 Hz, $^2J_{CP}$ = 30.6 Hz, $^1J_{CPt}$ = 218.5 Hz, CHCHCO), 59.7 (dd, $^2J_{CP}$ = 4.5 Hz, $^2J_{CP}$ = 32.5 Hz, C$_{Ar}$$C$HCH), 59.7 (ddd, $^2J_{CP}$ = 4.5 Hz, $^2J_{CP}$ = 32.5 Hz, $^1J_{CPt}$ = 188.1 Hz, C$_{Ar}$$C$HCH), 66.3 (O$C$H(CH$_3$)$_2$), 123.4 (d, $^5J_{CP}$ = 2.2 Hz, C$_{Ar}$H), 123.4 (dd, $^5J_{CP}$ = 2.2 Hz, $^4J_{CPt}$ = 9.1 Hz, C$_{Ar}$H), 125.3 (d, $^4J_{CP}$ = 3.0 Hz, C$_{Ar}$H), 125.3 (dd, $^4J_{CP}$ = 3.0 Hz, $^3J_{CPt}$ = 14.0 Hz, C$_{Ar}$H), 128.0 (C$_{Ar}$H), 128.1 (C$_{Ar}$H), 128.2 (C$_{Ar}$H), 128.3 (C$_{Ar}$H), 129.7 (d, $^1J_{CP}$ = 2.1 Hz, C$_{Ar}$H), 129.9 (d, $^1J_{CP}$ = 2.1 Hz, C$_{Ar}$H), 133.9 (C$_{Ar}$H), 133.9 (d, $^3J_{CPt}$ = 20.9 Hz, C$_{Ar}$H), 134.1 (C$_{Ar}$H), 134.1 (d, $^3J_{CPt}$ = 20.1 Hz, C$_{Ar}$H), 134.3 (C$_{Ar}$H), 134.3 (d, $^3J_{CPt}$ = 19.7 Hz,

$C_{Ar}H$), 134.5 ($C_{Ar}H$), 134.5 (d, $^3J_{CPt}$ = 19.6 Hz, $C_{Ar}H$), 134.7 (d, $^1J_{CP}$ = 2.3 Hz, $C_{Ar}P$), 134.7 (dd, $^1J_{CP}$ = 2.3 Hz, $^2J_{CPt}$ = 32.2 Hz, $C_{Ar}P$), 135.2 (d, $^1J_{CP}$ = 2.2 Hz, $C_{Ar}P$), 135.2 (dd, $^1J_{CP}$ = 2.2 Hz, $^2J_{CPt}$ = 30.0 Hz, $C_{Ar}P$), 136.4 (d, $^1J_{CP}$ = 2.5 Hz, $C_{Ar}P$), 136.4 (dd, $^1J_{CP}$ = 2.5 Hz, $^2J_{CPt}$ = 30.8 Hz, $C_{Ar}P$), 136.8 (d, $^1J_{CP}$ = 2.4 Hz, $C_{Ar}P$), 136.8 (dd, $^1J_{CP}$ = 2.4 Hz, $^2J_{CPt}$ = 28.4 Hz, $C_{Ar}P$), 144.4 (d, $^3J_{CP}$ = 3.2 Hz, $C_{Ar}N$), 152.7 (dd, $^3J_{CP}$ = 6.7 Hz, $^3J_{CP}$ = 1.8 Hz, $C_{Ar}CH$), 152.7 (ddd, $^3J_{CP}$ = 6.7 Hz, $^3J_{CP}$ = 1.8 Hz, $^2J_{CPt}$ = 38.0 Hz, $C_{Ar}CH$), 172.8 (dd, $^3J_{CP}$ 2.1 Hz, $^3J_{CP}$ = 5.2 Hz, CHCOO), 172.8 (ddd, $^3J_{CP}$ = 2.1 Hz, $^3J_{CP}$ = 5.1 Hz, $^2J_{CPt}$ = 43.2 Hz, CHCOO).

^{31}P-NMR (161.97 MHz; C_6D_6): δ = 27.2 (d, $^2J_{PP}$ = 35.8 Hz), 27.2 (dd, $^2J_{PP}$ = 35.8 Hz, $^1J_{PPt}$ = 3991.6 Hz), 28.5 (d, $^2J_{PP}$ = 35.7 Hz), 28.5 (dd, $^2J_{PP}$ = 35.7 Hz, $^1J_{PPt}$ = 3773.1 Hz).

^{195}Pt-NMR (85.48 MHz; C_6D_6): δ = -5051.3 (dd, $^1J_{PPt}$ = 3773.7 Hz, $^1J_{PPt}$ = 3992.9 Hz).

FAB-MS: m/z (%): 718.9 (100) [Pt(PPh$_3$)$_2^+$], 954.0 (1.59) [(C$_{12}$H$_{13}$NO$_4$)Pt(PPh$_3$)$_2^+$].

FT-IR (ATR, cm^{-1}): 3047 (w), 2895 (w), 1683 (m, CO), 1576 (m), 1504 (m), 1434 (m), 1332 (s), 1275 (m), 1159 (m), 1108 (s), 1095 (m), 1031 (w), 1001 (w), 849 (m), 743 (s), 690 (s), 538 (m), 509 (s), 499 (s), 417 (m).

Bis(triphenylphosphine)-(η^2-*tert*-butyl 4-nitrocinnamate)palladium(0)

PdPhNO$_2$-^{t}Bu: C$_{49}$H$_{45}$NO$_4$P$_2$Pd, M = 880.25 *g/mol*, orange solid, yield: 70%, **route**: *Pd-I*.

^{1}H-NMR (400.13 MHz; C_6D_6): δ = 1.30 (s, 9H, C(CH_3)$_3$), 4.53 (dpt, $^3J_{HH}$ = 10.0 Hz, $^3J_{HP}$ = 6.2 Hz, $^3J_{HP}$ = 6.2 Hz, 1H, CHCHCO), 5.10 (ddd, $^3J_{HH}$ = 10.0 Hz, $^3J_{HP}$ = 7.8 Hz, $^3J_{HP}$ = 5.7 Hz, 1H, C$_{Ar}$CHCH), 6.21 - 6.27 (m, 2H, H_{Ar}), 6.81 - 6.88

(m, 6H, H_{Ar}), 6.89 - 7.04 (m, 18H, H_{Ar}), 7.47 - 7.55 (m, 6H, H_{Ar}), 7.56 - 7.61 (m, 2H, H_{Ar}).

^{13}C-NMR (100.62 MHz; C_6D_6): δ = 28.3 (CH_3), 61.2 (dd, $^2J_{CP}$ = 4.9 Hz, $^2J_{CP}$ = 20.1 Hz, CHCHCO), 73.8 (dd, $^2J_{CP}$ = 3.1 Hz, $^2J_{CP}$ = 21.8 Hz, $C_{Ar}CH$CH), 78.5 (OC(CH$_3$)$_3$), 124.0 (bs, C_{Ar}H), 124.7 (d, $^3J_{CP}$ = 1.5 Hz, C_{Ar}H), 128.3 (C_{Ar}), 128.5 (C_{Ar}H), 128.6 (C_{Ar}H), 129.6 (dd, $^1J_{CP}$ = 17.6 Hz, $^3J_{CP}$ = 1.1 Hz, C_{Ar}P), 134.1 (d, $^2J_{CP}$ = 14.4 Hz, C_{Ar}H), 134.5 (d, $^2J_{CP}$ = 15.3 Hz, C_{Ar}H), 135.1 (dd, $^1J_{CP}$ = 28.5 Hz, $^3J_{CP}$ = 1.5 Hz, C_{Ar}P), 136.8 (dd, $^1J_{CP}$ = 28.4 Hz, $^3J_{CP}$ = 2.0 Hz, C_{Ar}P), 144.5 (dd, $^3J_{CP}$ = 2.6 Hz, $^3J_{CP}$ = 1.6 Hz, C_{Ar}CH), 150.2 (d, $^3J_{CP}$ = 5.4 Hz, C_{Ar}N), 171.1 - 171.2 (m, CHCOO).

^{31}P-NMR (161.97 MHz; C_6D_6): δ = 24.1 (d, $^2J_{PP}$ = 9.7 Hz), 26.7 (d, $^2J_{PP}$ = 9.7 Hz).

FAB-MS: m/z (%): 368.0 (6.59) [Pd(PPh$_3$)$^+$], 630.1 (24.35) [Pd(PPh$_3$)$_2$$^+$].

FT-IR (ATR, cm^{-1}): 3060 (w), 2984 (w), 1681 (s, CO), 1583 (s), 1504 (m), 1477 (w), 1432 (m), 1327 (s), 1305 (w), 1285 (s), 1217 (w), 1136 (s), 1109 (w), 1091 (m), 985 (w), 850 (m), 742 (s), 690 (s), 525 (w), 513 (s), 501 (s), 447 (w), 432 (w), 415 (m).

elemental analysis for $C_{49}H_{45}NO_4P_2Pd$ (880.25 g/mol):

calcd.: C 66.86, H 5.15, N 1.59, O 7.27, P 7.04, Pd 12.09

found: C 66.56, H 5.41, N 1.73

Bis(triphenylphosphine)-(η^2-*tert*-butyl 4-nitrocinnamate)platinum(0)

PtPhNO$_2$-^{t}Bu: $C_{49}H_{45}NO_4P_2Pt$, M = 968.91 g/mol, orange solid, yield: 80%, **route:** Pt-III.

^{1}H-NMR (400.13 MHz; C_6D_6): δ = 1.31 (s, 9H, C(CH_3)$_3$), 3.88 (dpt, $^3J_{HH}$ = 8.6 Hz, $^3J_{HP}$ = 4.7 Hz, $^3J_{HP}$ = 8.6 Hz, 1H, CHCHCO), 3.88 (ddpt, $^3J_{HH}$ = 8.6 Hz, $^3J_{HP}$ = 4.7 Hz,

$^3J_{HP} = 8.6$ Hz, $^2J_{HPt} = 59.4$ Hz, 1H, CHCHCO), 4.33 (dpt, $^3J_{HH} = 9.1$ Hz, $^3J_{HP} = 5.0$ Hz,

$^3J_{HP} = 9.1$ Hz, 1H, C$_{Ar}$CHCH), 4.33 (ddpt, $^3J_{HH} = 9.1$ Hz, $^3J_{HP} = 5.0$ Hz, $^3J_{HP} = 9.1$ Hz,

$^2J_{HPt} = 51.9$ Hz, 1H, C$_{Ar}$$C$HCH), 6.18 - 6.21 (m, 2H, H_{Ar}), 6.78 - 6.91 (m, 9H, H_{Ar}),

6.91 - 7.01 (m, 9H, H_{Ar}), 7.01 - 7.10 (m, 6H, H_{Ar}), 7.48 - 7.57 (m, 6H, H_{Ar}), 7.61 (d,

$^3J_{HH} = 8.8$ Hz, 2H, H_{Ar}).

^{13}C-NMR (100.62 MHz; C$_6$D$_6$): δ = 28.4 (CH$_3$), 50.7 (dd, $^2J_{CP}$ = 5.9 Hz,

$^2J_{CP} = 31.0$ Hz, CHCHCO), 50.7 (ddd, $^2J_{CP} = 5.9$ Hz, $^2J_{CP} = 31.0$ Hz, $^1J_{CPt} = 218.6$ Hz,

CHCHCO), 60.6 (dd, $^2J_{CP}$ = 4.9 Hz, $^2J_{CP}$ = 31.8 Hz, C$_{Ar}$$C$HCH), 60.6 (ddd,

$^2J_{CP} = 4.9$ Hz, $^2J_{CP} = 31.8$ Hz, $^1J_{CPt} = 175.8$ Hz, C$_{Ar}$$C$HCH), 78.2 (O$C$(CH$_3$)$_3$), 123.5

(d, $^5J_{CP} = 2.3$ Hz, C_{Ar}H), 123.5 (dd, $^5J_{CP} = 2.3$ Hz, $^4J_{CPt} = 9.1$ Hz, C_{Ar}H), 125.2 (d,

$^4J_{CP} = 3.0$ Hz, C_{Ar}H), 125.2 (dd, $^4J_{CP} = 3.0$ Hz, $^3J_{CPt} = 13.6$ Hz, C_{Ar}H), 128.0 (C_{Ar}H),

128.1 (C_{Ar}H), 128.3 (C_{Ar}H), 128.4 (C_{Ar}H), 129.7 (d, $^1J_{CP} = 2.1$ Hz, C_{Ar}H), 129.9 (d,

$^1J_{CP} = 2.1$ Hz, C_{Ar}H), 134.0 (C_{Ar}H), 134.0 (d, $^3J_{CPt} = 20.0$ Hz, C_{Ar}H), 134.1 (C_{Ar}H),

134.1 (d, $^3J_{CPt} = 19.9$ Hz, C_{Ar}H), 134.4 (C_{Ar}H), 134.4 (d, $^3J_{CPt} = 19.9$ Hz, C_{Ar}H), 134.6

(C_{Ar}H), 134.6 (d, $^3J_{CPt} = 19.8$ Hz, C_{Ar}H), 134.7 (d, $^1J_{CP} = 2.1$ Hz, C_{Ar}P), 134.7 (dd,

$^1J_{CP} = 2.1$ Hz, $^2J_{CPt} = 30.0$ Hz, C_{Ar}P), 135.2 (d, $^1J_{CP} = 2.1$ Hz, C_{Ar}P), 135.2 (dd,

$^1J_{CP} = 2.1$ Hz, $^2J_{CPt} = 30.3$ Hz, C_{Ar}P), 136.5 (d, $^1J_{CP} = 2.4$ Hz, C_{Ar}P), 136.5 (dd,

$^1J_{CP} = 2.4$ Hz, $^2J_{CPt} = 30.2$ Hz, C_{Ar}P), 136.9 (d, $^1J_{CP} = 2.5$ Hz, C_{Ar}P), 136.9 (dd,

$^1J_{CP} = 2.5$ Hz, $^2J_{CPt} = 30.7$ Hz, C_{Ar}P), 144.4 (d, $^3J_{CP} = 3.2$ Hz, C_{Ar}N), 152.7 (dd,

$^3J_{CP} = 6.7$ Hz, $^3J_{CP} = 1.8$ Hz, C_{Ar}CH), 152.7 (ddd, $^3J_{CP} = 6.7$ Hz, $^3J_{CP} = 1.8$ Hz,

$^2J_{CPt} = 38.4$ Hz, C_{Ar}CH), 173.0 (dd, $^3J_{CP}$ 2.1 Hz, $^3J_{CP} = 5.1$ Hz, CHCOO), 173.0 (ddd,

$^3J_{CP} = 2.1$ Hz, $^3J_{CP} = 5.1$ Hz, $^2J_{CPt} = 38.0$ Hz, CHCOO).

^{31}P-NMR (161.97 MHz; C$_6$D$_6$): δ = 26.6 (d, $^2J_{PP}$ = 36.5 Hz), 26.6 (dd,

$^2J_{PP} = 36.5$ Hz, $^1J_{PPt} = 3956.0$ Hz), 28.5 (d, $^2J_{PP} = 36.6$ Hz), 28.5 (dd, $^2J_{PP} = 36.6$ Hz,

$^1J_{PPt} = 3811.3$ Hz).

^{195}Pt-NMR (85.48 MHz; C$_6$D$_6$): δ = -5052.3 (dd, $^1J_{PPt}$ = 3811.1 Hz,

$^1J_{PPt} = 3957.7$ Hz).

FAB-MS: m/z (%): 719.1 (100) [Pt(PPh$_3$)$_2{}^+$], 969.1 (1.66) [(C$_{13}$H$_{15}$NO$_4$)Pt(PPh$_3$)$_2{}^+$].

FT-IR (ATR, cm^{-1}): 3032 (w), 2956 (w), 1683 (s, CO), 1583 (m), 1502 (m), 1433 (s),

1328 (s), 1282 (s), 1135 (s), 1091 (m), 971 (w), 910 (w), 849 (s), 744 (s), 691 (s), 577 (w), 538 (m), 517 (m), 509 (s), 440 (m), 410 (m).

elemental analysis for $C_{49}H_{45}NO_4P_2Pt$ (968.91 g/mol):

calcd.: C 60.74, H 4.68, N 1.45, O 6.61, P 6.39, Pt 20.13

found: C 58.57, H 4.73, N 1.34

Bis(triphenylphosphine)-(η^2-4-methoxyphenyl 4-methoxycinnamate)nickel(0)

NiPhOMe-PhOMe: $C_{53}H_{46}NiO_4P_2$, M = 867.57 g/mol, red solid, yield: >99%, **route**: *Ni*.

^{1}H-NMR (400.13 MHz; C_7D_8): δ = 3.34 (s, 3H, OCH_3), 3.36 (s, 3H, OCH_3), 3.71 (dpt, $^3J_{HH}$ = 8.8 Hz, $^3J_{HP}$ = 3.5 Hz, $^3J_{HP}$ = 8.8 Hz, 1H, CHCHCO), 4.36 (dpt, $^3J_{HH}$ = 8.8 Hz, $^3J_{HP}$ = 5.0 Hz, $^3J_{HP}$ = 8.8 Hz, 1H, CC$_{Ar}$CHCH), 6.49 (d, $^3J_{HH}$ = 8.7 Hz, 2H, H_{Ar}), 6.53 (d, $^3J_{HH}$ = 8.7 Hz, 2H, H_{Ar}), 6.63 (s, 4H, H_{Ar}), 6.89 - 7.05 (m, 18H, H_{Ar}), 7.06 - 7.26 (m, 6H, H_{Ar}), 7.52 - 7.67 (m, 6H, H_{Ar}).

^{13}C-NMR (100.53 MHz; C_7D_8): δ = 52.5 (dd, $^2J_{CP}$ = 1.6 Hz, $^2J_{CP}$ = 12.5 Hz, CHCHCO), 54.8 (OCH$_3$), 54.9 (OCH$_3$), 65.2 (d, $^2J_{CP}$ = 16.5 Hz, C$_{Ar}$CHCH), 114.0 (C_{Ar}H), 114.6 (C_{Ar}H), 123.2 (C_{Ar}H), 126.4 (d, $^4J_{CP}$ = 1.9 Hz, C_{Ar}H), 128.1 (C_{Ar}H), 128.1 (C_{Ar}H), 128.3 (C_{Ar}H), 128.4 (C_{Ar}H), 129.0 (C_{Ar}H), 129.3 (d, $^3J_{CP}$ = 1.1 Hz, C_{Ar}H), 132.4 (d, $^1J_{CP}$ = 9.8 Hz, C_{Ar}), 134.2 (d, $^2J_{CP}$ = 12.5 Hz, C_{Ar}H), 134.4 (d, $^2J_{CP}$ = 13.4 Hz, C_{Ar}H), 135.0 (d, $^1J_{CP}$ = 3.0 Hz, C_{Ar}), 135.3 (d, $^1J_{CP}$ = 3.0 Hz, C_{Ar}), 136.1 (d, $^1J_{CP}$ = 1.7 Hz, C_{Ar}), 136.1 (d, $^1J_{CP}$ = 2.6 Hz, C_{Ar}), 136.5 (d, $^1J_{CP}$ = 2.4 Hz, C_{Ar}), 146.3 (C_{Ar}), 156.8 (C_{Ar}), 157.7 (d, $^3J_{CP}$ = 1.4 Hz, C_{Ar}), 170.9 (dd, $^3J_{CP}$ = 1.1 Hz, $^3J_{CP}$ = 3.4 Hz, CHCOO).

^{31}P-NMR (161.97 MHz; C_7D_8): δ = 31.7 (d, $^2J_{PP}$ = 36.5 Hz), 33.3 (d, $^2J_{PP}$ = 36.5 Hz).

FT-IR (ATR, cm^{-1}): 1688 (m, CO).

elemental analysis for C$_{53}$H$_{46}$NiO$_4$P$_2$ (867.57 *g/mol*):

calcd.: C 73.37, H 5.34, Ni 6.77, O 7.38, P 7.14

found: C 71.94, H 5.46

Bis(triphenylphosphine)-(η^2-4-methoxyphenyl 4-methoxycinnamate)palladium(0)

PdPhOMe-PhOMe: C$_{53}$H$_{46}$O$_4$P$_2$Pd, M = 915.30 *g/mol*, yellow solid, yield: 94%, **route**: *Pd-I*.

^{1}H-NMR (400.13 MHz; C$_7$D$_8$): δ = 3.30 (s, 3H, OCH_3), 3.32 (s, 3H, OCH_3), 4.90 (bs, 1H, CHCHCO), 5.35 (bs, 1H, C$_{Ar}$CHCH), 6.41 - 6.51 (bm, 2H, H_{Ar}), 6.52 - 6.71 (bm, 6H, H_{Ar}), 6.84 - 7.03 (bm, 18H, H_{Ar}), 7.04 - 7.19 (bm, 6H, H_{Ar}), 7.47 - 7.68 (bm, 6H, H_{Ar}).

^{1}H-NMR (400.13 MHz; 246 K; C$_7$D$_8$): δ = 3.30 (s, 3H, OCH_3), 3.31 (s, 3H, OCH_3), 5.09 (dpt, $^3J_{HH}$ = 10.2 Hz, $^3J_{HP}$ = 6.0 Hz, $^3J_{HP}$ = 6.0 Hz, 1H, CHCHCO), 5.44 (dpt, $^3J_{HH}$ = 10.2 Hz, $^3J_{HP}$ = 7.1 Hz, $^3J_{HP}$ = 7.1 Hz, 1H, C$_{Ar}$CHCH), 6.43 - 6.48 (m, 2H, H_{Ar}), 6.55 - 6.59 (m, 2H, H_{Ar}), 6.69 (bs, 4H, H_{Ar}), 6.92 - 7.04 (m, 18H, H_{Ar}), 7.10 - 7.17 (m, 6H, H_{Ar}), 7.61 - 7.70 (m, 6H, H_{Ar}).

^{13}C-NMR (100.62 MHz; C$_7$D$_8$): δ = 54.7 (d, $^2J_{CP}$ = 4.2 Hz, OCH$_3$), 54.8 (d, $^2J_{CP}$ = 4.4 Hz, OCH$_3$), 60.0 (bs, CHCHCO), 76.7 (bs, C$_{Ar}$$C$HCH), 114.0 ($C_{Ar}$H), 114.1 ($C_{Ar}$H), 123.1 ($C_{Ar}$H), 126.4 - 127.0 (bm, C_{Ar}H), 128.3 (C_{Ar}H), 128.4 (C_{Ar}H), 128.4 (C_{Ar}H), 134.0 - 134.8 (bm, C_{Ar}H), 134.8 - 135.2 (bm, C_{Ar}H), 135.5 - 136.6 (bm, C_{Ar}H), 146.1 (C_{Ar}O), 156.8 (C_{Ar}O), 158.1 (bs, C_{Ar}O), 169.3 (bs, CHCOO).

^{13}C-NMR (100.53 MHz; 246 K; C$_7$D$_8$): δ = 54.5 (d, $^2J_{CP}$ = 1.6 Hz, OCH$_3$), 54.6 (d, $^2J_{CP}$ = 1.3 Hz, OCH$_3$), 59.4 (dd, $^2J_{CP}$ = 4.8 Hz, $^2J_{CP}$ = 18.8 Hz, CHCHCO), 76.5

(dd, $^2J_{CP} = 16.6$ Hz, $^2J_{CP} = 5.2$ Hz, $C_{Ar}CHCH$), 113.8 ($C_{Ar}H$), 113.8 ($C_{Ar}H$), 123.2 ($C_{Ar}H$), 126.5 (bs, C_{Ar}), 128.0 (C_{Ar}), 128.5 (C_{Ar}), 128.9 (C_{Ar}), 129.3 (C_{Ar}), 134.1 (d, $^2J_{CP} = 14.4$ Hz, $C_{Ar}H$), 134.4 (d, $^2J_{CP} = 15.5$ Hz, $C_{Ar}H$), 134.6(d, $^3J_{CP} = 6.7$ Hz, $C_{Ar}CH$), 135.6 (dd, $^1J_{CP} = 28.0$ Hz, $^3J_{CP} = 1.7$ Hz, $C_{Ar}P$), 137.1 (dd, $^3J_{CP} = 1.5$ Hz, $C_{Ar}P$), 145.8 (s, $C_{Ar}O$), 156.5 (s, $C_{Ar}O$), 157.8 (bs, $C_{Ar}O$), 169.5 - 169.6 (m, CHCOO).

^{31}P-NMR (161.97 MHz; C$_7$D$_8$): $\delta = 24.4$ (d, $^2J_{PP} = 18.5$ Hz), 25.5 (d, $^2J_{PP} = 18.5$ Hz).

^{31}P-NMR (161.97 MHz; 246 K; C$_7$D$_8$): $\delta = 23.8$ (d, $^2J_{PP} = 20.1$ Hz), 25.2 (d, $^2J_{PP} = 20.1$ Hz).

FT-IR (ATR, cm^{-1}): 3044 (m), 2923 (m), 1696 (s, CO), 1609 (w), 1504 (m), 1476 (s), 1432 (s), 1321 (w), 1252 (m), 1190 (s), 1122 (s), 1102 (s), 1039 (m), 972 (w), 821 (m), 745 (s), 699 (s), 527 (m), 517 (m), 500 (s), 427 (m).

elemental analysis for C$_{53}$H$_{46}$O$_4$P$_2$Pd (915.30 g/mol):

calcd.: C 69.55, H 5.07, O 6.99, P 6.77, Pd 11.63

found: C 69.41, H 5.19

Bis(triphenylphosphine)-(η^2-4-methoxyphenyl 4-methoxycinnamate)platinum(0)

PtPhOMe-PhOMe: C$_{53}$H$_{46}$O$_4$P$_2$Pt, M = 1003.96 g/mol, off-white solid, yield: 99%, route: *Pt-VI*.

Single crystals suitable for X-ray analysis were obtained by slow diffusion of pentane into a toluene solution.

^{1}H-NMR (400.13 MHz; C$_6$D$_6$): $\delta = 3.25$ (s, 3H, OCH_3), 3.32 (s, 3H, OCH_3), 4.30 (dpt, $^3J_{HH} = 8.6$ Hz, $^3J_{HP} = 4.4$ Hz, $^3J_{HP} = 8.6$ Hz, 1H, CHCHCO), 4.30 (ddpt, $^3J_{HH} = 8.6$ Hz, $^3J_{HP} = 4.4$ Hz, $^3J_{HP} = 8.6$ Hz, $^2J_{HPt} = 62.3$ Hz, 1H, CHCHCO), 4.57 (dpt, $^3J_{HH} = 8.4$ Hz, $^3J_{HP} = 5.7$ Hz, $^3J_{HP} = 8.4$ Hz, 1H, $C_{Ar}CH$CH), 4.57 (ddpt,

$^3J_{HH} = 8.4$ Hz, $^3J_{HP} = 5.7$ Hz, $^3J_{HP} = 8.4$ Hz, $^2J_{HPt} = 51.2$ Hz, 1H, $C_{Ar}CHCH$), 6.49 - 6.59 (m, 4H, H_{Ar}), 6.62 (d, $^3J_{HH} = 9.0$ Hz, 2H, H_{Ar}), 6.71 (d, $^3J_{HH} = 8.9$ Hz, 2H, H_{Ar}), 6.84 - 6.96 (m, 18H, H_{Ar}), 7.17 - 7.27 (m, 6H, H_{Ar}), 7.59 - 7.69 (m, 6H, H_{Ar}).

^{13}C-NMR (100.53 MHz; C_6D_6): $\delta = 49.4$(dd, $^2J_{CP} = 5.5$ Hz, $^2J_{CP} = 27.5$ Hz, CHCHCO), 49.4 (ddd, $^2J_{CP} = 5.5$ Hz, $^2J_{CP} = 27.5$ Hz, $^1J_{CPt} = 193.7$ Hz, CHCHCO), 54.8 (OCH_3), 55.0 (OCH_3), 61.7 (dd, $^2J_{CP} = 3.4$ Hz, $^2J_{CP} = 32.4$ Hz, $C_{Ar}CHCH$), 61.7 (ddd, $^2J_{CP} = 3.4$Hz, $^2J_{CP} = 32.4$ Hz, $^1J_{CPt} = 163.5$ Hz, $C_{Ar}CHCH$), 113.6 (d, $^5J_{CP} = 2.0$ Hz, $C_{Ar}H$), 114.2 ($C_{Ar}H$), 123.4 ($C_{Ar}H$), 127.1 (d, $^5J_{CP} = 3.1$ Hz, $C_{Ar}H$), 127.1 (dd, $^5J_{CP} = 3.1$ Hz, $^4J_{CPt} = 13.1$ Hz, $C_{Ar}H$), 128.0 ($C_{Ar}H$), 128.4 ($C_{Ar}H$), 128.4 ($C_{Ar}H$), 129.4 (d, $^1J_{CPt} = 1.9$ Hz, $C_{Ar}H$), 129.6 (d, $^1J_{CPt} = 1.8$ Hz, $C_{Ar}H$), 134.2 ($C_{Ar}H$), 134.2 (d, $^3J_{CPt} = 22.2$ Hz, $C_{Ar}H$), 134.3 ($C_{Ar}H$), 134.3 (d, $^3J_{CPt} = 22.2$ Hz, $C_{Ar}H$), 134.4 ($C_{Ar}H$), 134.4 (d, $^3J_{CPt} = 18.2$ Hz, $C_{Ar}H$), 134.5 ($C_{Ar}H$), 134.5 (d, $^3J_{CPt} = 19.2$ Hz, $C_{Ar}H$), 135.5 (d, $^1J_{CP} = 3.7$ Hz, $C_{Ar}P$), 135.5 (dd, $^1J_{CP} = 3.7$ Hz, $^2J_{CPt} = 33.1$ Hz, $C_{Ar}P$), 135.9 (d, $^1J_{CP} = 3.7$ Hz, $C_{Ar}P$), 135.9 (dd, $^1J_{CP} = 3.7$ Hz, $^2J_{CPt} = 32.1$ Hz, $C_{Ar}P$), 137.1 (d, $^1J_{CP} = 3.4$ Hz, $C_{Ar}P$), 137.1 (dd, $^1J_{CP} = 3.4$ Hz, $^2J_{CPt} = 28.2$ Hz, $C_{Ar}P$), 137.5 (d, $^1J_{CP} = 3.4$ Hz, $C_{Ar}P$), 137.5 (dd, $^1J_{CP} = 3.4$ Hz, $^2J_{CPt} = 27.3$ Hz, $C_{Ar}P$), 146.3 ($C_{Ar}O$), 156.8 ($C_{Ar}O$), 157.6 (d, $^3J_{CP} = 2.4$ Hz, $C_{Ar}CH$), 157.6 (dd, $^3J_{CP} = 2.4$ Hz, $^2J_{CPt} = 55.8$ Hz, $C_{Ar}CH$), 171.9 (dd, $^3J_{CP} = 2.3$ Hz, $^3J_{CP} = 4.8$ Hz, CHCOO), 171.9 (ddd, $^3J_{CP} = 2.3$ Hz, $^3J_{CP} = 4.8$ Hz, $^2J_{CPt} = 41.7$ Hz, CHCOO).

^{31}P-NMR (161.97 MHz; C_6D_6): $\delta = 27.6$ (d, $^2J_{PP} = 44.8$ Hz), 27.6 (dd, $^2J_{PP} = 44.8$ Hz, $^1J_{PPt} = 4217.8$ Hz), 28.3 (d, $^2J_{PP} = 44.8$ Hz), 28.3 (dd, $^2J_{PP} = 44.8$ Hz, $^1J_{PPt} = 3625.3$ Hz).

^{195}Pt-NMR (85.48 MHz; C_6D_6): $\delta = -5044.0$ (dd, $^1J_{PPt} = 3625.4$ Hz, $^1J_{PPt} = 4218.0$ Hz).

FAB-MS: m/z (%): 718.9 (100) $[Pt(PPh_3)_2^+]$, 1002.0 (1.64) $[(C_{17}H_{16}O_4)Pt(PPh_3)_2^+]$.

FT-IR (ATR, cm^{-1}): 3049 (w), 2928 (w), 2837 (w), 1701 (m, CO), 1605 (w), 1505 (s), 1434 (s), 1302 (m), 1246 (m), 1177 (m), 1122 (m), 1097 (s), 1044 (w), 829 (m), 741 (m), 693 (s), 544 (m), 507 (s), 423 (m).

elemental analysis for $C_{53}H_{46}O_4P_2Pt$ (1003.96 *g/mol*):

calcd.: C 63.41, H 4.62, O 6.37, P 6.17, Pt 19.43

found: C 63.07, H 4.84

Bis(triphenylphosphine)-(η^2-4-methylphenyl 4-methoxycinnamate)nickel(0)

NiPhOMe-PhMe: $C_{53}H_{46}NiO_3P_2$, M $= 851.57$ g/mol, orange red solid, yield: 90%, **route:** Ni.

^{1}H-NMR (400.13 MHz; C_7D_8): $\delta = 2.11$ (s, 3H, CH_3), 3.37 (s, 3H, OCH_3), 3.70 (dpt, $^3J_{HH} = 9.2$ Hz, $^3J_{HP} = 3.5$ Hz, $^3J_{HP} = 9.2$ Hz, 1H, CHCHCO), 4.36 (dpt, $^3J_{HH} = 9.2$ Hz, $^3J_{HP} = 5.0$ Hz, $^3J_{HP} = 9.2$ Hz, 1H, CC$_{Ar}$CHCH), 6.49 (d, $^3J_{HH} = 8.5$ Hz, 2H, H_{Ar}), 6.53 (d, $^3J_{HH} = 8.7$ Hz, 2H, H_{Ar}), 6.65 (d, $^3J_{HH} = 8.2$ Hz, 2H, H_{Ar}), 6.85 (d, $^3J_{HH} = 8.1$ Hz, 2H, H_{Ar}), 6.89 - 7.04 (m, 18H, H_{Ar}), 7.05 - 7.27 (m, 6H, H_{Ar}), 7.41 - 7.71 (m, 6H, H_{Ar}).

^{13}C-NMR (100.53 MHz; C_7D_8): $\delta = 20.8$ (CH_3), 52.5 (dd, $^2J_{CP} = 1.6$ Hz, $^2J_{CP} = 12.6$ Hz, CHCHCO), 54.8 (OCH$_3$), 65.1 (d, $^2J_{CP} = 16.4$ Hz, C$_{Ar}$$C$HCH), 114.6 ($C_{Ar}$H), 122.3 ($C_{Ar}$H), 126.4 (d, $^4J_{CP} = 1.7$ Hz, C_{Ar}H), 128.0 (C_{Ar}H), 128.1 (C_{Ar}H), 128.3 (C_{Ar}H), 128.4 (C_{Ar}H), 129.3 (bs, C_{Ar}H), 129.4 (C_{Ar}H), 132.5 (bs, C_{Ar}), 134.2 (d, $^2J_{CP} = 12.5$ Hz, C_{Ar}H), 134.4 (d, $^2J_{CP} = 13.4$ Hz, C_{Ar}H), 134.9 (d, $^1J_{CP} = 3.0$ Hz, C_{Ar}), 135.3 (d, $^1J_{CP} = 3.0$ Hz, C_{Ar}), 136.1 (d, $^1J_{CP} = 1.4$ Hz, C_{Ar}), 136.1 (d, $^1J_{CP} = 2.7$ Hz, C_{Ar}), 136.4 (d, $^1J_{CP} = 2.4$ Hz, C_{Ar}), 150.5 (C_{Ar}), 157.6 (d, $^3J_{CP} = 1.1$ Hz, C_{Ar}), 170.7 (dd, $^3J_{CP} = 1.1$ Hz, $^3J_{CP} = 3.2$ Hz, CHCOO).

^{31}P-NMR (161.97 MHz; C_7D_8): $\delta = 32.0$ (d, $^2J_{PP} = 36.5$ Hz), 33.6 (d, $^2J_{PP} = 36.5$ Hz).

FT-IR (ATR, cm^{-1}): 1692 (m, CO).

Bis(triphenylphosphine)-(η^2-4-methylphenyl 4-methoxycinnamate)palladium(0)

PdPhOMe-PhMe: $C_{53}H_{46}O_3P_2Pd$, M $= 899.30$ g/mol, light yellow solid, yield: $>99\%$, route: *Pd-I*.

^{1}H-NMR (400.13 MHz; C_7D_8): $\delta = 2.09$ (s, 3H, CH_3), 3.31 (s, 3H, OCH_3), 4.88 (bs, 1H, CHCHCO), 5.34 (bs, 1H, C_{Ar}CHCH), 6.41 - 6.48 (bm, 2H, H_{Ar}), 6.52 - 6.61 (bm, 2H, H_{Ar}), 6.63 - 6.73 (bm, 2H, H_{Ar}), 6.79 - 6.86 (bm, 2H, H_{Ar}), 6.88 - 7.05 (bm, 18H, H_{Ar}), 7.05 - 7.19 (bm, 6H, H_{Ar}), 7.50 - 7.65 (bm, 6H, H_{Ar}).

^{1}H-NMR (400.13 MHz; 246 K; C_7D_8): $\delta = 2.15$ (s, 3H, CH_3), 3.31 (s, 3H, OCH_3), 5.07 (dpt, $^3J_{HH} = 10.2$ Hz, $^3J_{HP} = 6.0$ Hz, $^3J_{HP} = 6.0$ Hz, 1H, CHCHCO), 5.43 (dpt, $^3J_{HH} = 10.2$ Hz, $^3J_{HP} = 7.1$ Hz, $^3J_{HP} = 7.1$ Hz, 1H, C_{Ar}CHCH), 6.43 - 6.48 (m, 2H, H_{Ar}), 6.54 - 6.59 (m, 2H, H_{Ar}), 6.59 - 6.65 (m, 2H, H_{Ar}), 6.83 - 6.87 (m, 2H, H_{Ar}), 6.91 - 7.04 (m, 18H, H_{Ar}), 7.10 - 7.16 (m, 6H, H_{Ar}), 7.62 - 7.69 (m, 6H, H_{Ar}).

^{13}C-NMR (100.62 MHz; C_7D_8): $\delta = 20.7$ (CH_3), 54.7 (d, $^2J_{CP} = 4.1$ Hz, OCH$_3$), 60.1 (bs, CHCHCO), 76.9 (bs, C_{Ar}CHCH), 114.1 (C_{Ar}H), 122.2 (C_{Ar}H), 126.8 (C_{Ar}H), 128.3 (C_{Ar}H), 128.5 (C_{Ar}H), 129.5 (C_{Ar}H), 133.9 - 135.0 (bm, C_{Ar}H), 135.5 - 136.6 (bm, C_{Ar}H), 150.4 (C_{Ar}O), 157.9 (bs, C_{Ar}O), 169.2 (bs, CHCOO).

^{13}C-NMR (100.53 MHz; 246 K; C_7D_8): $\delta = 20.8$ (CH_3), 54.5 (d, $^2J_{CP} = 1.4$ Hz, OCH$_3$), 59.5 (dd, $^2J_{CP} = 6.8$ Hz, $^2J_{CP} = 15.4$ Hz, CHCHCO), 76.6 (dd, $^2J_{CP} = 2.3$ Hz, $^2J_{CP} = 24.4$ Hz, C_{Ar}CHCH), 113.8 (C_{Ar}H), 122.3 (C_{Ar}H), 126.5 (bs, C_{Ar}), 127.4 (d, $^3J_{CP} = 1.5$ Hz, C_{Ar}H), 127.8 (d, $^3J_{CP} = 1.4$ Hz, C_{Ar}H), 128.0 (C_{Ar}), 128.5 (C_{Ar}), 129.3 (C_{Ar}), 129.4 (C_{Ar}), 133.4 (C_{Ar}CH$_3$), 134.1 (d, $^2J_{CP} = 14.3$ Hz, C_{Ar}H), 134.4 (d, $^2J_{CP} = 15.3$ Hz, C_{Ar}H), 134.6 (d, $^3J_{CP} = 6.7$ Hz, C_{Ar}CH), 135.6 (dd, $^1J_{CP} = 28.0$ Hz, $^3J_{CP} = 1.8$ Hz, C_{Ar}P), 137.1 (d, $^3J_{CP} = 1.6$ Hz, C_{Ar}P), 150.2 (s, C_{Ar}O), 157.7 - 157.8 (m, C_{Ar}O), 169.3 (dd, $^3J_{CP} = 3.8$ Hz, $^3J_{CP} = 2.3$ Hz, CHCOO).

^{31}P-NMR (161.97 MHz; C_7D_8): $\delta = 24.4$ (d, $^2J_{PP} = 18.1$ Hz), 25.5 (d, $^2J_{PP} = 18.1$ Hz).

^{31}P-NMR (161.97 MHz; 246 K; C_7D_8): $\delta = 23.8$ (d, $^2J_{PP} = 20.2$ Hz), 25.3 (d, $^2J_{PP} = 20.2$ Hz).

FAB-MS: m/z (%): 368.0 (12.27) [Pd(PPh$_3$)$^+$], 630.1 (29.13) [Pd(PPh$_3$)$_2^+$].

FT-IR (ATR, cm^{-1}): 3057 (w), 1693 (s, CO), 1594 (w), 1477 (m), 1433 (s), (w), 1247 (m), 1201 (m), 1116 (s), 1103 (s), 1091 (s), 1028 (m), 973 (w), 867 (w), 822 (w), 746 (s), 692 (s), 525 (m), 513 (m), 503 (s), 412 (s).

elemental analysis for $C_{53}H_{46}O_3P_2Pd$ (899.30 g/mol):

calcd.: C 70.78, H 5.16, O 5.34, P 6.89, Pd 11.83

found: C 70.66, H 5.22

Bis(triphenylphosphine)-(η^2-4-methylphenyl 4-methoxycinnamate)platinum(0)

PtPhOMe-PhMe: $C_{53}H_{46}O_3P_2Pt$, M = 987.96 g/mol, off-white solid, yield: 92%, **route**: *Pt-VI*.

^{1}H-NMR (400.13 MHz; C_6D_6): $\delta = 2.05$ (s, 3H, CH_3), 3.32 (s, 3H, OCH_3), 4.28 (dpt, $^3J_{HH} = 8.5$ Hz, $^3J_{HP} = 4.4$ Hz, $^3J_{HP} = 8.5$ Hz, 1H, CHCHCO), 4.28 (ddpt, $^3J_{HH} = 8.5$ Hz, $^3J_{HP} = 4.4$ Hz, $^3J_{HP} = 8.5$ Hz, $^2J_{HPt} = 62.0$ Hz, 1H, CHCHCO), 4.56 (dpt, $^3J_{HH} = 8.6$ Hz, $^3J_{HP} = 5.8$ Hz, $^3J_{HP} = 8.6$ Hz, 1H, C_{Ar}CHCH), 4.56 (ddpt, $^3J_{HH} = 8.6$ Hz, $^3J_{HP} = 5.8$ Hz, $^3J_{HP} = 8.6$ Hz, $^2J_{HPt} = 50.4$ Hz, 1H, C_{Ar}CHCH), 6.49 - 6.59 (m, 4H, H_{Ar}), 6.73 (d, $^3J_{HH} = 8.4$ Hz, 2H, H_{Ar}), 6.83 (d, $^3J_{HH} = 8.4$ Hz, 2H, H_{Ar}), 6.85 - 7.00 (m, 18H, H_{Ar}), 7.17 - 7.25 (m, 6H, H_{Ar}), 7.60 - 7.70 (m, 6H, H_{Ar}).

^{13}C-NMR (100.53 MHz; C_6D_6): $\delta = 20.8$ (CH_3), 49.4 (dd, $^2J_{CP} = 5.5$ Hz, $^2J_{CP} = 27.5$ Hz, CHCHCO), 49.4 (ddd, $^2J_{CP} = 5.5$ Hz, $^2J_{CP} = 27.5$ Hz, $^1J_{CPt} = 188.9$ Hz, CHCHCO), 54.8 (OCH$_3$), 61.7 (dd, $^2J_{CP} = 3.4$ Hz, $^2J_{CP} = 32.2$ Hz, $C_{Ar}$$C$HCH),

61.7 (ddd, $^2J_{CP}$ = 3.4 Hz, $^2J_{CP}$ = 32.2 Hz, $^1J_{CPt}$ = 193.1 Hz, $C_{Ar}CHCH$), 113.6 (d, $^5J_{CP}$ = 2.0 Hz, $C_{Ar}H$), 122.4 ($C_{Ar}H$), 127.1 (d, $^5J_{CP}$ = 3.1 Hz, $C_{Ar}H$), 127.1 (dd, $^5J_{CP}$ = 3.1 Hz, $^4J_{CPt}$ = 13.8 Hz, $C_{Ar}H$), 128.0 ($C_{Ar}H$), 128.4 ($C_{Ar}H$), 128.4 ($C_{Ar}H$), 129.4 (d, $^1J_{CPt}$ = 2.0 Hz, $C_{Ar}H$), 129.5 ($C_{Ar}H$), 129.6 (d, $^1J_{CPt}$ = 1.8 Hz, $C_{Ar}H$), 133.5 (C_{Ar}), 134.2 ($C_{Ar}H$), 134.2 (d, $^3J_{CPt}$ = 22.0 Hz, $C_{Ar}H$), 134.3 ($C_{Ar}H$), 134.3 (d, $^3J_{CPt}$ = 21.0 Hz, $C_{Ar}H$), 134.4 ($C_{Ar}H$), 134.4 (d, $^3J_{CPt}$ = 18.4 Hz, $C_{Ar}H$), 134.5 ($C_{Ar}H$), 134.5 (d, $^3J_{CPt}$ = 18.2 Hz, $C_{Ar}H$), 135.5 (d, $^1J_{CP}$ = 3.8 Hz, $C_{Ar}P$), 135.5 (dd, $^1J_{CP}$ = 3.8 Hz, $^2J_{CPt}$ = 34.7 Hz, $C_{Ar}P$), 135.9 (d, $^1J_{CP}$ = 3.8 Hz, $C_{Ar}P$), 135.9 (dd, $^1J_{CP}$ = 3.8 Hz, $^2J_{CPt}$ = 33.0 Hz, $C_{Ar}P$), 137.1 (d, $^1J_{CP}$ = 3.5 Hz, $C_{Ar}P$), 137.1 (dd, $^1J_{CP}$ = 3.5 Hz, $^2J_{CPt}$ = 28.4 Hz, $C_{Ar}P$), 137.5 (d, $^1J_{CP}$ = 3.4 Hz, $C_{Ar}P$), 137.5 (dd, $^1J_{CP}$ = 3.4 Hz, $^2J_{CPt}$ = 30.0 Hz, $C_{Ar}P$), 150.6 ($C_{Ar}O$), 157.6 (d, $^3J_{CP}$ = 2.4 Hz, $C_{Ar}CH$), 157.6 (dd, $^3J_{CP}$ = 2.4 Hz, $^2J_{CPt}$ = 51.3 Hz, $C_{Ar}CH$), 171.7 (dd, $^3J_{CP}$ = 2.4 Hz, $^3J_{CP}$ = 4.8 Hz, $CHCOO$), 171.7 (ddd, $^3J_{CP}$ = 2.4 Hz, $^3J_{CP}$ = 4.8 Hz, $^2J_{CPt}$ = 44.2 Hz, $CHCOO$).

^{31}P-NMR (161.97 MHz; C_6D_6): δ = 27.6 (d, $^2J_{PP}$ = 44.6 Hz), 27.6 (dd, $^2J_{PP}$ = 44.6 Hz, $^1J_{PPt}$ = 4219.2 Hz), 28.3 (d, $^2J_{PP}$ = 44.6 Hz), 28.3 (dd, $^2J_{PP}$ = 44.6 Hz, $^1J_{PPt}$ = 3625.5 Hz).

^{195}Pt-NMR (85.48 MHz; C_6D_6): δ = -5043.2 (dd, $^1J_{PPt}$ = 3624.9 Hz, $^1J_{PPt}$ = 4218.8 Hz).

FAB-MS: m/z (%): 719.2 (100) [$Pt(PPh_3)_2{}^+$], 987.3 (1.45) [$(C_{17}H_{16}O_3)Pt(PPh_3)_2{}^+$].

FT-IR (ATR, cm^{-1}): 3050 (w), 2974 (w), 1702 (s, CO), 1592 (w), 1509 (m), 1433 (m), 1304 (w), 1244 (m), 1191 (m), 1113 (s), 1094 (s), 1016 (m), 910 (w), 827 (m), 747 (m), 706 (s), 538 (m), 518 (m), 504 (s), 436 (m).

elemental analysis for $C_{53}H_{46}O_3P_2Pt$ (987.96 *g/mol*):

calcd.: C 64.43, H 4.69, O 4.86, P 6.27, Pt 19.75

found: C 63.91, H 4.62

Bis(triphenylphosphine)-(η^2-phenyl 4-methoxycinnamate)nickel(0)

NiPhOMe-Ph: $C_{52}H_{44}NiO_3P_2$, M $= 837.54\ g/mol$, orange red solid, yield: 85%, **route:** *Ni*.

^{1}H-NMR (400.13 MHz; C_7D_8): $\delta = 3.36$ (s, 3H, OCH_3), 3.70 (dpt, $^3J_{HH} = 9.1$ Hz, $^3J_{HP} = 3.5$ Hz, $^3J_{HP} = 9.1$ Hz, 1H, CHCHCO), 4.35 (dpt, $^3J_{HH} = 9.1$ Hz, $^3J_{HP} = 5.0$ Hz, $^3J_{HP} = 9.1$ Hz, 1H, CC$_{Ar}$CHCH), 6.49 (d, $^3J_{HH} = 8.5$ Hz, 2H, H_{Ar}), 6.53 (d, $^3J_{HH} = 8.6$ Hz, 2H, H_{Ar}), 6.73 (d, $^3J_{HH} = 7.9$ Hz, 2H, H_{Ar}), 6.88 - 7.19 (m, 27H, H_{Ar}), 7.54 - 7.64 (m, 6H, H_{Ar}).

^{13}C-NMR (100.53 MHz; C_7D_8): $\delta = 52.4$ (dd, $^2J_{CP} = 1.5$ Hz, $^2J_{CP} = 12.5$ Hz, CHCHCO), 54.8 (OCH$_3$), 65.0 (d, $^2J_{CP} = 16.5$ Hz, C_{Ar}CHCH), 114.6 (C_{Ar}H), 122.6 (C_{Ar}H), 126.4 (d, $^4J_{CP} = 1.7$ Hz, C_{Ar}H), 128.0 (C_{Ar}H), 128.1 (C_{Ar}H), 128.3 (C_{Ar}H), 128.4 (C_{Ar}H), 128.8 (C_{Ar}H), 129.3 (d, $^3J_{CP} = 0.7$ Hz, C_{Ar}H), 132.5 (bs, C_{Ar}), 134.2 (d, $^2J_{CP} = 12.5$ Hz, C_{Ar}H), 134.4 (d, $^2J_{CP} = 134.4$ Hz, C_{Ar}H), 134.9 (d, $^1J_{CP} = 3.0$ Hz, C_{Ar}), 135.2 (d, $^1J_{CP} = 3.0$ Hz, C_{Ar}), 136.0 (d, $^1J_{CP} = 2.7$ Hz, C_{Ar}), 136.1 (d, $^1J_{CP} = 2.4$ Hz, C_{Ar}), 136.4 (d, $^1J_{CP} = 2.4$ Hz, C_{Ar}), 152.8 (C_{Ar}), 157.7 (d, $^3J_{CP} = 1.3$ Hz, C_{Ar}), 170.5 (dd, $^3J_{CP} = 1.0$ Hz, $^3J_{CP} = 3.2$ Hz, CHCOO).

^{31}P-NMR (161.97 MHz; C_7D_8): $\delta = 32.1$ (d, $^2J_{PP} = 36.3$ Hz), 33.6 (d, $^2J_{PP} = 36.3$ Hz).

FT-IR (ATR, cm^{-1}): 1682 (m, CO).

elemental analysis for $C_{52}H_{44}NiO_3P_2$ ($837.54\ g/mol$):

calcd.: C 74.57, H 5.30, Ni 7.01, O 5.73, P 7.40

found: C 72.26, H 5.27

Bis(triphenylphosphine)-(η^2-phenyl 4-methoxycinnamate)palladium(0)

PdPhOMe-Ph: $C_{52}H_{44}O_3P_2Pd$, M = 885.27 g/mol, yellow solid, yield: >99%, **route**: *Pd-I*.

^{1}H-NMR (400.13 MHz; C_7D_8): $\delta = 3.31$ (s, 3H, OCH_3), 4.88 (bs, 1H, $CHCHCO$), 5.33 (bs, 1H, $C_{Ar}CHCH$), 6.40 - 6.50 (bm, 2H, H_{Ar}), 6.50 - 6.66 (bm, 2H, H_{Ar}), 6.69 - 6.84 (bm, 2H, H_{Ar}), 6.87 - 7.06 (bm, 21H, H_{Ar}), 7.06 - 7.21 (bm, 6H, H_{Ar}), 7.42 - 7.67 (bm, 6H, H_{Ar}).

^{1}H-NMR (400.13 MHz; 246 K; C_7D_8): $\delta = 3.27$ (s, 3H, OCH_3), 5.03 (dpt, $^3J_{HH} = 10.2$ Hz, $^3J_{HP} = 6.0$ Hz, $^3J_{HP} = 6.0$ Hz, 1H, $CHCHCO$), 5.37 (dpt, $^3J_{HH} = 10.2$ Hz, $^3J_{HP} = 7.1$ Hz, $^3J_{HP} = 7.1$ Hz, 1H, $C_{Ar}CHCH$), 6.38 - 6.43 (m, 2H, H_{Ar}), 6.49 - 6.54 (m, 2H, H_{Ar}), 6.62 - 6.70 (m, 2H, H_{Ar}), 6.87 - 7.02 (m, 21H, H_{Ar}), 7.04 - 7.12 (m, 6H, H_{Ar}), 7.57 - 7.64 (m, 6H, H_{Ar}).

^{13}C-NMR (100.62 MHz; C_7D_8): $\delta = 54.7$ (d, $^2J_{CP} = 4.3$ Hz, OCH_3), 60.2 (bs, $CHCHCO$), 76.9 (bs, $C_{Ar}CHCH$), 114.1 ($C_{Ar}H$), 122.5 ($C_{Ar}H$), 126.8 ($C_{Ar}H$), 128.3 ($C_{Ar}H$), 128.5 ($C_{Ar}H$), 133.6 - 135.2 (bm, $C_{Ar}H$), 135.6 - 136.7 (bm, $C_{Ar}H$), 152.7 (bs, $C_{Ar}O$), 158.2 (bs, $C_{Ar}O$), 168.9 (bs, $CHCOO$).

^{13}C-NMR (100.53 MHz; 246 K; C_7D_8): $\delta = 54.5$ (d, $^2J_{CP} = 1.3$ Hz, OCH_3), 59.3 (dd, $^2J_{CP} = 3.7$ Hz, $^2J_{CP} = 12.8$ Hz, $CHCHCO$), 76.7 (dd, $^2J_{CP} = 2.1$ Hz, $^2J_{CP} = 21.3$ Hz, $_{Ar}CHCH$), 113.8 ($C_{Ar}H$), 122.6 ($C_{Ar}H$), 126.5 (bs, $C_{Ar}CH$), 128.0 (C_{Ar}), 128.2 (C_{Ar}), 128.5 (C_{Ar}), 128.9 (C_{Ar}), 129.3 (C_{Ar}), 134.1 (d, $^2J_{CP} = 14.5$ Hz, $C_{Ar}H$), 134.4 (d, $^2J_{CP} = 15.2$ Hz, $C_{Ar}H$), 134.6 (d, $^3J_{CP} = 6.6$ Hz, $C_{Ar}CH$), 135.6 (dd, $^1J_{CP} = 28.0$ Hz, $^3J_{CP} = 1.8$ Hz, $C_{Ar}P$), 137.1 (d, $^3J_{CP} = 1.6$ Hz, $C_{Ar}P$), 152.4 ($C_{Ar}O$), 157.8 - 157.8 (m, $C_{Ar}O$), 169.1 (dd, $^3J_{CP} = 2.4$ Hz, $^3J_{CP} = 3.8$ Hz, $CHCOO$).

^{31}P-NMR (161.97 MHz; C_7D_8): $\delta = 24.4$ (d, $^2J_{PP} = 17.8$ Hz), 25.5 (d, $^2J_{PP} = 17.8$ Hz).

^{31}P-NMR (161.97 MHz; 246 K; C$_7$D$_8$): δ = 23.8 (d, $^2J_{PP}$ = 19.7 Hz), 25.2 (d, $^2J_{PP}$ = 19.7 Hz).

FT-IR (ATR, cm^{-1}): 3040 (m), 1692 (m, CO), 1605 (w), 1506 (w), 1476 (m), 1434 (m), 1317 (w), 1241 (m), 1194 (m), 1112 (s), 1101 (s), 1091 (s), 1022 (w), 961 (w), 809 (w), 741 (s), 691 (s), 496 (s), 408 (s).

elemental analysis for C$_{52}$H$_{44}$O$_3$P$_2$Pd (885.27 g/mol):

calcd.: C 70.55, H 5.01, O 5.42, P 7.00, Pd 12.02

found: C 70.11, H 5.03

Bis(triphenylphosphine)-(η^2-phenyl 4-methoxycinnamate)platinum(0)

PtPhOMe-Ph: C$_{52}$H$_{44}$O$_3$P$_2$Pt, M = 973.93 g/mol, off-white, yield: 96%, **route:** Pt-VI.

^{1}H-NMR (400.13 MHz; C$_6$D$_6$): δ = 3.32(s, 3H, OCH_3), 4.28 (dpt, $^3J_{HH}$ = 8.5 Hz, $^3J_{HP}$ = 4.3 Hz, $^3J_{HP}$ = 8.5 Hz, 1H, CHCHCO), 4.28 (ddpt, $^3J_{HH}$ = 8.5 Hz, $^3J_{HP}$ = 4.3 Hz, $^3J_{HP}$ = 8.5 Hz, $^2J_{HPt}$ = 60.9 Hz, 1H, CHCHCO), 4.54 (dpt, $^3J_{HH}$ = 8.5 Hz, $^3J_{HP}$ = 5.8 Hz, $^3J_{HP}$ = 8.5 Hz, 1H, C$_{Ar}$CHCH), 4.54 (ddpt, $^3J_{HH}$ = 8.5 Hz, $^3J_{HP}$ = 5.8 Hz, $^3J_{HP}$ = 8.5 Hz, $^2J_{HPt}$ = 51.3 Hz, 1H, C$_{Ar}$CHCH), 6.43 - 6.61 (m, 4H, H_{Ar}), 6.81 (d, $^3J_{HH}$ = 8.2 Hz, 2H, H_{Ar}), 6.85 - 6.96 (m, 19H, H_{Ar}), 7.01 (t, $^3J_{HH}$ = 7.8 Hz, 2H, H_{Ar}), 7.17 - 7.24 (m, 6H, H_{Ar}), 7.53 - 7.72 (m, 6H, H_{Ar}).

^{13}C-NMR (100.53 MHz; C$_6$D$_6$): δ = 49.3 (dd, $^2J_{CP}$ = 5.9 Hz, $^2J_{CP}$ = 27.7 Hz, CHCHCO), 49.3 (ddd, $^2J_{CP}$ = 5.9 Hz, $^2J_{CP}$ = 27.7 Hz, $^1J_{CPt}$ = 193.2 Hz, CHCHCO), 54.8 (OCH$_3$), 61.7 (dd, $^2J_{CP}$ = 3.5 Hz, $^2J_{CP}$ = 32.7 Hz, C$_{Ar}$ CHCH), 61.7 (ddd, $^2J_{CP}$ = 3.5 Hz, $^2J_{CP}$ = 32.7 Hz, $^1J_{CPt}$ = 197.2 Hz, C$_{Ar}$ CHCH), 113.6 (d, $^5J_{CP}$ = 2.0 Hz, C_{Ar}H), 122.7 (C_{Ar}H), 124.4 (C_{Ar}), 127.1 (d, $^5J_{CP}$ = 3.1 Hz, C_{Ar}H), 127.1 (dd, $^5J_{CP}$ = 3.1 Hz, $^4J_{CPt}$ = 13.7 Hz, C_{Ar}H), 128.0 (C_{Ar}H), 128.4 (C_{Ar}H), 128.4 (C_{Ar}H), 129.0 (C_{Ar}H), 129.4

(d, $^1J_{CPt}$ = 2.0 Hz, C_{Ar}H), 129.6 (d, $^1J_{CPt}$ = 1.8 Hz, C_{Ar}H), 134.1 (C_{Ar}H), 134.1 (d, $^3J_{CPt}$ = 21.8 Hz, C_{Ar}H), 134.3 (C_{Ar}H), 134.3 (d, $^3J_{CPt}$ = 25.2 Hz, C_{Ar}H), 134.4 (C_{Ar}H), 134.4 (d, $^3J_{CPt}$ = 18.4 Hz, C_{Ar}H), 134.5 (C_{Ar}H), 134.5 (d, $^3J_{CPt}$ = 18.6 Hz, C_{Ar}H), 135.5 (d, $^1J_{CP}$ = 3.5 Hz, C_{Ar}P), 135.5 (dd, $^1J_{CP}$ = 3.5 Hz, $^2J_{CPt}$ = 34.0 Hz, C_{Ar}P), 135.9 (d, $^1J_{CP}$ = 3.6 Hz, C_{Ar}P), 135.9 (dd, $^1J_{CP}$ = 3.6 Hz, $^2J_{CPt}$ = 30.8 Hz, C_{Ar}P), 137.1 (d, $^1J_{CP}$ = 3.2 Hz, C_{Ar}P), 137.1 (dd, $^1J_{CP}$ = 3.2 Hz, $^2J_{CPt}$ = 26.0 Hz, C_{Ar}P), 137.5 (d, $^1J_{CP}$ = 3.2 Hz, C_{Ar}P), 137.5 (dd, $^1J_{CP}$ = 3.2 Hz, $^2J_{CPt}$ = 29.0 Hz, C_{Ar}P), 152.8 (C_{Ar}O), 157.6 (d, $^3J_{CP}$ = 2.4 Hz, C_{Ar}CH), 157.6 (dd, $^3J_{CP}$ = 2.4 Hz, $^2J_{CPt}$ = 28.4 Hz, C_{Ar}CH), 171.5 (dd, $^3J_{CP}$ = 2.3 Hz, $^3J_{CP}$ = 4.9 Hz, CHCOO), 171.5 (ddd, $^3J_{CP}$ = 2.3 Hz, $^3J_{CP}$ = 4.9 Hz, $^2J_{CPt}$ = 51.2 Hz, CHCOO).

^{31}P-NMR (161.97 MHz; C_6D_6): δ = 27.5 (d, $^2J_{PP}$ = 44.5 Hz), 27.5 (dd, $^2J_{PP}$ = 44.5 Hz, $^1J_{PPt}$ = 4225.8 Hz), 28.3 (d, $^2J_{PP}$ = 44.5 Hz), 28.3 (dd, $^2J_{PP}$ = 44.5 Hz, $^1J_{PPt}$ = 3622.8 Hz).

^{195}Pt-NMR (85.48 MHz; C_6D_6): δ = -5042.1 (dd, $^1J_{PPt}$ = 3622.5 Hz, $^1J_{PPt}$ = 4225.6 Hz).

FAB-MS: m/z (%): 719.3 (100) [Pt(PPh$_3$)$_2^+$], 973.4 (1.25) [($C_{16}H_{14}O_3$)Pt(PPh$_3$)$_2^+$].

FT-IR (ATR, cm^{-1}): 3042 (w), 2996 (w), 2845 (w), 1708 (s, CO), 1595 (w), 1508 (s), 1473 (m), 1434 (s), 1398 (m), 1292 (w), 1243 (m), 1170 (w), 1140 (w), 1096 (s), 1034 (w), 973 (w), 898 (w), 822 (m), 746 (s), 685 (s), 538 (m), 517 (s), 508 (s), 438 (m).

elemental analysis for $C_{52}H_{44}O_3P_2Pt$ (973.93 *g/mol*):

calcd.: C 64.13, H 4.55, O 4.93, P 6.36, Pt 20.03

found: C 63.55, H 4.49

Bis(triphenylphosphine)-(η^2-4-nitrophenyl 4-methoxycinnamate)palladium(0)

PdPhOMe-PhNO$_2$: C$_{52}$H$_{43}$NO$_5$P$_2$Pd, M = 930.27 g/mol, yellow solid, yield: >99%, route: *Pd-I*.

^{1}H-NMR (400.13 MHz; C$_7$D$_8$): δ = 3.32 (s, 3H, OCH_3), 4.74 - 4.85 (bm, 1H, CHCHCO), 5.17 - 5.29 (bm, 1H, C$_{Ar}$CHCH), 6.41 - 6.50 (bm, 2H, H_{Ar}), 6.50 - 6.61 (bm, 4H, H_{Ar}), 6.85 - 7.01 (bm, 18H, H_{Ar}), 7.01 - 7.15 (bm, 6H, H_{Ar}), 7.39 - 7.54 (bm, 6H, H_{Ar}), 7.70 - 7.77 (bm, 2H, H_{Ar}).

^{1}H-NMR (400.13 MHz; 246 K; C$_7$D$_8$): δ = 3.27 (s, 3H, OCH_3), 4.94 (dpt, $^3J_{HH}$ = 10.1 Hz, $^3J_{HP}$ = 5.8 Hz, $^3J_{HP}$ = 5.8 Hz, 1H, CHCHCO), 5.26 (dpt, $^3J_{HH}$ = 10.1 Hz, $^3J_{HP}$ = 7.3 Hz, $^3J_{HP}$ = 7.3 Hz, 1H, C$_{Ar}$CHCH), 6.40 - 6.46 (m, 4H, H_{Ar}), 6.48 - 6.53 (m, 2H, H_{Ar}), 6.84 - 6.92 (m, 18H, H_{Ar}), 6.94 - 7.01 (m, 6H, H_{Ar}), 7.01 - 7.08 (m, 6H, H_{Ar}), 7.47 - 7.53 (m, 6H, H_{Ar}), 7.68 - 7.73 (m, 2H, H_{Ar}).

^{13}C-NMR (100.62 MHz; C$_7$D$_8$): δ = 54.7 (d, $^2J_{CP}$ = 4.4 Hz, OCH_3), 58.8 (bs, CHCHCO), 75.8 (bs, C$_{Ar}$CHCH), 114.2 (C$_{Ar}$H), 122.3 (C$_{Ar}$H), 126.8 (C$_{Ar}$H), 128.3 (C$_{Ar}$H), 128.5 (C$_{Ar}$H), 129.3 (C$_{Ar}$H), 133.8 - 134.7 (bm, C$_{Ar}$H), 135.3 - 135.9 (bm, C$_{Ar}$H), 144.5 (bs, C$_{Ar}$N), 157.3 (bs, C$_{Ar}$O), 158.4 (bs, C$_{Ar}$O), 167.9 (bs, CHCOO).

^{13}C-NMR (100.53 MHz; 246 K; C$_7$D$_8$): δ = 54.5 (d, $^2J_{CP}$ = 1.3 Hz, OCH_3), 58.1 (dd, $^2J_{CP}$ = 5.9 Hz, $^2J_{CP}$ = 17.5 Hz, CHCHCO), 75.6 (dd, $^2J_{CP}$ = 2.3 Hz, $^2J_{CP}$ = 22.9 Hz, C$_{Ar}$CHCH), 113.9 (C$_{Ar}$H), 122.3 (C$_{Ar}$H), 124.5 (C$_{Ar}$H), 126.5 (bs, C$_{Ar}$CH), 128.0 (C$_{Ar}$), 128.2 (C$_{Ar}$), 128.4 (C$_{Ar}$), 129.2 (C$_{Ar}$), 129.2 (C$_{Ar}$), 129.5 (C$_{Ar}$r), 134.0 (d, $^2J_{CP}$ = 14.8 Hz, C$_{Ar}$H), 134.2 (d, $^3J_{CP}$ = 6.5 Hz, C$_{Ar}$H), 134.3 (d, $^2J_{CP}$ = 15.4 Hz, C$_{Ar}$H), 135.2 (dd, $^1J_{CP}$ = 28.9 Hz, $^3J_{CP}$ = 1.7 Hz, C$_{Ar}$P), 136.8 (dd, $^1J_{CP}$ = 27.6 Hz, $^3J_{CP}$ = 1.2 Hz, C$_{Ar}$P), 144.2 (C$_{Ar}$N), 157.0 (C$_{Ar}$O), 158.0 - 158.1 (m, C$_{Ar}$O), 167.8 - 167.9 (m, CHCOO).

^{31}P-NMR (161.97 MHz; C$_7$D$_8$): δ = 24.6 (d, $^2J_{PP}$ = 17.4 Hz), 25.2 (d, $^2J_{PP}$ = 17.4 Hz).

^{31}P-NMR (161.97 MHz; 246 K; C$_7$D$_8$): δ = 23.9 (d, $^2J_{PP}$ = 16.2 Hz), 25.0 (d, $^2J_{PP}$ = 16.2 Hz).

FT-IR (ATR, cm^{-1}): 3050 (w), 2929 (w), 1702 (m, CO), 1587 (w), 1516 (m), 1474 (m), 1433 (m), 1344 (m), 1237 (m), 1163 (w), 1093 (s), 1042 (w), 966 (w), 860 (w), 799 (w), 739 (m), 692 (s), 501 (s), 421 (m).

elemental analysis for C$_{52}$H$_{43}$NO$_5$P$_2$Pd (930.27 g/mol):

calcd.: C 67.14, H 4.66, N 1.51, O 8.60, P 6.66, Pd 11.44

found: C 67.48, H 4.99, N 1.50

Bis(triphenylphosphine)-(η^2-4-nitrophenyl 4-methoxycinnamate)platinum(0)

PtPhOMe-PhNO$_2$: C$_{52}$H$_{43}$NO$_5$P$_2$Pt, M = 1018.93 g/mol, white solid, yield: 95%, **route**: *Pt-VI*.

^{1}H-NMR (400.13 MHz; C$_6$D$_6$): δ = 3.32 (s, 3H, OCH_3), 4.18 (dpt, $^3J_{HH}$ = 8.6 Hz, $^3J_{HP}$ = 3.9 Hz, $^3J_{HP}$ = 8.6 Hz, 1H, CHCHCO), 4.18 (ddpt, $^3J_{HH}$ = 8.6 Hz, $^3J_{HP}$ = 3.9 Hz, $^3J_{HP}$ = 8.6 Hz, $^2J_{HPt}$ = 61.2 Hz, 1H, CHCHCO), 4.43 (dpt, $^3J_{HH}$ = 8.6 Hz, $^3J_{HP}$ = 5.5 Hz, $^3J_{HP}$ = 8.6 Hz, 1H, C$_{Ar}$CHCH), 4.43 (ddpt, $^3J_{HH}$ = 8.6 Hz, $^3J_{HP}$ = 5.5 Hz, $^3J_{HP}$ = 8.6 Hz, $^2J_{HPt}$ = 50.2 Hz, 1H, C$_{Ar}$CHCH), 6.48 - 6.66 (m, 6H, H_{Ar}), 6.79 - 7.02 (m, 18H, H_{Ar}), 7.08 - 7.28 (m, 6H, H_{Ar}), 7.45 - 7.59 (m, 6H, H_{Ar}), 7.74 (d, $^3J_{HH}$ = 9.1 Hz, 2H, H_{Ar}).

^{13}C-NMR (100.53 MHz; C$_6$D$_6$): δ = 48.5 (dd, $^2J_{CP}$ = 5.9 Hz, $^2J_{CP}$ = 27.1 Hz, CHCHCO), 48.5 (ddd, $^2J_{CP}$ = 5.9 Hz, $^2J_{CP}$ = 27.1 Hz, $^1J_{CPt}$ = 157.0 Hz, CHCHCO), 54.9 (OCH$_3$), 60.9 (dd, $^2J_{CP}$ = 3.2 Hz, $^2J_{CP}$ = 33.3 Hz, C$_{Ar}$CHCH), 60.9 (ddd, $^2J_{CP}$ = 3.2 Hz, $^2J_{CP}$ = 33.3 Hz, $^1J_{CPt}$ = 200.8 Hz, C$_{Ar}$CHCH), 113.7 (d, $^5J_{CP}$ = 2.0 Hz, C_{Ar}H), 122.4 (C_{Ar}H), 124.7 (C_{Ar}H), 127.2 (d, $^5J_{CP}$ = 3.1 Hz, C_{Ar}H), 127.2 (dd,

$^5J_{CP} = 3.1$ Hz, $^4J_{CPt} = 12.4$ Hz, C_{Ar}H), 128.1 (C_{Ar}H), 128.4 (C_{Ar}H), 128.5 (C_{Ar}H), 129.5 (d, $^1J_{CPt} = 2.0$ Hz, C_{Ar}H), 129.8 (d, $^1J_{CPt} = 1.9$ Hz, C_{Ar}H), 134.1 (C_{Ar}H), 134.1 (d, $^3J_{CPt} = 22.0$ Hz, C_{Ar}H), 134.2 (C_{Ar}H), 134.2 (d, $^3J_{CPt} = 24.6$ Hz, C_{Ar}H), 134.2 (C_{Ar}H), 134.2 (d, $^3J_{CPt} = 17.6$ Hz, C_{Ar}H), 134.4 (C_{Ar}H), 134.4 (d, $^3J_{CPt} = 18.2$ Hz, C_{Ar}H), 135.1 (d, $^1J_{CP} = 3.5$ Hz, C_{Ar}P), 135.1 (dd, $^1J_{CP} = 3.5$ Hz, $^2J_{CPt} = 33.7$ Hz, C_{Ar}P), 135.5 (d, $^1J_{CP} = 3.4$ Hz, C_{Ar}P), 135.5 (dd, $^1J_{CP} = 3.4$ Hz, $^2J_{CPt} = 31.9$ Hz, C_{Ar}P), 136.7 (d, $^1J_{CP} = 3.0$ Hz, C_{Ar}P), 136.7 (dd, $^1J_{CP} = 3.0$ Hz, $^2J_{CPt} = 29.4$ Hz, C_{Ar}P), 137.1 (d, $^1J_{CP} = 3.0$ Hz, C_{Ar}P), 137.1 (dd, $^1J_{CP} = 3.0$ Hz, $^2J_{CPt} = 26.6$ Hz, C_{Ar}P), 144.5 (C_{Ar}O), 157.2 (C_{Ar}N), 157.8 (d, $^3J_{CP} = 2.4$ Hz, C_{Ar}CH), 157.8 (dd, $^3J_{CP} = 2.4$ Hz, $^2J_{CPt} = 43.9$ Hz, C_{Ar}CH), 170.5 (dd, $^3J_{CP} = 2.4$ Hz, $^3J_{CP} = 4.9$ Hz, CHCOO), 170.5 (ddd, $^3J_{CP} = 2.4$ Hz, $^3J_{CP} = 4.9$ Hz, $^2J_{CPt} = 45.6$ Hz, CHCOO).

^{31}P-NMR (161.97 MHz; C_6D_6): $\delta = 27.0$ (d, $^2J_{PP} = 41.4$ Hz), 27.0 (dd, $^2J_{PP} = 41.4$ Hz, $^1J_{PPt} = 4275.5$ Hz), 27.8 (d, $^2J_{PP} = 41.4$ Hz), 27.8 (dd, $^2J_{PP} = 41.4$ Hz, $^1J_{PPt} = 3581.5$ Hz).

^{195}Pt-NMR (85.48 MHz; C_6D_6): $\delta = $ -5047.6 (dd, $^1J_{PPt} = 3582.5$ Hz, $^1J_{PPt} = 4275.2$ Hz).

FAB-MS: m/z (%): 719.0 (100) [Pt(PPh$_3$)$_2{}^+$], 1019.1 (1.73) [(C$_{16}$H$_{13}$NO$_5$)Pt(PPh$_3$)$_2{}^+$].

FT-IR (ATR, cm^{-1}): 3053 (w), 2825 (w), 1704 (m, CO), 1582 (w), 1510 (s), 1434 (s), 1343 (m), 1222 (m), 1177 (w), 1092 (s), 1029 (w), 968 (w), 908 (w), 862 (w), 826 (m), 741 (m), 693 (s), 529 (m), 507 (s), 423 (m).

elemental analysis for C$_{52}$H$_{43}$NO$_5$P$_2$Pt (1018.93 g/mol):

calcd.: C 61.30, H 4.25, N 1.37, O 7.85, P 6.08, Pt 19.15

found: C 61.11, H 4.60, N 1.44

Bis(triphenylphosphine)-(η^2-4-methoxyphenyl 4-methylcinnamate)nickel(0)

NiPhMe-PhOMe: $C_{53}H_{46}NiO_3P_2$, M = 851.57 g/mol, red solid, yield: 99%, **route**: *Ni*.

^{1}H-NMR (400.13 MHz; C_7D_8): δ = 2.09 (s, 3H, CH_3), 3.33 (s, 3H, OCH_3), 3.73 (dpt, $^3J_{HH}$ = 8.9 Hz, $^3J_{HP}$ = 3.5 Hz, $^3J_{HP}$ = 8.9 Hz, 1H, CHCHCO), 4.38 (dpt, $^3J_{HH}$ = 8.9 Hz, $^3J_{HP}$ = 5.2 Hz, $^3J_{HP}$ = 8.9 Hz, 1H, $CC_{Ar}CH$CH), 6.48 (d, $^3J_{HH}$ = 7.7 Hz, 2H, H_{Ar}), 6.62 (s, 4H, H_{Ar}), 6.73 (d, $^3J_{HH}$ = 7.7 Hz, 2H, H_{Ar}), 6.89 - 7.04 (m, 18H, H_{Ar}), 7.05 - 7.21 (m, 6H, H_{Ar}), 7.51 - 7.71 (m, 6H, H_{Ar}).

^{13}C-NMR (100.53 MHz; C_7D_8): δ = 21.3 (CH_3), 52.4 (dd, $^2J_{CP}$ = 1.6 Hz, $^2J_{CP}$ = 12.6 Hz, CHCHCO), 54.9 (OCH_3), 65.5 (d, $^2J_{CP}$ = 16.3 Hz, $C_{Ar}CH$CH), 113.9 ($C_{Ar}H$), 123.2 ($C_{Ar}H$), 128.0 ($C_{Ar}H$), 128.1 ($C_{Ar}H$), 128.3 ($C_{Ar}H$), 128.4 ($C_{Ar}H$), 129.3 (d, $^3J_{CP}$ = 0.9 Hz, $C_{Ar}H$), 129.6 ($C_{Ar}H$), 132.4 (bs, C_{Ar}), 133.1 (d, $^1J_{CP}$ = 1.6 Hz, C_{Ar}), 134.2 (d, $^2J_{CP}$ = 12.5 Hz, $C_{Ar}H$), 134.4 (d, $^2J_{CP}$ = 13.4 Hz, $C_{Ar}H$), 134.8 (d, $^1J_{CP}$ = 2.9 Hz, C_{Ar}), 135.1 (d, $^1J_{CP}$ = 2.9 Hz, C_{Ar}), 136.0 (d, $^1J_{CP}$ = 2.3 Hz, C_{Ar}), 136.3 (d, $^1J_{CP}$ = 2.3 Hz, C_{Ar}), 140.5 (d, $^3J_{CP}$ = 5.5 Hz, C_{Ar}), 146.1 (C_{Ar}), 156.7 (C_{Ar}), 171.1 (dd, $^3J_{CP}$ = 1.1 Hz, $^3J_{CP}$ = 2.3 Hz, CHCOO).

^{31}P-NMR (161.97 MHz; C_7D_8): δ = 30.6 (d, $^2J_{PP}$ = 35.5 Hz), 32.9 (d, $^2J_{PP}$ = 35.5 Hz).

FT-IR (ATR, cm^{-1}): 1691 (m, CO).

Bis(triphenylphosphine)-(η^2-4-methoxyphenyl 4-methylcinnamate)palladium(0)

PdPhMe-PhOMe: $C_{53}H_{46}O_3P_2Pd$, M = 899.30 g/mol, light cream solid, yield: 93%, route: *Pd-I*.

^{1}H-NMR (400.13 MHz; C_7D_8): δ = 2.09 (s, 3H, CH_3), 3.30 (s, 3H, OCH_3), 4.91 (bs, 1H, CHCHCO), 5.32 (bs, 1H, $C_{Ar}CHCH$), 6.51 - 6.62 (bm, 4H, H_{Ar}), 6.62 - 6.73 (bm, 4H, H_{Ar}), 6.87 - 7.04 (bm, 18H, H_{Ar}), 7.04 - 7.20 (bm, 6H, H_{Ar}), 7.51 - 7.63 (bm, 6H, H_{Ar}).

^{1}H-NMR (400.13 MHz; 246 K; C_7D_8): δ = 2.09 (s, 3H, CH_3), 3.26 (s, 3H, OCH_3), 5.06 (dpt, $^3J_{HH}$ = 10.1 Hz, $^3J_{HP}$ = 6.1 Hz, $^3J_{HP}$ = 6.1 Hz, 1H, CHCHCO), 5.38 (dpt, $^3J_{HH}$ = 10.1 Hz, $^3J_{HP}$ = 7.1 Hz, $^3J_{HP}$ = 7.1 Hz, 1H, $C_{Ar}CHCH$), 6.51 - 6.55 (m, 2H, H_{Ar}), 6.56 (bs, 4H, H_{Ar}), 6.63 - 6.68 (m, 2H, H_{Ar}), 6.87 - 7.00 (m, 18H, H_{Ar}), 7.04 - 7.11 (m, 6H, H_{Ar}), 7.56 - 7.65 (m, 6H, H_{Ar}).

^{13}C-NMR (100.62 MHz; C_7D_8): δ = 21.1 (CH_3), 54.8 (OCH_3), 59.8 - 60.5 (bm, CHCHCO), 76.4 - 77.1 (bm, $C_{Ar}CHCH$), 114.0 ($C_{Ar}H$), 123.1 ($C_{Ar}H$), 128.2 - 128.3 (bm, $C_{Ar}H$), 128.4 ($C_{Ar}H$), 128.5 ($C_{Ar}H$), 129.2 ($C_{Ar}H$), 132.3 ($C_{Ar}H$), 132.4 ($C_{Ar}H$), 133.8 - 134.7 (bm, $C_{Ar}H$), 135.6 - 136.3 (bm, $C_{Ar}H$), 139.7 (bs, $C_{Ar}CH$), 146.0 ($C_{Ar}O$), 156.7 ($C_{Ar}O$), 169.5 (bs, CHCOO).

^{13}C-NMR (100.53 MHz; 246 K; C_7D_8): δ = 21.0 (CH_3), 54.6 (OCH_3), 59.5 (dd, $^2J_{CP}$ = 5.3 Hz, $^2J_{CP}$ = 17.0 Hz, CHCHCO), 76.7 (dd, $^2J_{CP}$ = 4.8 Hz, $^2J_{CP}$ = 19.7 Hz, $C_{Ar}CHCH$), 113.7 ($C_{Ar}H$), 123.2 ($C_{Ar}H$), 128.4 (C_{Ar}), 129.2 (C_{Ar}), 129.3 (C_{Ar}), 133.9 (d, $^3J_{CP}$ = 1.7 Hz, $C_{Ar}H$), 134.1 (d, $^2J_{CP}$ = 14.5 Hz, $C_{Ar}H$), 134.1 (d, $^2J_{CP}$ = 14.5 Hz, $C_{Ar}H$), 135.5 (dd, $^1J_{CP}$ = 28.2 Hz, $^3J_{CP}$ = 1.7 Hz, $C_{Ar}P$), 137.0 (d, $^3J_{CP}$ = 1.7 Hz, $C_{Ar}P$), 139.3 (d, $^3J_{CP}$ = 6.4 Hz, $C_{Ar}CH$), 145.8 ($C_{Ar}O$), 156.5 ($C_{Ar}O$), 169.6 (dd, $^3J_{CP}$ = 2.4 Hz, $^3J_{CP}$ = 3.8 Hz, CHCOO).

^{31}P-NMR (161.97 MHz; C$_7$D$_8$): $\delta = 24.3$ (d, $^2J_{PP} = 18.9$ Hz), 25.8 (d, $^2J_{PP} = 18.9$ Hz).

^{31}P-NMR (161.97 MHz; 246 K; C$_7$D$_8$): $\delta = 23.7$ (d, $^2J_{PP} = 18.2$ Hz), 25.6 (d, $^2J_{PP} = 18.2$ Hz).

FT-IR (ATR, cm^{-1}): 3051 (m), 1703 (s, CO), 1477 (m), 1433 (m), 1198 (m), 1118 (s), 1091 (m), 1034 (w), 974 (w), 898 (w), 822 (w), 747 (m), 693 (s), 524 (m), 514 (m), 504 (s), 428 (m).

elemental analysis for C$_{53}$H$_{46}$O$_3$P$_2$Pd (899.30 g/mol):

calcd.: C 70.78, H 5.16, O 5.34, P 6.89, Pd 11.83

found: C 70.36, H 5.21

Bis(triphenylphosphine)-(η^2-4-methoxyphenyl 4-methylcinnamate)platinum(0)

PtPhMe-PhOMe: C$_{53}$H$_{46}$O$_3$P$_2$Pt, M = 987.96 g/mol, white solid, yield: 78%, **route**: *Pt-VI*.

^{1}H-NMR (400.13 MHz; C$_6$D$_6$): $\delta = 2.13$ (s, 3H, CH_3), 3.26 (s, 3H, OCH_3), 4.30 (dpt, $^3J_{HH} = 8.6$ Hz, $^3J_{HP} = 4.5$ Hz, $^3J_{HP} = 8.6$ Hz, 1H, CHCHCO), 4.30 (ddpt, $^3J_{HH} = 8.6$ Hz, $^3J_{HP} = 4.5$ Hz, $^3J_{HP} = 8.6$ Hz, $^2J_{HPt} = 61.7$ Hz, 1H, CHCHCO), 4.57 (dpt, $^3J_{HH} = 8.5$ Hz, $^3J_{HP} = 5.4$ Hz, $^3J_{HP} = 8.5$ Hz, 1H, C$_{Ar}$CHCH), 4.57 (ddpt, $^3J_{HH} = 8.5$ Hz, $^3J_{HP} = 5.4$ Hz, $^3J_{HP} = 8.5$ Hz, $^2J_{HPt} = 51.7$ Hz, 1H, C$_{Ar}$CHCH), 6.61 (d, $^3J_{HH} = 6.7$ Hz, 4H, H_{Ar}), 6.75 (d, $^3J_{HH} = 9.0$ Hz, 4H, H_{Ar}), 6.85 - 6.96 (m, 18H, H_{Ar}), 7.17 - 7.25 (m, 6H, H_{Ar}), 7.58 - 7.66 (m, 6H, H_{Ar}).

^{13}C-NMR (100.53 MHz; C$_6$D$_6$): $\delta = 21.1$ (CH_3), 49.4 (dd, $^2J_{CP} = 5.4$ Hz, $^2J_{CP} = 28.8$ Hz, CHCHCO), 49.4 (ddd, $^2J_{CP} = 5.4$ Hz, $^2J_{CP} = 28.8$ Hz, $^1J_{CPt} = 186.0$ Hz, CHCHCO), 55.0 (OCH_3), 61.9 (dd, $^2J_{CP} = 3.6$ Hz, $^2J_{CP} = 33.3$ Hz, C$_{Ar}$CHCH), 61.9 (ddd, $^2J_{CP} = 3.6$ Hz, $^2J_{CP} = 33.3$ Hz, $^1J_{CPt} = 189.0$ Hz, C$_{Ar}$CHCH), 114.2 (C$_{Ar}$H),

123.3 (C_{Ar}H), 126.2 (d, $^4J_{CP}$ = 3.2 Hz, C_{Ar}H), 126.2 (dd, $^4J_{CP}$ = 3.2 Hz, $^3J_{CPt}$ = 7.2 Hz, C_{Ar}H), 127.9 (C_{Ar}H), 128.0 (C_{Ar}H), 128.2 (C_{Ar}H), 128.4 (C_{Ar}H), 128.7 (d, $^4J_{CP}$ = 2.2 Hz, C_{Ar}H), 128.7 (dd, $^4J_{CP}$ = 2.2 Hz, $^3J_{CPt}$ = 7.6 Hz, C_{Ar}H), 129.4 (d, $^1J_{CPt}$ = 2.0 Hz, C_{Ar}H), 129.6 (d, $^1J_{CPt}$ = 1.9 Hz, C_{Ar}H), 134.2 (C_{Ar}H), 134.2 (d, $^3J_{CPt}$ = 21.2 Hz, C_{Ar}H), 134.3 (C_{Ar}H), 134.3 (d, $^3J_{CPt}$ = 23.2 Hz, C_{Ar}H), 134.4 (C_{Ar}H), 134.4 (d, $^3J_{CPt}$ = 18.6 Hz, C_{Ar}H), 134.6 (C_{Ar}H), 134.6 (d, $^3J_{CPt}$ = 17.8 Hz, C_{Ar}H), 135.5 (d, $^1J_{CP}$ = 3.2 Hz, C_{Ar}P), 135.5 (dd, $^1J_{CP}$ = 3.2 Hz, $^2J_{CPt}$ = 34.6 Hz, C_{Ar}P), 135.9 (d, $^1J_{CP}$ = 3.2 Hz, C_{Ar}P), 135.9 (dd, $^1J_{CP}$ = 3.2 Hz, $^2J_{CPt}$ = 31.5 Hz, C_{Ar}P), 137.1 (d, $^1J_{CP}$ = 2.9 Hz, C_{Ar}P), 137.1 (dd, $^1J_{CP}$ = 2.9 Hz, $^2J_{CPt}$ = 28.5 Hz, C_{Ar}P), 137.5 (d, $^1J_{CP}$ = 3.0 Hz, C_{Ar}P), 137.5 (dd, $^1J_{CP}$ = 3.0 Hz, $^2J_{CPt}$ = 26.9 Hz, C_{Ar}P), 141.7 (dd, $^3J_{CP}$ = 1.3 Hz, $^3J_{CP}$ = 6.6 Hz, C_{Ar}CH), 141.7 (ddd, $^3J_{CP}$ = 1.3 Hz, $^3J_{CP}$ = 6.6 Hz, $^2J_{CPt}$ = 38.4 Hz, C_{Ar}CH), 146.3 (C_{Ar}O), 156.9 (C_{Ar}OCH$_3$), 172.0 (dd, $^3J_{CP}$ 2.2 Hz, $^3J_{CP}$ = 4.8 Hz, CHCOO), 172.0 (ddd, $^3J_{CP}$ = 2.2 Hz, $^3J_{CP}$ = 4.8 Hz, $^2J_{CPt}$ = 47.6 Hz, CHCOO).

^{31}P-NMR (161.97 MHz; C$_6$D$_6$): δ = 28.0 (d, $^2J_{PP}$ = 43.4 Hz), 28.0 (dd, $^2J_{PP}$ = 43.4 Hz, $^1J_{PPt}$ = 4196.8 Hz), 28.9 (d, $^2J_{PP}$ = 43.3 Hz), 28.9 (dd, $^2J_{PP}$ = 43.3 Hz, $^1J_{PPt}$ = 3641.7 Hz).

^{195}Pt-NMR (85.48 MHz; C$_6$D$_6$): δ = -5047.6 (dd, $^1J_{PPt}$ = 3642.2 Hz, $^1J_{PPt}$ = 4197.8 Hz).

FAB-MS: m/z (%): 719.0 (100) [Pt(PPh$_3$)$_2{}^+$], 987.1 (1.12) [(C$_{17}$H$_{16}$O$_3$)Pt(PPh$_3$)$_2{}^+$].

FT-IR (ATR, cm^{-1}): 3048 (w), 2942 (w), 1699 (m, CO), 1598 (w), 1505 (s), 1477 (w), 1434 (s), 1302 (w), 1242 (w), 1178 (m), 1095 (s), 1030 (m), 893 (w), 802 (m), 742 (m), 693 (s), 575 (w), 538 (m), 499 (s), 416 (m).

elemental analysis for C$_{53}$H$_{46}$O$_3$P$_2$Pt (987.96 g/mol):

calcd.: C 64.43, H 4.69, O 4.86, P 6.27, Pt 19.75

found: C 64.61, H 4.91

Bis(triphenylphosphine)-(η^2-4-methylphenyl 4-methylcinnamate)nickel(0)

NiPhMe-PhMe: $C_{53}H_{46}NiO_2P_2$, M = 835.57 *g/mol*, orange red solid, yield: >99%, **route:** *Ni.*

^{1}H-NMR (400.13 MHz; C_7D_8): δ = 2.09 (s, 3H, CH_3), 2.12 (s, 3H, CH_3), 3.64 - 3.76 (bm, 1H, CHCHCO), 4.30 - 4.41 (bm, 1H, CC$_{Ar}$CHCH), 6.47 (d, $^3J_{HH}$ = 6.9 Hz, 2H, H_{Ar}), 6.64 (d, $^3J_{HH}$ = 7.3 Hz, 2H, H_{Ar}), 6.72 (d, $^3J_{HH}$ = 6.9 Hz, 2H, H_{Ar}), 6.85 (d, $^3J_{HH}$ = 6.9 Hz, 2H, H_{Ar}), 6.89 - 7.04 (m, 17H, H_{Ar}), 7.05 - 7.15 (m, 7H, H_{Ar}), 7.56 - 7.66 (m, 6H, H_{Ar}).

^{13}C-NMR (100.62 MHz; C_7D_8): δ = 20.8 (CH$_3$), 21.3 (CH$_3$), 52.4 (dd, $^2J_{CP}$ = 1.1 Hz, $^2J_{CP}$ = 13.4 Hz, CHCHCO), 65.5 (d, $^2J_{CP}$ = 16.4 Hz, C$_{Ar}$$C$HCH), 122.3 ($C_{Ar}$H), 128.0 ($C_{Ar}$H), 128.1 ($C_{Ar}$H), 128.3 ($C_{Ar}$H), 128.4 ($C_{Ar}$H), 129.3 ($C_{Ar}$H), 129.4 ($C_{Ar}$H), 129.6 ($C_{Ar}$H), 132.4 (bs, C_{Ar}), 133.1 (bs, C_{Ar}), 134.2 (d, $^2J_{CP}$ = 12.5 Hz, C_{Ar}H), 134.4 (d, $^2J_{CP}$ = 13.4 Hz, C_{Ar}H), 134.8 (d, $^1J_{CP}$ = 2.8 Hz, C_{Ar}), 135.1 (d, $^1J_{CP}$ = 2.8 Hz, C_{Ar}), 136.0 (d, $^1J_{CP}$ = 2.1 Hz, C_{Ar}), 136.3 (d, $^1J_{CP}$ = 2.0 Hz, C_{Ar}), 140.5 (d, $^3J_{CP}$ = 5.4 Hz, C_{Ar}), 150.5 (C_{Ar}), 170.9 (d, $^3J_{CP}$ = 2.4 Hz, CHCOO).

^{31}P-NMR (161.97 MHz; C_7D_8): δ = 30.6 (d, $^2J_{PP}$ = 35.0 Hz), 32.9 (d, $^2J_{PP}$ = 35.0 Hz).

FT-IR (ATR, cm^{-1}): 1698 (m, CO).

Bis(triphenylphosphine)-(η^2-4-methylphenyl 4-methylcinnamate)palladium(0)

PdPhMe-PhMe: $C_{53}H_{46}O_2P_2Pd$, M = 883.30 g/mol, yellow solid, yield: >99%, **route**: *Pd-I*.

^{1}H-NMR (400.13 MHz; C_7D_8): $\delta = 2.09$ (s, 6H, CH_3), 4.89 (bs, 1H, CHCHCO), 5.31 (bs, 1H, C_{Ar}CHCH), 6.48 - 6.61 (bm, 2H, H_{Ar}), 6.62 - 6.72 (bm, 4H, H_{Ar}), 6.77 - 6.85 (bm, 2H, H_{Ar}), 6.86 - 7.04 (bm, 18H, H_{Ar}), 7.04 - 7.21 (bm, 6H, H_{Ar}), 7.51 - 7.65 (bm, 6H, H_{Ar}).

^{1}H-NMR (400.13 MHz; 246 K; C_7D_8): $\delta = 2.09$ (s, 3H, CH_3), 2.10 (s, 3H, CH_3), 5.04 (dpt, $^3J_{HH} = 10.1$ Hz, $^3J_{HP} = 6.1$ Hz, $^3J_{HP} = 6.1$ Hz, 1H, CHCHCO), 5.37 (dpt, $^3J_{HH} = 10.1$ Hz, $^3J_{HP} = 7.1$ Hz, $^3J_{HP} = 7.1$ Hz, 1H, C_{Ar}CHCH), 6.49 - 6.55 (m, 2H, H_{Ar}), 6.55 - 6.62 (m, 2H, H_{Ar}), 6.63 - 6.68 (m, 2H, H_{Ar}), 6.77 - 6.83 (m, 2H, H_{Ar}), 6.87 - 7.00 (m, 18H, H_{Ar}), 7.03 - 7.10 (m, 6H, H_{Ar}), 7.57 - 7.64 (m, 6H, H_{Ar}).

^{13}C-NMR (100.62 MHz; C_7D_8): $\delta = 20.7$ (CH_3), 21.1 (CH_3), 59.9 - 60.7 (bm, CHCHCO), 76.4 - 77.2 (bm, C_{Ar}CHCH), 125.6 (C_{Ar}H), 128.2 - 128.3 (bm, C_{Ar}H), 128.4 (C_{Ar}H), 128.5 (C_{Ar}H), 129.2 (C_{Ar}H), 129.2 (C_{Ar}H), 129.4 (C_{Ar}H), 132.3 (C_{Ar}H), 132.4 (C_{Ar}H), 134.0 - 134.7 (bm, C_{Ar}H), 135.7 - 136.4 (bm, C_{Ar}H), 150.4 (C_{Ar}O).

^{13}C-NMR (100.62 MHz; 246 K; C_7D_8): $\delta = 20.6$ (CH_3), 20.8 (CH_3), 59.5 (dd, $^2J_{CP} = 4.5$ Hz, $^2J_{CP} = 16.4$ Hz, CHCHCO), 76.6 (dd, $^2J_{CP} = 2.2$ Hz, $^2J_{CP} = 22.9$ Hz, C_{Ar}CHCH), 122.3 (C_{Ar}H), 125.2 (d, $^3J_{CP} = 1.4$ Hz, C_{Ar}H), 125.5 (d, $^3J_{CP} = 1.8$ Hz, C_{Ar}H), 128.2 (C_{Ar}), 128.4 (C_{Ar}), 129.2 (C_{Ar}), 129.3 (C_{Ar}), 129.4 (C_{Ar}), 133.4 (C_{Ar}CH$_3$), 133.9 (d, $^3J_{CP} = 1.5$ Hz, C_{Ar}H), 134.1 (d, $^2J_{CP} = 14.7$ Hz, C_{Ar}H), 134.4 (d, $^2J_{CP} = 15.4$ Hz, C_{Ar}H), 135.5 (dd, $^1J_{CP} = 28.3$ Hz, $^3J_{CP} = 1.5$ Hz, C_{Ar}P), 137.0 (d, $^3J_{CP} = 1.4$ Hz, C_{Ar}P), 139.3 (d, $^3J_{CP} = 6.6$ Hz, C_{Ar}CH), 150.2 (C_{Ar}O), 169.4 - 169.5 (m, CHCOO)

^{31}P-NMR (161.97 MHz; C$_7$D$_8$): $\delta = 24.4$ (d, $^2J_{PP} = 17.6$ Hz), 26.0 (d, $^2J_{PP} = 17.6$ Hz).

^{31}P-NMR (161.97 MHz; 246 K; C$_7$D$_8$): $\delta = 23.7$ (d, $^2J_{PP} = 18.1$ Hz), 25.6 (d, $^2J_{PP} = 18.1$ Hz).

FAB-MS: m/z (%): 368.0 (17.79) [Pd(PPh$_3$)$^+$], 630.0 (37.67) [Pd(PPh$_3$)$_2$$^+$].

FT-IR (ATR, cm^{-1}): 3043 (w), 1694 (m, CO), 1591 (w), 1500 (w), 1476 (m), 1432 (s), 1198 (m), 1115 (s), 1104 (m), 1092 (m), 1015 (w), 962 (w), 803 (w), 742 (s), 683 (s), 530 (m), 515 (m), 501 (s), 409 (s).

elemental analysis for C$_{53}$H$_{46}$O$_2$P$_2$Pd (883.30 *g/mol*):

calcd.: C 72.07, H 5.25, O 3.62, P 7.01, Pd 12.05

found: C 71.41, H 5.31

Bis(triphenylphosphine)-(η^2-4-methylphenyl 4-methylcinnamate)platinum(0)

PtPhMe-PhMe: C$_{53}$H$_{46}$O$_2$P$_2$Pt, M = 971.96 *g/mol*, white solid, yield: 98%, **route**: *Pt-VI*.

^{1}H-NMR (400.13 MHz; C$_6$D$_6$): $\delta = 2.06$ (s, 3H, CH_3), 2.13 (s, 3H, CH_3), 4.29 (dpt, $^3J_{HH} = 8.6$ Hz, $^3J_{HP} = 4.5$ Hz, $^3J_{HP} = 8.6$ Hz, 1H, CHCHCO), 4.29 (ddpt, $^3J_{HH} = 8.6$ Hz, $^3J_{HP} = 4.5$ Hz, $^3J_{HP} = 8.6$ Hz, $^2J_{HPt} = 61.5$ Hz, 1H, CHCHCO), 4.56 (dpt, $^3J_{HH} = 8.7$ Hz, $^3J_{HP} = 5.7$ Hz, $^3J_{HP} = 8.7$ Hz, 1H, C$_{Ar}$CHCH), 4.56 (ddpt, $^3J_{HH} = 8.7$ Hz, $^3J_{HP} = 5.7$ Hz, $^3J_{HP} = 8.7$ Hz, $^2J_{HPt} = 51.4$ Hz, 1H, C$_{Ar}$CHCH), 6.58 (d, $^3J_{HH} = 8.0$ Hz, 2H, H_{Ar}), 6.74 (d, $^3J_{HH} = 8.4$ Hz, 4H, H_{Ar}), 6.81 - 6.98 (m, 10H, H_{Ar}), 7.17 - 7.23 (m, 6H, H_{Ar}), 7.48 (d, $^3J_{HH} = 6.9$ Hz, 10H, H_{Ar}) 7.59 - 7.67 (m, 6H, H_{Ar}).

^{13}C-NMR (100.62 MHz; C$_6$D$_6$): $\delta = 20.8$ (CH_3), 21.1 (CH_3), 49.4 (dd, $^2J_{CP} = 6.1$ Hz, $^2J_{CP} = 28.1$ Hz, CHCHCO), 49.4 (ddd, $^2J_{CP} = 6.1$ Hz, $^2J_{CP} = 28.1$ Hz, $^1J_{CPt} = 198.9$ Hz, CHCHCO), 61.9 (dd, $^2J_{CP} = 3.9$ Hz, $^2J_{CP} = 32.7$ Hz, C$_{Ar}$CHCH), 61.9 (ddd,

$^2J_{CP}$ = 3.9 Hz, $^2J_{CP}$ = 32.7 Hz, $^1J_{CPt}$ = 195.7 Hz, $C_{Ar}CHCH$), 122.4 ($C_{Ar}H$), 126.2 (d,

$^4J_{CP}$ = 3.1 Hz, $C_{Ar}H$), 126.2 (dd, $^4J_{CP}$ = 3.1 Hz, $^3J_{CPt}$ = 13.4 Hz, $C_{Ar}H$), 127.9 ($C_{Ar}H$),

128.4 ($C_{Ar}H$), 128.4 ($C_{Ar}H$), 128.7 (d, $^4J_{CP}$ = 2.2 Hz, $C_{Ar}H$), 129.3 (d, $^4J_{CP}$ = 2.0 Hz,

$C_{Ar}H$), 129.5 ($C_{Ar}H$), 129.6 (d, $^4J_{CP}$ = 1.9 Hz, $C_{Ar}H$), 133.2 (d, $^4J_{CP}$ = 2.7 Hz, $C_{Ar}H$),

133.5 (C_{Ar}), 134.2 ($C_{Ar}H$), 134.2 (d, $^3J_{CPt}$ = 21.4 Hz, $C_{Ar}H$), 134.3 ($C_{Ar}H$), 134.3

(d, $^3J_{CPt}$ = 25.0 Hz, $C_{Ar}H$), 134.4 ($C_{Ar}H$), 134.4 (d, $^3J_{CPt}$ = 18.4 Hz, $C_{Ar}H$), 134.5

($C_{Ar}H$), 134.5 (d, $^3J_{CPt}$ = 18.8 Hz, $C_{Ar}H$), 135.4 (d, $^1J_{CP}$ = 3.1 Hz, $C_{Ar}P$), 135.4 (dd,

$^1J_{CP}$ = 3.1 Hz, $^2J_{CPt}$ = 34.1 Hz, $C_{Ar}P$), 135.9 (d, $^1J_{CP}$ = 3.1 Hz, $C_{Ar}P$), 135.9 (dd,

$^1J_{CP}$ = 3.1 Hz, $^2J_{CPt}$ = 31.0 Hz, $C_{Ar}P$), 137.0 (d, $^1J_{CP}$ = 2.9 Hz, $C_{Ar}P$), 137.0 (dd,

$^1J_{CP}$ = 2.9 Hz, $^2J_{CPt}$ = 28.4 Hz, $C_{Ar}P$), 137.4 (d, $^1J_{CP}$ = 2.8 Hz, $C_{Ar}P$), 137.4 (dd,

$^1J_{CP}$ = 2.8 Hz, $^2J_{CPt}$ = 28.7 Hz, $C_{Ar}P$), 141.7 (dd, $^3J_{CP}$ = 1.4 Hz, $^3J_{CP}$ = 6.6 Hz,

$C_{Ar}CH$), 141.7 (ddd, $^3J_{CP}$ = 1.4 Hz, $^3J_{CP}$ = 6.6 Hz, $^2J_{CPt}$ = 32.4 Hz, $C_{Ar}CH$), 150.6

($C_{Ar}O$), 171.8 (dd, $^3J_{CP}$ 2.4 Hz, $^3J_{CP}$ = 5.0 Hz, $CHCOO$), 171.8 (ddd, $^3J_{CP}$ = 2.4 Hz,

$^3J_{CP}$ = 5.0 Hz, $^2J_{CPt}$ = 50.1 Hz, $CHCOO$).

^{31}P-NMR (161.97 MHz; C_6D_6): δ = 27.3 (d, $^2J_{PP}$ = 43.3 Hz), 27.3 (dd,

$^2J_{PP}$ = 43.3 Hz, $^1J_{PPt}$ = 4197.8 Hz), 28.2 (d, $^2J_{PP}$ = 43.4Hz), 28.2 (dd, $^2J_{PP}$ = 43.4 Hz,

$^1J_{PPt}$ = 3641.7 Hz).

^{195}Pt-NMR (85.48 MHz; C_6D_6): δ = -5046.4 (dd, $^1J_{PPt}$ = 3641.1 Hz,

$^1J_{PPt}$ = 4197.0 Hz).

FAB-MS: m/z (%): 719.0 (100) [Pt(PPh$_3$)$_2^+$], 971.2 (1.20) [(C$_{17}$H$_{16}$O$_2$)Pt(PPh$_3$)$_2^+$].

FT-IR (ATR, cm^{-1}): 3041 (w), 1707 (s, CO), 1493 (w), 1432 (m), 1258 (w), 1205 (w),
1167 (w), 1124 (s), 1095 (m), 970 (w), 803 (w), 743 (m), 694 (s), 538 (m), 521 (m), 500
(s), 417 (m).

elemental analysis for C$_{53}$H$_{46}$O$_2$P$_2$Pt (971.96 g/mol):

calcd.: C 65.49, H 4.77, O 3.29, P 6.37, Pt 20.07

found: C 65.24, H 4.84

Bis(triphenylphosphine)-(η^2-phenyl 4-methylcinnamate)nickel(0)

NiPhMe-Ph: $C_{52}H_{44}NiO_2P_2$, M = 821.55 g/mol, yellow orange solid, yield: 82%, **route**: *Ni*.

^{1}H-NMR (400.13 MHz; C_7D_8): δ = 2.08 (s, 3H, CH_3), 3.69 (ddd, $^3J_{HH}$ = 8.3 Hz, $^3J_{HP}$ = 3.7 Hz, $^3J_{HP}$ = 9.5 Hz, 1H, CHCHCO), 4.34 (ddd, $^3J_{HH}$ = 8.3 Hz, $^3J_{HP}$ = 5.1 Hz, $^3J_{HP}$ = 9.2 Hz, 1H, CC$_{Ar}$CHCH), 6.46 (d, $^3J_{HH}$ = 7.7 Hz, 2H, H_{Ar}), 6.73 (dd, $^3J_{HH}$ = 5.1 Hz, $^3J_{HH}$ = 7.4 Hz, 4H, H_{Ar}), 6.86 - 7.19 (m, 27H, H_{Ar}), 7.49 - 7.75 (m, 6H, H_{Ar}).

^{13}C-NMR (100.53 MHz; C_7D_8): δ = 21.3 (CH_3), 52.3 (dd, $^2J_{CP}$ = 1.8 Hz, $^2J_{CP}$ = 12.7 Hz, CHCHCO), 65.3 (d, $^2J_{CP}$ = 16.3 Hz, $C_{Ar}CHCH$), 122.5 ($C_{Ar}H$), 124.2 ($C_{Ar}H$), 125.4 ($C_{Ar}H$), 128.0 ($C_{Ar}H$), 128.1 ($C_{Ar}H$), 128.3 ($C_{Ar}H$), 138.4 ($C_{Ar}H$), 128.8 ($C_{Ar}H$), 129.3 (d, $^3J_{CP}$ = 1.3 Hz, $C_{Ar}H$), 129.6 (d, $^3J_{CP}$ = 0.6 Hz, $C_{Ar}H$), 132.5 (bs, C_{Ar}), 133.1 (d, $^1J_{CP}$ = 1.4 Hz, C_{Ar}), 134.2 (d, $^2J_{CP}$ = 12.5 Hz, $C_{Ar}H$), 134.4 (d, $^2J_{CP}$ = 13.4 Hz, $C_{Ar}H$), 134.8 (d, $^1J_{CP}$ = 2.9 Hz, C_{Ar}), 135.2 (d, $^1J_{CP}$ = 2.9 Hz, C_{Ar}), 136.0 (d, $^1J_{CP}$ = 2.3 Hz, C_{Ar}), 136.3 (d, $^1J_{CP}$ = 2.3 Hz, C_{Ar}), 140.5 (d, $^3J_{CP}$ = 5.5 Hz, C_{Ar}), 152.7 (C_{Ar}), 170.7 (dd, $^3J_{CP}$ = 1.2 Hz, $^3J_{CP}$ = 3.4 Hz, CHCOO).

^{31}P-NMR (161.97 MHz; C_7D_8): δ = 31.6 (d, $^2J_{PP}$ = 35.3 Hz), 33.8 (d, $^2J_{PP}$ = 35.3 Hz).

FT-IR (ATR, cm^{-1}): 1696 (m, CO).

Bis(triphenylphosphine)-(η^2-phenyl 4-methylcinnamate)palladium(0)

PdPhMe-Ph: $C_{52}H_{44}O_2P_2Pd$, M $= 869.27$ g/mol, light yellow solid, yield: $>99\%$, **route**: *Pd-I*.

^{1}H-NMR (400.13 MHz; C_7D_8): $\delta = 2.09$ (s, 3H, CH_3), 4.89 (bs, 1H, CHCHCO), 5.30 (bs, 1H, C_{Ar}CHCH), 6.48 - 6.59 (bm, 2H, H_{Ar}), 6.63 - 6.72 (bm, 2H, H_{Ar}), 6.72 - 6.82 (bm, 2H, H_{Ar}), 6.83 - 7.04 (bm, 21H, H_{Ar}), 7.04 - 7.20 (bm, 6H, H_{Ar}), 7.46 - 7.62 (bm, 6H, H_{Ar}).

^{1}H-NMR (400.13 MHz; 246 K; C_7D_8): $\delta = 2.09$ (s, 3H, CH_3), 5.03 (dpt, $^3J_{HH} = 10.2$ Hz, $^3J_{HP} = 6.0$ Hz, $^3J_{HP} = 6.0$ Hz, 1H, CHCHCO), 5.36 (dpt, $^3J_{HH} = 10.2$ Hz, $^3J_{HP} = 7.1$ Hz, $^3J_{HP} = 7.1$ Hz, 1H, C_{Ar}CHCH), 6.48 - 6.55 (m, 2H, H_{Ar}), 6.61 - 6.70 (m, 4H, H_{Ar}), 6.86 - 7.02 (m, 21H, H_{Ar}), 7.03 - 7.10 (m, 6H, H_{Ar}), 7.55 - 7.63 (m, 6H, H_{Ar}).

^{13}C-NMR (100.62 MHz; C_7D_8): $\delta = 21.1$ (CH_3), 59.8 - 60.4 (bm, CHCHCO), 76.4 - 77.1 (bm, C_{Ar}CHCH), 122.5 (C_{Ar}H), 124.3 (C_{Ar}H), 125.6 (C_{Ar}H), 128.2 - 128.3 (bm, C_{Ar}H), 128.3 - 128.4 (bm, C_{Ar}H), 128.4 (C_{Ar}H), 128.5 (C_{Ar}H), 128.5 (C_{Ar}H), 129.2 (C_{Ar}H), 129.2 (C_{Ar}H), 129.3 (C_{Ar}H), 132.3 (C_{Ar}H), 132.4 (C_{Ar}H), 134.0 - 134.7 (bm, C_{Ar}H), 135.6 - 136.2 (bm, C_{Ar}H), 139.7 (bs, C_{Ar}CH), 152.6 (C_{Ar}O), 169.2 (bs, CHCOO).

^{13}C-NMR (100.53 MHz; 246 K; C_7D_8): $\delta = 20.8$ (CH_3), 59.3 (dd, $^3J_{CP} = 3.6$ Hz, $^3J_{CP} = 17.6$ Hz, CHCHCO), 76.6 (dd, $^3J_{CP} = 4.1$ Hz, $^3J_{CP} = 24.1$ Hz, C_{Ar}CHCH), 122.6 (C_{Ar}H), 128.0 (C_{Ar}), 128.2 (C_{Ar}), 128.4 (C_{Ar}), 129.2 (C_{Ar}), 129.3 (C_{Ar}), 134.0 (d, $^3J_{CP} = 3.1$ Hz, C_{Ar}H), 134.1 (d, $^2J_{CP} = 14.8$ Hz, C_{Ar}H), 134.4 (d, $^2J_{CP} = 15.4$ Hz, C_{Ar}H), 135.4 (dd, $^1J_{CP} = 28.4$ Hz, $^3J_{CP} = 1.8$ Hz, C_{Ar}P), 137.1 (dd, $^1J_{CP} = 27.4$ Hz, $^3J_{CP} = 1.6$ Hz, C_{Ar}P), 139.3 (d, $^3J_{CP} = 6.2$ Hz, C_{Ar}CH), 152.4 (C_{Ar}O), 169.2 (dd, $^3J_{CP} = 2.0$ Hz, $^3J_{CP} = 3.7$ Hz, CHCOO).

^{31}P-NMR (161.97 MHz; C$_7$D$_8$): $\delta = 24.4$ (d, $^2J_{PP} = 18.7$ Hz), 26.0 (d, $^2J_{PP} = 18.7$ Hz).

^{31}P-NMR (161.97 MHz; 246 K; C$_7$D$_8$): $\delta = 23.7$ (d, $^2J_{PP} = 17.7$ Hz), 25.6 (d, $^2J_{PP} = 17.7$ Hz).

FT-IR (ATR, cm^{-1}): 3045 (m), 1697 (s, CO), 1594 (w), 1475 (s), 1433 (s), 1306 (w), 1261 (w), 1200 (m), 1186 (m), 1109 (s), 1091 (s), 1034 (w), 973 (w), 928 (w), 814 (m), 746 (s), 692 (s), 526 (m), 514 (s), 412 (m).

elemental analysis for C$_{52}$H$_{44}$O$_2$P$_2$Pd (869.27 *g/mol*):

calcd.: C 71.85, H 5.10, O 3.68, P 7.13, Pd 12.24

found: C 72.19, H 5.17

Bis(triphenylphosphine)-(η^2-phenyl 4-methylcinnamate)platinum(0)

PtPhMe-Ph: C$_{52}$H$_{44}$O$_2$P$_2$Pt, M = 957.93 *g/mol*, white solid, yield: 85%, **route**: *Pt-VI*.

Single crystals suitable for X-ray analysis were obtained by slow diffusion of pentane into a toluene solution.

^{1}H-NMR (400.13 MHz; C$_7$D$_8$): $\delta = 2.13$ (s, 3H, CH_3), 4.29 (dpt, $^3J_{HH} = 8.5$ Hz, $^3J_{HP} = 4.4$ Hz, $^3J_{HP} = 8.5$ Hz, 1H, CHCHCO), 4.29 (ddpt, $^3J_{HH} = 8.5$ Hz, $^3J_{HP} = 4.4$ Hz, $^3J_{HP} = 8.5$ Hz, $^2J_{HPt} = 61.3$ Hz, 1H, CHCHCO), 4.56 (dpt, $^3J_{HH} = 8.6$ Hz, $^3J_{HP} = 5.8$ Hz, $^3J_{HP} = 8.6$ Hz, 1H, C$_{Ar}$CHCH), 4.56 (ddpt, $^3J_{HH} = 8.6$ Hz, $^3J_{HP} = 5.8$ Hz, $^3J_{HP} = 8.6$ Hz, $^2J_{HPt} = 51.7$ Hz, 1H, C$_{Ar}$CHCH), 6.57 (d, $^3J_{HH} = 7.4$ Hz, 2H, H_{Ar}), 6.75 (d, $^3J_{HH} = 7.7$ Hz, 2H, H_{Ar}), 6.78 - 6.98 (m, 21H, H_{Ar}), 7.48 (t, $^3J_{HH} = 7.7$ Hz, 2H, H_{Ar}), 7.16 - 7.24 (m, 6H, H_{Ar}), 7.55 - 7.67 (m, 6H, H_{Ar}).

^{13}C-NMR (100.53 MHz; C$_7$D$_8$): $\delta = 21.0$ (CH_3), 49.4 (dd, $^2J_{CP} = 6.2$ Hz, $^2J_{CP} = 28.2$ Hz, CHCHCO), 49.4 (ddd, $^2J_{CP} = 6.2$ Hz, $^2J_{CP} = 28.2$ Hz, $^1J_{CPt} = 197.6$ Hz, CHCHCO), 61.9 (dd, $^2J_{CP} = 4.0$ Hz, $^2J_{CP} = 32.8$ Hz, C$_{Ar}$CHCH), 61.9 (ddd,

$^2J_{CP}$ = 4.0 Hz, $^2J_{CP}$ = 32.8 Hz, $^1J_{CPt}$ = 196.8 Hz, $C_{Ar}CHCH$), 122.5 ($C_{Ar}H$), 124.2 (C_{Ar}), 126.1 (d, $^5J_{CP}$ = 3.2 Hz, $C_{Ar}H$), 126.1 (dd, $^5J_{CP}$ = 3.2 Hz, $^4J_{CPt}$ = 13.1 Hz, $C_{Ar}H$), 127.9 ($C_{Ar}H$), 128.0 ($C_{Ar}H$), 128.2 ($C_{Ar}H$), 128.3 ($C_{Ar}H$), 128.8 ($C_{Ar}H$), 129.3 (d, $^1J_{CPt}$ = 2.0 Hz, $C_{Ar}H$), 129.6 (d, $^1J_{CPt}$ = 1.9 Hz, $C_{Ar}H$), 133.1 (d, $^4J_{CP}$ = 2.7 Hz, $C_{Ar}H$), 133.1 (dd, $^4J_{CP}$ = 2.7 Hz, $^3J_{CPt}$ = 9.4 Hz, $C_{Ar}H$), 134.1 ($C_{Ar}H$), 134.1 (d, $^3J_{CPt}$ = 21.4 Hz, $C_{Ar}H$), 134.2 ($C_{Ar}H$), 134.2 (d, $^3J_{CPt}$ = 21.6 Hz, $C_{Ar}H$), 134.3 ($C_{Ar}H$), 134.3 (d, $^3J_{CPt}$ = 18.8 Hz, $C_{Ar}H$), 134.5 ($C_{Ar}H$), 134.5 (d, $^3J_{CPt}$ = 18.4 Hz, $C_{Ar}H$), 135.4 (d, $^1J_{CP}$ = 3.0 Hz, $C_{Ar}P$), 135.4 (dd, $^1J_{CP}$ = 3.0 Hz, $^2J_{CPt}$ = 34.9 Hz, $C_{Ar}P$), 135.9 (d, $^1J_{CP}$ = 3.1 Hz, $C_{Ar}P$), 135.9 (dd, $^1J_{CP}$ = 3.1 Hz, $^2J_{CPt}$ = 33.2 Hz, $C_{Ar}P$), 137.0 (d, $^1J_{CP}$ = 2.7 Hz, $C_{Ar}P$), 137.0 (dd, $^1J_{CP}$ = 2.7 Hz, $^2J_{CPt}$ = 28.2 Hz, $C_{Ar}P$), 137.4 (d, $^1J_{CP}$ = 2.7 Hz, $C_{Ar}P$), 137.4 (dd, $^1J_{CP}$ = 2.7 Hz, $^2J_{CPt}$ = 27.2 Hz, $C_{Ar}P$), 141.6 (dd, $^3J_{CP}$ = 1.4 Hz, $^3J_{CP}$ = 6.7 Hz, $C_{Ar}CH$), 141.6 (ddd, $^3J_{CP}$ = 1.4 Hz, $^3J_{CP}$ = 6.7 Hz, $^2J_{CPt}$ = 33.0 Hz, $C_{Ar}CH$), 152.8 ($C_{Ar}O$), 171.2 (dd, $^3J_{CP}$ 2.3 Hz, $^3J_{CP}$ = 5.0 Hz, CHCOO), 171.2 (ddd, $^3J_{CP}$ = 2.3 Hz, $^3J_{CP}$ = 5.0 Hz, $^2J_{CPt}$ = 47.7 Hz, CHCOO).

^{31}P-NMR (161.97 MHz; C_7D_8): δ = 27.9 (d, $^2J_{PP}$ = 43.3 Hz), 27.9 (dd, $^2J_{PP}$ = 43.3 Hz, $^1J_{PPt}$ = 4206.0 Hz), 29.0 (d, $^2J_{PP}$ = 43.4 Hz), 29.0 (dd, $^2J_{PP}$ = 43.4 Hz, $^1J_{PPt}$ = 3643.7 Hz).

^{195}Pt-NMR (85.48 MHz; C_7D_8): δ = -5052.0 (dd, $^1J_{PPt}$ = 3643.7 Hz, $^1J_{PPt}$ = 4206.7 Hz).

FAB-MS: m/z (%): 719.0 (100) [Pt(PPh$_3$)$_2^+$], 957.1 (1.62) [(C$_{16}$H$_{14}$O$_2$)Pt(PPh$_3$)$_2^+$].

FT-IR (ATR, cm^{-1}): 3052 (w), 1703 (m, CO), 1470 (w), 1433 (s), 1304 (w), 1258 (w), 1205 (w), 1145 (w), 1093 (s), 985 (w), 917 (w), 811 (w), 743 (m), 692 (s), 538 (m), 517 (s), 501 (s), 440 (m).

elemental analysis for C$_{52}$H$_{44}$O$_2$P$_2$Pt (957.93 g/mol):

calcd.: C 65.20, H 4.63, O 3.34, P 6.47, Pt 20.36

found: C 64.80, H 4.84

Bis(triphenylphosphine)-(η^2-4-nitrophenyl 4-methylcinnamate)palladium(0)

PdPhMe-PhNO$_2$: C$_{52}$H$_{43}$NO$_4$P$_2$Pd, M = 914.27 *g/mol*, deep yellow solid, yield: >99%, **route**: *Pd-I*.

^{1}H-NMR (400.13 MHz; C$_7$D$_8$): δ = 2.09 (s, 3H, CH_3), 4.74 - 4.86 (bm, 1H, CHCHCO), 5.15 - 5.27 (bm, 1H, C$_{Ar}$CHCH), 6.49 - 6.60 (bm, 4H, H_{Ar}), 6.63 - 6.70 (bm, 2H, H_{Ar}), 6.85 - 7.17 (bm, 24H, H_{Ar}), 7.40 - 7.53 (bm, 6H, H_{Ar}), 7.70 - 7.80 (bm, 2H, H_{Ar}).

^{1}H-NMR (400.13 MHz; 246 K; C$_7$D$_8$): δ = 2.09 (s, 3H, CH_3), 4.95 (dpt, $^3J_{HH}$ = 10.2 Hz, $^3J_{HP}$ = 5.9 Hz, $^3J_{HP}$ = 5.9 Hz, 1H, CHCHCO), 5.26 (dpt, $^3J_{HH}$ = 10.2 Hz, $^3J_{HP}$ = 7.2 Hz, $^3J_{HP}$ = 7.2 Hz, 1H, C$_{Ar}$CHCH), 6.40 - 6.45 (m, 2H, H_{Ar}), 6.48 - 6.53 (m, 2H, H_{Ar}), 6.64 - 6.68 (m, 2H, H_{Ar}), 6.84 - 6.91 (m, 12H, H_{Ar}), 6.94 - 7.06 (m, 12H, H_{Ar}), 7.45 - 7.52 (m, 6H, H_{Ar}), 7.68 - 7.73 (m, 2H, H_{Ar}).

^{13}C-NMR (100.62 MHz; C$_7$D$_8$): δ = 21.1 (CH_3), 58.3 - 59.3 (bm, CHCHCO), 75.3 - 76.7 (bm, C$_{Ar}C$HCH), 122.3 (C_{Ar}H), 124.5 (C_{Ar}H), 125.6 (C_{Ar}H), 128.2 - 128.3 (bm, C_{Ar}H), 128.3 - 128.4 (bm, C_{Ar}H), 128.5 (C_{Ar}H), 128.5 (C_{Ar}H), 129.2 (C_{Ar}H), 129.3 (C_{Ar}H), 129.3 (C_{Ar}H), 129.5 - 129.6 (bm, C_{Ar}H), 132.3 (C_{Ar}H), 132.4 (C_{Ar}H), 134.0 - 134.5 (bm, C_{Ar}H), 135.2 - 135.8 (bm, C_{Ar}H).

^{13}C-NMR (100.62 MHz; 246 K; C$_7$D$_8$): δ = 22.9 (CH_3), 58.2 (dd, $^2J_{CP}$ = 3.6 Hz, $^2J_{CP}$ = 21.1 Hz, CHCHCO), 75.6 (dd, $^2J_{CP}$ = 3.3 Hz, $^2J_{CP}$ = 22.6 Hz, C$_{Ar}C$HCH), 122.3 (C_{Ar}H), 124.5 (C_{Ar}H), 125.4 (bs, C_{Ar}H), 128.0 (C_{Ar}), 128.3 (C_{Ar}), 128.4 (C_{Ar}), 129.2 (C_{Ar}), 129.2 (C_{Ar}), 129.5 (C_{Ar}), 134.0 (d, $^2J_{CP}$ = 15.9 Hz, C_{Ar}H), 134.2 (d, $^2J_{CP}$ = 16.1 Hz, C_{Ar}H), 134.4 (d, $^3J_{CP}$ = 1.2 Hz, C_{Ar}H), 135.1 (dd, $^1J_{CP}$ = 29.1 Hz, $^3J_{CP}$ = 1.7 Hz, C_{Ar}P), 136.7 (dd, $^1J_{CP}$ = 27.7 Hz, $^3J_{CP}$ = 1.3 Hz, C_{Ar}P), 138.9 (d, $^3J_{CP}$ = 6.7 Hz, C_{Ar}CH), 144.2 (C_{Ar}N), 156.9 (C_{Ar}O), 167.9 - 168.0 (m, CHCOO).

^{31}P-NMR (161.97 MHz; C$_7$D$_8$): δ = 24.5 (d, $^2J_{PP}$ = 15.3 Hz), 25.5 (d, $^2J_{PP}$ = 15.3 Hz).

^{31}P-NMR (161.97 MHz; 246 K; C$_7$D$_8$): δ = 23.9 (d, $^2J_{PP}$ = 14.4 Hz), 25.3 (d, $^2J_{PP}$ = 14.4 Hz).

FT-IR (ATR, cm^{-1}): 3063 (m), 2926 (m), 2851 (m), 1699 (m, CO), 1585 (w), 1519 (m), 1475 (m), 1433 (s), 1344 (m), 1206 (m), 1092 (s), 964 (m), 904 (m), 858 (m), 798 (m), 745 (m), 737 (s), 692 (s), 528 (m), 513 (m), 500 (s), 419 (m).

elemental analysis for C$_{52}$H$_{43}$NO$_4$P$_2$Pd (914.27 g/mol):

calcd.: C 68.31, H 4.74, N 1.53, O 7.00, P 6.78, Pd 11.64

found: C 69.43, H 5.32, N 1.35

Bis(triphenylphosphine)-(η^2-4-nitrophenyl 4-methylcinnamate)platinum(0)

PtPhMe-PhNO$_2$: C$_{52}$H$_{43}$NO$_4$P$_2$Pt, M = 1002.93 g/mol, off-white solid, yield: 90%, route: *Pt-VI*.

^{1}H-NMR (400.13 MHz; C$_6$D$_6$): δ = 2.12 (s, 3H, CH_3), 4.21 (dpt, $^3J_{HH}$ = 8.7 Hz, $^3J_{HP}$ = 4.0 Hz, $^3J_{HP}$ = 8.7 Hz, 1H, CHCHCO), 4.21 (ddpt, $^3J_{HH}$ = 8.7 Hz, $^3J_{HP}$ = 4.0 Hz, $^3J_{HP}$ = 8.7 Hz, $^2J_{HPt}$ = 60.6 Hz, 1H, CHCHCO), 4.45 (dpt, $^3J_{HH}$ = 8.7 Hz, $^3J_{HP}$ = 5.5 Hz, $^3J_{HP}$ = 8.7 Hz, 1H, C$_{Ar}$CHCH), 4.45 (ddpt, $^3J_{HH}$ = 8.7 Hz, $^3J_{HP}$ = 5.5 Hz, $^3J_{HP}$ = 8.7 Hz, $^2J_{HPt}$ = 51.4 Hz, 1H, C$_{Ar}$CHCH), 6.51 - 6.62 (m, 4H, H_{Ar}), 6.75 (d, $^3J_{HH}$ = 7.9 Hz, 2H, H_{Ar}), 6.81 - 6.96 (m, 18H, H_{Ar}), 7.10 - 7.22 (m, 6H, H_{Ar}), 7.46 - 7.58 (m, 6H, H_{Ar}), 7.73 (d, $^3J_{HH}$ = 9.0 Hz, 2H, H_{Ar}).

^{13}C-NMR (100.62 MHz; C$_6$D$_6$): δ = 20.7 (CH$_3$), 48.1 (dd, $^2J_{CP}$ = 6.2 Hz, $^2J_{CP}$ = 27.7 Hz, CHCHCO), 48.1 (ddd, $^2J_{CP}$ = 6.2 Hz, $^2J_{CP}$ = 27.7 Hz, $^1J_{CPt}$ = 186.0 Hz, CHCHCO), 60.8 (dd, $^2J_{CP}$ = 3.6 Hz, $^2J_{CP}$ = 33.4 Hz, C$_{Ar}$$C$HCH), 60.8 (ddd, $^2J_{CP}$ = 3.6 Hz, $^2J_{CP}$ = 33.4 Hz, $^1J_{CPt}$ = 200.3 Hz, C$_{Ar}$$C$HCH), 122.1 (C$_{Ar}$H), 124.3 (C$_{Ar}$H), 125.8 (d, $^4J_{CP}$ = 3.1 Hz, C$_{Ar}$H), 125.8 (dd, $^4J_{CP}$ = 3.1 Hz, $^3J_{CPt}$ = 12.8 Hz,

C_{Ar}H), 127.6 (C_{Ar}H), 128.0 (C_{Ar}H), 128.1 (C_{Ar}H), 128.4 (d, $^4J_{CP}$ = 2.2 Hz, C_{Ar}H), 129.2
(d, $^4J_{CP}$ = 2.1 Hz, C_{Ar}H), 129.5 (d, $^4J_{CP}$ = 1.9 Hz, C_{Ar}H), 133.3 (d, $^4J_{CP}$ = 2.7 Hz, C_{Ar}H),
133.7 (C_{Ar}H), 133.7 (d, $^3J_{CPt}$ = 21.8 Hz, C_{Ar}H), 133.9 (C_{Ar}H), 133.9 (d, $^3J_{CPt}$ = 13.6 Hz,
C_{Ar}H), 133.9 (C_{Ar}H), 133.9 (d, $^3J_{CPt}$ = 17.4 Hz, C_{Ar}H), 134.0 (C_{Ar}H), 134.0 (d,
$^3J_{CPt}$ = 18.6 Hz, C_{Ar}H), 134.6 (d, $^1J_{CP}$ = 3.0 Hz, C_{Ar}P), 134.6 (dd, $^1J_{CP}$ = 3.0 Hz,
$^2J_{CPt}$ = 33.8 Hz, C_{Ar}P), 135.1 (d, $^1J_{CP}$ = 3.0 Hz, C_{Ar}P), 135.1 (dd, $^1J_{CP}$ = 3.0 Hz,
$^2J_{CPt}$ = 33.4 Hz, C_{Ar}P), 136.3 (d, $^1J_{CP}$ = 2.6 Hz, C_{Ar}P), 136.3 (dd, $^1J_{CP}$ = 2.6 Hz,
$^2J_{CPt}$ = 28.2 Hz, C_{Ar}P), 136.7 (d, $^1J_{CP}$ = 2.6 Hz, C_{Ar}P), 136.7 (dd, $^1J_{CP}$ = 2.6 Hz,
$^2J_{CPt}$ = 27.9 Hz, C_{Ar}P), 140.9 (dd, $^3J_{CP}$ = 1.2 Hz, $^3J_{CP}$ = 6.6 Hz, C_{Ar}CH), 140.9
(ddd, $^3J_{CP}$ = 1.2 Hz, $^3J_{CP}$ = 6.6 Hz, $^2J_{CPt}$ = 32.0 Hz, C_{Ar}CH), 144.1 (C_{Ar}O), 156.9
(C_{Ar}N), 170.0 (dd, $^3J_{CP}$ 2.5 Hz, $^3J_{CP}$ = 5.0 Hz, CHCOO), 170.0 (ddd, $^3J_{CP}$ = 2.5 Hz,
$^3J_{CP}$ = 5.0 Hz, $^2J_{CPt}$ = 44.9 Hz, CHCOO).

^{31}P-NMR (161.97 MHz; C_6D_6): δ = 26.8 (d, $^2J_{PP}$ = 40.2 Hz), 26.8 (dd,
$^2J_{PP}$ = 40.2 Hz, $^1J_{PPt}$ = 4248.6 Hz), 27.7 (d, $^2J_{PP}$ = 40.2 Hz), 27.7 (dd, $^2J_{PP}$ = 40.2 Hz,
$^1J_{PPt}$ = 3599.1 Hz).

^{195}Pt-NMR (85.48 MHz; C_6D_6): δ = -5045.1 (dd, $^1J_{PPt}$ = 3598.7 Hz,
$^1J_{PPt}$ = 4246.2 Hz).

FAB-MS: m/z (%): 718.9 (100) [Pt(PPh$_3$)$_2^+$], 1002.0 (1.72) [($C_{16}H_{13}NO_4$)Pt(PPh$_3$)$_2^+$].

FT-IR (ATR, cm^{-1}): 3047 (w), 1713 (s, CO), 1592 (w), 1524 (m), 1486 (w), 1434 (m),
1345 (m), 1228 (w), 1152 (w), 1092 (s), 1001 (w), 925 (w), 804 (w), 744 (m), 692 (s), 577
(w), 540 (m), 516 (s), 499 (s), 430 (m).

elemental analysis for $C_{52}H_{43}NO_4P_2Pt$ (1002.93 *g/mol*):

calcd.: C 62.27, H 4.32, N 1.40, O 6.38, P 6.18, Pt 19.45

found: C 61.53, H 4.52, N 1.24

Bis(triphenylphosphine)-(η^2-4-methoxyphenyl cinnamate)nickel(0)

NiPh-PhOMe: $C_{52}H_{44}NiO_3P_2$, M = 837.54 g/mol, orange red solid, yield: >99%, route: Ni.

^{1}H-NMR (399.78 MHz; C_7D_8): δ = 3.33 (s, 3H, OCH_3), 3.72 (dpt, $^3J_{HH}$ = 9.0 Hz, $^3J_{HP}$ = 3.7 Hz, $^3J_{HP}$ = 9.0 Hz, 1H, CHCHCO), 4.34 (dpt, $^3J_{HH}$ = 9.0 Hz, $^3J_{HP}$ = 5.2 Hz, $^3J_{HP}$ = 9.0 Hz, 1H, CC$_{Ar}$CHCH), 6.55 (d, $^3J_{HH}$ = 7.5 Hz, 2H, H_{Ar}), 6.62 (s, 4H, H_{Ar}), 6.87 - 7.14 (m, 27H, H_{Ar}), 7.53 - 7.65 (m, 6H, H_{Ar}).

^{13}C-NMR (100.53 MHz; C_7D_8): δ = 52.3 (dd, $^2J_{CP}$ = 1.7 Hz, $^2J_{CP}$ = 12.9 Hz, CHCHCO), 54.9 (OCH$_3$), 65.3 (d, $^2J_{CP}$ = 16.3 Hz, C_{Ar}CHCH), 113.9 (C_{Ar}H), 123.2 (C_{Ar}H), 125.4 (d, $^4J_{CP}$ = 2.1 Hz, C_{Ar}H), 128.1 (C_{Ar}H), 128.3 (C_{Ar}H), 128.4 (C_{Ar}H), 129.3 (d, $^3J_{CP}$ = 1.0 Hz, C_{Ar}H), 131.5 (bs, C_{Ar}), 132.4 (d, $^1J_{CP}$ = 9.4 Hz, C_{Ar}), 134.2 (d, $^2J_{CP}$ = 12.4 Hz, C_{Ar}H), 134.4 (d, $^2J_{CP}$ = 13.4 Hz, C_{Ar}H), 134.6 (d, $^1J_{CP}$ = 2.8 Hz, C_{Ar}), 135.0 (d, $^1J_{CP}$ = 2.9 Hz, C_{Ar}), 135.9 (d, $^1J_{CP}$ = 2.3 Hz, C_{Ar}), 136.2 (d, $^1J_{CP}$ = 2.3 Hz, C_{Ar}), 143.6 (d, $^3J_{CP}$ = 5.4 Hz, C_{Ar}), 146.1 (C_{Ar}), 156.8 (C_{Ar}), 171.1 (dd, $^3J_{CP}$ = 1.0 Hz, $^3J_{CP}$ = 3.3 Hz, CHCOO).

^{31}P-NMR (161.97 MHz; C_7D_8): δ = 31.5 (d, $^2J_{PP}$ = 34.4 Hz), 33.3 (d, $^2J_{PP}$ = 34.4 Hz).

FT-IR (ATR, cm^{-1}): 1697 (m, CO).

elemental analysis for $C_{52}H_{44}NiO_3P_2$ (837.54 g/mol):

calcd.: C 74.57, H 5.30, Ni 7.01, O 5.73, P 7.40

found: C 72.27, H 5.21

Bis(triphenylphosphine)-(η^2-4-methoxyphenyl cinnamate)palladium(0)

PdPh-PhOMe: $C_{52}H_{44}O_3P_2Pd$, M $= 885.27$ g/mol, yellow solid, yield: $>99\%$, **route**: Pd-I.

^{1}H-NMR (400.13 MHz; C_7D_8): $\delta = 3.31$ (s, 3H, OCH_3), 4.88 (bs, 1H, CHCHCO), 5.26 (bs, 1H, $C_{Ar}CH$CH), 6.54 - 6.71 (bm, 6H, H_{Ar}), 6.81 - 7.01 (bm, 21H, H_{Ar}), 7.02 - 7.17 (bm, 6H, H_{Ar}), 7.48 - 7.63 (bm, 6H, H_{Ar}).

^{1}H-NMR (399.78 MHz; 246 K; C_7D_8): $\delta = 3.25$ (s, 3H, OCH_3), 5.04 (dpt, $^3J_{HH} = 10.2$ Hz, $^3J_{HP} = 6.0$ Hz, $^3J_{HP} = 6.0$ Hz, 1H, CHCHCO), 5.32 (dpt, $^3J_{HH} = 10.2$ Hz, $^3J_{HP} = 7.1$ Hz, $^3J_{HP} = 7.1$ Hz, 1H, $C_{Ar}CH$CH), 6.56 (bs, 4H, H_{Ar}), 6.58 - 6.62 (m, 2H, H_{Ar}), 6.85 - 6.97 (m, 21H, H_{Ar}), 7.03 - 7.10 (m, 6H, H_{Ar}), 7.55 - 7.62 (m, 6H, H_{Ar}).

^{13}C-NMR (100.62 MHz; C_7D_8): $\delta = 54.8$ (OCH_3), 59.9 - 60.5 (bm, CHCHCO), 75.9 - 76.7 (bm, $C_{Ar}CH$CH), 114.0 ($C_{Ar}H$), 114.8 ($C_{Ar}H$), 123.1 ($C_{Ar}H$), 125.6 - 125.8 (bm, $C_{Ar}H$), 128.2 - 128.3 (bm, $C_{Ar}H$), 128.5 ($C_{Ar}H$), 129.2 ($C_{Ar}H$), 129.4 ($C_{Ar}H$), 133.9 - 134.7 (bm, $C_{Ar}H$), 135.5 - 136.2 (bm, $C_{Ar}H$), 142.7 (bs, C_{Ar}CH), 146.1 (C_{Ar}O), 156.8 (C_{Ar}O), 169.7 (bs, CHCOO).

^{13}C-NMR (100.53 MHz; 246 K; C_7D_8): $\delta = 54.6$ (OCH_3), 59.4 (dd, $^2J_{CP} = 5.9$ Hz, $^2J_{CP} = 17.7$ Hz, CHCHCO), 76.1 (dd, $^2J_{CP} = 4.1$ Hz, $^2J_{CP} = 20.6$ Hz, $C_{Ar}CH$CH), 113.8 ($C_{Ar}H$) 123.2 ($C_{Ar}H$), 125.5 (bs, $C_{Ar}H$), 128.2 (C_{Ar}), 128.5 (C_{Ar}), 129.4 (C_{Ar}), 134.1 (d, $^2J_{CP} = 14.3$ Hz, $C_{Ar}H$), 134.4 (d, $^2J_{CP} = 15.5$ Hz, $C_{Ar}H$), 135.3 (dd, $^1J_{CP} = 28.6$ Hz, $^3J_{CP} = 1.7$ Hz, C_{Ar}P), 137.0 (dd, $^1J_{CP} = 27.5$ Hz, $^3J_{CP} = 1.6$ Hz, C_{Ar}P), 142.3 (d, $^3J_{CP} = 7.0$ Hz, C_{Ar}CH), 145.7 (C_{Ar}O), 156.6 (C_{Ar}O), 169.9 (dd, $^3J_{CP} = 2.1$ Hz, $^3J_{CP} = 3.8$ Hz, CHCOO).

^{31}P-NMR (161.97 MHz; C_7D_8): $\delta = 24.8$ (d, $^2J_{PP} = 16.9$ Hz), 25.9 (d, $^2J_{PP} = 16.9$ Hz).

^{31}P-NMR (161.97 MHz; 246 K; C$_7$D$_8$): δ = 24.1 (d, $^2J_{PP}$ = 15.8 Hz), 25.6 (d, $^2J_{PP}$ = 15.8 Hz).

FT-IR (ATR, cm^{-1}): 3041 (w), 1693 (s, CO), 1606 (w), 1504 (m), 1476 (s), 1433 (s), 1272 (w), 1242 (w), 1196 (m), 1181 (w), 1122 (s), 1092 (s), 1245 (w), 969 (w), 848 (w), 818 (w), 742 (s), 693 (s), 530 (m), 515 (m), 500 (s), 419 (m).

elemental analysis for C$_{52}$H$_{44}$O$_3$P$_2$Pd (885.27 g/mol):

calcd.: C 70.55, H 5.01, O 5.42, P 7.00, Pd 12.02

found: C 70.34, H 5.00

Bis(triphenylphosphine)-(η^2-4-methoxyphenyl cinnamate)platinum(0)

PtPh-PhOMe: C$_{52}$H$_{44}$O$_3$P$_2$Pt, M = 973.93 g/mol, white solid, yield: 71%, **route:** Pt-VI.

^{1}H-NMR (399.78 MHz; C$_6$D$_6$): δ = 3.26 (s, 3H, OCH$_3$), 4.29 (dpt, $^3J_{HH}$ = 8.6 Hz, $^3J_{HP}$ = 4.3 Hz, $^3J_{HP}$ = 8.6 Hz, 1H, CHCHCO), 4.29 (ddpt, $^3J_{HH}$ = 8.6 Hz, $^3J_{HP}$ = 4.3 Hz, $^3J_{HP}$ = 8.6 Hz, $^2J_{HPt}$ = 60.5 Hz, 1H, CHCHCO), 4.52 (dpt, $^3J_{HH}$ = 8.7 Hz, $^3J_{HP}$ = 5.6 Hz, $^3J_{HP}$ = 8.7 Hz, 1H, C$_{Ar}$CHCH), 4.52 (ddpt, $^3J_{HH}$ = 8.7 Hz, $^3J_{HP}$ = 5.6 Hz, $^3J_{HP}$ = 8.7 Hz, $^2J_{HPt}$ = 51.1 Hz, 1H, C$_{Ar}$CHCH), 6.62 (d, $^3J_{HH}$ = 9.2 Hz, 2H, H_{Ar}), 6.64 - 6.69 (m, 2H, H_{Ar}), 6.72 (d, $^3J_{HH}$ = 9.2 Hz, 2H, H_{Ar}), 6.83 - 7.00 (m, 15H, H_{Ar}), 7.16 - 7.24 (m, 6H, H_{Ar}), 7.42 - 7.52 (m, 6H, H_{Ar}), 7.57 - 7.68 (m, 6H, H_{Ar}).

^{13}C-NMR (100.53 MHz; C$_6$D$_6$): δ = 49.6 (dd, $^2J_{CP}$ = 5.7 Hz, $^2J_{CP}$ = 28.4 Hz, CHCHCO), 49.6 (ddd, $^2J_{CP}$ = 5.7 Hz, $^2J_{CP}$ = 28.4 Hz, $^1J_{CPt}$ = 198.3 Hz, CHCHCO), 55.0 (OCH$_3$), 61.7 (dd, $^2J_{CP}$ = 3.7 Hz, $^2J_{CP}$ = 32.6 Hz, C$_{Ar}$CHCH), 61.7 (ddd, $^2J_{CP}$ = 3.7 Hz, $^2J_{CP}$ = 32.6 Hz, $^1J_{CPt}$ = 197.6 Hz, C$_{Ar}$CHCH), 114.2 (C$_{Ar}$H), 123.3 (C$_{Ar}$H), 124.2 (d, $^5J_{CP}$ = 2.5 Hz, C$_{Ar}$H), 124.2 (dd, $^5J_{CP}$ = 2.5 Hz, $^4J_{CPt}$ = 7.4 Hz, C$_{Ar}$H), 126.3 (d,

$^4J_{CP} = 3.1$ Hz, $C_{Ar}H$), 126.3 (dd, $^4J_{CP} = 3.1$ Hz, $^3J_{CPt} = 13.6$ Hz, $C_{Ar}H$), 128.0 ($C_{Ar}H$), 128.1 ($C_{Ar}H$), 128.2 ($C_{Ar}H$), 128.4 $C_{Ar}H$), 128.6 $C_{Ar}H$), 129.4 (d, $^1J_{CPt} = 2.0$ Hz, $C_{Ar}H$), 129.7 (d, $^1J_{CPt} = 1.9$ Hz, $C_{Ar}H$), 134.1 ($C_{Ar}H$), 134.1 (d, $^3J_{CPt} = 21.4$ Hz, $C_{Ar}H$), 134.3 ($C_{Ar}H$), 134.3 (d, $^3J_{CPt} = 20.4$ Hz, $C_{Ar}H$), 134.5 ($C_{Ar}H$), 134.5 (d, $^3J_{CPt} = 18.8$ Hz, $C_{Ar}H$), 134.6 ($C_{Ar}H$), 134.6 (d, $^3J_{CPt} = 18.6$ Hz, $C_{Ar}H$), 135.4 (d, $^1J_{CP} = 3.3$ Hz, $C_{Ar}P$), 135.4 (dd, $^1J_{CP} = 3.3$ Hz, $^2J_{CPt} = 33.2$ Hz, $C_{Ar}P$), 135.8 (d, $^1J_{CP} = 3.2$ Hz, $C_{Ar}P$), 135.8 (dd, $^1J_{CP} = 3.2$ Hz, $^2J_{CPt} = 31.8$ Hz, $C_{Ar}P$), 137.0 (d, $^1J_{CP} = 3.0$ Hz, $C_{Ar}P$), 137.0 (dd, $^1J_{CP} = 3.0$ Hz, $^2J_{CPt} = 29.3$ Hz, $C_{Ar}P$), 137.4 (d, $^1J_{CP} = 3.0$ Hz, $C_{Ar}P$), 137.4 (dd, $^1J_{CP} = 3.0$ Hz, $^2J_{CPt} = 27.1$ Hz, $C_{Ar}P$), 140.3 (bd, $^5J_{CP} = 22.7$ Hz, $C_{Ar}H$), 144.8 (dd, $^3J_{CP} = 1.5$ Hz, $^3J_{CP} = 6.5$ Hz, $C_{Ar}CH$), 144.8 (ddd, $^3J_{CP} = 1.5$ Hz, $^3J_{CP} = 6.5$ Hz, $^2J_{CPt} = 33.1$ Hz, $C_{Ar}CH$), 146.3 ($C_{Ar}O$), 156.9 ($C_{Ar}OCH_3$), 172.0 (dd, $^3J_{CP}$ 2.4 Hz, $^3J_{CP} = 4.9$ Hz, $CHCOO$), 172.0 (ddd, $^3J_{CP} = 2.4$ Hz, $^3J_{CP} = 4.9$ Hz, $^2J_{CPt} = 45.4$ Hz, $CHCOO$).

^{31}P-NMR (161.97 MHz; C_6D_6): $\delta = 28.0$ (d, $^2J_{PP} = 41.8$ Hz), 28.0 (dd, $^2J_{PP} = 41.8$ Hz, $^1J_{PPt} = 4167.6$ Hz), 28.8 (d, $^2J_{PP} = 41.8$ Hz), 28.8 (dd, $^2J_{PP} = 41.8$ Hz, $^1J_{PPt} = 3644.6$ Hz).

^{195}Pt-NMR (85.48 MHz; C_6D_6): $\delta = -5054.2$ (dd, $^1J_{PPt} = 3645.0$ Hz, $^1J_{PPt} = 4168.4$ Hz).

FAB-MS: m/z (%): 718.3 (100) [Pt(PPh$_3$)(PPh$_2$C$_6$H$_4$)$^+$], 972.2 (1.06) [(C$_{16}$H$_{14}$O$_3$)Pt(PPh$_3$)(PPh$_2$C$_6$H$_4$)$^+$].

ESI-MS: m/z (%): 759.1 (100) [Pt(PPh$_3$)(PPh$_2$C$_6$H$_4$)CH$_3$CN$^+$].

FT-IR (ATR, cm^{-1}): 3047 (w), 1698 (m CO), 1577 (w), 1505 (m), 1433 (m), 1320 (w), 1244 (w), 1183 (w), 1121 (s), 1097 (m), 1032 (w), 971 (w), 835 (w), 744 (m), 695 (s), 539 (m), 517 (m), 504 (s), 418 (m).

elemental analysis for C$_{52}$H$_{44}$O$_3$P$_2$Pt (973.93 *g/mol*):

calcd.: C 64.13, H 4.55, O 4.93, P 6.36, Pt 20.03

found: C 62.92, H 4.70

Bis(triphenylphosphine)-(η^2-4-methylphenyl cinnamate)nickel(0)

NiPh-PhMe: $C_{52}H_{44}NiO_2P_2$, M = 821.55 g/mol, orange red solid, yield: 95%, **route:** *Ni.*

^{1}H-NMR (399.78 MHz; C_7D_8): δ = 2.12 (s, 3H, CH_3), 3.71 (dpt, $^3J_{HH}$ = 8.8 Hz, $^3J_{HP}$ = 3.7 Hz, $^3J_{HP}$ = 8.8 Hz, 1H, CHCHCO), 4.34 (dpt, $^3J_{HH}$ = 8.8 Hz, $^3J_{HP}$ = 5.2 Hz, $^3J_{HP}$ = 8.8 Hz, 1H, CC$_{Ar}$CHCH), 6.55 (d, $^3J_{HH}$ = 7.5 Hz, 2H, H_{Ar}), 6.62 (d, $^3J_{HH}$ = 8.2 Hz, 2H, H_{Ar}), 6.85 (d, $^3J_{HH}$ = 8.2 Hz, 2H, H_{Ar}), 6.88 - 7.04 (m, 21H, H_{Ar}), 7.05 - 7.15 (m, 6H, H_{Ar}), 7.54 - 7.65 (m, 6H, H_{Ar}).

^{13}C-NMR (100.53 MHz; C_7D_8): δ = 20.8 (CH$_3$), 52.4 (dd, $^2J_{CP}$ = 1.8 Hz, $^2J_{CP}$ = 12.9 Hz, CHCHCO), 65.2 (d, $^2J_{CP}$ = 16.3 Hz, $C_{Ar}C$HCH), 122.3 (C_{Ar}H), 125.4 (d, $^4J_{CP}$ = 2.1 Hz, C_{Ar}H), 128.1 (C_{Ar}H), 128.3 (C_{Ar}H), 128.4 (C_{Ar}H), 129.3 (d, $^3J_{CP}$ = 1.2 Hz, C_{Ar}H), 129.4 (C_{Ar}H), 131.5 (bs, C_{Ar}), 132.4 (bs, C_{Ar}), 134.2 (d, $^2J_{CP}$ = 12.4 Hz, C_{Ar}H), 134.4 (d, $^2J_{CP}$ = 13.4 Hz, C_{Ar}H), 134.6 (d, $^1J_{CP}$ = 2.9 Hz, C_{Ar}), 135.0 (d, $^1J_{CP}$ = 2.9 Hz, C_{Ar}), 135.9 (d, $^1J_{CP}$ = 2.3 Hz, C_{Ar}), 136.2 (d, $^1J_{CP}$ = 2.3 Hz, C_{Ar}), 143.6 (d, $^3J_{CP}$ = 5.4 Hz, C_{Ar}), 150.5 (C_{Ar}), 170.9 (dd, $^3J_{CP}$ = 1.2 Hz, $^3J_{CP}$ = 3.5 Hz, CHCOO).

^{31}P-NMR (161.97 MHz; C_7D_8): δ = 32.0 (d, $^2J_{PP}$ = 34.3 Hz), 33.8 (d, $^2J_{PP}$ = 34.3 Hz).

FT-IR (ATR, cm^{-1}): 1692 (m, CO).

Bis(triphenylphosphine)-(η^2-4-methylphenyl cinnamate)palladium(0)

PdPh-PhMe: $C_{52}H_{44}O_2P_2Pd$, M $= 869.27$ g/mol, yellow solid, yield: 98%, **route**: *Pd-I*.

^{1}H-NMR (400.13 MHz; C_7D_8): $\delta = 2.09$ (s, 3H, CH_3), 4.86 (bs, 1H, CHCHCO), 5.25 (bs, 1H, C_{Ar}CHCH), 6.57 - 6.72 (bm, 4H, H_{Ar}), 6.77 - 7.02 (bm, 23H, H_{Ar}), 7.02 - 7.16 (bm, 6H, H_{Ar}), 7.49 - 7.65 (bm, 6H, H_{Ar}).

^{1}H-NMR (399.78 MHz; 246 K; C_7D_8): $\delta = 2.11$ (s, 3H, CH_3), 5.03 (dpt, $^3J_{HH} = 10.2$ Hz, $^3J_{HP} = 6.1$ Hz, $^3J_{HP} = 6.1$ Hz, 1H, CHCHCO), 5.32 (dpt, $^3J_{HH} = 10.2$ Hz, $^3J_{HP} = 7.0$ Hz, $^3J_{HP} = 7.0$ Hz, 1H, C_{Ar}CHCH), 6.55 - 6.63 (m, 4H, H_{Ar}), 6.78 - 6.83 (m, 2H, H_{Ar}), 6.85 - 6.99 (m, 21H, H_{Ar}), 7.03 - 7.10 (m, 6H, H_{Ar}), 7.56 - 7.63 (m, 6H, H_{Ar}).

^{13}C-NMR (100.62 MHz; C_7D_8): $\delta = 20.7$ (CH_3), 59.9 - 60.5 (bm, CHCHCO), 75.9 - 76.8 (bm, C_{Ar}CHCH), 122.2 (C_{Ar}H), 125.7 (C_{Ar}H), 128.2 - 128.3 (bm, C_{Ar}H), 128.5 (C_{Ar}H), 129.2 (C_{Ar}H), 129.3 (C_{Ar}H), 129.4 (C_{Ar}H), 133.9 - 134.8 (bm, C_{Ar}H), 135.5 - 136.2 (bm, C_{Ar}H), 142.7 (bs, C_{Ar}CH), 150.6 (C_{Ar}O), 169.5 (bs, CHCOO).

^{13}C-NMR (100.53 MHz; 246 K; C_7D_8): $\delta = 20.8$ (CH_3), 59.5 (dd, $^2J_{CP} = 5.1$ Hz, $^2J_{CP} = 17.3$ Hz, CHCHCO), 76.2 (dd, $^2J_{CP} = 1.5$ Hz, $^2J_{CP} = 21.9$ Hz, C_{Ar}CHCH), 122.3 (C_{Ar}H), 124.7 (bs, C_{Ar}H), 125.5 (bs C_{Ar}H), 128.2 (C_{Ar}), 128.4 (C_{Ar}), 129.3 (C_{Ar}), 129.4 (C_{Ar}), 133.4 (C_{Ar}CH$_3$), 134.1 (d, $^2J_{CP} = 14.3$ Hz, C_{Ar}H), 134.4 (d, $^2J_{CP} = 15.3$ Hz, C_{Ar}H), 135.1 - 135.5 (m, C_{Ar}P), 136.9 - 137.2 (m, C_{Ar}P), 142.3 (d, $^3J_{CP} = 6.4$ Hz, C_{Ar}CH), 150.1 (C_{Ar}O), 169.5 (dd, $^3J_{CP} = 2.6$ Hz, $^3J_{CP} = 3.9$ Hz, CHCOO).

^{31}P-NMR (161.97 MHz; C_7D_8): $\delta = 24.8$ (d, $^2J_{PP} = 16.5$ Hz), 25.9 (d, $^2J_{PP} = 16.5$ Hz).

^{31}P-NMR (161.97 MHz; 246 K; C_7D_8): $\delta = 24.1$ (d, $^2J_{PP} = 15.8$ Hz), 25.6 (d, $^2J_{PP} = 15.8$ Hz).

FAB-MS: m/z (%): 367.9 (17.74) [Pd(PPh$_3$)$^+$], 630.0 (34.29) [Pd(PPh$_3$)$_2$$^+$].

FT-IR (ATR, cm^{-1}): 3046 (m), 1698 (s, CO), 1591 (w), 1500 (w), 1477 (m), 1432 (s), 1318 (w), 1273 (w), 1199 (m), 1123 (s), 1092 (m), 970 (m), 849 (w), 743 (s), 692 (s), 530 (m), 515 (m), 500 (s), 417 (m).

elemental analysis for C$_{52}$H$_{44}$O$_2$P$_2$Pd (869.27 g/mol):

calcd.: C 71.85, H 5.10, O 3.68, P 7.13, Pd 12.24

found: C 71.91, H 5.16

Bis(triphenylphosphine)-(η^2-4-methylphenyl cinnamate)platinum(0)

PtPh-PhMe: C$_{52}$H$_{44}$O$_2$P$_2$Pt, M = 957.93 g/mol, white solid, yield: 89%, **route:** *Pt-VI.*

Single crystals suitable for X-ray analysis were obtained by slow diffusion of pentane into a toluene solution.

^{1}H-NMR (399.78 MHz; C$_6$D$_6$): δ = 2.06 (s, 3H, CH$_3$), 4.28 (dpt, $^3J_{HH}$ = 8.7 Hz, $^3J_{HP}$ = 4.2 Hz, $^3J_{HP}$ = 8.7 Hz, 1H, CHCHCO), 4.28 (ddpt, $^3J_{HH}$ = 8.7 Hz, $^3J_{HP}$ = 4.2Hz, $^3J_{HP}$ = 8.7 Hz, $^2J_{HPt}$ = 61.0 Hz, 1H, CHCHCO), 4.52 (dpt, $^3J_{HH}$ = 8.7 Hz, $^3J_{HP}$ = 5.6 Hz, $^3J_{HP}$ = 8.7 Hz, 1H, C$_{Ar}$CHCH), 4.52 (ddpt, $^3J_{HH}$ = 8.7 Hz, $^3J_{HP}$ = 5.6 Hz, $^3J_{HP}$ = 8.7 Hz, $^2J_{HPt}$ = 51.7 Hz, 1H, C$_{Ar}$CHCH), 6.65 (d, $^3J_{HH}$ = 8.2 Hz, 2H, H$_{Ar}$), 6.73 (d, $^3J_{HH}$ = 8.2 Hz, 2H, H$_{Ar}$), 6.80 - 6.98 (m, 17H, H$_{Ar}$), 7.15 - 7.23 (m, 6H, H$_{Ar}$), 7.44 - 7.52 (m, 6H, H$_{Ar}$), 7.58 - 767 (m, 6H, H$_{Ar}$).

^{13}C-NMR (100.53 MHz; C$_6$D$_6$): δ = 20.8 (CH$_3$), 49.6 (dd, $^2J_{CP}$ = 5.7 Hz, $^2J_{CP}$ = 28.2 Hz, CHCHCO), 49.6 (ddd, $^2J_{CP}$ = 5.7 Hz, $^2J_{CP}$ = 28.2 Hz, $^1J_{CPt}$ = 199.1 Hz, CHCHCO), 61.7 (dd, $^2J_{CP}$ = 3.7 Hz, $^2J_{CP}$ = 32.7 Hz, C$_{Ar}$CHCH), 61.7 (ddd, $^2J_{CP}$ = 3.7 Hz, $^2J_{CP}$ = 32.7 Hz, $^1J_{CPt}$ = 198.4 Hz, C$_{Ar}$CHCH), 122.4 (C$_{Ar}$H), 124.2 (d, $^5J_{CP}$ = 2.5 Hz, C$_{Ar}$H), 124.2 (dd, $^5J_{CP}$ = 2.5 Hz, $^4J_{CPt}$ = 7.0 Hz, C$_{Ar}$H), 126.3 (d,

$^4J_{CP}$ = 3.2 Hz, C_{Ar}H), 126.3 (dd, $^4J_{CP}$ = 3.2 Hz, $^3J_{CPt}$ = 13.2 Hz, C_{Ar}H), 128.0 (C_{Ar}H), 128.1 (C_{Ar}H), 128.4 (C_{Ar}H), 128.6 (C_{Ar}H), 129.4 (d, $^1J_{CPt}$ = 2.0 Hz, C_{Ar}H), 129.7 (d, $^1J_{CPt}$ = 1.9 Hz, C_{Ar}H), 129.5 (C_{Ar}H), 134.1 (C_{Ar}H), 134.1 (d, $^3J_{CPt}$ = 21.4 Hz, C_{Ar}H), 134.3 (C_{Ar}H), 134.3 (d, $^3J_{CPt}$ = 21.0 Hz, C_{Ar}H), 134.4 (C_{Ar}H), 134.4 (d, $^3J_{CPt}$ = 18.6 Hz, C_{Ar}H), 134.5 (C_{Ar}H), 134.5 (d, $^3J_{CPt}$ = 18.8 Hz, C_{Ar}H), 135.4 (d, $^1J_{CP}$ = 3.3 Hz, C_{Ar}P), 135.4 (dd, $^1J_{CP}$ = 3.3 Hz, $^2J_{CPt}$ = 33.9 Hz, C_{Ar}P), 135.8 (d, $^1J_{CP}$ = 3.3 Hz, C_{Ar}P), 135.8 (dd, $^1J_{CP}$ = 3.3 Hz, $^2J_{CPt}$ = 32.2 Hz, C_{Ar}P), 137.0 (d, $^1J_{CP}$ = 3.0 Hz, C_{Ar}P), 137.0 (dd, $^1J_{CP}$ = 3.0 Hz, $^2J_{CPt}$ = 29.3 Hz, C_{Ar}P), 137.4 (d, $^1J_{CP}$ = 3.0 Hz, C_{Ar}P), 137.4 (dd, $^1J_{CP}$ = 3.0 Hz, $^2J_{CPt}$ = 28.7 Hz, C_{Ar}P), 140.4 (bd, $^5J_{CP}$ = 21.2 Hz, C_{Ar}H), 144.8 (dd, $^3J_{CP}$ = 1.4 Hz, $^3J_{CP}$ = 6.5 Hz, C_{Ar}CH), 144.8 (ddd, $^3J_{CP}$ = 1.4 Hz, $^3J_{CP}$ = 6.5 Hz, $^2J_{CPt}$ = 32.3 Hz, C_{Ar}CH), 150.6 (C_{Ar}O), 171.8 (dd, $^3J_{CP}$ 2.3 Hz, $^3J_{CP}$ = 5.0 Hz, CHCOO), 171.8 (ddd, $^3J_{CP}$ = 2.3 Hz, $^3J_{CP}$ = 5.0 Hz, $^2J_{CPt}$ = 48.4 Hz, CHCOO).

^{31}P-NMR (161.97 MHz; C_6D_6): δ = 28.0 (d, $^2J_{PP}$ = 41.6 Hz), 28.0 (dd, $^2J_{PP}$ = 41.6 Hz, $^1J_{PPt}$ = 4168.4 Hz), 28.8 (d, $^2J_{PP}$ = 41.6 Hz), 28.8 (dd, $^2J_{PP}$ = 41.6 Hz, $^1J_{PPt}$ = 3644.7 Hz).

^{195}Pt-NMR (85.48 MHz; C_6D_6): δ = -5053.4 (dd, $^1J_{PPt}$ = 3645.1 Hz, $^1J_{PPt}$ = 4169.8 Hz).

FAB-MS: m/z (%): 718.3 (100) [Pt(PPh$_3$)(PPh$_2$C$_6$H$_4$)$^+$], 956.2 (0.84) [(C$_{16}$H$_{14}$O$_2$)Pt(PPh$_3$)(PPh$_2$C$_6$H$_4$)$^+$].

ESI-MS: m/z (%): 759.1 (100) [Pt(PPh$_3$)(PPh$_2$C$_6$H$_4$)CH$_3$CN$^+$].

FT-IR (ATR, cm^{-1}): 3052 (w), 1703 (s, CO), 1592 (w), 1501 (w), 1470 (w), 1433 (m), 1304 (w), 1273 (w), 1213 (w), 1182 (w), 1121 (s), 1094 (m), 970 (w), 895 (w), 819 (w), 743 (m), 694 (s), 535 (m), 516 (m), 503 (s), 438 (w), 417 (m).

elemental analysis for C$_{52}$H$_{44}$O$_2$P$_2$Pt (957.93 *g/mol*):

calcd.: C 65.20, H 4.63, O 3.34, P 6.47, Pt 20.36

found: C 64.23, H 4.71

Bis(triphenylphosphine)-(η^2-phenyl cinnamate)nickel(0)

NiPh-Ph: $C_{51}H_{42}NiO_2P_2$, M = 807.52 g/mol, yellow orange solid, yield: >99%, **route**: *Ni*.

^{1}H-NMR (399.78 MHz; C_7D_8): δ = 3.71 (dpt, $^3J_{HH}$ = 8.9 Hz, $^3J_{HP}$ = 3.7 Hz, $^3J_{HP}$ = 8.9 Hz, 1H, CH*CH*CO), 4.32 (dpt, $^3J_{HH}$ = 8.9 Hz, $^3J_{HP}$ = 5.3 Hz, $^3J_{HP}$ = 8.9 Hz, 1H, CC$_{Ar}$*CH*CH), 6.55 (d, $^3J_{HH}$ = 7.5 Hz, 2H, H_{Ar}), 6.71 (d, $^3J_{HH}$ = 7.9 Hz, 2H, H_{Ar}), 6.82 - 7.20 (m, 29H, H_{Ar}), 7.49 - 7.65 (m, 6H, H_{Ar}).

^{13}C-NMR (100.53 MHz; C_7D_8): δ = 52.2 (dd, $^2J_{CP}$ = 1.7 Hz, $^2J_{CP}$ = 12.8 Hz, CH*CH*CO), 65.1 (d, $^2J_{CP}$ = 16.4 Hz, C$_{Ar}$*CH*CH), 122.5 (C_{Ar}H), 125.4 (d, $^4J_{CP}$ = 2.0 Hz, C_{Ar}H), 128.1 (C_{Ar}H), 128.3 (C_{Ar}H), 128.4 (C_{Ar}H), 128.8 (C_{Ar}H), 129.3 (d, $^3J_{CP}$ = 1.1 Hz, C_{Ar}H), 131.5 (bs, C_{Ar}), 132.4 (bs, C_{Ar}), 134.2 (d, $^2J_{CP}$ = 12.4 Hz, C_{Ar}H), 134.4 (d, $^2J_{CP}$ = 13.3 Hz, C_{Ar}H), 134.6 (d, $^1J_{CP}$ = 2.9 Hz, C_{Ar}), 134.9 (d, $^1J_{CP}$ = 2.9 Hz, C_{Ar}), 135.8 (d, $^1J_{CP}$ = 2.3 Hz, C_{Ar}), 136.2 (d, $^1J_{CP}$ = 2.3 Hz, C_{Ar}), 143.5 (d, $^3J_{CP}$ = 5.4 Hz, C_{Ar}), 152.6 (C_{Ar}), 170.7 (dd, $^3J_{CP}$ = 1.1 Hz, $^3J_{CP}$ = 3.5 Hz, CH*C*OO).

^{31}P-NMR (161.97 MHz; C_7D_8): δ = 31.2 (d, $^2J_{PP}$ = 34.1 Hz), 32.9 (d, $^2J_{PP}$ = 34.1 Hz).

FT-IR (ATR, cm^{-1}): 1708 (m, CO).

Bis(triphenylphosphine)-(η^2-phenyl cinnamate)palladium(0)

PdPh-Ph: $C_{51}H_{42}O_2P_2Pd$, M = 855.25 g/mol, light yellow solid, yield: 94%, **route**: *Pd-I*.

^{1}H-NMR (400.13 MHz; C_7D_8): $\delta = 4.85$ (bs, 1H, CHCHCO), 5.24 (bs, 1H, C_{Ar}CHCH), 6.54 - 6.67 (bm, 2H, H_{Ar}), 6.70 - 6.83 (bm, 2H, H_{Ar}), 6.83 - 7.03 (bm, 24H, H_{Ar}), 7.03 - 7.17 (bm, 6H, H_{Ar}), 7.46 - 7.64 (bm, 6H, H_{Ar}).

^{1}H-NMR (399.78 MHz; 246 K; C_7D_8): $\delta = 5.02$ (dpt, $^3J_{HH} = 10.2$ Hz, $^3J_{HP} = 6.0$ Hz, $^3J_{HP} = 6.0$ Hz, 1H, CHCHCO), 5.30 (dpt, $^3J_{HH} = 10.2$ Hz, $^3J_{HP} = 7.0$ Hz, $^3J_{HP} = 7.0$ Hz, 1H, C_{Ar}CHCH), 6.56 - 6.62 (m, 2H, H_{Ar}), 6.63 - 6.68 (m, 2H, H_{Ar}), 6.85 - 7.02 (m, 24H, H_{Ar}), 7.03 - 7.09 (m, 6H, H_{Ar}), 7.55 - 7.62 (m, 6H, H_{Ar}).

^{13}C-NMR (100.62 MHz; C_7D_8): $\delta = 59.7 - 60.3$ (bm, CHCHCO), 75.9 - 76.5 (bm, $C_{Ar}$$C$HCH), 122.4 ($C_{Ar}$H), 124.3 ($C_{Ar}$H), 125.7 ($C_{Ar}$H), 128.1 - 128.2 (bm, C_{Ar}H), 128.2 - 128.3 (bm, C_{Ar}H), 128.5 (C_{Ar}H), 129.1 (C_{Ar}H), 129.4 (C_{Ar}H), 134.0 - 134.7 (bm, C_{Ar}H), 135.5 - 136.1 (bm, C_{Ar}H), 142.5 (bs, C_{Ar}CH), 152.5 (C_{Ar}O), 169.2 (bs, CHCOO).

^{13}C-NMR (100.53 MHz; 246 K; C_7D_8): $\delta = 59.3$ (dd, $^2J_{CP} = 5.7$ Hz, $^2J_{CP} = 18.0$ Hz, CHCHCO), 76.1 (dd, $^2J_{CP} = 3.2$ Hz, $^2J_{CP} = 23.7$ Hz, $C_{Ar}$$C$HCH), 122.5 ($C_{Ar}$H), 124.4 (bs, C_{Ar}H), 125.5 (C_{Ar}H), 128.2 (C_{Ar}), 128.5 (C_{Ar}), 128.9 (C_{Ar}), 129.4 (C_{Ar}), 134.1 (d, $^2J_{CP} = 14.4$ Hz, C_{Ar}H), 134.3 (d, $^2J_{CP} = 15.4$ Hz, C_{Ar}H), 135.3 (dd, $^1J_{CP} = 28.6$ Hz, $^3J_{CP} = 1.5$ Hz, C_{Ar}P), 137.0 (dd, $^1J_{CP} = 27.6$ Hz, $^3J_{CP} = 1.5$ Hz, C_{Ar}P), 142.3 (d, $^3J_{CP} = 6.5$ Hz, C_{Ar}CH), 152.3 (C_{Ar}O), 169.3 (dd, $^3J_{CP} = 1.6$ Hz, $^3J_{CP} = 3.5$ Hz, CHCOO).

^{31}P-NMR (161.97 MHz; C_7D_8): $\delta = 24.8$ (d, $^2J_{PP} = 17.0$ Hz), 25.9 (d, $^2J_{PP} = 17.0$ Hz).

^{31}P-NMR (161.97 MHz; 246 K; C_7D_8): $\delta = 24.1$ (d, $^2J_{PP} = 15.5$ Hz), 25.6 (d, $^2J_{PP} = 15.5$ Hz).

FT-IR (ATR, cm^{-1}): 3059 (m), 1710 (s, CO), 1588 (w), 1471 (m), 1433 (s), 1316 (w), 1255 (w), 1197 (m), 1097 (s), 952 (w), 922 (w), 740 (m), 692 (s), 527 (m), 514 (m), 495 (s), 407 (m).

elemental analysis for $C_{51}H_{42}O_2P_2Pd$ (855.25 g/mol):

calcd.: C 71.62, H 4.95, O 3.74, P 7.24, Pd 12.44

found: C 71.58, H 5.08

Bis(triphenylphosphine)-(η^2-phenyl cinnamate)platinum(0)

PtPh-Ph: $C_{51}H_{42}O_2P_2Pt$, M = 943.90 g/mol, white solid, yield: 77%, **route**: *Pt-VI*.

Single crystals suitable for X-ray analysis were obtained by slow diffusion of pentane into a toluene solution.

^{1}H-NMR (399.78 MHz; C_6D_6): δ = 4.27 (dpt, $^3J_{HH}$ = 8.7 Hz, $^3J_{HP}$ = 4.3 Hz, $^3J_{HP}$ = 8.7 Hz, 1H, CHCHCO), 4.27 (ddpt, $^3J_{HH}$ = 8.7 Hz, $^3J_{HP}$ = 4.3 Hz, $^3J_{HP}$ = 8.7 Hz, $^2J_{HPt}$ = 60.6 Hz, 1H, CHCHCO), 4.50 (dpt, $^3J_{HH}$ = 8.7 Hz, $^3J_{HP}$ = 5.5 Hz, $^3J_{HP}$ = 8.7 Hz, 1H, $C_{Ar}C$HCH), 4.50 (ddpt, $^3J_{HH}$ = 8.7 Hz, $^3J_{HP}$ = 5.5 Hz, $^3J_{HP}$ = 8.7 Hz, $^2J_{HPt}$ = 51.1 Hz, 1H, $C_{Ar}C$HCH), 6.64 (d, $^3J_{HH}$ = 8.2 Hz, 2H, H_{Ar}), 6.79 - 7.04 (m, 18H, H_{Ar}), 7.14 - 7.22 (m, 6H, H_{Ar}), 7.41 - 7.52 (m, 8H, H_{Ar}), 7.55 - 7.66 (m, 6H, H_{Ar}).

^{13}C-NMR (100.53 MHz; C_6D_6): δ = 49.5 (dd, $^2J_{CP}$ = 6.0 Hz, $^2J_{CP}$ = 28.3 Hz, CHCHCO), 49.5 (ddd, $^2J_{CP}$ = 6.0 Hz, $^2J_{CP}$ = 28.3 Hz, $^1J_{CPt}$ = 196.0 Hz, CHCHCO), 61.7 (dd, $^2J_{CP}$ = 3.8 Hz, $^2J_{CP}$ = 32.8 Hz, C_{Ar}CHCH), 61.7 (ddd, $^2J_{CP}$ = 3.8 Hz, $^2J_{CP}$ = 32.8 Hz, $^1J_{CPt}$ = 194.5 Hz, C_{Ar}CHCH), 122.6 (C_{Ar}H), 124.2 (d, $^5J_{CP}$ = 2.5 Hz, C_{Ar}H), 124.2 (dd, $^5J_{CP}$ = 2.5 Hz, $^4J_{CPt}$ = 6.8 Hz, C_{Ar}H), 124.4 (C_{Ar}), 126.3 (d, $^4J_{CP}$ = 3.2 Hz, C_{Ar}H), 126.3 (dd, $^4J_{CP}$ = 3.2 Hz, $^3J_{CPt}$ = 13.5 Hz, C_{Ar}H), 128.0 (C_{Ar}H), 128.2 (C_{Ar}H), 128.4 (C_{Ar}H), 128.6 (C_{Ar}H), 129.4 (d, $^1J_{CPt}$ = 2.0 Hz, C_{Ar}H), 129.7 (d, $^1J_{CPt}$ = 1.9 Hz, C_{Ar}H), 134.1 (C_{Ar}H), 134.1 (d, $^3J_{CPt}$ = 21.6 Hz, C_{Ar}H), 134.3 (C_{Ar}H), 134.3 (d, $^3J_{CPt}$ = 21.2 Hz, C_{Ar}H), 134.4 (C_{Ar}H), 134.4 (d, $^3J_{CPt}$ = 18.8 Hz, C_{Ar}H), 134.5 (C_{Ar}H), 134.5 (d, $^3J_{CPt}$ = 18.4 Hz, C_{Ar}H), 135.3 (d, $^1J_{CP}$ = 3.1 Hz, C_{Ar}P), 135.3 (dd, $^1J_{CP}$ = 3.1 Hz, $^2J_{CPt}$ = 34.5 Hz, C_{Ar}P), 135.7 (d, $^1J_{CP}$ = 3.1 Hz, C_{Ar}P), 135.7 (dd, $^1J_{CP}$ = 3.1 Hz, $^2J_{CPt}$ = 33.2 Hz, C_{Ar}P), 136.9 (d, $^1J_{CP}$ = 2.9 Hz, C_{Ar}P), 136.9 (dd, $^1J_{CP}$ = 2.9 Hz, $^2J_{CPt}$ = 29.3 Hz, C_{Ar}P), 137.3 (d, $^1J_{CP}$ = 2.9 Hz, C_{Ar}P), 137.3 (dd, $^1J_{CP}$ = 2.9 Hz, $^2J_{CPt}$ = 28.3 Hz, C_{Ar}P), 140.3 (bd, $^5J_{CP}$ = 17.7 Hz, C_{Ar}H), 144.7 (dd, $^3J_{CP}$ = 1.4 Hz, $^3J_{CP}$ = 6.5 Hz, C_{Ar}CH), 144.7 (ddd, $^3J_{CP}$ = 1.4 Hz, $^3J_{CP}$ = 6.5 Hz,

$^2J_{CPt} = 33.1$ Hz, C_{Ar}CH), 152.8 (C_{Ar}O), 171.5 (dd, $^3J_{CP}$ 2.3 Hz, $^3J_{CP} = 5.0$ Hz, CHCOO),
171.5 (ddd, $^3J_{CP} = 2.4$ Hz, $^3J_{CP} = 5.0$ Hz, $^2J_{CPt} = 47.7$ Hz, CHCOO).

^{31}P-NMR (161.97 MHz; C$_6$D$_6$): δ = 27.9 (d, $^2J_{PP}$ = 41.3 Hz), 27.9 (dd, $^2J_{PP}$ = 41.3 Hz, $^1J_{PPt}$ = 4175.0 Hz), 28.8 (d, $^2J_{PP}$ = 41.4 Hz), 28.8 (dd, $^2J_{PP}$ = 41.4 Hz, $^1J_{PPt}$ = 3641.6 Hz).

^{195}Pt-NMR (85.48 MHz; C$_6$D$_6$): δ = -5052.5 (dd, $^1J_{PPt}$ = 3642.0 Hz, $^1J_{PPt}$ = 4176.1 Hz).

FAB-MS: m/z (%): 718.3 (100) [Pt(PPh$_3$)(PPh$_2$C$_6$H$_4$)$^+$], 942.2 (1.08) [(C$_{15}$H$_{12}$O$_2$)Pt(PPh$_3$)(PPh$_2$C$_6$H$_4$)$^+$].

ESI-MS: m/z (%): 759.1 (100) [Pt(PPh$_3$)(PPh$_2$C$_6$H$_4$)CH$_3$CN$^+$].

FT-IR (ATR, cm^{-1}): 3050 (w), 1717 (s, CO), 1582 (w), 1468 (m), 1434 (s), 1408 (m), 1317 (w), 1256 (w), 1196 (m), 1143 (m), 1090 (s), 976 (w), 923 (m), 893 (w), 741 (s), 681 (s), 574 (w), 539 (m), 517 (m), 507 (s), 498 (s), 423 (s).

elemental analysis for C$_{51}$H$_{42}$O$_2$P$_2$Pt (943.90 *g/mol*):

calcd.: C 64.89, H 4.48, O 3.39, P 6.56, Pt 20.67

found: C 63.94, H 4.57

Bis(triphenylphosphine)-(η^2-4-nitrophenyl cinnamate)palladium(0)

PdPh-PhNO$_2$: C$_{51}$H$_{41}$NO$_4$P$_2$Pd, M = 900.24 *g/mol*, deep yellow solid, yield: 93%, **route**: *Pd-I.*

^{1}H-NMR (400.13 MHz; C$_7$D$_8$): δ = 4.70 - 4.82 (bm, 1H, CHCHHCO), 5.08 - 5.20 (bm, 1H, C$_{Ar}$CHCH), 6.54 - 6.65 (bm, 4H, H_{Ar}), 6.83 - 7.10 (bm, 27H, H_{Ar}), 7.40 - 7.51 (bm, 6H, H_{Ar}), 7.70 - 7.78 (bm, 2H, H_{Ar}).

^{1}H-NMR (399.78 MHz; 246 K; C_7D_8): $\delta = 4.93$ (ddd, $^3J_{HH} = 10.2$ Hz, $^3J_{HP} = 4.5$ Hz, $^3J_{HP} = 6.3$ Hz, 1H, CHCHCO), 5.19 (dpt, $^3J_{HH} = 10.2$ Hz, $^3J_{HP} = 7.0$ Hz, $^3J_{HP} = 7.0$ Hz, 1H, $C_{Ar}CH$CH), 6.41 - 6.45 (m, 2H, H_{Ar}), 6.55 - 6.60 (m, 2H, H_{Ar}), 6.84 - 7.05 (m, 27H, H_{Ar}), 7.45 - 7.51 (m, 6H, H_{Ar}), 7.68 - 7.72 (m, 2H, H_{Ar}).

^{13}C-NMR (100.62 MHz; C_7D_8): $\delta = 58.7$ (d, $^2J_{CP} = 10.7$ Hz, CHCHCO), 75.3 (d, $^2J_{CP} = 22.6$ Hz, $C_{Ar}CH$CH), 122.2 (C_{Ar}H), 124.5 (C_{Ar}H), 125.8 (C_{Ar}H), 128.2 (C_{Ar}H), 128.3 (C_{Ar}H), 128.5 (C_{Ar}H), 129.2 (C_{Ar}H), 129.2 (C_{Ar}H), 129.6 (C_{Ar}H), 134.0 - 134.6 (bm, C_{Ar}H), 135.4 (d, $^1J_{CP} = 30.5$ Hz, C_{Ar}H), 136.8 (d, $^1J_{CP} = 27.8$ Hz, C_{Ar}H), 142.2 (d, $^3J_{CP} = 6.0$ Hz, C_{Ar}CH), 144.5 (C_{Ar}N), 157.1 (C_{Ar}O), 168.0 (bs, CHCOO).

^{13}C-NMR (100.53 MHz; 246 K; C_7D_8): $\delta = 58.1$ (dd, $^2J_{CP} = 5.9$ Hz, $^2J_{CP} = 17.8$ Hz, CHCHCO), 75.1 (dd, $^2J_{CP} = 1.0$ Hz, $^2J_{CP} = 22.9$ Hz, $C_{Ar}CH$CH), 122.3 (C_{Ar}H), 124.5 (C_{Ar}H), 125.5 (bs, C_{Ar}H), 128.3 (C_{Ar}), 129.2 (C_{Ar}), 129.6 (C_{Ar}), 134.0 (d, $^2J_{CP} = 14.2$ Hz, C_{Ar}H), 134.2 (d, $^2J_{CP} = 15.3$ Hz, C_{Ar}H), 134.7 - 135.1 (m, C_{Ar}P), 136.4 - 136.8 (m, C_{Ar}P), 141.9 (d, $^3J_{CP} = 6.4$ Hz, C_{Ar}CH), 144.2 (C_{Ar}N), 156.9 (C_{Ar}O), 168.0 - 168.1 (m, CHCOO).

^{31}P-NMR (161.97 MHz; C_7D_8): $\delta = 25.0$ (d, $^2J_{PP} = 13.4$ Hz), 25.6 (d, $^2J_{PP} = 13.4$ Hz).

^{31}P-NMR (161.83 MHz; 246 K; C_7D_8): $\delta = 24.2$ (d, $^2J_{PP} = 12.1$ Hz), 25.3 (d, $^2J_{PP} = 12.1$ Hz).

FT-IR (ATR, cm^{-1}): 3043 (m), 1695 (s, CO), 1605 (m), 1519 (s), 1476 (s), 1433 (s), 1214 (s), 1089 (s), 983 (m), 938 (m), 862 (m), 741 (s), 692 (s), 526 (m), 499 (s), 423 (m).

elemental analysis for $C_{51}H_{41}NO_4P_2Pd$ (900.24 g/mol):

calcd.: C 68.04, H 4.59, N 1.56, O 7.11, P 6.88, Pd 11.82

found: C 67.91, H 4.59, N 1.38

Bis(triphenylphosphine)-(η^2-4-nitrophenyl cinnamate)platinum(0)

PtPh-PhNO$_2$: C$_{51}$H$_{41}$NO$_4$P$_2$Pt, M $=$ 988.90 g/mol, white solid, yield: 91%, **route:** *Pt-VI.*

^{1}H-NMR (399.78 MHz; C$_6$D$_6$): δ $=$ 4.18 (dpt, $^3J_{HH}$ $=$ 8.7 Hz, $^3J_{HP}$ $=$ 3.9 Hz, $^3J_{HP}$ $=$ 8.7 Hz, 1H, CHC*H*CO), 4.18 (ddpt, $^3J_{HH}$ $=$ 8.7 Hz, $^3J_{HP}$ $=$ 3.9 Hz, $^3J_{HP}$ $=$ 8.7 Hz, $^2J_{HPt}$ $=$ 60.4 Hz, 1H, CHC*H*CO), 4.40 (dpt, $^3J_{HH}$ $=$ 8.7 Hz, $^3J_{HP}$ $=$ 5.5 Hz, $^3J_{HP}$ $=$ 8.7 Hz, 1H, C$_{Ar}$C*H*CH), 4.40 (ddpt, $^3J_{HH}$ $=$ 8.7 Hz, $^3J_{HP}$ $=$ 5.5 Hz, $^3J_{HP}$ $=$ 8.7 Hz, $^2J_{HPt}$ $=$ 51.9 Hz, 1H, C$_{Ar}$C*H*CH), 6.57(d, $^3J_{HH}$ $=$ 8.9 Hz, 2H, H_{Ar}), 6.64 (d, $^3J_{HH}$ $=$ 7.1 Hz, 2H, H_{Ar}), 6.81 - 7.00 (m, 22H, H_{Ar}), 7.15 (t, $^3J_{HH}$ $=$ 9.0 Hz, 5H, H_{Ar}), 7.40 - 7.48 (m, 3H, H_{Ar}), 7.48 - 7.56 (m, 3H, H_{Ar}), 7.74 (d, $^3J_{HH}$ $=$ 9.0 Hz, 2H, H_{Ar}).

^{13}C-NMR (100.53 MHz; C$_6$D$_6$): δ $=$ 49.0 (dd, $^2J_{CP}$ $=$ 5.4 Hz, $^2J_{CP}$ $=$ 27.7 Hz, CHC*H*CO), 49.0 (ddd, $^2J_{CP}$ $=$ 5.4 Hz, $^2J_{CP}$ $=$ 27.7 Hz, $^1J_{CPt}$ $=$ 188.3 Hz, CHC*H*CO), 61.3 (dd, $^2J_{CP}$ $=$ 3.1 Hz, $^2J_{CP}$ $=$ 33.0 Hz, C$_{Ar}$C*H*CH), 61.3 (ddd, $^2J_{CP}$ $=$ 3.1 Hz, $^2J_{CP}$ $=$ 33.0 Hz, $^1J_{CPt}$ $=$ 199.8 Hz, C$_{Ar}$C*H*CH), 122.7 (C_{Ar}H), 124.8 (C_{Ar}), 125.0 (C_{Ar}H), 126.6 (C_{Ar}H), 128.5 (C_{Ar}H), 128.7 (C_{Ar}H), 128.8 (C_{Ar}H), 129.0 (C_{Ar}H), 129.9 (C_{Ar}H), 130.2 (C_{Ar}H), 134.4 (C_{Ar}H), 134.5(C_{Ar}H), 134.5 (C_{Ar}H), 134.6 (C_{Ar}H), 134.7 (C_{Ar}H), 135.2 (d, $^1J_{CP}$ $=$ 3.1 Hz, C_{Ar}P), 135.2 (dd, $^1J_{CP}$ $=$ 3.1 Hz, $^2J_{CPt}$ $=$ 34.6 Hz, C_{Ar}P), 135.7 (d, $^1J_{CP}$ $=$ 3.0 Hz, C_{Ar}P), 135.7 (dd, $^1J_{CP}$ $=$ 3.0 Hz, $^2J_{CPt}$ $=$ 31.5 Hz, C_{Ar}P), 136.9 (d, $^1J_{CP}$ $=$ 2.3 Hz, C_{Ar}P), 136.9 (dd, $^1J_{CP}$ $=$ 2.3 Hz, $^2J_{CPt}$ $=$ 26.9 Hz, C_{Ar}P), 137.3 (d, $^1J_{CP}$ $=$ 3.1 Hz, C_{Ar}P), 137.3 (dd, $^1J_{CP}$ $=$ 3.1 Hz, $^2J_{CPt}$ $=$ 27.0 Hz, C_{Ar}P), 139.8 (bs, C_{Ar}H), 144.7 (d, $^3J_{CP}$ $=$ 7.0 Hz, C_{Ar}CH), 144.7 (dd, $^3J_{CP}$ $=$ 7.0 Hz, $^2J_{CPt}$ $=$ 32.2 Hz, C_{Ar}CH), 144.9 (C_{Ar}N), 157.6 (C_{Ar}O), 170.7 (dd, $^3J_{CP}$ $=$ 2.3 Hz, $^3J_{CP}$ $=$ 4.6 Hz, CHCOO), 170.7 (ddd, $^3J_{CP}$ $=$ 2.3 Hz, $^3J_{CP}$ $=$ 4.6 Hz, $^2J_{CPt}$ $=$ 50.7 Hz, CHCOO).

^{31}P-NMR (161.83 MHz; C$_6$D$_6$): δ = 27.5 (d, $^2J_{PP}$ = 38.6 Hz), 27.5 (dd, $^2J_{PP}$ = 38.6 Hz, $^1J_{PPt}$ = 4218.0 Hz), 28.3 (d, $^2J_{PP}$ = 38.6 Hz), 28.3 (dd, $^2J_{PP}$ = 38.6 Hz, $^1J_{PPt}$ = 3603.7 Hz).

^{195}Pt-NMR (85.48 MHz; C$_6$D$_6$): δ = -5051.0 (dd, $^1J_{PPt}$ = 3604.4 Hz, $^1J_{PPt}$ = 4219.6 Hz).

FAB-MS: m/z (%): 718.2 (100) [Pt(PPh$_3$)(PPh$_2$C$_6$H$_4$)$^+$], 987.1 (0.82) [(C$_{15}$H$_{11}$NO$_4$)Pt(PPh$_3$)(PPh$_2$C$_6$H$_4$)$^+$].

ESI-MS: m/z (%): 459.0 (100) [Pt(PPh$_3$)(PPh$_2$C$_6$H$_4$)CH$_3$CN$^+$].

FT-IR (ATR, cm^{-1}): 3053 (w), 1705 (m, CO), 1589 (w), 1519 (m), 1433 (m), 1345 (m), 1226 (m), 1150 (w), 1074 (s), 983 (w), 915 (m), 938 (w), 786 (w), 741 (m), 688 (s), 566 (w), 538 (m), 517 (m), 498 (s), 422 (m).

elemental analysis for C$_{51}$H$_{41}$NO$_4$P$_2$Pt (988.90 g/mol):

calcd.: C 61.94, H 4.18, N 1.42, O 6.47, P 6.26, Pt 19.73

found: C 60.90, H 4.20, N 1.36

Bis(triphenylphosphine)-(η^2-4-methoxyphenyl 4-chlorocinnamate)palladium(0)

PdPhCl-PhOMe: C$_{52}$H$_{43}$ClO$_3$P$_2$Pd, M = 919.72 g/mol, light yellow solid, yield: >99%, **route:** Pd-I.

^{1}H-NMR (400.13 MHz; C$_7$D$_8$): δ = 3.31 (s, 3H, OCH$_3$), 4.74 (bs, 1H, CHCHCO), 5.13 (bs, 1H, C$_{Ar}$CHCH), 6.33 - 6.41 (bm, 2H, H$_{Ar}$), 6.55 - 6.62 (bm, 2H, H$_{Ar}$), 6.62 - 6.68 (bm, 2H, H$_{Ar}$), 6.73 - 6.80 (bm, 2H, H$_{Ar}$), 6.86 - 7.01 (bm, 18H, H$_{Ar}$), 7.02 - 7.12 (bm, 6H, H$_{Ar}$), 7.47 - 7.60 (bm, 6H, H$_{Ar}$).

^{1}H-NMR (400.13 MHz; 246 K; C$_7$D$_8$): δ = 3.25 (s, 3H, OCH$_3$), 4.90 (dpt, $^3J_{HH}$ = 10.1 Hz, $^3J_{HP}$ = 6.1 Hz, $^3J_{HP}$ = 6.1 Hz, 1H, CHCHCO), 5.18 (dpt,

$^3J_{HH} = 10.1$ Hz, $^3J_{HP} = 7.1$ Hz, $^3J_{HP} = 7.1$ Hz, 1H, $C_{Ar}CHCH$), 6.31 (d, $^3J_{HH} = 7.7$ Hz, 2H, H_{Ar}), 6.50 - 6.58 (m, 4H, H_{Ar}), 6.72 (d, $^3J_{HH} = 8.6$ Hz, 2H, H_{Ar}), 6.84 - 7.00 (m, 18H, H_{Ar}), 7.00 - 7.07 (m, 6H, H_{Ar}), 7.52 - 7.62 (m, 6H, H_{Ar}).

^{13}C-NMR (100.62 MHz; C_7D_8): δ = 54.9 (OCH_3), 59.2 - 59.8 (bm, $CHCHCO$), 74.2 - 74.6 (bm, $C_{Ar}CHCH$), 114.1 ($C_{Ar}H$), 123.0 ($C_{Ar}H$), 126.6 ($C_{Ar}H$), 128.3 ($C_{Ar}H$), 128.5 ($C_{Ar}H$), 129.4 ($C_{Ar}H$), 133.8 - 134.8 (bm, $C_{Ar}H$), 135.2 - 136.0 (bm, $C_{Ar}H$), 141.3 (bs, $C_{Ar}CH$), 146.0 ($C_{Ar}O$), 156.9 ($C_{Ar}O$), 169.5 (bs, $CHCOO$).

^{13}C-NMR (100.53 MHz; 246 K; C_7D_8): δ = 54.6 (OCH_3), 58.7 (dd, $^2J_{CP} = 6.0$ Hz, $^2J_{CP} = 19.0$ Hz, $CHCHCO$), 74.3 (d, $^2J_{CP} = 22.6$ Hz, $C_{Ar}CHCH$), 113.8 ($C_{Ar}H$), 123.1 ($C_{Ar}H$), 126.4 (bs, $C_{Ar}H$), 128.2 ($C_{Ar}H$), 128.4 ($C_{Ar}H$), 129.3 (d, $^3J_{CP} = 25.4$ Hz, $C_{Ar}H$), 134.0 (d, $^2J_{CP} = 14.3$ Hz, $C_{Ar}H$), 134.3 (d, $^2J_{CP} = 15.4$ Hz, $C_{Ar}H$), 135.0 (dd, $^1J_{CP} = 29.0$ Hz, $^3J_{CP} = 1.7$ Hz, $C_{Ar}P$), 136.8 (dd, $^1J_{CP} = 28.0$ Hz, $^3J_{CP} = 1.4$ Hz, $C_{Ar}P$), 140.9 (d, $^3J_{CP} = 6.8$ Hz, CHC_{Ar}), 145.6 ($C_{Ar}O$), 156.6 ($C_{Ar}O$), 169.7 (dd, $^3J_{CP} = 1.8$ Hz, $^3J_{CP} = 2.5$ Hz, $CHCOO$).

^{31}P-NMR (161.97 MHz; C_7D_8): δ = 23.8 (d, $^2J_{PP} = 15.1$ Hz), 25.3 (d, $^2J_{PP} = 15.1$ Hz).

^{31}P-NMR (161.97 MHz; 246 K; C_7D_8): δ = 23.8 (d, $^2J_{PP} = 14.0$ Hz), 25.7 (d, $^2J_{PP} = 14.0$ Hz).

FT-IR (ATR, cm^{-1}): 3057 (w), 2911 (w), 1698 (m, CO), 1605 (w), 1505 (m), 1476 (w), 1433 (m), 1320 (m), 1190 (m), 1117 (s), 1092 (m), 1064 (m), 1037 (w), 969 (w), 817 (w), 741 (m), 693 (s), 525 (w), 504 (s), 410 (m).

Bis(triphenylphosphine)-(η^2-4-methoxyphenyl 4-chlorocinnamate)platinum(0)

PtPhCl-PhOMe: $C_{52}H_{43}ClO_3P_2Pt$, M = 1008.37 g/mol, white solid, yield: >99%, **route:** *Pt-VI.*

^{1}H-NMR (400.13 MHz; C$_6$D$_6$): δ = 3.26 (s, 3H, OCH_3), 4.15 (dpt, $^3J_{HH}$ = 8.6 Hz, $^3J_{HP}$ = 4.3 Hz, $^3J_{HP}$ = 8.6 Hz, 1H, CHCHCO), 4.15 (ddpt, $^3J_{HH}$ = 8.6 Hz, $^3J_{HP}$ = 4.3 Hz, $^3J_{HP}$ = 8.6 Hz, $^2J_{HPt}$ = 59.9 Hz, 1H, CHCHCO), 4.41 (dpt, $^3J_{HH}$ = 8.6 Hz, $^3J_{HP}$ = 5.8 Hz, $^3J_{HP}$ = 8.6 Hz, 1H, C$_{Ar}$CHCH), 4.41 (ddpt, $^3J_{HH}$ = 8.6 Hz, $^3J_{HP}$ = 5.8 Hz, $^3J_{HP}$ = 8.6 Hz, $^2J_{HPt}$ = 50.8 Hz, 1H, C$_{Ar}$CHCH), 6.38 (d, $^3J_{HH}$ = 8.3 Hz, 2H, H_{Ar}), 6.62 (d, $^3J_{HH}$ = 8.6 Hz, 2H, H_{Ar}), 6.72 (d, $^3J_{HH}$ = 7.4 Hz, 2H, H_{Ar}), 6.80 - 7.00 (m, 19H, H_{Ar}), 7.11 - 7.21 (m, 7H, H_{Ar}), 7.54 - 7.65 (m, 6H, H_{Ar}).

^{13}C-NMR (100.53 MHz; C$_6$D$_6$): δ = 48.9 (dd, $^2J_{CP}$ = 6.5 Hz, $^2J_{CP}$ = 29.6 Hz, CHCHCO), 48.9 (ddd, $^2J_{CP}$ = 6.5 Hz, $^2J_{CP}$ = 29.6 Hz, $^1J_{CPt}$ = 206.3 Hz, CHCHCO), 55.0 (OCH$_3$), 60.1 (dd, $^2J_{CP}$ = 4.0 Hz, $^2J_{CP}$ = 33.7 Hz, C$_{Ar}$$C$HCH), 60.1 (ddd, $^2J_{CP}$ = 4.0 Hz, $^2J_{CP}$ = 33.7 Hz, $^1J_{CPt}$ = 201.0 Hz, C$_{Ar}$$C$HCH), 114.2 ($C_{Ar}$H), 123.2 ($C_{Ar}$H), 127.3 (d, $^5J_{CP}$ = 3.1 Hz, C_{Ar}H), 127.3 (dd, $^5J_{CP}$ = 3.1 Hz, $^4J_{CPt}$ = 13.8 Hz, C_{Ar}H), 128.0 (C_{Ar}H), 128.1 (C_{Ar}H), 128.4 (C_{Ar}H), 128.5 (C_{Ar}H), 128.6 (C_{Ar}), 129.5 (d, $^5J_{CP}$ = 1.2 Hz, C_{Ar}H), 129.8 (d, $^5J_{CP}$ = 1.4 Hz, C_{Ar}H), 134.0 (C_{Ar}P), 134.0 (d, $^3J_{CPt}$ = 21.2 Hz, C_{Ar}P), 134.2 (C_{Ar}P), 134.2 (d, $^3J_{CPt}$ = 25.0 Hz, C_{Ar}P), 134.4 (C_{Ar}P), 134.4 (d, $^3J_{CPt}$ = 26.4 Hz, C_{Ar}P), 134.5 (C_{Ar}P), 134.5 (d, $^3J_{CPt}$ = 18.0 Hz, C_{Ar}P), 135.1 (d, $^2J_{CP}$ = 2.6 Hz, C_{Ar}P), 135.1 (dd, $^2J_{CP}$ = 2.6 Hz, $^3J_{CPt}$ = 34.0 Hz, C_{Ar}P), 135.5 (d, $^2J_{CP}$ = 2.7 Hz, C_{Ar}P), 135.5 (dd, $^2J_{CP}$ = 2.7 Hz, $^3J_{CPt}$ = 38.1 Hz, C_{Ar}P), 136.7 (d, $^2J_{CP}$ = 2.5 Hz, C_{Ar}P), 136.7 (dd, $^2J_{CP}$ = 2.5 Hz, $^3J_{CPt}$ = 30.5 Hz, C_{Ar}P), 137.1 (d, $^2J_{CP}$ = 2.5 Hz, C_{Ar}P), 137.1 (dd, $^2J_{CP}$ = 2.5 Hz, $^3J_{CPt}$ = 29.4 Hz, C_{Ar}P), 143.4 (dd, $^3J_{CP}$ = 1.2 Hz, $^3J_{CP}$ = 6.9 Hz, C_{Ar}CH), 143.4 (ddd, $^3J_{CP}$ = 1.2 Hz, $^3J_{CP}$ = 6.9 Hz, $^2J_{CPt}$ = 35.6 Hz, C_{Ar}CH), 146.2 (C_{Ar}O), 156.9 (C_{Ar}O), 171.9 (dd, $^3J_{CP}$ 2.1 Hz, $^3J_{CP}$ = 4.9 Hz, CHCOO), 171.9 (ddd, $^3J_{CP}$ = 2.1 Hz, $^3J_{CP}$ = 4.9 Hz, $^2J_{CPt}$ = 37.9 Hz, CHCOO).

^{31}P-NMR (161.97 MHz; C$_6$D$_6$): δ = 26.8 (d, $^2J_{PP}$ = 39.7 Hz), 26.8 (dd, $^2J_{PP}$ = 39.7 Hz, $^1J_{PPt}$ = 4139.4 Hz), 27.9 (d, $^2J_{PP}$ = 39.7 Hz), 27.9 (dd, $^2J_{PP}$ = 39.7 Hz, $^1J_{PPt}$ = 3651.2 Hz).

^{195}Pt-NMR (85.48 MHz; C$_6$D$_6$): δ = -5055.6 (dd, $^1J_{PPt}$ = 3650.1 Hz, $^1J_{PPt}$ = 4138.5 Hz).

FAB-MS: m/z (%): 719.0 (100) [Pt(PPh$_3$)(PPh$_2$C$_6$H$_4$)$^+$].

FT-IR (ATR, cm^{-1}): 1701 (m, CO), 1504 (m), 1433 (m), 1394 (w), 1303 (w), 1242 (w), 1193 (m), 1176 (m), 1112 (m), 1092 (s), 1030 (w), 818 (m), 742 (m), 693 (s), 545 (m), 508 (s), 424 (m).

Bis(triphenylphosphine)-(η^2-4-methylphenyl 4-chlorocinnamate)palladium(0)

PdPhCl-PhMe: $C_{52}H_{43}ClO_2P_2Pd$, M $= 903.72$ g/mol, light yellow solid, yield: 90%, **route**: *Pd-I*.

^{1}H-NMR (400.13 MHz; C_7D_8): $\delta = 2.09$ (s, 3H, CH_3), 4.72 (bs, 1H, CHCHCO), 5.13 (bs, 1H, $C_{Ar}CH$CH), 6.33 - 6.42 (bm, 2H, H_{Ar}), 6.62 - 6.71 (bm, 2H, H_{Ar}), 6.72 - 6.79 (bm, 2H, H_{Ar}), 6.79 - 6.86 (bm, 2H, H_{Ar}), 6.87 - 7.01 (bm, 18H, H_{Ar}), 7.01 - 7.12 (bm, 6H, H_{Ar}), 7.47 - 7.62 (bm, 6H, H_{Ar}).

^{1}H-NMR (400.13 MHz; 246 K; C_7D_8): $\delta = 2.10$ (s, 3H, CH_3), 4.88 (dpt, $^3J_{HH} = 10.1$ Hz, $^3J_{HP} = 6.1$ Hz, $^3J_{HP} = 6.1$ Hz, 1H, CHCHCO), 5.17 (dpt, $^3J_{HH} = 10.1$ Hz, $^3J_{HP} = 7.1$ Hz, $^3J_{HP} = 7.1$ Hz, 1H, $C_{Ar}CH$CH), 6.30 (d, $^3J_{HH} = 7.7$ Hz, 2H, H_{Ar}), 6.56 (d, $^3J_{HH} = 8.3$ Hz, 2H, H_{Ar}), 6.72 (d, $^3J_{HH} = 8.6$ Hz, 2H, H_{Ar}), 6.80 (d, $^3J_{HH} = 8.1$ Hz, 2H, H_{Ar}), 6.85 - 6.98 (m, 18H, H_{Ar}), 6.99 - 7.09 (m, 6H, H_{Ar}), 7.52 - 7.62 (m, 6H, H_{Ar}).

^{13}C-NMR (100.62 MHz; C_7D_8): $\delta = 20.7$ (CH_3), 59.4 - 60.0 (bm, CHCHCO), 74.0 - 74.7 (bm, C_{Ar}CHCH), 122.1 (C_{Ar}H), 126.7 (C_{Ar}H), 128.3 (C_{Ar}H), 128.5 (C_{Ar}H), 129.2 (C_{Ar}H), 129.2 (C_{Ar}H), 129.4 (C_{Ar}H), 129.5 (C_{Ar}H), 133.9 - 134.8 (bm, C_{Ar}H), 135.4 - 135.9 (bm, C_{Ar}H), 141.1 (bs, C_{Ar}CH), 150.4 (C_{Ar}O), 169.2 (bs, CHCOO).

^{13}C-NMR (100.53 MHz; 246 K; C_7D_8): $\delta = 20.6$ (CH_3), 55.8 (dd, $^2J_{CP} = 5.1$ Hz, $^2J_{CP} = 17.8$ Hz, CHCHCO), 74.3 (d, $^2J_{CP} = 22.8$ Hz, C_{Ar}CHCH), 122.2 (C_{Ar}H), 126.4 (bs, C_{Ar}H), 128.2 (C_{Ar}H), 128.5 (C_{Ar}H), 129.3 (d, $^3J_{CP} = 24.4$ Hz, C_{Ar}H), 129.5 (C_{Ar}H), 134.0 (d, $^2J_{CP} = 14.3$ Hz, C_{Ar}H), 134.3 (d, $^2J_{CP} = 15.3$ Hz, C_{Ar}H), 135.0 (dd,

$^1J_{CP} = 28.9$ Hz, $^3J_{CP} = 1.6$ Hz, C_{Ar}P), 136.8 (dd, $^1J_{CP} = 28.0$ Hz, $^3J_{CP} = 1.7$ Hz, C_{Ar}P), 140.9 (d, $^3J_{CP} = 6.5$ Hz, C_{Ar}CH), 150.0 (C_{Ar}O), 169.4 (dd, $^3J_{CP} = 2.7$ Hz, $^3J_{CP} = 4.4$ Hz, CHCOO).

^{31}P-NMR (161.97 MHz; C$_7$D$_8$): $\delta = 23.8$ (d, $^2J_{PP} = 15.0$ Hz), 25.3 (d, $^2J_{PP} = 15.0$ Hz).

^{31}P-NMR (161.97 MHz; 246 K; C$_7$D$_8$): $\delta = 23.8$ (d, $^2J_{PP} = 13.9$ Hz), 25.7 (d, $^2J_{PP} = 13.9$ Hz).

FT-IR (ATR, cm^{-1}): 3047 (w), 1698 (m , CO), 1471 (w), 1433 (m), 1309 (w), 1197 (m), 1164 (m), 1117 (s), 1090 (s), 968 (w), 908 (w), 825 (w), 741 (m), 692 (s), 523 (m), 503 (s), 411 (m).

Bis(triphenylphosphine)-(η^2-4-methylphenyl 4-chlorocinnamate)platinum(0)

PtPhCl-PhMe: C$_{52}$H$_{43}$ClO$_2$P$_2$Pt, M = 992.38 g/mol, white solid, yield: >99%, **route:** *Pt-VI.*

^{1}H-NMR (400.13 MHz; C$_6$D$_6$): $\delta = 2.06$ (s, 3H, CH_3), 4.13 (dpt, $^3J_{HH} = 8.6$ Hz, $^3J_{HP} = 4.3$ Hz, $^3J_{HP} = 8.6$ Hz, 1H, CHCHCO), 4.13 (ddpt, $^3J_{HH} = 8.6$ Hz, $^3J_{HP} = 4.3$ Hz, $^3J_{HP} = 8.6$ Hz, $^2J_{HPt} = 59.8$ Hz, 1H, CHCHCO), 4.41 (dpt, $^3J_{HH} = 8.6$ Hz, $^3J_{HP} = 5.8$ Hz, $^3J_{HP} = 8.6$ Hz, 1H, $C_{Ar}CH$CH), 4.41 (ddpt, $^3J_{HH} = 8.6$ Hz, $^3J_{HP} = 5.8$ Hz, $^3J_{HP} = 8.6$ Hz, $^2J_{HPt} = 52.2$ Hz, 1H, $C_{Ar}CH$CH), 6.37 (d, $^3J_{HH} = 8.1$ Hz, 2H, H_{Ar}), 6.73 (d, $^3J_{HH} = 7.9$ Hz, 2H, H_{Ar}), 6.80 - 6.98 (m, 21H, H_{Ar}), 7.10 - 7.21 (m, 7H, H_{Ar}), 7.53 - 7.65 (m, 6H, H_{Ar}).

^{13}C-NMR (100.53 MHz; C$_6$D$_6$): $\delta = 20.8$ (CH$_3$), 48.9 (dd, $^2J_{CP} = 6.0$ Hz, $^2J_{CP} = 28.6$ Hz, CHCHCO), 48.9 (ddd, $^2J_{CP} = 6.0$ Hz, $^2J_{CP} = 28.6$ Hz, $^1J_{CPt} = 205.1$ Hz, CHCHCO), 60.2 (dd, $^2J_{CP} = 4.1$ Hz, $^2J_{CP} = 33.5$ Hz, C_{Ar}CHCH), 60.2 (ddd, $^2J_{CP} = 4.1$ Hz, $^2J_{CP} = 33.5$ Hz, $^1J_{CPt} = 199.5$ Hz, C_{Ar}CHCH), 114.2 (C_{Ar}H), 127.3

(d, $^5J_{CP}$ = 3.1 Hz, C_{Ar}H), 127.3 (dd, $^5J_{CP}$ = 3.1 Hz, $^4J_{CPt}$ = 13.8 Hz, C_{Ar}H), 128.1 (C_{Ar}H), 128.4 (C_{Ar}H), 128.5 (C_{Ar}H), 128.6 (C_{Ar}), 129.6 (d, $^5J_{CP}$ = 3.5 Hz, C_{Ar}H), 129.8 (d, $^5J_{CP}$ = 1.4 Hz, C_{Ar}H), 133.7 (C_{Ar}H), 134.0 (C_{Ar}P), 134.0 (d, $^3J_{CPt}$ = 21.2 Hz, C_{Ar}P), 134.2 (C_{Ar}P), 134.2 (d, $^3J_{CPt}$ = 19.2 Hz, C_{Ar}P), 134.4 (C_{Ar}P), 134.4 (d, $^3J_{CPt}$ = 26.4 Hz, C_{Ar}P), 134.5 (C_{Ar}P), 134.5 (d, $^3J_{CPt}$ = 18.0 Hz, C_{Ar}P), 135.1 (d, $^2J_{CP}$ = 2.7 Hz, C_{Ar}P), 135.1 (dd, $^2J_{CP}$ = 2.7 Hz, $^3J_{CPt}$ = 33.5 Hz, C_{Ar}P), 135.5 (d, $^2J_{CP}$ = 2.7 Hz, C_{Ar}P), 135.5 (dd, $^2J_{CP}$ = 2.7 Hz, $^3J_{CPt}$ = 30.6 Hz, C_{Ar}P), 136.7 (d, $^2J_{CP}$ = 2.4 Hz, C_{Ar}P), 136.7 (dd, $^2J_{CP}$ = 2.4 Hz, $^3J_{CPt}$ = 30.1 Hz, C_{Ar}P), 137.1 (d, $^2J_{CP}$ = 2.5 Hz, C_{Ar}P), 137.1 (dd, $^2J_{CP}$ = 2.5 Hz, $^3J_{CPt}$ = 28.2 Hz, C_{Ar}P), 143.4 (dd, $^3J_{CP}$ = 0.9 Hz, $^3J_{CP}$ = 6.7 Hz, C_{Ar}CH), 143.4 (ddd, $^3J_{CP}$ = 0.9 Hz, $^3J_{CP}$ = 6.7 Hz, $^2J_{CPt}$ = 34.1 Hz, C_{Ar}CH), 150.5 (C_{Ar}O), 171.7 (dd, $^3J_{CP}$ 2.3 Hz, $^3J_{CP}$ = 6.7 Hz, CHCOO), 171.7 (ddd, $^3J_{CP}$ = 2.3 Hz, $^3J_{CP}$ = 6.7 Hz, $^2J_{CPt}$ = 45.2 Hz, CHCOO).

^{31}P-NMR (161.97 MHz; C$_6$D$_6$): δ = 26.8 (d, $^2J_{PP}$ = 39.7 Hz), 26.8 (dd, $^2J_{PP}$ = 39.7 Hz, $^1J_{PPt}$ = 4140.7 Hz), 27.9 (d, $^2J_{PP}$ = 39.7 Hz), 27.9 (dd, $^2J_{PP}$ = 39.7 Hz, $^1J_{PPt}$ = 3651.2 Hz).

^{195}Pt-NMR (85.48 MHz; C$_6$D$_6$): δ = -5054.9 (dd, $^1J_{PPt}$ = 3650.0 Hz, $^1J_{PPt}$ = 4139.6 Hz).

FAB-MS: m/z (%): 719.1 (100) [Pt(PPh$_3$)(PPh$_2$C$_6$H$_4$)$^+$].

FT-IR (ATR, cm^{-1}): 3050 (w), 1701 (s, CO), 1574 (w), 1477 (m), 1432 (s), 1392 (w), 1301 (w), 1179 (m), 1105 (m), 1090 (s), 998 (w), 892 (w), 822 (m), 741 (s), 692 (s), 534 (m), 498 (s), 423 (m).

Bis(triphenylphosphine)-(η^2-phenyl 4-chlorocinnamate)palladium(0)

PdPhCl-Ph: C$_{51}$H$_{41}$ClO$_2$P$_2$Pd, M = 889.69 *g/mol*, light yellow solid, yield: 94%, **route:** *Pd-I.*

^{1}H-NMR (400.13 MHz; C$_7$D$_8$): $\delta = 4.72$ (bs, 1H, CHCHCO), 5.11 (bs, 1H, C$_{Ar}$CHCH), 6.30 - 6.41 (bm, 2H, H_{Ar}), 6.69 - 6.83 (bm, 4H, H_{Ar}), 6.86 - 7.13 (bm, 28H, H_{Ar}), 7.46 - 7.63 (bm, 6H, H_{Ar}).

^{1}H-NMR (400.13 MHz; 246 K; C$_7$D$_8$): $\delta = 4.87$ (dpt, $^3J_{HH} = 10.1$ Hz, $^3J_{HP} = 76.1$ Hz, $^3J_{HP} = 6.1$ Hz, 1H, CHCHCO), 5.16 (dpt, $^3J_{HH} = 10.1$ Hz, $^3J_{HP} = 7.1$ Hz, $^3J_{HP} = 7.1$ Hz, 1H, C$_{Ar}$CHCH), 6.30 (d, $^3J_{HH} = 7.0$ Hz, 2H, H_{Ar}), 6.64 (d, $^3J_{HH} = 7.7$ Hz, 2H, H_{Ar}), 6.72 (d, $^3J_{HH} = 8.6$ Hz, 2H, H_{Ar}), 6.83 - 7.07 (m, 26H, H_{Ar}), 7.40 (t, $^3J_{HH} = 7.6$ Hz, 2H, H_{Ar}), 7.49 - 7.60 (m, 6H, H_{Ar}).

^{13}C-NMR (100.62 MHz; C$_7$D$_8$): $\delta = 58.0$ - 58.4 (bm, CHCHCO), 73.2 - 73.7 (bm, C$_{Ar}$$C$HCH), 122.2 ($C_{Ar}$H), 124.6 ($C_{Ar}$H), 126.7 ($C_{Ar}$H), 128.3 ($C_{Ar}$H), 128.4 ($C_{Ar}$H), 128.5 ($C_{Ar}$H), 129.4 ($C_{Ar}$H), 129.7 ($C_{Ar}$H), 133.8 - 134.6 (bm, C_{Ar}H), 134.9 - 135.4 (bm, C_{Ar}H), 136.4 - 136.9 (bm, C_{Ar}H), 140.7 (bs, C_{Ar}CH), 157.0 (C_{Ar}O), 167.9 (bs, CHCOO).

^{13}C-NMR (100.62 MHz; 246 K; C$_7$D$_8$): $\delta = 58.6$ (dd, $^2J_{CP} = 5.4$ Hz, $^2J_{CP} = 17.6$ Hz, CHCHCO), 74.3 (dd, $^2J_{CP} = 2.4$ Hz, $^2J_{CP} = 24.1$ Hz, C$_{Ar}$$C$HCH), 122.5 ($C_{Ar}$H), 126.4 (bs, C_{Ar}H), 128.2 (C_{Ar}H), 128.5 (C_{Ar}H), 128.9 (C_{Ar}H), 129.3 (d, $^3J_{CP} = 23.8$ Hz, C_{Ar}H), 134.0 (d, $^2J_{CP} = 14.2$ Hz, C_{Ar}H), 134.3 (d, $^2J_{CP} = 15.4$ Hz, C_{Ar}H), 135.0 (dd, $^1J_{CP} = 29.0$ Hz, $^3J_{CP} = 1.5$ Hz, C_{Ar}P), 136.8 (dd, $^1J_{CP} = 28.3$ Hz, $^3J_{CP} = 1.5$ Hz, C_{Ar}P), 140.9 (d, $^3J_{CP} = 6.4$ Hz, C_{Ar}CH), 152.2 (C_{Ar}O), 169.2 (dd, $^3J_{CP} = 2.3$ Hz, $^3J_{CP} = 4.2$ Hz, CHCOO).

^{31}P-NMR (161.97 MHz; C$_7$D$_8$): $\delta = 23.7$ (d, $^2J_{PP} = 14.6$ Hz), 25.2 (d, $^2J_{PP} = 14.6$ Hz).

^{31}P-NMR (161.97 MHz; 246 K; C$_7$D$_8$): $\delta = 23.8$ (d, $^2J_{PP} = 13.5$ Hz), 25.7 (d, $^2J_{PP} = 13.5$ Hz).

FT-IR (ATR, cm^{-1}): 3045 (w), 1697 (s, CO), 1595 (w), 1473 (m), 1433 (s), 1307 (m), 1198 (m), 1116 (s), 1089 (s), 973 (w), 822 (w), 746 (s), 693 (s), 525 (w), 514 (m), 505 (s), 415 (m).

Bis(triphenylphosphine)-(η^2-phenyl 4-chlorocinnamate)platinum(0)

PtPhCl-Ph: $C_{51}H_{41}ClO_2P_2Pt$, $M = 978.35$ *g/mol*, white solid, yield: >99%, **route**: *Pt-VI*.

^{1}H-NMR (400.13 MHz; C_6D_6): $\delta = 4.13$ (dpt, $^3J_{HH} = 8.4$ Hz, $^3J_{HP} = 4.1$ Hz, $^3J_{HP} = 8.4$ Hz, 1H, CHC*H*CO), 4.13 (ddpt, $^3J_{HH} = 8.4$ Hz, $^3J_{HP} = 4.1$ Hz, $^3J_{HP} = 8.4$ Hz, $^2J_{HPt} = 60.4$ Hz, 1H, CHC*H*CO), 4.39 (dpt, $^3J_{HH} = 8.4$ Hz, $^3J_{HP} = 5.9$ Hz, $^3J_{HP} = 8.4$ Hz, 1H, $C_{Ar}C$HCH), 4.39 (ddpt, $^3J_{HH} = 8.4$ Hz, $^3J_{HP} = 5.9$ Hz, $^3J_{HP} = 8.4$ Hz, $^2J_{HPt} = 51.4$ Hz, 1H, $C_{Ar}C$HCH), 6.36 (d, $^3J_{HH} = 7.8$ Hz, 2H, H_{Ar}), 6.79 - 6.97 (m, 22H, H_{Ar}), 7.01 (t, $^3J_{HH} = 7.3$ Hz, 2H, H_{Ar}), 7.10 - 7.19 (m, 7H, H_{Ar}), 7.53 - 7.64 (m, 6H, H_{Ar}).

^{13}C-NMR (100.62 MHz; C_6D_6): $\delta = 48.8$ (dd, $^2J_{CP} = 6.0$ Hz, $^2J_{CP} = 29.9$ Hz, CHCHCO), 48.8 (ddd, $^2J_{CP} = 6.0$ Hz, $^2J_{CP} = 29.9$ Hz, $^1J_{CPt} = 199.9$ Hz, CHCHCO), 60.1 (dd, $^2J_{CP} = 4.3$ Hz, $^2J_{CP} = 33.6$ Hz, C_{Ar}CHCH), 60.1 (ddd, $^2J_{CP} = 4.3$ Hz, $^2J_{CP} = 33.6$ Hz, $^1J_{CPt} = 199.7$ Hz, C_{Ar}CHCH), 122.6 (C_{Ar}H), 124.5 (C_{Ar}H), 127.3 (d, $^5J_{CP} = 3.1$ Hz, C_{Ar}H), 127.3 (dd, $^5J_{CP} = 3.1$ Hz, $^4J_{CPt} = 14.1$ Hz, C_{Ar}H), 128.0 (C_{Ar}H), 128.1 (C_{Ar}H), 128.2 (C_{Ar}H), 128.5 (C_{Ar}H), 128.6 (C_{Ar}), 129.0 (C_{Ar}H), 129.6 (d, $^5J_{CP} = 1.9$ Hz, C_{Ar}H), 129.7 (C_{Ar}), 129.8 (d, $^5J_{CP} = 1.5$ Hz, C_{Ar}H), 134.0 (C_{Ar}P), 134.0 (d, $^3J_{CPt} = 21.4$ Hz, C_{Ar}P), 134.2 (C_{Ar}P), 134.2 (d, $^3J_{CPt} = 16.0$ Hz, C_{Ar}P), 134.3 (C_{Ar}P), 134.3 (d, $^3J_{CPt} = 26.4$ Hz, C_{Ar}P), 134.5 (C_{Ar}P), 134.5 (d, $^3J_{CPt} = 18.2$ Hz, C_{Ar}P), 135.0 (d, $^2J_{CP} = 2.7$ Hz, C_{Ar}P), 135.0 (dd, $^2J_{CP} = 2.7$ Hz, $^3J_{CPt} = 34.5$ Hz, C_{Ar}P), 135.5 (d, $^2J_{CP} = 2.7$ Hz, C_{Ar}P), 135.5 (dd, $^2J_{CP} = 2.7$ Hz, $^3J_{CPt} = 30.9$ Hz, C_{Ar}P), 136.7 (d, $^2J_{CP} = 2.4$ Hz, C_{Ar}P), 136.7 (dd, $^2J_{CP} = 2.4$ Hz, $^3J_{CPt} = 29.3$ Hz, C_{Ar}P), 137.1 (d, $^2J_{CP} = 2.4$ Hz, C_{Ar}P), 137.1 (dd, $^2J_{CP} = 2.4$ Hz, $^3J_{CPt} = 27.1$ Hz, C_{Ar}P), 143.3 (dd, $^3J_{CP} = 1.3$ Hz, $^3J_{CP} = 6.4$ Hz, C_{Ar}CH), 143.3 (ddd, $^3J_{CP} = 1.3$ Hz, $^3J_{CP} = 6.4$ Hz,

$^2J_{CPt} = 32.4$ Hz, C_{Ar}CH), 152.7 (C_{Ar}O), 171.4 (dd, $^3J_{CP}$ 2.2 Hz, $^3J_{CP} = 5.3$ Hz, CHCOO),
171.4 (ddd, $^3J_{CP} = 2.2$ Hz, $^3J_{CP} = 5.3$ Hz, $^2J_{CPt} = 42.7$ Hz, CHCOO).

^{31}P-NMR (161.97 MHz; C$_6$D$_6$): $\delta = 26.8$ (d, $^2J_{PP} = 39.6$ Hz), 26.8 (dd, $^2J_{PP} = 39.6$ Hz, $^1J_{PPt} = 4147.0$ Hz), 27.9 (d, $^2J_{PP} = 39.6$ Hz), 27.9 (dd, $^2J_{PP} = 39.6$ Hz, $^1J_{PPt} = 3647.7$ Hz).

^{195}Pt-NMR (85.48 MHz; C$_6$D$_6$): $\delta = $ -5053.8 (dd, $^1J_{PPt} = 3647.0$ Hz, $^1J_{PPt} = 4146.2$ Hz).

FAB-MS: m/z (%): 719.1 (100) [Pt(PPh$_3$)(PPh$_2$C$_6$H$_4$)$^+$], 977.8 (1.63) [(C$_{15}$H$_{11}$ClO$_2$)Pt(PPh$_3$)(PPh$_2$C$_6$H$_4$)$^+$].

FT-IR (ATR, cm^{-1}): 3038 (w), 1705 (s, CO), 1576 (w), 1479 (m), 1433 (s), 1394 (w), 1288 (w), 1182 (m), 1111 (m), 1092 (s), 985 (w), 909 (w), 818 (m), 742 (s), 692 (s), 508 (s), 424 (m).

Bis(triphenylphosphine)-(η^2-4-nitrophenyl 4-chlorocinnamate)palladium(0)

PdPhCl-PhNO$_2$: C$_{51}$H$_{40}$ClNO$_4$P$_2$Pd, M $= 934.69$ g/mol, light yellow solid, yield: 96%, **route**: Pd-I.

^{1}H-NMR (400.13 MHz; C$_7$D$_8$): $\delta = 4.58$ - 4.69 (bm, 1H, CHCHCO), 4.96 - 5.09 (bm, 1H, C$_{Ar}$CHCH), 6.35 (bd, $^3J_{HH} = 7.7$ Hz, 2H, H_{Ar}), 6.57 (bd, $^3J_{HH} = 8.3$ Hz, 2H, H_{Ar}), 6.76 (bd, $^3J_{HH} = 7.7$ Hz, 2H, H_{Ar}), 6.85 - 7.09 (bm, 24H, H_{Ar}), 7.37 - 7.50 (bm, 6H, H_{Ar}), 7.73 (bd, $^3J_{HH} = 8.3$ Hz, 2H, H_{Ar}).

^{1}H-NMR (400.13 MHz; 246 K; C$_7$D$_8$): $\delta = 4.79$ (dpt, $^3J_{HH} = 10.1$ Hz, $^3J_{HP} = 6.0$ Hz, $^3J_{HP} = 6.0$ Hz, 1H, CHCHCO), 5.05 (dpt, $^3J_{HH} = 10.1$ Hz, $^3J_{HP} = 7.3$ Hz, $^3J_{HP} = 7.3$ Hz, 1H, C$_{Ar}$CHCH), 6.28 (d, $^3J_{HH} = 7.3$ Hz, 2H, H_{Ar}), 6.41 (d, $^3J_{HH} = 8.9$ Hz,

2H, H_{Ar}), 6.73 (d, $^3J_{HH}$ = 8.3 Hz, 2H, H_{Ar}), 6.78 - 7.09 (m, 24H, H_{Ar}), 7.38 - 7.52 (m, 6H, H_{Ar}), 7.69 (d, $^3J_{HH}$ = 9.0 Hz, 2H, H_{Ar}).

^{13}C-NMR (100.62 MHz; C_7D_8): δ = 59.4 (d, $^2J_{CP}$ = 16.8 Hz, CHCHCO), 74.4 (d, $^2J_{CP}$ = 22.0 Hz, C_{Ar}CHCH), 122.4 (C_{Ar}H), 124.4 (C_{Ar}H), 126.6 (C_{Ar}H), 128.3 (C_{Ar}H), 128.5 (C_{Ar}H), 129.2 (C_{Ar}H), 129.4 (C_{Ar}H), 133.8 - 134.7 (bm, C_{Ar}H), 135.6 (d, $^1J_{CP}$ = 27.5 Hz, C_{Ar}H), 136.6 - 137.4 (bm, C_{Ar}H), 139.2 (bs, C_{Ar}CH), 140.9 (C_{Ar}N), 152.6 (C_{Ar}O), 168.9 (bs, CHCOO).

^{13}C-NMR (100.53 MHz; 246 K; C_7D_8): δ = 57.5 (dd, $^2J_{CP}$ = 7.3 Hz, $^2J_{CP}$ = 19.9 Hz, CHCHCO), 73.3 (dd, $^2J_{CP}$ = 1.8 Hz, $^2J_{CP}$ = 23.7 Hz, C_{Ar}CHCH), 122.2 (C_{Ar}H), 124.5 (C_{Ar}H), 126.4 (bs, C_{Ar}H), 128.2 (C_{Ar}H), 128.3 (C_{Ar}H), 128.5 (C_{Ar}H), 128.6 (C_{Ar}H), 129.5 (d, $^3J_{CP}$ = 27.8 Hz, C_{Ar}H), 133.9 (d, $^2J_{CP}$ = 14.3 Hz, C_{Ar}H), 134.1 (d, $^2J_{CP}$ = 15.3 Hz, C_{Ar}H), 134.6 (dd, $^1J_{CP}$ = 29.8 Hz, $^3J_{CP}$ = 1.5 Hz, C_{Ar}P), 136.4 (dd, $^1J_{CP}$ = 28.6 Hz, $^3J_{CP}$ = 1.4 Hz, C_{Ar}P), 140.5 (d, $^3J_{CP}$ = 6.9 Hz, C_{Ar}CH), 144.3 (C_{Ar}N), 156.7 (C_{Ar}O), 168.0 (dd, $^3J_{CP}$ = 2.6 Hz, $^3J_{CP}$ = 4.2 Hz, CHCOO).

^{31}P-NMR (161.97 MHz; C_7D_8): δ = 24.0 (d, $^2J_{PP}$ = 11.3 Hz), 25.0 (d, $^2J_{PP}$ = 11.3 Hz).

^{31}P-NMR (161.97 MHz; 246 K; C_7D_8): δ = 24.0 (d, $^2J_{PP}$ = 10.1 Hz), 25.4 (d, $^2J_{PP}$ = 10.1 Hz).

FT-IR (ATR, cm^{-1}): 3062 (w), 1699 (m, CO), 1590 (w), 1518 (m), 1476 (m), 1433 (s), 1343 (m), 1307 (m), 1205 (w), 1181 (w), 1152 (m), 1087 (s), 965 (w), 893 (w), 847 (w), 818 (m), 741 (s), 692 (s), 503 (s), 412 (m).

Bis(triphenylphosphine)-(η^2-4-nitrophenyl 4-chlorocinnamate)platinum(0)

PtPhCl-PhNO$_2$: $C_{51}H_{40}ClNO_4P_2Pt$, M = 1023.35 g/mol, white solid, yield: >99%, **route**: *Pt-VI*.

^{1}H-NMR (400.13 MHz; C$_6$D$_6$): δ = 4.03 (dpt, $^3J_{HH}$ = 8.7 Hz, $^3J_{HP}$ = 3.8 Hz, $^3J_{HP}$ = 8.7 Hz, 1H, CHCHCO), 4.03 (ddpt, $^3J_{HH}$ = 8.7 Hz, $^3J_{HP}$ = 3.8 Hz, $^3J_{HP}$ = 8.7 Hz, $^2J_{HPt}$ = 60.5 Hz, 1H, CHCHCO), 4.29 (dpt, $^3J_{HH}$ = 8.7 Hz, $^3J_{HP}$ = 5.4 Hz, $^3J_{HP}$ = 8.7 Hz, 1H, C$_{Ar}$CHCH), 4.29 (ddpt, $^3J_{HH}$ = 8.7 Hz, $^3J_{HP}$ = 5.4 Hz, $^3J_{HP}$ = 8.7 Hz, $^2J_{HPt}$ = 51.4 Hz, 1H, C$_{Ar}$CHCH), 6.37 (d, $^3J_{HH}$ = 7.9 Hz, 2H, H_{Ar}), 6.57 (d, $^3J_{HH}$ = 8.6 Hz, 2H, H_{Ar}), 6.82 - 6.95 (m, 19H, H_{Ar}), 7.07 - 7.19 (m, 7H, H_{Ar}), 7.43 - 7.54 (m, 6H, H_{Ar}), 7.74 (d, $^3J_{HH}$ = 8.6 Hz, 2H, H_{Ar}).

^{13}C-NMR (100.53 MHz; C$_6$D$_6$): δ = 48.0 (dd, $^2J_{CP}$ = 5.3 Hz, $^2J_{CP}$ = 27.5 Hz, CHCHCO), 48.0 (ddd, $^2J_{CP}$ = 5.3 Hz, $^2J_{CP}$ = 27.5 Hz, $^1J_{CPt}$ = 175.0 Hz, CHCHCO), 59.4 (dd, $^2J_{CP}$ = 3.8 Hz, $^2J_{CP}$ = 34.9 Hz, C$_{Ar}$$C$HCH), 59.4 (ddd, $^2J_{CP}$ = 3.8 Hz, $^2J_{CP}$ = 34.9 Hz, $^1J_{CPt}$ = 206.16 Hz, C$_{Ar}$$C$HCH), 122.4 ($C_{Ar}$H), 124.7 ($C_{Ar}$H), 127.3 (d, $^5J_{CP}$ = 3.2 Hz, C_{Ar}H), 127.3 (dd, $^5J_{CP}$ = 3.2 Hz, $^4J_{CPt}$ = 11.6 Hz, C_{Ar}H), 127.9 (C_{Ar}H), 128.2 (C_{Ar}H), 128.4 (C_{Ar}H), 128.5 (C_{Ar}H), 128.6 (C_{Ar}), 129.7 (d, $^5J_{CP}$ = 1.8 Hz, C_{Ar}H), 130.0 (d, $^5J_{CP}$ = 1.5 Hz, C_{Ar}H), 134.0 (C_{Ar}P), 134.0 (d, $^3J_{CPt}$ = 21.4 Hz, C_{Ar}P), 134.1 (C_{Ar}P), 134.1 (d, $^3J_{CPt}$ = 17.6 Hz, C_{Ar}P), 134.2 (C_{Ar}P), 134.2 (d, $^3J_{CPt}$ = 18.8 Hz, C_{Ar}P), 134.3 (C_{Ar}P), 134.3 (d, $^3J_{CPt}$ = 18.2 Hz, C_{Ar}P), 134.7 (d, $^2J_{CP}$ = 2.6 Hz, C_{Ar}P), 134.7 (dd, $^2J_{CP}$ = 2.6 Hz, $^3J_{CPt}$ = 31.2 Hz, C_{Ar}P), 135.1 (d, $^2J_{CP}$ = 2.6 Hz, C_{Ar}P), 135.1 (dd, $^2J_{CP}$ = 2.6 Hz, $^3J_{CPt}$ = 32.5 Hz, C_{Ar}P), 136.3 (d, $^2J_{CP}$ = 2.3 Hz, C_{Ar}P), 136.3 (dd, $^2J_{CP}$ = 2.3 Hz, $^3J_{CPt}$ = 27.8 Hz, C_{Ar}P), 136.7 (d, $^2J_{CP}$ = 2.3 Hz, C_{Ar}P), 136.7 (dd, $^2J_{CP}$ = 2.3 Hz, $^3J_{CPt}$ = 28.7 Hz, C_{Ar}P), 143.0 (dd, $^3J_{CP}$ = 1.2 Hz, $^3J_{CP}$ = 7.7 Hz, C_{Ar}CH), 143.0 (ddd, $^3J_{CP}$ = 1.2 Hz, $^3J_{CP}$ = 7.7 Hz, $^2J_{CPt}$ = 33.1 Hz, C_{Ar}CH), 144.6 (C_{Ar}O), 157.1 (C_{Ar}NO$_2$), 170.3 (dd, $^3J_{CP}$ 2.0 Hz, $^3J_{CP}$ = 5.1 Hz, CHCOO), 170.3 (ddd, $^3J_{CP}$ = 2.0 Hz, $^3J_{CP}$ = 5.1 Hz, $^2J_{CPt}$ = 47.6 Hz, CHCOO).

^{31}P-NMR (161.97 MHz; C$_6$D$_6$): δ = 26.3 (d, $^2J_{PP}$ = 36.9 Hz), 26.3 (dd, $^2J_{PP}$ = 36.9 Hz, $^1J_{PPt}$ = 4188.7 Hz), 27.3 (d, $^2J_{PP}$ = 36.9 Hz), 27.3 (dd, $^2J_{PP}$ = 36.9 Hz, $^1J_{PPt}$ = 3611.9 Hz).

^{195}Pt-NMR (85.48 MHz; C$_6$D$_6$): δ = -5053.0 (dd, $^1J_{PPt}$ = 3610.4 Hz, $^1J_{PPt}$ = 4187.7 Hz).

FAB-MS: m/z (%): 719.1 (100) [Pt(PPh$_3$)(PPh$_2$C$_6$H$_4$)$^+$], 1021.8 (1.68)

$[(C_{15}H_{10}ClNO_4)Pt(PPh_3)(PPh_2C_6H_4)^+]$

FT-IR (ATR, cm^{-1}): 3053 (w), 1705 (s, CO), 1589 (m), 1517 (s), 1488 (m), 1433 (s), 1392 (w), 1342 (s), 1305 (m), 1218 (m), 1148 (m), 1083 (s), 997 (m), 910 (m), 822 (m), 741 (s), 691 (s), 574 (w), 544 (m), 506 (s), 422 (m).

Bis(triphenylphosphine)-(η^2-4-methoxyphenyl 4-trifluoromethylcinnamate)-palladium(0)

PdPhCF$_3$-PhOMe: $C_{53}H_{43}F_3O_3P_2Pd$, M = 953.27 *g/mol*, deep yellow solid, yield: >99%, **route**: *Pd-I*.

^{1}H-NMR (400.13 MHz; C$_6$D$_6$): δ = 3.27 (s, 3H, OCH_3), 4.75 - 4.87 (bm, 1H, CHCHCO), 5.18 - 5.28 (bm, 1H, C$_{Ar}$CHCH), 6.49 (bd, $^3J_{HH}$ = 7.7 Hz, 2H, H_{Ar}), 6.62 (bd, $^3J_{HH}$ = 7.9 Hz, 2H, H_{Ar}), 6.71 (bd, $^3J_{HH}$ = 7.8 Hz, 2H, H_{Ar}), 6.84 - 6.98 (bm, 18H, H_{Ar}), 7.01 (bd, $^3J_{HH}$ = 7.8 Hz, 2H, H_{Ar}), 7.03 - 7.10 (bm, 6H, H_{Ar}), 7.53 - 7.63 (bm, 6H, H_{Ar}).

^{1}H-NMR (400.13 MHz; 278 K; C$_6$D$_6$): δ = 3.24 (s, 3H, OCH_3), 4.89 (dpt, $^3J_{HH}$ = 10.1 Hz, $^3J_{HP}$ = 6.2 Hz, $^3J_{HP}$ = 6.2 Hz, 1H, CHCHCO), 5.25 (dpt, $^3J_{HH}$ = 10.1 Hz, $^3J_{HP}$ = 7.4 Hz, $^3J_{HP}$ = 7.4 Hz, 1H, C$_{Ar}$CHCH), 6.46 (d, $^3J_{HH}$ = 7.9 Hz, 2H, H_{Ar}), 6.61 (d, $^3J_{HH}$ = 9.1 Hz, 2H, H_{Ar}), 6.68 (d, $^3J_{HH}$ = 9.0 Hz, 2H, H_{Ar}), 6.83 - 6.96 (m, 18H, H_{Ar}), 6.99 (d, $^3J_{HH}$ = 8.2 Hz, 2H, H_{Ar}), 7.02 - 7.10 (m, 6H, H_{Ar}), 7.55 - 7.63 (m, 6H, H_{Ar}).

^{13}C-NMR (100.53 MHz; C$_6$D$_6$): δ = 55.0 (OCH$_3$), 59.3 (dd, $^2J_{CP}$ = 5.6 Hz, $^2J_{CP}$ = 18.7 Hz, CHCHCO), 73.6 (d, $^2J_{CP}$ = 23.2 Hz, C$_{Ar}$$C$HCH), 114.3 ($C_{Ar}$H), 123.2 ($C_{Ar}$H), 125.3 ($C_{Ar}$H), 125.4 (bs, C_{Ar}H), 127.9 (C_{Ar}H), 128.2 (C_{Ar}H), 128.4 (C_{Ar}H), 128.6 (C_{Ar}H), 128.7 (C_{Ar}H), 129.5 (d, $^3J_{CP}$ = 23.0 Hz, C_{Ar}H), 134.1 (d, $^2J_{CP}$ = 14.3 Hz,

C_{Ar}H), 134.5 (d, $^2J_{CP}$ = 15.2 Hz, C_{Ar}H), 135.3 (d, $^1J_{CP}$ = 27.9 Hz, C_{Ar}P), 136.8 (d, $^1J_{CP}$ = 27.4 Hz, C_{Ar}P), 146.0 (C_{Ar}O), 146.7 (d, $^3J_{CP}$ = 6.6 Hz, CHC_{Ar}), 157.0 (C_{Ar}O), 170.0 (dd, $^3J_{CP}$ = 2.1 Hz, $^3J_{CP}$ = 2.4 Hz, CHCOO).

^{19}F-NMR (376.50 MHz; C$_6$D$_6$): δ = -61.4.

^{31}P-NMR (161.97 MHz; C$_6$D$_6$): δ = 25.0 (d, $^2J_{PP}$ = 11.0 Hz), 26.4 (d, $^2J_{PP}$ = 11.0 Hz).

^{31}P-NMR (161.97 MHz; 278 K; C$_6$D$_6$): δ = 24.7 (d, $^2J_{PP}$ = 10.6 Hz), 26.3 (dd, $^1J_{PF}$ = 2.9 Hz, $^2J_{PP}$ = 10.6 Hz).

FT-IR (ATR, cm^{-1}): 3047 (w), 2911 (w), 1699 (s, CO), 1609 (m), 1505 (s), 1477 (m), 1433 (s), 1320 (s), 1273 (w), 1188 (m), 1163 (s), 1116 (s), 1102 (s), 1064 (s), 1035 (m), 1012 (w), 961 (w), 830 (m), 741 (s), 692 (s), 503 (s), 411 (s).

Bis(triphenylphosphine)-(η^2-4-methoxyphenyl 4-trifluoromethylcinnamate)-platinum(0)

PtPhCF$_3$-PhOMe: C$_{53}$H$_{43}$F$_3$O$_3$P$_2$Pt, M = 1041.93 g/mol, white solid, yield: 90%, route: *Pt-VI*.

^{1}H-NMR (400.13 MHz; C$_6$D$_6$): δ = 3.27 (s, 3H, OCH_3), 4.15 (dpt, $^3J_{HH}$ = 8.6 Hz, $^3J_{HP}$ = 4.2 Hz, $^3J_{HP}$ = 8.6 Hz, 1H, CHCHCO), 4.15 (ddpt, $^3J_{HH}$ = 8.6 Hz, $^3J_{HP}$ = 4.2 Hz, $^3J_{HP}$ = 8.6 Hz, $^2J_{HPt}$ = 59.6 Hz, 1H, CHCHCO), 4.44 (dpt, $^3J_{HH}$ = 8.6 Hz, $^3J_{HP}$ = 5.5 Hz, $^3J_{HP}$ = 8.6 Hz, 1H, C$_{Ar}$CHCH), 4.44 (ddpt, $^3J_{HH}$ = 8.6 Hz, $^3J_{HP}$ = 5.5 Hz, $^3J_{HP}$ = 8.6 Hz, $^2J_{HPt}$ = 51.4 Hz, 1H, C$_{Ar}$CHCH), 6.45 (d, $^3J_{HH}$ = 7.7 Hz, 2H, H_{Ar}), 6.63 (d, $^3J_{HH}$ = 8.7 Hz, 2H, H_{Ar}), 6.73 (d, $^3J_{HH}$ = 7.3 Hz, 2H, H_{Ar}), 6.81 - 6.99 (m, 17H, H_{Ar}), 7.05 (d, $^3J_{HH}$ = 7.9 Hz, 2H, H_{Ar}), 7.08 - 7.19 (m, 7H, H_{Ar}), 7.54 - 7.64 (m, 6H, H_{Ar}).

^{13}C-NMR (100.53 MHz; C$_6$D$_6$): δ = 48.9 (dd, $^2J_{CP}$ = 5.8 Hz, $^2J_{CP}$ = 28.6 Hz, CH*CH*CO), 48.9 (ddd, $^2J_{CP}$ = 5.8 Hz, $^2J_{CP}$ = 28.6 Hz, $^1J_{CPt}$ = 197.5 Hz, CH*CH*CO), 55.1 (O*CH$_3$*), 59.8 (dd, $^2J_{CP}$ = 4.0 Hz, $^2J_{CP}$ = 33.2 Hz, C$_{Ar}$*CH*CH), 59.8 (ddd, $^2J_{CP}$ = 4.0 Hz, $^2J_{CP}$ = 33.2 Hz, $^1J_{CPt}$ = 195.4 Hz, C$_{Ar}$*CH*CH), 114.3 (*C$_{Ar}$*H), 123.2 (*C$_{Ar}$*H), 124.9 (bm, *C$_{Ar}$*H), 125.8 (d, $^5J_{CP}$ = 3.0 Hz, *C$_{Ar}$*H), 125.8 (dd, $^5J_{CP}$ = 3.0 Hz, $^4J_{CPt}$ = 13.5 Hz, *C$_{Ar}$*H), 127.9 (*C$_{Ar}$*H), 128.2 (*C$_{Ar}$*H), 128.4 (*C$_{Ar}$*H), 128.5 (*C$_{Ar}$*H), 128.6 (*C$_{Ar}$*), 129.6 (d, $^5J_{CP}$ = 1.6 Hz, *C$_{Ar}$*H), 129.9 (d, $^5J_{CP}$ = 1.6 Hz, *C$_{Ar}$*H), 134.0 (*C$_{Ar}$*P), 134.0 (d, $^3J_{CPt}$ = 20.8 Hz, *C$_{Ar}$*P), 134.1 (*C$_{Ar}$*P), 134.1 (d, $^3J_{CPt}$ = 25.0 Hz, *C$_{Ar}$*P), 134.3 (*C$_{Ar}$*P), 134.3 (d, $^3J_{CPt}$ = 19.0 Hz, *C$_{Ar}$*P), 134.5 (*C$_{Ar}$*P), 134.5 (d, $^3J_{CPt}$ = 18.4 Hz, *C$_{Ar}$*P), 134.8 (d, $^2J_{CP}$ = 2.5 Hz, *C$_{Ar}$*P), 134.8 (dd, $^2J_{CP}$ = 2.5 Hz, $^3J_{CPt}$ = 34.4 Hz, *C$_{Ar}$*P), 135.3 (d, $^2J_{CP}$ = 2.5 Hz, *C$_{Ar}$*P), 135.3 (dd, $^2J_{CP}$ = 2.5 Hz, $^3J_{CPt}$ = 36.0 Hz, *C$_{Ar}$*P), 136.5 (d, $^2J_{CP}$ = 2.4 Hz, *C$_{Ar}$*P), 136.5 (dd, $^2J_{CP}$ = 2.4 Hz, $^3J_{CPt}$ = 27.8 Hz, *C$_{Ar}$*P), 136.9 (d, $^2J_{CP}$ = 2.4 Hz, *C$_{Ar}$*P), 136.9 (dd, $^2J_{CP}$ = 2.4 Hz, $^3J_{CPt}$ = 28.5 Hz, *C$_{Ar}$*P), 146.1 (*C$_{Ar}$*O), 149.0 (dd, $^3J_{CP}$ = 2.3 Hz, $^3J_{CP}$ = 7.5 Hz, *C$_{Ar}$*CH), 149.0 (ddd, $^3J_{CP}$ = 2.3 Hz, $^3J_{CP}$ = 7.5 Hz, $^2J_{CPt}$ = 35.0 Hz, *C$_{Ar}$*CH), 157.0 (*C$_{Ar}$*O), 171.9 (dd, $^3J_{CP}$ 2.1 Hz, $^3J_{CP}$ = 5.3 Hz, CH*C*OO), 171.9 (ddd, $^3J_{CP}$ = 2.1 Hz, $^3J_{CP}$ = 5.3 Hz, $^2J_{CPt}$ = 46.7 Hz, CH*C*OO).

^{19}F-NMR (376.50 MHz; C$_6$D$_6$): δ = -61.28.

^{31}P-NMR (161.97 MHz; C$_6$D$_6$): δ = 26.7 (d, $^2J_{PP}$ = 37.0 Hz), 26.7 (dd, $^2J_{PP}$ = 37.0 Hz, $^1J_{PPt}$ = 4097.1 Hz), 27.8 (dd, $^1J_{PF}$ = 2.8 Hz, $^2J_{PP}$ = 37.0 Hz), 27.8 (ddd, $^1J_{PF}$ = 2.8 Hz, $^2J_{PP}$ = 37.0 Hz, $^1J_{PPt}$ = 3695.6 Hz).

^{195}Pt-NMR (85.48 MHz; C$_6$D$_6$): δ = -5058.6 (dd, $^1J_{PPt}$ = 3683.4 Hz, $^1J_{PPt}$ = 4088.8 Hz).

FAB-MS: m/z (%): 719.0 (100) [Pt(PPh$_3$)(PPh$_2$C$_6$H$_4$)$^+$].

FT-IR (ATR, cm^{-1}): 3050 (w), 1702 (m, CO), 1609 (w), 1504 (m), 1433 (m), 1322 (s), 1245 (m), 1156 (m), 1105 (s), 1095 (s), 1063 (s), 1029 (w), 983 (m), 834 (m), 741 (m), 692 (s), 605 (w), 537 (m), 507 (s), 423 (m).

Bis(triphenylphosphine)-(η^2-4-methylphenyl 4-trifluoromethylcinnamate)-palladium(0)

PdPhCF$_3$-PhMe: C$_{53}$H$_{43}$F$_3$O$_2$P$_2$Pd, M $= 937.27$ g/mol, deep yellow solid, yield: $>99\%$, route: *Pd-I*.

^{1}H-NMR (400.13 MHz; C$_7$D$_8$): $\delta = 2.09$ (s, 3H, CH_3), 4.65 - 4.78 (bm, 1H, CHCHCO), 5.06 - 5.18 (bm, 1H, C$_{Ar}$CHCH), 6.44 (bd, $^3J_{HH} = 7.8$ Hz, 2H, H_{Ar}), 6.65 (bd, $^3J_{HH} = 7.6$ Hz, 2H, H_{Ar}), 6.81 (bd, $^3J_{HH} = 7.7$ Hz, 2H, H_{Ar}), 6.86 - 7.09 (bm, 26H, H_{Ar}), 7.49 - 7.62 (bm, 6H, H_{Ar}).

^{1}H-NMR (400.13 MHz; 278 K; C$_7$D$_8$): $\delta = 2.10$ (s, 3H, CH_3), 4.77 (dpt, $^3J_{HH} = 10.1$ Hz, $^3J_{HP} = 6.2$ Hz, $^3J_{HP} = 6.2$ Hz, 1H, CHCHCO), 5.14 (dpt, $^3J_{HH} = 10.1$ Hz, $^3J_{HP} = 7.5$ Hz, $^3J_{HP} = 7.5$ Hz, 1H, C$_{Ar}$CHCH), 6.42 (d, $^3J_{HH} = 8.1$ Hz, 2H, H_{Ar}), 6.63 (d, $^3J_{HH} = 8.4$ Hz, 2H, H_{Ar}), 6.82 (d, $^3J_{HH} = 8.1$ Hz, 2H, H_{Ar}), 6.86 - 7.08 (m, 26H, H_{Ar}), 7.51 - 7.59 (m, 6H, H_{Ar}).

^{1}H-NMR (400.13 MHz; 246 K; C$_7$D$_8$): $\delta = 2.11$ (s, 3H, CH_3), 4.88 (dpt, $^3J_{HH} = 10.0$ Hz, $^3J_{HP} = 6.2$ Hz, $^3J_{HP} = 6.2$ Hz, 1H, CHCHCO), 5.16 (dpt, $^3J_{HH} = 10.0$ Hz, $^3J_{HP} = 7.4$ Hz, $^3J_{HP} = 7.4$ Hz, 1H, C$_{Ar}$CHCH), 6.37 (d, $^3J_{HH} = 6.8$ Hz, 2H, H_{Ar}), 6.56 (d, $^3J_{HH} = 8.3$ Hz, 2H, H_{Ar}), 6.81 (d, $^3J_{HH} = 8.2$ Hz, 2H, H_{Ar}), 6.84 - 7.05 (m, 26H, H_{Ar}), 7.52 - 7.61 (m, 6H, H_{Ar}).

^{13}C-NMR (100.53 MHz; C$_7$D$_8$): $\delta = 20.7$ (CH$_3$), 59.4 (dd, $^2J_{CP} = 5.2$ Hz, $^2J_{CP} = 19.6$ Hz, CHCHCO), 73.5 (d, $^2J_{CP} = 23.2$ Hz, C$_{Ar}$$C$HCH), 122.1 ($C_{Ar}$H), 128.2 ($C_{Ar}$H), 128.3 ($C_{Ar}$H), 128.5 ($C_{Ar}$H), 128.5 ($C_{Ar}$H), 129.4 (d, $^3J_{CP} = 20.5$ Hz, C_{Ar}H), 134.1 (d, $^2J_{CP} = 14.3$ Hz, C_{Ar}H), 134.3 (C_{Ar}), 134.4 (d, $^2J_{CP} = 15.2$ Hz, C_{Ar}H), 135.3 (dd, $^1J_{CP} = 28.7$ Hz, $^3J_{CP} = 1.1$ Hz, C_{Ar}P), 136.8 (d, $^1J_{CP} = 27.9$ Hz, C_{Ar}P), 146.7 (d, $^3J_{CP} = 5.6$ Hz, CHC_{Ar}), 150.3 (C_{Ar}O), 169.4 (dd, $^3J_{CP} = 1.9$ Hz, $^3J_{CP} = 4.2$ Hz,

CH$CHOO$).

^{19}F-NMR (376.50 MHz; C$_7$D$_8$): δ = -61.4.

^{31}P-NMR (161.97 MHz; C$_7$D$_8$): δ = 25.0 (d, $^2J_{PP}$ = 11.1 Hz), 26.4 (d, $^2J_{PP}$ = 11.1 Hz).

^{31}P-NMR (161.97 MHz; 278 K; C$_7$D$_8$): δ = 24.7 (d, $^2J_{PP}$ = 10.8 Hz), 26.3 (dd, $^1J_{PF}$ = 2.9 Hz, $^2J_{PP}$ = 10.8 Hz).

^{31}P-NMR (161.97 MHz; 246 K; C$_7$D$_8$): δ = 24.3 (d, $^2J_{PP}$ = 10.1 Hz), 26.1 (dd, $^1J_{PF}$ = 2.9 Hz, $^2J_{PP}$ = 10.1 Hz).

FT-IR (ATR, cm^{-1}): 3053 (w), 1674 (m, CO), 1597 (w), 1476 (m), 1433 (s), 1307 (m), 1268 (w), 1178 (w), 1155 (m), 1092 (s), 1064 (m), 997 (w), 820 (w), 740 (s), 692 (s), 500 (s), 410 (m).

Bis(triphenylphosphine)-(η^2-4-methylphenyl 4-trifluoromethylcinnamate)-platinum(0)

PtPhCF$_3$-PhMe: C$_{53}$H$_{43}$F$_3$O$_2$P$_2$Pt, M = 1025.93 *g/mol*, white solid, yield: >99%, **route**: *Pt-VI.*

^{1}H-NMR (400.13 MHz; C$_6$D$_6$): δ = 2.06 (s, 3H, CH_3), 4.13 (dpt, $^3J_{HH}$ = 8.6 Hz, $^3J_{HP}$ = 4.2 Hz, $^3J_{HP}$ = 8.6 Hz, 1H, CHCHCO), 4.13 (ddpt, $^3J_{HH}$ = 8.6 Hz, $^3J_{HP}$ = 4.2 Hz, $^3J_{HP}$ = 8.6 Hz, $^2J_{HPt}$ = 59.0 Hz, 1H, CHCHCO), 4.43 (dpt, $^3J_{HH}$ = 8.6 Hz, $^3J_{HP}$ = 5.5 Hz, $^3J_{HP}$ = 8.6 Hz, 1H, C$_{Ar}$CHCH), 4.43 (ddpt, $^3J_{HH}$ = 8.6 Hz, $^3J_{HP}$ = 5.5 Hz, $^3J_{HP}$ = 8.6 Hz, $^2J_{HPt}$ = 51.8 Hz, 1H, C$_{Ar}$CHCH), 6.44 (d, $^3J_{HH}$ = 7.8 Hz, 2H, H_{Ar}), 6.74 (d, $^3J_{HH}$ = 7.3 Hz, 2H, H_{Ar}), 6.80 - 6.99 (m, 19H, H_{Ar}), 7.04 (d, $^3J_{HH}$ = 7.8 Hz, 2H, H_{Ar}), 7.08 - 7.19 (m, 7H, H_{Ar}), 7.53 - 7.66 (m, 6H, H_{Ar}).

^{13}C-NMR (100.53 MHz; C$_6$D$_6$): δ = 20.8 (CH_3), 48.9 (dd, $^2J_{CP}$ = 5.7 Hz, $^2J_{CP}$ = 28.6 Hz, CHCHCO), 48.9 (ddd, $^2J_{CP}$ = 5.7 Hz, $^2J_{CP}$ = 28.6 Hz, $^1J_{CPt}$ = 201.0 Hz,

CH$CHCO$), 59.8 (dd, $^2J_{CP}$ = 4.6 Hz, $^2J_{CP}$ = 33.9 Hz, C$_{Ar}$$CHCH$), 59.8 (ddd, $^2J_{CP}$ = 4.6 Hz, $^2J_{CP}$ = 33.9 Hz, $^1J_{CPt}$ = 194.2 Hz, C$_{Ar}$$CHCH$), 122.3 ($C_{Ar}$H), 124.9 (bm, C_{Ar}H), 125.8 (d, $^5J_{CP}$ = 3.0 Hz, C_{Ar}H), 125.8 (dd, $^5J_{CP}$ = 3.0 Hz, $^4J_{CPt}$ = 14.2 Hz, C_{Ar}H), 127.9 (C_{Ar}H), 128.2 (C_{Ar}H), 128.4 (C_{Ar}H), 128.5 (C_{Ar}H), 128.6 (C_{Ar}), 129.6 (d, $^5J_{CP}$ = 3.5 Hz, C_{Ar}H), 129.9 (d, $^5J_{CP}$ = 1.3 Hz, C_{Ar}H), 133.8 (C_{Ar}H), 134.0 (C_{Ar}P), 134.0 (d, $^3J_{CPt}$ = 21.0 Hz, C_{Ar}P), 134.1 (C_{Ar}P), 134.1 (d, $^3J_{CPt}$ = 21.0 Hz, C_{Ar}P), 134.3 (C_{Ar}P), 134.3 (d, $^3J_{CPt}$ = 19.0 Hz, C_{Ar}P), 134.5 (C_{Ar}P), 134.5 (d, $^3J_{CPt}$ = 19.0 Hz, C_{Ar}P), 134.8 (d, $^2J_{CP}$ = 2.5 Hz, C_{Ar}P), 134.8 (dd, $^2J_{CP}$ = 2.5 Hz, $^3J_{CPt}$ = 33.2 Hz, C_{Ar}P), 135.3 (d, $^2J_{CP}$ = 2.6 Hz, C_{Ar}P), 135.3 (dd, $^2J_{CP}$ = 2.6 Hz, $^3J_{CPt}$ = 29.0 Hz, C_{Ar}P), 136.5 (d, $^2J_{CP}$ = 2.3 Hz, C_{Ar}P), 136.5 (dd, $^2J_{CP}$ = 2.3 Hz, $^3J_{CPt}$ = 29.7 Hz, C_{Ar}P), 136.9 (d, $^2J_{CP}$ = 2.4 Hz, C_{Ar}P), 136.9 (dd, $^2J_{CP}$ = 2.4 Hz, $^3J_{CPt}$ = 29.0 Hz, C_{Ar}P), 149.0 (d, $^3J_{CP}$ = 7.5 Hz, C_{Ar}CH), 149.0 (dd, $^3J_{CP}$ = 7.5 Hz, $^2J_{CPt}$ = 34.7 Hz, C_{Ar}CH), 150.4 (C_{Ar}O), 171.7 (dd, $^3J_{CP}$ 1.8 Hz, $^3J_{CP}$ = 5.2 Hz, CHCOO), 171.7 (ddd, $^3J_{CP}$ = 1.8 Hz, $^3J_{CP}$ = 5.2 Hz, $^2J_{CPt}$ = 44.3 Hz, CHCOO).

^{19}F-NMR (376.50 MHz; C$_6$D$_6$): δ = -61.18.

^{31}P-NMR (161.97 MHz; C$_6$D$_6$): δ = 26.7 (d, $^2J_{PP}$ = 36.9 Hz), 26.7 (dd, $^2J_{PP}$ = 36.9 Hz, $^1J_{PPt}$ = 4097.6 Hz), 27.8 (dd, $^1J_{PF}$ = 2.8 Hz, $^2J_{PP}$ = 36.9 Hz), 27.8 (ddd, $^1J_{PF}$ = 2.8 Hz, $^2J_{PP}$ = 36.9 Hz, $^1J_{PPt}$ = 3695.0 Hz).

^{195}Pt-NMR (85.48 MHz; C$_6$D$_6$): δ = -5057.8 (dd, $^1J_{PPt}$ = 3678.6 Hz, $^1J_{PPt}$ = 4104.7 Hz).

FAB-MS: m/z (%): 719.1 (100) [Pt(PPh$_3$)(PPh$_2$C$_6$H$_4$)$^+$].

FT-IR (ATR, cm^{-1}): 3050 (w), 1702 (m, CO), 1590 (w), 1433 (m), 1322 (s), 1257 (w), 1182 (w), 1157 (m), 1113 (s), 1096 (s), 1064 (s), 1013 (w), 833 (m), 742 (m), 692 (s), 590 (w), 530 (m), 507 (s), 424 (m).

Bis(triphenylphosphine)-(η^2-phenyl 4-trifluoromethylcinnamate)palladium(0)

PdPhCF$_3$-Ph: C$_{52}$H$_{41}$F$_3$O$_2$P$_2$Pd, M = 923.24 *g/mol*, deep yellow solid, yield: 98%, **route**: *Pd-I*.

^{1}H-NMR (400.13 MHz; C$_6$D$_6$): δ = 4.76 - 4.85 (bm, 1H, CHCHCO), 5.15 - 5.28 (bm, 1H, C$_{Ar}$CHCH), 6.47 (bd, $^3J_{HH}$ = 7.9 Hz, 2H, H_{Ar}), 6.81 (bd, $^3J_{HH}$ = 8.0 Hz, 2H, H_{Ar}), 6.84 - 7.11 (bm, 29H, H_{Ar}), 7.52 - 7.63 (bm, 6H, H_{Ar}).

^{1}H-NMR (400.13 MHz; 278 K; C$_6$D$_6$): δ = 4.86 (dpt, $^3J_{HH}$ = 10.1 Hz, $^3J_{HP}$ = 6.2 Hz, $^3J_{HP}$ = 6.2 Hz, 1H, CHCHCO), 5.22 (dpt, $^3J_{HH}$ = 10.1 Hz, $^3J_{HP}$ = 7.5 Hz, $^3J_{HP}$ = 7.5 Hz, 1H, C$_{Ar}$CHCH), 6.44 (d, $^3J_{HH}$ = 7.9 Hz, 2H, H_{Ar}), 6.77 (d, $^3J_{HH}$ = 7.6 Hz, 2H, H_{Ar}), 6.83 - 7.12 (m, 29H, H_{Ar}), 7.54 - 7.62 (m, 6H, H_{Ar}).

^{13}C-NMR (100.62 MHz; C$_6$D$_6$): δ = 59.2 (dd, $^2J_{CP}$ = 5.3 Hz, $^2J_{CP}$ = 19.0 Hz, CHCHCO), 73.6 (dd, $^2J_{CP}$ = 2.6 Hz, $^2J_{CP}$ = 22.7 Hz, C$_{Ar}$CHCH), 122.5 (C_{Ar}H), 125.4 (bs, C_{Ar}H), 127.9 (C_{Ar}H), 128.2 (C_{Ar}H), 128.4 (C_{Ar}H), 128.7 (C_{Ar}H), 129.1 (C_{Ar}H), 129.5 (d, $^3J_{CP}$ = 21.8 Hz, C_{Ar}H), 134.1 (d, $^2J_{CP}$ = 14.4 Hz, C_{Ar}H), 134.4 (C_{Ar}), 134.4 (d, $^2J_{CP}$ = 15.6 Hz, C_{Ar}H), 135.2 (d, $^1J_{CP}$ = 28.8 Hz, C_{Ar}P), 136.8 (d, $^1J_{CP}$ = 27.5 Hz, C_{Ar}P), 146.6 (d, $^3J_{CP}$ = 5.8 Hz, CHC$_{Ar}$), 152.5 (C_{Ar}O), 169.5 (dd, $^3J_{CP}$ = 2.2 Hz, $^3J_{CP}$ = 3.9 Hz, CHCOO).

^{19}F-NMR (376.50 MHz; C$_6$D$_6$): δ = -61.4.

^{31}P-NMR (161.97 MHz; C$_6$D$_6$): δ = 25.0 (d, $^2J_{PP}$ = 10.6 Hz), 26.4 (d, $^2J_{PP}$ = 10.6 Hz).

^{31}P-NMR (161.97 MHz; 278 K; C$_6$D$_6$): δ = 24.7 (d, $^2J_{PP}$ = 10.1 Hz), 26.3 (dd, $^1J_{PF}$ = 2.9 Hz, $^2J_{PP}$ = 10.1 Hz).

FT-IR (ATR, cm^{-1}): 3047 (w), 1699 (m, CO), 1609 (w), 1477 (w), 1433 (m), 1320 (s), 1272 (w), 1188 (w), 1161 (m), 1116 (s), 1102 (s), 1064 (s), 1026 (w), 968 (w), 831 (w), 740 (s), 692 (s), 524 (m), 503 (s), 405 (m).

Bis(triphenylphosphine)-(η^2-phenyl 4-trifluoromethylcinnamate)platinum(0)

PtPhCF$_3$-Ph: C$_{52}$H$_{41}$F$_3$O$_2$P$_2$Pt, M = 1011.90 *g/mol*, white solid, yield: >99%, **route**: *Pt-VI.*

^{1}H-NMR (400.13 MHz; C$_6$D$_6$): δ = 4.13 (dpt, $^3J_{HH}$ = 8.7 Hz, $^3J_{HP}$ = 4.2 Hz, $^3J_{HP}$ = 8.7 Hz, 1H, CHC*H*CO), 4.13 (ddpt, $^3J_{HH}$ = 8.7 Hz, $^3J_{HP}$ = 4.2 Hz, $^3J_{HP}$ = 8.7 Hz, $^2J_{HPt}$ = 59.2 Hz, 1H, CHC*H*CO), 4.41 (dpt, $^3J_{HH}$ = 8.7 Hz, $^3J_{HP}$ = 5.5 Hz, $^3J_{HP}$ = 8.7 Hz, 1H, C$_{Ar}$C*H*CH), 4.41 (ddpt, $^3J_{HH}$ = 8.7 Hz, $^3J_{HP}$ = 5.5 Hz, $^3J_{HP}$ = 8.7 Hz, $^2J_{HPt}$ = 52.0 Hz, 1H, C$_{Ar}$C*H*CH), 6.43 (d, $^3J_{HH}$ = 7.8 Hz, 2H, H_{Ar}), 6.79 - 6.98 (m, 20H, H_{Ar}), 6.98 - 7.07 (m, 4H, H_{Ar}), 7.07 - 7.18 (m, 7H, H_{Ar}), 7.53 - 7.64 (m, 6H, H_{Ar}).

^{13}C-NMR (100.62 MHz; C$_6$D$_6$): δ = 48.7 (dd, $^2J_{CP}$ = 6.0 Hz, $^2J_{CP}$ = 30.3 Hz, CH*C*HCO), 48.7 (ddd, $^2J_{CP}$ = 6.0 Hz, $^2J_{CP}$ = 30.3 Hz, $^1J_{CPt}$ = 209.8 Hz, CH*C*HCO), 59.8 (dd, $^2J_{CP}$ = 4.3 Hz, $^2J_{CP}$ = 33.9 Hz, C$_{Ar}$*C*HCH), 59.8 (ddd, $^2J_{CP}$ = 4.3 Hz, $^2J_{CP}$ = 33.9 Hz, $^1J_{CPt}$ = 211.3 Hz, C$_{Ar}$*C*HCH), 115.7 (C_{Ar}H), 122.5 (bm, C_{Ar}H), 125.8 (d, $^5J_{CP}$ = 3.1 Hz, C_{Ar}H), 125.8 (dd, $^5J_{CP}$ = 3.1 Hz, $^4J_{CPt}$ = 13.6 Hz, C_{Ar}H), 127.9 (C_{Ar}H), 128.2 (C_{Ar}H), 128.4 (C_{Ar}H), 128.5 (C_{Ar}H), 128.6 (C_{Ar}H), 129.1 (C_{Ar}H), 129.7 (d, $^5J_{CP}$ = 1.8 Hz, C_{Ar}H), 129.8 (C_{Ar}), 129.9 (d, $^5J_{CP}$ = 1.6 Hz, C_{Ar}H), 134.0 (C_{Ar}P), 134.0 (d, $^3J_{CPt}$ = 21.2 Hz, C_{Ar}P), 134.1 (C_{Ar}P), 134.1 (d, $^3J_{CPt}$ = 15.6 Hz, C_{Ar}P), 134.3 (C_{Ar}P), 134.3 (d, $^3J_{CPt}$ = 19.0 Hz, C_{Ar}P), 134.4 (C_{Ar}P), 134.4 (d, $^3J_{CPt}$ = 19.2 Hz, C_{Ar}P), 134.8 (d, $^2J_{CP}$ = 2.5 Hz, C_{Ar}P), 134.8 (dd, $^2J_{CP}$ = 2.5 Hz, $^3J_{CPt}$ = 34.4 Hz, C_{Ar}P), 135.2 (d, $^2J_{CP}$ = 2.4 Hz, C_{Ar}P), 135.2 (dd, $^2J_{CP}$ = 2.4 Hz, $^3J_{CPt}$ = 35.3 Hz, C_{Ar}P), 136.4 (d, $^2J_{CP}$ = 2.3 Hz, C_{Ar}P), 136.4 (dd, $^2J_{CP}$ = 2.3 Hz, $^3J_{CPt}$ = 29.2 Hz, C_{Ar}P), 136.8 (d, $^2J_{CP}$ = 2.3 Hz, C_{Ar}P), 136.8 (dd, $^2J_{CP}$ = 2.3 Hz, $^3J_{CPt}$ = 29.6 Hz, C_{Ar}P), 148.9 (d, $^3J_{CP}$ = 7.3 Hz, C_{Ar}CH), 148.9 (dd, $^3J_{CP}$ = 7.3 Hz, $^2J_{CPt}$ = 38.1 Hz, C_{Ar}CH), 152.6 (C_{Ar}O), 171.7 (dd, $^3J_{CP}$ 2.3 Hz, $^3J_{CP}$ = 5.3 Hz, CH*C*OO), 171.7 (ddd, $^3J_{CP}$ = 2.3 Hz, $^3J_{CP}$ = 5.3 Hz, $^2J_{CPt}$ = 47.4 Hz, CH*C*OO).

^{19}F-NMR (376.50 MHz; C$_6$D$_6$): δ = -61.28.

^{31}P-NMR (161.97 MHz; C$_6$D$_6$): δ = 26.6 (d, $^2J_{PP}$ = 36.5 Hz), 26.6 (dd, $^2J_{PP}$ = 36.5 Hz, $^1J_{PPt}$ = 4107.8 Hz), 27.7 (dd, $^1J_{PF}$ = 2.8 Hz, $^2J_{PP}$ = 36.5 Hz), 27.7 (ddd, $^1J_{PF}$ = 2.8 Hz, $^2J_{PP}$ = 36.5 Hz, $^1J_{PPt}$ = 3685.8 Hz).

^{195}Pt-NMR (85.48 MHz; C$_6$D$_6$): δ = -5056.6 (dd, $^1J_{PPt}$ = 3675.1 Hz, $^1J_{PPt}$ = 4109.6 Hz).

FAB-MS: m/z (%): 719.1 (100) [Pt(PPh$_3$)(PPh$_2$C$_6$H$_4$)$^+$].

FT-IR (ATR, cm^{-1}): 3056 (w), 1707 (m, CO), 1590 (w), 1478 (m), 1433 (s), 1322 (s), 1265 (m), 1182 (w), 1156 (m), 1113 (s), 1095 (s), 1064 (s), 1013 (w), 735 (m), 742 (s), 692 (s), 529 (m), 507 (s), 423 (m).

Bis(triphenylphosphine)-(η^2-4-nitrophenyl 4-trifluoromethylcinnamate)palladium(0)

PdPhCF$_3$-PhNO$_2$: C$_{52}$H$_{40}$F$_3$NO$_4$P$_2$Pd, M = 968.24 *g/mol*, deep yellow solid, yield: >99%, **route**: *Pd-I*.

^{1}H-NMR (400.13 MHz; C$_6$D$_6$): δ = 4.66 - 4.76 (bm, 1H, CHC*H*CO), 5.06 - 5.17 (bm, 1H, C$_{Ar}$C*H*CH), 6.47 (bd, $^3J_{HH}$ = 7.8 Hz, 2H, H_{Ar}), 6.58 (bd, $^3J_{HH}$ = 7.8 Hz, 2H, H_{Ar}), 6.81 - 6.97 (bm, 18H, H_{Ar}), 6.97 - 7.06 (bm, 8H, H_{Ar}), 7.41 - 7.52 (bm, 6H, H_{Ar}), 7.75 (bd, $^3J_{HH}$ = 7.8 Hz, 2H, H_{Ar}).

^{1}H-NMR (400.13 MHz; 278 K; C$_6$D$_6$): δ = 4.77 (ddd, $^3J_{HH}$ = 10.1 Hz, $^3J_{HP}$ = 5.4 Hz, $^3J_{HP}$ = 6.6 Hz, 1H, CHC*H*CO), 5.12 (ddd, $^3J_{HH}$ = 10.1 Hz, $^3J_{HP}$ = 6.3 Hz, $^3J_{HP}$ = 10.1 Hz, 1H, C$_{Ar}$C*H*CH), 6.44 (d, $^3J_{HH}$ = 7.9 Hz, 2H, H_{Ar}), 6.52 (d, $^3J_{HH}$ = 9.1 Hz, 2H, H_{Ar}), 6.80 - 6.96 (m, 18H, H_{Ar}), 6.97 - 7.06 (m, 8H, H_{Ar}), 7.42 - 7.53 (m, 6H, H_{Ar}), 7.73 (d, $^3J_{HH}$ = 9.1 Hz, 2H, H_{Ar}).

^{13}C-NMR (100.53 MHz; C$_6$D$_6$): δ = 58.0 (dd, $^2J_{CP}$ = 5.5 Hz, $^2J_{CP}$ = 18.8 Hz, CHCHCO), 72.6 (dd, $^2J_{CP}$ = 1.9 Hz, $^2J_{CP}$ = 24.0 Hz, C$_{Ar}$CHCH), 122.3 (C_{Ar}H), 124.7 (C_{Ar}H), 125.4 (C_{Ar}H), 125.5 (bs, C_{Ar}H), 127.9 (C_{Ar}H), 128.2 (C_{Ar}H), 128.5 (C_{Ar}H), 128.6 (C_{Ar}H), 128.7 (C_{Ar}H), 128.8 (C_{Ar}H), 129.7 (d, $^3J_{CP}$ = 25.6 Hz, C_{Ar}H), 134.0 (d, $^2J_{CP}$ = 14.2 Hz, C_{Ar}H), 134.3 (d, $^2J_{CP}$ = 15.2 Hz, C_{Ar}H), 134.9 (dd, $^1J_{CP}$ = 29.8 Hz, $^3J_{CP}$ = 1.4 Hz, C_{Ar}P), 136.4 (dd, $^1J_{CP}$ = 28.6 Hz, $^3J_{CP}$ = 1.2 Hz, C_{Ar}P), 144.7 (C_{Ar}O), 146.3 (d, $^3J_{CP}$ = 7.1 Hz, CHC_{Ar}), 156.9 (C_{Ar}N), 168.4 (dd, $^3J_{CP}$ = 2.1 Hz, $^3J_{CP}$ = 4.3 Hz, CHCOO).

^{19}F-NMR (376.50 MHz; C$_6$D$_6$): δ = -61.5.

^{31}P-NMR (161.97 MHz; C$_6$D$_6$): δ = 25.1 (d, $^2J_{PP}$ = 7.1 Hz), 26.1 (dd, $^1J_{PF}$ = 2.9 Hz, $^2J_{PP}$ = 7.1 Hz).

^{31}P-NMR (161.97 MHz; 278 K; C$_6$D$_6$): δ = 24.8 (d, $^2J_{PP}$ = 6.8 Hz), 26.0 (dd, $^1J_{PF}$ = 3.2 Hz, $^2J_{PP}$ = 6.8 Hz).

FT-IR (ATR, cm^{-1}): 3047 (w), 1699 (s, CO), 1609 (m), 1518 (m), 1477 (m), 1433 (m), 1343 (w), 1319 (s), 1271 (w), 1198 (w), 1163 (m), 1113 (s), 1091 (s), 1063 (s), 1012 (w), 966 (w), 897 (w), 829 (w), 741 (m), 692 (s), 596 (w), 525 (m), 504 (s), 406 (m).

Bis(triphenylphosphine)-(η^2-4-nitrophenyl 4-trifluoromethylcinnamate)platinum(0)

PtPhCF$_3$-PhNO$_2$: C$_{52}$H$_{40}$F$_3$NO$_4$P$_2$Pt, M = 1056.90 g/mol, white solid, yield: >99%, route: *Pt-VI*.

^{1}H-NMR (400.13 MHz; C$_6$D$_6$): δ = 4.04 (dpt, $^3J_{HH}$ = 8.9 Hz, $^3J_{HP}$ = 3.4 Hz, $^3J_{HP}$ = 8.9 Hz, 1H, CHCHCO), 4.04 (ddpt, $^3J_{HH}$ = 8.9 Hz, $^3J_{HP}$ = 3.4 Hz, $^3J_{HP}$ = 8.9 Hz, $^2J_{HPt}$ = 58.6 Hz, 1H, CHCHCO), 4.32 (dpt, $^3J_{HH}$ = 8.9 Hz, $^3J_{HP}$ = 5.0 Hz, $^3J_{HP}$ = 8.9 Hz, 1H, C$_{Ar}$CHCH), 4.32 (ddpt, $^3J_{HH}$ = 8.9 Hz, $^3J_{HP}$ = 5.0 Hz, $^3J_{HP}$ = 8.9 Hz,

$^2J_{HPt}$ = 52.1 Hz, 1H, $C_{Ar}CHCH$), 6.44 (d, $^3J_{HH}$ = 7.8 Hz, 2H, H_{Ar}), 6.58 (d,
$^3J_{HH}$ = 7.8 Hz, 2H, H_{Ar}), 6.80 - 6.94 (m, 19H, H_{Ar}), 7.03 - 7.12 (m, 7H, H_{Ar}), 7.41 - 7.54
(m, 6H, H_{Ar}), 7.75 (d, $^3J_{HH}$ = 7.8 Hz, 2H, H_{Ar}).

^{13}C-NMR (100.53 MHz; C_6D_6): δ = 47.9 (dd, $^2J_{CP}$ = 5.5 Hz, $^2J_{CP}$ = 28.4 Hz,
CHCHCO), 47.9 (ddd, $^2J_{CP}$ = 5.5 Hz, $^2J_{CP}$ = 28.4 Hz, $^1J_{CPt}$ = 193.8 Hz, CHCHCO),
59.1 (dd, $^2J_{CP}$ = 3.5 Hz, $^2J_{CP}$ = 32.6 Hz, $C_{Ar}CHCH$), 59.1 (ddd, $^2J_{CP}$ = 3.5 Hz,
$^2J_{CP}$ = 32.6 Hz, $^1J_{CPt}$ = 206.5 Hz, $C_{Ar}CHCH$), 122.4 ($C_{Ar}H$), 124.7 ($C_{Ar}H$), 125.0 (bm,
$C_{Ar}H$), 125.8 (d, $^5J_{CP}$ = 3.1 Hz, $C_{Ar}H$), 125.8 (dd, $^5J_{CP}$ = 3.1 Hz, $^4J_{CPt}$ = 13.0 Hz,
$C_{Ar}H$), 127.9 ($C_{Ar}H$), 128.2 ($C_{Ar}H$), 128.4 ($C_{Ar}H$), 128.5 ($C_{Ar}H$), 128.6 (C_{Ar}), 129.8
(d, $^5J_{CP}$ = 1.8 Hz, $C_{Ar}H$), 130.1 (d, $^5J_{CP}$ = 1.6 Hz, $C_{Ar}H$), 133.9 ($C_{Ar}P$), 133.9 (d,
$^3J_{CPt}$ = 21.2 Hz, $C_{Ar}P$), 134.0 ($C_{Ar}P$), 134.0 (d, $^3J_{CPt}$ = 24.8 Hz, $C_{Ar}P$), 134.2 ($C_{Ar}P$),
134.2 (d, $^3J_{CPt}$ = 20.0 Hz, $C_{Ar}P$), 134.3 ($C_{Ar}P$), 134.3 (d, $^3J_{CPt}$ = 18.8 Hz, $C_{Ar}P$), 134.4
(d, $^2J_{CP}$ = 2.6 Hz, $C_{Ar}P$), 134.4 (dd, $^2J_{CP}$ = 2.6 Hz, $^3J_{CPt}$ = 31.7 Hz, $C_{Ar}P$), 134.9
(d, $^2J_{CP}$ = 2.5 Hz, $C_{Ar}P$), 134.9 (dd, $^2J_{CP}$ = 2.5 Hz, $^3J_{CPt}$ = 31.2 Hz, $C_{Ar}P$), 136.1
(d, $^2J_{CP}$ = 2.1 Hz, $C_{Ar}P$), 136.1 (dd, $^2J_{CP}$ = 2.1 Hz, $^3J_{CPt}$ = 29.4 Hz, $C_{Ar}P$), 136.5
(d, $^2J_{CP}$ = 2.3 Hz, $C_{Ar}P$), 136.5 (dd, $^2J_{CP}$ = 2.3 Hz, $^3J_{CPt}$ = 29.9 Hz, $C_{Ar}P$), 144.7
($C_{Ar}O$), 148.6 (d, $^3J_{CP}$ = 6.0 Hz, $C_{Ar}CH$), 148.6 (dd, $^3J_{CP}$ = 6.0 Hz, $^2J_{CPt}$ = 42-7 Hz,
$C_{Ar}CH$), 157.0 ($C_{Ar}NO_2$), 170.4 (dd, $^3J_{CP}$ 2.1 Hz, $^3J_{CP}$ = 4.8 Hz, CHCOO), 170.4 (ddd,
$^3J_{CP}$ = 2.1 Hz, $^3J_{CP}$ = 4.8 Hz, $^2J_{CPt}$ = 52.8 Hz, CHCOO).

^{19}F-NMR (376.50 MHz; C_6D_6): δ = -61.36.

^{31}P-NMR (161.97 MHz; C_6D_6): δ = 26.2 (d, $^2J_{PP}$ = 34.1 Hz), 26.2 (dd, $^2J_{PP}$ = 34.1 Hz,
$^1J_{PPt}$ = 4142.9 Hz), 27.2 (dd, $^1J_{PF}$ = 2.7 Hz, $^2J_{PP}$ = 34.1 Hz), 27.2 (ddd, $^1J_{PF}$ = 2.7 Hz,
$^2J_{PP}$ = 34.1 Hz, $^1J_{PPt}$ = 3648.7 Hz).

^{195}Pt-NMR (85.48 MHz; C_6D_6): δ = -5055.9 (dd, $^1J_{PPt}$ = 3642.3 Hz,
$^1J_{PPt}$ = 4145.5 Hz).

FAB-MS: m/z (%): 719.0 (100) [Pt(PPh$_3$)(PPh$_2$C$_6$H$_4$)$^+$].

FT-IR (ATR, cm^{-1}): 3043 (w), 1710 (m, CO), 1610 (w), 1518 (m), 1479 (w), 1434 (m),
1343 (m), 1321 (s), 1265 (w), 1219 (w), 1154 (m), 1090 (s), 1063 (s), 1012 (w), 984 (w),
834 (m), 741 (m), 691 (s), 605 (w), 544 (m), 507 (s), 423 (m).

Bis(triphenylphosphine)-(η^2-4-methoxyphenyl 4-nitrocinnamate)palladium(0)

PdPhNO$_2$-PhOMe: C$_{52}$H$_{43}$NO$_5$P$_2$Pd, M $=$ 930.27 g/mol, orange solid, yield: 75%, **route**: *Pd-I.*

Single crystals suitable for x-ray analysis were obtained by slow diffusion of pentane into a toluene solution.

^{1}H-NMR (400.13 MHz; C$_6$D$_6$): $\delta = 3.27$ (s, 3H, OCH_3), 4.69 (dpt, $^3J_{HH} = 10.1$ Hz, $^3J_{HP} = 6.6$ Hz, $^3J_{HP} = 6.6$ Hz, 1H, CHCHCO), 5.18 (ddd, $^3J_{HH} = 10.1$ Hz, $^3J_{HP} = 5.6$ Hz, $^3J_{HP} = 8.0$ Hz, 1H, C$_{Ar}$CHCH), 6.29 - 6.34 (m, 2H, H_{Ar}), 6.60 - 6.64 (m, 2H, H_{Ar}), 6.68 - 6.73 (m, 2H, H_{Ar}), 6.79 - 6.85 (m, 6H, H_{Ar}), 6.87 - 6.96 (m, 12H, H_{Ar}), 7.00 - 7.07 (m, 6H, H_{Ar}), 7.50 - 7.57 (m, 6H, H_{Ar}), 7.59 - 7.63 (m, 2H, H_{Ar}).

^{13}C-NMR (100.53 MHz; C$_6$D$_6$): $\delta = 55.1$ (OCH$_3$), 58.6 (dd, $^2J_{CP} = 5.1$ Hz, $^2J_{CP} = 19.9$ Hz, CHCHCO), 72.7 (dd, $^2J_{CP} = 2.4$ Hz, $^2J_{CP} = 22.8$ Hz, C$_{Ar}$$C$HCH), 114.3 ($C_{Ar}$H), 123.1 ($C_{Ar}$H), 124.0 (bs, C_{Ar}H), 124.9 (d, $^3J_{CP} = 2.0$ Hz, C_{Ar}H), 128.4 (C_{Ar}), 128.6 (C_{Ar}H), 128.7 (C_{Ar}H), 129.7 (dd, $^1J_{CP} = 18.8$ Hz, $^3J_{CP} = 0.8$ Hz, C_{Ar}P), 134.0 (d, $^2J_{CP} = 14.4$ Hz, C_{Ar}H), 134.4 (d, $^2J_{CP} = 15.1$ Hz, C_{Ar}H), 134.9 (dd, $^1J_{CP} = 29.5$ Hz, $^3J_{CP} = 1.3$ Hz, C_{Ar}P), 136.4 (dd, $^1J_{CP} = 29.1$ Hz, $^3J_{CP} = 1.5$ Hz, C_{Ar}P), 144.7 (dd, $^3J_{CP} = 1.0$ Hz, $^3J_{CP} = 2.0$ Hz, C_{Ar}CH), 145.9 (C_{Ar}O), 149.9 (d, $^3J_{CP} = 6.3$ Hz, C_{Ar}N), 157.1 (C_{Ar}O), 170.0 (dd, $^3J_{CP} = 1.9$ Hz, $^3J_{CP} = 4.4$ Hz, CHCOO).

^{31}P-NMR (161.97 MHz; C$_6$D$_6$): $\delta = 25.0$ (d, $^2J_{PP} = 5.2$ Hz), 26.9 (d, $^2J_{PP} = 5.2$ Hz).

FAB-MS: m/z (%): 368.0 (17.24) [Pd(PPh$_3$)$^+$], 630.0 (38.46) [Pd(PPh$_3$)$_2$$^+$].

FT-IR (ATR, cm^{-1}): 1708 (s, CO), 1585 (m), 1506 (s), 1433 (s), 1334 (s), 1305 (w), 1273 (w), 1251 (w), 1197 (w), 1180 (m), 1119 (s), 1092 (m), 1025 (w), 964 (w), 889 (w), 820 (m), 743 (s), 692 (s), 528 (m), 517 (s), 498 (s), 445 (w), 419 (m).

elemental analysis for C$_{52}$H$_{43}$NO$_5$P$_2$Pd (930.27 g/mol):

calcd.: C 67.14, H 4.66, N 1.51, O 8.60, P 6.66, Pd 11.44

found: C 66.82, H 4.89, N 1.59

Bis(triphenylphosphine)-(η^2-4-methoxyphenyl 4-nitrocinnamate)platinum(0)

PtPhNO$_2$-PhOMe: C$_{52}$H$_{43}$NO$_5$P$_2$Pt, M = 1018.93 *g/mol*, yellow solid, yield: 99%, **route**: *Pt-III*.

Single crystals suitable for X-ray analysis were obtained by slow diffusion of pentane into a toluene solution.

^{1}H-NMR (400.13 MHz; C$_6$D$_6$): δ = 3.27 (s, 3H, OCH$_3$), 4.06 (dpt, $^3J_{HH}$ = 8.8 Hz, $^3J_{HP}$ = 4.3 Hz, $^3J_{HP}$ = 8.8 Hz, 1H, CHC*H*CO), 4.06 (ddpt, $^3J_{HH}$ = 8.8 Hz, $^3J_{HP}$ = 4.3 Hz, $^3J_{HP}$ = 8.8 Hz, $^2J_{HPt}$ = 59.2 Hz, 1H, CHC*H*CO), 4.44 (dpt, $^3J_{HH}$ = 9.1 Hz, $^3J_{HP}$ = 4.9 Hz, $^3J_{HP}$ = 9.1 Hz, 1H, C$_{Ar}$C*H*CH), 4.44 (ddpt, $^3J_{HH}$ = 9.1 Hz, $^3J_{HP}$ = 4.9 Hz, $^3J_{HP}$ = 9.1 Hz, $^2J_{HPt}$ = 51.2 Hz, 1H, C$_{Ar}$C*H*CH), 6.29 (d, $^3J_{HH}$ = 7.3 Hz, 2H, H$_{Ar}$), 6.64 (d, $^3J_{HH}$ = 9.1 Hz, 2H, H$_{Ar}$), 6.72 (d, $^3J_{HH}$ = 5.9 Hz, 2H, H$_{Ar}$), 6.75 - 6.99 (m, 17H, H$_{Ar}$), 7.01 - 7.14 (m, 7H, H$_{Ar}$), 7.48 - 7.60 (m, 6H, H$_{Ar}$), 7.66 (d, $^3J_{HH}$ = 8.7 Hz, 2H, H$_{Ar}$).

^{13}C-NMR (100.53 MHz; C$_6$D$_6$): δ = 48.6 (dd, $^2J_{CP}$ = 5.6 Hz, $^2J_{CP}$ = 30.4 Hz, CH*C*HCO), 48.6 (ddd, $^2J_{CP}$ = 5.6 Hz, $^2J_{CP}$ = 30.4 Hz, $^1J_{CPt}$ = 206.0 Hz, CH*C*HCO), 55.0 (O*C*H$_3$), 59.5 (dd, $^2J_{CP}$ = 4.3 Hz, $^2J_{CP}$ = 32.4 Hz, C$_{Ar}$*C*HCH), 59.5 (ddd, $^2J_{CP}$ = 4.3 Hz, $^2J_{CP}$ = 32.4 Hz, $^1J_{CPt}$ = 187.6 Hz, C$_{Ar}$*C*HCH), 114.3 (*C*$_{Ar}$H), 123.1 (*C*$_{Ar}$H), 123.5 (d, $^5J_{CP}$ = 2.2 Hz, *C*$_{Ar}$H), 123.5 (dd, $^5J_{CP}$ = 2.2 Hz, $^4J_{CPt}$ = 9.8 Hz, *C*$_{Ar}$H), 125.4 (d, $^4J_{CP}$ = 3.0 Hz, *C*$_{Ar}$H), 125.4 (dd, $^4J_{CP}$ = 3.0 Hz, $^3J_{CPt}$ = 11.6 Hz, *C*$_{Ar}$H), 128.2 (*C*$_{Ar}$H), 128.4 (*C*$_{Ar}$H), 128.5 (*C*$_{Ar}$H), 129.8 (d, $^1J_{CPt}$ = 2.1 Hz, *C*$_{Ar}$H), 130.0 (d, $^1J_{CPt}$ = 2.1 Hz, *C*$_{Ar}$H), 133.9 (*C*$_{Ar}$H), 133.9 (d, $^3J_{CPt}$ = 20.8 Hz, *C*$_{Ar}$H), 134.1 (*C*$_{Ar}$H), 134.1 (d, $^3J_{CPt}$ = 21.0 Hz, *C*$_{Ar}$H), 134.3 (*C*$_{Ar}$H), 134.3 (d, $^3J_{CPt}$ = 20.2 Hz,

C_{Ar}H), 134.4 (C_{Ar}H), 134.4 (d, $^3J_{CPt}$ = 17.0 Hz, C_{Ar}H), 134.5 (d, $^1J_{CP}$ = 2.1 Hz, C_{Ar}P),

134.5 (dd, $^1J_{CP}$ = 2.1 Hz, $^2J_{CPt}$ = 27.0 Hz, C_{Ar}P), 134.9 (d, $^1J_{CP}$ = 2.2 Hz, C_{Ar}P),

134.9 (dd, $^1J_{CP}$ = 2.2 Hz, $^2J_{CPt}$ = 27.7 Hz, C_{Ar}P), 136.1 (d, $^1J_{CP}$ = 2.3 Hz, C_{Ar}P),

136.1 (dd, $^1J_{CP}$ = 2.3 Hz, $^2J_{CPt}$ = 28.5 Hz, C_{Ar}P), 136.5 (d, $^1J_{CP}$ = 2.2 Hz, C_{Ar}P), 136.5

(dd, $^1J_{CP}$ = 2.2 Hz, $^2J_{CPt}$ = 30.0 Hz, C_{Ar}P), 144.5 (d, $^3J_{CP}$ = 3.1 Hz, C_{Ar}N), 146.0

(C_{Ar}O), 152.7 (dd, $^3J_{CP}$ = 1.6 Hz, $^3J_{CP}$ = 6.7 Hz, C_{Ar}CH), 152.7 (ddd, $^3J_{CP}$ = 1.6 Hz,

$^3J_{CP}$ = 6.7 Hz, $^2J_{CPt}$ = 37.0 Hz, C_{Ar}CH), 157.1 (C_{Ar}OCH$_3$), 171.8 (dd, $^3J_{CP}$ 2.1 Hz,

$^3J_{CP}$ = 5.3 Hz, CHCOO), 171.8 (ddd, $^3J_{CP}$ = 2.1 Hz, $^3J_{CP}$ = 5.3 Hz, $^2J_{CPt}$ = 46.0 Hz,

CHCOO).

^{31}P-NMR (161.97 MHz; C$_6$D$_6$): δ = 26.7 (d, $^2J_{PP}$ = 32.9 Hz), 26.7 (dd, $^2J_{PP}$ = 32.9 Hz, $^1J_{PPt}$ = 4031.3 Hz), 28.0 (d, $^2J_{PP}$ = 32.9 Hz), 28.0 (dd, $^2J_{PP}$ = 32.9 Hz, $^1J_{PPt}$ = 3741.4 Hz).

^{195}Pt-NMR (85.48 MHz; C$_6$D$_6$): δ = -5048.8 (dd, $^1J_{PPt}$ = 3741.8 Hz, $^1J_{PPt}$ = 4032.3 Hz).

FAB-MS: m/z (%): 718.3 (100) [Pt(PPh$_3$)(PPh$_2$C$_6$H$_4$)$^+$], 1017.1 (0.73) [(C$_{16}$H$_{13}$NO$_5$)Pt(PPh$_3$)(PPh$_2$C$_6$H$_4$)$^+$].

ESI-MS: m/z (%): 759.1 (100) [Pt(PPh$_3$)(PPh$_2$C$_6$H$_4$)CH$_3$CN$^+$].

FT-IR (ATR, cm^{-1}): 3048 (w), 2987 (w), 1714 (m, CO), 1577 (m), 1506 (s), 1434 (m), 1395 (w), 1333 (s), 1251 (w), 1183 (m), 1115 (s), 1095 (m), 1017 (w), 979 (w), 895 (w), 835 (m), 744 (m), 693 (s), 577 (w), 539 (m), 524 (s), 508 (s), 418 (m).

elemental analysis for C$_{52}$H$_{43}$NO$_5$P$_2$Pt (1018.93 g/mol):

calcd.: C 61.30, H 4.25, N 1.37, O 7.85, P 6.08, Pt 19.15

found: C 58.66, H 4.17, N 1.36

Bis(triphenylphosphine)-(η^2-4-methylphenyl 4-nitrocinnamate)palladium(0)

PdPhNO$_2$-PhMe: C$_{52}$H$_{43}$NO$_4$P$_2$Pd, M = 914.27 g/mol, red solid, yield: 85%, **route**: *Pd-I*.

^{1}H-NMR (400.13 MHz; C$_6$D$_6$): δ = 2.07 (s, 3H, CH_3), 4.68 (ddd, $^3J_{HH}$ = 10.1 Hz, $^3J_{HP}$ = 6.6 Hz, $^3J_{HP}$ = 6.0 Hz, 1H, CHCHCO), 5.18 (ddd, $^3J_{HH}$ = 10.1 Hz, $^3J_{HP}$ = 8.0 Hz, $^3J_{HP}$ = 5.6 Hz, 1H, C$_{Ar}$CHCH), 6.29 - 6.34 (m, 2H, H_{Ar}), 6.71 - 6.76 (m, 2H, H_{Ar}), 6.80 - 6.86 (m, 8H, H_{Ar}), 6.87 - 6.99 (m, 12H, H_{Ar}), 7.00 - 7.08 (m, 6H, H_{Ar}), 7.51 - 7.59 (m, 6H, H_{Ar}), 7.60 - 7.64 (m, 2H, H_{Ar}).

^{13}C-NMR (100.53 MHz; C$_6$D$_6$): δ = 20.4 (CH_3), 58.2 (dd, $^2J_{CP}$ = 5.1 Hz, $^2J_{CP}$ = 19.9 Hz, CHCHCO), 72.4 (dd, $^2J_{CP}$ = 2.6 Hz, $^2J_{CP}$ = 22.9 Hz, C$_{Ar}$$C$HCH), 121.7 ($C_{Ar}$H), 123.6 (bs, C_{Ar}H), 124.6 (d, $^3J_{CP}$ = 1.9 Hz, C_{Ar}H), 128.0 (C_{Ar}), 128.2 (C_{Ar}H), 128.3 (C_{Ar}H), 129.3 (dd, $^1J_{CP}$ = 17.1 Hz, $^3J_{CP}$ = 0.9 Hz, C_{Ar}P), 129.3 (C_{Ar}H), 133.6 (C_{Ar}CH$_3$), 133.7 (d, $^2J_{CP}$ = 14.4 Hz, C_{Ar}H), 134.0 (d, $^2J_{CP}$ = 15.1 Hz, C_{Ar}H), 134.5 (dd, $^1J_{CP}$ = 29.5 Hz, $^3J_{CP}$ = 1.3 Hz, C_{Ar}P), 136.1 (dd, $^1J_{CP}$ = 29.0 Hz, $^3J_{CP}$ = 1.4 Hz, C_{Ar}P), 144.3 (dd, $^3J_{CP}$ = 1.1 Hz, $^3J_{CP}$ = 1.9 Hz, C_{Ar}CH), 149.5 (d, $^3J_{CP}$ = 6.3 Hz, C_{Ar}N), 149.8 (C_{Ar}O), 169.4 (dd, $^3J_{CP}$ = 1.8 Hz, $^3J_{CP}$ = 4.4 Hz, CHCOO).

^{31}P-NMR (161.97 MHz; C$_6$D$_6$): δ = 25.0 (d, $^2J_{PP}$ = 5.2 Hz), 26.9 (d, $^2J_{PP}$ = 5.2 Hz).

FAB-MS: m/z (%): 368.0 (15.76) [Pd(PPh$_3$)$^+$], 630.0 (40.88) [Pd(PPh$_3$)$_2$$^+$].

FT-IR (ATR, cm^{-1}): 3050 (w), 1701 (s, CO), 1583 (m), 1504 (m), 1433 (s), 1330 (s), 1278 (m), 1204 (m), 1126 (s), 1109 (s), 1092 (s), 971 (w), 849 (m), 738 (m), 692 (s), 531 (m), 516 (m), 501 (s), (m).

elemental analysis for C$_{52}$H$_{43}$NO$_4$P$_2$Pd (914.27 g/mol):

calcd.: C 68.31, H 4.74, N 1.53, O 7.00, P 6.78, Pd 11.64

found: C 70.15, 5.13, N 1.46

Bis(triphenylphosphine)-(η^2-4-methylphenyl 4-nitrocinnamate)platinum(0)

PtPhNO$_2$-PhMe: C$_{52}$H$_{43}$NO$_4$P$_2$Pt, M $=$ 1002.93 g/mol, yellow solid, yield: 99%, route: *Pt-III*.

Single crystals suitable for X-ray analysis were obtained by slow diffusion of pentane into a toluene solution.

^{1}H-NMR (400.13 MHz; C$_6$D$_6$): $\delta = 2.07$ (s, 3H, CH$_3$), 4.04 (dpt, $^3J_{HH} = 8.7$ Hz, $^3J_{HP} = 4.1$ Hz, $^3J_{HP} = 8.7$ Hz, 1H, CHCHCO), 4.04 (ddpt, $^3J_{HH} = 8.7$ Hz, $^3J_{HP} = 4.1$ Hz, $^3J_{HP} = 8.7$ Hz, $^2J_{HPt} = 57.8$ Hz, 1H, CHCHCO), 4.43 (dpt, $^3J_{HH} = 8.9$ Hz, $^3J_{HP} = 4.8$ Hz, $^3J_{HP} = 8.9$ Hz, 1H, C$_{Ar}$CHCH), 4.43 (ddpt, $^3J_{HH} = 8.9$ Hz, $^3J_{HP} = 4.8$ Hz, $^3J_{HP} = 8.9$ Hz, $^2J_{HPt} = 52.0$ Hz, 1H, C$_{Ar}$CHCH), 6.28 (d, $^3J_{HH} = 8.6$ Hz, 2H, H_{Ar}), 6.62 - 7.01 (m, 21H, H_{Ar}), 7.01 - 7.16 (m, 7H, H_{Ar}), 7.50 - 7.62 (m, 6H, H_{Ar}), 7.66 (d, $^3J_{HH} = 8.0$ Hz, 2H, H_{Ar}).

^{13}C-NMR (100.53 MHz; C$_6$D$_6$): $\delta = 20.8$ (CH$_3$), 48.7 (dd, $^2J_{CP} = 5.7$ Hz, $^2J_{CP} = 30.3$ Hz, CHCHCO), 48.7 (ddd, $^2J_{CP} = 5.7$ Hz, $^2J_{CP} = 30.3$ Hz, $^1J_{CPt} = 209.0$ Hz, CHCHCO), 59.6 (dd, $^2J_{CP} = 4.5$ Hz, $^2J_{CP} = 32.5$ Hz, C$_{Ar}$$C$HCH), 59.6 (ddd, $^2J_{CP} = 4.5$ Hz, $^2J_{CP} = 32.5$ Hz, $^1J_{CPt} = 184.4$ Hz, C$_{Ar}$$C$HCH), 122.2 ($C_{Ar}$H), 123.5 (d, $^5J_{CP} = 2.2$ Hz, C_{Ar}H), 123.5 (dd, $^5J_{CP} = 2.2$ Hz, $^4J_{CPt} = 6.6$ Hz, C_{Ar}H), 125.4 (d, $^4J_{CP} = 3.0$ Hz, C_{Ar}H), 125.4 (dd, $^4J_{CP} = 3.0$ Hz, $^3J_{CPt} = 12.0$ Hz, C_{Ar}H), 128.2 (C_{Ar}H), 128.4 (C_{Ar}H), 128.5 (C_{Ar}H), 129.6 (C_{Ar}H), 129.8 (d, $^1J_{CPt} = 2.1$ Hz, C_{Ar}H), 130.0 (d, $^1J_{CPt} = 2.0$ Hz, C_{Ar}H), 133.9 (C_{Ar}H), 133.9 (d, $^3J_{CPt} = 20.8$ Hz, C_{Ar}H), 134.1 (C_{Ar}H), 134.1 (d, $^3J_{CPt} = 20.2$ Hz, C_{Ar}H), 134.3 (C_{Ar}H), 134.3 (d, $^3J_{CPt} = 19.4$ Hz, C_{Ar}H), 134.4 (C_{Ar}H), 134.4 (d, $^3J_{CPt} = 20.6$ Hz, C_{Ar}H), 134.5 (d, $^1J_{CP} = 2.0$ Hz, C_{Ar}P), 134.5 (dd, $^1J_{CP} = 2.0$ Hz, $^2J_{CPt} = 29.4$ Hz, C_{Ar}P), 135.9 (d, $^1J_{CP} = 2.2$ Hz, C_{Ar}P), 135.9 (dd, $^1J_{CP} = 2.2$ Hz, $^2J_{CPt} = 29.2$ Hz, C_{Ar}P), 136.1 (d, $^1J_{CP} = 2.3$ Hz, C_{Ar}P), 136.1 (dd, $^1J_{CP} = 2.3$ Hz, $^2J_{CPt} = 29.2$ Hz, C_{Ar}P), 136.5 (d, $^1J_{CP} = 2.3$ Hz, C_{Ar}P), 136.5

(dd, $^1J_{CP}$ = 2.3 Hz, $^2J_{CPt}$ = 26.0 Hz, C_{Ar}P), 144.5 (d, $^3J_{CP}$ = 3.4 Hz, C_{Ar}N), 150.3 (C_{Ar}O), 152.7 (dd, $^3J_{CP}$ = 1.6 Hz, $^3J_{CP}$ = 6.8 Hz, C_{Ar}CH), 152.7 (ddd, $^3J_{CP}$ = 1.6 Hz, $^3J_{CP}$ = 6.8 Hz, $^2J_{CPt}$ = 40.0 Hz, C_{Ar}CH), 171.6 (dd, $^3J_{CP}$ 2.0 Hz, $^3J_{CP}$ = 5.3 Hz, CHCOO), 171.6 (ddd, $^3J_{CP}$ = 2.0 Hz, $^3J_{CP}$ = 5.3 Hz, $^2J_{CPt}$ = 41.6 Hz, CHCOO).

^{31}P-NMR (161.97 MHz; C_6D_6): δ = 26.1 (d, $^2J_{PP}$ = 32.9 Hz), 26.1 (dd, $^2J_{PP}$ = 32.9 Hz, $^1J_{PPt}$ = 4032.8 Hz), 27.3 (d, $^2J_{PP}$ = 33.0 Hz), 27.3 (dd, $^2J_{PP}$ = 33.0 Hz, $^1J_{PPt}$ = 3741.5 Hz).

^{195}Pt-NMR (85.48 MHz; C_6D_6): δ = -5048.2 (dd, $^1J_{PPt}$ = 3741.9 Hz, $^1J_{PPt}$ = 4033.1 Hz).

FAB-MS: m/z (%): 718.3 (100) [Pt(PPh$_3$)(PPh$_2$C$_6$H$_4$)$^+$], 1001.2 (0.84) [(C$_{16}$H$_{13}$NO$_4$)Pt(PPh$_3$)(PPh$_2$C$_6$H$_4$)$^+$].

ESI-MS: m/z (%): 760.1 (100) [Pt(PPh$_3$)(PPh$_3$)CH$_3$CN$^+$].

FT-IR (ATR, cm^{-1}): 3032 (w), 1713 (s, CO), 1585 (m), 1506 (s), 1434 (m), 1390 (w), 1334 (s), 1267 (m), 1206 (w), 1152 (m), 1113 (s), 1094 (s), 981 (w), 905 (w), 844 (m), 753 (s), 691 (s), 572 (w), 541 (m), 519 (s), 502 (s), 451 (m), 432 (m).

elemental analysis for C$_{52}$H$_{43}$NO$_4$P$_2$Pt (1002.93 g/mol):

calcd.: C 62.27, H 4.32, N 1.40, O 6.38, P 6.18, Pt 19.45

found: C 59.87, H 4.16, N 1.63

Bis(triphenylphosphine)-(η^2-phenyl 4-nitrocinnamate)palladium(0)

PdPhNO$_2$-Ph: C$_{51}$H$_{41}$NO$_4$P$_2$Pd, M = 900.24 g/mol, red solid, yield: 78%, **route**: *Pd-I*.

Single crystals suitable for X-ray analysis were obtained by slow diffusion of pentane into a toluene solution.

¹H-NMR (400.13 MHz; C₆D₆): δ = 4.67 (dpt, $^3J_{HH}$ = 10.0 Hz, $^3J_{HP}$ = 6.2 Hz, $^3J_{HP}$ = 6.2 Hz, 1H, CHCHCO), 5.16 (ddd, $^3J_{HH}$ = 10.0 Hz, $^3J_{HP}$ = 8.0 Hz, $^3J_{HP}$ = 5.7 Hz, 1H, C$_{Ar}$CHCH), 6.27 - 6.35 (m, 2H, H_{Ar}), 6.79 - 6.86 (m, 8H, H_{Ar}), 6.87 - 6.98 (m, 12H, H_{Ar}), 7.00 - 7.08 (m, 8H, H_{Ar}), 7.47 - 7.59 (m, 6H, H_{Ar}), 7.58 - 7.66 (m, 2H, H_{Ar}).

¹³C-NMR (100.62 MHz; C₆D₆): δ = 58.5 (dd, $^2J_{CP}$ = 5.1 Hz, $^2J_{CP}$ = 19.9 Hz, CHCHCO), 72.7 (dd, $^2J_{CP}$ = 2.6 Hz, $^2J_{CP}$ = 22.1 Hz, C$_{Ar}$CHCH), 122.4 (C$_{Ar}$H), 124.0 (d, $^3J_{CP}$ = 1.4 Hz, C$_{Ar}$H), 124.8 (C$_{Ar}$H), 125.0 (d, $^3J_{CP}$ = 2.3 Hz, C$_{Ar}$H), 128.4 (C$_{Ar}$), 128.6 (C$_{Ar}$H), 128.7 (C$_{Ar}$H), 129.1 (C$_{Ar}$H), 129.7 (dd, $^1J_{CP}$ = 17.2 Hz, $^3J_{CP}$ = 1.6 Hz, C$_{Ar}$P), 134.0 (d, $^2J_{CP}$ = 14.4 Hz, C$_{Ar}$H), 134.0 (d, $^2J_{CP}$ = 15.1 Hz, C$_{Ar}$H), 134.9 (dd, $^1J_{CP}$ = 29.6 Hz, $^3J_{CP}$ = 1.6 Hz, C$_{Ar}$P), 136.4 (dd, $^1J_{CP}$ = 29.1 Hz, $^3J_{CP}$ = 1.7 Hz, C$_{Ar}$P), 144.7 (dd, $^3J_{CP}$ = 3.3 Hz, $^3J_{CP}$ = 1.3 Hz, C$_{Ar}$CH), 149.9 (d, $^3J_{CP}$ = 6.0 Hz, C$_{Ar}$N), 152.4 (C$_{Ar}$O), 169.6 (dd, $^3J_{CP}$ = 2.0 Hz, $^3J_{CP}$ = 4.6 Hz, CHCOO).

³¹P-NMR (161.97 MHz; C₆D₆): δ = 25.0 (d, $^2J_{PP}$ = 4.8 Hz), 26.9 (d, $^2J_{PP}$ = 4.8 Hz).

FAB-MS: m/z (%): 368.0 (16.55) [Pd(PPh₃)⁺], 630.1 (39.71) [Pd(PPh₃)₂⁺].

FT-IR (ATR, cm⁻¹): 3063 (m), 1699 (m, CO), 1583 (m), 1504 (m), 1494 (w), 1468 (m), 1434 (m), 1343 (s), 1306 (w), 1277 (m), 1220 (w), 1200 (w), 1184 (w), 1123 (s), 1110 (s), 1092 (s), 1032 (w), 971 (w), 858 (w), 812 (w), 737 (s), 692 (s), (m), 500 (s), 468 (w), 418 (m).

elemental analysis for C₅₁H₄₁NO₄P₂Pd (900.24 g/mol):

calcd.: C 68.04, H 4.59, N 1.56, O 7.11, P 6.88, Pd 11.82

found: C 69.95, H 5.00, N 1.35

Bis(triphenylphosphine)-(η^2-phenyl 4-nitrocinnamate)platinum(0)

PtPhNO$_2$-Ph: $C_{51}H_{41}NO_4P_2Pt$, M = 988.90 *g/mol*, yellow solid, yield: 99%, **route:** *Pt-III.*

Single crystals suitable for X-ray analysis were obtained by slow diffusion of pentane into a toluene solution.

^{1}H-NMR (400.13 MHz; C_6D_6): δ = 4.03 (dpt, $^3J_{HH}$ = 8.3 Hz, $^3J_{HP}$ = 4.6 Hz, $^3J_{HP}$ = 8.3 Hz, 1H, CHC*H*CO), 4.03 (ddpt, $^3J_{HH}$ = 8.3 Hz, $^3J_{HP}$ = 4.6 Hz, $^3J_{HP}$ = 8.3 Hz, $^2J_{HPt}$ = 59.8 Hz, 1H, CHC*H*CO), 4.40 (dpt, $^3J_{HH}$ = 9.1 Hz, $^3J_{HP}$ = 4.7 Hz, $^3J_{HP}$ = 9.1 Hz, 1H, C_{Ar}C*H*CH), 4.40 (ddpt, $^3J_{HH}$ = 9.1 Hz, $^3J_{HP}$ = 4.7 Hz, $^3J_{HP}$ = 9.1 Hz, $^2J_{HPt}$ = 31.6 Hz, 1H, C_{Ar}C*H*CH), 6.27 (d, $^3J_{HH}$ = 8.3 Hz, 2H, H_{Ar}), 6.63 - 6.97 (m, 20H, H_{Ar}), 6.97 - 7.14 (m, 9H, H_{Ar}), 7.47 - 7.60 (m, 6H, H_{Ar}), 7.65 (d, $^3J_{HH}$ = 7.8 Hz, 2H, H_{Ar}).

^{13}C-NMR (100.62 MHz; C_6D_6): δ = 48.5 (dd, $^2J_{CP}$ = 5.8 Hz, $^2J_{CP}$ = 30.3 Hz, CH*C*HCO), 48.5 (ddd, $^2J_{CP}$ = 5.8 Hz, $^2J_{CP}$ = 30.3 Hz, $^1J_{CPt}$ = 201.0 Hz, CH*C*HCO), 59.5 (dd, $^2J_{CP}$ = 4.5 Hz, $^2J_{CP}$ = 32.5 Hz, C_{Ar}*C*HCH), 59.5 (ddd, $^2J_{CP}$ = 4.5 Hz, $^2J_{CP}$ = 32.5 Hz, $^1J_{CPt}$ = 180.2 Hz, C_{Ar}*C*HCH), 122.4 (C_{Ar}H), 123.5 (d, $^5J_{CP}$ = 2.2 Hz, C_{Ar}H), 123.5 (dd, $^5J_{CP}$ = 2.2 Hz, $^4J_{CPt}$ = 7.0 Hz, C_{Ar}H), 124.7 (C_{Ar}H), 125.4 (d, $^4J_{CP}$ = 3.0 Hz, C_{Ar}H), 125.4 (dd, $^4J_{CP}$ = 3.0 Hz, $^3J_{CPt}$ = 13.7 Hz, C_{Ar}H), 128.2 (C_{Ar}H), 128.4 (C_{Ar}H), 128.5 (C_{Ar}H), 129.1 (C_{Ar}H), 129.8 (d, $^1J_{CPt}$ = 2.0 Hz, C_{Ar}H), 130.0 (d, $^1J_{CPt}$ = 2.0 Hz, C_{Ar}H), 133.9 (C_{Ar}H), 133.9 (d, $^3J_{CPt}$ = 20.8 Hz, C_{Ar}H), 134.1 (C_{Ar}H), 134.1 (d, $^3J_{CPt}$ = 19.8 Hz, C_{Ar}H), 134.2 (C_{Ar}H), 134.2 (d, $^3J_{CPt}$ = 19.6 Hz, C_{Ar}H), 134.4 (C_{Ar}H), 134.4 (d, $^3J_{CPt}$ = 19.4 Hz, C_{Ar}H), 134.4 (d, $^1J_{CP}$ = 2.0 Hz, C_{Ar}P), 134.4 (dd, $^1J_{CP}$ = 2.0 Hz, $^2J_{CPt}$ = 31.4 Hz, C_{Ar}P), 134.9 (d, $^1J_{CP}$ = 2.1 Hz, C_{Ar}P), 134.9 (dd, $^1J_{CP}$ = 2.1 Hz, $^2J_{CPt}$ = 30.8 Hz, C_{Ar}P), 136.0 (d, $^1J_{CP}$ = 2.2 Hz, C_{Ar}P), 136.0 (dd, $^1J_{CP}$ = 2.2 Hz, $^2J_{CPt}$ = 25.6 Hz, C_{Ar}P), 136.4 (d, $^1J_{CP}$ = 2.2 Hz, C_{Ar}P), 136.4

(dd, $^1J_{CP}$ = 2.2 Hz, $^2J_{CPt}$ = 30.2 Hz, C_{Ar}P), 144.6 (d, $^3J_{CP}$ = 3.2 Hz, C_{Ar}N), 152.5 (C_{Ar}O), 152.7 (dd, $^3J_{CP}$ = 1.6 Hz, $^3J_{CP}$ = 6.8 Hz, C_{Ar}CH), 152.7 (ddd, $^3J_{CP}$ = 1.6 Hz, $^3J_{CP}$ = 6.8 Hz, $^2J_{CPt}$ = 36.6 Hz, C_{Ar}CH), 171.4 (dd, $^3J_{CP}$ 2.0 Hz, $^3J_{CP}$ = 5.4 Hz, CHCOO), 171.4 (ddd, $^3J_{CP}$ = 2.0 Hz, $^3J_{CP}$ = 5.4 Hz, $^2J_{CPt}$ = 43.8 Hz, CHCOO).

^{31}P-NMR (161.97 MHz; C$_6$D$_6$): δ = 26.0 (d, $^2J_{PP}$ = 32.5 Hz), 26.0 (dd, $^2J_{PP}$ = 32.5 Hz, $^1J_{PPt}$ = 4038.3 Hz), 27.3 (d, $^2J_{PP}$ = 32.6 Hz), 27.3 (dd, $^2J_{PP}$ = 32.6 Hz, $^1J_{PPt}$ = 3736.9 Hz).

^{195}Pt-NMR (85.48 MHz; C$_6$D$_6$): δ = -5047.1 (dd, $^1J_{PPt}$ = 3737.1 Hz, $^1J_{PPt}$ = 4038.7 Hz).

FAB-MS: m/z (%): 718.2 (100) [Pt(PPh$_3$)(PPh$_2$C$_6$H$_4$)$^+$], 987.1 (0.64) [(C$_{15}$H$_{11}$NO$_4$)Pt(PPh$_3$)(PPh$_2$C$_6$H$_4$)$^+$].

ESI-MS: m/z (%): 760.1 (100) [Pt(PPh$_3$)(PPh$_3$)CH$_3$CN$^+$].

FT-IR (ATR, cm^{-1}): 3082 (w), 3062 (w), 3015 (w), 2995 (w), 1715 (s, CO), 1595 (m), 1515 (m), 1495 (m), 1482 (m), 1434 (s), 1408 (w), 1335 (s), 1304 (m), 1266 (m), 1198 (m), 1155 (m), 1111 (s), 1093 (s), 1024 (w), 985 (w), 931 (w), 838 (m), 745 (s), 691 (s), 540 (s), 519 (s), 500 (s), 450 (w), 437 (m).

elemental analysis for C$_{51}$H$_{41}$NO$_4$P$_2$Pt (988.90 g/mol):

calcd.: C 61.94, H 4.18, N 1.42, O 6.47, P 6.26, Pt 19.73

found: C 60.13, H 3.80, N 1.33

Bis(triphenylphosphine)-(η^2-4-nitrophenyl 4-nitrocinnamate)palladium(0)

PdPhNO$_2$-PhNO$_2$: C$_{51}$H$_{40}$N$_2$O$_6$P$_2$Pd, M = 945.24 g/mol, orange solid, yield: 79%, route: *Pd-I*.

Single crystals suitable for X-ray analysis were obtained by slow diffusion of pentane into a toluene solution.

^{1}H-NMR (400.13 MHz; C$_6$D$_6$): δ = 4.59 (ddd, $^3J_{HH}$ = 10.0 Hz, $^3J_{HP}$ = 6.9 Hz, $^3J_{HP}$ = 5.2 Hz, 1H, CHCHCO), 5.06 (ddd, $^3J_{HH}$ = 10.0 Hz, $^3J_{HP}$ = 8.1 Hz, $^3J_{HP}$ = 5.5 Hz, 1H, C$_{Ar}$CHCH), 6.29 - 6.34 (m, 2H, H_{Ar}), 6.57 - 6.62 (m, 2H, H_{Ar}), 6.79 - 6.86 (m, 6H, H_{Ar}), 6.86 - 6.95 (m, 12H, H_{Ar}), 6.96 - 7.04 (m, 6H, H_{Ar}), 7.40 - 7.48 (m, 6H, H_{Ar}), 7.61 - 7.65 (m, 2H, H_{Ar}), 7.74 - 7.78 (m, 2H, H_{Ar}).

^{13}C-NMR (100.53 MHz; C$_6$D$_6$): δ = 57.3 (dd, $^2J_{CP}$ = 5.4 Hz, $^2J_{CP}$ = 19.6 Hz, CHCHCO), 71.7 (dd, $^2J_{CP}$ = 2.0 Hz, $^2J_{CP}$ = 24.2 Hz, C$_{Ar}$CHCH), 122.2 (C_{Ar}H), 124.0 (bs, C_{Ar}H), 124.8 (C_{Ar}H), 125.0 (d, $^3J_{CP}$ = 2.0 Hz, C_{Ar}H), 128.4 (C_{Ar}), 128.5 (C_{Ar}), 128.7 (C_{Ar}H), 128.8 (C_{Ar}H), 129.8 (dd, $^1J_{CP}$ = 21.2 Hz, $^3J_{CP}$ = 1.5 Hz, C_{Ar}P), 134.0 (d, $^2J_{CP}$ = 14.2 Hz, C_{Ar}H), 134.2 (d, $^2J_{CP}$ = 15.0 Hz, C_{Ar}H), 134.5 (dd, $^1J_{CP}$ = 30.5 Hz, $^3J_{CP}$ = 1.5 Hz, C_{Ar}P), 136.0 (dd, $^1J_{CP}$ = 29.5 Hz, $^3J_{CP}$ = 1.6 Hz, C_{Ar}P), 144.77 (C_{Ar}N), 144.9 (dd, $^3J_{CP}$ = 1.5 Hz, $^3J_{CP}$ = 3.2 Hz, C_{Ar}CH), 149.5 (d, $^3J_{CP}$ = 6.7 Hz, C_{Ar}N), 156.8 (C_{Ar}O), 168.5 (dd, $^3J_{CP}$ = 2.1 Hz, $^3J_{CP}$ = 4.5 Hz, CHCOO).

^{31}P-NMR (161.97 MHz; C$_6$D$_6$): δ = 25.2 (d, $^2J_{PP}$ = 1.4 Hz), 26.5 (d, $^2J_{PP}$ = 1.4 Hz).

FAB-MS: m/z (%): 368.0 (14.24) [Pd(PPh$_3$)$^+$], 630.0 (39.90) [Pd(PPh$_3$)$_2$$^+$].

FT-IR (ATR, cm^{-1}): 3051 (m), 1702 (s, CO), 1585 (m), 1519(s), 1467 (m), 1434 (s), 1326 (s), 1278 (m), 1229 (m), 1210 (m), 1095 (s), 962 (m), 902 (w), 849 (m), 788 (m), 735 (s), 692 (s), 527 (m), 514 (m), 500 (s), 466 (w), 418 (m).

elemental analysis for C$_{51}$H$_{40}$N$_2$O$_6$P$_2$Pd (945.24 g/mol):

calcd.: C 64.80, H 4.27, N 2.96, O 10.16, P 6.55, Pd 11.26

found: C 66.95, H 4.64, N 2.71

Bis(triphenylphosphine)-(η^2-4-nitrophenyl 4-nitrocinnamate)platinum(0)

PtPhNO$_2$-PhNO$_2$: $C_{51}H_{40}N_2O_6P_2Pt$, M = 1033.90 g/mol, orange solid, yield: 99%, route: *Pt-III*.

Single crystals suitable for X-ray analysis were obtained by slow diffusion of pentane into a toluene solution.

^{1}H-NMR (400.13 MHz; C_6D_6): δ = 3.96 (dpt, $^3J_{HH}$ = 8.8 Hz, $^3J_{HP}$ = 3.5 Hz, $^3J_{HP}$ = 8.8 Hz, 1H, CHCHCO), 3.96 (ddpt, $^3J_{HH}$ = 8.8 Hz, $^3J_{HP}$ = 3.5 Hz, $^3J_{HP}$ = 8.8 Hz, $^2J_{HPt}$ = 57.8 Hz, 1H, CHCHCO), 4.31 (dpt, $^3J_{HH}$ = 9.1 Hz, $^3J_{HP}$ = 4.9 Hz, $^3J_{HP}$ = 9.1 Hz, 1H, C_{Ar}CHCH), 4.31 (ddpt, $^3J_{HH}$ = 9.1 Hz, $^3J_{HP}$ = 4.9 Hz, $^3J_{HP}$ = 9.1 Hz, $^2J_{HPt}$ = 50.2 Hz, 1H, C_{Ar}CHCH), 6.29 (d, $^3J_{HH}$ = 8.4 Hz, 2H, H_{Ar}), 6.59 (d, $^3J_{HH}$ = 7.9 Hz, 2H, H_{Ar}), 6.64 - 7.14 (m, 24H, H_{Ar}), 7.36 - 7.53 (m, 6H, H_{Ar}), 7.68 (d, $^3J_{HH}$ = 7.8 Hz, 2H, H_{Ar}), 7.76 (d, $^3J_{HH}$ = 8.0 Hz, 2H, H_{Ar}).

^{13}C-NMR (100.53 MHz; C_6D_6): δ = 47.7 (dd, $^2J_{CP}$ = 5.8 Hz, $^2J_{CP}$ = 29.7 Hz, CHCHCO), 47.7 (ddd, $^2J_{CP}$ = 5.8 Hz, $^2J_{CP}$ = 29.7 Hz, $^1J_{CPt}$ = 201.4 Hz, CHCHCO), 58.7 (dd, $^2J_{CP}$ = 4.2 Hz, $^2J_{CP}$ = 33.6 Hz, $C_{Ar}$$C$HCH), 58.7 (ddd, $^2J_{CP}$ = 4.2 Hz, $^2J_{CP}$ = 33.6 Hz, $^1J_{CPt}$ = 187.6 Hz, $C_{Ar}$$C$HCH), 122.3 ($C_{Ar}$H), 123.5 (d, $^5J_{CP}$ = 2.3 Hz, C_{Ar}H), 123.5 (dd, $^5J_{CP}$ = 2.3 Hz, $^4J_{CPt}$ = 6.6 Hz, C_{Ar}H), 124.7 (C_{Ar}H), 125.5 (d, $^4J_{CP}$ = 3.2 Hz, C_{Ar}H), 125.5 (dd, $^4J_{CP}$ = 3.2 Hz, $^3J_{CPt}$ = 14.0 Hz, C_{Ar}H), 128.1 (C_{Ar}H), 128.2 (C_{Ar}H), 128.5 (C_{Ar}H), 128.6 (C_{Ar}H), 130.0 (d, $^1J_{CPt}$ = 2.2 Hz, C_{Ar}H), 130.2 (d, $^1J_{CPt}$ = 2.0 Hz, C_{Ar}H), 133.9 (C_{Ar}H), 133.9 (d, $^3J_{CPt}$ = 21.4 Hz, C_{Ar}H), 134.0 (C_{Ar}H), 134.0 (d, $^3J_{CPt}$ = 25.2 Hz, C_{Ar}H), 134.1 (C_{Ar}H), 134.1 (d, $^3J_{CPt}$ = 22.4 Hz, C_{Ar}H), 134.2 (C_{Ar}H), 134.2 (d, $^3J_{CPt}$ = 26.2 Hz, C_{Ar}H), 134.6 (d, $^1J_{CP}$ = 2.1 Hz, C_{Ar}P), 134.6 (dd, $^1J_{CP}$ = 2.1 Hz, $^2J_{CPt}$ = 33.8 Hz, C_{Ar}P), 135.7 (d, $^1J_{CP}$ = 2.1 Hz, C_{Ar}P), 135.7 (dd, $^1J_{CP}$ = 2.1 Hz, $^2J_{CPt}$ = 29.6 Hz, C_{Ar}P), 136.1 (d, $^1J_{CP}$ = 2.0 Hz, C_{Ar}P), 136.1 (dd, $^1J_{CP}$ = 2.0 Hz, $^2J_{CPt}$ = 28.4 Hz, C_{Ar}P), 144.7 (d, $^3J_{CP}$ = 3.0 Hz,

$C_{Ar}N$), 144.7 ($C_{Ar}N$), 152.2 (dd, $^3J_{CP}$ = 1.4 Hz, $^3J_{CP}$ = 6.7 Hz, $C_{Ar}CH$), 152.2 (ddd, $^3J_{CP}$ = 1.4 Hz, $^3J_{CP}$ = 6.7 Hz, $^2J_{CPt}$ = 33.8 Hz, $C_{Ar}CH$), 156.9 ($C_{Ar}O$), 170.3 (dd, $^3J_{CP}$ 2.5 Hz, $^3J_{CP}$ = 5.1 Hz, CHCOO), 170.3 (ddd, $^3J_{CP}$ = 2.5 Hz, $^3J_{CP}$ = 5.1 Hz, $^2J_{CPt}$ = 35.4 Hz, CHCOO).

^{31}P-NMR (161.97 MHz; C_6D_6): δ = 25.6 (d, $^2J_{PP}$ = 30.1 Hz), 25.6 (dd, $^2J_{PP}$ = 30.1 Hz, $^1J_{PPt}$ = 4092.8 Hz), 26.8 (d, $^2J_{PP}$ = 30.2 Hz), 26.8 (dd, $^2J_{PP}$ = 30.2 Hz, $^1J_{PPt}$ = 3682.9 Hz).

^{195}Pt-NMR (85.48 MHz; C_6D_6): δ = -5047.0 (dd, $^1J_{PPt}$ = 3681.7 Hz, $^1J_{PPt}$ = 4095.1 Hz).

FAB-MS: m/z (%): 718.1 (100) [Pt(PPh$_3$)(PPh$_2$C$_6$H$_4$)$^+$], 1032.8 (1.54) [(C$_{15}$H$_{10}$N$_2$O$_6$)Pt(PPh$_3$)(PPh$_2$C$_6$H$_4$)$^+$].

ESI-MS: m/z (%): 760.0 (100) [Pt(PPh$_3$)(PPh$_2$C$_6$H$_4$)CH$_3$CN$^+$].

FT-IR (ATR, cm^{-1}): 3048 (w), 1715 (m, CO), 1589 (m), 1524 (m), 1508 (m), 1491 (m), 1434 (m), 1394 (w), 1334 (s), 1258 (w), 1220(m), 1152 (m), 1093 (s), 985 (m), 909 (m), 845 (m), 743 (s), 691 (s), 576 (w), 546 (m), 517 (s), 501 (s), 451 (w), 432 (m).

elemental analysis for C$_{51}$H$_{40}$N$_2$O$_6$P$_2$Pt (1033.90 *g/mol*):

calcd.: C 59.25, H 3.90, N 2.71, O 9.28, P 5.99, Pt 18.87

found: C 57.87, H 3.91, N 3.11

Bis(triphenylphosphine)-(η^2-1-methoxypropan-2-yl 4-nitrocinnamate)palladium(0)

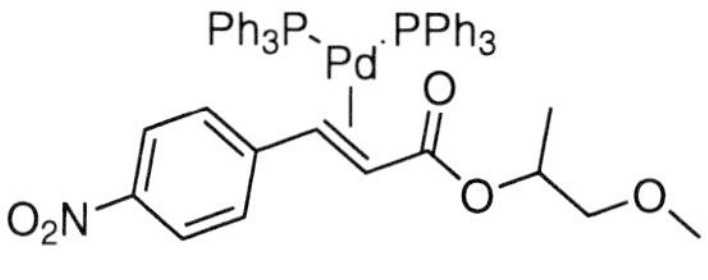

PdPhNO$_2$-MeO^{i}Pr: C$_{49}$H$_{45}$NO$_5$P$_2$Pd, M = 896.25 *g/mol*, red solid, **route**: *Pd-I*.

major diastereoisomer:

^{1}H-NMR (400.13 MHz; C_6D_6): δ = 0.85 (d, $^3J_{HH}$ = 6.4 Hz, 3H, CHCH_3), 3.06 (s, 3H, OCH_3), 3.20 (dd, $^2J_{HH}$ = 9.9 Hz, $^3J_{HH}$ = 4.8 Hz, 1H, CHCH_2O), 3.32 (dd,

$^2J_{HH} = 9.9$ Hz, $^3J_{HH} = 5.4$ Hz, 1H, CHCH_2O), 4.56 (dpt, $^3J_{HH} = 11.3$ Hz, $^3J_{HP} = 6.0$ Hz,

$^3J_{HP} = 11.3$ Hz, 1H, CHCHCO), 5.00 - 5.18 (m, 2H, OCH, C$_{Ar}$CHCH), 6.21 (d,

$^3J_{HH} = 8.5$ Hz, 2H, H_{Ar}), 6.78 - 6.87 (m, 6H, H_{Ar}), 6.87 - 6.93 (m, 3H, H_{Ar}), 6.93 -

7.04 (m, 15H, H_{Ar}), 7.47 - 7.61 (m, 8H, H_{Ar}).

^{13}C-NMR (100.62 MHz; C$_6$D$_6$): δ = 16.2 (CHCH_3), 58.4 (OCH_3), 59.4 (dd,

$^2J_{CP} = 5.3$ Hz, $^2J_{CP} = 20.1$ Hz, CHCHCO), 68.1 (OCH), 72.9 (dd, $^2J_{CP} = 3.1$ Hz,

$^2J_{CP} = 21.9$ Hz, C$_{Ar}$$CH$CH), 75.3 (CH$CH_2$O), 123.6($C_{Ar}$H), 124.4 (d, $^4J_{CP} = 2.5$ Hz,

C_{Ar}), 128.2 (d, $^3J_{CP} = 9.4$ Hz, C_{Ar}H), 129.1 (d, $^4J_{CP} = 1.4$ Hz, C_{Ar}H), 129.3 (d,

$^4J_{CP} = 1.4$ Hz, C_{Ar}H), 133.7 (d, $^2J_{CP} = 14.4$ Hz, C_{Ar}H), 134.1 (d, $^2J_{CP} = 15.2$ Hz,

C_{Ar}H), 134.6 (dd, $^1J_{CP} = 29.0$ Hz, $^3J_{CP} = 1.7$ Hz, C_{Ar}), 136.3 (dd, $^1J_{CP} = 28.7$ Hz,

$^3J_{CP} = 1.9$ Hz, C_{Ar}), 144.1 (dd, $^3J_{CP} = 1.3$ Hz, $^3J_{CP} = 3.1$ Hz, C_{Ar}CH), 149.6 (d,

$^3J_{CP} = 6.5$ Hz, C_{Ar}N), 170.5 (dd, $^3J_{CP} = 1.9$ Hz, $^3J_{CP} = 4.5$ Hz, CHCOO).

^{31}P-NMR (161.97 MHz; C$_6$D$_6$): δ = 24.3 (d, $^2J_{PP} = 8.1$ Hz), 26.9 (d, $^2J_{PP} = 8.1$ Hz).

minor diastereoisomer:

^{1}H-NMR (400.13 MHz; C$_6$D$_6$): δ = 1.31 (d, $^3J_{HH} = 6.4$ Hz, 3H, CHCH_3), 2.75(dd,

$^2J_{HH} = 9.6$ Hz, $^3J_{HH} = 5.8$ Hz, 1H, CHCH_2O), 2.91 - 2.96 (m, 1H, CHCH_2O) 2.94(s, 3H,

OCH_3), 4.56 (dpt, $^3J_{HH} = 11.3$ Hz, $^3J_{HP} = 6.0$ Hz, $^3J_{HP} = 11.3$ Hz, 1H, CHCHCO),

5.00 - 5.18 (m, 2H, OCH, C$_{Ar}$CHCH), 6.21 (d, $^3J_{HH} = 8.5$ Hz, 2H, H_{Ar}), 6.78 - 6.87 (m,

6H, H_{Ar}), 6.87 - 6.93 (m, 3H, H_{Ar}), 6.93 - 7.04 (m, 15H, H_{Ar}), 7.47 - 7.61 (m, 8H, H_{Ar}).

^{13}C-NMR (100.62 MHz; C$_6$D$_6$): δ = 17.2 (CHCH_3), 58.3 (OCH_3), 59.2 (dd,

$^2J_{CP} = 5.1$ Hz, $^2J_{CP} = 19.9$ Hz, CHCHCO), 68.0 (OCH), 72.8 (dd, $^2J_{CP} = 2.7$ Hz,

$^2J_{CP} = 22.3$ Hz, C$_{Ar}$$CH$CH), 74.4 (CH$CH_2$O), 123.6($C_{Ar}$H), 124.4 (d, $^4J_{CP} = 2.5$ Hz,

C_{Ar}), 128.2 (d, $^3J_{CP} = 9.4$ Hz, C_{Ar}H), 129.1 (d, $^4J_{CP} = 1.4$ Hz, C_{Ar}H), 129.3 (d,

$^4J_{CP} = 1.4$ Hz, C_{Ar}H), 133.7 (d, $^2J_{CP} = 14.4$ Hz, C_{Ar}H), 134.1 (d, $^2J_{CP} = 15.2$ Hz,

C_{Ar}H), 134.6 (dd, $^1J_{CP} = 28.8$ Hz, $^3J_{CP} = 2.3$ Hz, C_{Ar}), 136.4 (dd, $^1J_{CP} = 28.6$ Hz,

$^3J_{CP} = 2.0$ Hz, C_{Ar}), 144.1 (dd, $^3J_{CP} = 1.3$ Hz, $^3J_{CP} = 3.1$ Hz, C_{Ar}CH), 149.6 (d,

$^3J_{CP} = 6.5$ Hz, C_{Ar}N), 170.5 (dd, $^3J_{CP} = 1.8$ Hz, $^3J_{CP} = 4.6$ Hz, CHCOO).

^{31}P-NMR (161.97 MHz; C$_6$D$_6$): δ = 24.4 (d, $^2J_{PP} = 8.4$ Hz), 27.0 (d, $^2J_{PP} = 8.4$ Hz).

Bis(triphenylphosphine)-(η^2-1-methoxypropan-2-yl 4-nitrocinnamate)platinum(0)

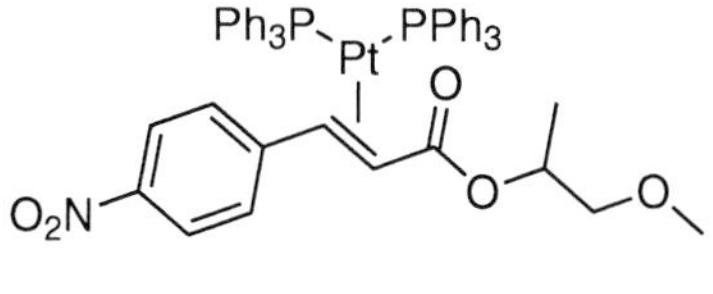

PtPhNO$_2$-MeO^{i}Pr: $C_{49}H_{45}NO_5P_2Pt$, M = 984.91 *g/mol*, yellow solid, NMR scale, **route**: *Pt-III*.

major diastereoisomer:

^{1}H-NMR (400.13 MHz; C_6D_6): δ = 0.85 (d, $^3J_{HH}$ = 6.4 Hz, 3H, CHCH_3), 3.08 (s, 3H, OCH_3), 3.20 (dd, $^2J_{HH}$ = 9.8 Hz, $^3J_{HH}$ = 5.0 Hz, 1H, CHCH_2O), 3.33 (dd, $^2J_{HH}$ = 9.8 Hz, $^3J_{HH}$ = 5.3 Hz, 1H, CHCH_2O), 3.93 (dpt, $^3J_{HH}$ = 4.7 Hz, $^3J_{HP}$ = 8.7 Hz, $^3J_{HP}$ = 8.7 Hz, 1H, CHCHCO), 3.93 (ddpt, $^3J_{HH}$ = 4.7 Hz, $^3J_{HP}$ = 8.7 Hz, $^3J_{HP}$ = 8.7 Hz, $^2J_{HPt}$ = 59.2 Hz, 1H, CHCHCO), 4.35 (dpt, $^3J_{HH}$ = 4.7 Hz, $^3J_{HP}$ = 9.2 Hz, $^3J_{HP}$ = 9.2 Hz, 1H $C_{Ar}CH$CH), 4.35 (ddpt, $^3J_{HH}$ = 4.7 Hz, $^3J_{HP}$ = 9.2 Hz, $^3J_{HP}$ = 9.2 Hz, $^2J_{HPt}$ = 51.8 Hz, 1H $C_{Ar}CH$CH), 5.14 (pdq, $^3J_{HH}$ = 4.8 Hz, $^3J_{HH}$ = 11.0 Hz, 1H, OCH), 6.18 (d, $^3J_{HP}$ = 8.7 Hz, 2H, H_{Ar}), 6.76 - 6.89 (m, 2H, H_{Ar}), 6.92 - 7.01 (m, 20H, H_{Ar}), 7.38 - 7.47 (m, 10H, H_{Ar}).

^{31}P-NMR (161.97 MHz; C_6D_6): δ = 26.2 (d, $^2J_{PP}$ = 35.2 Hz), 26.2 (dd, $^2J_{PP}$ = 35.2 Hz, $^1J_{PPt}$ = 4003.5 Hz), 27.7 (d, $^2J_{PP}$ = 35.2 Hz), 27.7 (dd, $^2J_{PP}$ = 35.2 Hz, $^1J_{PPt}$ = 3769.0 Hz).

minor diastereoisomer:

^{1}H-NMR (400.13 MHz; C_6D_6): δ = 1.31 (d, $^3J_{HH}$ = 6.4 Hz, 3H, CHCH_3), 2.79 (dd, $^2J_{HH}$ = 9.6 Hz, $^3J_{HH}$ = 6.0 Hz, 1H, CHCH_2O), 2.94 (s, 3H, OCH_3), 2.97 (dd, $^2J_{HH}$ = 9.6 Hz, $^3J_{HH}$ = 4.4 Hz, 1H, CHCH_2O), 3.92 (dpt, $^3J_{HH}$ = 9.2 Hz, $^3J_{HP}$ = 4.4 Hz, $^3J_{HP}$ = 9.2 Hz, 1H, CHCHCO), 3.92 (ddpt, $^3J_{HH}$ = 9.2 Hz, $^3J_{HP}$ = 4.4 Hz, $^3J_{HP}$ = 9.2 Hz, $^2J_{HPt}$ = 59.2 Hz, 1H, CHCHCO), 4.36 (dpt, $^3J_{HH}$ = 9.2 Hz, $^3J_{HP}$ = 4.6 Hz, $^3J_{HP}$ = 9.2 Hz, 1H $C_{Ar}CH$CH), 4.36 (ddpt, $^3J_{HH}$ = 9.2 Hz, $^3J_{HP}$ = 4.6 Hz, $^3J_{HP}$ = 9.2 Hz, $^2J_{HPt}$ = 51.8 Hz, 1H $C_{Ar}CH$CH),

5.11 (pdq, $^3J_{HH}$ = 4.4 Hz, $^3J_{HH}$ = 10.2 Hz, 1H, OCH), 6.18 (d, $^3J_{HP}$ = 8.7 Hz, 2H, H_{Ar}), 6.76 - 6.89 (m, 2H, H_{Ar}), 6.92 - 7.01 (m, 20H, H_{Ar}), 7.38 - 7.47 (m, 10H, H_{Ar}).

^{31}P-NMR (161.97 MHz; C$_6$D$_6$): δ = 26.2 (d, $^2J_{PP}$ = 35.2 Hz), 26.2 (dd, $^2J_{PP}$ = 35.2 Hz, $^1J_{PPt}$ = 3996.3 Hz), 27.8 (d, $^2J_{PP}$ = 35.2 Hz), 27.8 (dd, $^2J_{PP}$ = 35.2 Hz, $^1J_{PPt}$ = 3768.9 Hz).

8.3.9. Cinnamate complexes with chelating phosphines

(1,2-Bis(diphenylphosphino)ethane)-(η^2-ethyl 4-methoxycinnamate)palladium(0)

Pd(dppe)PhOMe-Et: C$_{38}$H$_{38}$O$_3$P$_2$Pd, M = 711.07 g/mol, yellow solid, yield: >99%, **route**: *Pd-IV*.

^{1}H-NMR (400.13 MHz; C$_6$D$_6$): δ = 1.01 (t, $^3J_{HH}$ = 7.1 Hz, 3H, CH$_2$CH_3), 1.57 - 1.75 (m, 1H, PCH_2CH$_2$P), 1.75 - 1.92 (m, 1H, PCH_2CH$_2$P), 1.93 - 2.18 (m, 2H, PCH_2CH$_2$P), 3.27 (s, 3H, OCH_3), 3.97 - 4.13 (m, 1H, CH_2CH$_3$), 4.16 - 4.29 (m, 1H, CH_2CH$_3$), 4.97 - 5.13 (m, 1H, CHCHCO), 6.04 - 6.17 (m, 1H, C$_{Ar}$CHCH), 6.44 - 6.58 (m, 2H, H_{Ar}), 6.81 - 7.27 (m, 16H, H_{Ar}), 7.60 - 7.78 (m, 4H, H_{Ar}), 7.78 - 7.90 (m, 2H, H_{Ar}).

^{13}C-NMR (100.62 MHz; C$_6$D$_6$): δ = 14.8 (CH$_2$$CH_3$), 26.8 - 27.6 (m, P$CH_2$), 54.7 - 54.9 (m, CH$C$HCO), 54.8 (O$CH_3$), 58.9 ($CH_2CH_3$), 66.9 (d, $^2J_{CP}$ = 24.3 Hz, C$_{Ar}$$C$HCH), 114.3 ($C_{Ar}$H), 126.7 ($C_{Ar}$H), 127.9 ($C_{Ar}$H), 128.2 ($C_{Ar}$H), 128.6 (d, $^3J_{CP}$ = 8.2 Hz, C_{Ar}H), 128.8 (d, $^3J_{CP}$ = 8.2 Hz, C_{Ar}H), 129.0 (d, $^3J_{CP}$ = 8.2 Hz, C_{Ar}H), 129.4 (C_{Ar}H), 129.8 (C_{Ar}H), 130.0 (C_{Ar}H), 132.8 (d, $^2J_{CP}$ = 13.1 Hz, C_{Ar}H), 132.8 (d, $^2J_{CP}$ = 13.1 Hz, C_{Ar}H), 133.1 (d, $^2J_{CP}$ = 13.9 Hz, C_{Ar}H), 133.1 (d, $^2J_{CP}$ = 13.9 Hz, C_{Ar}H), 133.4 (d, $^2J_{CP}$ = 14.5 Hz, C_{Ar}H), 133.4 (d, $^2J_{CP}$ = 14.5 Hz, C_{Ar}H), 133.7 (d, $^2J_{CP}$ = 13.9 Hz, C_{Ar}H), 133.7 (d, $^2J_{CP}$ = 13.9 Hz, C_{Ar}H), 136.2 - 137.1 (m, C_{Ar}), 137.9 (d, $^3J_{CP}$ = 6.3 Hz, C_{Ar}CH), 156.9 - 157.0 (m, C_{Ar}O), 172.7 - 172.8 (m, CHCOO).

^{31}P-NMR (161.97 MHz; C$_6$D$_6$): $\delta = 32.2$ (d, $^2J_{PP} = 42.9$ Hz), 32.6 (d, $^2J_{PP} = 42.9$ Hz).

(1,3-Bis(diphenylphosphino)propane)-(η^2-ethyl 4-methoxycinnamate)palladium(0)

Pd(dppp)PhOMe-Et: C$_{39}$H$_{40}$O$_3$P$_2$Pd, M = 725.10 *g/mol*, yellow solid, yield: 93%, **route**: *Pd-IV*.

^{1}H-NMR (400.13 MHz; C$_6$D$_6$): $\delta = 0.94$ (t, $^3J_{HH} = 7.1$ Hz, 3H, CH$_2$CH_3), 1.28 - 1.57 (m, 2H, CH$_2$CH_2CH$_2$), 1.80 - 2.12 (m, 4H, PCH_2CH$_2$), 3.28 (s, 3H, OCH_3), 3.80 (dq, $^2J_{HH} = $ Hz, $^3J_{HH} = $ Hz, 1H, CH_2CH$_3$), 4.14 (dq, $^2J_{HH} = $ Hz, $^3J_{HH} = $ Hz, 1H, CH_2CH$_3$), 4.80 (ddd, $^3J_{HH} = 10.8$ Hz, $^3J_{HP} = 4.4$ Hz, $^3J_{HP} = 7.2$ Hz, 1H, CHCHCO), 5.69 (ddd, $^3J_{HH} = 10.8$ Hz, $^3J_{HP} = 5.3$ Hz, $^3J_{HP} = 8.2$ Hz, 1H, C$_{Ar}$CHCH), 6.41 (d, $^3J_{HP} = 8.6$ Hz, 2H, H_{Ar}), 6.77 - 7.21 (m, 16H, H_{Ar}), 7.53 - 7.62 (m, 2H, H_{Ar}), 7.79 - 7.89 (m, 4H, H_{Ar}).

^{13}C-NMR (100.62 MHz; C$_6$D$_6$): $\delta = 14.7$ (CH$_2$$CH_3$), 19.5 (t, $^2J_{CP} = 5.1$ Hz, PCH$_2$$CH_2$), 29.6 (d, $^1J_{CP} = 17.2$ Hz, PCH$_2$CH$_2$), 30.0 (d, $^1J_{CP} = 18.2$ Hz, PCH$_2$CH$_2$), 54.7 (dd, $^2J_{CP} = 6.4$ Hz, $^2J_{CP} = 19.1$ Hz, CHCHCO), 54.8 (OCH$_3$), 58.8 (CH$_2$CH$_3$), 67.2 (dd, $^2J_{CP} = 0.9$ Hz, $^2J_{CP} = 26.0$ Hz, C$_{Ar}$$C$HCH), 114.0 ($C_{Ar}$H), 126.5 (dd, $^3J_{CP} = 1.4$ Hz, $^3J_{CP} = 3.4$ Hz, C_{Ar}H), 127.9 (C_{Ar}H), 128.2 (C_{Ar}H), 128.4 (d, $^3J_{CP} = 9.1$ Hz, C_{Ar}H), 128.5 (d, $^4J_{CP} = 1.5$ Hz, C_{Ar}H), 128.6 (d, $^3J_{CP} = 9.0$ Hz, C_{Ar}H), 128.7 (d, $^3J_{CP} = 9.5$ Hz, C_{Ar}H), 129.5 (d, $^4J_{CP} = 1.4$ Hz, C_{Ar}H), 129.5 (d, $^4J_{CP} = 1.4$ Hz, C_{Ar}H), 130.0 (d, $^4J_{CP} = 1.5$ Hz, C_{Ar}H), 132.2 (d, $^2J_{CP} = 13.9$ Hz, C_{Ar}H), 133.4 (d, $^2J_{CP} = 14.9$ Hz, C_{Ar}H), 133.5 (d, $^2J_{CP} = 15.2$ Hz, C_{Ar}H), 134.2 (d, $^2J_{CP} = 16.3$ Hz, C_{Ar}H), 135.9 (dd, $^1J_{CP} = 23.7$ Hz, $^3J_{CP} = 5.2$ Hz, C_{Ar}), 136.2 (dd, $^1J_{CP} = 25.6$ Hz, $^3J_{CP} = 3.3$ Hz, C_{Ar}), 137.0 (d, $^3J_{CP} = 6.5$ Hz, C_{Ar}CH), 137.9 (dd, $^1J_{CP} = 22.8$ Hz, $^3J_{CP} = 4.0$ Hz, C_{Ar}), 138.2

(dd, $^1J_{CP}$ = 23.0 Hz, $^3J_{CP}$ = 4.0 Hz, C_{Ar}), 157.1 (dd, $^3J_{CP}$ = 1.3 Hz, $^3J_{CP}$ = 2.7 Hz, $C_{Ar}O$), 172.4 (dd, $^3J_{CP}$ = 3.5 Hz, $^3J_{CP}$ = 3.5 Hz, CHCOO).

^{31}P-NMR (161.97 MHz; C$_6$D$_6$): δ = 7.6 (d, $^2J_{PP}$ = 12.6 Hz), 10.4 (d, $^2J_{PP}$ = 12.6 Hz).

(1,3-Bis(di-*iso*-propylphosphino)propane)-(η^2-ethyl 4-methoxycinnamate)palladium(0)

Pd(dippp)PhOMe-Et: C$_{27}$H$_{48}$O$_3$P$_2$Pd, M = 589.04 g/mol, yellow solid, **route**: *Pd-III*.

^{31}P-NMR (161.97 MHz; C$_6$H$_6$): δ = 26.8 (d, $^2J_{PP}$ = 10.4 Hz), 27.9 (d, $^2J_{PP}$ = 10.4 Hz).

(1,4-Bis(diphenylphosphino)butane)-(η^2-ethyl 4-methoxycinnamate)palladium(0)

Pd(dppb)PhOMe-Et: C$_{40}$H$_{42}$O$_3$P$_2$Pd, M = 739.13 g/mol, yellow solid, NMR scale, **route**: *Pd-II*.

^{1}H-NMR (400.13 MHz; C$_6$D$_6$): δ = 0.92 (t, $^3J_{HH}$ = 7.1 Hz, 3H, CH$_2$CH_3), 1.37 - 1.75 (m, 4H, PCH$_2$CH_2), 1.87 - 2.17 (m, 4H, PCH_2CH$_2$), 3.25 (s, 3H, OCH_3), 3.75 (dq, $^2J_{HH}$ = 10.7 Hz, $^3J_{HH}$ = 7.1 Hz, 1H, CH_2CH$_3$), 4.16 (dq, $^2J_{HH}$ = 10.7 Hz, $^3J_{HH}$ = 7.1 Hz, 1H, CH_2CH$_3$), 4.60 (ddd, $^3J_{HH}$ = 10.8 Hz, $^3J_{HP}$ = 4.7 Hz, $^3J_{HP}$ = 7.0 Hz, 1H, CHCHCO), 5.43 (ddd, $^3J_{HH}$ = 10.8 Hz, $^3J_{HP}$ = 5.4 Hz, $^3J_{HP}$ = 8.0 Hz, 1H, C$_{Ar}$CHCH),

6.35 (d, $^3J_{HP} = 8.6$ Hz, 2H, H_{Ar}), 6.73 (d, $^3J_{HP} = 7.2$ Hz, 2H, H_{Ar}), 6.86 - 7.00 (m, 5H, H_{Ar}), 7.01 - 7.26 (m, 9H, H_{Ar}), 7.58 - 7.71 (m, 4H, H_{Ar}), 7.71 - 7.85 (m, 2H, H_{Ar}).

^{13}C-NMR (100.62 MHz; C_6D_6): δ = 14.6 (CH_2CH_3), 23.7 (dd, $^2J_{CP} = 7.6$ Hz, $^3J_{CP} = 4.1$ Hz, PCH_2CH_2), 24.3 (dd, $^2J_{CP} = 7.2$ Hz, $^3J_{CP} = 4.7$ Hz, PCH_2CH_2), 27.6 (d, $^1J_{CP} = 15.5$ Hz, PCH_2CH_2), 28.3 (d, $^1J_{CP} = 16.3$ Hz, PCH_2CH_2), 54.7 (OCH_3), 56.9 (dd, $^2J_{CP} = 5.8$ Hz, $^2J_{CP} = 18.8$ Hz, $CHCHCO$), 58.7 (CH_2CH_3), 69.3 (dd, $^2J_{CP} = 1.5$ Hz, $^2J_{CP} = 25.2$ Hz, $C_{Ar}CHCH$), 113.9 (d, $^3J_{CP} = 1.5$ Hz, $C_{Ar}H$), 126.5 (dd, $^3J_{CP} = 1.1$ Hz, $^3J_{CP} = 2.2$ Hz, $C_{Ar}H$), 127.9 ($C_{Ar}H$), 128.2 ($C_{Ar}H$), 128.6 (d, $^3J_{CP} = 6.0$ Hz, $C_{Ar}H$), 128.6 (d, $^3J_{CP} = 6.0$ Hz, $C_{Ar}H$), 128.7 (d, $^3J_{CP} = 3.6$ Hz, $C_{Ar}H$), 129.0 (d, $^4J_{CP} = 0.8$ Hz, $C_{Ar}H$), 129.5 (d, $^4J_{CP} = 1.5$ Hz, $C_{Ar}H$), 129.7 (d, $^4J_{CP} = 1.6$ Hz, $C_{Ar}H$), 131.0 - 131.4 (m, C_{Ar}), 132.5 (d, $^2J_{CP} = 13.8$ Hz, $C_{Ar}H$), 133.0 (d, $^2J_{CP} = 14.2$ Hz, $C_{Ar}H$), 133.6 (d, $^2J_{CP} = 15.3$ Hz, $C_{Ar}H$), 134.0 (d, $^2J_{CP} = 15.6$ Hz, $C_{Ar}H$), 136.0 (dd, $^1J_{CP} = 24.0$ Hz, $^3J_{CP} = 3.6$ Hz, C_{Ar}), 136.8 (dd, $^3J_{CP} = 1.0$ Hz, $^3J_{CP} = 6.6$ Hz, $C_{Ar}CH$), 138.1 (dd, $^1J_{CP} = 21.8$ Hz, $^3J_{CP} = 3.1$ Hz, C_{Ar}), 138.1 (dd, $^1J_{CP} = 25.7$ Hz, $^3J_{CP} = 3.4$ Hz, C_{Ar}), 139.6 (dd, $^1J_{CP} = 23.2$ Hz, $^3J_{CP} = 3.7$ Hz, C_{Ar}), 157.1 (dd, $^3J_{CP} = 1.1$ Hz, $^3J_{CP} = 2.6$ Hz, $C_{Ar}O$), 172.1 (dd, $^3J_{CP} = 3.5$ Hz, $^3J_{CP} = 3.5$ Hz, $CHCOO$).

^{31}P-NMR (161.97 MHz; C_6D_6): $\delta = 17.3$ (d, $^2J_{PP} = 25.0$ Hz), 18.2 (d, $^2J_{PP} = 25.0$ Hz).

(1,2-Bis(diphenylphosphino)ethane)-(η^2-*iso*-propyl 4-methoxycinnamate)palladium(0)

Pd(dppe)PhOMe-^{i}Pr: $C_{39}H_{40}O_3P_2Pd$, M = 725.10 *g/mol*, yellow solid, NMR scale, **route**: *Pd-II*.

^{31}P-NMR (161.97 MHz; C_6H_6): $\delta = 31.9$ (d, $^2J_{PP} = 42.9$ Hz), 32.3 (d, $^2J_{PP} = 42.9$ Hz).

(1,3-Bis(diphenylphosphino)propane)-(η^2-*iso*-propyl 4-methoxycinnamate)palladium(0)

Pd(dppp)PhOMe-^{i}Pr: $C_{40}H_{42}O_3P_2Pd$, M $= 739.13$ g/mol, yellow solid, NMR scale, route: *Pd-II*.

^{31}P-NMR (161.97 MHz; C_6H_6): $\delta = 7.1$ (d, $^2J_{PP} = 13.7$ Hz), 11.0 (d, $^2J_{PP} = 13.7$ Hz).

(1,2-Bis(diphenylphosphino)ethane)-(η^2-ethyl 4-methylcinnamate)palladium(0)

Pd(dppe)PhMe-Et: $C_{38}H_{38}O_2P_2Pd$, M $= 695.07$ g/mol, yellow solid, NMR scale **route**: *Pd-III*, 8 eq. olefin.

^{31}P-NMR (161.97 MHz; C_6D_6): $\delta = 32.3$ (d, $^2J_{PP} = 39.9$ Hz), 32.7 (d, $^2J_{PP} = 39.9$ Hz).

(1,3-Bis(diphenylphosphino)propane)-(η^2-ethyl 4-methylcinnamate)palladium(0)

Pd(dppp)PhMe-Et: $C_{39}H_{40}O_2P_2Pd$, M $=$ 709.10 g/mol, yellow solid, NMR scale, route: *Pd-III*, 10 eq. olefin.

^{31}P-NMR (161.97 MHz; C_6D_6): $\delta = 8.0$ (d, $^2J_{PP} = 9.3$ Hz), 10.0 (d, $^2J_{PP} = 9.3$ Hz).

(1,3-Bis(di-*iso*-propylphosphino)propane)-(η^2-ethyl 4-methylcinnamate)palladium(0)

Pd(dippp)PhMe-Et: $C_{27}H_{48}O_2P_2Pd$, M $=$ 573.04 g/mol, yellow solid, route: *Pd-III*.

^{31}P-NMR (161.97 MHz; C_6H_6): $\delta = 26.8$ (d, $^2J_{PP} = 12.5$ Hz), 28.2 (d, $^2J_{PP} = 12.5$ Hz).

(1,4-Bis(diphenylphosphino)butane)-(η^2-ethyl 4-methylcinnamate)palladium(0)

Pd(dppb)PhMe-Et: $C_{40}H_{42}O_2P_2Pd$, M $=$ 723.13 g/mol, yellow solid, NMR scale, route: *Pd-II*.

^{1}H-NMR (400.13 MHz; C$_6$D$_6$): δ = 0.91 (t, $^3J_{HH}$ = 7.1 Hz, 3H, CH$_2$CH_3), 1.36 - 1.75 (m, 4H, PCH$_2$CH_2), 2.01(s, 3H, CH_3), 1.87 - 2.10 (m, 4H, PCH_2CH$_2$), 3.73 (dq, $^2J_{HH}$ = 10.7 Hz, $^3J_{HH}$ = 7.1 Hz, 1H, CH_2CH$_3$), 4.14 (dq, $^2J_{HH}$ = 10.7 Hz, $^3J_{HH}$ = 7.1 Hz, 1H, CH_2CH$_3$), 4.63 (ddd, $^3J_{HH}$ = 10.6 Hz, $^3J_{HP}$ = 2.4 Hz, $^3J_{HP}$ = 4.1 Hz, 1H, CHCHCO), 5.41 (ddd, $^3J_{HH}$ = 10.6 Hz, $^3J_{HP}$ = 2.9 Hz, $^3J_{HP}$ = 4.2 Hz, 1H, C$_{Ar}$CHCH), 6.56 (d, $^3J_{HP}$ = 7.9 Hz, 2H, H_{Ar}), 6.75 (d, $^3J_{HP}$ = 7.7 Hz, 2H, H_{Ar}), 6.81 - 7.00 (m, 4H, H_{Ar}), 7.00 - 7.27 (m, 10H, H_{Ar}), 7.63 (t, $^3J_{HP}$ = 8.4 Hz, 2H, H_{Ar}), 7.69 - 7.83 (m, 4H, H_{Ar}).

^{13}C-NMR (100.62 MHz; C$_6$D$_6$): δ = 14.6 (CH$_2$CH_3), 21.2 (CH$_3$), 23.7 (dd, $^2J_{CP}$ = 6.8 Hz, $^3J_{CP}$ = 4.8 Hz, PCH$_2$CH_2), 24.3 (dd, $^2J_{CP}$ = 5.5 Hz, $^3J_{CP}$ = 5.5 Hz, PCH$_2$CH_2), 27.6 (d, $^1J_{CP}$ = 13.9 Hz, $^3J_{CP}$ = 2.2 Hz, PCH_2CH$_2$), 28.3 (d, $^1J_{CP}$ = 12.8 Hz, $^3J_{CP}$ = 4.7 Hz, PCH_2CH$_2$), 56.9 (d, $^2J_{CP}$ = 13.6 Hz, CHCHCO), 58.7 (CH$_2$$CH_3$), 69.5 (dd, $^2J_{CP}$ = 3.2 Hz, $^2J_{CP}$ = 19.8 Hz, C$_{Ar}$$C$HCH), 125.5 - 125.6 (m, C_{Ar}H), 127.9 (C_{Ar}H), 128.2 (C_{Ar}H), 128.6 (d, $^3J_{CP}$ = 7.9 Hz, C_{Ar}H), 128.6 (d, $^3J_{CP}$ = 9.5 Hz, C_{Ar}H), 128.9 - 129.1 (m, C_{Ar}H), 129.5 (d, $^4J_{CP}$ = 0.5 Hz, C_{Ar}H), 129.7 (d, $^4J_{CP}$ = 0.6 Hz, C_{Ar}H), 131.0 - 131.4 (m, C_{Ar}), 132.5 (dd, $^2J_{CP}$ = 11.6 Hz, $^4J_{CP}$ = 2.6 Hz, C_{Ar}H), 133.1 (dd, $^2J_{CP}$ = 11.6 Hz, $^4J_{CP}$ = 2.7 Hz, C_{Ar}H), 133.4 (dd, $^2J_{CP}$ = 12.5 Hz, $^4J_{CP}$ = 2.9 Hz, C_{Ar}H), 134.0 (dd, $^2J_{CP}$ = 13.0 Hz, $^4J_{CP}$ = 2.8 Hz, C_{Ar}H), 135.8 (dd, $^1J_{CP}$ = 20.1 Hz, $^3J_{CP}$ = 8.5 Hz, C_{Ar}), 138.0 (dd, $^1J_{CP}$ = 27.7 Hz, $^3J_{CP}$ = 7.3 Hz, C_{Ar}), 138.1 (dd, $^1J_{CP}$ = 16.6 Hz, $^3J_{CP}$ = 8.8 Hz, C_{Ar}), 139.4 (dd, $^1J_{CP}$ = 20.6 Hz, $^3J_{CP}$ = 6.6 Hz, C_{Ar}), 141.5 (d, $^3J_{CP}$ = 5.3 Hz, C_{Ar}CH), 172.2 (dd, $^3J_{CP}$ = 2.1 Hz, $^3J_{CP}$ = 2.1 Hz, CHCOO).

^{31}P-NMR (161.97 MHz; C$_6$D$_6$): δ = 17.7 (d, $^2J_{PP}$ = 22.6 Hz), 17.9 (d, $^2J_{PP}$ = 22.6 Hz).

(1,2-Bis(diphenylphosphino)ethane)-(η^2-ethyl cinnamate)palladium(0)

Pd(dppe)Ph-Et: $C_{37}H_{36}O_2P_2Pd$, M $= 681.05$ g/mol, yellow solid, yield: 96%, **route**: *Pd-IV*.

^{1}H-NMR (400.13 MHz; C_6D_6): $\delta = 0.99$ (t, $^3J_{HH} = 7.1$ Hz, 3H, CH_2CH_3), 1.57 - 1.72 (m, 1H, PCH_2CH_2P), 1.73 - 1.88 (m, 1H, PCH_2CH_2P), 1.94 - 2.14 (m, 2H, PCH_2CH_2P), 4.01 (dq, $^2J_{HH} = 10.8$ Hz, $^3J_{HH} = 7.1$ Hz, 1H, CH_2CH_3), 4.20 (dq, $^2J_{HH} = 10.8$ Hz, $^3J_{HH} = 7.1$ Hz, 1H, CH_2CH_3), 5.06 (ddd, $^3J_{HH} = 11.2$ Hz, $^3J_{HP} = 3.0$ Hz, $^3J_{HP} = 7.5$ Hz, 1H, $CHCHCO$), 6.03 (ddd, $^3J_{HH} = 11.2$ Hz, $^3J_{HP} = 8.4$ Hz, $^3J_{HP} = 3.4$ Hz, 1H, $C_{Ar}CHCH$), 6.77 - 7.14 (m, 17H, H_{Ar}), 7.23 (d, $^3J_{HP} = 7.6$ Hz, 2H, H_{Ar}), 7.59 - 7.75 (m, 4H, H_{Ar}), 7.77 - 7.86 (m, 2H, H_{Ar}).

^{13}C-NMR (100.62 MHz; C_6D_6): $\delta = 14.8$ (CH_2CH_3), 27.0 (dd, $^1J_{CP} = 25.4$ Hz, $^2J_{CP} = 18.1$ Hz, PCH_2), 27.2 (dd, $^1J_{CP} = 23.3$ Hz, $^2J_{CP} = 18.1$ Hz, PCH_2), 54.7 (dd, $^2J_{CP} = 4.4$ Hz, $^2J_{CP} = 19.7$ Hz, $CHCHCO$), 58.9 (CH_2CH_3), 66.9 (d, $^2J_{CP} = 26.3$ Hz, $C_{Ar}CHCH$), 123.5 (dd, $^3J_{CP} = 1.6$ Hz, $^3J_{CP} = 2.9$ Hz, $C_{Ar}H$), 125.8 (dd, $^3J_{CP} = 1.6$ Hz, $^3J_{CP} = 3.7$ Hz, $C_{Ar}H$), 127.9 ($C_{Ar}H$), 128.2 ($C_{Ar}H$), 128.6 - 128.7 (m, $C_{Ar}H$), 128.8 (d, $^3J_{CP} = 3.0$ Hz, $C_{Ar}H$), 128.9 (d, $^3J_{CP} = 7.6$ Hz, $C_{Ar}H$), 129.0 (d, $^3J_{CP} = 8.4$ Hz, $C_{Ar}H$), 129.4 (d, $^4J_{CP} = 1.6$ Hz, $C_{Ar}H$), 129.9 (d, $^4J_{CP} = 1.4$ Hz, $C_{Ar}H$), 130.0 - 130.1 (m, $C_{Ar}H$), 132.8 (d, $^2J_{CP} = 14.4$ Hz, $C_{Ar}H$), 133.1 (d, $^2J_{CP} = 11.5$ Hz, $C_{Ar}H$), 133.2 (d, $^2J_{CP} = 11.8$ Hz, $C_{Ar}H$), 133.7 (d, $^2J_{CP} = 15.1$ Hz, $C_{Ar}H$), 134.5 (dd, $^1J_{CP} = 22.7$ Hz, $^3J_{CP} = 4.0$ Hz, C_{Ar}), 136.2 - 136.8 (m, C_{Ar}), 145.6 (d, $^3J_{CP} = 7.0$ Hz, $C_{Ar}CH$), 172.8 (dd, $^3J_{CP} = 2.6$ Hz, $^3J_{CP} = 4.1$ Hz, $CHCOO$).

^{31}P-NMR (161.97 MHz; C_6D_6): $\delta = 33.0$ (d, $^2J_{PP} = 37.4$ Hz), 33.5 (d, $^2J_{PP} = 37.4$ Hz).

(1,3-Bis(diphenylphosphino)propane)-(η^2-ethyl cinnamate)palladium(0)

Pd(dppp)Ph-Et: $C_{38}H_{38}O_2P_2Pd$, M = 695.07 g/mol, yellow solid, yield: 83%, **route:** Pd-IV.

^{1}H-NMR (400.13 MHz; C_6D_6): δ = 0.93 (t, $^3J_{HH}$ = 7.1 Hz, 3H, CH_2CH_3), 1.29 - 1.53 (m, 2H, $CH_2CH_2CH_2$), 1.77 - 2.11 (m, 4H, PCH_2CH_2), 3.76 (dq, $^2J_{HH}$ = 10.8 Hz, $^3J_{HH}$ = 7.1 Hz, 1H, CH_2CH_3), 4.13 (dq, $^2J_{HH}$ = 10.8 Hz, $^3J_{HH}$ = 7.1 Hz, 1H, CH_2CH_3), 4.81 (ddd, $^3J_{HH}$ = 10.8 Hz, $^3J_{HP}$ = 4.3 Hz, $^3J_{HP}$ = 7.4 Hz, 1H, $CHCHCO$), 5.63 (ddd, $^3J_{HH}$ = 10.8 Hz, $^3J_{HP}$ = 5.0 Hz, $^3J_{HP}$ = 8.4 Hz, 1H, $C_{Ar}CHCH$), 6.73 - 6.90 (m, 7H, H_{Ar}), 6.90 - 7.20 (m, 12H, H_{Ar}), 7.49 - 7.60 (m, 2H, H_{Ar}), 7.73 - 7.87 (m, 4H, H_{Ar}).

^{13}C-NMR (100.62 MHz; C_6D_6): δ = 14.7 (CH_2CH_3), 19.5 (t, $^2J_{CP}$ = 4.9 Hz, PCH_2CH_2), 29.5 (d, $^1J_{CP}$ = 17.5 Hz, PCH_2CH_2), 29.8 (d, $^1J_{CP}$ = 18.2 Hz, PCH_2CH_2), 54.8 (dd, $^2J_{CP}$ = 5.9 Hz, $^2J_{CP}$ = 19.6 Hz, $CHCHCO$), 58.8 (CH_2CH_3), 67.3 (d, $^2J_{CP}$ = 25.9 Hz, $C_{Ar}CHCH$), 123.5 - 123.6 (m, $C_{Ar}H$), 125.6 - 125.8 (m, $C_{Ar}H$), 128.4 (d, $^3J_{CP}$ = 9.1 Hz, $C_{Ar}H$), 128.6 (d, $^3J_{CP}$ = 9.0 Hz, $C_{Ar}H$), 128.6 (d, $^4J_{CP}$ = 1.5 Hz, $C_{Ar}H$), 128.7 (d, $^3J_{CP}$ = 9.6 Hz, $C_{Ar}H$), 129.5 (d, $^4J_{CP}$ = 1.4 Hz, $C_{Ar}H$), 129.6 (d, $^4J_{CP}$ = 1.4 Hz, $C_{Ar}H$), 130.0 (d, $^4J_{CP}$ = 1.5 Hz, $C_{Ar}H$), 132.2 (d, $^2J_{CP}$ = 13.9 Hz, $C_{Ar}H$), 133.4 (d, $^2J_{CP}$ = 15.5 Hz, $C_{Ar}H$), 133.5 (d, $^2J_{CP}$ = 15.4 Hz, $C_{Ar}H$), 134.1 (d, $^2J_{CP}$ = 16.0 Hz, $C_{Ar}H$), 135.6 (dd, $^1J_{CP}$ = 24.3 Hz, $^3J_{CP}$ = 4.9 Hz, C_{Ar}), 136.1 (dd, $^1J_{CP}$ = 26.1 Hz, $^3J_{CP}$ = 3.2 Hz, C_{Ar}), 137.7 (dd, $^1J_{CP}$ = 19.0 Hz, $^3J_{CP}$ = 4.0 Hz, C_{Ar}), 137.9 (dd, $^1J_{CP}$ = 19.3 Hz, $^3J_{CP}$ = 4.0 Hz, C_{Ar}), 144.6 (d, $^3J_{CP}$ = 6.9 Hz, $C_{Ar}CH$), 172.4 (dd, $^3J_{CP}$ = 3.1 Hz, $^3J_{CP}$ = 3.9 Hz, $CHCOO$).

^{31}P-NMR (161.97 MHz; C_6D_6): δ = 8.1 (d, $^2J_{PP}$ = 7.8 Hz), 10.5 (d, $^2J_{PP}$ = 7.8 Hz).

(1,3-Bis(di-*iso*-propylphosphino)propane)-(η^2-ethyl cinnamate)palladium(0)

Pd(dippp)Ph-Et: $C_{26}H_{46}O_2P_2Pd$, M = 559.01 g/mol, orange solid, **route**: *Pd-III*.

^{31}P-NMR (161.97 MHz; C_6H_6): δ = 26.4 (d, $^2J_{PP}$ = 14.4 Hz), 28.5 (d, $^2J_{PP}$ = 14.4 Hz).

(1,4-Bis(diphenylphosphino)butane)-(η^2-ethyl cinnamate)palladium(0)

Pd(dppb)Ph-Et: $C_{39}H_{40}O_2P_2P$, M = 709.10 g/mol, yellow solid, NMR scale, **route**: *Pd-III*.

^{1}H-NMR (400.13 MHz; C_6D_6): δ = 0.90 (t, $^3J_{HH}$ = 7.1 Hz, 3H, CH_2CH_3), 1.35 - 1.77 (m, 4H, PCH_2CH_2), 1.86 - 2.15 (m, 4H, PCH_2CH_2), 3.73 (dq, $^2J_{HH}$ = 10.6 Hz, $^3J_{HH}$ = 7.1 Hz, 1H, CH_2CH_3), 4.14 (dq, $^2J_{HH}$ = 10.6 Hz, $^3J_{HH}$ = 7.1 Hz, 1H, CH_2CH_3), 4.60 (ddd, $^3J_{HH}$ = 11.0 Hz, $^3J_{HP}$ = 4.5 Hz, $^3J_{HP}$ = 7.2 Hz, 1H, $CHCHCO$), 5.35 (ddd, $^3J_{HH}$ = 11.0 Hz, $^3J_{HP}$ = 5.1 Hz, $^3J_{HP}$ = 8.1 Hz, 1H, $C_{Ar}CHCH$), 6.71 - 6.99 (m, 19H, H_{Ar}), 7.01 - 7.24 (m, 10H, H_{Ar}), 7.58 - 7.66 (m, 2H, H_{Ar}), 7.66 - 7.74 (m, 2H, H_{Ar}), 7.74 - 7.83 (m, 2H, H_{Ar}).

^{13}C-NMR (100.62 MHz; C_6D_6): δ = 14.6 (CH_2CH_3), 23.7 (dd, $^2J_{CP}$ = 7.3 Hz, $^3J_{CP}$ = 4.0 Hz, PCH_2CH_2), 24.3 (dd, $^2J_{CP}$ = 6.5 Hz, $^3J_{CP}$ = 4.4 Hz, PCH_2CH_2), 27.7 (d, $^1J_{CP}$ = 15.8 Hz, PCH_2CH_2), 28.3 (d, $^1J_{CP}$ = 16.4 Hz, PCH_2CH_2), 57.2 (dd, $^2J_{CP}$ = 5.8 Hz, $^2J_{CP}$ = 19.5 Hz, $CHCHCO$), 58.7 (CH_2CH_3), 69.2 (dd,

$^2J_{CP}$ = 1.8 Hz, $^2J_{CP}$ = 25.2 Hz, $C_{Ar}CHCH$), 123.6 - 123.7 (m, $C_{Ar}H$), 125.7 (dd, $^3J_{CP}$ = 1.1 Hz, $^3J_{CP}$ = 3.3 Hz, $C_{Ar}H$), 127.9 ($C_{Ar}H$), 128.2 ($C_{Ar}H$), 128.5 ($C_{Ar}H$), 128.6 (d, $^3J_{CP}$ = 8.9 Hz, $C_{Ar}H$), 128.8 (d, $^4J_{CP}$ = 1.4 Hz, $C_{Ar}H$), 129.0 (d, $^4J_{CP}$ = 1.1 Hz, $C_{Ar}H$), 129.5 (d, $^4J_{CP}$ = 1.5 Hz, $C_{Ar}H$), 129.7 (d, $^4J_{CP}$ = 1.6 Hz, $C_{Ar}H$), 131.0 - 131.3 (m, C_{Ar}), 132.4 (d, $^2J_{CP}$ = 13.8 Hz, $C_{Ar}H$), 133.0 (d, $^2J_{CP}$ = 14.2 Hz, $C_{Ar}H$), 133.3 (d, $^2J_{CP}$ = 15.0 Hz, $C_{Ar}H$), 134.0 (d, $^2J_{CP}$ = 15.5 Hz, $C_{Ar}H$), 135.6 (dd, $^1J_{CP}$ = 24.5 Hz, $^3J_{CP}$ = 3.5 Hz, C_{Ar}), 137.9 (dd, $^1J_{CP}$ = 23.3 Hz, $^3J_{CP}$ = 3.5 Hz, C_{Ar}), 138.2 (dd, $^1J_{CP}$ = 25.2 Hz, $^3J_{CP}$ = 4.2 Hz, C_{Ar}), 139.3 (dd, $^1J_{CP}$ = 23.9 Hz, $^3J_{CP}$ = 3.6 Hz, C_{Ar}), 144.5 (dd, $^3J_{CP}$ = 1.0 Hz, $^3J_{CP}$ = 6.6 Hz, $C_{Ar}CH$), 172.2 (dd, $^3J_{CP}$ = 3.5 Hz, $^3J_{CP}$ = 3.5 Hz, $CHCOO$).

^{31}P-NMR (161.97 MHz; C_6D_6): δ = 17.8 (d, $^2J_{PP}$ = 20.2 Hz), 18.5 (d, $^2J_{PP}$ = 20.2 Hz).

(1,2-Bis(diphenylphosphino)ethane)-(η^2-methyl 4-nitrocinnamate)palladium(0)

Pd(dppe)PhNO$_2$-Me: $C_{36}H_{33}NO_4P_2Pd$, M = 712.02 g/mol, red solid, yield: >99%, **route:** *Pd-III*.

^{1}H-NMR (400.13 MHz; C_6D_6): δ = 1.60 - 1.93 (m, 4H, PCH_2CH_2P), 3.47 (s, 3H, OCH_3), 4.75 (dpt, $^3J_{HH}$ = 8.4 Hz, $^3J_{HP}$ = 2.5 Hz, $^3J_{HP}$ = 8.4 Hz, 1H, $CHCHCO$), 5.75 (pt, $^3J_{HH}$ = 8.4 Hz, $^3J_{HP}$ = 8.4 Hz, $^3J_{HP}$ = 8.4 Hz, 1H, $C_{Ar}CHCH$), 6.26 (d, $^3J_{HP}$ = 8.5 Hz, 2H, H_{Ar}), 6.58 - 6.86 (m, 6H, H_{Ar}), 6.93 - 7.17 (m, 8H, H_{Ar}), 7.46 - 7.66 (m, 6H, H_{Ar}), 7.66 - 7.75 (m, 2H, H_{Ar}).

^{31}P-NMR (161.97 MHz; C_6D_6): δ = 36.8 (d, $^2J_{PP}$ = 17.2 Hz), 36.5 (d, $^2J_{PP}$ = 17.2 Hz).

(1,3-Bis(diphenylphosphino)propane)-(η^2-methyl 4-nitrocinnamate)palladium(0)

Pd(dppp)PhNO$_2$-Me: C$_{37}$H$_{35}$NO$_4$P$_2$Pd, M = 726.05 g/mol, deep orange solid, yield: 98%, **route:** *Pd-III.*

^{1}H-NMR (400.13 MHz; C$_6$D$_6$): δ = 1.13 - 1.49 (m, 2H, CH$_2$C*H*$_2$CH$_2$), 1.66 - 1.77 (m, 1H, PC*H*$_2$CH$_2$), 1.82 - 2.09 (m, 3H, PC*H*$_2$CH$_2$), 3.32 (s, 3H, OC*H*$_3$), 4.51 (ddd, $^3J_{HH}$ = 10.8 Hz, $^3J_{HP}$ = 3.7 Hz, $^3J_{HP}$ = 8.1 Hz, 1H, CHC*H*CO), 5.37 (dpt, $^3J_{HH}$ = 10.8 Hz, $^3J_{HP}$ = 3.3 Hz, $^3J_{HP}$ = 10.8 Hz, 1H, C$_{Ar}$C*H*CH), 6.27 (d, $^3J_{HP}$ = 8.7 Hz, 1H, *H$_{Ar}$*), 6.45 (d, $^3J_{HP}$ = 7.9 Hz, 2H, *H$_{Ar}$*), 6.47 - 6.58 (m, 2H, *H$_{Ar}$*), 6.67 - 6.76 (m, 2H, *H$_{Ar}$*), 6.84 (t, $^3J_{HP}$ = 7.4 Hz, 1H, *H$_{Ar}$*), 6.93 - 7.24 (m, 8H, *H$_{Ar}$*), 7.41 - 7.56 (m, 4H, *H$_{Ar}$*), 7.56 - 7.76 (m, 4H, *H$_{Ar}$*).

^{13}C-NMR (100.62 MHz; C$_6$D$_6$): δ = 19.3 (t, $^2J_{CP}$ = 4.8 Hz, PCH$_2$*C*H$_2$), 29.3 (d, $^1J_{CP}$ = 19.4 Hz, P*C*H$_2$CH$_2$), 29.8 (d, $^1J_{CP}$ = 19.0 Hz, P*C*H$_2$CH$_2$), 50.3 (O*C*H$_3$), 53.1 (dd, $^2J_{CP}$ = 4.8 Hz, $^2J_{CP}$ = 21.5 Hz, CH*C*HCO), 65.4 (dd, $^2J_{CP}$ = 1.7 Hz, $^2J_{CP}$ = 24.7 Hz, C$_{Ar}$*C*HCH), 123.4 (*C$_{Ar}$*H), 124.0 - 124.1 (m, *C$_{Ar}$*H), 124.2 (dd, $^3J_{CP}$ = 1.6 Hz, $^3J_{CP}$ = 3.2 Hz, *C$_{Ar}$*H), 127.9 (*C$_{Ar}$*H), 128.2 (*C$_{Ar}$*H), 128.5 (d, $^3J_{CP}$ = 9.4 Hz, *C$_{Ar}$*H), 128.6 (d, $^3J_{CP}$ = 9.3 Hz, *C$_{Ar}$*H), 128.8 (d, $^3J_{CP}$ = 9.1 Hz, *C$_{Ar}$*H), 128.9 (d, $^3J_{CP}$ = 9.5 Hz, *C$_{Ar}$*H), 129.2 (d, $^4J_{CP}$ = 1.5 Hz, *C$_{Ar}$*H), 129.8 (d, $^4J_{CP}$ = 1.7 Hz, *C$_{Ar}$*H), 129.9 (d, $^4J_{CP}$ = 1.6 Hz, *C$_{Ar}$*H), 130.5 (d, $^4J_{CP}$ = 1.8 Hz, *C$_{Ar}$*H), 131.3 (d, $^2J_{CP}$ = 12.9 Hz, *C$_{Ar}$*H), 133.1 (d, $^2J_{CP}$ = 15.0 Hz, *C$_{Ar}$*H), 133.4 (d, $^2J_{CP}$ = 14.9 Hz, *C$_{Ar}$*H), 134.2 (dd, $^1J_{CP}$ = 28.6 Hz, $^3J_{CP}$ = 2.4 Hz, *C$_{Ar}$*), 134.5 (dd, $^1J_{CP}$ = 27.4 Hz, $^3J_{CP}$ = 3.1 Hz, *C$_{Ar}$*), 134.5 (d, $^2J_{CP}$ = 16.0 Hz, *C$_{Ar}$*H), 136.7 (dd, $^1J_{CP}$ = 26.4 Hz, $^3J_{CP}$ = 3.0 Hz, *C$_{Ar}$*), 137.3 (dd, $^1J_{CP}$ = 26.1 Hz, $^3J_{CP}$ = 3.3 Hz, *C$_{Ar}$*), 143.4 (dd, $^3J_{CP}$ = 2.1 Hz, $^3J_{CP}$ = 4.0 Hz, *C$_{Ar}$*CH), 151.8 (dd, $^3J_{CP}$ = 0.6 Hz, $^3J_{CP}$ = 7.0 Hz, *C$_{Ar}$*N), 172.6 (dd, $^3J_{CP}$ = 2.2 Hz, $^3J_{CP}$ = 4.7 Hz, CH*C*OO).

^{31}P-NMR (161.97 MHz; C_6D_6): $\delta = 9.8$ (d, $^2J_{PP} = 12.6$ Hz), 11.5 (d, $^2J_{PP} = 12.6$ Hz).

(1,4-Bis(diphenylphosphino)butane)-(η^2-methyl 4-nitrocinnamate)palladium(0)

Pd(dppb)PhNO$_2$-Me: $C_{38}H_{37}NO_4P_2Pd$, M $= 740.07$ g/mol, red solid, yield: 83%, route: *Pd-III*.

^{1}H-NMR (400.13 MHz; C_6D_6): $\delta = 1.19 - 1.51$ (m, 4H, PCH$_2$CH$_2$), 1.77 - 2.07 (m, 4H, PCH$_2$CH$_2$), 3.28 (s, 3H, OCH$_3$), 4.28 (ddd, $^3J_{HH} = 11.9$ Hz, $^3J_{HP} = 4.5$ Hz, $^3J_{HP} = 7.8$ Hz, 1H, CHCHCO), 5.19 (dpt, $^3J_{HH} = 11.9$ Hz, $^3J_{HP} = 3.5$ Hz, $^3J_{HP} = 11.9$ Hz, 1H, C$_{Ar}$CHCH), 6.33 (d, $^3J_{HP} = 8.5$Hz, 2H, H_{Ar}), 6.64 (t, $^3J_{HP} = 8.9$Hz, 2H, H_{Ar}), 6.73 (t, $^3J_{HP} = 7.5$Hz, 2H, H_{Ar}), 6.85 (t, $^3J_{HP} = 7.3$Hz, 1H, H_{Ar}), 6.98 - 7.30 (m, 7H, H_{Ar}), 7.39 - 7.56 (m, 6H, H_{Ar}), 7.59 - 7.72 (m, 4H, H_{Ar}).

^{13}C-NMR (100.62 MHz; C_6D_6): $\delta = 23.2$ (dd, $^2J_{CP} = 6.4$ Hz, $^3J_{CP} = 4.1$ Hz, PCH$_2$CH$_2$), 23.9 (dd, $^2J_{CP} = 5.2$ Hz, $^3J_{CP} = 5.2$ Hz, PCH$_2$CH$_2$), 27.1 (d, $^1J_{CP} = 17.1$ Hz, PCH$_2$CH$_2$), 27.8 (d, $^1J_{CP} = 17.4$ Hz, PCH$_2$CH$_2$), 50.2 (OCH$_3$), 55.0 (dd, $^2J_{CP} = 4.9$ Hz, $^2J_{CP} = 21.4$ Hz, CHCHCO), 66.6 (dd, $^2J_{CP} = 1.6$ Hz, $^2J_{CP} = 24.9$ Hz, C$_{Ar}$CHCH), 123.4 (C_{Ar}), 124.0 (d, $^3J_{CP} = 1.6$ Hz, C_{Ar}H), 124.3 (d, $^3J_{CP} = 2.2$ Hz, C_{Ar}H), 127.9 (C_{Ar}H), 128.2 (C_{Ar}H), 128.5 (C_{Ar}H), 128.8 (d, $^3J_{CP} = 9.0$ Hz, C_{Ar}H), 128.8 (d, $^3J_{CP} = 8.2$ Hz, C_{Ar}H), 128.9 (d, $^3J_{CP} = 9.5$ Hz, C_{Ar}H), 129.2 (d, $^4J_{CP} = 1.2$ Hz, C_{Ar}H), 129.5 (d, $^4J_{CP} = 0.8$ Hz, C_{Ar}H), 130.0 (d, $^4J_{CP} = 1.3$ Hz, C_{Ar}H), 130.2 (d, $^4J_{CP} = 1.3$ Hz, C_{Ar}H), 131.9 (d, $^2J_{CP} = 13.3$ Hz, C_{Ar}H), 132.2 (d, $^2J_{CP} = 13.4$ Hz, C_{Ar}H), 133.8 (d, $^2J_{CP} = 12.9$ Hz, C_{Ar}H), 134.0 (d, $^2J_{CP} = 13.2$ Hz, C_{Ar}H), 135.0 (dd, $^1J_{CP} = 26.9$ Hz, $^3J_{CP} = 2.7$ Hz, C_{Ar}), 135.5 (dd, $^1J_{CP} = 27.5$ Hz, $^3J_{CP} = 2.8$ Hz, C_{Ar}), 136.8 (dd, $^1J_{CP} = 26.1$ Hz, $^3J_{CP} = 2.9$ Hz, C_{Ar}), 138.6 (dd, $^1J_{CP} = 26.7$ Hz, $^3J_{CP} = 2.7$ Hz,

C_{Ar}), 143.5 (dd, $^3J_{CP}$ = 1.6 Hz, $^3J_{CP}$ = 3.6 Hz, C_{Ar}CH), 151.6 (dd, $^3J_{CP}$ = 1.2 Hz, $^3J_{CP}$ = 6.9 Hz, C_{Ar}N), 172.4 (dd, $^3J_{CP}$ = 2.5 Hz, $^3J_{CP}$ = 4.4 Hz, CHCOO).

^{31}P-NMR (161.97 MHz; C$_6$D$_6$): δ = 19.7 (d, $^2J_{PP}$ = 3.1 Hz), 20.2 (d, $^2J_{PP}$ = 3.1 Hz).

(1,2-Bis(diphenylphosphino)ethane)-(η^2-ethyl 4-nitrocinnamate)palladium(0)

Pd(dppe)PhNO$_2$-Et: C$_{37}$H$_{35}$NO$_4$P$_2$Pd, M = 726.05 *g/mol*, red solid, yield: >99%, **route**: *Pd-III*.

^{1}H-NMR (400.13 MHz; C$_6$D$_6$): δ = 0.98 (t, $^3J_{HH}$ = 7.1 Hz, 3H, CH$_2$CH$_3$), 1.56 - 1.72 (m, 1H, PCH$_2$CH$_2$P), 1.75 - 1.98 (m, 3H, PCH$_2$CH$_2$P), 3.97 (dq, $^2J_{HH}$ = 10.8 Hz, $^3J_{HH}$ = 7.1 Hz, 1H, CH$_2$CH$_3$), 4.18 (dq, $^2J_{HH}$ = 10.8 Hz, $^3J_{HH}$ = 7.1 Hz, 1H, CH$_2$CH$_3$), ´ 4.77 (ddd, $^3J_{HH}$ = 11.1 Hz, $^3J_{HP}$ = 2.7 Hz, $^3J_{HP}$ = 8.2 Hz, 1H, CHCHCO), 5.77 (pt, $^3J_{HH}$ = 10.4 Hz, $^3J_{HP}$ = 10.4 Hz, $^3J_{HP}$ = 10.4 Hz, 1H, C$_{Ar}$CHCH), 6.59 - 6.86 (m, 8H, H$_{Ar}$), 6.93 - 7.13 (m, 8H, H$_{Ar}$), 7.48 - 7.65 (m, 6H, H$_{Ar}$), 7.72 - 7.79 (m, 2H, H$_{Ar}$).

^{13}C-NMR (100.62 MHz; C$_6$D$_6$): δ = 14.7 (CH$_2$CH$_3$), 27.1 (dd, $^1J_{CP}$ = 17.7 Hz, $^2J_{CP}$ = 7.7 Hz, PCH$_2$), 27.4 (dd, $^1J_{CP}$ = 18.1 Hz, $^2J_{CP}$ = 8.1 Hz, PCH$_2$), 53.4 (dd, $^2J_{CP}$ = 3.5 Hz, $^2J_{CP}$ = 21.9 Hz, CHCHCO), 59.2 (CH$_2$CH$_3$), 65.2 (d, $^2J_{CP}$ = 24.9 Hz, C$_{Ar}$CHCH), 124.4 (dd, $^3J_{CP}$ = 1.4 Hz, $^3J_{CP}$ = 9.8 Hz, C$_{Ar}$H), 127.9 (C$_{Ar}$H), 128.2 (C$_{Ar}$H), 128.6 (d, $^3J_{CP}$ = 9.0 Hz, C$_{Ar}$H), 128.9 (d, $^3J_{CP}$ = 9.4 Hz, C$_{Ar}$H), 129.2 (d, $^3J_{CP}$ = 9.3 Hz, C$_{Ar}$H), 130.0 (d, $^4J_{CP}$ = 1.3 Hz, C$_{Ar}$H), 130.4 (C$_{Ar}$H), 130.6 (d, $^4J_{CP}$ = 1.4 Hz, C$_{Ar}$H), 131.9 (d, $^2J_{CP}$ = 13.6 Hz, C$_{Ar}$H), 133.0 (d, $^2J_{CP}$ = 14.8 Hz, C$_{Ar}$H), 133.5 (d, $^2J_{CP}$ = 15.0 Hz, C$_{Ar}$H), 133.6 (d, $^2J_{CP}$ = 15.8 Hz, C$_{Ar}$H), 134.7 (dd, $^1J_{CP}$ = 25.8 Hz, $^3J_{CP}$ = 3.8 Hz, C$_{Ar}$), 135.5 (dd, $^1J_{CP}$ = 24.2 Hz, $^3J_{CP}$ = 4.4 Hz, C$_{Ar}$), 135.8 (dd, $^1J_{CP}$ = 24.8 Hz, $^3J_{CP}$ = 3.2 Hz, C$_{Ar}$), 143.2 (dd, $^3J_{CP}$ = 2.2 Hz, $^3J_{CP}$ = 4.2 Hz,

C_{Ar}CH), 152.8 (d, $^3J_{CP}$ = 7.3 Hz, C_{Ar}N), 172.6 (dd, $^3J_{CP}$ = 1.6 Hz, $^3J_{CP}$ = 4.8 Hz, CHCOO).

^{31}P-NMR (161.97 MHz; C$_6$D$_6$): δ = 36.3 (d, $^2J_{PP}$ = 17.3 Hz), 36.6 (d, $^2J_{PP}$ = 17.3 Hz).

(1,3-Bis(diphenylphosphino)propane)-(η^2-ethyl 4-nitrocinnamate)palladium(0)

Pd(dppp)PhNO$_2$-Et: C$_{38}$H$_{37}$NO$_4$P$_2$Pd, M = 740.07 g/mol, red solid, yield: 83%, **route**: *Pd-III*.

Single crystals suitable for X-ray analysis were obtained by slow diffusion of pentane into a toluene solution.

^{1}H-NMR (400.13 MHz; C$_6$D$_6$): δ = 0.91 (t, $^3J_{HH}$ = 7.1 Hz, 3H, CH$_2$CH_3), 1.08 - 1.31 (m, 1H, CH$_2$CH_2CH$_2$), 1.31 - 1.56 (m, 1H, CH$_2$CH_2CH$_2$), 1.65 (pt, $^2J_{HH}$ = 12.9 Hz, $^3J_{HH}$ = 12.9 Hz, 1H, PCH_2CH$_2$), 1.86 (pt, $^2J_{HH}$ = 12.9 Hz, $^3J_{HH}$ = 12.9 Hz, 1H, PCH_2CH$_2$), 1.93 - 2.16 (m, 2H, PCH_2CH$_2$), 3.78 (dq, $^2J_{HH}$ = 10.8 Hz, $^3J_{HH}$ = 7.1 Hz, 1H, CH_2CH$_3$), 4.09 (dq, $^2J_{HH}$ = 10.8 Hz, $^3J_{HH}$ = 7.1 Hz, 1H, CH_2CH$_3$), 4.50 (ddd, $^3J_{HH}$ = 10.9 Hz, $^3J_{HP}$ = 3.8 Hz, $^3J_{HP}$ = 8.1 Hz, 1H, CHCHCO), 5.39 (dpt, $^3J_{HH}$ = 9.6 Hz, $^3J_{HP}$ = 3.4 Hz, $^3J_{HP}$ = 9.6 Hz, 1H, C$_{Ar}$CHCH), 6.43 (d, $^3J_{HP}$ = 7.9 Hz, 2H, H_{Ar}), 6.48 - 6.56 (m, 2H, H_{Ar}), 6.66 - 6.74 (m, 2H, H_{Ar}), 6.84 (t, $^3J_{HP}$ = 7.4 Hz, 1H, H_{Ar}), 6.92 - 7.26 (m, 11H, H_{Ar}), 7.41 - 7.52 (m, 4H, H_{Ar}), 7.64 - 7.74 (m, 2H, H_{Ar}), 7.76 - 7.85 (m, 2H, H_{Ar}).

^{13}C-NMR (100.62 MHz; C$_6$D$_6$): δ = 14.6 (CH$_2$CH_3), 19.2 (t, $^2J_{CP}$ = 4.8 Hz, PCH$_2$CH_2), 29.3 (d, $^1J_{CP}$ = 18.9 Hz, PCH$_2$CH$_2$), 29.8 (d, $^1J_{CP}$ = 19.2 Hz, PCH$_2$CH$_2$), 53.4 (dd, $^2J_{CP}$ = 4.8 Hz, $^2J_{CP}$ = 21.5 Hz, CHCHCO), 59.1 (CH_2CH$_3$), 65.4 (dd, $^2J_{CP}$ = 1.8 Hz, $^2J_{CP}$ = 24.5 Hz, C$_{Ar}$$C$HCH), 124.0 - 124.1 (m, C_{Ar}H), 124.2 (dd, $^3J_{CP}$ = 1.4 Hz, $^3J_{CP}$ = 3.1 Hz, C_{Ar}H), 127.9 (C_{Ar}H), 128.2 (C_{Ar}H), 128.2 (C_{Ar}H), 128.6

(d, $^3J_{CP}$ = 9.5 Hz, C_{Ar}H), 128.8 (d, $^3J_{CP}$ = 9.3 Hz, C_{Ar}H), 128.9 (d, $^3J_{CP}$ = 10.1 Hz, C_{Ar}H), 129.0 (d, $^4J_{CP}$ = 1.8 Hz, C_{Ar}H), 129.9 (d, $^4J_{CP}$ = 1.2 Hz, C_{Ar}H), 130.6 (d, $^4J_{CP}$ = 1.7 Hz, C_{Ar}H), 131.1 (d, $^2J_{CP}$ = 12.8 Hz, C_{Ar}H), 133.0 (d, $^2J_{CP}$ = 14.9 Hz, C_{Ar}H), 133.6 (d, $^2J_{CP}$ = 15.2 Hz, C_{Ar}H), 133.8 (dd, $^1J_{CP}$ = 29.0 Hz, $^3J_{CP}$ = 1.8 Hz, C_{Ar}), 134.7 (d, $^2J_{CP}$ = 16.1 Hz, C_{Ar}H), 134.8 (dd, $^1J_{CP}$ = 26.0 Hz, $^3J_{CP}$ = 4.1 Hz, C_{Ar}), 136.7 (dd, $^1J_{CP}$ = 26.7 Hz, $^3J_{CP}$ = 2.4 Hz, C_{Ar}), 137.5 (dd, $^1J_{CP}$ = 25.7 Hz, $^3J_{CP}$ = 4.0 Hz, C_{Ar}), 143.3 (dd, $^3J_{CP}$ = 2.0 Hz, $^3J_{CP}$ = 3.8 Hz, C_{Ar}CH), 151.8 (dd, $^3J_{CP}$ = 0.6 Hz, $^3J_{CP}$ = 7.0 Hz, C_{Ar}N), 172.4 (dd, $^3J_{CP}$ = 2.2 Hz, $^3J_{CP}$ = 4.6 Hz, CHCOO).

^{31}P-NMR (161.97 MHz; C$_6$D$_6$): δ = 9.2 (d, $^2J_{PP}$ = 12.0 Hz), 12.0 (d, $^2J_{PP}$ = 12.0 Hz).

(1,3-Bis(di-*iso*-propylphosphino)propane)-(η^2-ethyl 4-nitrocinnamate)palladium(0)

Pd(dippp)PhNO$_2$-Et: C$_{26}$H$_{45}$NO$_4$P$_2$Pd, M = 604.01 *g/mol*, violet solid, **route**: *Pd-III*.

^{31}P-NMR (161.97 MHz; C$_6$H$_6$): δ = 26.2 (d, $^2J_{PP}$ = 30.5 Hz), 30.7 (d, $^2J_{PP}$ = 30.5 Hz).

(1,4-Bis(diphenylphosphino)butane)-(η^2-ethyl 4-nitrocinnamate)palladium(0)

Pd(dppb)PhNO$_2$-Et: C$_{39}$H$_{39}$NO$_4$P$_2$Pd, M = 754.10 *g/mol*, red solid, NMR scale, route: *Pd-II*.

^{1}H-NMR (400.13 MHz; C_6D_6): $\delta = 0.88$ (t, $^3J_{HH} = 7.1$ Hz, 3H, CH_2CH_3), 1.23 - 1.58 (m, 4H, PCH_2CH_2), 1.78 - 2.06 (m, 4H, PCH_2CH_2), 3.67 (dq, $^2J_{HH} = 10.8$ Hz, $^3J_{HH} = 7.1$ Hz, 1H, CH_2CH_3), 4.09 (dq, $^2J_{HH} = 10.8$ Hz, $^3J_{HH} = 7.1$ Hz, 1H, CH_2CH_3), 4.28 (ddd, $^3J_{HH} = 10.0$ Hz, $^3J_{HP} = 4.5$ Hz, $^3J_{HP} = 7.9$ Hz, 1H, $CHCHCO$), 5.21 (dpt, $^3J_{HH} = 10.0$ Hz, $^3J_{HP} = 3.8$ Hz, $^3J_{HP} = 10.0$ Hz, 1H, $C_{Ar}CHCH$), 6.32 (d, $^3J_{HP} = 7.6$ Hz, 2H, H_{Ar}), 6.58 - 6.67 (m, 2H, H_{Ar}), 6.72 (t, $^3J_{HP} = 6.9$ Hz, 2H, H_{Ar}), 6.84 (t, $^3J_{HP} = 7.3$ Hz, 1H, H_{Ar}), 6.99 - 7.21 (m, 9H, H_{Ar}), 7.43 (d, $^3J_{HP} = 8.7$ Hz, 2H, H_{Ar}), 7.48 - 7.57 (m, 2H, H_{Ar}), 7.60 - 7.68 (m, 2H, H_{Ar}), 7.68 - 7.75 (m, 2H, H_{Ar}).

^{13}C-NMR (100.62 MHz; C_6D_6): $\delta = 14.5$ (CH_2CH_3), 23.2 (dd, $^2J_{CP} = 6.5$ Hz, $^3J_{CP} = 4.2$ Hz, PCH_2CH_2), 23.8 (dd, $^2J_{CP} = 5.3$ Hz, $^3J_{CP} = 5.3$ Hz, PCH_2CH_2), 27.0 (d, $^1J_{CP} = 17.1$ Hz, PCH_2CH_2), 27.8 (d, $^1J_{CP} = 17.3$ Hz, PCH_2CH_2), 55.5 (dd, $^2J_{CP} = 5.0$ Hz, $^2J_{CP} = 21.7$ Hz, $CHCHCO$), 59.0 (CH_2CH_3), 66.7 (dd, $^2J_{CP} = 2.4$ Hz, $^2J_{CP} = 24.2$ Hz, $C_{Ar}CHCH$), 124.0 (d, $^3J_{CP} = 1.7$ Hz, $C_{Ar}H$), 124.3 (d, $^3J_{CP} = 2.1$ Hz, $C_{Ar}H$), 127.9 ($C_{Ar}H$), 128.2 ($C_{Ar}H$), 128.7 (d, $^3J_{CP} = 9.7$ Hz, $C_{Ar}H$), 128.8 (d, $^3J_{CP} = 7.5$ Hz, $C_{Ar}H$), 128.9 (d, $^3J_{CP} = 8.1$ Hz, $C_{Ar}H$), 129.2 (d, $^4J_{CP} = 1.5$ Hz, $C_{Ar}H$), 129.4 (d, $^4J_{CP} = 1.2$ Hz, $C_{Ar}H$), 130.1 (d, $^4J_{CP} = 1.7$ Hz, $C_{Ar}H$), 130.2 (d, $^4J_{CP} = 1.6$ Hz, $C_{Ar}H$), 131.9 (d, $^2J_{CP} = 13.3$ Hz, $C_{Ar}H$), 132.1 (d, $^2J_{CP} = 13.3$ Hz, $C_{Ar}H$), 134.0 (d, $^2J_{CP} = 15.3$ Hz, $C_{Ar}H$), 134.0 (d, $^2J_{CP} = 15.5$ Hz, $C_{Ar}H$), 135.1 (dd, $^1J_{CP} = 26.7$ Hz, $^3J_{CP} = 2.8$ Hz, C_{Ar}), 135.5 (dd, $^1J_{CP} = 27.3$ Hz, $^3J_{CP} = 2.8$ Hz, C_{Ar}), 136.8 (dd, $^1J_{CP} = 26.1$ Hz, $^3J_{CP} = 2.7$ Hz, C_{Ar}), 138.8 (dd, $^1J_{CP} = 26.5$ Hz, $^3J_{CP} = 3.0$ Hz, C_{Ar}), 143.5 (dd, $^3J_{CP} = 1.5$ Hz, $^3J_{CP} = 3.7$ Hz, $C_{Ar}CH$), 151.6 (dd, $^3J_{CP} = 1.2$ Hz, $^3J_{CP} = 6.9$ Hz, $C_{Ar}N$), 172.1 (dd, $^3J_{CP} = 2.5$ Hz, $^3J_{CP} = 4.4$ Hz, $CHCOO$).

^{31}P-NMR (161.97 MHz; C_6D_6): $\delta = 19.5$ (d, $^2J_{PP} = 4.2$ Hz), 20.2 (d, $^2J_{PP} = 4.2$ Hz).

(1,2-Bis(diphenylphosphino)ethane)-(η^2-*iso*-propyl 4-nitrocinnamate)palladium(0)

Pd(dppe)PhNO$_2$-^{i}Pr: C$_{38}$H$_{37}$NO$_4$P$_2$Pd, M = 740.07 g/mol, dark red solid, yield: 98%, route: *Pd-III.*

Single crystals suitable for X-ray analysis were obtained by slow diffusion of pentane into a toluene solution.

^{1}H-NMR (400.13 MHz; C$_6$D$_6$): $\delta = 0.88$ (d, $^3J_{HH} = 6.2$ Hz, 3H, CHCH_3), 1.21 (d, $^3J_{HH} = 6.2$ Hz, 3H, CHCH_3), 1.49 - 1.68 (m, 1H, PCH_2CH$_2$P), 1.73 - 2.02 (m, 3H, PCH_2CH$_2$P), 4.64 - 4.82 (m, 1H, CHCHCO), 5.15 (hp, $^3J_{HH} = 6.2$ Hz, 1H, CH(CH$_3$)$_2$), 5.79 (pt, $^3J_{HH} = 8.5$ Hz, $^3J_{HP} = 8.5$ Hz, $^3J_{HP} = 8.5$ Hz, 1H, C$_{Ar}$CHCH), 6.59 - 6.86 (m, 6H, H_{Ar}), 6.91 - 7.22 (m, 10H, H_{Ar}), 7.47 - 7.68 (m, 6H, H_{Ar}), 7.78 - 7.88 (m, 2H, H_{Ar}).

^{13}C-NMR (100.62 MHz; C$_6$D$_6$): $\delta = 22.0$ (CHCH$_3$), 22.7 (CHCH$_3$), 27.3 (dd, $^1J_{CP} = 18.1$ Hz, $^2J_{CP} = 8.0$ Hz, PCH$_2$), 27.5 (dd, $^1J_{CP} = 17.8$ Hz, $^2J_{CP} = 7.8$ Hz, PCH$_2$), 54.0 (dd, $^2J_{CP} = 3.3$ Hz, $^2J_{CP} = 21.9$ Hz, CHCHCO), 65.4 (d, $^2J_{CP} = 25.1$ Hz, C$_{Ar}$$C$HCH), 66.0 (O$C$H(CH$_3$)$_2$), 124.4 (d, $^3J_{CP} = 14.5$ Hz, C$_{Ar}$H), 127.9 (C_{Ar}H), 128.2 (C_{Ar}H), 128.6 (d, $^3J_{CP} = 9.0$ Hz, C$_{Ar}$H), 128.9 (d, $^3J_{CP} = 9.5$ Hz, C$_{Ar}$H), 129.1 (d, $^3J_{CP} = 9.3$ Hz, C$_{Ar}$H), 129.1 (d, $^3J_{CP} = 9.3$ Hz, C$_{Ar}$H), 130.0 (d, $^4J_{CP} = 1.6$ Hz, C$_{Ar}$H), 130.3 - 130.5 (m, C_{Ar}H), 130.6 (d, $^4J_{CP} = 1.6$ Hz, C$_{Ar}$H), 133.1 (d, $^2J_{CP} = 14.8$ Hz, C$_{Ar}$H), 133.6 (d, $^2J_{CP} = 14.9$ Hz, C$_{Ar}$H), 133.6 (d, $^2J_{CP} = 15.8$ Hz, C$_{Ar}$H), 134.9 (dd, $^1J_{CP} = 25.9$ Hz, $^3J_{CP} = 4.0$ Hz, C$_{Ar}$), 135.4 (dd, $^1J_{CP} = 24.4$ Hz, $^3J_{CP} = 5.0$ Hz, C$_{Ar}$), 135.9 (dd, $^1J_{CP} = 24.5$ Hz, $^3J_{CP} = 2.8$ Hz, C$_{Ar}$), 143.2 (dd, $^3J_{CP} = 2.2$ Hz, $^3J_{CP} = 4.2$ Hz, C$_{Ar}$CH), 152.8 (d, $^3J_{CP} = 7.4$ Hz, C$_{Ar}$NO$_2$), 172.2 (dd, $^3J_{CP} = 1.7$ Hz, $^3J_{CP} = 4.6$ Hz, CHCOO).

^{31}P-NMR (161.97 MHz; C$_6$D$_6$): $\delta = 36.5$ (d, $^2J_{PP} = 17.6$ Hz), 36.1 (d, $^2J_{PP} = 17.6$ Hz).

(1,3-Bis(diphenylphosphino)propane)-(η^2-*iso*-propyl 4-nitrocinnamate)palladium(0)

Pd(dppp)PhNO$_2$-^{i}Pr: C$_{39}$H$_{39}$NO$_4$P$_2$Pd, M = 754.10 g/mol, dark red solid, yield: 90%, route: *Pd-III*.

^{1}H-NMR (400.13 MHz; C$_6$D$_6$): δ = 0.76 (d, $^3J_{HH}$ = 6.3 Hz, 3H, CHCH_3), 1.19 (d, $^3J_{HH}$ = 6.3 Hz, 3H, CHCH_3), 1.32 - 1.69 (m, 2H, CH$_2$CH_2CH$_2$), 1.58 (pt, $^2J_{HH}$ = 13.0 Hz, $^3J_{HH}$ = 13.0 Hz, 1H, PCH_2CH$_2$), 1.81 (pt, $^2J_{HH}$ = 13.0 Hz, $^3J_{HH}$ = 13.0 Hz, 1H, PCH_2CH$_2$), 1.97 - 2.23 (m, 2H, PCH_2CH$_2$), 4.50 (ddd, $^3J_{HH}$ = 11.7 Hz, $^3J_{HP}$ = 3.9 Hz, $^3J_{HP}$ = 8.1 Hz, 1H, CHCHCO), 5.03 (hp, $^3J_{HH}$ = 6.3 Hz, 1H, CH(CH$_3$)$_2$), 5.40 (dpt, $^3J_{HH}$ = 11.7 Hz, $^3J_{HP}$ = 3.4 Hz, $^3J_{HP}$ = 11.7 Hz, 1H, C$_{Ar}$CHCH), 6.41 (d, $^3J_{HP}$ = 7.9 Hz, 2H, H_{Ar}), 6.47 - 6.56 (m, 2H, H_{Ar}), 6.66 - 6.74 (m, 2H, H_{Ar}), 6.84 (t, $^3J_{HP}$ = 7.4 Hz, 1H, H_{Ar}), 6.91 - 7.20 (m, 7H, H_{Ar}), 7.27 (t, $^3J_{HP}$ = 7.0 Hz, 2H, H_{Ar}), 7.39 - 7.54 (m, 4H, H_{Ar}), 7.69 - 7.78 (m, 2H, H_{Ar}), 7.87 - 7.95 (m, 2H, H_{Ar}).

^{13}C-NMR (100.62 MHz; C$_6$D$_6$): δ = 19.1 (t, $^2J_{CP}$ = 5.0 Hz, PCH$_2$CH_2), 21.8 (CHCH$_3$), 22.6 (CHCH$_3$), 29.4 (d, $^1J_{CP}$ = 18.7 Hz, PCH$_2$CH$_2$), 29.9 (d, $^1J_{CP}$ = 19.9 Hz, PCH$_2$CH$_2$), 53.8 (dd, $^2J_{CP}$ = 4.8 Hz, $^2J_{CP}$ = 21.6 Hz, CHCHCO), 65.5 (dd, $^2J_{CP}$ = 1.7 Hz, $^2J_{CP}$ = 24.5 Hz, C$_{Ar}$$C$HCH), 66.0 (O$C$H(CH$_3$)$_2$), 124.0 - 124.1 (m, C_{Ar}H), 124.2 (dd, $^3J_{CP}$ = 1.3 Hz, $^3J_{CP}$ = 2.9 Hz, C_{Ar}H), 127.9 (C_{Ar}H), 128.2 (C_{Ar}H), 128.2 (C_{Ar}H), 128.5 (d, $^3J_{CP}$ = 9.6 Hz, C_{Ar}H), 128.8 (d, $^3J_{CP}$ = 9.3 Hz, C_{Ar}H), 129.0 (d, $^3J_{CP}$ = 9.8 Hz, C_{Ar}H), 129.9 (d, $^4J_{CP}$ = 1.6 Hz, C_{Ar}H), 130.0 (d, $^4J_{CP}$ = 1.8 Hz, C_{Ar}H), 130.2 - 130.4 (m, C_{Ar}H), 130.7 (d, $^4J_{CP}$ = 1.8 Hz, C_{Ar}H), 130.9 (d, $^2J_{CP}$ = 12.6 Hz, C_{Ar}H), 132.9 (d, $^2J_{CP}$ = 15.0 Hz, C_{Ar}H), 133.3 - 133.7 (m, C_{Ar}), 133.9 (d, $^2J_{CP}$ = 15.5 Hz, C_{Ar}H), 135.0 (d, $^2J_{CP}$ = 16.3 Hz, C_{Ar}H), 135.1 (dd, $^1J_{CP}$ = 23.1 Hz, $^3J_{CP}$ = 4.6 Hz, C_{Ar}), 136.6 (dd, $^1J_{CP}$ = 26.8 Hz, $^3J_{CP}$ = 1.8 Hz, C_{Ar}), 137.7 (dd, $^1J_{CP}$ = 25.4 Hz, $^3J_{CP}$ = 4.6 Hz, C_{Ar}),

143.3 (dd, $^3J_{CP}$ = 1.9 Hz, $^3J_{CP}$ = 3.8 Hz, C_{Ar}CH), 151.7 (d, $^3J_{CP}$ = 7.0 Hz, C_{Ar}N), 172.2 (dd, $^3J_{CP}$ = 2.2 Hz, $^3J_{CP}$ = 4.4 Hz, CHCOO).

^{31}P-NMR (161.97 MHz; C$_6$D$_6$): δ = 8.9 (d, $^2J_{PP}$ = 10.6 Hz), 12.5 (d, $^2J_{PP}$ = 10.6 Hz).

(1,4-Bis(diphenylphosphino)butane)-(η^2-*iso*-propyl 4-nitrocinnamate)palladium(0)

Pd(dppb)PhNO$_2$-^{i}Pr: C$_{40}$H$_{41}$NO$_4$P$_2$Pd, M = 768.13 g/mol, red solid, yield: 89%, **route**: *Pd-III*.

^{1}H-NMR (400.13 MHz; C$_6$D$_6$): δ = 0.73 (d, $^3J_{HH}$ = 6.2 Hz, 3H, CHCH_3), 1.20 (d, $^3J_{HH}$ = 6.2 Hz, 3H, CHCH_3), 1.26 - 1.46 (m, 3H, PCH$_2$CH_2), 1.46 - 1.64 (m, 1H, PCH$_2$CH_2), 1.79 - 2.06 (m, 4H, PCH$_2$CH_2), 4.25 (ddd, $^3J_{HH}$ = 12.2 Hz, $^3J_{HP}$ = 4.5 Hz, $^3J_{HP}$ = 7.8 Hz, 1H, CHCHCO), 4.95 (hp, $^3J_{HH}$ = 6.2 Hz, 1H, CH(CH$_3$)$_2$), 5.20 (dpt, $^3J_{HH}$ = 12.2 Hz, $^3J_{HP}$ = 4.0 Hz, $^3J_{HP}$ = 12.2 Hz, 1H, C$_{Ar}$CHCH), 6.31 (d, $^3J_{HP}$ = 7.6 Hz, 2H, H_{Ar}), 6.58 - 6.69 (m, 2H, H_{Ar}), 6.69 - 6.76 (m, 2H, H_{Ar}), 6.81 - 6.88 (m, 1H, H_{Ar}), 7.01 - 7.30 (m, 9H, H_{Ar}), 7.40 (d, $^3J_{HP}$ = 8.7 Hz, 2H, H_{Ar}), 7.52 - 7.61 (m, 2H, H_{Ar}), 7.61 - 7.72 (m, 2H, H_{Ar}), 7.77 (t, $^3J_{HP}$ = 8.6 Hz, 2H, H_{Ar}).

^{13}C-NMR (100.62 MHz; C$_6$D$_6$): δ = 21.7 (CHCH$_3$), 22.6 (CHCH$_3$), 23.1 (dd, $^2J_{CP}$ = 6.4 Hz, $^3J_{CP}$ = 4.4 Hz, PCH$_2$$CH_2$), 23.7 (dd, $^2J_{CP}$ = 4.9 Hz, $^3J_{CP}$ = 4.9 Hz, PCH$_2$$CH_2$), 26.8 (d, $^1J_{CP}$ = 17.2 Hz, PCH$_2$CH$_2$), 27.7 (d, $^1J_{CP}$ = 17.1 Hz, PCH$_2$CH$_2$), 56.0 (dd, $^2J_{CP}$ = 4.6 Hz, $^2J_{CP}$ = 21.6 Hz, CHCHCO), 66.1 (OCH(CH$_3$)$_2$), 66.9 (dd, $^2J_{CP}$ = 1.0 Hz, $^2J_{CP}$ = 24.2 Hz, C$_{Ar}$$C$HCH), 123.8 - 124.0 (m, C_{Ar}, C_{Ar}H), 124.3 (d, $^3J_{CP}$ = 9.5 Hz, C_{Ar}H), 127.9 (C_{Ar}H), 128.2 (C_{Ar}H), 128.7 (d, $^3J_{CP}$ = 9.5 Hz, C_{Ar}H), 128.8 (d, $^3J_{CP}$ = 8.8 Hz, C_{Ar}H), 128.9 (d, $^3J_{CP}$ = 10.2 Hz, C_{Ar}H), 129.1 - 129.2 (m, C_{Ar}H), 129.2 - 129.4 (m, C_{Ar}H), 130.2 - 130.3 (m, C_{Ar}H), 131.8 (d, $^2J_{CP}$ = 13.3 Hz, C_{Ar}H), 132.1 (d, $^2J_{CP}$ = 13.2 Hz, C_{Ar}H), 134.1 (d, $^2J_{CP}$ = 15.5 Hz, C_{Ar}H), 134.3 (d,

$^2J_{CP}$ = 15.6 Hz, $C_{Ar}H$), 135.2 (dd, $^1J_{CP}$ = 26.3 Hz, $^3J_{CP}$ = 2.8 Hz, C_{Ar}), 135.4 (dd, $^1J_{CP}$ = 27.1 Hz, $^3J_{CP}$ = 2.7 Hz, C_{Ar}), 136.8 (dd, $^1J_{CP}$ = 26.2 Hz, $^3J_{CP}$ = 2.6 Hz, C_{Ar}), 139.1 (dd, $^1J_{CP}$ = 26.3 Hz, $^3J_{CP}$ = 3.2 Hz, C_{Ar}), 143.4 (dd, $^3J_{CP}$ = 1.5 Hz, $^3J_{CP}$ = 3.6 Hz, $C_{Ar}CH$), 151.7 (dd, $^3J_{CP}$ = 1.4 Hz, $^3J_{CP}$ = 6.9 Hz, $C_{Ar}N$), 171.9 (dd, $^3J_{CP}$ = 2.4 Hz, $^3J_{CP}$ = 4.4 Hz, CHCOO).

^{31}P-NMR (161.97 MHz; C$_6$D$_6$): δ = 19.3 (d, $^2J_{PP}$ = 4.5 Hz), 20.1 (d, $^2J_{PP}$ = 4.5 Hz).

(1,2-Bis(diphenylphosphino)ethane)-(η^2-*tert*-butyl 4-nitrocinnamate)palladium(0)

Pd(dppe)PhNO$_2$-^{t}Bu: C$_{39}$H$_{39}$NO$_4$P$_2$Pd, M = 754.10 *g/mol*, red solid, yield: >99%, **route:** *Pd-III*, low solubility in aromatic solvents.

^{1}H-NMR (400.13 MHz; C$_6$D$_6$): δ = 1.41 (s, 9H, C(CH$_3$)$_3$), 1.41 - 1.60 (m, 2H, PCH$_2$CH$_2$P), 1.84 - 2.14 (m, 2H, PCH$_2$CH$_2$P), 4.69 (dpt, $^3J_{HH}$ = 9.4 Hz, $^3J_{HP}$ = 3.6 Hz, $^3J_{HP}$ = 9.4 Hz, 1H, CHCHCO), 5.75 (dpt, $^3J_{HH}$ = 9.4 Hz, $^3J_{HP}$ = 2.6 Hz, $^3J_{HP}$ = 9.4 Hz, 1H, C$_{Ar}$CHCH), 6.54 - 6.85 (m, 6H, H_{Ar}), 6.85 - 7.26 (m, 10H, H_{Ar}), 7.44 - 7.69 (m, 6H, H_{Ar}), 7.82 - 7.95 (m, 2H, H_{Ar}).

^{31}P-NMR (161.97 MHz; C$_6$D$_6$): δ = 35.3 (d, $^2J_{PP}$ = 18.4 Hz), 36.4 (d, $^2J_{PP}$ = 18.4 Hz).

(1,3-Bis(diphenylphosphino)propane)-(η^2-*tert*-butyl 4-nitrocinnamate)palladium(0)

Pd(dppp)PhNO$_2$-^{t}Bu: C$_{40}$H$_{41}$NO$_4$P$_2$Pd, M = 768.13 *g/mol*, red solid, yield: 86%, **route**: *Pd-III*.

^{1}H-NMR (400.13 MHz; C$_6$D$_6$): δ = 1.06 - 1.29 (m, 1H, CH$_2$C*H*$_2$CH$_2$), 1.33 (s, 9H, C(C*H*$_3$)$_3$), 1.35 - 1.51 (m, 1H, CH$_2$C*H*$_2$CH$_2$), 1.56 (pt, $^2J_{HH}$ = 12.5 Hz, $^3J_{HH}$ = 12.5 Hz, 1H, PC*H*$_2$CH$_2$), 1.80 (pt, $^2J_{HH}$ = 12.5 Hz, $^3J_{HH}$ = 12.5 Hz, 1H, PC*H*$_2$CH$_2$), 2.06 (dd, $^2J_{HH}$ = 20.9 Hz, $^3J_{HH}$ = 14.5 Hz, 1H, PC*H*$_2$CH$_2$), 2.19 (dd, $^2J_{HH}$ = 20.9 Hz, $^3J_{HH}$ = 13.0 Hz, 1H, PC*H*$_2$CH$_2$), 4.42 (ddd, $^3J_{HH}$ = 11.1 Hz, $^3J_{HP}$ = 3.9 Hz, $^3J_{HP}$ = 7.9 Hz, 1H, CHC*H*CO), 5.34 (dpt, $^3J_{HH}$ = 11.1 Hz, $^3J_{HP}$ = 3.4 Hz, $^3J_{HP}$ = 11.1 Hz, 1H, C$_{Ar}$C*H*CH), 6.39 (d, $^3J_{HP}$ = 8.1 Hz, 2H, *H*$_{Ar}$), 6.44 - 6.51 (m, 2H, *H*$_{Ar}$), 6.68 (t, $^3J_{HP}$ = 6.9 Hz, 2H, *H*$_{Ar}$), 6.84 (t, $^3J_{HP}$ = 7.4 Hz, 1H, *H*$_{Ar}$), 6.91 - 7.06 (m, 3H, *H*$_{Ar}$), 7.07 - 7.22 (m, 4H, *H*$_{Ar}$), 7.29 (t, $^3J_{HP}$ = 7.1 Hz, 2H, *H*$_{Ar}$), 7.45 (d, $^3J_{HP}$ = 8.7 Hz, 2H, *H*$_{Ar}$), 7.47 - 7.56 (m, 2H, *H*$_{Ar}$), 7.67 - 7.76 (m, 2H, *H*$_{Ar}$), 7.88 - 7.99 (m, 2H, *H*$_{Ar}$).

^{13}C-NMR (100.62 MHz; C$_6$D$_6$): δ = 18.9 (t, $^2J_{CP}$ = 4.6 Hz, PCH$_2$*C*H$_2$), 28.6 (*C*H$_3$), 29.4 (d, $^1J_{CP}$ = 18.9 Hz, P*C*H$_2$CH$_2$), 29.8 (d, $^1J_{CP}$ = 19.2 Hz, P*C*H$_2$CH$_2$), 54.8 (dd, $^2J_{CP}$ = 4.8 Hz, $^2J_{CP}$ = 22.2 Hz, CH*C*HCO), 66.0 (dd, $^2J_{CP}$ = 2.0 Hz, $^2J_{CP}$ = 24.4 Hz, C$_{Ar}$*C*HCH), 78.0 (O*C*(CH$_3$)$_3$), 124.0 - 124.1 (m, *C*$_{Ar}$H), 124.1 - 124.2 (m, *C*$_{Ar}$H), 127.9 (*C*$_{Ar}$H), 128.2 (*C*$_{Ar}$H), 128.5 (d, $^3J_{CP}$ = 9.6 Hz, *C*$_{Ar}$H), 128.8 (d, $^3J_{CP}$ = 9.3 Hz, *C*$_{Ar}$H), 128.9 (d, $^3J_{CP}$ = 9.8 Hz, *C*$_{Ar}$H), 129.9 (d, $^4J_{CP}$ = 1.6 Hz, *C*$_{Ar}$H), 130.0 (d, $^4J_{CP}$ = 1.7 Hz, *C*$_{Ar}$H), 130.7 (d, $^4J_{CP}$ = 1.7 Hz, *C*$_{Ar}$H), 130.8 (d, $^2J_{CP}$ = 12.4 Hz, *C*$_{Ar}$H), 133.0 (d, $^2J_{CP}$ = 15.2 Hz, *C*$_{Ar}$H), 134.1 (d, $^2J_{CP}$ = 15.4 Hz, *C*$_{Ar}$H), 135.0 (d, $^2J_{CP}$ = 16.2 Hz, *C*$_{Ar}$H), 135.1 (dd, $^1J_{CP}$ = 26.0 Hz, $^3J_{CP}$ = 4.8 Hz, *C*$_{Ar}$), 136.3 (dd, $^1J_{CP}$ = 26.6 Hz,

$^3J_{CP} = 1.5$ Hz, C_{Ar}), 137.8 (dd, $^1J_{CP} = 24.9$ Hz, $^3J_{CP} = 5.1$ Hz, C_{Ar}), 143.2 (dd, $^3J_{CP} = 2.0$ Hz, $^3J_{CP} = 3.7$ Hz, $C_{Ar}CH$), 152.0 (dd, $^3J_{CP} = 1.2$ Hz, $^3J_{CP} = 6.9$ Hz, $C_{Ar}N$), 172.5 (dd, $^3J_{CP} = 2.4$ Hz, $^3J_{CP} = 4.2$ Hz, CHCOO).

^{31}P-NMR (161.97 MHz; C$_6$D$_6$): $\delta = 8.5$ (d, $^2J_{PP} = 10.3$ Hz), 11.9 (d, $^2J_{PP} = 10.3$ Hz).

(1,4-Bis(diphenylphosphino)butane)-(η^2-*tert*-butyl 4-nitrocinnamate)palladium(0)

Pd(dppb)PhNO$_2$-^{t}Bu: C$_{41}$H$_{43}$NO$_4$P$_2$Pd, M $= 782.15$ *g/mol*, red solid, yield: 82%, **route:** *Pd-III.*

^{1}H-NMR (400.13 MHz; C$_6$D$_6$): $\delta = 1.34$ (s, 9H, C(CH_3)$_3$), 1.18 - 1.57 (m, 4H, PCH$_2$CH_2), 1.72 - 1.84 (m, 1H, PCH$_2$CH_2), 1.84 - 2.12 (m, 3H, PCH$_2$CH_2), 4.16 (ddd, $^3J_{HH} = 11.6$ Hz, $^3J_{HP} = 4.8$ Hz, $^3J_{HP} = 7.6$ Hz, 1H, CHCHCO), 5.17 (ddd, $^3J_{HH} = 11.6$ Hz, $^3J_{HP} = 3.9$ Hz, $^3J_{HP} = 9.0$ Hz, 1H, C$_{Ar}$CHCH), 6.33(d, $^3J_{HP} = 7.6$ Hz, 2H, H_{Ar}), 6.54 - 6.65 (m, 2H, H_{Ar}), 6.68 - 6.79 (m, 2H, H_{Ar}), 6.80 - 6.88 (m, 1H, H_{Ar}), 6.98 - 7.30 (m, 9H, H_{Ar}), 7.45 (d, $^3J_{HP} = 8.7$ Hz, 2H, H_{Ar}), 7.58 - 7.78 (m, 6H, H_{Ar}).

^{13}C-NMR (100.62 MHz; C$_6$D$_6$): $\delta = 23.5$ (dd, $^2J_{CP} = 5.8$ Hz, $^3J_{CP} = 4.3$ Hz, PCH$_2$ CH$_2$), 23.9 (dd, $^2J_{CP} = 6.6$ Hz, $^3J_{CP} = 4.4$ Hz, PCH$_2$ CH$_2$), 27.3 (d, $^1J_{CP} = 16.8$ Hz, PCH$_2$CH$_2$), 27.7 (d, $^1J_{CP} = 17.5$ Hz, PCH$_2$CH$_2$), 28.6 (CH_3), 57.1 (dd, $^2J_{CP} = 4.9$ Hz, $^2J_{CP} = 22.5$ Hz, CHCHCO), 67.5 (dd, $^2J_{CP} = 1.1$ Hz, $^2J_{CP} = 23.6$ Hz, C$_{Ar}$ CHCH), 78.0 (OC(CH$_3$)$_3$), 123.9 (d, $^3J_{CP} = 1.3$ Hz, C_{Ar}H), 124.3 (d, $^3J_{CP} = 2.0$ Hz, C_{Ar}H), 127.9 (C_{Ar}H), 128.2 (C_{Ar}H), 128.2 (C_{Ar}H), 128.7 (d, $^3J_{CP} = 9.2$ Hz, C_{Ar}H), 128.8 (d, $^3J_{CP} = 10.1$ Hz, C_{Ar}H), 128.8 (d, $^3J_{CP} = 9.4$ Hz, C_{Ar}H), 129.1 (d, $^4J_{CP} = 1.1$ Hz, C_{Ar}H), 129.3 (d, $^4J_{CP} = 0.5$ Hz, C_{Ar}H), 130.1 (d, $^4J_{CP} = 1.2$ Hz, C_{Ar}H), 130.2 (d, $^4J_{CP} = 1.1$ Hz, C_{Ar}H), 131.7 (d, $^2J_{CP} = 13.1$ Hz, C_{Ar}H), 132.2 (d, $^2J_{CP} = 13.7$ Hz, C_{Ar}H), 134.2 (d, $^2J_{CP} = 10.2$ Hz, C_{Ar}H), 134.3 (d, $^2J_{CP} = 10.2$ Hz, C_{Ar}H), 135.1 (dd,

$^1J_{CP} = 21.3$ Hz, $^3J_{CP} = 2.8$ Hz, C_{Ar}), 135.4 (dd, $^1J_{CP} = 22.5$ Hz, $^3J_{CP} = 2.9$ Hz, C_{Ar}), 136.6 (dd, $^1J_{CP} = 25.7$ Hz, $^3J_{CP} = 2.6$ Hz, C_{Ar}), 139.6 (dd, $^1J_{CP} = 25.9$ Hz, $^3J_{CP} = 3.7$ Hz, C_{Ar}), 143.4 (dd, $^3J_{CP} = 1.5$ Hz, $^3J_{CP} = 3.5$ Hz, $C_{Ar}CH$), 151.9 (dd, $^3J_{CP} = 1.5$ Hz, $^3J_{CP} = 6.7$ Hz, $C_{Ar}N$), 172.2 (dd, $^3J_{CP} = 2.5$ Hz, $^3J_{CP} = 4.1$ Hz, CHCOO).

^{31}P-NMR (161.97 MHz; C$_6$D$_6$): $\delta = 19.1$ (d, $^2J_{PP} = 5.1$ Hz), 19.6 (d, $^2J_{PP} = 5.1$ Hz).

(1,2-Bis(diphenylphosphino)ethane)-(η^2-4-methoxyphenyl 4-nitrocinnamate) palladium(0)

Pd(dppe)PhNO$_2$-PhOMe: C$_{42}$H$_{37}$NO$_5$P$_2$Pd, M = 804.11 g/mol, deep orange solid, yield: >99%, **route**: *Pd-III*.

^{1}H-NMR (400.13 MHz; C$_6$D$_6$): $\delta = 2.04 - 2.53$ (m, 4H, PCH_2CH_2P), 3.77 (s, 3H, OCH_3), 4.71 (ddd, $^3J_{HH} = 10.8$ Hz, $^3J_{HP} = 2.3$ Hz, $^3J_{HP} = 8.7$ Hz, 1H, CHCHCO), 5.44 (pt, $^3J_{HH} = 10.8$ Hz, $^3J_{HP} = 9.6$ Hz, $^3J_{HP} = 9.6$ Hz, 1H, $C_{Ar}CH$CH), 6.54 (d, $^3J_{HP} = 8.9$ Hz, 2H, H_{Ar}), 6.68 (d, $^3J_{HP} = 8.9$ Hz, 2H, H_{Ar}), 6.77 - 6.88 (m, 2H, H_{Ar}), 7.03 (t, $^3J_{HP} = 7.3$ Hz, 2H, H_{Ar}), 7.13 - 7.22 (m, 3H, H_{Ar}), 7.28 - 7.37 (m, 2H, H_{Ar}), 7.37 - 7.44 (m, 1H, H_{Ar}), 7.45 - 7.59 (m, 8H, H_{Ar}), 7.65 - 7.76 (m, 2H, H_{Ar}), 7.84 (d, $^3J_{HP} = 8.7$ Hz, 2H, H_{Ar}), 7.88 - 7.97 (m, 2H, H_{Ar}).

^{31}P-NMR (161.97 MHz; C$_6$D$_6$): $\delta = 37.8$ (d, $^2J_{PP} = 18.2$ Hz), 37.9 (d, $^2J_{PP} = 18.2$ Hz).

(1,3-Bis(diphenylphosphino)propane)-(η^2-4-methoxyphenyl 4-nitrocinnamate) palladium(0)

Pd(dppp)PhNO$_2$-PhOMe: C$_{43}$H$_{39}$NO$_5$P$_2$Pd, M = 818.14 *g/mol*, deep orange solid, yield: >99%, **route:** *Pd-III*.

^{1}H-NMR (400.13 MHz; C$_6$D$_6$): δ = 1.07 - 1.33 (m, 1H, CH$_2$C*H*$_2$CH$_2$), 1.33 - 1.57 (m, 1H, CH$_2$C*H*$_2$CH$_2$), 1.63 (pt, $^2J_{HH}$ = 12.9 Hz, $^3J_{HH}$ = 12.9 Hz, 1H, PC*H*$_2$CH$_2$), 1.84 (pt, $^2J_{HH}$ = 12.9 Hz, $^3J_{HH}$ = 12.9 Hz, 1H, PC*H*$_2$CH$_2$), 2.03 (dd, $^2J_{HH}$ = 20.6 Hz, $^3J_{HH}$ = 13.5 Hz, 1H, PC*H*$_2$CH$_2$), 2.18 (dd, $^2J_{HH}$ = 20.6 Hz, $^3J_{HH}$ = 12.7 Hz, 1H, PC*H*$_2$CH$_2$), 3.24 (s, 3H, OC*H*$_3$), 4.68 (ddd, $^3J_{HH}$ = 10.4 Hz, $^3J_{HP}$ = 3.5 Hz, $^3J_{HP}$ = 8.2 Hz, 1H, CHC*H*CO), 5.40 (dpt, $^3J_{HH}$ = 10.4 Hz, $^3J_{HP}$ = 3.5 Hz, $^3J_{HP}$ = 3.5 Hz, 1H, C$_{Ar}$C*H*CH), 6.45 (d, $^3J_{HP}$ = 8.1 Hz, 2H, H_{Ar}), 6.48 - 6.57 (m, 2H, H_{Ar}), 6.59 (d, $^3J_{HP}$ = 8.9 Hz, 2H, H_{Ar}), 6.67 - 6.79 (m, 2H, H_{Ar}), 6.82 - 7.01 (m, 4H, H_{Ar}), 7.01 - 7.14 (m, 4H, H_{Ar}), 7.22 (t, $^3J_{HP}$ = 7.1 Hz, 2H, H_{Ar}), 7.44 - 7.52 (m, 4H, H_{Ar}), 7.64 - 7.73 (m, 2H, H_{Ar}), 7.89 - 7.97 (m, 2H, H_{Ar}).

^{31}P-NMR (161.97 MHz; C$_6$D$_6$): δ = 8.8 (d, $^2J_{PP}$ = 16.4 Hz), 12.7 (d, $^2J_{PP}$ = 16.4 Hz).

(1,4-Bis(diphenylphosphino)butane)-(η^2-4-methoxyphenyl 4-nitrocinnamate) palladium(0)

Pd(dppb)PhNO$_2$-PhOMe: C$_{44}$H$_{41}$NO$_5$P$_2$Pd, M $= 832.17\ g/mol$, red solid, NMR scale, **route:** *Pd-II.*

Single crystals suitable for X-ray analysis were obtained from a saturated benzene solution.

^{1}H-NMR (400.13 MHz; C$_6$D$_6$): δ = 1.24 - 1.58 (m, 4H, PCH$_2$CH_2), 1.79 - 2.10 (m, 4H, PCH_2CH$_2$), 3.28 (s, 3H, OCH_3), 4.41 (ddd, $^3J_{HH}$ = 10.7 Hz, $^3J_{HP}$ = 4.3 Hz, $^3J_{HP}$ = 8.0 Hz, 1H, CHCHCO), 5.22 (dpt, $^3J_{HH}$ = 10.7 Hz, $^3J_{HP}$ = 3.9 Hz, $^3J_{HP}$ = 3.9 Hz, 1H, C$_{Ar}$CHCH), 6.36 (d, $^3J_{HP}$ = 8.0 Hz, 2H, H_{Ar}), 6.64 (d, $^3J_{HP}$ = 9.0 Hz, 4H, H_{Ar}), 6.69 - 6.82 (m, 4H, H_{Ar}), 6.94 - 7.15 (m, 8H, H_{Ar}), 7.18 - 7.26 (m, 2H, H_{Ar}), 7.46 (d, $^3J_{HP}$ = 8.6 Hz, 2H, H_{Ar}), 7.52 - 7.61 (m, 2H, H_{Ar}), 7.61 - 7.69 (m, 2H, H_{Ar}), 7.79 (t, $^3J_{HP}$ = 8.7 Hz, 2H, H_{Ar}).

^{13}C-NMR (100.53 MHz; C$_6$D$_6$): δ = 23.3 (dd, $^2J_{CP}$ = 6.2 Hz, $^3J_{CP}$ = 3.9 Hz, PCH$_2$CH_2), 23.7 (dd, $^2J_{CP}$ = 5.7 Hz, $^3J_{CP}$ = 5.7 Hz, PCH$_2$CH_2), 26.9 (d, $^1J_{CP}$ = 17.5 Hz, PCH_2CH$_2$), 27.7 (d, $^1J_{CP}$ = 17.5 Hz, PCH_2CH$_2$), 54.6 (dd, $^2J_{CP}$ = 5.1 Hz, $^2J_{CP}$ = 21.5 Hz, CHCHCO), 55.1 (OCH_3), 66.6 (d, $^2J_{CP}$ = 25.2 Hz, C$_{Ar}$CHCH), 114.2 (C$_{Ar}$H), 123.1 (C$_{Ar}$H), 123.9 - 124.0 (m, C$_{Ar}$H), 124.5 (d, $^3J_{CP}$ = 1.8 Hz, C$_{Ar}$H), 127.9 (C$_{Ar}$H), 128.2 (C$_{Ar}$H), 128.5 - 128.7 (m, C$_{Ar}$H), 128.9 (d, $^3J_{CP}$ = 9.0 Hz, C$_{Ar}$H), 129.1 (d, $^3J_{CP}$ = 9.5 Hz, C$_{Ar}$H), 129.2 - 129.3 (m, C$_{Ar}$H), 129.5 - 129.6 (m, C$_{Ar}$H), 130.1 - 130.3 (m, C$_{Ar}$H), 130.3 - 130.4 (m, C$_{Ar}$H), 131.8 (d, $^2J_{CP}$ = 13.1 Hz, C$_{Ar}$H), 132.1 (d, $^2J_{CP}$ = 13.4 Hz, C$_{Ar}$H), 134.1 (d, $^2J_{CP}$ = 15.2 Hz, C$_{Ar}$H), 134.1 (d, $^2J_{CP}$ = 15.4 Hz, C$_{Ar}$H), 135.0 (dd, $^1J_{CP}$ = 27.2 Hz, $^3J_{CP}$ = 2.7 Hz, C$_{Ar}$), 135.1 (dd, $^1J_{CP}$ = 28.1 Hz, $^3J_{CP}$ = 1.8 Hz, C$_{Ar}$), 136.2 (dd, $^1J_{CP}$ = 26.7 Hz, $^3J_{CP}$ = 2.4 Hz, C$_{Ar}$), 139.0 (dd,

$^1J_{CP} = 27.0$ Hz, $^3J_{CP} = 3.1$ Hz, C_{Ar}), 143.7 (dd, $^3J_{CP} = 1.3$ Hz, $^3J_{CP} = 3.4$ Hz, C_{Ar}CH), 146.1 (C_{Ar}O), 151.4 (d, $^3J_{CP} = 6.7$ Hz, C_{Ar}N), 157.0 (C_{Ar}O), 170.7 (dd, $^3J_{CP} = 2.4$ Hz, $^3J_{CP} = 4.5$ Hz, CHCOO).

^{31}P-NMR (161.97 MHz; C$_6$D$_6$): $\delta = 19.3$ (s), 20.7 (s).

8.3.10. Further complexes

Di(η^2-bicyclo[2.2.1]hept-2-ene)-(η^2-4-methylphenyl nitrocinnamate)platinum(0)

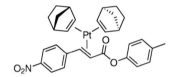

Pt(nbe)PhNO$_2$-PhMe: C$_{30}$H$_{33}$NO$_4$Pt, M = 666.67 g/mol, yellow solid, yield: >99%, route: *Pt-II*.

^{1}H-NMR (400.13 MHz; C$_6$D$_6$): $\delta = -0.04$ (d, $^2J_{HH} = 8.5$ Hz, 2H, CHCH_3CH), 0.17 (d, $^2J_{HH} = 8.5$ Hz, 2H, CHCH_2CH), 1.03 (bs, 4H, CHCH_2CH_2CH), 1.43 (bs, 4H, CHCH_2CH_2CH), 2.03 (s, 3H, CH_3), 2.47 (s, 2H, CHCH$_2$CH$_2$CH), 2.88 (s, 2H, CHCH$_2$CH$_2$CH), 3.27 (s, 2H, CHCHCHCH), 3.27 (d, $^2J_{HPt} = 63.2$ Hz, 2H, CHCHCHCH), 3.72 (s, 2H, CHCHCHCH), 3.72 (d, $^2J_{HPt} = 63.4$ Hz, 2H, CHCHCHCH), 4.18 (d, $^3J_{HH} = 9.8$ Hz, 1H, CHCHCO), 4.18 (dd, $^3J_{HH} = 9.8$ Hz, $^2J_{HPt} = 62.0$ Hz, 1H, CHCHCO), 4.77 (d, $^3J_{HH} = 9.8$ Hz, 1H, C$_{Ar}$CHCH), 4.77 (dd, $^3J_{HH} = 9.8$ Hz, $^2J_{HPt} = 55.9$ Hz, 1H, C$_{Ar}$CHCH), 6.63 (d, $^3J_{HH} = 8.0$ Hz, 2H, H_{Ar}), 6.93 (d, $^3J_{HH} = 7.5$ Hz, 2H, H_{Ar}), 7.14 (d, $^3J_{HH} = 9.0$ Hz, 2H, H_{Ar}), 7.82 (d, $^3J_{HH} = 7.9$ Hz, 2H, H_{Ar}).

^{13}C-NMR (100.62 MHz; C$_6$D$_6$): $\delta = 20.7$ (CH$_3$), 27.3 (CHCH$_2$CH), 27.3 (d, $^2J_{CPt} = 42.3$ Hz, CHCH$_2$CH), 27.4 (CHCH$_2$CH), 27.4 (d, $^2J_{CPt} = 43.2$ Hz, CCHCH$_2$CH), 40.4 (CHCH$_2$CH), 40.4 (d, $^3J_{CPt} = 38.4$ Hz, CHCH$_2$CH), 42.5 (CHCH$_2$CH$_2$CH), 42.5 (d, $^3J_{CPt} = 13.2$ Hz, CHCH$_2$CH$_2$CH), 43.4 (CHCH$_2$CH$_2$CH),

43.4 (d, $^3J_{CPt}$ = 13.2 Hz, CHCH_2CH_2CH), 50.4 (CHCHCO), 50.4 (d, $^1J_{CPt}$ = 174.0 Hz, CHCHCO), 59.6 (C$_{Ar}CH$CH), 59.6 (d, $^1J_{CPt}$ = 161.2 Hz, C$_{Ar}CH$CHCH), 68.3 (CHCHCHCH), 68.3 (d, $^1J_{CPt}$ = 191.5 Hz, CHCHCHCH), 79.3 (CHCHCHCH), 79.3 (d, $^1J_{CPt}$ = 155.9 Hz, CHCHCHCH), 121.9 (C$_{Ar}$H), 123.4 (C$_{Ar}$H), 126.4 (C$_{Ar}$H), 126.4 (d, $^3J_{CPt}$ = 14.7 Hz, C$_{Ar}$H), 130.1 (C$_{Ar}$H), 134.9 (C$_{Ar}$), 146.0 (C$_{Ar}$N), 149.8 (C$_{Ar}$O), 150.0 (C$_{Ar}$CH), 150.0 (d, $^2J_{CPt}$ = 37.7 Hz, C$_{Ar}$CH), 169.1 (CHCOO), 169.1 (d, $^2J_{CPt}$ = 47.5 Hz, CHCOO)

^{195}Pt-NMR (85.48 MHz; C$_6$D$_6$): δ = -6160.9.

Bis(triphenylphosphine)palladium(0)

Ph$_3$P-Pd-PPh$_3$

6.2: C$_{36}$H$_{30}$P$_2$Pd, M = 630.99 *g/mol*.

476 mg (2.24 mmol) (η^3-allyl)(η^5-cyclopentadienyl)palladium(II) was dissolved in pentane, 1.18 g (4.48 mmol) PPh$_3$ dissolved in pentane was added and the reaction mixture stirred over night. The solvent was decanted off and the residue washed with pentane to give 1.39 g (2.20 mmol, 98%) crude Pd(PPh$_3$)$_2$ as a light yellow solid. A pure sample for NMR analysis was obtained by *Schlenck flash* column chromatography (degased Al$_2$O$_3$; toluene).

^{1}H-NMR (400.13 MHz; C$_7$D$_8$): δ = 6.86 - 6.94 (m, 12H, H_{Ar}), 6.94 - 7.01 (m, 6H, H_{Ar}), 7.44 - 7.54 (m, 12H, H_{Ar})

^{13}C-NMR (100.62 MHz; C$_7$D$_8$): δ = 127.9 (s, C$_{Ar}$H), 128.3 (d, $^3J_{CP}$ = 17.7 Hz, C$_{Ar}$H), 134.3 (d, $^2J_{CP}$ = 17.3 Hz, C$_{Ar}$H), 139.7 (d, $^1J_{CP}$ = 16.0 Hz, C$_{Ar}$)

^{31}P-NMR (161.97 MHz; C$_7$D$_8$): δ = 20.6 (s).

elemental analysis for C$_{36}$H$_{30}$P$_2$Pd (630.99 *g/mol*):

calcd.: C 68.52, H 4.79, P 9.82, Pd 16.87

found: C 70.39, H 5.37

8.4. Reactivity studies & catalysis

8.4.1. NMR reactivity studies

Light induced ligand exchange

30.0 μmol diazo complex and 30.0 μmol triphenylphosphine were weighed into a *J.-Young* NMR tube, dissolved in 0.5 ml benzene-d$_6$ and irradiated at 300 nm for the given amount of time.

Photo-activated oxidative addition

9.00 μmol diazo complex was weighed into a *J.-Young* NMR tube, dissolved in 0.5 ml C$_6$D$_6$, 9.90 μmol silane or aryl halide was added via *Hamilton*-syringe and irradiated at 300 nm.

Ligand exchange experiment

5.00 mg (5.40 μmol) **PtPhOMe-Et** was weighed into a septum NMR tube, dissolved in 0.2 ml benzene-d$_6$ and 5.40 μmol cinnamic acid ethyl ester dissolved in 0.3 ml C$_6$D$_6$ added via syringe through the septum, the tube shaken and spectra recording started immediately.

Oxidative addition procedure: silane

10.0 μmol (1.0 eq) complex was dissolved in 0.5 ml C$_6$D$_6$ and 2.00 μl (1.99 mg, 10.8 μmol, 1.1 eq.) diphenylsilane added by means of a *Hamilton*-syringe and spectra recording commenced.

Oxidative addition procedure: aryl halid

5.00 mg complex was weighed into a *J.-Young* NMR tube dissolved in toluene, aryl halide was added via *Hamilton*-syringe and heated for the stated time at the given temperature.

8.4.2. Catalysis

Hydrosilylation of *p*-fluoroacetophenone

5.00 μmol catalyst and 65.0 μl (74.3 mg, 0.54 mmol, 1.0 eq.) *p*-fluoroacetophenone were dissolved in 0.25 ml CD_2Cl_2 placed in a *J.-Young* NMR tube, 0.15 ml (149.0 mg, 0.81 mmol, 1.5 eq.) Ph_2SiH_2 added and spectra recording commenced.

Hydrosilylation of acetophenone

6.7 μmol (5 mol%) catalyst was weighed into a septum NMR tube, 27.0 μl (26.8 mg, 145.0 μmol, 1.1 eq.) Ph_2SiH_2 and 16.6 μl (16.1 mg, 134.0 μmol, 1.0 eq.) acetophenone dissolved in 0.4 ml benzene-d_6 were added via syringe through the septum, the tube shaken and spectra recording started immediately.

Photo-triggered catalysis

7.39 mg (6.7 μmol, 5 mol%) complex was weighed into a septum NMR tube, 27.0 μl (26.8 mg, 145.0 μmol, 1.1 eq.) Ph_2SiH_2 and 16.6 μl (16.1 mg, 134.0 μmol, 1.0 eq.) acetophenone dissolved in 0.4 ml benzene-d_6 were added via syringe through the septum, the tube shaken and spectra recording started immediately. After the stated amount of time the NMR tube was irradiated at 300 nm for the described time and spectra recording continued afterwards.

Homocoupling of diphenylsilane

10.0 μmol complex was weighed into a *J.-Young* NMR tube, 0.5 ml (0.50 g, 2.69 mmol) Ph_2SiH_2 added and heated to 75 °C for the described amount of time.

Karstedt inhibition assay

0.40 ml (327.6 mg, 1.47 mmol, 2.0 eq.) 1,1,1,3,5,5,5-heptamethyltrisiloxane, 0.17 ml (137.5 mg, 0.74 mmol, 1.0 eq.) 1,3-divinyltetramethyldisiloxane and 1.80 μl *Karstedt*-solution (toluene, 3% Pt) were mixed, added to 24.0 μmol inhibitor in a NMR tube and heated to 100 °C for the given amount of time.

9. Bibliography

[1] B. Marciniec, H. Maciejewski, C. Pietraszuk, P. Pawluc, *Hydrosilylation: A Comprehensive Review on Recent Advances*, Springer Science+Business Media B. V., **2009**.

[2] L. N. Lewis, R. E. Colborn, H. Grade, G. L. Bryant, C. A. Sumpter, R. A. Scott, *Organometallics* **1995**, *14*(5), 2202–2213.

[3] L. N. Lewis, J. Stein, Y. Gao, R. E. Colborn, G. Hutchins, *Plat. Met. Rev.* **1997**, *41*, 66–75.

[4] M. L. Clarke, *Polyhedron* **2001**, *20*(3-4), 151 – 164.

[5] L. N. Lewis, J. Stein, R. E. Colborn, Y. Gao, J. Dong, *J. Organomet. Chem.* **1996**, *521*(1-2), 221 – 227.

[6] P. Steffanut, J. A. Osborn, A. DeCian, J. Fisher, *Chem. Eur. J.* **1998**, *4*(10), 2008–2017.

[7] A. W. Norris, M. Bahadur, M. Yoshitake, in *Proc. SPIE*, volume 5941, SPIE, San Diego, CA, USA, volume 5941, 594115–7.

[8] T. Luce, *LEDs Magazine* **2008**, *4*, 22.

[9] G. Geoffroy, *J. Chem. Ed.* **1983**, *60*(10), 861–866.

[10] R. Larciprete, *App. Surf. Sci.* **1990**, *46*(1-4), 19–26.

[11] M. Wrighton, *Chem. Rev.* **1974**, *74*(4), 401–430.

[12] M. L. H. Green, *Pure Appl. Chem.* **1978**, *50*, 27–35.

[13] J. D. Feldman, J. C. Peters, T. D. Tilley, *Organometallics* **2002**, *21*(20), 4050–4064.

[14] M. Herberhold, F. Wehrmann, D. Neugebauer, G. Huttner, *J. Organomet. Chem.* **1978**, *152*(3), 329–336.

[15] A. E. Stiegman, D. R. Tyler, *Coord. Chem. Rev.* **1985**, *63*, 217–240.

[16] D. M. Haddleton, R. N. Perutz, *J. Chem. Soc., Chem. Commun.* **1985**, (20), 1372–1374.

[17] S. T. Belt, S. B. Duckett, M. Helliwell, R. N. Perutz, *J. Chem. Soc., Chem. Commun.* **1989**, (14), 928–930.

[18] J. Müller, T. Akhnoukh, P. E. Gaede, A.-l. Guo, P. Moran, K. Qiao, *J. Organomet. Chem.* **1997**, *541*(1-2), 207–217.

[19] J. Cooke, D. E. Berry, K. L. Fawkes, *J. Chem. Ed.* **2007**, *84*(1), 115–118.

[20] V. Jakubek, A. J. Lees, *Inorg. Chem.* **2004**, *43*(22), 6869–6871.

[21] D. Wrobel, U. Irmer, Y.-F. Wang, **2009**, US2009062417.

[22] T. Naumann, S. Bosshammer, M. Putzer, U. Irmer, **2006**, DE102004036573.

[23] R. Meuser, G. Mignani, **1995**, AU680129B2.

[24] L. J. Farrugia, *J. Appl. Crystallogr.* **1997**, *30*, 565.

[25] C. Eaborn, A. Pidcock, B. Ratcliff, *J. Organomet. Chem.* **1972**, *43*(1), C2 – C4.

[26] C. Eaborn, B. Ratcliff, A. Pidcock, *J. Organomet. Chem.* **1974**, *65*(2), 181 – 186.

[27] J. Braddock-Wilking, J. Y. Corey, K. A. Trankler, H. Xu, L. M. French, N. Praingam, C. White, N. P. Rath, *Organometallics* **2006**, *25*(11), 2859–2871.

[28] H. Arii, M. Takahashi, A. Noda, M. Nanjo, K. Mochida, *Organometallics* **2008**, *27*(8), 1929–1935.

[29] G. Beuter, O. Heyke, I.-P. Lorenz, *Z. Naturforsch. B* **1991**, *46*, 1694–1698.

[30] P. B. Hitchcock, M. F. Lappert, N. J. W. Warhurst, *Angew. Chem. Int. Ed.* **1991**, *30*(4), 438–440.

[31] H. Tomioka, K. Kimoto, H. Murata, Y. Izawa, *J. Chem. Soc., Perkin Trans. 1* **1991**, 471–477.

[32] W. Kirmse, L. Horner, *Just. Lieb. Ann. Chem.* **1959**, *625*(1), 34–43.

[33] H. Staudinger, E. Anthes, F. Pfenninger, *Ber. dtsch. Chem. Ges.* **1916**, *49*(2), 1928–1941.

[34] R. M. Letcher, T.-Y. Yue, K.-F. Chiu, A. S. Kelkar, K.-K. Cheung, *J. Chem. Soc., Perkin Trans. 1* **1998**, (19), 3267–3276.

[35] A. Mustafa, O. H. Hishmat, *J. Am. Chem. Soc.* **1957**, *79*(9), 2225–2230.

[36] M. P. Cava, A. A. Deana, K. Muth, M. J. Mitchell, *Org. Synth.* **1961**, *41*, 93–95.

[37] A. Schwarzer, E. Weber, *Cryst. Growth Des.* **2008**, *8*(8), 2862–2874.

[38] B. Pal, P. K. Pradhan, P. Jaisankar, V. S. Giri, *Synthesis* **2003**, 1549–1552.

[39] J. R. Dimmock, N. M. Kandepu, A. J. Nazarali, N. L. Motaganahalli, T. P. Kowalchuk, U. Pugazhenthi, J. S. Prisciak, J. W. Quail, T. M. Allen, R. LeClerc, C. L. Santos, E. De Clercq, J. Balzarini, *J. Med. Chem.* **2000**, *43*(21), 3933–3940.

[40] M. Mahendra, B. H. Doreswamy, M. A. Sridhar, J. S. Prasad, S. A. Khanum, S. Shashikanth, T. D. Venu, *J. Chem. Cryst.* **2005**, *35*, 463–467.

[41] H. Tomioka, K. Nakanishi, Y. Izawa, *J. Chem. Soc., Perkin Trans. 1* **1991**, *1991*, 465–470.

[42] W.-H. Cheung, S.-L. Zheng, W.-Y. Yu, G.-C. Zhou, C.-M. Che, *Org. Lett.* **2003**, *5*(14), 2535–2538.

[43] B. M. Adger, S. Bradbury, M. Keating, C. W. Rees, R. C. Storr, M. T. Williams, *J. Chem. Soc., Perkin Trans. 1* **1975**, 31–40.

[44] R. J. McMahon, C. J. Abelt, O. L. Chapman, J. W. Johnson, C. L. Kreil, J. P. LeRoux, A. M. Mooring, P. R. West, *J. Am. Chem. Soc.* **1987**, *109*(8), 2456–2469.

[45] H.-J. Teuber, R. Braun, *Chem. Ber.* **1967**, *100*(4), 1353–1366.

[46] R. E. Ireland, D. M. Obrecht, *Helv. Chim. Acta* **1986**, *69*(6), 1273–1286.

[47] K. Inamoto, T. Saito, M. Katsuno, T. Sakamoto, K. Hiroya, *Org. Lett.* **2007**, *9*(15), 2931–2934.

[48] R. Gomez, J. L. Segura, N. Martin, *Org. Lett.* **2005**, *7*(4), 717–720.

[49] S. Roy, A. Nangia, *Cryst. Growth Des.* **2007**, *7*(10), 2047–2058.

[50] C. C. Dudman, C. B. Reese, *Synthesis* **1982**, *5*, 419–421.

[51] K. E. Eichstadt, J. C. Reepmeyer, R. B. Cook, P. G. Riley, D. P. Davis, R. A. Wiley, *J. Med. Chem.* **1976**, *19*(1), 47–51.

[52] G. W. Jones, K. T. Chang, H. Shechter, *J. Am. Chem. Soc.* **1979**, *101*(14), 3906–3916.

[53] D. M. Guldi, B. Nuber, P. J. Bracher, C. A. Alabi, S. MacMahon, J. W. Kukol, S. R. Wilson, D. I. Schuster, *J. Phys. Chem. A* **2003**, *107*(18), 3215–3221.

[54] H. Tomioka, N. Kobayashi, S. Murata, Y. Ohtawa, *J. Am. Chem. Soc.* **1991**, *113*(23), 8771–8778.

[55] A. Bondi, *J. Phys. Chem.* **1964**, *68*(3), 441–451.

[56] J. Firl, W. Runge, W. Hartmann, *Angew. Chem.* **1974**, *86*(7), 274–275.

[57] R. K. Huff, E. G. Savins, *Chem. Commun.* **1980**, 742–743.

[58] V. Sander, P. Weyerstahl, *Chem. Ber.* **1978**, *111*(12), 3879–3891.

[59] R. M. Silverstein, G. C. Bassler, T. C. Morrill, *Spectroscopic Identification of Organic Compounds*, 4th ed., John Wiley and Sons, New York, **1981**.

[60] K. Nakanishi, P. H. Solomon, *Infrared Absorption Spectroscopy*, 2nd ed., Emerson-Adams Press, **1998**.

[61] M. Yamashita, K. Okuyama, T. Kawajiri, A. Takada, Y. Inagaki, H. Nakano, M. Tomiyama, A. Ohnaka, I. Terayama, I. Kawasaki, S. Ohta, *Tetrahedron* **2002**, *58*(8), 1497–1505.

[62] M. Karplus, *J. Am. Chem. Soc.* **1963**, *85*(18), 2870–2871.

[63] A. Padwa, M. J. Chughtai, J. Boonsombat, P. Rashatasakhon, *Tetrahedron* **2008**, *64*(21), 4758–4767.

[64] H. Suga, S. Higuchi, M. Ohtsuka, D. Ishimoto, T. Arikawa, Y. Hashimoto, S. Misawa, T. Tsuchida, A. Kakehi, T. Baba, *Tetrahedron* **2010**, *66*(16), 3070–3089.

[65] R. Brückner, *Reaktionsmechanismen*, 2nd ed., Spektrum Akademischer Verlag, **2003**.

[66] J. T. Sharp, R. H. Findlay, P. B. Thorogood, *J. Chem. Soc., Perkin Trans. 1* **1975**, 102–113.

[67] H. Staudinger, S. Schotz, *Ber. dtsch. Chem. Ges. A/B* **1920**, *53*(6), 1105–1124.

[68] S. Koda, Y. Ohnuma, T. Ohkawa, S. Tsuchiya, *Bull. Chem. Soc. Jpn.* **1980**, *53*(12), 3447–3456.

[69] T. Majima, C. Pac, H. Sakurai, *J. Am. Chem. Soc.* **1980**, *102*(16), 5265–5273.

[70] T. Bach, *Synthesis* **1998**, 683–703.

[71] A. G. Griesbeck, S. Stadtmueller, *J. Am. Chem. Soc.* **1991**, *113*(18), 6923–6928.

[72] H. Staudinger, *Helv. Chim. Acta* **1925**, *8*(1), 306–332.

[73] F. O. Rice, J. Greenberg, *J. Am. Chem. Soc.* **1934**, *56*(10), 2132–2134.

[74] R. Oda, S. Munemiya, M. Okano, *Makromol. Chem.* **1961**, *43*(1), 149–151.

[75] K. C. Khemani, F. Wudl, *J. Am. Chem. Soc.* **1989**, *111*(25), 9124–9125.

[76] W. Davies, S. Middleton, *J. Chem. Soc.* **1958**, 822–825.

[77] I. A. Kashulin, I. E. Nifant'ev, *J. Org. Chem.* **2004**, *69*(16), 5476–5479.

[78] C. Hansch, A. Leo, R. W. Taft, *Chem. Rev.* **1991**, *91*(2), 165–195.

[79] J.-P. Djukic, C. Michon, D. Heiser, N. Kyritsakas-Gruber, A. de Cian, K. Dötz, M. Pfeffer, *Eur. J. Inorg. Chem.* **2004**, 2107–2122.

[80] M. Okubo, Y. Uematsu, *Bull. Chem. Soc. Jpn.* **1982**, *55*(4), 1121–1126.

[81] L. F. Fieser, A. M. Seligman, *J. Am. Chem. Soc.* **1935**, *57*(11), 2174–2176.

[82] E. Bergmann, M. Orchin, *J. Am. Chem. Soc.* **1949**, *71*(3), 1111–1112.

[83] H. E. Bronstein, N. Choi, L. T. Scott, *J. Am. Chem. Soc.* **2002**, *124*(30), 8870–8875.

[84] E. H. Huntress, K. Pfister, K. H. T. Pfister, *J. Am. Chem. Soc.* **1942**, *64*(12), 2845–2849.

[85] N. Kharasch, T. C. Bruice, *J. Am. Chem. Soc.* **1951**, *73*(7), 3240–3244.

[86] L. Horner, D. W. Baston, *Just. Lieb. Ann. Chem.* **1973**, 910–935.

[87] A. K. Lala, R. R. Dixit, *J. Chem. Soc., Chem. Commun.* **1989**, *1989*, 636–638.

[88] G. Just, Z. Y. Wang, L. Chan, *J. Org. Chem.* **1988**, *53*(5), 1030–1033.

[89] W. Kirmse, B. Krzossa, *Tetrahedron Lett.* **1998**, *39*(8), 799 – 802.

[90] X. M. Cherian, S. A. Van Arman, A. W. Czarnik, *J. Am. Chem. Soc.* **1990**, *112*(11), 4490–4498.

[91] D. M. A. Grieve, G. E. Lewis, M. D. Ravenscroft, P. Skrabal, T. Sonoda, I. Szele, H. Zollinger, *Helv. Chim. Acta* **1985**, *68*(5), 1427–1443.

[92] W. Zeng, T. E. Ballard, A. G. Tkachenko, V. A. Burns, D. L. Feldheim, C. Melander, *Bioorg. Med. Chem. Lett.* **2006**, *16*(19), 5148 – 5151.

[93] Y.-B. Zhao, B. Mariampillai, D. Candito, B. Laleu, M. Li, M. Lautens, *Angew. Chem. Int. Ed.* **2009**, *48*(10), 1849–1852.

[94] D. Tilly, S. S. Samanta, F. Faigl, J. Mortier, *Tetrahedron Lett.* **2002**, *43*(46), 8347–8350.

[95] D. Tilly, S. S. Samanta, A. De, A.-S. Castanet, J. Mortier, *Org. Lett.* **2005**, *7*(5), 827–830.

[96] D. Tilly, S. S. Samanta, A.-S. Castanet, A. De, J. Mortier, *Eur. J. Org. Chem.* **2006**, 174–182.

[97] M. A. Campo, R. C. Larock, *Org. Lett.* **2000**, *2*(23), 3675–3677.

[98] M. A. Campo, R. C. Larock, *J. Org. Chem.* **2002**, *67*(16), 5616–5620.

[99] P. J. Coelho, L. M. Carvalho, S. Rodrigues, A. M. F. Oliveira-Campos, R. Dubest, J. Aubard, A. Samat, R. Guglielmetti, *Tetrahedron* **2002**, *58*(5), 925 – 931.

[100] T.-H. Nguyen, A.-S. Castanet, J. Mortier, *Org. Lett.* **2006**, *8*(4), 765–768.

[101] J. A. Hirsch, *Topics in Stereochemistry, Volume 1*, (editors N. L. Allinger, E. L. Eliel), John Wiley & Sons, Inc., **1967**.

[102] K. M. Gillespie, C. J. Sanders, P. O'Shaughnessy, I. Westmoreland, C. P. Thickitt, P. Scott, *J. Org. Chem.* **2002**, *67*(10), 3450–3458.

[103] S. W. Wright, D. L. Hageman, A. S. Wright, L. D. McClure, *Tetrahedron Lett.* **1997**, *38*(42), 7345 – 7348.

[104] H. G. O. Becker, Autorenkollektiv, *Organikum*, 21st ed., Wiley-VCH Verlag, **2001**.

[105] R. Ugo, F. Cariati, G. L. Monica, *Inorg. Synth.* **1990**, *28*, 123–126.

[106] P. G. Gassman, I. G. Cesa, *Organometallics* **1984**, *3*(1), 119–128.

[107] U. Nagel, *Chem. Ber.* **1982**, *115*(5), 1998–1999.

[108] Y. Tatsuno, T. Yoshida, S. Otsuka, *Inorg. Synth.* **1990**, *28*, 343–346.

[109] T. Yoshida, S. Otsuka, *Inorg. Synth.* **1990**, *28*, 113–119.

[110] H. Maciejewski, A. Sydor, B. Marciniec, M. Kubicki, P. B. Hitchcock, *Inorg. Chim. Acta* **2006**, *359*(9), 2989 – 2997.

[111] B. M. Still, P. G. A. Kumar, J. R. Aldrich-Wright, W. S. Price, *Chem. Soc. Rev.* **2007**, *36*, 665 – 686.

[112] R. H. Crabtree, *The organometallic chemistry of the transition metals*, 4th ed., John Wiley & Sons Limited, **2005**.

[113] J. Ishizu, T. Yamamoto, A. Yamamoto, *Bull. Chem. Soc. Jpn.* **1978**, *51*, 2646–2650.

[114] P.-T. Cheng, S. C. Nyburg, *Can. J. Chem.* **1972**, *50*, 912–916.

[115] V. V. Grushin, *Organometallics* **2000**, *19*(10), 1888–1900.

[116] V. P. W. Böhm, W. A. Herrmann, *Chem. Eur. J.* **2001**, *7*(19), 4191–4197.

[117] W. J. Marshall, R. J. Young, V. V. Grushin, *Organometallics* **2001**, *20*(3), 523–533.

[118] S. L. Fraser, M. Y. Antipin, V. N. Khroustalyov, V. V. Grushin, *J. Am. Chem. Soc.* **1997**, *119*(20), 4769–4770.

[119] G. T. L. Broadwood-Strong, P. A. Chaloner, P. B. Hitchcock, *Polyhedron* **1993**, *12*(7), 721–729.

[120] R. U. Kirss, R. Eisenberg, *Inorg. Chem.* **1989**, *28*(17), 3372–3378.

[121] M. R. Mason, J. G. Verkade, *Organometallics* **1992**, *11*(6), 2212–2220.

[122] P. A. McLaughlin, J. G. Verkade, *Organometallics* **1998**, *17*(26), 5937–5940.

[123] C. A. Tolman, *J. Am. Chem. Soc.* **1970**, *92*(10), 2953–2956.

[124] S. Hietkamp, D. J. Stufkens, K. Vrieze, *J. Organomet. Chem.* **1979**, *169*(1), 107–113.

[125] P. E. Garrou, *Chem. Rev.* **1981**, *81*(3), 229–266.

[126] H. C. Clark, P. N. Kapoor, I. J. Mcmahon, *J. Organomet. Chem.* **1984**, *265*(1), 107 – 115.

[127] A. Scrivanti, R. Campostrini, G. Carturan, *Inorg. Chim. Acta* **1988**, *142*(2), 187 – 189.

[128] J. M. Brown, J. J. Perez-Torrente, N. W. Alcock, *Organometallics* **1995**, *14*(3), 1195–1203.

[129] H. Petzold, H. Görls, W. Weigand, *J. Organomet. Chem.* **2007**, *692*(13), 2736 – 2742.

[130] T. G. Appleton, M. A. Bennett, I. B. Tomkins, *J. Chem. Soc., Dalton Trans.* **1976**, 439–446.

[131] M. P. Brown, R. J. Puddephatt, M. Rashidi, K. R. Seddon, *J. Chem. Soc., Dalton Trans.* **1977**, 951–955.

[132] D. A. Slack, M. C. Baird, *Inorg. Chim. Acta* **1977**, *24*, 277 – 280.

[133] G. Banditelli, A. L. Bandini, F. Bonati, G. Minghetti, *Inorg. Chim. Acta* **1982**, *60*, 93 – 98.

[134] H. C. Clark, L. E. Manzer, *J. Organomet. Chem.* **1973**, *59*, 411 – 428.

[135] D. Drew, J. R. Doyle, *Inorg. Synth.* **1990**, *28*, 346–349.

[136] M. Green, J. A. K. Howard, J. L. Spencer, F. G. A. Stone, *J. Chem. Soc., Dalton Trans.* **1977**, 271–277.

[137] M. Green, J. A. Howard, J. L. Spencer, F. G. A. Stone, *J. Chem. Soc., Chem. Commun.* **1975**, 3–4.

[138] M. T. Chicote, M. Green, J. L. Spencer, F. Gordon, A. Stone, J. Vicente, *J. Organomet. Chem.* **1977**, *137*(1), C8–C10.

[139] M. T. Chicote, M. Green, J. L. Spencer, F. G. A. Stone, J. Vicente, *J. Chem. Soc., Dalton Trans.* **1979**, (3), 536–541.

[140] F. Gordon, A. Stone, *Acc. Chem. Res.* **1981**, *14*(10), 318–325.

[141] J. W. Sprengers, M. J. Agerbeek, C. J. Elsevier, H. Kooijman, A. L. Spek, *Organometallics* **2004**, *23*(13), 3117–3125.

[142] S. D. Ittel, *Inorg. Chem.* **1977**, *16*(10), 2589–2597.

[143] P. Sehnal, H. Taghzouti, I. J. S. Fairlamb, A. Jutand, A. F. Lee, A. C. Whitwood, *Organometallics* **2009**, *28*(3), 824–829.

[144] R. Freeman, H. D. W. Hill, R. Kaptein, *J. Magn. Reson.* **1972**, *7*, 327–329.

[145] M. L. Martin, J.-J. Delpuech, G. J. Martin, *Practical NMR Spectroscopy*, Heyden, London, **1980**, 231–235.

[146] B. Neises, W. Steglich, *Angew. Chem. Int. Ed.* **1978**, *17*(7), 522–524.

[147] J. W. Schoenecker, A. E. Takemori, P. S. Portoghese, *J. Med. Chem.* **1986**, *29*(10), 1868–1871.

[148] U. Aeberhard, R. Keese, E. Stamm, U.-C. Vögeli, W. Lau, J. K. Kochi, *Helv. Chim. Acta* **1983**, *66*(8), 2740–2759.

[149] S. Patai, Z. Rappoport, *The chemistry of organic silicon compounds*, John Wiley & Sons Limited, **1989**.

[150] N. S. Fedotov, G. E. Évert, E. V. Bolosova, Z. V. Gorislavskaya, G. B. Belyakova, L. V. Sobolevskaya, E. K. Dobrovinskaya, V. F. Mironov, *Russ. J. gen. Chem.* **1982**, *52*, 1628–1633.

[151] M. Nasim, A. K. Saxena, indu P. Pal, L. M. Pande, *Synth. React. Inorg. Met.-Org. Chem.* **1987**, *17*, 1003–1009.

[152] M. D. Healy, A. R. Barron, *J. Organomet. Chem.* **1990**, *381*(2), 165 – 172.

[153] A. Pla-Quintana, A. Roglans, J. V. de Julián-Ortiz, M. Moreno-Mañas, T. Parella, J. Benet-Buchholz, X. Solans, *Chem. Eur. J.* **2005**, *11*(9), 2689–2697.

[154] A. Pla-Quintana, A. Torrent, A. Dachs, A. Roglans, R. Pleixats, M. Moreno-Manas, T. Parella, J. Benet-Buchholz, *Organometallics* **2006**, *25*(23), 5612–5620.

[155] A. Pla-Quintana, A. Roglans, T. Parella, J. Benet-Buchholz, *J. Organomet. Chem.* **2007**, *692*(14), 2997 – 3004.

[156] A. Dachs, J. Masllorens, A. Pla-Quintana, A. Roglans, J. Farjas, T. Parella, *Organometallics* **2008**, *27*(22), 5768–5776.

[157] J. Cortés, M. Moreno-Mañas, R. Pleixats, *Eur. J. Org. Chem.* **2000**, 239–243.

[158] S. Cerezo, J. Cortès, E. Lago, E. Molins, M. Moreno-Mañas, T. Parella, R. Pleixats, J. Torrejón, A. Vallribera, *Eur. J. Inorg. Chem.* **2001**, 1999–2006.

[159] S. Cerezo, J. Cortès, D. Galvan, E. Lago, C. Marchi, E. Molins, M. Moreno-Mañas, R. Pleixats, J. Torrejón, A. Vallribera, *Eur. J. Org. Chem.* **2001**, 329–337.

[160] J. Krause, K.-J. Haack, G. Cestaric, R. Goddard, K.-R. Pörschke, *Chem. Commun.* **1998**, 1291–1292.

[161] J. Krause, G. Cestaric, K.-J. Haack, K. Seevogel, W. Storm, K.-R. Porschke, *J. Am. Chem. Soc.* **1999**, *121*(42), 9807–9823.

[162] K. Blum, E. S. Chernyshova, R. Goddard, K. Jonas, K.-R. Porschke, *Organometallics* **2007**, *26*(21), 5174–5178.

[163] H. Werner, D. Tune, G. Parker, C. Krüger, D. J. Brauer, *Angew. Chem.* **1975**, *87*(6), 205–206.

[164] D. M. Norton, E. A. Mitchell, N. R. Botros, P. G. Jessop, M. C. Baird, *J. Org. Chem.* **2009**, *74*(17), 6674–6680.

[165] H. Urata, H. Suzuki, Y. Moro-oka, T. Ikawa, *J. Organomet. Chem.* **1989**, *364*(1-2), 235 – 244.

[166] B. E. Mann, A. Musco, *J. Chem. Soc., Dalton Trans.* **1975**, (16-17), 1673–1677.

[167] E. O. Fischer, H. Werner, *Chem. Ber.* **1962**, *95*(3), 703–708.

[168] H. Werner, *Angew. Chem.* **1977**, *89*(1), 1–10.

[169] E. Ye, H. Tan, S. Li, W. Y. Fan, *Angew. Chem. Int. Ed.* **2006**, *45*(7), 1120–1123.

[170] V. Harder, H. Werner, *Helv. Chim. Acta* **1973**, *56*(1), 549–553.

[171] Z. Csákai, R. Skoda-Földes, L. Kollár, *Inorg. Chim. Acta* **1999**, *286*(1), 93–97.

[172] H. Werner, A. Kühn, *Angew. Chem.* **1977**, *89*(6), 427–428.

[173] S. Berger, S. Braun, *200 and More NMR Experiments*, WILEY-VCH Verlag, **2004**.

[174] J. Jeener, B. H. Meier, P. Bachmann, R. R. Ernst, *J. Chem. Phys.* **1979**, *71*, 4546–4553.

[175] C. L. Perrin, T. J. Dwyer, *Chem. Rev.* **1990**, *90*(6), 935–967.

[176] V. S. Sergienko, M. A. Porai-Koshits, *J. Struct. Chem.* **1988**, *28*, 548–552.

[177] V. Albano, P. L. Bellon, V. Scatturin, *Chem. Commun.* **1966**, (15), 507–507.

[178] V. G. Andrianov, I. S. Akhrem, N. M. Chistovalova, Y. T. Struchkov, *J. Struct. Chem.* **1976**, *17*, 111–116.

[179] P. A. Chaloner, P. B. Hitchcock, G. T. L. Broadwood-Strong, *Acta Cryst. C* **1989**, *45*(9), 1309–1311.

[180] R. van der Linde, R. O. de Jongh, *J. Chem. Soc., Dalton Trans.* **1971**, 563–563.

[181] A. Visser, R. van der Linde, R. O. de Jongh, *Inorg. Synth.* **1976**, *16*, 127–130.

[182] J. J. M. de Pater, D. S. Tromp, D. M. Tooke, A. L. Spek, B.-J. Deelman, G. van Koten, C. J. Elsevier, *Organometallics* **2005**, *24*(26), 6411–6419.

[183] D. M. Fenton, *J. Org. Chem.* **1973**, *38*(18), 3192–3198.

[184] J. F. Hartwig, S. Richards, D. Baranano, F. Paul, *J. Am. Chem. Soc.* **1996**, *118*(15), 3626–3633.

[185] L. M. Alcazar-Roman, J. F. Hartwig, A. L. Rheingold, L. M. Liable-Sands, I. A. Guzei, *J. Am. Chem. Soc.* **2000**, *122*(19), 4618–4630.

[186] C. Azerraf, A. Shpruhman, D. Gelman, *Chem. Commun.* **2009**, (4), 466–468.

[187] M. Yamashita, I. Takamiya, K. Jin, K. Nozaki, *Organometallics* **2006**, *25*(19), 4588–4595.

[188] W. Lesueur, E. Solari, C. Floriani, A. Chiesi-Villa, C. Rizzoli, *Inorg. Chem.* **1997**, *36*(15), 3354–3362.

[189] A. B. Pangborn, M. A. Giardello, R. H. Grubbs, R. K. Rosen, F. J. Timmers, *Organometallics* **1996**, *15*(5), 1518–1520.

[190] W. L. F. Armarego, C. L. L. Chai, *Purification of laboratory chemicals*, 5th ed., Butterworth-Heinemann, **2003**.

[191] D. W. Later, B. W. Wilson, M. L. Lee, *Anal. Chem.* **1985**, *57*(14), 2979–2984.

[192] W. C. Still, M. Kahn, A. Mitra, *J. Org. Chem.* **1978**, *43*(14), 2923–2925.

[193] H. E. Gottlieb, V. Kotlyar, A. Nudelman, *J. Org. Chem.* **1997**, *62*(21), 7512–7515.

[194] G. R. Fulmer, A. J. M. Miller, N. H. Sherden, H. E. Gottlieb, A. Nudelman, B. M. Stoltz, J. E. Bercaw, K. I. Goldberg, *Organometallics* **2010**, *29*(9), 2176–2179.

[195] G. M. Sheldrick, *Acta Cryst. A* **2008**, *64*(1), 112–122.

[196] H. Fuess, T. Hahn, H. Wondratschek, U. Müller, U. Shmueli, E. Prince, A. Authier, V. Kopsky, D. B. Litvin, M. G. Rossmann, S. H. E. Arnold, B. McMahon., *Complete online set of International tables for crystallography Volume A-G*, Springer Verlag, **2007**.

[197] T. E. Ready, J. C. Chien, M. D. Rausch, *J. Organomet. Chem.* **1999**, *583*(1-2), 11–27.

[198] H. Erlenmeyer, W. Schoenauer, *Helv. Chim. Acta* **1937**, *20*(1), 1008–1012.

[199] L. E. Crascall, J. L. Spencer, *Inorg. Synth.* **1990**, *28*, 126–132.

[200] N. Kamigata, M. Satoh, T. Fukushima, *Bull. Chem. Soc. Jpn.* **1990**, *63*(7), 2118–2120.

[201] O. Tsuge, K. Sone, S. Urano, K. Matsuda, *J. Org. Chem.* **1982**, *47*(26), 5171–5177.

[202] W. M. Phillips, D. J. Currie, *Can. J. Chem.* **1969**, *47*(17), 3137–3146.

[203] Z.-Z. Huang, Y. Tang, *J. Org. Chem.* **2002**, *67*(15), 5320–5326.

[204] T. Hashimoto, T. Shiomi, J.-i. Ito, H. Nishiyama, *Tetrahedron* **2007**, *63*(52), 12883–12887.

[205] J. P. Parrish, E. E. Dueno, S.-I. Kim, K. W. Jung, *Synth. Commun.* **2000**, *30*(15), 2687–2700.

[206] R. Imashiro, M. Seki, *J. Org. Chem.* **2004**, *69*(12), 4216–4226.

[207] V. D. Novokreshchennykh, S. S. Mochalov, Y. S. Shabarov, *J. Org. Chem. USSR* **1982**, *18*, 262–268.

[208] J. Zheng, Y. Shen, *Synth. Commun.* **1994**, *24*(14), 2069–2073.

[209] H. García, S. Iborra, M. A. Miranda, J. Primo, *Heterocycles* **1985**, *23*(8), 1983–1989.

[210] R. Walter, *Ber. dtsch. Chem. Ges. A/B* **1925**, *58*(10), 2303–2310.

[211] Belozwetow, *J. Gen. Chem. USSR* **1966**, *36*, 1212.

[212] G. Cevasco, S. Thea, *J. Org. Chem.* **1994**, *59*(21), 6274–6278.

[213] J. M. Lohar, J. S. Dave, *Mol. Cryst. Liq. Cryst.* **1983**, *103*, 143–153.

[214] O. B. Nagy, V. Reuliaux, N. Bertrand, A. V. D. Mensbrugghe, J. Leseul, J. B. Nagy, *Bull. Soc. Chim. Bel.* **1985**, *94*, 1055–1074.

[215] C. Pardin, J. N. Pelletier, W. D. Lubell, J. W. Keillor, *J. Org. Chem.* **2008**, *73*(15), 5766–5775.

[216] A. B. Charette, M. K. Janes, H. Lebel, *Tetrahedron: Asymmetry* **2003**, *14*(7), 867–872.

[217] R. Bairwa, M. Kakwani, N. R. Tawari, J. Lalchandani, M. Ray, M. Rajan, M. S. Degani, *Bioorg. Med. Chem. Lett.* **2010**, *20*(5), 1623–1625.

[218] T. Okutome, H. Kawamura, S. Taira, T. Nakayama, S. Nunomura, M. Kurumi, Y. Sakurai, T. Aoyama, S. Fujii, *Chem. Pharm. Bull.* **1984**, *32*(5), 1854–865.

[219] J. Masllorens, M. Moreno-Manas, A. Pla-Quintana, A. Roglans, *Org. Lett.* **2003**, *5*(9), 1559–1561.

[220] W. Adam, M. Ahrweiler, M. Sauter, *Chem. Ber.* **1994**, *127*(5), 941–946.

[221] J. T. Gerig, R. S. McLeod, *Can. J. Chem.* **1975**, *53*(4), 513–518.

[222] S. Skraup, E. Beng, *Ber. dtsch. Chem. Ges. A/B* **1927**, *60*(4), 942–950.

[223] J. M. Concellón, J. A. Pérez-Andrés, H. Rodríguez-Solla, *Angew. Chem. Int. Ed.* **2000**, *39*(15), 2773–2775.

[224] J. P. Parrish, Y. C. Jung, S. I. Shin, K. W. Jung, *J. Org. Chem.* **2002**, *67*(20), 7127–7130.

[225] C. S. Rondestvedt, C. D. Ver Nooy, *J. Am. Chem. Soc.* **1955**, *77*(18), 4878–4883.

[226] F. Bohlmann, J. Jacob, *Chem. Ber.* **1974**, *107*(8), 2578–2584.

[227] J. A. Kampmeier, S. H. Harris, R. M. Rodehorst, *J. Am. Chem. Soc.* **1981**, *103*(6), 1478–1485.

[228] R. Anschütz, *Ber. dtsch. Chem. Ges. A/B* **1927**, *60*(6), 1320–1322.

[229] Z. Wu, G. S. Minhas, D. Wen, H. Jiang, K. Chen, P. Zimniak, J. Zheng, *J. Med. Chem.* **2004**, *47*(12), 3282–3294.

A. Appendix

A.1. X-ray structure parameters

PtPhNO$_2$–bp

Table A.1.: Structural parameters of **PtPhNO$_2$–bp**:

molecular formula	$C_{65}H_{53}NO_5P_2Pt$
molecular weight [a.m.u.]	1185.11
colour/habit	yellow fragment
crystal dimensions [mm^3]	$0.40 \times 0.18 \times 0.15$
crystal system	monoclinic
space group	P2$_1$/n
unit cell dimensions	a = 13.5749(3) Å $\quad$ $\alpha = 90.00°$
	b = 9.6451(3) Å $\quad$ $\beta = 97.4740(10)°$
	c = 41.5616(10) Å $\quad$ $\gamma = 90.00°$
cell volume [Å^3]	5397.2(2)
Z	4
T [K]	173(2)
calculated density [gcm^{-3}]	1.458
absorption coefficient μ [mm^{-1}]	2.711
F(000)	2392
Θ range [°]	0.99 - 25.34
index ranges [h, k, l]	±16, $+10/-9$, ±50

continued on next page

Table A.1 – continued from previous page

no. of independent reflections / R_{int}	9301
no. of absorbed reflections	8776
no. of data	9301
no. of restraints	0
no. of parameters	848
R_1 $(I > 2\sigma(I))$	0.0246
wR_2 $(I > 2\sigma(I))$	0.0519
R_1 (all data)	0.0271
wR_2 (all data)	0.0527
goodness of fit (on F^2)	1.229
largest difference peak [$e\text{Å}^3$]	0.858
largest difference hole [$e\text{Å}^3$]	-1.579

Table A.2.: Bond lengths of $\mathbf{PtPhNO_2-bp}$ [Å]:

C1–C2	1.391(5)	C25–C26	1.384(5)	C46–C47	1.379(5)
C1–C6	1.394(5)	C25–C30	1.387(5)	C46–C51	1.390(5)
C1–P1	1.841(3)	C25–P2	1.840(3)	C46–O4	1.403(4)
C2–C3	1.389(5)	C26–C27	1.389(5)	C47–C48	1.384(5)
C3–C4	1.373(6)	C27–C28	1.378(6)	C48–C49	1.371(6)
C4–C5	1.382(6)	C28–C29	1.375(5)	C49–C50	1.373(6)
C5–C6	1.387(5)	C29–C30	1.392(5)	C50–C51	1.405(5)
C7–C8	1.392(5)	C31–C32	1.388(4)	C51–C52	1.503(5)
C7–C12	1.399(5)	C31–C36	1.392(5)	C52–O5	1.218(4)
C7–P1	1.830(3)	C31–P2	1.840(3)	C52–C53	1.497(6)
C8–C9	1.383(6)	C32–C33	1.391(5)	C53–C58	1.393(6)
C9–C10	1.360(6)	C33–C34	1.376(6)	C53–C54	1.396(6)
C10–C11	1.377(6)	C34–C35	1.378(6)	C54–C55	1.377(8)
C11–C12	1.386(6)	C35–C36	1.385(5)	C55–C56	1.373(9)

continued on next page

Table A.2 – continued from previous page

C13–C18	1.383(5)	C37–C38	1.449(4)	C56–C57	1.370(7)
C13–C14	1.392(5)	C37–C39	1.478(4)	C57–C58	1.383(7)
C13–P1	1.827(3)	C37–Pt1	2.119(3)	C59–C64	1.317(11)
C14–C15	1.392(5)	C38–C45	1.459(4)	C59–C60	1.343(10)
C15–C16	1.374(7)	C38–Pt1	2.138(3)	C59–C65	1.558(10)
C16–C17	1.376(6)	C39–C44	1.398(5)	C60–C61	1.423(11)
C17–C18	1.384(5)	C39–C40	1.404(5)	C61–C62	1.417(12)
C19–C24	1.390(4)	C40–C41	1.385(5)	C62–C63	1.283(11)
C19–C20	1.391(5)	C41–C42	1.382(5)	C63–C64	1.450(12)
C19–P2	1.827(3)	C42–C43	1.383(5)	N1–O2	1.221(4)
C20–C21	1.385(5)	C42–N1	1.468(4)	N1–O1	1.225(5)
C21–C22	1.382(5)	C43–C44	1.383(5)	P1–Pt1	2.2959(8)
C22–C23	1.382(6)	C45–O3	1.207(4)	P2–Pt1	2.2844(8)
C23–C24	1.392(5)	C45–O4	1.387(4)		

Table A.3.: Bond angles of **PtPhNO$_2$–bp** [°]:

C2–C1–C6	118.1(3)	C32–C31–C36	118.9(3)	C58–C53–C54	117.1(5)
C2–C1–P1	123.6(3)	C32–C31–P2	123.3(3)	C58–C53–C52	123.7(4)
C6–C1–P1	118.2(2)	C36–C31–P2	117.8(2)	C54–C53–C52	119.1(4)
C3–C2–C1	120.3(4)	C31–C32–C33	120.4(3)	C55–C54–C53	121.5(6)
C4–C3–C2	120.9(4)	C34–C33–C32	120.3(3)	C56–C55–C54	120.1(6)
C3–C4–C5	119.7(3)	C33–C34–C35	119.6(4)	C57–C56–C55	119.7(6)
C4–C5–C6	119.8(4)	C34–C35–C36	120.7(4)	C56–C57–C58	120.4(6)
C5–C6–C1	121.3(3)	C35–C36–C31	120.1(3)	C57–C58–C53	121.1(5)
C8–C7–C12	117.5(3)	C38–C37–C39	120.2(3)	C64–C59–C60	120.2(9)
C8–C7–P1	123.6(2)	C38–C37–Pt1	70.82(17)	C64–C59–C65	121.0(8)
C12–C7–P1	118.9(3)	C39–C37–Pt1	111.3(2)	C60–C59–C65	118.8(8)
C9–C8–C7	120.8(3)	C37–C38–C45	119.7(3)	C59–C60–C61	120.0(9)

continued on next page

Table A.3 – continued from previous page

C10–C9–C8	120.9(4)	C37–C38–Pt1	69.39(16)	C62–C61–C60	117.7(9)
C9–C10–C11	119.7(4)	C45–C38–Pt1	113.3(2)	C63–C62–C61	121.9(11)
C10–C11–C12	120.1(4)	C44–C39–C40	117.5(3)	C62–C63–C64	118.1(10)
C11–C12–C7	120.9(4)	C44–C39–C37	122.1(3)	C59–C64–C63	122.1(9)
C18–C13–C14	119.2(3)	C40–C39–C37	120.5(3)	O2–N1–O1	123.7(3)
C18–C13–P1	117.5(3)	C41–C40–C39	121.5(4)	O2–N1–C42	117.9(4)
C14–C13–P1	123.1(3)	C42–C41–C40	118.6(3)	O1–N1–C42	118.4(3)
C15–C14–C13	119.8(4)	C41–C42–C43	122.0(3)	C45–O4–C46	118.8(3)
C16–C15–C14	120.5(4)	C41–C42–N1	119.5(3)	C13–P1–C7	104.04(14)
C15–C16–C17	119.7(4)	C43–C42–N1	118.5(3)	C13–P1–C1	102.03(14)
C16–C17–C18	120.5(4)	C42–C43–C44	118.4(4)	C7–P1–C1	102.06(15)
C13–C18–C17	120.3(4)	C43–C44–C39	122.0(3)	C13–P1–Pt1	112.12(11)
C24–C19–C20	119.1(3)	O3–C45–O4	121.2(3)	C7–P1–Pt1	113.59(10)
C24–C19–P2	123.1(3)	O3–C45–C38	128.9(3)	C1–P1–Pt1	121.03(10)
C20–C19–P2	117.8(2)	O4–C45–C38	109.9(3)	C19–P2–C31	103.45(14)
C21–C20–C19	120.6(3)	C47–C46–C51	121.2(3)	C19–P2–C25	105.48(14)
C22–C21–C20	119.9(4)	C47–C46–O4	115.7(3)	C31–P2–C25	100.61(15)
C23–C22–C21	120.1(4)	C51–C46–O4	123.1(3)	C19–P2–Pt1	113.27(11)
C22–C23–C24	120.1(4)	C46–C47–C48	120.1(4)	C31–P2–Pt1	113.78(10)
C19–C24–C23	120.1(4)	C49–C48–C47	120.1(4)	C25–P2–Pt1	118.48(10)
C26–C25–C30	118.4(3)	C48–C49–C50	119.8(4)	C37–Pt1–C38	39.79(12)
C26–C25–P2	119.6(3)	C49–C50–C51	121.6(4)	C37–Pt1–P2	102.82(9)
C30–C25–P2	122.0(2)	C46–C51–C50	117.2(3)	C38–Pt1–P2	142.55(9)
C25–C26–C27	121.2(4)	C46–C51–C52	122.1(3)	C37–Pt1–P1	148.44(9)
C28–C27–C26	119.9(4)	C50–C51–C52	120.6(3)	C38–Pt1–P1	109.04(9)
C29–C28–C27	119.5(4)	O5–C52–C53	121.1(4)	P2–Pt1–P1	108.38(3)
C28–C29–C30	120.6(4)	O5–C52–C51	119.2(4)		
C25–C30–C29	120.4(3)	C53–C52–C51	119.7(3)		

PhOMe−bp

Table A.4.: Structural parameters of **PhOMe−bp**:

molecular formula	$C_{23}H_{18}O_4$
molecular weight [a.m.u.]	258.37
colour/habit	coulorless block
crystal dimensions [mm^3]	$0.30 \times 0.10 \times 0.10$
crystal system	monoclinic
space group	$P2_1/c$
unit cell dimensions	a = 13.941(2) Å α = 90.00°
	b = 9.6365(16) Å β = 108.517(7)°
	c = 14.264(2) Å γ = 90.00°
cell volume [Å^3]	1817.1(5)
Z	4
T [K]	173(2)
calculated density [gcm^{-3}]	1.310
absorption coefficient μ [mm^{-1}]	0.089
F(000)	752
Θ range [°]	1.54 - 30.49
index ranges [h, k, l]	$\pm19, \pm13, \pm20$
no. of independent reflections / R_{int}	5508
no. of absorbed reflections	4659
no. of data	5508
no. of restraints	0
no. of parameters	316
R_1 ($I > 2\sigma(I)$)	0.0395
wR_2 ($I > 2\sigma(I)$)	0.1108

continued on next page

Table A.4 – continued from previous page

R_1 (all data)	0.0478
wR_2 (all data)	0.1200
goodness of fit (on F^2)	1.004

Table A.5.: Bond lengths of **PhOMe−bp** [Å]:

O1–C11	1.3766(11)	C3–C8	1.4917(13)	C9–C16	1.3843(14)
O1–C4	1.4035(11)	C3–C5	1.4999(13)	C9–C10	1.3972(14)
C1–C15	1.3406(13)	C4–C13	1.3883(14)	C12–C18	1.3856(15)
C1–C11	1.4640(13)	C4–C5	1.3933(13)	C13–C19	1.3909(16)
O2–C10	1.3633(12)	O3–C11	1.2033(12)	C14–C22	1.3947(15)
O2–C20	1.4355(13)	C5–C12	1.4032(13)	C17–C21	1.3907(15)
C2–C6	1.4012(13)	C6–C7	1.3868(14)	C18–C19	1.3904(17)
C2–C16	1.4065(13)	C7–C10	1.3954(13)	C21–C23	1.3846(19)
C2–C15	1.4602(13)	C8–C14	1.3934(14)	C22–C23	1.3905(19)
C3–O4	1.2174(12)	C8–C17	1.3960(14)		

Table A.6.: Bond angles of **PhOMe−bp** [°]:

C11–O1–C4	115.95(7)	C4–C5–C3	123.99(8)	O1–C11–C1	110.53(8)
C15–C1–C11	120.33(9)	C12–C5–C3	117.82(8)	C18–C12–C5	121.14(9)
C10–O2–C20	117.79(8)	C7–C6–C2	121.87(9)	C4–C13–C19	119.3(1)
C6–C2–C16	117.58(9)	C6–C7–C10	119.43(9)	C8–C14–C22	119.63(10)
C6–C2–C15	122.98(8)	C14–C8–C17	120.15(9)	C1–C15–C2	126.58(9)
C16–C2–C15	119.44(9)	C14–C8–C3	121.37(9)	C9–C16–C2	121.14(9)
O4–C3–C8	120.10(9)	C17–C8–C3	118.31(9)	C21–C17–C8	119.75(11)
O4–C3–C5	119.54(8)	C16–C9–C10	120.11(9)	C12–C18–C19	119.53(9)
C8–C3–C5	120.30(8)	O2–C10–C7	124.27(9)	C18–C19–C13	120.48(10)
C13–C4–C5	121.48(9)	O2–C10–C9	115.90(8)	C23–C21–C17	120.03(11)

continued on next page

Table A.6 – continued from previous page

C13–C4–O1	116.40(8)	C7–C10–C9	119.82(9)	C23–C22–C14	119.88(11)
C5–C4–O1	122.11(8)	O3–C11–O1	121.83(9)	C21–C23–C22	120.39(10)
C4–C5–C12	118.06(9)	O3–C11–C1	127.63(9)		

Me−bpNNHTs

Table A.7.: Structural parameters of **Me−bpNNHTs**:

molecular formula	$C_{96}H_{88}N_8O_{16}S_4$
molecular weight [a.m.u.]	1738.06
crystal system	monoclinic
space group	$P2_1/n$
unit cell dimensions	$a = 15.1191(50)$ Å $\quad \alpha = 90.00°$
	$b = 9.5164(50)$ Å $\quad \beta = 99.179(5)°$
	$c = 15.7294(50)$ Å $\quad \gamma = 90.00°$
cell volume [Å^3]	2234.16(156)
Z	4
calculated density [gcm^{-3}]	1.29174
absorption coefficient μ [mm^{-1}]	0.18
F(000)	908
no. of independent reflections / R_{int}	4709
no. of absorbed reflections	4238
no. of data	4709
no. of restraints	0
no. of parameters	367
R_1 ($I > 2\sigma(I)$)	0.0515
R_1 (all data)	0.0579

Table A.8.: Bond lengths of **Me−bpNNHTs** [Å]:

S1–C8	1.7553(4)	C5–C6	1.3797(3)	C20–C7	1.3833(3)
O4–C23	1.2077(4)	C4–C5	1.3853(4)	C2–C1	1.3835(4)
O3–C23	1.3699(4)	C4–O3	1.4022(3)	C19–C20	1.3721(4)
O2–S1	1.4309(3)	C3–C2	1.3947(3)	C19–C10	1.3536(4)
O1–S1	1.4240(4)	C3–C4	1.3834(5)	C18–C13	1.3815(3)
N2–C22	1.2847(4)	C25–C26	1.4901(4)	C17–C18	1.3797(3)
N1–N2	1.4039(3)	C24–C25	1.3178(3)	C16–C15	1.3938(3)
N1–S1	1.6539(4)	C23–C24	1.4580(4)	C16–C17	1.3993(4)
C9–C10	1.3786(3)	C22–C3	1.5001(5)	C15–C14	1.3793(3)
C8–C7	1.3674(4)	C22–C16	1.4845(3)	C13–C14	1.3817(4)
C8–C9	1.3737(4)	C21–C19	1.5065(4)	C1–C6	1.3772(5)

Table A.9.: Bond angles of **Me−bpNNHTs** [°]:

S1–C8–C9	120.175(19)	C8–C9–C10	120.029(21)	C22–C16–C17	120.607(19)
S1–C8–C7	120.303(18)	C5–C4–O3	118.845(20)	C21–C19–C20	121.802(20)
O4–C23–C24	127.401(23)	C5–C6–C1	120.284(22)	C21–C19–C10	120.599(20)
O3–C23–O4	122.363(23)	C4–C5–C6	119.064(20)	C20–C19–C10	117.599(21)
O3–C23–C24	110.233(20)	C4–C3–C2	118.064(22)	C2–C1–C6	120.283(21)
O2–S1–N1	103.809(18)	C4–O3–C23	116.386(20)	C19–C10–C9	121.591(20)
O2–S1–C8	109.453(17)	C3–C4–O3	119.306(22)	C19–C20–C7	122.271(21)
O2–S1–O1	120.096(20)	C3–C22–C16	119.455(19)	C18–C13–C14	119.822(20)
O1–S1–C8	107.803(18)	C3–C4–C5	121.827(21)	C17–C16–C15	118.471(20)
N2–N1–S1	111.725(19)	C3–C2–C1	120.471(20)	C17–C18–C13	120.410(21)
N2–C22–C16	116.649(21)	C24–C25–C26	124.048(21)	C16–C17–C18	120.368(20)
N1–N2–C22	115.973(22)	C23–C24–C25	121.793(21)	C16–C15–C14	120.771(21)

continued on next page

Table A.9 – continued from previous page

N1–S1–O1	106.800(19)	C22–C3–C2	119.567(19)	C15–C14–C13	120.15(2)
C9–C8–C7	119.515(21)	C22–C3–C4	122.366(20)		
C8–C7–C20	118.836(20)	C22–C16–C15	120.91(2)		

3.1-Ph

Table A.10.: Structural parameters of 3.1-Ph:

molecular formula	$C_{22}H_{16}O_2$
molecular weight [a.m.u.]	312.35
colour/habit	colourless block
crystal dimensions [mm^3]	$0.45 \times 0.35 \times 0.25$
crystal system	triclinic
space group	P-1
unit cell dimensions	a = 5.6455(7) Å $\alpha = 106.515(9)°$
	b = 9.7737(12) Å $\beta = 93.748(9)°$
	c = 15.2205(17) Å $\gamma = 96.692(10)°$
cell volume [$Å^3$]	795.41(17)
Z	2
T [K]	150(2)
calculated density [gcm^{-3}]	1.304
absorption coefficient μ [mm^{-1}]	0.082
F(000)	328
Θ range [°]	5.12 - 27.10
index ranges [h, k, l]	$\pm7, \pm12, \pm19$
no. of independent reflections / R_{int}	3446
no. of absorbed reflections	2961
no. of data	3446
no. of restraints	0

continued on next page

Table A.10 – continued from previous page

no. of parameters	281
R_1 $(I > 2\sigma(I))$	0.0466
wR_2 $(I > 2\sigma(I))$	0.1108
R_1 (all data)	0.0550
wR_2 (all data)	0.1137
goodness of fit (on F^2)	1.139
largest difference peak [e$\mathring{A}^3$]	0.238
largest difference hole [e$\mathring{A}^3$]	-0.218

Table A.11.: Bond lengths of **3.1-Ph** [$\mathring{A}$]:

O1–C9	1.2016(19)	C7–C17	1.499(2)	C14–C15	1.376(3)
O2–C9	1.3656(19)	C7–C8	1.517(2)	C15–C16	1.385(2)
O2–C1	1.3996(19)	C7–C10	1.530(2)	C17–C18	1.388(2)
C1–C2	1.382(2)	C8–C9	1.472(2)	C17–C22	1.393(2)
C1–C6	1.388(2)	C8–C10	1.522(2)	C18–C19	1.389(2)
C2–C3	1.381(3)	C10–C11	1.488(2)	C19–C20	1.380(2)
C3–C4	1.381(3)	C11–C12	1.391(2)	C20–C21	1.382(2)
C4–C5	1.386(2)	C11–C16	1.391(2)	C21–C22	1.387(2)
C5–C6	1.393(2)	C12–C13	1.385(3)		
C6–C7	1.490(2)	C13–C14	1.380(3)		

Table A.12.: Bond angles of **3.1-Ph** [°]:

C9–O2–C1	121.84(12)	C6–C7–C10	114.54(12)	C16–C11–C10	117.65(14)
C2–C1–C6	121.87(15)	C17–C7–C10	119.68(12)	C13–C12–C11	120.38(18)
C2–C1–O2	114.91(14)	C8–C7–C10	59.94(10)	C14–C13–C12	120.52(18)
C6–C1–O2	123.22(14)	C9–C8–C7	121.07(13)	C15–C14–C13	119.75(17)
C3–C2–C1	119.60(16)	C9–C8–C10	116.86(13)	C14–C15–C16	119.97(18)

continued on next page

Table A.12 – continued from previous page

C4–C3–C2	119.99(16)	C7–C8–C10	60.46(10)	C15–C16–C11	121.01(17)
C3–C4–C5	119.71(17)	O1–C9–O2	117.72(14)	C18–C17–C22	118.53(14)
C4–C5–C6	121.47(15)	O1–C9–C8	124.32(15)	C18–C17–C7	120.92(13)
C1–C6–C5	117.34(14)	O2–C9–C8	117.89(13)	C22–C17–C7	120.49(14)
C1–C6–C7	120.09(14)	C11–C10–C8	121.63(13)	C17–C18–C19	120.77(15)
C5–C6–C7	122.51(13)	C11–C10–C7	122.96(13)	C20–C19–C18	120.08(16)
C6–C7–C17	116.82(12)	C8–C10–C7	59.6(1)	C19–C20–C21	119.80(15)
C6–C7–C8	113.99(12)	C12–C11–C16	118.36(15)	C20–C21–C22	120.19(15)
C17–C7–C8	119.67(13)	C12–C11–C10	123.98(15)	C21–C22–C17	120.63(15)

PtPhOMe–PhOMe

Table A.13.: Structural parameters of **PtPhOMe–PhOMe**:

molecular formula	$C_{53}H_{46}O_4P_2Pt$	
molecular weight [a.m.u.]	1003.96	
crystal system	monoclinic	
space group	$P2_1/c$	
unit cell dimensions	a = 9.3173(3) Å	$\alpha = 90.00°$
	b = 10.1344(3) Å	$\beta = 101.32°$
	c = 47.4630(11) Å	$\gamma = 90.00°$
cell volume [Å^3]	4394.5(2)	
Z	4	
T [K]	293(2)	
calculated density [gcm^{-3}]	1.482	
absorption coefficient μ [mm^{-1}]	3.320	
F(000)	1876	
Θ range [°]	2.97 - 32.79	
index ranges [h, k, l]	+12/-13, +14/-15, +70/-66	

continued on next page

Table A.13 – continued from previous page

no. of independent reflections / R_{int}	14838
no. of absorbed reflections	12336
no. of data	14838
no. of restraints	0
no. of parameters	543
R_1 $(I > 2\sigma(I))$	0.0661
wR_2 $(I > 2\sigma(I))$	0.0817
R_1 (all data)	0.1151
wR_2 (all data)	0.1187
goodness of fit (on F^2)	1.320

Table A.14.: Bond lengths of **PtPhOMe–PhOMe** [Å]:

C1–C6	1.385(8)	C19–P2	1.828(6)	C38–C39	1.457(8)
C1–C2	1.390(8)	C20–C21	1.389(9)	C38–Pt1	2.140(5)
C1–P1	1.829(5)	C20–H20	0.9300	C38–H38	0.9300
C2–C3	1.384(9)	C21–C22	1.387(10)	C39–O1	1.200(7)
C2–H2	0.9300	C21–H21	0.9300	C39–O2	1.385(7)
C3–C4	1.378(9)	C22–C23	1.372(10)	C40–C41	1.374(8)
C3–H3	0.9300	C22–H22	0.9300	C40–C45	1.378(9)
C4–C5	1.387(9)	C23–C24	1.398(9)	C40–O2	1.399(7)
C4–H4	0.9300	C23–H23	0.9300	C41–C42	1.385(8)
C5–C6	1.381(8)	C24–H24	0.9300	C41–H41	0.9300
C5–H5	0.9300	C25–C26	1.388(8)	C42–C43	1.381(8)
C6–H6	0.9300	C25–C30	1.395(8)	C42–H42	0.9300
C7–C8	1.391(8)	C25–P2	1.828(5)	C43–O3	1.371(7)
C7–C12	1.392(8)	C26–C27	1.394(8)	C43–C44	1.399(8)
C7–P1	1.826(5)	C26–H26	0.9300	C44–C45	1.371(9)
C8–C9	1.387(8)	C27–C28	1.379(9)	C44–H44	0.9300

continued on next page

Table A.14 – continued from previous page

C8–H8	0.9300	C27–H27	0.9300	C45–H45	0.9300
C9–C10	1.366(10)	C28–C29	1.378(10)	C46–O3	1.425(8)
C9–H9	0.9300	C28–H28	0.9300	C46–H46A	0.9600
C10–C11	1.384(10)	C29–C30	1.377(8)	C46–H46B	0.9600
C10–H10	0.9300	C29–H29	0.9300	C46–H46C	0.9600
C11–C12	1.380(8)	C30–H30	0.9300	C47–C48	1.390(7)
C11–H11	0.9300	C31–C32	1.389(7)	C47–C52	1.397(8)
C12–H12	0.9300	C31–C36	1.392(7)	C48–C49	1.394(8)
C13–C18	1.393(8)	C31–P2	1.841(5)	C48–H48	0.9300
C13–C14	1.395(8)	C32–C33	1.388(8)	C49–C50	1.381(8)
C13–P1	1.823(5)	C32–H32	0.9300	C49–H49	0.9300
C14–C15	1.385(8)	C33–C34	1.377(9)	C50–O4	1.377(7)
C14–H14	0.9300	C33–H33	0.9300	C50–C51	1.385(8)
C15–C16	1.376(10)	C34–C35	1.389(9)	C51–C52	1.381(8)
C15–H15	0.9300	C34–H34	0.9300	C51–H51	0.9300
C16–C17	1.388(10)	C35–C36	1.393(9)	C52–H52	0.9300
C16–H16	0.9300	C35–H35	0.9300	C53–O4	1.404(8)
C17–C18	1.381(8)	C36–H36	0.9300	C53–H53A	0.9600
C17–H17	0.9300	C37–C38	1.439(7)	C53–H53B	0.9600
C18–H18	0.9300	C37–C47	1.480(7)	C53–H53C	0.9600
C19–C24	1.379(8)	C37–Pt1	2.132(5)	P1–Pt1	2.2793(13)
C19–C20	1.389(8)	C37–H37	0.9300	P2–Pt1	2.2909(13)

Table A.15.: Bond angles of **PtPhOMe–PhOMe** [°]:

C6–C1–C2	118.1(5)	C23–C22–H22	119.800	C42–C41–H41	119.800
C6–C1–P1	120.7(4)	C21–C22–H22	119.800	C43–C42–C41	119.7(5)
C2–C1–P1	121.2(4)	C22–C23–C24	119.6(6)	C43–C42–H42	120.100
C3–C2–C1	120.7(6)	C22–C23–H23	120.200	C41–C42–H42	120.100

continued on next page

Table A.15 – continued from previous page

C3–C2–H2	119.700	C24–C23–H23	120.200	O3–C43–C42	124.4(5)
C1–C2–H2	119.700	C19–C24–C23	120.9(6)	O3–C43–C44	116.1(5)
C4–C3–C2	120.6(6)	C19–C24–H24	119.500	C42–C43–C44	119.5(5)
C4–C3–H3	119.700	C23–C24–H24	119.500	C45–C44–C43	120.0(6)
C2–C3–H3	119.700	C26–C25–C30	118.3(5)	C45–C44–H44	120.000
C3–C4–C5	119.2(6)	C26–C25–P2	122.1(4)	C43–C44–H44	120.000
C3–C4–H4	120.400	C30–C25–P2	119.6(4)	C44–C45–C40	120.3(6)
C5–C4–H4	120.400	C25–C26–C27	120.3(6)	C44–C45–H45	119.900
C6–C5–C4	119.9(6)	C25–C26–H26	119.800	C40–C45–H45	119.900
C6–C5–H5	120.000	C27–C26–H26	119.800	O3–C46–H46A	109.500
C4–C5–H5	120.000	C28–C27–C26	120.4(6)	O3–C46–H46B	109.500
C5–C6–C1	121.5(5)	C28–C27–H27	119.800	H46A–C46–H46B	109.500
C5–C6–H6	119.300	C26–C27–H27	119.800	O3–C46–H46C	109.500
C1–C6–H6	119.300	C29–C28–C27	119.6(6)	H46A–C46–H46C	109.500
C8–C7–C12	118.3(5)	C29–C28–H28	120.200	H46B–C46–H46C	109.500
C8–C7–P1	124.8(4)	C27–C28–H28	120.200	C48–C47–C52	117.1(5)
C12–C7–P1	116.9(4)	C30–C29–C28	120.3(6)	C48–C47–C37	119.9(5)
C9–C8–C7	120.6(6)	C30–C29–H29	119.900	C52–C47–C37	123.0(5)
C9–C8–H8	119.700	C28–C29–H29	119.900	C47–C48–C49	122.0(5)
C7–C8–H8	119.700	C29–C30–C25	121.1(6)	C47–C48–H48	119.000
C10–C9–C8	120.3(6)	C29–C30–H30	119.400	C49–C48–H48	119.000
C10–C9–H9	119.900	C25–C30–H30	119.400	C50–C49–C48	119.3(5)
C8–C9–H9	119.900	C32–C31–C36	119.3(5)	C50–C49–H49	120.300
C9–C10–C11	120.0(6)	C32–C31–P2	123.6(4)	C48–C49–H49	120.300
C9–C10–H10	120.000	C36–C31–P2	117.1(4)	O4–C50–C49	124.1(5)
C11–C10–H10	120.000	C33–C32–C31	120.8(5)	O4–C50–C51	116.1(5)
C12–C11–C10	120.0(6)	C33–C32–H32	119.600	C49–C50–C51	119.8(5)
C12–C11–H11	120.000	C31–C32–H32	119.600	C52–C51–C50	120.3(5)

continued on next page

Table A.15 – continued from previous page

C10–C11–H11	120.000	C34–C33–C32	119.7(5)	C52–C51–H51	119.900
C11–C12–C7	120.7(6)	C34–C33–H33	120.200	C50–C51–H51	119.900
C11–C12–H12	119.700	C32–C33–H33	120.200	C51–C52–C47	121.4(5)
C7–C12–H12	119.700	C33–C34–C35	120.4(6)	C51–C52–H52	119.300
C18–C13–C14	118.4(5)	C33–C34–H34	119.800	C47–C52–H52	119.300
C18–C13–P1	116.9(4)	C35–C34–H34	119.800	O4–C53–H53A	109.500
C14–C13–P1	124.5(4)	C34–C35–C36	119.9(6)	O4–C53–H53B	109.500
C15–C14–C13	120.2(6)	C34–C35–H35	120.000	H53A–C53–H53B	109.500
C15–C14–H14	119.900	C36–C35–H35	120.000	O4–C53–H53C	109.500
C13–C14–H14	119.900	C31–C36–C35	120.0(5)	H53A–C53–H53C	109.500
C16–C15–C14	120.7(6)	C31–C36–H36	120.000	H53B–C53–H53C	109.500
C16–C15–H15	119.700	C35–C36–H36	120.000	C39–O2–C40	120.6(5)
C14–C15–H15	119.700	C38–C37–C47	121.6(5)	C43–O3–C46	117.6(5)
C15–C16–C17	120.0(6)	C38–C37–Pt1	70.6(3)	C50–O4–C53	117.8(5)
C15–C16–H16	120.000	C47–C37–Pt1	114.8(4)	C13–P1–C7	103.3(2)
C17–C16–H16	120.000	C38–C37–H37	119.200	C13–P1–C1	104.6(3)
C18–C17–C16	119.4(6)	C47–C37–H37	119.200	C7–P1–C1	99.7(2)
C18–C17–H17	120.300	Pt1–C37–H37	84.900	C13–P1–Pt1	114.21(18)
C16–C17–H17	120.300	C37–C38–C39	120.2(5)	C7–P1–Pt1	112.02(18)
C17–C18–C13	121.4(6)	C37–C38–Pt1	70.0(3)	C1–P1–Pt1	120.83(19)
C17–C18–H18	119.300	C39–C38–Pt1	111.4(4)	C25–P2–C19	100.3(2)
C13–C18–H18	119.300	C37–C38–H38	119.900	C25–P2–C31	102.1(2)
C24–C19–C20	118.6(5)	C39–C38–H38	119.900	C19–P2–C31	105.7(3)
C24–C19–P2	123.8(5)	Pt1–C38–H38	88.600	C25–P2–Pt1	119.65(18)
C20–C19–P2	117.6(4)	O1–C39–O2	121.5(5)	C19–P2–Pt1	117.44(18)
C21–C20–C19	121.1(6)	O1–C39–C38	129.3(5)	C31–P2–Pt1	109.76(17)
C21–C20–H20	119.400	O2–C39–C38	109.1(5)	C37–Pt1–C38	39.37(19)
C19–C20–H20	119.400	C41–C40–C45	120.1(6)	C37–Pt1–P1	100.65(15)

continued on next page

Table A.15 – continued from previous page

C22–C21–C20	119.3(6)	C41–C40–O2	118.1(6)	C38–Pt1–P1	139.95(14)
C22–C21–H21	120.300	C45–C40–O2	121.4(6)	C37–Pt1–P2	148.85(15)
C20–C21–H21	120.300	C40–C41–C42	120.4(6)	C38–Pt1–P2	109.47(14)
C23–C22–C21	120.4(6)	C40–C41–H41	119.800	P1–Pt1–P2	110.47(5)

PtPhMe−Ph

Table A.16.: Structural parameters of **PtPhMe−Ph**:

molecular formula	$C_{52}H_{44}O_2P_2Pt$
molecular weight [a.m.u.]	957.90
colour/habit	yellow fragment
crystal system	triclinic
space group	P-1
unit cell dimensions	a = 13.1504(19) Å $\alpha = 83.120(7)°$
	b = 13.584(2) Å $\beta = 83.981(7)°$
	c = 24.442(4) Å $\gamma = 85.467(6)°$
cell volume [Å^3]	4301.1(11)
Z	4
T [K]	173(2)
calculated density [gcm^{-3}]	1.479
absorption coefficient μ [mm^{-1}]	3.377
F(000)	1920
Θ range [°]	0.84 - 29.51
index ranges [h, k, l]	±18, ±18, ±33
no. of independent reflections / R_{int}	23692
no. of absorbed reflections	18689
no. of data	23692
no. of restraints	0

continued on next page

Table A.16 – continued from previous page

no. of parameters	1379
R_1 $(I > 2\sigma(I))$	0.0324
wR_2 $(I > 2\sigma(I))$	0.0650
R_1 (all data)	0.0509
wR_2 (all data)	0.0716
goodness of fit (on F^2)	1.000

Table A.17.: Bond lengths of **PtPhMe−Ph** [Å]:

Pt1–C45	2.118(3)	C26–H26	0.93(4)	C66–C67	1.386(5)
Pt1–C44	2.139(3)	C27–C28	1.383(7)	C66–H66	0.90(4)
Pt1–P1	2.2771(8)	C27–H27	0.88(4)	C67–C68	1.385(6)
Pt1–P2	2.2986(8)	C28–C29	1.370(7)	C67–H67	0.95(4)
P2–C31	1.823(3)	C28–H28	0.97(4)	C68–C69	1.373(6)
P2–C19	1.839(3)	C29–C30	1.388(5)	C68–H68	0.88(4)
P2–C25	1.845(3)	C29–H29	0.97(4)	C69–C70	1.393(5)
P1–C1	1.823(3)	C30–H30	0.94(4)	C69–H69	0.85(5)
P1–C7	1.834(3)	C31–C36	1.389(5)	C70–H70	0.89(4)
P1–C13	1.834(3)	C31–C32	1.400(5)	C71–C72	1.397(5)
Pt2–C97	2.120(3)	C32–C33	1.390(5)	C71–C76	1.400(5)
Pt2–C96	2.133(3)	C32–H32	0.99(4)	C72–C73	1.386(5)
Pt2–P3	2.2776(8)	C33–C34	1.383(6)	C72–H72	0.97(3)
Pt2–P4	2.2915(8)	C33–H33	0.93(4)	C73–C74	1.383(6)
P4–C77	1.829(3)	C34–C35	1.373(6)	C73–H73	0.95(4)
P4–C71	1.829(3)	C34–H34	0.93(5)	C74–C75	1.387(6)
P4–C83	1.835(3)	C35–C36	1.390(5)	C74–H74	0.91(4)
P3–C53	1.833(3)	C35–H35	0.87(4)	C75–C76	1.389(5)
P3–C59	1.836(3)	C36–H36	0.92(4)	C75–H75	0.92(5)
P3–C65	1.839(3)	C37–C42	1.377(5)	C76–H76	0.93(4)

continued on next page

Table A.17 – continued from previous page

O3–C95	1.382(4)	C37–C38	1.384(5)	C77–C78	1.393(5)
O3–C89	1.407(4)	C38–C39	1.396(5)	C77–C82	1.396(5)
O4–C95	1.211(4)	C38–H38	0.92(4)	C78–C79	1.408(6)
O1–C43	1.206(4)	C39–C40	1.375(6)	C78–H78	0.86(4)
O2–C43	1.390(4)	C39–H39	0.90(4)	C79–C80	1.379(7)
O2–C37	1.399(4)	C40–C41	1.384(6)	C79–H79	0.91(5)
C1–C2	1.402(5)	C40–H40	0.90(4)	C80–C81	1.384(6)
C1–C6	1.403(5)	C41–C42	1.390(5)	C80–H80	0.92(5)
C2–C3	1.395(5)	C41–H41	0.91(5)	C81–C82	1.389(5)
C2–H2	0.92(4)	C42–H42	0.92(4)	C81–H81	0.97(5)
C3–C4	1.381(6)	C43–C44	1.467(5)	C82–H82	0.97(4)
C3–H3	0.91(4)	C44–C45	1.452(4)	C83–C88	1.387(4)
C4–C5	1.378(6)	C44–H44	0.94(3)	C83–C84	1.400(4)
C4–H4	0.91(4)	C45–C46	1.483(4)	C84–C85	1.391(5)
C5–C6	1.387(5)	C45–H45	0.95(4)	C84–H84	0.90(4)
C5–H5	0.89(4)	C46–C51	1.399(4)	C85–C86	1.386(5)
C6–H6	0.91(3)	C46–C47	1.401(5)	C85–H85	0.92(4)
C7–C8	1.392(5)	C47–C48	1.393(5)	C86–C87	1.368(5)
C7–C12	1.399(5)	C47–H47	0.89(4)	C86–H86	0.94(4)
C8–C9	1.398(5)	C48–C49	1.393(5)	C87–C88	1.401(5)
C8–H8	0.97(4)	C48–H48	0.94(3)	C87–H87	0.91(4)
C9–C10	1.382(6)	C49–C50	1.385(5)	C88–H88	0.94(4)
C9–H9	0.95(4)	C49–C52	1.506(5)	C89–C94	1.382(4)
C10–C11	1.387(6)	C50–C51	1.394(5)	C89–C90	1.384(5)
C10–H10	0.92(4)	C50–H50	0.97(4)	C90–C91	1.390(5)
C11–C12	1.393(5)	C51–H51	0.93(4)	C90–H90	0.95(4)
C11–H11	1.00(4)	C52–H52A	0.92(5)	C91–C92	1.381(5)
C12–H12	0.96(3)	C52–H52B	0.97(5)	C91–H91	0.93(4)

continued on next page

Table A.17 – continued from previous page

C13–C18	1.375(5)	C52–H52C	0.97(6)	C92–C93	1.384(5)
C13–C14	1.390(5)	C53–C58	1.395(5)	C92–H92	0.89(4)
C14–C15	1.393(5)	C53–C54	1.401(5)	C93–C94	1.390(5)
C14–H14	0.93(4)	C54–C55	1.388(5)	C93–H93	0.95(4)
C15–C16	1.363(6)	C54–H54	0.97(4)	C94–H94	0.95(4)
C15–H15	0.92(4)	C55–C56	1.385(5)	C95–C96	1.463(4)
C16–C17	1.382(6)	C55–H55	0.90(4)	C96–C97	1.443(4)
C16–H16	0.95(4)	C56–C57	1.376(6)	C96–H96	0.89(3)
C17–C18	1.389(5)	C56–H56	0.88(4)	C97–C98	1.472(4)
C17–H17	0.94(5)	C57–C58	1.397(5)	C97–H97	0.90(3)
C18–H18	0.96(5)	C57–H57	0.88(5)	C98–C103	1.402(4)
C19–C20	1.389(5)	C58–H58	0.87(4)	C98–C99	1.409(4)
C19–C24	1.395(5)	C59–C64	1.395(4)	C99–C100	1.388(5)
C20–C21	1.390(6)	C59–C60	1.399(4)	C99–H99	0.99(3)
C20–H20	0.93(4)	C60–C61	1.384(5)	C100–C101	1.398(5)
C21–C22	1.381(6)	C60–H60	0.95(4)	C100–H100	0.89(4)
C21–H21	0.98(5)	C61–C62	1.389(5)	C101–C102	1.394(5)
C22–C23	1.378(6)	C61–H61	0.91(4)	C101–C104	1.511(5)
C22–H22	0.97(4)	C62–C63	1.386(5)	C102–C103	1.394(5)
C23–C24	1.383(5)	C62–H62	0.87(4)	C102–H102	0.91(3)
C23–H23	0.90(4)	C63–C64	1.399(5)	C103–H103	0.92(3)
C24–H24	0.92(4)	C63–H63	0.90(4)	C104–H104	0.96(5)
C25–C30	1.388(5)	C64–H64	0.99(3)	C104–H105	0.91(6)
C25–C26	1.398(5)	C65–C70	1.389(4)	C104–H106	0.87(6)
C26–C27	1.390(5)	C65–C66	1.398(5)		

Table A.18.: Bond angles of **PtPhMe−Ph** [°]:

C45–Pt1–C44	39.87(12)	C46–C47–H47	122.(3)
C45–Pt1–P1	100.99(9)	C47–C48–C49	121.3(3)
C44–Pt1–P1	140.73(9)	C47–C48–H48	122.(2)
C45–Pt1–P2	147.09(9)	C49–C48–H48	116.(2)
C44–Pt1–P2	107.96(9)	C50–C49–C48	117.7(3)
P1–Pt1–P2	111.27(3)	C50–C49–C52	121.5(3)
C31–P2–C19	105.77(15)	C48–C49–C52	120.7(4)
C31–P2–C25	99.47(15)	C49–C50–C51	121.5(3)
C19–P2–C25	102.01(15)	C49–C50–H50	121.(2)
C31–P2–Pt1	116.87(11)	C51–C50–H50	118.(2)
C19–P2–Pt1	118.65(10)	C50–C51–C46	121.1(3)
C25–P2–Pt1	111.44(10)	C50–C51–H51	120.(2)
C1–P1–C7	102.18(15)	C46–C51–H51	119.(2)
C1–P1–C13	104.43(15)	C49–C52–H52A	114.(3)
C7–P1–C13	100.22(14)	C49–C52–H52B	107.(3)
C1–P1–Pt1	114.78(10)	H52A–C52–H52B	111.(4)
C7–P1–Pt1	115.35(10)	C49–C52–H52C	116.(3)
C13–P1–Pt1	117.69(10)	H52A–C52–H52C	107.(4)
C97–Pt2–C96	39.67(11)	H52B–C52–H52C	102.(4)
C97–Pt2–P3	100.17(8)	C58–C53–C54	118.1(3)
C96–Pt2–P3	139.40(8)	C58–C53–P3	123.3(3)
C97–Pt2–P4	144.43(8)	C54–C53–P3	118.6(2)
C96–Pt2–P4	105.57(8)	C55–C54–C53	120.7(3)
P3–Pt2–P4	115.02(3)	C55–C54–H54	121.(2)
C77–P4–C71	106.04(15)	C53–C54–H54	118.(2)
C77–P4–C83	101.59(14)	C56–C55–C54	120.6(4)
C71–P4–C83	100.25(14)	C56–C55–H55	125.(2)

continued on next page

Table A.18 – continued from previous page

C77–P4–Pt2	117.95(11)	C54–C55–H55	114.(2)
C71–P4–Pt2	116.13(11)	C57–C56–C55	119.5(3)
C83–P4–Pt2	112.52(10)	C57–C56–H56	116.(3)
C53–P3–C59	101.52(14)	C55–C56–H56	125.(3)
C53–P3–C65	102.25(14)	C56–C57–C58	120.5(4)
C59–P3–C65	105.00(14)	C56–C57–H57	122.(3)
C53–P3–Pt2	113.32(10)	C58–C57–H57	117.(3)
C59–P3–Pt2	116.13(10)	C53–C58–C57	120.7(4)
C65–P3–Pt2	116.65(10)	C53–C58–H58	122.(2)
C95–O3–C89	116.6(3)	C57–C58–H58	117.(2)
C43–O2–C37	119.1(3)	C64–C59–C60	118.6(3)
C2–C1–C6	118.6(3)	C64–C59–P3	122.6(2)
C2–C1–P1	118.0(2)	C60–C59–P3	118.7(2)
C6–C1–P1	123.4(3)	C61–C60–C59	121.1(3)
C3–C2–C1	120.2(3)	C61–C60–H60	123.(2)
C3–C2–H2	122.(2)	C59–C60–H60	116.(2)
C1–C2–H2	118.(2)	C60–C61–C62	120.0(3)
C4–C3–C2	120.2(4)	C60–C61–H61	119.(2)
C4–C3–H3	123.(2)	C62–C61–H61	121.(2)
C2–C3–H3	116.(2)	C63–C62–C61	119.8(3)
C5–C4–C3	120.1(4)	C63–C62–H62	121.(2)
C5–C4–H4	118.(3)	C61–C62–H62	119.(2)
C3–C4–H4	122.(3)	C62–C63–C64	120.3(3)
C4–C5–C6	120.6(4)	C62–C63–H63	122.(2)
C4–C5–H5	118.(2)	C64–C63–H63	118.(2)
C6–C5–H5	121.(2)	C59–C64–C63	120.3(3)
C5–C6–C1	120.3(3)	C59–C64–H64	120.(2)
C5–C6–H6	120.(2)	C63–C64–H64	120.(2)

continued on next page

Table A.18 – continued from previous page

C1–C6–H6	119.(2)	C70–C65–C66	118.7(3)
C8–C7–C12	118.9(3)	C70–C65–P3	123.0(3)
C8–C7–P1	123.2(3)	C66–C65–P3	118.3(2)
C12–C7–P1	117.8(2)	C67–C66–C65	120.4(3)
C7–C8–C9	120.2(4)	C67–C66–H66	121.(2)
C7–C8–H8	116.(2)	C65–C66–H66	119.(2)
C9–C8–H8	124.(2)	C68–C67–C66	120.2(4)
C10–C9–C8	120.3(4)	C68–C67–H67	123.(3)
C10–C9–H9	121.(3)	C66–C67–H67	117.(3)
C8–C9–H9	119.(3)	C69–C68–C67	119.8(4)
C9–C10–C11	120.1(4)	C69–C68–H68	122.(3)
C9–C10–H10	123.(3)	C67–C68–H68	118.(3)
C11–C10–H10	117.(3)	C68–C69–C70	120.5(4)
C10–C11–C12	119.8(4)	C68–C69–H69	119.(3)
C10–C11–H11	121.(2)	C70–C69–H69	121.(3)
C12–C11–H11	119.(2)	C65–C70–C69	120.4(4)
C11–C12–C7	120.7(3)	C65–C70–H70	118.(3)
C11–C12–H12	120.(2)	C69–C70–H70	122.(3)
C7–C12–H12	119.(2)	C72–C71–C76	118.3(3)
C18–C13–C14	118.3(3)	C72–C71–P4	118.9(2)
C18–C13–P1	118.7(3)	C76–C71–P4	122.7(3)
C14–C13–P1	123.0(3)	C73–C72–C71	121.2(3)
C13–C14–C15	120.6(3)	C73–C72–H72	118.(2)
C13–C14–H14	121.(2)	C71–C72–H72	120.(2)
C15–C14–H14	119.(2)	C74–C73–C72	119.6(4)
C16–C15–C14	120.7(4)	C74–C73–H73	124.(2)
C16–C15–H15	119.(3)	C72–C73–H73	116.(2)
C14–C15–H15	121.(3)	C73–C74–C75	120.3(4)

continued on next page

Table A.18 – continued from previous page

C15–C16–C17	119.1(4)	C73–C74–H74	117.(2)
C15–C16–H16	122.(3)	C75–C74–H74	122.(2)
C17–C16–H16	118.(3)	C74–C75–C76	119.9(4)
C16–C17–C18	120.4(4)	C74–C75–H75	120.(3)
C16–C17–H17	119.(3)	C76–C75–H75	120.(3)
C18–C17–H17	120.(3)	C75–C76–C71	120.6(4)
C13–C18–C17	120.9(4)	C75–C76–H76	119.(3)
C13–C18–H18	119.(3)	C71–C76–H76	121.(3)
C17–C18–H18	120.(3)	C78–C77–C82	118.3(3)
C20–C19–C24	117.5(3)	C78–C77–P4	124.4(3)
C20–C19–P2	125.4(3)	C82–C77–P4	117.2(2)
C24–C19–P2	117.0(3)	C77–C78–C79	120.1(4)
C19–C20–C21	121.4(4)	C77–C78–H78	123.(3)
C19–C20–H20	122.(3)	C79–C78–H78	117.(3)
C21–C20–H20	117.(3)	C80–C79–C78	120.2(4)
C22–C21–C20	120.0(4)	C80–C79–H79	121.(3)
C22–C21–H21	115.(3)	C78–C79–H79	119.(3)
C20–C21–H21	125.(3)	C79–C80–C81	120.3(4)
C23–C22–C21	119.4(4)	C79–C80–H80	121.(3)
C23–C22–H22	119.(3)	C81–C80–H80	119.(3)
C21–C22–H22	121.(3)	C80–C81–C82	119.4(4)
C22–C23–C24	120.5(4)	C80–C81–H81	120.(3)
C22–C23–H23	120.(3)	C82–C81–H81	121.(3)
C24–C23–H23	120.(3)	C81–C82–C77	121.6(4)
C23–C24–C19	121.1(3)	C81–C82–H82	120.(2)
C23–C24–H24	118.(2)	C77–C82–H82	118.(2)
C19–C24–H24	121.(2)	C88–C83–C84	118.3(3)
C30–C25–C26	118.5(3)	C88–C83–P4	123.9(2)

continued on next page

Table A.18 – continued from previous page

C30–C25–P2	124.6(3)	C84–C83–P4	117.7(2)
C26–C25–P2	116.9(3)	C85–C84–C83	120.7(3)
C27–C26–C25	120.7(4)	C85–C84–H84	120.(2)
C27–C26–H26	121.(3)	C83–C84–H84	119.(2)
C25–C26–H26	119.(3)	C86–C85–C84	120.1(3)
C28–C27–C26	119.8(4)	C86–C85–H85	121.(2)
C28–C27–H27	121.(3)	C84–C85–H85	119.(2)
C26–C27–H27	120.(3)	C87–C86–C85	119.9(3)
C29–C28–C27	120.0(4)	C87–C86–H86	119.(3)
C29–C28–H28	119.(3)	C85–C86–H86	121.(3)
C27–C28–H28	121.(3)	C86–C87–C88	120.4(3)
C28–C29–C30	120.6(4)	C86–C87–H87	123.(2)
C28–C29–H29	122.(3)	C88–C87–H87	117.(2)
C30–C29–H29	117.(3)	C83–C88–C87	120.7(3)
C25–C30–C29	120.5(4)	C83–C88–H88	122.(2)
C25–C30–H30	120.(3)	C87–C88–H88	118.(2)
C29–C30–H30	120.(3)	C94–C89–C90	121.4(3)
C36–C31–C32	118.5(3)	C94–C89–O3	117.3(3)
C36–C31–P2	119.3(2)	C90–C89–O3	121.2(3)
C32–C31–P2	122.0(3)	C89–C90–C91	119.0(3)
C33–C32–C31	120.3(4)	C89–C90–H90	119.(2)
C33–C32–H32	119.(2)	C91–C90–H90	122.(3)
C31–C32–H32	120.(2)	C92–C91–C90	120.4(4)
C34–C33–C32	120.0(4)	C92–C91–H91	125.(3)
C34–C33–H33	121.(2)	C90–C91–H91	114.(3)
C32–C33–H33	119.(2)	C91–C92–C93	120.0(4)
C35–C34–C33	120.4(3)	C91–C92–H92	118.(3)
C35–C34–H34	123.(3)	C93–C92–H92	122.(3)

continued on next page

Table A.18 – continued from previous page

C33–C34–H34	117.(3)	C92–C93–C94	120.4(3)
C34–C35–C36	119.9(4)	C92–C93–H93	118.(3)
C34–C35–H35	119.(3)	C94–C93–H93	122.(3)
C36–C35–H35	121.(3)	C89–C94–C93	119.0(3)
C31–C36–C35	120.9(4)	C89–C94–H94	121.(2)
C31–C36–H36	123.(2)	C93–C94–H94	120.(2)
C35–C36–H36	116.(2)	O4–C95–O3	121.7(3)
C42–C37–C38	121.3(3)	O4–C95–C96	127.8(3)
C42–C37–O2	116.7(3)	O3–C95–C96	110.5(3)
C38–C37–O2	121.8(3)	C97–C96–C95	118.8(3)
C37–C38–C39	118.7(4)	C97–C96–Pt2	69.70(16)
C37–C38–H38	119.(3)	C95–C96–Pt2	110.6(2)
C39–C38–H38	122.(3)	C97–C96–H96	122.(2)
C40–C39–C38	120.4(4)	C95–C96–H96	113.(2)
C40–C39–H39	122.(3)	Pt2–C96–H96	114.(2)
C38–C39–H39	118.(3)	C96–C97–C98	122.2(3)
C39–C40–C41	120.2(4)	C96–C97–Pt2	70.63(16)
C39–C40–H40	123.(3)	C98–C97–Pt2	112.9(2)
C41–C40–H40	117.(3)	C96–C97–H97	116.(2)
C40–C41–C42	120.0(4)	C98–C97–H97	114.(2)
C40–C41–H41	120.(3)	Pt2–C97–H97	112.5(19)
C42–C41–H41	120.(3)	C103–C98–C99	116.9(3)
C37–C42–C41	119.4(4)	C103–C98–C97	120.9(3)
C37–C42–H42	122.(3)	C99–C98–C97	122.2(3)
C41–C42–H42	119.(3)	C100–C99–C98	121.0(3)
O1–C43–O2	121.8(3)	C100–C99–H99	120.8(19)
O1–C43–C44	129.0(3)	C98–C99–H99	118.2(19)
O2–C43–C44	109.1(3)	C99–C100–C101	121.9(3)

continued on next page

Table A.18 – continued from previous page

C45–C44–C43	118.3(3)	C99–C100–H100	119.(2)
C45–C44–Pt1	69.27(18)	C101–C100–H100	119.(2)
C43–C44–Pt1	115.2(2)	C102–C101–C100	117.2(3)
C45–C44–H44	118.1(19)	C102–C101–C104	121.5(3)
C43–C44–H44	115.5(19)	C100–C101–C104	121.3(4)
Pt1–C44–H44	112.(2)	C101–C102–C103	121.4(3)
C44–C45–C46	121.3(3)	C101–C102–H102	124.(2)
C44–C45–Pt1	70.85(18)	C103–C102–H102	115.(2)
C46–C45–Pt1	115.5(2)	C102–C103–C98	121.5(3)
C44–C45–H45	116.(2)	C102–C103–H103	121.(2)
C46–C45–H45	113.(2)	C98–C103–H103	118.(2)
Pt1–C45–H45	112.(2)	C101–C104–H104	109.(3)
C51–C46–C47	117.2(3)	C101–C104–H105	111.(3)
C51–C46–C45	119.9(3)	H104–C104–H105	108.(4)
C47–C46–C45	122.8(3)	C101–C104–H106	113.(4)
C48–C47–C46	121.1(3)	H104–C104–H106	115.(5)
C48–C47–H47	117.(3)	H105–C104–H106	101.(5)

PtPh–PhMe

Table A.19.: Structural parameters of **PtPh–PhMe**:

molecular formula	$C_{52}H_{44}O_2P_2Pt$
molecular weight [a.m.u.]	957.90
colour/habit	yellow fragment
crystal dimensions [mm^3]	$0.20 \times 0.10 \times 0.08$
crystal system	monoclinic
space group	$P2_1/c$
unit cell dimensions	a = 9.6547(5) Å $\alpha = 90.00°$

continued on next page

Table A.19 – continued from previous page

	b = 18.4881(8) Å	$\beta = 98.323(2)°$
	c = 24.5659(10) Å	$\gamma = 90.00°$
cell volume [Å³]	4338.8(3)	
Z	4	
T [K]	173(2)	
calculated density [gcm^{-3}]	1.466	
absorption coefficient μ [mm^{-1}]	3.348	
F(000)	1920	
Θ range [°]	1.38 - 25.35	
index ranges [h, k, l]	±11, ±22, +29/-27	
no. of independent reflections / R_{int}	7870	
no. of absorbed reflections	7282	
no. of data	7870	
no. of restraints	0	
no. of parameters	678	
R_1 ($I > 2\sigma(I)$)	0.0277	
wR_2 ($I > 2\sigma(I)$)	0.0622	
R_1 (all data)	0.0314	
wR_2 (all data)	0.0634	
goodness of fit (on F^2)	1.268	
largest difference peak [e$\mathring{A}^3$]	1.497	
largest difference hole [e$\mathring{A}^3$]	-0.969	

Table A.20.: Bond lengths of **PtPh−PhMe** [Å]:

Pt1–C38	2.127(4)	C14–C15	1.386(6)	C34–C35	1.379(6)
Pt1–C37	2.128(4)	C14–H11	0.93(4)	C34–H28	0.98(4)
Pt1–P2	2.2782(9)	C15–C16	1.377(7)	C35–C36	1.386(6)
Pt1–P1	2.2826(9)	C15–H12	0.91(5)	C35–H29	0.88(4)

continued on next page

Table A.20 – continued from previous page

P1–C1	1.828(4)	C16–C17	1.390(7)	C36–H30	0.90(4)
P1–C7	1.831(4)	C16–H13	0.89(5)	C37–C38	1.430(5)
P1–C13	1.832(4)	C17–C18	1.378(6)	C37–C39	1.469(5)
P2–C31	1.826(4)	C17–H14	0.89(5)	C37–H31	0.91(4)
P2–C19	1.838(4)	C18–H15	0.85(4)	C38–C45	1.466(5)
P2–C25	1.844(4)	C19–C24	1.386(5)	C38–H32	0.87(4)
O1–C45	1.379(4)	C19–C20	1.399(5)	C39–C40	1.397(5)
O1–C46	1.401(5)	C20–C21	1.389(6)	C39–C44	1.400(5)
O2–C45	1.214(4)	C20–H16	0.93(4)	C40–C41	1.381(6)
C1–C6	1.395(5)	C21–C22	1.374(7)	C40–H33	0.94(4)
C1–C2	1.397(6)	C21–H17	0.89(4)	C41–C42	1.393(7)
C2–C3	1.381(6)	C22–C23	1.380(7)	C41–H34	0.94(5)
C2–H1	0.86(4)	C22–H18	0.97(5)	C42–C43	1.375(6)
C3–C4	1.374(6)	C23–C24	1.383(6)	C42–H35	0.94(4)
C3–H2	0.89(4)	C23–H19	0.91(4)	C43–C44	1.386(6)
C4–C5	1.378(7)	C24–H20	0.88(5)	C43–H36	0.80(5)
C4–H3	0.91(5)	C25–C30	1.392(6)	C44–H37	0.92(4)
C5–C6	1.390(6)	C25–C26	1.398(6)	C46–C51	1.377(6)
C5–H4	0.85(5)	C26–C27	1.395(6)	C46–C47	1.380(6)
C6–H5	0.92(5)	C26–H21	0.90(4)	C47–C48	1.382(6)
C7–C12	1.392(6)	C27–C28	1.378(7)	C47–H38	0.86(5)
C7–C8	1.395(6)	C27–H22	0.91(5)	C48–C49	1.378(6)
C8–C9	1.371(6)	C28–C29	1.377(7)	C48–H39	0.88(5)
C8–H6	0.86(4)	C28–H23	0.97(5)	C49–C50	1.387(7)
C9–C10	1.380(7)	C29–C30	1.377(6)	C49–C52	1.511(6)
C9–H7	0.94(5)	C29–H24	0.98(5)	C50–C51	1.374(7)
C10–C11	1.373(7)	C30–H25	1.01(5)	C50–H40	0.87(6)
C10–H8	0.91(5)	C31–C36	1.386(5)	C51–H41	0.87(4)

continued on next page

Table A.20 – continued from previous page

C11–C12	1.388(6)	C31–C32	1.400(5)	C52–H42	0.9800
C11–H9	0.95(5)	C32–C33	1.380(6)	C52–H43	0.9800
C12–H10	0.90(4)	C32–H26	0.89(5)	C52–H44	0.9800
C13–C14	1.379(5)	C33–C34	1.382(6)		
C13–C18	1.400(5)	C33–H27	0.90(5)		

Table A.21.: Bond angles of **PtPh–PhMe** [°]:

C38–Pt1–C37	39.28(14)	C16–C15–H12	120.(3)	C35–C36–C31	120.5(4)
C38–Pt1–P2	139.4(1)	C14–C15–H12	120.(3)	C35–C36–H30	120.(2)
C37–Pt1–P2	101.74(10)	C15–C16–C17	119.7(4)	C31–C36–H30	119.(2)
C38–Pt1–P1	112.47(11)	C15–C16–H13	122.(4)	C38–C37–C39	122.7(3)
C37–Pt1–P1	150.18(10)	C17–C16–H13	118.(4)	C38–C37–Pt1	70.3(2)
P2–Pt1–P1	107.69(3)	C18–C17–C16	120.0(4)	C39–C37–Pt1	113.5(2)
C1–P1–C7	102.58(17)	C18–C17–H14	119.(3)	C38–C37–H31	116.(3)
C1–P1–C13	101.50(16)	C16–C17–H14	121.(3)	C39–C37–H31	116.(3)
C7–P1–C13	105.07(17)	C17–C18–C13	120.5(4)	Pt1–C37–H31	108.(3)
C1–P1–Pt1	116.06(12)	C17–C18–H15	121.(3)	C37–C38–C45	117.1(3)
C7–P1–Pt1	116.34(12)	C13–C18–H15	118.(3)	C37–C38–Pt1	70.4(2)
C13–P1–Pt1	113.50(12)	C24–C19–C20	118.7(4)	C45–C38–Pt1	106.1(2)
C31–P2–C19	103.93(16)	C24–C19–P2	123.5(3)	C37–C38–H32	121.(3)
C31–P2–C25	104.01(17)	C20–C19–P2	117.7(3)	C45–C38–H32	115.(3)
C19–P2–C25	101.08(16)	C21–C20–C19	120.3(4)	Pt1–C38–H32	118.(3)
C31–P2–Pt1	117.46(12)	C21–C20–H16	116.(3)	C40–C39–C44	117.2(4)
C19–P2–Pt1	109.63(12)	C19–C20–H16	123.(3)	C40–C39–C37	120.1(3)
C25–P2–Pt1	118.63(12)	C22–C21–C20	120.3(4)	C44–C39–C37	122.6(3)
C45–O1–C46	116.1(3)	C22–C21–H17	124.(3)	C41–C40–C39	121.6(4)
C6–C1–C2	118.0(4)	C20–C21–H17	116.(3)	C41–C40–H33	121.(3)
C6–C1–P1	123.0(3)	C21–C22–C23	119.6(4)	C39–C40–H33	118.(3)

continued on next page

Table A.21 – continued from previous page

C2–C1–P1	119.0(3)	C21–C22–H18	123.(3)	C40–C41–C42	120.5(4)
C3–C2–C1	120.6(4)	C23–C22–H18	117.(3)	C40–C41–H34	122.(3)
C3–C2–H1	119.(3)	C22–C23–C24	120.8(4)	C42–C41–H34	117.(3)
C1–C2–H1	121.(3)	C22–C23–H19	120.(3)	C43–C42–C41	118.5(4)
C4–C3–C2	121.1(4)	C24–C23–H19	119.(3)	C43–C42–H35	125.(2)
C4–C3–H2	119.(3)	C23–C24–C19	120.3(4)	C41–C42–H35	117.(2)
C2–C3–H2	120.(3)	C23–C24–H20	120.(3)	C42–C43–C44	121.4(4)
C3–C4–C5	119.0(4)	C19–C24–H20	120.(3)	C42–C43–H36	122.(4)
C3–C4–H3	121.(3)	C30–C25–C26	117.7(4)	C44–C43–H36	116.(4)
C5–C4–H3	120.(3)	C30–C25–P2	120.5(3)	C43–C44–C39	120.8(4)
C4–C5–C6	120.8(4)	C26–C25–P2	121.6(3)	C43–C44–H37	123.(2)
C4–C5–H4	124.(3)	C27–C26–C25	120.3(5)	C39–C44–H37	116.(2)
C6–C5–H4	116.(3)	C27–C26–H21	120.(3)	O2–C45–O1	121.3(3)
C5–C6–C1	120.4(4)	C25–C26–H21	120.(3)	O2–C45–C38	126.9(3)
C5–C6–H5	118.(3)	C28–C27–C26	120.5(4)	O1–C45–C38	111.8(3)
C1–C6–H5	121.(3)	C28–C27–H22	124.(3)	C51–C46–C47	120.6(4)
C12–C7–C8	117.8(4)	C26–C27–H22	115.(3)	C51–C46–O1	118.6(4)
C12–C7–P1	123.4(3)	C29–C28–C27	119.6(4)	C47–C46–O1	120.8(4)
C8–C7–P1	118.8(3)	C29–C28–H23	119.(3)	C46–C47–C48	118.7(4)
C9–C8–C7	121.2(4)	C27–C28–H23	121.(3)	C46–C47–H38	122.(3)
C9–C8–H6	120.(3)	C28–C29–C30	120.2(4)	C48–C47–H38	120.(3)
C7–C8–H6	119.(3)	C28–C29–H24	118.(3)	C49–C48–C47	122.1(4)
C8–C9–C10	120.4(4)	C30–C29–H24	121.(3)	C49–C48–H39	117.(3)
C8–C9–H7	120.(3)	C29–C30–C25	121.7(4)	C47–C48–H39	121.(3)
C10–C9–H7	120.(3)	C29–C30–H25	119.(3)	C48–C49–C50	117.7(4)
C11–C10–C9	119.7(4)	C25–C30–H25	119.(3)	C48–C49–C52	122.2(4)
C11–C10–H8	119.(3)	C36–C31–C32	118.3(3)	C50–C49–C52	120.1(4)
C9–C10–H8	122.(3)	C36–C31–P2	118.6(3)	C51–C50–C49	121.4(5)

continued on next page

Table A.21 – continued from previous page

C10–C11–C12	120.1(4)	C32–C31–P2	123.1(3)	C51–C50–H40	119.(4)
C10–C11–H9	120.(3)	C33–C32–C31	121.0(4)	C49–C50–H40	119.(4)
C12–C11–H9	120.(3)	C33–C32–H26	119.(3)	C50–C51–C46	119.6(4)
C11–C12–C7	120.8(4)	C31–C32–H26	120.(3)	C50–C51–H41	120.(3)
C11–C12–H10	118.(3)	C32–C33–C34	120.1(4)	C46–C51–H41	120.(3)
C7–C12–H10	121.(3)	C32–C33–H27	121.(3)	C49–C52–H42	109.500
C14–C13–C18	118.8(4)	C34–C33–H27	119.(3)	C49–C52–H43	109.500
C14–C13–P1	125.3(3)	C35–C34–C33	119.4(4)	H42–C52–H43	109.500
C18–C13–P1	115.9(3)	C35–C34–H28	118.(2)	C49–C52–H44	109.500
C13–C14–C15	120.7(4)	C33–C34–H28	123.(2)	H42–C52–H44	109.500
C13–C14–H11	120.(2)	C34–C35–C36	120.8(4)	H43–C52–H44	109.500
C15–C14–H11	120.(2)	C34–C35–H29	120.(3)		
C16–C15–C14	120.2(4)	C36–C35–H29	119.(3)		

PtPh–Ph

Table A.22.: Structural parameters of **PtPh–Ph**:

molecular formula	$C_{51}H_{42}O_2P_2Pt$
molecular weight [a.m.u.]	943.88
colour/habit	yellow fragment
crystal system	monoclinic
space group	$P2_1/c$
unit cell dimensions	a = 10.2103(9) Å $\alpha = 90.00°$
	b = 17.8507(13) Å $\beta = 94.566(3)°$
	c = 22.6104(19) Å $\gamma = 90.00°$
cell volume [Å^3]	4107.9(6)
Z	4
T [K]	173(2)

continued on next page

Table A.22 – continued from previous page

calculated density $[gcm^{-3}]$	1.526
absorption coefficient μ $[mm^{-1}]$	3.534
F(000)	1888
Θ range $[°]$	1.46 - 28.28
index ranges [h, k, l]	$\pm 13, +17/-23, \pm 30$
no. of independent reflections / R_{int}	10099
no. of absorbed reflections	9111
no. of data	10099
no. of restraints	0
no. of parameters	673
R_1 $(I > 2\sigma(I))$	0.0205
wR_2 $(I > 2\sigma(I))$	0.0485
R_1 (all data)	0.0256
wR_2 (all data)	0.0513
goodness of fit (on F^2)	0.985

Table A.23.: Bond lengths of **PtPh−Ph** $[\mathring{A}]$:

Pt1–C43	2.108(2)	C9–C10	1.376(5)	C31–C36	1.397(3)
Pt1–C44	2.135(2)	C10–C11	1.390(4)	C32–C33	1.383(4)
Pt1–P1	2.2715(5)	C11–C12	1.385(3)	C33–C34	1.375(4)
Pt1–P2	2.2844(6)	C13–C14	1.389(3)	C34–C35	1.391(3)
P1–C36	1.826(2)	C13–C18	1.391(3)	C35–C36	1.395(3)
P1–C30	1.837(2)	C14–C15	1.381(4)	C37–C38	1.387(3)
P1–C24	1.838(2)	C15–C16	1.385(4)	C37–C42	1.397(3)
P2–C12	1.827(2)	C16–C17	1.385(3)	C38–C39	1.389(4)
P2–C18	1.828(2)	C17–C18	1.392(3)	C39–C40	1.383(4)
P2–C6	1.843(2)	C19–C24	1.391(3)	C40–C41	1.389(3)
O1–C45	1.201(3)	C19–C20	1.397(3)	C41–C42	1.403(3)

continued on next page

Table A.23 – continued from previous page

O2–C45	1.390(3)	C20–C21	1.383(4)	C42–C43	1.480(3)
O2–C51	1.402(3)	C21–C22	1.381(4)	C43–C44	1.443(3)
C1–C2	1.386(3)	C22–C23	1.396(3)	C44–C45	1.470(3)
C1–C6	1.388(3)	C23–C24	1.397(3)	C46–C51	1.378(4)
C2–C3	1.384(4)	C25–C26	1.385(3)	C46–C47	1.406(4)
C3–C4	1.378(4)	C25–C30	1.402(3)	C47–C48	1.371(5)
C4–C5	1.390(4)	C26–C27	1.381(4)	C48–C49	1.365(5)
C5–C6	1.390(3)	C27–C28	1.382(4)	C49–C50	1.392(4)
C7–C12	1.384(4)	C28–C29	1.383(4)	C50–C51	1.386(4)
C7–C8	1.388(3)	C29–C30	1.392(3)		
C8–C9	1.368(4)	C31–C32	1.395(3)		

Table A.24.: Bond angles of **PtPh−Ph** [°]:

C43–Pt1–C44	39.76(8)	C12–C11–C10	120.2(3)	C33–C34–C35	119.6(2)
C43–Pt1–P1	102.58(6)	C7–C12–C11	118.5(2)	C34–C35–C36	120.7(2)
C44–Pt1–P1	141.87(6)	C7–C12–P2	122.66(17)	C35–C36–C31	118.9(2)
C43–Pt1–P2	148.26(6)	C11–C12–P2	118.80(19)	C35–C36–P1	117.10(16)
C44–Pt1–P2	108.99(6)	C14–C13–C18	120.2(2)	C31–C36–P1	123.82(17)
P1–Pt1–P2	109.024(19)	C15–C14–C13	120.4(2)	C38–C37–C42	121.1(2)
C36–P1–C30	101.48(10)	C14–C15–C16	119.8(2)	C37–C38–C39	120.2(2)
C36–P1–C24	105.84(9)	C17–C16–C15	119.8(2)	C40–C39–C38	119.5(2)
C30–P1–C24	103.38(10)	C16–C17–C18	121.0(2)	C39–C40–C41	120.4(2)
C36–P1–Pt1	115.59(7)	C13–C18–C17	118.8(2)	C40–C41–C42	120.9(2)
C30–P1–Pt1	113.57(6)	C13–C18–P2	124.14(17)	C37–C42–C41	117.8(2)
C24–P1–Pt1	115.37(7)	C17–C18–P2	117.08(16)	C37–C42–C43	120.0(2)
C12–P2–C18	104.47(10)	C24–C19–C20	121.2(2)	C41–C42–C43	122.2(2)
C12–P2–C6	102.49(10)	C21–C20–C19	120.0(2)	C44–C43–C42	120.96(19)
C18–P2–C6	100.85(10)	C22–C21–C20	119.6(2)	C44–C43–Pt1	71.16(12)

continued on next page

Table A.24 – continued from previous page

C12–P2–Pt1	112.48(7)	C21–C22–C23	120.5(2)	C42–C43–Pt1	114.47(14)
C18–P2–Pt1	119.89(8)	C22–C23–C24	120.7(2)	C43–C44–C45	117.63(19)
C6–P2–Pt1	114.58(7)	C19–C24–C23	118.0(2)	C43–C44–Pt1	69.09(11)
C45–O2–C51	118.11(17)	C19–C24–P1	120.93(16)	C45–C44–Pt1	111.75(14)
C2–C1–C6	120.8(2)	C23–C24–P1	120.78(16)	O1–C45–O2	122.0(2)
C3–C2–C1	120.2(3)	C26–C25–C30	120.9(2)	O1–C45–C44	128.0(2)
C4–C3–C2	119.5(2)	C27–C26–C25	120.2(2)	O2–C45–C44	109.94(18)
C3–C4–C5	120.5(3)	C26–C27–C28	119.6(2)	C51–C46–C47	118.3(3)
C6–C5–C4	120.4(2)	C27–C28–C29	120.4(3)	C48–C47–C46	120.5(3)
C1–C6–C5	118.6(2)	C28–C29–C30	121.0(2)	C49–C48–C47	120.5(3)
C1–C6–P2	118.03(17)	C29–C30–C25	117.9(2)	C48–C49–C50	120.3(3)
C5–C6–P2	123.34(17)	C29–C30–P1	122.56(16)	C51–C50–C49	119.1(3)
C12–C7–C8	120.9(2)	C25–C30–P1	119.53(17)	C46–C51–C50	121.3(2)
C9–C8–C7	120.1(3)	C32–C31–C36	120.0(2)	C46–C51–O2	121.9(2)
C8–C9–C10	119.6(3)	C33–C32–C31	119.9(2)	C50–C51–O2	116.7(2)
C9–C10–C11	120.5(3)	C34–C33–C32	120.8(2)		

PtPhNO$_2$–PhOMe

Table A.25.: Structural parameters of **PtPhNO$_2$–PhOMe**:

molecular formula	C$_{52}$H$_{43}$NO$_5$P$_2$Pt
molecular weight [a.m.u.]	1018.90
colour/habit	yellow fragment
crystal dimensions [mm^3]	0.66 × 0.51 × 0.25
crystal system	monoclinic
space group	P2$_1$/c
unit cell dimensions	a = 23.2571(10) Å $\alpha = 90.00°$
	b = 9.6685(4) Å $\beta = 92.352(2)°$

continued on next page

Table A.25 – continued from previous page

	c = 19.6859(9) Å	$\gamma = 90.00°$
cell volume [Å^3]	4422.9(3)	
Z	4	
T [K]	172(2)	
calculated density [gcm^{-3}]	1.530	
absorption coefficient μ [mm^{-1}]	3.294	
F(000)	2040	
Θ range [°]	0.88 - 25.32	
index ranges [h, k, l]	±27, ±11, ±23	
no. of independent reflections / R$_{int}$	8070	
no. of absorbed reflections	7705	
no. of data	8070	
no. of restraints	0	
no. of parameters	722	
R$_1$ ($I > 2\sigma(I)$)	0.0205	
wR$_2$ ($I > 2\sigma(I)$)	0.0509	
R$_1$ (all data)	0.0224	
wR$_2$ (all data)	0.0527	
goodness of fit (on F^2)	1.106	
largest difference peak [eÅ^3]	1.322	
largest difference hole [eÅ^3]	-1.119	

Table A.26.: Bond lengths of **PtPhNO$_2$–PhOMe** [Å]:

C1–C2	1.394(4)	C19–P2	1.826(3)	C38–C45	1.474(4)
C1–C6	1.394(4)	C20–C21	1.384(5)	C38–Pt1	2.117(2)
C1–P1	1.825(3)	C20–H16	0.91(3)	C38–H32	0.95(3)
C2–C3	1.387(4)	C21–C22	1.380(5)	C39–C44	1.401(4)
C2–H1	1.00(3)	C21–H17	0.89(4)	C39–C40	1.404(4)

continued on next page

Table A.26 – continued from previous page

C3–C4	1.371(5)	C22–C23	1.376(5)	C40–C41	1.381(4)
C3–H2	0.93(5)	C22–H18	0.88(4)	C40–H33	0.86(3)
C4–C5	1.381(6)	C23–C24	1.388(4)	C41–C42	1.388(4)
C4–H3	0.92(4)	C23–H19	0.94(4)	C41–H34	0.90(3)
C5–C6	1.386(5)	C24–H20	0.97(3)	C42–C43	1.376(4)
C5–H4	0.85(4)	C25–C26	1.383(4)	C42–N1	1.458(4)
C6–H5	0.94(3)	C25–C30	1.396(4)	C43–C44	1.380(4)
C7–C8	1.391(4)	C25–P2	1.831(3)	C43–H35	0.88(4)
C7–C12	1.397(4)	C26–C27	1.388(4)	C44–H36	0.97(3)
C7–P1	1.833(3)	C26–H21	0.91(3)	C45–O3	1.201(3)
C8–C9	1.392(5)	C27–C28	1.374(4)	C45–O4	1.373(3)
C8–H6	0.94(3)	C27–H22	0.95(4)	C46–C51	1.371(4)
C9–C10	1.373(6)	C28–C29	1.380(4)	C46–C47	1.373(4)
C9–H7	1.03(4)	C28–H23	0.90(4)	C46–O4	1.411(3)
C10–C11	1.375(6)	C29–C30	1.382(4)	C47–C48	1.376(5)
C10–H8	0.93(4)	C29–H24	0.93(4)	C47–H37	0.95(3)
C11–C12	1.388(4)	C30–H25	0.90(3)	C48–C49	1.386(5)
C11–H9	0.86(4)	C31–C36	1.384(4)	C48–H38	0.95(5)
C12–H10	0.94(4)	C31–C32	1.398(4)	C49–O5	1.370(3)
C13–C18	1.380(4)	C31–P2	1.840(3)	C49–C50	1.381(5)
C13–C14	1.396(4)	C32–C33	1.386(5)	C50–C51	1.392(4)
C13–P1	1.837(2)	C32–H26	0.97(3)	C50–H39	0.93(3)
C14–C15	1.381(4)	C33–C34	1.385(6)	C51–H40	0.91(4)
C14–H11	0.91(3)	C33–H27	0.95(4)	C52–O5	1.415(5)
C15–C16	1.381(5)	C34–C35	1.364(6)	C52–H41	0.90(4)
C15–H12	0.92(3)	C34–H28	0.97(4)	C52–H42	1.06(4)
C16–C17	1.370(5)	C35–C36	1.405(5)	C52–H43	0.94(4)
C16–H13	0.93(4)	C35–H29	0.88(4)	N1–O2	1.232(4)

continued on next page

Table A.26 – continued from previous page

C17–C18	1.390(4)	C36–H30	0.91(3)	N1–O1	1.232(4)
C17–H14	0.92(4)	C37–C38	1.445(4)	P1–Pt1	2.2926(6)
C18–H15	0.96(3)	C37–C39	1.481(4)	P2–Pt1	2.2739(6)
C19–C24	1.396(4)	C37–Pt1	2.110(2)		
C19–C20	1.398(4)	C37–H31	0.97(3)		

Table A.27.: Bond angles of **PtPhNO₂−PhOMe** [°]:

C2–C1–C6	119.0(3)	C23–C22–C21	120.2(3)	C40–C41–C42	118.7(3)
C2–C1–P1	118.6(2)	C23–C22–H18	119.(3)	C40–C41–H34	120.(2)
C6–C1–P1	122.4(2)	C21–C22–H18	121.(3)	C42–C41–H34	121.(2)
C3–C2–C1	120.0(3)	C22–C23–C24	120.5(3)	C43–C42–C41	121.8(3)
C3–C2–H1	123.2(19)	C22–C23–H19	121.(2)	C43–C42–N1	118.9(3)
C1–C2–H1	116.8(19)	C24–C23–H19	119.(2)	C41–C42–N1	119.3(3)
C4–C3–C2	120.7(3)	C23–C24–C19	120.0(3)	C42–C43–C44	118.9(3)
C4–C3–H2	119.(3)	C23–C24–H20	122.0(19)	C42–C43–H35	118.(2)
C2–C3–H2	120.(3)	C19–C24–H20	118.0(19)	C44–C43–H35	123.(2)
C3–C4–C5	119.8(3)	C26–C25–C30	118.3(2)	C43–C44–C39	121.6(3)
C3–C4–H3	119.(2)	C26–C25–P2	119.8(2)	C43–C44–H36	118.9(19)
C5–C4–H3	121.(2)	C30–C25–P2	121.9(2)	C39–C44–H36	120.(2)
C4–C5–C6	120.4(3)	C25–C26–C27	120.9(3)	O3–C45–O4	122.2(2)
C4–C5–H4	123.(2)	C25–C26–H21	118.(2)	O3–C45–C38	128.1(2)
C6–C5–H4	116.(2)	C27–C26–H21	121.(2)	O4–C45–C38	109.7(2)
C5–C6–C1	120.1(3)	C28–C27–C26	120.2(3)	C51–C46–C47	121.1(3)
C5–C6–H5	122.(2)	C28–C27–H22	120.(2)	C51–C46–O4	119.7(3)
C1–C6–H5	118.(2)	C26–C27–H22	120.(2)	C47–C46–O4	119.0(3)
C8–C7–C12	118.5(3)	C27–C28–C29	119.7(3)	C46–C47–C48	119.4(3)
C8–C7–P1	122.1(2)	C27–C28–H23	123.(2)	C46–C47–H37	122.(2)
C12–C7–P1	119.0(2)	C29–C28–H23	117.(2)	C48–C47–H37	118.(2)

continued on next page

Table A.27 – continued from previous page

C7–C8–C9	120.3(3)	C28–C29–C30	120.3(3)	C47–C48–C49	120.3(3)
C7–C8–H6	115.4(19)	C28–C29–H24	119.(2)	C47–C48–H38	125.(3)
C9–C8–H6	124.3(19)	C30–C29–H24	120.(2)	C49–C48–H38	115.(3)
C10–C9–C8	120.6(3)	C29–C30–C25	120.6(3)	O5–C49–C50	124.3(3)
C10–C9–H7	125.(2)	C29–C30–H25	118.(2)	O5–C49–C48	115.7(3)
C8–C9–H7	114.(2)	C25–C30–H25	121.(2)	C50–C49–C48	120.0(3)
C9–C10–C11	119.7(3)	C36–C31–C32	119.2(3)	C49–C50–C51	119.3(3)
C9–C10–H8	122.(3)	C36–C31–P2	123.4(2)	C49–C50–H39	122.(2)
C11–C10–H8	119.(3)	C32–C31–P2	117.4(2)	C51–C50–H39	119.(2)
C10–C11–C12	120.5(4)	C33–C32–C31	120.4(3)	C46–C51–C50	119.8(3)
C10–C11–H9	121.(3)	C33–C32–H26	122.(2)	C46–C51–H40	121.(2)
C12–C11–H9	119.(3)	C31–C32–H26	118.(2)	C50–C51–H40	119.(2)
C11–C12–C7	120.4(3)	C34–C33–C32	120.2(4)	O5–C52–H41	112.(3)
C11–C12–H10	121.(2)	C34–C33–H27	118.(2)	O5–C52–H42	111.(2)
C7–C12–H10	119.(2)	C32–C33–H27	122.(2)	H41–C52–H42	110.(3)
C18–C13–C14	118.7(2)	C35–C34–C33	119.6(3)	O5–C52–H43	103.(3)
C18–C13–P1	124.2(2)	C35–C34–H28	122.(3)	H41–C52–H43	109.(3)
C14–C13–P1	117.11(19)	C33–C34–H28	118.(3)	H42–C52–H43	113.(3)
C15–C14–C13	120.7(3)	C34–C35–C36	121.2(4)	O2–N1–O1	123.6(2)
C15–C14–H11	119.(2)	C34–C35–H29	122.(3)	O2–N1–C42	118.4(3)
C13–C14–H11	121.(2)	C36–C35–H29	116.(3)	O1–N1–C42	118.0(3)
C16–C15–C14	120.3(3)	C31–C36–C35	119.4(4)	C45–O4–C46	117.2(2)
C16–C15–H12	120.(2)	C31–C36–H30	119.3(19)	C49–O5–C52	117.0(3)
C14–C15–H12	120.(2)	C35–C36–H30	121.3(19)	C1–P1–C7	103.30(12)
C17–C16–C15	119.1(3)	C38–C37–C39	120.5(2)	C1–P1–C13	102.77(12)
C17–C16–H13	121.(2)	C38–C37–Pt1	70.29(13)	C7–P1–C13	104.76(12)
C15–C16–H13	120.(2)	C39–C37–Pt1	114.84(17)	C1–P1–Pt1	112.14(9)
C16–C17–C18	121.2(3)	C38–C37–H31	117.0(16)	C7–P1–Pt1	119.84(9)

continued on next page

Table A.27 – continued from previous page

C16–C17–H14	122.(2)	C39–C37–H31	113.3(16)	C13–P1–Pt1	112.31(8)
C18–C17–H14	117.(2)	Pt1–C37–H31	114.0(16)	C19–P2–C25	104.71(12)
C13–C18–C17	119.9(3)	C37–C38–C45	118.3(2)	C19–P2–C31	104.73(13)
C13–C18–H15	121.7(19)	C37–C38–Pt1	69.74(13)	C25–P2–C31	100.53(12)
C17–C18–H15	118.4(19)	C45–C38–Pt1	113.26(17)	C19–P2–Pt1	113.12(9)
C24–C19–C20	118.6(3)	C37–C38–H32	120.3(15)	C25–P2–Pt1	119.52(8)
C24–C19–P2	122.8(2)	C45–C38–H32	111.7(15)	C31–P2–Pt1	112.54(8)
C20–C19–P2	118.5(2)	Pt1–C38–H32	117.3(15)	C37–Pt1–C38	39.97(9)
C21–C20–C19	120.8(3)	C44–C39–C40	117.7(3)	C37–Pt1–P2	103.45(7)
C21–C20–H16	121.(2)	C44–C39–C37	119.8(2)	C38–Pt1–P2	143.41(7)
C19–C20–H16	118.(2)	C40–C39–C37	122.5(2)	C37–Pt1–P1	145.89(7)
C22–C21–C20	119.8(3)	C41–C40–C39	121.3(3)	C38–Pt1–P1	105.99(7)
C22–C21–H17	120.(2)	C41–C40–H33	115.(2)	P2–Pt1–P1	110.51(2)
C20–C21–H17	121.(2)	C39–C40–H33	123.(2)		

PtPhNO$_2$–PhMe

Table A.28.: Structural parameters of **PtPhNO$_2$–PhMe**:

molecular formula	C$_{52}$H$_{43}$NO$_4$P$_2$Pt
molecular weight [a.m.u.]	1002.90
colour/habit	yellow fragment
crystal dimensions [mm^3]	0.18 × 0.13 × 0.03
crystal system	monoclinic
space group	P2$_1$/c
unit cell dimensions	a = 23.2464(13) Å $\alpha = 90.00°$
	b = 9.6633(4) Å $\beta = 93.877(2)°$
	c = 19.5892(10) Å $\gamma = 90.00°$
cell volume [Å^3]	4390.4(4)

continued on next page

Table A.28 – continued from previous page

Z	4
T [K]	173(2)
calculated density $[gcm^{-3}]$	1.517
absorption coefficient μ $[mm^{-1}]$	3.316
F(000)	2008
Θ range [°]	0.88 - 25.34
index ranges [h, k, l]	±27, ±11, ±23
no. of independent reflections / R_{int}	8039
no. of absorbed reflections	7228
no. of data	8039
no. of restraints	0
no. of parameters	701
R_1 $(I > 2\sigma(I))$	0.0243
wR_2 $(I > 2\sigma(I))$	0.0464
R_1 (all data)	0.0310
wR_2 (all data)	0.0492
goodness of fit (on F^2)	1.156
largest difference peak $[e\mathring{A}^3]$	1.042
largest difference hole $[e\mathring{A}^3]$	-0.755

Table A.29.: Bond lengths of **PtPhNO$_2$–PhMe** $[\mathring{A}]$:

C1–C2	1.392(5)	C19–C20	1.398(5)	C37–Pt1	2.122(3)
C1–C6	1.396(5)	C19–P1	1.835(3)	C37–H31	0.93(3)
C1–P2	1.835(3)	C20–C21	1.386(5)	C38–C45	1.468(4)
C2–C3	1.392(5)	C20–H16	0.97(4)	C38–Pt1	2.115(3)
C2–H1	0.91(3)	C21–C22	1.378(7)	C38–H32	0.93(3)
C3–C4	1.379(6)	C21–H17	0.97(4)	C39–C40	1.398(5)
C3–H2	0.91(4)	C22–C23	1.361(7)	C39–C44	1.404(4)

continued on next page

Table A.29 – continued from previous page

C4–C5	1.372(6)	C22–H18	0.90(5)	C40–C41	1.373(5)
C4–H3	0.88(4)	C23–C24	1.400(6)	C40–H33	0.89(3)
C5–C6	1.388(5)	C23–H19	0.83(4)	C41–C42	1.388(5)
C5–H4	0.90(4)	C24–H20	0.94(4)	C41–H34	0.90(3)
C6–H5	0.90(3)	C25–C30	1.390(5)	C42–C43	1.377(5)
C7–C8	1.393(5)	C25–C26	1.395(5)	C42–N1	1.461(4)
C7–C12	1.395(5)	C25–P1	1.825(3)	C43–C44	1.384(5)
C7–P2	1.828(3)	C26–C27	1.382(5)	C43–H35	0.95(4)
C8–C9	1.375(5)	C26–H21	0.88(3)	C44–H36	0.93(3)
C8–H6	0.93(4)	C27–C28	1.388(6)	C45–O2	1.204(4)
C9–C10	1.374(6)	C27–H22	0.89(4)	C45–O1	1.376(4)
C9–H7	0.89(4)	C28–C29	1.369(6)	C46–C47	1.369(5)
C10–C11	1.381(6)	C28–H23	0.90(4)	C46–C51	1.378(5)
C10–H8	0.94(4)	C29–C30	1.387(5)	C46–O1	1.412(4)
C11–C12	1.390(5)	C29–H24	0.93(4)	C47–C48	1.386(5)
C11–H9	0.89(4)	C30–H25	0.90(4)	C47–H37	0.95(4)
C12–H10	0.88(3)	C31–C36	1.387(5)	C48–C49	1.378(5)
C13–C18	1.379(4)	C31–C32	1.391(4)	C48–H38	0.91(4)
C13–C14	1.402(4)	C31–P1	1.833(3)	C49–C50	1.378(6)
C13–P2	1.833(3)	C32–C33	1.388(5)	C49–C52	1.507(5)
C14–C15	1.381(5)	C32–H26	0.88(4)	C50–C51	1.389(5)
C14–H11	0.87(3)	C33–C34	1.369(5)	C50–H39	0.93(4)
C15–C16	1.374(5)	C33–H27	0.97(4)	C51–H40	0.95(4)
C15–H12	0.92(4)	C34–C35	1.380(5)	C52–H52A	0.9800
C16–C17	1.374(5)	C34–H28	0.90(3)	C52–H52B	0.9800
C16–H13	0.85(4)	C35–C36	1.384(5)	C52–H52C	0.9800
C17–C18	1.388(5)	C35–H29	0.92(4)	N1–O3	1.229(4)
C17–H14	0.85(4)	C36–H30	0.88(4)	N1–O4	1.234(4)

continued on next page

Table A.29 – continued from previous page

C18–H15	0.94(3)	C37–C38	1.444(4)	P1–Pt1	2.2728(8)
C19–C24	1.382(5)	C37–C39	1.470(5)	P2–Pt1	2.2903(8)

Table A.30.: Bond angles of **PtPhNO$_2$–PhMe** [°]:

C2–C1–C6	118.6(3)	C23–C22–C21	119.8(4)	C40–C41–C42	119.1(3)
C2–C1–P2	119.3(2)	C23–C22–H18	121.(3)	C40–C41–H34	121.(2)
C6–C1–P2	121.9(3)	C21–C22–H18	119.(3)	C42–C41–H34	120.(2)
C1–C2–C3	120.7(3)	C22–C23–C24	121.0(4)	C43–C42–C41	121.5(3)
C1–C2–H1	121.(2)	C22–C23–H19	123.(3)	C43–C42–N1	119.1(3)
C3–C2–H1	118.(2)	C24–C23–H19	116.(3)	C41–C42–N1	119.4(3)
C4–C3–C2	119.7(4)	C19–C24–C23	119.8(4)	C42–C43–C44	118.8(3)
C4–C3–H2	122.(2)	C19–C24–H20	119.(2)	C42–C43–H35	122.(2)
C2–C3–H2	118.(3)	C23–C24–H20	121.(2)	C44–C43–H35	119.(2)
C5–C4–C3	120.3(3)	C30–C25–C26	118.5(3)	C43–C44–C39	121.5(3)
C5–C4–H3	122.(3)	C30–C25–P1	122.8(3)	C43–C44–H36	120.3(19)
C3–C4–H3	118.(3)	C26–C25–P1	118.4(2)	C39–C44–H36	118.2(19)
C4–C5–C6	120.4(4)	C27–C26–C25	120.7(3)	O2–C45–O1	121.7(3)
C4–C5–H4	123.(2)	C27–C26–H21	120.(2)	O2–C45–C38	128.0(3)
C6–C5–H4	116.(2)	C25–C26–H21	120.(2)	O1–C45–C38	110.3(3)
C5–C6–C1	120.3(4)	C26–C27–C28	120.1(4)	C47–C46–C51	120.8(3)
C5–C6–H5	120.(2)	C26–C27–H22	120.(2)	C47–C46–O1	119.0(3)
C1–C6–H5	119.(2)	C28–C27–H22	120.(2)	C51–C46–O1	119.9(3)
C8–C7–C12	119.0(3)	C29–C28–C27	119.5(4)	C46–C47–C48	119.5(3)
C8–C7–P2	118.3(3)	C29–C28–H23	122.(2)	C46–C47–H37	119.(2)
C12–C7–P2	122.7(3)	C27–C28–H23	118.(2)	C48–C47–H37	122.(2)
C9–C8–C7	120.3(4)	C28–C29–C30	120.9(4)	C49–C48–C47	121.1(4)
C9–C8–H6	123.(2)	C28–C29–H24	119.(2)	C49–C48–H38	117.(3)
C7–C8–H6	116.(2)	C30–C29–H24	120.(2)	C47–C48–H38	122.(3)

continued on next page

Table A.30 – continued from previous page

C10–C9–C8	120.8(4)	C29–C30–C25	120.2(4)	C50–C49–C48	118.2(3)
C10–C9–H7	120.(2)	C29–C30–H25	120.(2)	C50–C49–C52	120.6(4)
C8–C9–H7	119.(2)	C25–C30–H25	120.(2)	C48–C49–C52	121.1(4)
C9–C10–C11	119.8(4)	C36–C31–C32	118.4(3)	C49–C50–C51	121.6(4)
C9–C10–H8	122.(3)	C36–C31–P1	122.3(2)	C49–C50–H39	119.(3)
C11–C10–H8	119.(3)	C32–C31–P1	119.3(2)	C51–C50–H39	119.(3)
C10–C11–C12	120.2(4)	C33–C32–C31	120.5(3)	C46–C51–C50	118.7(4)
C10–C11–H9	120.(3)	C33–C32–H26	120.(2)	C46–C51–H40	119.(2)
C12–C11–H9	120.(3)	C31–C32–H26	119.(2)	C50–C51–H40	122.(2)
C11–C12–C7	119.9(4)	C34–C33–C32	120.4(3)	C49–C52–H52A	109.500
C11–C12–H10	121.(2)	C34–C33–H27	122.(2)	C49–C52–H52B	109.500
C7–C12–H10	119.(2)	C32–C33–H27	117.(2)	H52A–C52–H52B	109.500
C18–C13–C14	118.4(3)	C33–C34–C35	119.6(3)	C49–C52–H52C	109.500
C18–C13–P2	124.4(2)	C33–C34–H28	123.(2)	H52A–C52–H52C	109.500
C14–C13–P2	117.2(2)	C35–C34–H28	117.(2)	H52B–C52–H52C	109.500
C15–C14–C13	120.7(3)	C34–C35–C36	120.4(3)	O3–N1–O4	123.4(3)
C15–C14–H11	121.(2)	C34–C35–H29	122.(2)	O3–N1–C42	118.4(3)
C13–C14–H11	118.(2)	C36–C35–H29	118.(2)	O4–N1–C42	118.2(3)
C16–C15–C14	120.3(3)	C35–C36–C31	120.6(3)	C45–O1–C46	117.8(2)
C16–C15–H12	121.(2)	C35–C36–H30	119.(2)	C25–P1–C31	104.95(14)
C14–C15–H12	119.(2)	C31–C36–H30	120.(2)	C25–P1–C19	104.74(15)
C17–C16–C15	119.5(4)	C38–C37–C39	121.7(3)	C31–P1–C19	100.86(15)
C17–C16–H13	121.(2)	C38–C37–Pt1	69.79(16)	C25–P1–Pt1	113.15(11)
C15–C16–H13	120.(2)	C39–C37–Pt1	113.9(2)	C31–P1–Pt1	118.71(10)
C16–C17–C18	120.9(4)	C38–C37–H31	115.5(19)	C19–P1–Pt1	112.84(10)
C16–C17–H14	118.(3)	C39–C37–H31	115.2(19)	C7–P2–C13	103.14(14)
C18–C17–H14	121.(3)	Pt1–C37–H31	112.0(18)	C7–P2–C1	103.29(14)
C13–C18–C17	120.3(3)	C37–C38–C45	118.4(3)	C13–P2–C1	104.51(14)

continued on next page

Table A.30 – continued from previous page

C13–C18–H15	122.(2)	C37–C38–Pt1	70.35(16)	C7–P2–Pt1	111.83(10)
C17–C18–H15	118.(2)	C45–C38–Pt1	113.3(2)	C13–P2–Pt1	111.66(9)
C24–C19–C20	118.9(3)	C37–C38–H32	121.(2)	C1–P2–Pt1	120.68(10)
C24–C19–P1	123.6(3)	C45–C38–H32	110.(2)	C38–Pt1–C37	39.85(12)
C20–C19–P1	117.5(3)	Pt1–C38–H32	118.9(19)	C38–Pt1–P1	144.03(9)
C21–C20–C19	120.4(4)	C40–C39–C44	117.5(3)	C37–Pt1–P1	104.18(9)
C21–C20–H16	121.(2)	C40–C39–C37	123.2(3)	C38–Pt1–P2	105.55(9)
C19–C20–H16	118.(2)	C44–C39–C37	119.3(3)	C37–Pt1–P2	145.30(9)
C22–C21–C20	120.1(5)	C41–C40–C39	121.6(3)	P1–Pt1–P2	110.35(3)
C22–C21–H17	124.(3)	C41–C40–H33	120.(2)		
C20–C21–H17	116.(3)	C39–C40–H33	118.(2)		

PtPhNO$_2$–Ph

Table A.31.: Structural parameters of **PtPhNO$_2$–Ph**:

molecular formula	$C_{51}H_{41}NO_4P_2Pt$
molecular weight [a.m.u.]	988.88
colour/habit	yellow fragment
crystal dimensions [mm^3]	$0.40 \times 0.10 \times 0.01$
crystal system	monoclinic
space group	P2$_1$/c
unit cell dimensions	a = 22.9412(9) Å $\quad \alpha = 90.00°$
	b = 9.6754(3) Å $\quad \beta = 96.099(2)°$
	c = 19.4574(7) Å $\quad \gamma = 90.00°$
cell volume [$Å^3$]	4294.4(3)
Z	4
T [K]	173(2)
calculated density [gcm^{-3}]	1.529

continued on next page

Table A.31 – continued from previous page

absorption coefficient μ $[mm^{-1}]$	3.389
F(000)	1976
Θ range $[°]$	0.89 - 26.35
index ranges [h, k, l]	±28, $+12/-11$, ±24
no. of independent reflections / R_{int}	8745
no. of absorbed reflections	7064
no. of data	8745
no. of restraints	0
no. of parameters	696
R_1 $(I > 2\sigma(I))$	0.0196
wR_2 $(I > 2\sigma(I))$	0.0385
R_1 (all data)	0.0334
wR_2 (all data)	0.0437
goodness of fit (on F^2)	1.069
largest difference peak $[e\text{Å}^3]$	0.680
largest difference hole $[e\text{Å}^3]$	-0.533

Table A.32.: Bond lengths of **PtPhNO$_2$–Ph** [Å]:

C1–C2	1.393(4)	C19–C20	1.393(4)	C38–C37	1.458(4)
C1–C6	1.395(4)	C19–C24	1.399(4)	C38–C45	1.467(4)
C1–P1	1.826(3)	C19–P2	1.836(3)	C38–Pt1	2.133(3)
C2–C3	1.393(4)	C20–C21	1.388(4)	C38–H32	0.90(3)
C2–H1	0.90(3)	C20–H16	0.91(3)	C37–C39	1.470(4)
C3–C4	1.377(5)	C21–C22	1.377(5)	C37–Pt1	2.116(2)
C3–H2	0.90(3)	C21–H17	0.90(3)	C37–H31	0.93(3)
C4–C5	1.381(5)	C22–C23	1.383(5)	C39–C40	1.401(4)
C4–H3	0.93(3)	C22–H18	0.92(3)	C39–C44	1.406(4)
C5–C6	1.385(4)	C23–C24	1.391(4)	C40–C41	1.384(4)

continued on next page

Table A.32 – continued from previous page

C5–H4	0.86(3)	C23–H19	0.94(3)	C40–H33	0.94(3)
C6–H5	0.90(3)	C24–H20	0.95(3)	C41–C42	1.377(4)
C7–C8	1.390(4)	C25–C26	1.387(4)	C41–H34	0.88(3)
C7–C12	1.391(4)	C25–C30	1.396(4)	C42–C43	1.391(4)
C7–P1	1.833(3)	C25–P2	1.827(3)	C42–N1	1.462(3)
C8–C9	1.393(4)	C26–C27	1.385(4)	C43–C44	1.381(4)
C8–H6	0.96(3)	C26–H21	0.92(3)	C43–H35	0.94(3)
C9–C10	1.381(4)	C27–C28	1.380(5)	C44–H36	0.92(3)
C9–H7	0.92(3)	C27–H22	0.92(3)	C45–O4	1.202(3)
C10–C11	1.375(4)	C28–C29	1.380(5)	C45–O3	1.382(3)
C10–H8	0.93(3)	C28–H23	0.98(3)	C46–C51	1.368(4)
C11–C12	1.386(4)	C29–C30	1.389(4)	C46–C47	1.373(4)
C11–H9	0.92(3)	C29–H24	0.91(3)	C46–O3	1.417(3)
C12–H10	0.88(3)	C30–H25	0.87(3)	C47–C48	1.390(5)
C13–C18	1.384(4)	C31–C36	1.383(4)	C47–H37	0.87(3)
C13–C14	1.395(4)	C31–C32	1.400(4)	C48–C49	1.371(6)
C13–P1	1.837(3)	C31–P2	1.830(2)	C48–H38	0.90(4)
C14–C15	1.393(4)	C32–C33	1.380(4)	C49–C50	1.374(5)
C14–H11	0.94(3)	C32–H26	0.92(3)	C49–H39	0.94(3)
C15–C16	1.376(6)	C33–C34	1.384(4)	C50–C51	1.387(4)
C15–H12	0.92(4)	C33–H27	0.93(3)	C50–H40	0.87(4)
C16–C17	1.369(6)	C34–C35	1.367(4)	C51–H41	0.96(3)
C16–H13	0.93(4)	C34–H28	0.90(3)	N1–O2	1.227(3)
C17–C18	1.399(4)	C35–C36	1.394(4)	N1–O1	1.236(3)
C17–H14	0.90(4)	C35–H29	0.91(3)	P1–Pt1	2.2726(6)
C18–H15	0.91(3)	C36–H30	0.92(3)	P2–Pt1	2.2923(6)

Table A.33.: Bond angles of **PtPhNO$_2$–Ph** [°]:

C2–C1–C6	118.7(3)	C22–C21–H17	120.1(19)	C44–C39–C37	122.8(2)
C2–C1–P1	122.6(2)	C20–C21–H17	120.(2)	C41–C40–C39	122.0(3)
C6–C1–P1	118.4(2)	C21–C22–C23	120.2(3)	C41–C40–H33	121.0(17)
C1–C2–C3	120.0(3)	C21–C22–H18	124.(2)	C39–C40–H33	117.0(17)
C1–C2–H1	118.2(19)	C23–C22–H18	116.(2)	C42–C41–C40	118.8(3)
C3–C2–H1	121.7(19)	C22–C23–C24	120.1(3)	C42–C41–H34	119.(2)
C4–C3–C2	120.7(3)	C22–C23–H19	123.(2)	C40–C41–H34	122.(2)
C4–C3–H2	122.(2)	C24–C23–H19	116.(2)	C41–C42–C43	121.6(3)
C2–C3–H2	118.(2)	C23–C24–C19	120.4(3)	C41–C42–N1	119.1(3)
C3–C4–C5	119.6(3)	C23–C24–H20	118.6(17)	C43–C42–N1	119.3(3)
C3–C4–H3	120.9(19)	C19–C24–H20	121.0(17)	C44–C43–C42	118.8(3)
C5–C4–H3	119.5(19)	C26–C25–C30	119.1(3)	C44–C43–H35	122.4(17)
C4–C5–C6	120.3(3)	C26–C25–P2	117.7(2)	C42–C43–H35	118.7(17)
C4–C5–H4	122.(2)	C30–C25–P2	123.1(2)	C43–C44–C39	121.6(3)
C6–C5–H4	118.(2)	C27–C26–C25	120.4(3)	C43–C44–H36	118.1(17)
C5–C6–C1	120.6(3)	C27–C26–H21	119.5(17)	C39–C44–H36	120.3(17)
C5–C6–H5	121.4(19)	C25–C26–H21	120.1(17)	O4–C45–O3	121.4(2)
C1–C6–H5	118.0(19)	C28–C27–C26	120.2(3)	O4–C45–C38	128.0(2)
C8–C7–C12	118.6(2)	C28–C27–H22	118.(2)	O3–C45–C38	110.6(2)
C8–C7–P1	119.5(2)	C26–C27–H22	122.(2)	C51–C46–C47	121.6(3)
C12–C7–P1	121.9(2)	C29–C28–C27	120.0(3)	C51–C46–O3	118.4(3)
C7–C8–C9	120.7(3)	C29–C28–H23	121.7(18)	C47–C46–O3	119.9(3)
C7–C8–H6	121.9(17)	C27–C28–H23	118.3(18)	C46–C47–C48	118.6(3)
C9–C8–H6	117.4(17)	C28–C29–C30	120.1(3)	C46–C47–H37	118.(2)
C10–C9–C8	119.8(3)	C28–C29–H24	123.(2)	C48–C47–H37	123.(2)
C10–C9–H7	122.2(17)	C30–C29–H24	117.(2)	C49–C48–C47	120.8(4)
C8–C9–H7	118.1(17)	C29–C30–C25	120.1(3)	C49–C48–H38	123.(3)

continued on next page

Table A.33 – continued from previous page

C11–C10–C9	120.1(3)	C29–C30–H25	120.7(18)	C47–C48–H38	116.(3)
C11–C10–H8	123.3(19)	C25–C30–H25	119.2(18)	C48–C49–C50	119.5(3)
C9–C10–H8	116.6(19)	C36–C31–C32	118.3(2)	C48–C49–H39	120.(2)
C10–C11–C12	120.3(3)	C36–C31–P2	124.0(2)	C50–C49–H39	121.(2)
C10–C11–H9	120.9(18)	C32–C31–P2	117.50(19)	C49–C50–C51	120.6(3)
C12–C11–H9	118.7(18)	C33–C32–C31	120.9(3)	C49–C50–H40	118.(3)
C11–C12–C7	120.6(3)	C33–C32–H26	120.0(17)	C51–C50–H40	121.(3)
C11–C12–H10	120.7(19)	C31–C32–H26	119.1(17)	C46–C51–C50	119.0(3)
C7–C12–H10	118.7(19)	C32–C33–C34	120.2(3)	C46–C51–H41	118.9(19)
C18–C13–C14	118.8(3)	C32–C33–H27	120.6(19)	C50–C51–H41	122.1(19)
C18–C13–P1	123.6(2)	C34–C33–H27	119.2(19)	O2–N1–O1	123.5(3)
C14–C13–P1	117.5(2)	C35–C34–C33	119.3(3)	O2–N1–C42	118.4(3)
C15–C14–C13	120.3(3)	C35–C34–H28	119.(2)	O1–N1–C42	118.1(3)
C15–C14–H11	118.4(18)	C33–C34–H28	121.(2)	C45–O3–C46	116.8(2)
C13–C14–H11	121.3(17)	C34–C35–C36	121.1(3)	C1–P1–C7	105.36(12)
C16–C15–C14	120.2(4)	C34–C35–H29	120.(2)	C1–P1–C13	104.77(12)
C16–C15–H12	122.(3)	C36–C35–H29	118.(2)	C7–P1–C13	101.02(12)
C14–C15–H12	118.(3)	C31–C36–C35	120.1(3)	C1–P1–Pt1	113.19(9)
C17–C16–C15	120.0(3)	C31–C36–H30	121.2(17)	C7–P1–Pt1	117.89(8)
C17–C16–H13	123.(2)	C35–C36–H30	118.8(17)	C13–P1–Pt1	113.12(8)
C15–C16–H13	117.(2)	C37–C38–C45	118.2(2)	C25–P2–C31	104.46(11)
C16–C17–C18	120.5(3)	C37–C38–Pt1	69.30(14)	C25–P2–C19	102.62(12)
C16–C17–H14	124.(2)	C45–C38–Pt1	113.06(17)	C31–P2–C19	103.68(11)
C18–C17–H14	115.(2)	C37–C38–H32	119.2(17)	C25–P2–Pt1	110.46(8)
C13–C18–C17	120.2(3)	C45–C38–H32	114.5(17)	C31–P2–Pt1	111.64(8)
C13–C18–H15	119.7(19)	Pt1–C38–H32	114.4(17)	C19–P2–Pt1	122.28(8)
C17–C18–H15	120.0(19)	C38–C37–C39	121.7(2)	C37–Pt1–C38	40.12(10)
C20–C19–C24	118.4(2)	C38–C37–Pt1	70.58(14)	C37–Pt1–P1	104.29(7)

continued on next page

Table A.33 – continued from previous page

C20–C19–P2	119.3(2)	C39–C37–Pt1	113.44(18)	C38–Pt1–P1	144.40(7)
C24–C19–P2	122.2(2)	C38–C37–H31	115.7(15)	C37–Pt1–P2	146.35(7)
C21–C20–C19	121.0(3)	C39–C37–H31	115.1(15)	C38–Pt1–P2	106.33(7)
C21–C20–H16	120.7(17)	Pt1–C37–H31	112.0(15)	P1–Pt1–P2	109.25(2)
C19–C20–H16	118.3(17)	C40–C39–C44	117.2(2)		
C22–C21–C20	119.9(3)	C40–C39–C37	120.0(2)		

PtPhNO$_2$–PhNO$_2$

Table A.34.: Structural parameters of **PtPhNO$_2$–PhNO$_2$**:

molecular formula	C$_{51}$H$_{40}$N$_2$O$_6$P$_2$Pt
molecular weight [a.m.u.]	1033.88
colour/habit	yellow fragment
crystal dimensions [mm^3]	0.40 × 0.30 × 0.20
crystal system	monoclinic
space group	P2$_1$/c
unit cell dimensions	a = 23.0336(13) Å $\alpha = 90.00°$
	b = 9.5895(6) Å $\beta = 93.581(2)°$
	c = 19.7733(12) Å $\gamma = 90.00°$
cell volume [Å^3]	4359.0(5)
Z	4
T [K]	173(2)
calculated density [gcm^{-3}]	1.575
absorption coefficient μ [mm^{-1}]	3.346
F(000)	2064
Θ range [°]	0.89 - 25.39
index ranges [h, k, l]	±27, ±11, ±23
no. of independent reflections / R$_{int}$	7955

continued on next page

Table A.34 – continued from previous page

no. of absorbed reflections	7665
no. of data	7955
no. of restraints	0
no. of parameters	719
R_1 $(I > 2\sigma(I))$	0.0149
wR_2 $(I > 2\sigma(I))$	0.0343
R_1 (all data)	0.0164
wR_2 (all data)	0.0360
goodness of fit (on F^2)	1.119
largest difference peak [e$\mathring{A}^3$]	0.433
largest difference hole [e$\mathring{A}^3$]	-0.637

Table A.35.: Bond lengths of **PtPhNO$_2$−PhNO$_2$** [$\mathring{A}$]:

Pt1–C37	2.1064(19)	C47–H33	0.93(3)	C3–C2	1.379(3)
Pt1–C38	2.116(2)	C48–C49	1.374(3)	C3–H2	0.90(3)
Pt1–P2	2.2738(5)	C48–H34	0.90(3)	C2–H1	0.89(2)
Pt1–P1	2.2920(5)	C49–C50	1.372(3)	C25–C30	1.386(3)
P1–C7	1.827(2)	C50–C51	1.386(3)	C25–C26	1.392(3)
P1–C13	1.828(2)	C50–H35	0.93(3)	C30–C29	1.385(3)
P1–C1	1.829(2)	C51–H36	0.91(3)	C30–H25	0.88(3)
P2–C31	1.823(2)	C7–C8	1.389(3)	C29–C28	1.373(3)
P2–C25	1.827(2)	C7–C12	1.390(3)	C29–H24	0.93(3)
P2–C19	1.834(2)	C8–C9	1.389(4)	C28–C27	1.376(3)
O1–N1	1.225(3)	C8–H6	0.89(3)	C28–H23	0.92(3)
O2–N1	1.227(3)	C9–C10	1.379(4)	C27–C26	1.385(3)
O3–C45	1.200(3)	C9–H7	0.90(3)	C27–H22	0.93(3)
O4–C46	1.390(3)	C10–C11	1.371(4)	C26–H21	0.90(3)
O4–C45	1.390(2)	C10–H8	0.92(3)	C31–C36	1.391(3)

continued on next page

Table A.35 – continued from previous page

O5–N2	1.219(3)	C11–C12	1.383(3)	C31–C32	1.392(3)
O6–N2	1.215(3)	C11–H9	0.94(3)	C36–C35	1.384(3)
N1–C42	1.462(3)	C12–H10	0.93(3)	C36–H30	0.91(3)
N2–C49	1.472(3)	C13–C18	1.393(3)	C35–C34	1.380(4)
C37–C38	1.445(3)	C13–C14	1.394(3)	C35–H29	0.92(3)
C37–C39	1.474(3)	C18–C17	1.384(3)	C34–C33	1.376(4)
C37–H31	0.92(2)	C18–H15	0.94(3)	C34–H28	0.91(3)
C38–C45	1.460(3)	C17–C16	1.375(4)	C33–C32	1.392(3)
C38–H32	0.90(2)	C17–H14	0.90(3)	C33–H27	0.95(3)
C39–C44	1.399(3)	C16–C15	1.377(4)	C32–H26	0.93(3)
C39–C40	1.401(3)	C16–H13	0.96(3)	C19–C20	1.382(3)
C40–C41	1.378(3)	C15–C14	1.387(3)	C19–C24	1.395(3)
C40–H37	0.94(3)	C15–H12	0.88(3)	C20–C21	1.397(4)
C41–C42	1.388(3)	C14–H11	0.94(2)	C20–H16	0.92(2)
C41–H38	0.89(3)	C1–C6	1.384(3)	C21–C22	1.367(5)
C42–C43	1.375(3)	C1–C2	1.396(3)	C21–H17	0.89(3)
C43–C44	1.381(3)	C6–C5	1.387(3)	C22–C23	1.379(5)
C43–H39	0.94(3)	C6–H5	0.95(2)	C22–H18	0.92(3)
C44–H40	0.98(3)	C5–C4	1.372(3)	C23–C24	1.385(4)
C46–C51	1.373(3)	C5–H4	0.90(3)	C23–H19	0.93(3)
C46–C47	1.379(3)	C4–C3	1.378(3)	C24–H20	0.96(3)
C47–C48	1.382(3)	C4–H3	0.91(3)	C38–H31	2.0186(198)

Table A.36.: Bond angles of **PtPhNO$_2$–PhNO$_2$** [°]:

C37–Pt1–C38	40.03(8)	C46–C47–H33	120.0(16)	C4–C3–C2	120.2(2)
C37–Pt1–P2	103.06(6)	C48–C47–H33	120.7(16)	C4–C3–H2	120.3(16)
C38–Pt1–P2	143.09(6)	C49–C48–C47	118.8(2)	C2–C3–H2	119.4(16)
C37–Pt1–P1	145.96(6)	C49–C48–H34	119.9(16)	C3–C2–C1	120.5(2)

continued on next page

Table A.36 – continued from previous page

C38–Pt1–P1	106.07(6)	C47–C48–H34	121.2(16)	C3–C2–H1	120.1(15)
P2–Pt1–P1	110.779(18)	C50–C49–C48	122.3(2)	C1–C2–H1	119.3(15)
C7–P1–C13	103.37(9)	C50–C49–N2	119.3(2)	C30–C25–C26	118.4(2)
C7–P1–C1	103.07(9)	C48–C49–N2	118.4(2)	C30–C25–P2	119.65(16)
C13–P1–C1	103.56(9)	C49–C50–C51	118.6(2)	C26–C25–P2	121.93(16)
C7–P1–Pt1	111.21(7)	C49–C50–H35	120.0(17)	C29–C30–C25	120.6(2)
C13–P1–Pt1	121.24(7)	C51–C50–H35	121.4(17)	C29–C30–H25	120.0(17)
C1–P1–Pt1	112.51(6)	C46–C51–C50	119.5(2)	C25–C30–H25	119.4(17)
C31–P2–C25	104.83(10)	C46–C51–H36	118.5(15)	C28–C29–C30	120.5(2)
C31–P2–C19	105.04(10)	C50–C51–H36	121.9(15)	C28–C29–H24	120.0(18)
C25–P2–C19	101.01(10)	C8–C7–C12	118.8(2)	C30–C29–H24	119.4(18)
C31–P2–Pt1	112.93(7)	C8–C7–P1	123.46(17)	C29–C28–C27	119.6(2)
C25–P2–Pt1	119.18(7)	C12–C7–P1	117.68(16)	C29–C28–H23	123.5(15)
C19–P2–Pt1	112.32(7)	C7–C8–C9	120.1(2)	C27–C28–H23	116.9(15)
C46–O4–C45	118.02(16)	C7–C8–H6	119.1(17)	C28–C27–C26	120.4(2)
O1–N1–O2	123.7(2)	C9–C8–H6	120.8(17)	C28–C27–H22	120.6(16)
O1–N1–C42	118.2(2)	C10–C9–C8	120.2(3)	C26–C27–H22	119.0(16)
O2–N1–C42	118.1(2)	C10–C9–H7	121.8(19)	C27–C26–C25	120.5(2)
O6–N2–O5	123.8(2)	C8–C9–H7	118.0(19)	C27–C26–H21	120.2(16)
O6–N2–C49	118.3(2)	C11–C10–C9	120.1(2)	C25–C26–H21	119.3(16)
O5–N2–C49	117.9(2)	C11–C10–H8	120.1(19)	C36–C31–C32	119.1(2)
C38–C37–C39	121.28(19)	C9–C10–H8	119.9(19)	C36–C31–P2	118.57(17)
C38–C37–Pt1	70.36(11)	C10–C11–C12	120.1(3)	C32–C31–P2	122.16(18)
C39–C37–Pt1	115.30(14)	C10–C11–H9	120.0(19)	C35–C36–C31	120.7(2)
C38–C37–H31	115.7(13)	C12–C11–H9	119.9(19)	C35–C36–H30	120.6(16)
C39–C37–H31	114.1(13)	C11–C12–C7	120.7(2)	C31–C36–H30	118.7(16)
Pt1–C37–H31	112.7(13)	C11–C12–H10	122.1(15)	C34–C35–C36	119.9(3)
C37–C38–C45	117.61(19)	C7–C12–H10	117.2(15)	C34–C35–H29	120.7(18)

continued on next page

Table A.36 – continued from previous page

C37–C38–Pt1	69.61(11)	C18–C13–C14	118.8(2)	C36–C35–H29	119.5(18)
C45–C38–Pt1	112.76(14)	C18–C13–P1	122.63(17)	C33–C34–C35	120.2(2)
C37–C38–H32	118.7(13)	C14–C13–P1	118.37(16)	C33–C34–H28	119.2(19)
C45–C38–H32	115.1(13)	C17–C18–C13	120.2(2)	C35–C34–H28	120.5(19)
Pt1–C38–H32	114.8(13)	C17–C18–H15	120.7(15)	C34–C33–C32	120.3(2)
C44–C39–C40	118.0(2)	C13–C18–H15	119.0(15)	C34–C33–H27	119.8(17)
C44–C39–C37	119.5(2)	C16–C17–C18	120.5(2)	C32–C33–H27	119.9(17)
C40–C39–C37	122.51(19)	C16–C17–H14	121.3(18)	C33–C32–C31	119.9(2)
C41–C40–C39	121.3(2)	C18–C17–H14	118.1(18)	C33–C32–H26	120.8(16)
C41–C40–H37	118.0(16)	C17–C16–C15	119.8(2)	C31–C32–H26	119.3(16)
C39–C40–H37	120.6(16)	C17–C16–H13	121.2(17)	C20–C19–C24	119.0(2)
C40–C41–C42	118.7(2)	C15–C16–H13	119.0(17)	C20–C19–P2	123.53(19)
C40–C41–H38	121.2(16)	C16–C15–C14	120.4(2)	C24–C19–P2	117.44(17)
C42–C41–H38	120.1(16)	C16–C15–H12	121.3(17)	C19–C20–C21	119.8(3)
C43–C42–C41	121.9(2)	C14–C15–H12	118.2(17)	C19–C20–H16	119.1(15)
C43–C42–N1	119.0(2)	C15–C14–C13	120.2(2)	C21–C20–H16	121.0(15)
C41–C42–N1	119.1(2)	C15–C14–H11	120.1(15)	C22–C21–C20	120.8(3)
C42–C43–C44	118.8(2)	C13–C14–H11	119.7(15)	C22–C21–H17	121.(2)
C42–C43–H39	120.9(16)	C6–C1–C2	118.82(19)	C20–C21–H17	118.(2)
C44–C43–H39	120.2(16)	C6–C1–P1	123.36(16)	C21–C22–C23	119.9(3)
C43–C44–C39	121.3(2)	C2–C1–P1	117.81(15)	C21–C22–H18	121.(2)
C43–C44–H40	119.7(14)	C1–C6–C5	119.9(2)	C23–C22–H18	119.(2)
C39–C44–H40	118.9(14)	C1–C6–H5	120.4(15)	C22–C23–C24	120.0(3)
O3–C45–O4	120.86(19)	C5–C6–H5	119.7(15)	C22–C23–H19	119.(2)
O3–C45–C38	128.85(19)	C4–C5–C6	120.9(2)	C24–C23–H19	121.(2)
O4–C45–C38	110.29(17)	C4–C5–H4	119.9(17)	C23–C24–C19	120.5(3)
C51–C46–C47	121.4(2)	C6–C5–H4	119.1(17)	C23–C24–H20	119.2(17)
C51–C46–O4	117.48(19)	C5–C4–C3	119.5(2)	C19–C24–H20	120.3(17)

continued on next page

Table A.36 – continued from previous page

| C47–C46–O4 | 120.8(2) | C5–C4–H3 | 119.9(16) |
| C46–C47–C48 | 119.2(2) | C3–C4–H3 | 120.6(16) |

NiPh−Me

Table A.37.: Structural parameters of **NiPh−Me**:

molecular formula	$C_{46}H_{40}NiO_2P_2$
molecular weight [a.m.u.]	745.43
colour/habit	red block
crystal dimensions [mm^3]	$0.25 \times 0.20 \times 0.10$
crystal system	triclinic
space group	P-1
unit cell dimensions	a = 10.3744(3) Å $\alpha = 102.425(2)°$
	b = 10.4307(3) Å $\beta = 93.986(2)°$
	c = 19.8392(7) Å $\gamma = 113.7190(10)°$
cell volume [$Å^3$]	1890.00(10)
Z	2
T [K]	150(2)
calculated density [gcm^{-3}]	1.310
absorption coefficient μ [mm^{-1}]	0.636
F(000)	780
Θ range [°]	1.07 - 25.38
index ranges [h, k, l]	$\pm12, \pm12, +22/-23$
no. of independent reflections / R_{int}	6944
no. of absorbed reflections	6121
no. of data	6944
no. of restraints	0
no. of parameters	620

continued on next page

Table A.37 – continued from previous page

R_1 $(I > 2\sigma(I))$	0.0263
wR_2 $(I > 2\sigma(I))$	0.0613
R_1 (all data)	0.0329
wR_2 (all data)	0.0655
goodness of fit (on F^2)	1.024
largest difference peak [e$\mathring{A}^3$]	0.330
largest difference hole [e$\mathring{A}^3$]	-0.314

Table A.38.: Bond lengths of **NiPh−Me** [$\mathring{A}$]:

C1–C6	1.390(2)	C17–C18	1.385(3)	C33–H27	0.97(2)
C1–C2	1.398(2)	C17–H14	0.962(19)	C34–C35	1.381(3)
C1–P1	1.8330(17)	C18–H15	0.968(19)	C34–H28	0.95(2)
C2–C3	1.387(2)	C19–C20	1.394(2)	C35–C36	1.384(3)
C2–H1	0.959(18)	C19–C24	1.400(2)	C35–H29	0.91(2)
C3–C4	1.385(3)	C19–P2	1.8389(17)	C36–H30	0.952(19)
C3–H2	0.96(2)	C20–C21	1.391(3)	C37–C38	1.417(2)
C4–C5	1.378(3)	C20–H16	0.912(19)	C37–C39	1.457(2)
C4–H3	0.95(2)	C21–C22	1.380(3)	C37–Ni1	1.9939(16)
C5–C6	1.392(3)	C21–H17	0.96(2)	C37–H31	0.949(19)
C5–H4	0.94(2)	C22–C23	1.387(3)	C38–C41	1.474(2)
C6–H5	0.93(2)	C22–H18	0.97(2)	C38–Ni1	1.9939(16)
C7–C12	1.400(2)	C23–C24	1.385(3)	C38–H32	0.948(18)
C7–C8	1.401(2)	C23–H19	0.94(2)	C39–O1	1.220(2)
C7–P1	1.8346(16)	C24–H20	0.973(18)	C39–O2	1.358(2)
C8–C9	1.389(3)	C25–C26	1.395(2)	C40–O2	1.439(2)
C8–H6	0.956(18)	C25–C30	1.396(2)	C40–H33	0.96(2)
C9–C10	1.385(3)	C25–P2	1.8296(16)	C40–H35	1.02(2)
C9–H7	0.96(2)	C26–C27	1.391(2)	C40–H34	0.93(2)

continued on next page

Table A.38 – continued from previous page

C10–C11	1.381(3)	C26–H21	0.961(19)	C41–C46	1.399(2)
C10–H8	0.93(2)	C27–C28	1.385(3)	C41–C42	1.403(2)
C11–C12	1.390(3)	C27–H22	0.94(2)	C42–C43	1.389(3)
C11–H9	0.94(2)	C28–C29	1.385(3)	C42–H36	0.964(19)
C12–H10	0.950(19)	C28–H23	0.952(19)	C43–C44	1.388(3)
C13–C14	1.393(2)	C29–C30	1.387(2)	C43–H37	0.97(2)
C13–C18	1.398(2)	C29–H24	0.97(2)	C44–C45	1.384(3)
C13–P1	1.8463(16)	C30–H25	0.957(18)	C44–H38	0.97(2)
C14–C15	1.388(3)	C31–C32	1.389(2)	C45–C46	1.384(3)
C14–H11	0.899(19)	C31–C36	1.395(2)	C45–H39	0.95(2)
C15–C16	1.380(3)	C31–P2	1.8441(16)	C46–H40	0.937(19)
C15–H12	0.95(2)	C32–C33	1.397(3)	P1–Ni1	2.1809(5)
C16–C17	1.389(3)	C32–H26	0.96(2)	P2–Ni1	2.1707(5)
C16–H13	0.967(19)	C33–C34	1.376(3)		

Table A.39.: Bond angles of **NiPh−Me** [°]:

C6–C1–C2	118.52(16)	C24–C19–P2	123.94(13)	C39–C37–H31	113.2(11)
C6–C1–P1	123.68(13)	C21–C20–C19	120.96(17)	Ni1–C37–H31	116.5(11)
C2–C1–P1	117.80(12)	C21–C20–H16	118.5(12)	C37–C38–C41	122.87(15)
C3–C2–C1	120.96(17)	C19–C20–H16	120.5(12)	C37–C38–Ni1	69.18(9)
C3–C2–H1	119.2(10)	C22–C21–C20	119.99(18)	C41–C38–Ni1	113.56(11)
C1–C2–H1	119.9(10)	C22–C21–H17	121.2(12)	C37–C38–H32	116.5(10)
C4–C3–C2	119.72(18)	C20–C21–H17	118.8(12)	C41–C38–H32	115.6(10)
C4–C3–H2	121.1(12)	C21–C22–C23	119.75(17)	Ni1–C38–H32	108.5(10)
C2–C3–H2	119.2(12)	C21–C22–H18	119.7(12)	O1–C39–O2	121.84(16)
C5–C4–C3	119.95(18)	C23–C22–H18	120.5(12)	O1–C39–C37	126.25(16)
C5–C4–H3	120.0(12)	C24–C23–C22	120.51(18)	O2–C39–C37	111.90(14)
C3–C4–H3	120.1(12)	C24–C23–H19	118.5(12)	O2–C40–H33	104.9(13)

continued on next page

Table A.39 – continued from previous page

C4–C5–C6	120.49(19)	C22–C23–H19	121.0(12)	O2–C40–H35	112.0(13)
C4–C5–H4	120.3(14)	C23–C24–C19	120.39(17)	H33–C40–H35	111.3(18)
C6–C5–H4	119.1(14)	C23–C24–H20	120.3(11)	O2–C40–H34	110.0(14)
C1–C6–C5	120.31(18)	C19–C24–H20	119.3(10)	H33–C40–H34	113.5(19)
C1–C6–H5	121.6(12)	C26–C25–C30	118.94(16)	H35–C40–H34	105.3(18)
C5–C6–H5	118.1(12)	C26–C25–P2	125.50(13)	C46–C41–C42	117.64(16)
C12–C7–C8	118.70(15)	C30–C25–P2	115.28(12)	C46–C41–C38	119.57(16)
C12–C7–P1	122.29(13)	C27–C26–C25	120.26(17)	C42–C41–C38	122.79(15)
C8–C7–P1	119.00(13)	C27–C26–H21	120.0(11)	C43–C42–C41	121.11(17)
C9–C8–C7	120.26(17)	C25–C26–H21	119.8(11)	C43–C42–H36	121.2(11)
C9–C8–H6	120.8(10)	C28–C27–C26	120.27(17)	C41–C42–H36	117.7(11)
C7–C8–H6	118.9(10)	C28–C27–H22	120.1(12)	C44–C43–C42	120.13(19)
C10–C9–C8	120.17(18)	C26–C27–H22	119.6(12)	C44–C43–H37	121.5(12)
C10–C9–H7	121.1(13)	C29–C28–C27	119.80(17)	C42–C43–H37	118.4(12)
C8–C9–H7	118.7(13)	C29–C28–H23	120.6(12)	C45–C44–C43	119.38(18)
C11–C10–C9	120.34(18)	C27–C28–H23	119.6(12)	C45–C44–H38	120.9(13)
C11–C10–H8	118.8(13)	C28–C29–C30	120.21(18)	C43–C44–H38	119.7(13)
C9–C10–H8	120.9(13)	C28–C29–H24	121.6(11)	C46–C45–C44	120.65(18)
C10–C11–C12	119.89(18)	C30–C29–H24	118.2(11)	C46–C45–H39	118.1(13)
C10–C11–H9	121.9(13)	C29–C30–C25	120.45(17)	C44–C45–H39	121.2(13)
C12–C11–H9	118.2(12)	C29–C30–H25	119.5(10)	C45–C46–C41	121.07(18)
C11–C12–C7	120.62(17)	C25–C30–H25	120.(1)	C45–C46–H40	120.4(11)
C11–C12–H10	118.8(12)	C32–C31–C36	118.40(16)	C41–C46–H40	118.5(11)
C7–C12–H10	120.6(11)	C32–C31–P2	124.79(14)	C39–O2–C40	115.29(15)
C14–C13–C18	117.78(16)	C36–C31–P2	116.76(13)	C1–P1–C7	103.07(7)
C14–C13–P1	123.77(13)	C31–C32–C33	120.38(19)	C1–P1–C13	101.66(7)
C18–C13–P1	118.46(13)	C31–C32–H26	119.8(12)	C7–P1–C13	101.03(7)
C15–C14–C13	120.86(17)	C33–C32–H26	119.8(12)	C1–P1–Ni1	113.67(5)

continued on next page

Table A.39 – continued from previous page

C15–C14–H11	120.3(12)	C34–C33–C32	120.3(2)	C7–P1–Ni1	118.19(6)
C13–C14–H11	118.9(12)	C34–C33–H27	122.9(13)	C13–P1–Ni1	116.87(5)
C16–C15–C14	120.80(18)	C32–C33–H27	116.7(13)	C25–P2–C19	108.06(7)
C16–C15–H12	120.2(13)	C33–C34–C35	119.83(19)	C25–P2–C31	99.77(8)
C14–C15–H12	119.0(13)	C33–C34–H28	119.5(14)	C19–P2–C31	99.53(7)
C15–C16–C17	119.07(17)	C35–C34–H28	120.6(14)	C25–P2–Ni1	114.84(5)
C15–C16–H13	121.3(11)	C34–C35–C36	120.1(2)	C19–P2–Ni1	115.75(6)
C17–C16–H13	119.7(11)	C34–C35–H29	121.6(13)	C31–P2–Ni1	116.74(5)
C18–C17–C16	120.26(18)	C36–C35–H29	118.3(13)	C37–Ni1–C38	41.64(7)
C18–C17–H14	118.5(11)	C35–C36–C31	120.99(19)	C37–Ni1–P2	105.95(5)
C16–C17–H14	121.2(11)	C35–C36–H30	120.9(11)	C38–Ni1–P2	147.09(5)
C17–C18–C13	121.23(17)	C31–C36–H30	118.1(11)	C37–Ni1–P1	141.12(5)
C17–C18–H15	119.9(11)	C38–C37–C39	119.14(15)	C38–Ni1–P1	101.05(5)
C13–C18–H15	118.8(11)	C38–C37–Ni1	69.18(9)	P2–Ni1–P1	111.863(18)
C20–C19–C24	118.37(16)	C39–C37–Ni1	104.53(11)		
C20–C19–P2	117.49(13)	C38–C37–H31	123.7(11)		

PtPhNO$_2$–^{i}Pr

Table A.40.: Structural parameters of **PtPhNO$_2$–^{i}Pr**:

molecular formula	C$_{51}$ · 50 H$_{47}$NO$_4$P$_2$Pt
molecular weight [a.m.u.]	1000.93
colour/habit	yellow block
crystal dimensions [mm^3]	0.20 × 0.19 × 0.14
crystal system	monoclinic
space group	P2$_1$/c
unit cell dimensions	a = 12.1942(3) Å α = 90.00°
	b = 20.9616(6) Å β = 116.6680(10)°

continued on next page

Table A.40 – continued from previous page

	$c = 19.1668(5)$ Å	$\gamma = 90.00°$
cell volume [Å^3]	4378.1(2)	
Z	4	
T [K]	150(2)	
calculated density [gcm^{-3}]	1.519	
absorption coefficient μ [mm^{-1}]	3.325	
F(000)	2012	
Θ range [°]	1.54 - 25.43	
index ranges [h, k, l]	$\pm14, +23/-25, \pm23$	
no. of independent reflections / R$_{int}$	8048	
no. of absorbed reflections	7428	
no. of data	8048	
no. of restraints	6	
no. of parameters	524	
R$_1$ $(I > 2\sigma(I))$	0.0180	
wR$_2$ $(I > 2\sigma(I))$	0.0419	
R$_1$ (all data)	0.0209	
wR$_2$ (all data)	0.0431	
goodness of fit (on F^2)	1.041	
largest difference peak [eÅ^3]	0.952	
largest difference hole [eÅ^3]	-0.565	

Table A.41.: Bond lengths of **PtPhNO$_2$–^{i}Pr** [Å]:

Pt1–C38	2.122(2)	C14–H14	0.9500	C37–C38	1.445(3)
Pt1–C37	2.123(2)	C15–C16	1.378(4)	C37–C39	1.477(3)
Pt1–P2	2.2839(6)	C15–H15	0.9500	C37–H37	0.8922
Pt1–P1	2.2877(6)	C16–C17	1.381(4)	C38–C45	1.472(3)
P1–C13	1.831(2)	C16–H16	0.9500	C38–H38	0.9126

continued on next page

Table A.41 – continued from previous page

P1–C7	1.835(2)	C17–C18	1.392(3)	C46–C47	1.514(4)
P1–C1	1.838(2)	C17–H17	0.9500	C46–C48	1.515(3)
P2–C19	1.833(2)	C18–H18	0.9500	C46–H46	1.3900
P2–C25	1.835(2)	C19–C20	1.386(3)	C47–H47A	0.9800
P2–C31	1.841(2)	C19–C24	1.399(3)	C47–H47B	0.9800
N1–O2	1.224(3)	C20–C21	1.392(4)	C47–H47C	0.9800
N1–O1	1.236(3)	C20–H20	0.9500	C48–H48A	0.9800
N1–C42	1.469(3)	C21–C22	1.380(4)	C48–H48B	0.9800
O3–C45	1.216(3)	C21–H21	0.9500	C48–H48C	0.9800
O4–C45	1.356(3)	C22–C23	1.380(4)	C39–C40	1.402(3)
O4–C46	1.462(3)	C22–H22	0.9500	C39–C44	1.406(3)
C1–C2	1.398(3)	C23–C24	1.386(4)	C40–C41	1.384(3)
C1–C6	1.401(3)	C23–H23	0.9500	C40–H40	0.9500
C2–C3	1.389(4)	C24–H24	0.9500	C41–C42	1.385(4)
C2–H2	0.9500	C25–C26	1.392(3)	C41–H41	0.9500
C3–C4	1.383(4)	C25–C30	1.403(3)	C42–C43	1.379(4)
C3–H3	0.9500	C26–C27	1.389(4)	C43–C44	1.384(3)
C4–C5	1.385(4)	C26–H26	0.9500	C43–H43	0.9500
C4–H4	0.9500	C27–C28	1.378(4)	C44–H44	0.9500
C5–C6	1.389(3)	C27–H27	0.9500	C51–C52	1.3900
C5–H5	0.9500	C28–C29	1.382(4)	C51–C56	1.3900
C6–H6	0.9500	C28–H28	0.9500	C51–H51	0.9500
C7–C12	1.392(3)	C29–C30	1.386(4)	C52–C53	1.3900
C7–C8	1.400(3)	C29–H29	0.9500	C52–H52	0.9500
C8–C9	1.386(3)	C30–H30	0.9500	C53–C54	1.3900
C8–H8	0.9500	C31–C32	1.395(3)	C53–H53	0.9500
C9–C10	1.380(4)	C31–C36	1.400(3)	C54–C55	1.3900
C9–H9	0.9500	C32–C33	1.395(4)	C54–H54	0.9500

continued on next page

Table A.41 – continued from previous page

C10–C11	1.377(4)	C32–H32	0.9500	C55–C56	1.3900
C10–H10	0.9500	C33–C34	1.378(4)	C55–H55	0.9500
C11–C12	1.396(3)	C33–H33	0.9500	C56–C57	1.423(14)
C11–H11	0.9500	C34–C35	1.390(4)	C57–H57A	0.9800
C12–H12	0.9500	C34–H34	0.9500	C57–H57B	0.9800
C13–C14	1.392(3)	C35–C36	1.383(3)	C57–H57C	0.9800
C13–C18	1.395(3)	C35–H35	0.9500		
C14–C15	1.391(4)	C36–H36	0.9500		

Table A.42.: Bond angles of **PtPhNO$_2$–^{i}Pr** [°]:

C38–Pt1–C37	39.80(8)	C15–C16–C17	119.5(2)	C39–C37–H37	113.000
C38–Pt1–P2	142.55(6)	C15–C16–H16	120.300	Pt1–C37–H37	112.900
C37–Pt1–P2	102.80(6)	C17–C16–H16	120.300	C37–C38–C45	117.3(2)
C38–Pt1–P1	109.70(6)	C16–C17–C18	120.0(2)	C37–C38–Pt1	70.14(12)
C37–Pt1–P1	149.49(6)	C16–C17–H17	120.000	C45–C38–Pt1	110.03(15)
P2–Pt1–P1	107.70(2)	C18–C17–H17	120.000	C37–C38–H38	119.400
C13–P1–C7	103.7(1)	C17–C18–C13	121.0(2)	C45–C38–H38	115.000
C13–P1–C1	103.51(10)	C17–C18–H18	119.500	Pt1–C38–H38	116.800
C7–P1–C1	98.73(10)	C13–C18–H18	119.500	O3–C45–O4	123.1(2)
C13–P1–Pt1	116.55(7)	C20–C19–C24	118.7(2)	O3–C45–C38	126.2(2)
C7–P1–Pt1	114.18(8)	C20–C19–P2	117.35(18)	O4–C45–C38	110.7(2)
C1–P1–Pt1	117.76(7)	C24–C19–P2	123.81(19)	O4–C46–C47	109.0(2)
C19–P2–C25	106.73(11)	C19–C20–C21	120.9(2)	O4–C46–C48	104.7(2)
C19–P2–C31	100.0(1)	C19–C20–H20	119.500	C47–C46–C48	113.9(2)
C25–P2–C31	104.46(11)	C21–C20–H20	119.500	O4–C46–H46	109.700
C19–P2–Pt1	118.47(8)	C22–C21–C20	119.9(3)	C47–C46–H46	109.700
C25–P2–Pt1	114.40(8)	C22–C21–H21	120.100	C48–C46–H46	109.700
C31–P2–Pt1	111.01(8)	C20–C21–H21	120.100	C46–C47–H47A	109.500

continued on next page

Table A.42 – continued from previous page

O2–N1–O1	123.8(3)	C21–C22–C23	119.7(2)	C46–C47–H47B	109.500
O2–N1–C42	118.0(3)	C21–C22–H22	120.200	H47A–C47–H47B	109.500
O1–N1–C42	118.1(3)	C23–C22–H22	120.200	C46–C47–H47C	109.500
C45–O4–C46	117.48(18)	C22–C23–C24	120.8(2)	H47A–C47–H47C	109.500
C2–C1–C6	118.5(2)	C22–C23–H23	119.600	H47B–C47–H47C	109.500
C2–C1–P1	123.67(18)	C24–C23–H23	119.600	C46–C48–H48A	109.500
C6–C1–P1	117.25(18)	C23–C24–C19	120.0(2)	C46–C48–H48B	109.500
C3–C2–C1	120.5(2)	C23–C24–H24	120.000	H48A–C48–H48B	109.500
C3–C2–H2	119.700	C19–C24–H24	120.000	C46–C48–H48C	109.500
C1–C2–H2	119.700	C26–C25–C30	118.4(2)	H48A–C48–H48C	109.500
C4–C3–C2	120.5(2)	C26–C25–P2	122.71(19)	H48B–C48–H48C	109.500
C4–C3–H3	119.800	C30–C25–P2	118.56(19)	C40–C39–C44	118.1(2)
C2–C3–H3	119.800	C27–C26–C25	120.4(3)	C40–C39–C37	123.2(2)
C3–C4–C5	119.6(2)	C27–C26–H26	119.800	C44–C39–C37	118.7(2)
C3–C4–H4	120.200	C25–C26–H26	119.800	C41–C40–C39	121.1(2)
C5–C4–H4	120.200	C28–C27–C26	120.6(3)	C41–C40–H40	119.400
C4–C5–C6	120.5(2)	C28–C27–H27	119.700	C39–C40–H40	119.400
C4–C5–H5	119.800	C26–C27–H27	119.700	C40–C41–C42	118.7(2)
C6–C5–H5	119.800	C27–C28–C29	119.8(3)	C40–C41–H41	120.700
C5–C6–C1	120.4(2)	C27–C28–H28	120.100	C42–C41–H41	120.700
C5–C6–H6	119.800	C29–C28–H28	120.100	C43–C42–C41	122.3(2)
C1–C6–H6	119.800	C28–C29–C30	120.2(3)	C43–C42–N1	118.4(2)
C12–C7–C8	118.8(2)	C28–C29–H29	119.900	C41–C42–N1	119.3(3)
C12–C7–P1	120.39(18)	C30–C29–H29	119.900	C42–C43–C44	118.5(2)
C8–C7–P1	120.83(18)	C29–C30–C25	120.6(3)	C42–C43–H43	120.700
C9–C8–C7	120.6(2)	C29–C30–H30	119.700	C44–C43–H43	120.700
C9–C8–H8	119.700	C25–C30–H30	119.700	C43–C44–C39	121.3(2)
C7–C8–H8	119.700	C32–C31–C36	118.5(2)	C43–C44–H44	119.300

continued on next page

Table A.42 – continued from previous page

C10–C9–C8	119.9(2)	C32–C31–P2	124.29(19)	C39–C44–H44	119.300
C10–C9–H9	120.000	C36–C31–P2	117.24(17)	C52–C51–C56	120.000
C8–C9–H9	120.000	C31–C32–C33	120.3(2)	C52–C51–H51	120.000
C11–C10–C9	120.4(2)	C31–C32–H32	119.900	C56–C51–H51	120.000
C11–C10–H10	119.800	C33–C32–H32	119.900	C53–C52–C51	120.000
C9–C10–H10	119.800	C34–C33–C32	120.5(2)	C53–C52–H52	120.000
C10–C11–C12	120.1(2)	C34–C33–H33	119.800	C51–C52–H52	120.000
C10–C11–H11	119.900	C32–C33–H33	119.800	C54–C53–C52	120.000
C12–C11–H11	119.900	C33–C34–C35	119.8(2)	C54–C53–H53	120.000
C7–C12–C11	120.1(2)	C33–C34–H34	120.100	C52–C53–H53	120.000
C7–C12–H12	119.900	C35–C34–H34	120.100	C55–C54–C53	120.000
C11–C12–H12	119.900	C36–C35–C34	120.0(3)	C55–C54–H54	120.000
C14–C13–C18	118.3(2)	C36–C35–H35	120.000	C53–C54–H54	120.000
C14–C13–P1	123.37(18)	C34–C35–H35	120.000	C54–C55–C56	120.000
C18–C13–P1	118.12(17)	C35–C36–C31	120.9(2)	C54–C55–H55	120.000
C15–C14–C13	120.2(2)	C35–C36–H36	119.500	C56–C55–H55	120.000
C15–C14–H14	119.900	C31–C36–H36	119.500	C55–C56–C51	120.000
C13–C14–H14	119.900	C38–C37–C39	123.3(2)	C55–C56–C57	118.3(6)
C16–C15–C14	120.9(2)	C38–C37–Pt1	70.06(12)	C51–C56–C57	121.7(6)
C16–C15–H15	119.500	C39–C37–Pt1	114.77(15)		
C14–C15–H15	119.500	C38–C37–H37	115.600		

PdPhCl–PhNO$_2$

Table A.43.: Structural parameters of **PdPhCl–PhNO$_2$**:

molecular formula	$C_{58}H_{49}ClNO_4P_2Pd$
molecular weight [a.m.u.]	1027.77
colour/habit	yellow block

continued on next page

Table A.43 – continued from previous page

crystal system	triclinic
space group	P-1
unit cell dimensions	a = 9.4951(3) Å　　$\alpha = 84.034(2)°$
	b = 14.3559(5) Å　　$\beta = 80.540(2)°$
	c = 18.8085(7) Å　　$\gamma = 73.842(2)°$
cell volume [Å^3]	2424.45(15)
Z	2
T [K]	296(2)
calculated density [gcm^{-3}]	1.408
absorption coefficient μ [mm^{-1}]	0.553
F(000)	1058
Θ range [°]	4.17 - 25.50
index ranges [h, k, l]	±11, ±17, ±22
no. of independent reflections / R$_{int}$	8953
no. of absorbed reflections	7842
no. of data	8953
no. of restraints	0
no. of parameters	605
R$_1$ $(I > 2\sigma(I))$	0.0290
wR$_2$ $(I > 2\sigma(I))$	0.0870
R$_1$ (all data)	0.0364
wR$_2$ (all data)	0.0933
goodness of fit (on F^2)	1.052

Table A.44.: Bond lengths of **PdPhCl–PhNO$_2$** [Å]:

Pd1–C9	2.137(2)	C9–C10	1.473(3)	C34–C35	1.391(3)
Pd1–C8	2.166(2)	C10–C11	1.400(3)	C34–C39	1.391(3)
Pd1–P1	2.3250(6)	C10–C15	1.402(3)	C35–C36	1.387(4)

continued on next page

Table A.44 – continued from previous page

Pd1–P2	2.3449(6)	C11–C12	1.385(3)	C36–C37	1.386(4)
P1–C28	1.828(2)	C12–C13	1.381(3)	C37–C38	1.381(4)
P1–C22	1.831(2)	C13–C14	1.379(4)	C38–C39	1.397(3)
P1–C16	1.835(2)	C14–C15	1.387(3)	C40–C41	1.387(3)
P2–C34	1.831(2)	C16–C21	1.393(3)	C40–C45	1.399(3)
P2–C40	1.833(2)	C16–C17	1.398(3)	C41–C42	1.388(4)
P2–C46	1.836(2)	C17–C18	1.388(3)	C42–C43	1.381(4)
Cl1–C13	1.748(2)	C18–C19	1.383(4)	C43–C44	1.388(4)
N1–O2	1.217(3)	C19–C20	1.378(4)	C44–C45	1.387(4)
N1–O1	1.228(3)	C20–C21	1.388(4)	C46–C47	1.385(4)
N1–C1	1.473(3)	C22–C23	1.397(3)	C46–C51	1.388(3)
O3–C4	1.385(3)	C22–C27	1.398(3)	C47–C48	1.386(4)
O3–C7	1.398(3)	C23–C24	1.391(4)	C48–C49	1.379(4)
O4–C7	1.205(3)	C24–C25	1.381(4)	C49–C50	1.377(4)
C1–C2	1.379(4)	C25–C26	1.386(4)	C50–C51	1.396(4)
C1–C6	1.390(4)	C26–C27	1.386(3)	C52–C53	1.515(4)
C2–C3	1.380(4)	C28–C33	1.393(3)	C53–C58	1.384(4)
C3–C4	1.384(3)	C28–C29	1.393(3)	C53–C54	1.387(4)
C4–C5	1.382(4)	C29–C30	1.386(3)	C54–C55	1.388(4)
C5–C6	1.386(4)	C30–C31	1.382(4)	C55–C56	1.383(4)
C7–C8	1.442(3)	C31–C32	1.383(4)	C56–C57	1.377(4)
C8–C9	1.427(3)	C32–C33	1.386(3)	C57–C58	1.386(4)

Table A.45.: Bond angles of **PdPhCl–PhNO$_2$** [°]:

C9–Pd1–C8	38.73(9)	C9–C8–Pd1	69.54(12)	C30–C31–C32	120.0(2)
C9–Pd1–P1	99.34(6)	C7–C8–Pd1	106.15(15)	C31–C32–C33	120.3(2)
C8–Pd1–P1	137.57(6)	C8–C9–C10	123.6(2)	C32–C33–C28	120.1(2)
C9–Pd1–P2	153.02(6)	C8–C9–Pd1	71.73(13)	C35–C34–C39	119.1(2)

continued on next page

Table A.45 – continued from previous page

C8–Pd1–P2	114.84(6)	C10–C9–Pd1	114.14(15)	C35–C34–P2	117.48(18)
P1–Pd1–P2	107.47(2)	C11–C10–C15	117.5(2)	C39–C34–P2	123.46(18)
C28–P1–C22	103.01(10)	C11–C10–C9	123.0(2)	C36–C35–C34	120.6(2)
C28–P1–C16	104.25(10)	C15–C10–C9	119.4(2)	C37–C36–C35	120.2(2)
C22–P1–C16	102.28(10)	C12–C11–C10	121.5(2)	C38–C37–C36	119.8(2)
C28–P1–Pd1	114.92(8)	C13–C12–C11	119.0(2)	C37–C38–C39	120.3(2)
C22–P1–Pd1	119.23(7)	C14–C13–C12	121.4(2)	C34–C39–C38	120.1(2)
C16–P1–Pd1	111.40(7)	C14–C13–Cl1	119.19(19)	C41–C40–C45	118.6(2)
C34–P2–C40	104.56(11)	C12–C13–Cl1	119.37(19)	C41–C40–P2	118.40(18)
C34–P2–C46	102.16(10)	C13–C14–C15	119.1(2)	C45–C40–P2	123.03(18)
C40–P2–C46	101.57(10)	C14–C15–C10	121.4(2)	C40–C41–C42	120.9(2)
C34–P2–Pd1	113.69(8)	C21–C16–C17	118.9(2)	C43–C42–C41	120.2(3)
C40–P2–Pd1	114.71(7)	C21–C16–P1	123.37(18)	C42–C43–C44	119.6(2)
C46–P2–Pd1	118.30(8)	C17–C16–P1	117.73(18)	C45–C44–C43	120.3(2)
O2–N1–O1	124.3(3)	C18–C17–C16	120.5(2)	C44–C45–C40	120.4(2)
O2–N1–C1	118.5(2)	C19–C18–C17	119.9(2)	C47–C46–C51	119.1(2)
O1–N1–C1	117.3(3)	C20–C19–C18	120.1(2)	C47–C46–P2	117.83(18)
C4–O3–C7	118.61(18)	C19–C20–C21	120.5(3)	C51–C46–P2	122.96(19)
C2–C1–C6	122.7(2)	C20–C21–C16	120.1(2)	C46–C47–C48	121.0(2)
C2–C1–N1	119.0(2)	C23–C22–C27	118.7(2)	C49–C48–C47	119.7(3)
C6–C1–N1	118.3(2)	C23–C22–P1	122.06(17)	C50–C49–C48	120.0(2)
C3–C2–C1	118.7(2)	C27–C22–P1	118.97(17)	C49–C50–C51	120.4(2)
C2–C3–C4	119.2(2)	C24–C23–C22	120.6(2)	C46–C51–C50	119.8(2)
C5–C4–C3	121.9(2)	C25–C24–C23	119.8(2)	C58–C53–C54	118.2(3)
C5–C4–O3	122.1(2)	C24–C25–C26	120.3(2)	C58–C53–C52	121.7(3)
C3–C4–O3	115.8(2)	C25–C26–C27	120.1(2)	C54–C53–C52	120.1(3)
C4–C5–C6	119.4(2)	C26–C27–C22	120.4(2)	C53–C54–C55	121.0(3)
C5–C6–C1	118.1(2)	C33–C28–C29	119.0(2)	C56–C55–C54	119.9(3)

continued on next page

Table A.45 – continued from previous page

O4–C7–O3	121.3(2)	C33–C28–P1	118.17(18)	C57–C56–C55	119.8(3)
O4–C7–C8	129.1(2)	C29–C28–P1	122.80(18)	C56–C57–C58	120.0(3)
O3–C7–C8	109.57(19)	C30–C29–C28	120.6(2)	C53–C58–C57	121.2(3)
C9–C8–C7	117.7(2)	C31–C30–C29	119.9(2)		

PdPhNO$_2$–PhOMe

Table A.46.: Structural parameters of **PdPhNO$_2$–PhOMe**:

molecular formula	C$_{52}$H$_{43}$NO$_5$Pd
molecular weight [a.m.u.]	930.21
colour/habit	orange fragment
crystal dimensions [mm^3]	0.15 × 0.10 × 0.05
crystal system	monoclinic
space group	P2$_1$7c
unit cell dimensions	a = 23.3448(9) Å $\alpha = 90.00°$
	b = 9.6937(3) Å $\beta = 92.460(2)°$
	c = 19.6595(7) Å $\gamma = 90.00°$
cell volume [Å^3]	4444.8(3)
Z	4
T [K]	173(2)
calculated density [gcm^{-3}]	1.390
absorption coefficient μ [mm^{-1}]	0.539
F(000)	1912
Θ range [°]	0.87 - 23.23
index ranges [h, k, l]	±25, ±10, ±21
no. of independent reflections / R$_{int}$	6140
no. of absorbed reflections	5463
no. of data	6140

continued on next page

Table A.46 – continued from previous page

no. of restraints	0
no. of parameters	722
R_1 $(I > 2\sigma(I))$	0.0274
wR_2 $(I > 2\sigma(I))$	0.0628
R_1 (all data)	0.0331
wR_2 (all data)	0.0653
goodness of fit (on F^2)	0.955
largest difference peak [eÅ^3]	0.352
largest difference hole [eÅ^3]	-0.375

Table A.47.: Bond lengths of **PdPhNO$_2$–PhOMe** [Å]:

C1–C6	1.382(5)	C19–P2	1.835(3)	C38–C45	1.460(4)
C1–C2	1.392(4)	C20–C21	1.382(5)	C38–Pd1	2.140(3)
C1–P1	1.835(3)	C20–H16	0.89(3)	C38–H32	0.92(3)
C2–C3	1.370(5)	C21–C22	1.371(5)	C39–C44	1.397(5)
C2–H1	0.85(3)	C21–H17	0.90(4)	C39–C40	1.399(5)
C3–C4	1.375(5)	C22–C23	1.373(5)	C40–C41	1.375(5)
C3–H2	0.93(4)	C22–H18	0.91(4)	C40–H33	0.89(3)
C4–C5	1.377(5)	C23–C24	1.381(5)	C41–C42	1.382(5)
C4–H3	0.94(4)	C23–H19	0.90(4)	C41–H34	0.92(4)
C5–C6	1.383(5)	C24–H20	0.89(3)	C42–C43	1.376(5)
C5–H4	0.89(4)	C25–C26	1.390(5)	C42–N1	1.461(4)
C6–H5	0.93(3)	C25–C30	1.389(5)	C43–C44	1.375(5)
C7–C8	1.390(5)	C25–P2	1.832(3)	C43–H35	0.84(3)
C7–C12	1.393(5)	C26–C27	1.384(5)	C44–H36	0.87(3)
C7–P1	1.829(3)	C26–H21	0.90(3)	C45–O3	1.202(4)
C8–C9	1.387(6)	C27–C28	1.372(6)	C45–O4	1.377(4)
C8–H6	0.86(3)	C27–H22	0.92(4)	C46–C47	1.368(5)

continued on next page

Table A.47 – continued from previous page

C9–C10	1.377(6)	C28–C29	1.380(6)	C46–C51	1.372(5)
C9–H7	0.92(4)	C28–H23	0.90(4)	C46–O4	1.412(4)
C10–C11	1.372(6)	C29–C30	1.383(5)	C47–C48	1.389(5)
C10–H8	0.90(4)	C29–H24	0.90(4)	C47–H37	0.90(4)
C11–C12	1.377(5)	C30–H25	0.89(3)	C48–C49	1.373(5)
C11–H9	0.91(4)	C31–C32	1.384(5)	C48–H38	0.88(4)
C12–H10	0.87(3)	C31–C36	1.395(5)	C49–O5	1.375(4)
C13–C14	1.391(5)	C31–P2	1.835(3)	C49–C50	1.386(5)
C13–C18	1.392(5)	C32–C33	1.395(6)	C50–C51	1.377(5)
C13–P1	1.836(3)	C32–H26	0.88(3)	C50–H39	0.90(4)
C14–C15	1.394(5)	C33–C34	1.371(7)	C51–H40	0.93(4)
C14–H11	0.92(3)	C33–H27	0.89(4)	C52–O5	1.419(5)
C15–C16	1.370(6)	C34–C35	1.376(7)	C52–H41	0.95(4)
C15–H12	0.88(4)	C34–H28	0.92(5)	C52–H42	1.00(4)
C16–C17	1.372(6)	C35–C36	1.384(5)	C52–H43	0.98(4)
C16–H13	0.93(4)	C35–H29	0.94(5)	N1–O2	1.228(4)
C17–C18	1.388(5)	C36–H30	0.95(3)	N1–O1	1.234(4)
C17–H14	0.91(4)	C37–C38	1.413(5)	P1–Pd1	2.3334(8)
C18–H15	0.91(3)	C37–C39	1.471(5)	P2–Pd1	2.3216(8)
C19–C20	1.381(5)	C37–Pd1	2.132(3)		
C19–C24	1.387(5)	C37–H31	0.86(3)		

Table A.48.: Bond angles of **PdPhNO$_2$–PhOMe** [°]:

C6–C1–C2	117.8(3)	C21–C22–C23	119.3(3)	C40–C41–C42	119.0(3)
C6–C1–P1	124.5(2)	C21–C22–H18	121.(2)	C40–C41–H34	124.(2)
C2–C1–P1	117.7(2)	C23–C22–H18	119.(2)	C42–C41–H34	117.(2)
C3–C2–C1	121.6(3)	C22–C23–C24	120.3(4)	C43–C42–C41	121.2(3)
C3–C2–H1	119.(2)	C22–C23–H19	120.(2)	C43–C42–N1	119.1(3)

continued on next page

Table A.48 – continued from previous page

C1–C2–H1	119.(2)	C24–C23–H19	119.(2)	C41–C42–N1	119.7(3)
C2–C3–C4	120.4(3)	C23–C24–C19	120.8(3)	C42–C43–C44	119.2(4)
C2–C3–H2	119.(2)	C23–C24–H20	120.(2)	C42–C43–H35	121.(2)
C4–C3–H2	120.(2)	C19–C24–H20	119.(2)	C44–C43–H35	120.(2)
C5–C4–C3	118.7(3)	C26–C25–C30	118.4(3)	C43–C44–C39	121.7(4)
C5–C4–H3	120.(2)	C26–C25–P2	123.1(3)	C43–C44–H36	121.(2)
C3–C4–H3	122.(2)	C30–C25–P2	118.4(3)	C39–C44–H36	117.(2)
C4–C5–C6	121.2(4)	C27–C26–C25	120.4(4)	O3–C45–O4	121.7(3)
C4–C5–H4	119.(2)	C27–C26–H21	120.(2)	O3–C45–C38	128.3(3)
C6–C5–H4	120.(2)	C25–C26–H21	120.(2)	O4–C45–C38	109.9(3)
C1–C6–C5	120.4(3)	C28–C27–C26	120.6(4)	C47–C46–C51	120.9(3)
C1–C6–H5	119.(2)	C28–C27–H22	120.(2)	C47–C46–O4	120.0(3)
C5–C6–H5	121.(2)	C26–C27–H22	119.(2)	C51–C46–O4	118.8(3)
C8–C7–C12	118.4(3)	C27–C28–C29	119.8(4)	C46–C47–C48	119.9(4)
C8–C7–P1	123.1(3)	C27–C28–H23	120.(2)	C46–C47–H37	120.(2)
C12–C7–P1	118.5(3)	C29–C28–H23	120.(2)	C48–C47–H37	120.(2)
C9–C8–C7	120.5(4)	C28–C29–C30	119.9(4)	C49–C48–C47	119.6(4)
C9–C8–H6	120.(2)	C28–C29–H24	122.(2)	C49–C48–H38	122.(2)
C7–C8–H6	119.(2)	C30–C29–H24	118.(2)	C47–C48–H38	118.(2)
C10–C9–C8	119.8(4)	C29–C30–C25	120.9(4)	C48–C49–O5	124.7(3)
C10–C9–H7	122.(2)	C29–C30–H25	119.(2)	C48–C49–C50	119.9(3)
C8–C9–H7	118.(2)	C25–C30–H25	120.(2)	O5–C49–C50	115.4(3)
C11–C10–C9	120.4(4)	C32–C31–C36	118.5(3)	C51–C50–C49	120.4(4)
C11–C10–H8	120.(3)	C32–C31–P2	124.0(3)	C51–C50–H39	123.(2)
C9–C10–H8	120.(3)	C36–C31–P2	117.5(3)	C49–C50–H39	117.(2)
C10–C11–C12	119.9(4)	C31–C32–C33	120.0(4)	C46–C51–C50	119.3(3)
C10–C11–H9	123.(3)	C31–C32–H26	117.(2)	C46–C51–H40	118.(2)
C12–C11–H9	117.(3)	C33–C32–H26	123.(2)	C50–C51–H40	123.(2)

continued on next page

Table A.48 – continued from previous page

C11–C12–C7	120.9(4)	C34–C33–C32	120.9(5)	O5–C52–H41	111.(3)
C11–C12–H10	123.(2)	C34–C33–H27	124.(3)	O5–C52–H42	111.(2)
C7–C12–H10	117.(2)	C32–C33–H27	115.(3)	H41–C52–H42	107.(3)
C14–C13–C18	118.5(3)	C35–C34–C33	119.5(4)	O5–C52–H43	100.(2)
C14–C13–P1	122.1(3)	C35–C34–H28	119.(3)	H41–C52–H43	114.(4)
C18–C13–P1	118.9(2)	C33–C34–H28	121.(3)	H42–C52–H43	114.(3)
C13–C14–C15	119.9(4)	C34–C35–C36	120.1(5)	O2–N1–O1	123.6(3)
C13–C14–H11	120.(2)	C34–C35–H29	119.(3)	O2–N1–C42	118.1(3)
C15–C14–H11	120.(2)	C36–C35–H29	121.(3)	O1–N1–C42	118.3(3)
C16–C15–C14	120.8(4)	C35–C36–C31	120.9(4)	C45–O4–C46	117.6(2)
C16–C15–H12	124.(2)	C35–C36–H30	119.(2)	C49–O5–C52	117.0(3)
C14–C15–H12	115.(2)	C31–C36–H30	120.(2)	C7–P1–C1	102.79(14)
C15–C16–C17	119.8(4)	C38–C37–C39	122.3(3)	C7–P1–C13	103.07(15)
C15–C16–H13	120.(2)	C38–C37–Pd1	70.99(17)	C1–P1–C13	104.80(14)
C17–C16–H13	120.(3)	C39–C37–Pd1	114.0(2)	C7–P1–Pd1	112.70(11)
C16–C17–C18	120.2(4)	C38–C37–H31	116.2(19)	C1–P1–Pd1	113.24(10)
C16–C17–H14	121.(2)	C39–C37–H31	114.7(19)	C13–P1–Pd1	118.56(10)
C18–C17–H14	119.(2)	Pd1–C37–H31	109.9(18)	C25–P2–C19	104.43(14)
C17–C18–C13	120.8(4)	C37–C38–C45	119.5(3)	C25–P2–C31	104.88(15)
C17–C18–H15	119.(2)	C37–C38–Pd1	70.40(17)	C19–P2–C31	100.88(14)
C13–C18–H15	120.(2)	C45–C38–Pd1	110.9(2)	C25–P2–Pd1	112.03(11)
C20–C19–C24	118.2(3)	C37–C38–H32	119.4(17)	C19–P2–Pd1	119.3(1)
C20–C19–P2	119.1(2)	C45–C38–H32	115.7(17)	C31–P2–Pd1	113.75(10)
C24–C19–P2	122.6(2)	Pd1–C38–H32	111.0(17)	C37–Pd1–C38	38.61(12)
C19–C20–C21	120.7(3)	C44–C39–C40	117.3(3)	C37–Pd1–P2	103.13(9)
C19–C20–H16	120.(2)	C44–C39–C37	120.1(3)	C38–Pd1–P2	141.74(9)
C21–C20–H16	119.(2)	C40–C39–C37	122.6(3)	C37–Pd1–P1	144.36(9)
C22–C21–C20	120.6(4)	C41–C40–C39	121.7(3)	C38–Pd1–P1	105.87(9)

continued on next page

Table A.48 – continued from previous page

C22–C21–H17	118.(2)	C41–C40–H33	119.(2)	P2–Pd1–P1	112.31(3)
C20–C21–H17	121.(2)	C39–C40–H33	120.(2)		

PdPhNO$_2$–Ph

Table A.49.: Structural parameters of **PdPhNO$_2$–Ph**:

molecular formula	C$_{51}$H$_{41}$NO$_4$P$_2$Pd	
molecular weight [a.m.u.]	900.19	
colour/habit	orange block	
crystal system	monoclinic	
space group	C1$_2$/c	
unit cell dimensions	a = 47.8785(16) Å	$\alpha = 90.00°$
	b = 9.8204(4) Å	$\beta = 108.5150(10)°$
	c = 19.4453(7) Å	$\gamma = 90.00°$
cell volume [Å^3]	8669.7(6)	
Z	8	
T [K]	173(2)	
calculated density [gcm^{-3}]	1.379	
absorption coefficient μ [mm^{-1}]	0.548	
F(000)	3696	
Θ range [°]	0.90 - 26.64	
index ranges [h, k, l]	±60, +11/-12, +24/-23	
no. of independent reflections / R$_{int}$	8951	
no. of absorbed reflections	8173	
no. of data	8951	
no. of restraints	0	
no. of parameters	696	
R$_1$ $(I > 2\sigma(I))$	0.0215	

continued on next page

Table A.49 – continued from previous page

wR$_2$ $(I > 2\sigma(I))$	0.0523
R$_1$ (all data)	0.0251
wR$_2$ (all data)	0.0546
goodness of fit (on F^2)	1.061

Table A.50.: Bond lengths of **PdPhNO$_2$–Ph** [Å]:

Pd1–C45	2.1360(15)	C4–C5	1.379(3)	C27–C28	1.378(3)
Pd1–C44	2.1514(15)	C5–C6	1.388(2)	C28–C29	1.386(3)
Pd1–P2	2.3184(4)	C7–C8	1.387(2)	C29–C30	1.391(2)
Pd1–P1	2.3311(4)	C7–C12	1.396(2)	C31–C36	1.391(2)
P1–C13	1.8258(15)	C8–C9	1.390(2)	C31–C32	1.395(2)
P1–C7	1.8337(15)	C9–C10	1.377(3)	C32–C33	1.387(3)
P1–C1	1.8358(15)	C10–C11	1.386(3)	C33–C34	1.388(3)
P2–C25	1.8271(16)	C11–C12	1.383(2)	C34–C35	1.377(3)
P2–C19	1.8331(16)	C13–C18	1.393(2)	C35–C36	1.396(3)
P2–C31	1.8331(16)	C13–C14	1.393(2)	C37–C42	1.365(3)
C44–C45	1.422(2)	C14–C15	1.393(2)	C37–C38	1.366(3)
C44–C43	1.467(2)	C15–C16	1.384(3)	C38–C39	1.389(3)
C45–C46	1.473(2)	C16–C17	1.376(3)	C39–C40	1.365(3)
O1–C43	1.3824(19)	C17–C18	1.387(2)	C40–C41	1.369(4)
O1–C37	1.4085(19)	C19–C24	1.390(2)	C41–C42	1.390(3)
O2–C43	1.2022(19)	C19–C20	1.398(2)	C46–C51	1.400(2)
O3–N1	1.226(2)	C20–C21	1.385(3)	C46–C47	1.406(2)
O4–N1	1.221(2)	C21–C22	1.378(3)	C47–C48	1.379(2)
N1–C49	1.466(2)	C22–C23	1.383(3)	C48–C49	1.386(3)
C1–C2	1.394(2)	C23–C24	1.388(2)	C49–C50	1.379(3)

continued on next page

Table A.50 – continued from previous page

C1–C6	1.397(2)	C25–C30	1.388(2)	C50–C51	1.384(2)
C2–C3	1.385(2)	C25–C26	1.399(2)		
C3–C4	1.383(3)	C26–C27	1.388(2)		

Table A.51.: Bond angles of **PdPhNO$_2$–Ph** [°]:

C45–Pd1–C44	38.73(6)	C5–C4–C3	119.89(17)	C28–C29–C30	120.38(17)
C45–Pd1–P2	103.86(4)	C4–C5–C6	120.27(18)	C25–C30–C29	120.07(17)
C44–Pd1–P2	142.52(4)	C5–C6–C1	120.54(17)	C36–C31–C32	118.88(15)
C45–Pd1–P1	144.70(4)	C8–C7–C12	118.43(14)	C36–C31–P2	123.86(13)
C44–Pd1–P1	106.20(4)	C8–C7–P1	123.70(12)	C32–C31–P2	117.26(12)
P2–Pd1–P1	111.288(14)	C12–C7–P1	117.77(12)	C33–C32–C31	120.53(16)
C13–P1–C7	104.22(7)	C7–C8–C9	120.45(16)	C32–C33–C34	120.12(19)
C13–P1–C1	102.46(7)	C10–C9–C8	120.81(17)	C35–C34–C33	119.88(18)
C7–P1–C1	102.83(7)	C9–C10–C11	119.13(16)	C34–C35–C36	120.23(18)
C13–P1–Pd1	110.11(5)	C12–C11–C10	120.46(17)	C31–C36–C35	120.36(17)
C7–P1–Pd1	113.51(5)	C11–C12–C7	120.71(16)	C42–C37–C38	121.38(17)
C1–P1–Pd1	121.84(5)	C18–C13–C14	119.25(15)	C42–C37–O1	119.58(17)
C25–P2–C19	105.09(7)	C18–C13–P1	116.96(12)	C38–C37–O1	118.86(16)
C25–P2–C31	105.00(7)	C14–C13–P1	123.67(13)	C37–C38–C39	119.1(2)
C19–P2–C31	101.04(7)	C13–C14–C15	120.03(17)	C40–C39–C38	120.5(2)
C25–P2–Pd1	111.04(5)	C16–C15–C14	119.83(18)	C39–C40–C41	119.5(2)
C19–P2–Pd1	118.50(5)	C17–C16–C15	120.54(17)	C40–C41–C42	120.8(2)
C31–P2–Pd1	114.77(5)	C16–C17–C18	119.88(19)	C37–C42–C41	118.7(2)
C45–C44–C43	119.35(15)	C17–C18–C13	120.45(17)	O2–C43–O1	121.90(14)
C45–C44–Pd1	70.05(9)	C24–C19–C20	118.72(15)	O2–C43–C44	128.16(15)
C43–C44–Pd1	110.37(10)	C24–C19–P2	119.24(12)	O1–C43–C44	109.94(13)
C44–C45–C46	121.55(15)	C20–C19–P2	121.98(13)	C51–C46–C47	117.52(15)
C44–C45–Pd1	71.22(9)	C21–C20–C19	120.03(17)	C51–C46–C45	120.21(15)

continued on next page

Table A.51 – continued from previous page

C46–C45–Pd1	109.97(10)	C22–C21–C20	120.74(18)	C47–C46–C45	122.27(14)
C43–O1–C37	117.24(12)	C21–C22–C23	119.76(17)	C48–C47–C46	121.38(16)
O4–N1–O3	123.49(17)	C22–C23–C24	119.92(17)	C47–C48–C49	119.03(17)
O4–N1–C49	118.55(17)	C23–C24–C19	120.79(16)	C50–C49–C48	121.59(16)
O3–N1–C49	117.96(17)	C30–C25–C26	119.03(15)	C50–C49–N1	119.29(16)
C2–C1–C6	118.40(15)	C30–C25–P2	123.20(12)	C48–C49–N1	119.12(16)
C2–C1–P1	118.74(12)	C26–C25–P2	117.48(12)	C49–C50–C51	118.77(16)
C6–C1–P1	122.83(12)	C27–C26–C25	120.54(16)	C50–C51–C46	121.71(16)
C3–C2–C1	120.77(17)	C28–C27–C26	119.99(17)		
C4–C3–C2	120.13(18)	C27–C28–C29	119.98(17)		

PdPhNO$_2$–PhNO$_2$

Table A.52.: Structural parameters of **PdPhNO$_2$–PhNO$_2$**:

molecular formula	C$_{58}$H$_{48}$N$_2$O$_6$P$_2$Pd
molecular weight [a.m.u.]	1037.32
colour/habit	orange fragment
crystal dimensions [mm^3]	0.25 × 0.08 × 0.03
crystal system	triclinic
space group	P-1
unit cell dimensions	a = 9.6523(4) Å $\quad$ $\alpha = 84.013(2)°$
	b = 14.5248(7) Å $\quad$ $\beta = 81.214(2)°$
	c = 18.5210(9) Å $\quad$ $\gamma = 74.7620(10)°$
cell volume [$\overset{\circ}{A}{}^3$]	2470.3(2)
Z	2
T [K]	123(2)
calculated density [gcm^{-3}]	1.395
absorption coefficient μ [mm^{-1}]	0.495

continued on next page

Table A.52 – continued from previous page

F(000)	1068
Θ range [°]	1.11 - 25.39
index ranges [h, k, l]	+10/-9, ±17, ±22
no. of independent reflections / R_{int}	8725
no. of absorbed reflections	7734
no. of data	8725
no. of restraints	0
no. of parameters	802
R_1 $(I > 2\sigma(I))$	0.0367
wR_2 $(I > 2\sigma(I))$	0.0955
R_1 (all data)	0.0446
wR_2 (all data)	0.1018
goodness of fit (on F^2)	1.047
largest difference peak [e$\mathring{A}^3$]	0.985
largest difference hole [e$\mathring{A}^3$]	-1.134

Table A.53.: Bond lengths of **PdPhNO$_2$–PhNO$_2$** [$\mathring{A}$]:

Pd1–C37	2.125(3)	C13–C14	1.405(5)	C36–H30	0.89(4)
Pd1–C38	2.157(3)	C14–C15	1.386(5)	C37–C38	1.411(5)
Pd1–P2	2.3288(8)	C14–H11	0.95(4)	C37–C39	1.467(5)
Pd1–P1	2.3415(8)	C15–C16	1.391(6)	C37–H31	0.94(4)
P1–C13	1.825(3)	C15–H12	0.94(5)	C38–C45	1.448(5)
P1–C7	1.831(3)	C16–C17	1.385(6)	C38–H32	0.88(4)
P1–C1	1.834(3)	C16–H13	0.97(4)	C39–C44	1.408(5)
P2–C31	1.827(3)	C17–C18	1.389(5)	C39–C40	1.409(5)
P2–C19	1.835(3)	C17–H14	0.95(4)	C40–C41	1.381(5)
P2–C25	1.850(3)	C18–H15	0.93(4)	C40–H33	0.93(4)
N1–O1	1.226(4)	C19–C24	1.392(5)	C41–C42	1.378(5)

continued on next page

Table A.53 – continued from previous page

N1–O2	1.233(4)	C19–C20	1.395(5)	C41–H34	0.90(4)
N1–C42	1.464(4)	C20–C21	1.389(5)	C42–C43	1.393(5)
N2–O6	1.221(4)	C20–H16	0.94(4)	C43–C44	1.385(5)
N2–O5	1.224(4)	C21–C22	1.379(6)	C43–H35	0.94(4)
N2–C49	1.475(4)	C21–H17	0.95(4)	C44–H36	0.98(4)
O3–C45	1.208(4)	C22–C23	1.385(6)	C46–C51	1.379(5)
O4–C46	1.384(4)	C22–H18	0.93(4)	C46–C47	1.396(5)
O4–C45	1.404(4)	C23–C24	1.399(5)	C47–C48	1.379(5)
C1–C6	1.391(5)	C23–H19	0.94(4)	C47–H37	0.93(4)
C1–C2	1.394(5)	C24–H20	0.93(4)	C48–C49	1.383(5)
C2–C3	1.393(5)	C25–C30	1.380(5)	C48–H38	0.96(4)
C2–H1	0.93(4)	C25–C26	1.394(5)	C49–C50	1.386(5)
C3–C4	1.377(6)	C26–C27	1.384(5)	C50–C51	1.386(5)
C3–H2	0.93(5)	C26–H21	0.85(3)	C50–H39	0.93(4)
C4–C5	1.378(6)	C27–C28	1.392(6)	C51–H40	0.91(4)
C4–H3	0.85(4)	C27–H22	0.97(4)	C52–C53	1.508(6)
C5–C6	1.390(5)	C28–C29	1.384(6)	C52–H52A	0.9800
C5–H4	0.99(4)	C28–H23	0.95(4)	C52–H52B	0.9800
C6–H5	0.91(4)	C29–C30	1.394(5)	C52–H52C	0.9800
C7–C8	1.393(5)	C29–H24	0.92(5)	C53–C58	1.377(6)
C7–C12	1.395(5)	C30–H25	0.96(4)	C53–C54	1.399(6)
C8–C9	1.391(5)	C31–C36	1.388(5)	C54–C55	1.383(7)
C8–H6	0.99(4)	C31–C32	1.398(5)	C54–H44	0.96(5)
C9–C10	1.385(6)	C32–C33	1.384(5)	C55–C56	1.380(7)
C9–H7	0.89(5)	C32–H26	0.93(4)	C55–H45	0.96(5)
C10–C11	1.372(6)	C33–C34	1.386(6)	C56–C57	1.384(7)
C10–H8	1.00(4)	C33–H27	0.93(4)	C56–H46	0.94(5)
C11–C12	1.393(5)	C34–C35	1.385(6)	C57–C58	1.380(7)

continued on next page

Table A.53 – continued from previous page

C11–H9	0.95(4)	C34–H28	0.92(4)	C57–H47	0.92(5)
C12–H10	0.91(4)	C35–C36	1.392(5)	C58–H48	0.91(4)
C13–C18	1.393(5)	C35–H29	0.99(4)		

Table A.54.: Bond angles of **PdPhNO$_2$–PhNO$_2$** [°]:

C37–Pd1–C38	38.47(12)	C17–C16–C15	120.0(4)	C37–C38–Pd1	69.53(18)
C37–Pd1–P2	100.69(9)	C17–C16–H13	120.(3)	C45–C38–Pd1	106.8(2)
C38–Pd1–P2	138.52(9)	C15–C16–H13	120.(3)	C37–C38–H32	125.(2)
C37–Pd1–P1	152.33(9)	C16–C17–C18	120.3(4)	C45–C38–H32	115.(2)
C38–Pd1–P1	114.48(9)	C16–C17–H14	119.(2)	Pd1–C38–H32	109.(2)
P2–Pd1–P1	106.83(3)	C18–C17–H14	121.(2)	C44–C39–C40	118.2(3)
C13–P1–C7	104.73(15)	C17–C18–C13	120.4(4)	C44–C39–C37	122.2(3)
C13–P1–C1	102.05(15)	C17–C18–H15	120.(2)	C40–C39–C37	119.6(3)
C7–P1–C1	102.18(14)	C13–C18–H15	120.(2)	C41–C40–C39	120.9(3)
C13–P1–Pd1	114.68(10)	C24–C19–C20	119.0(3)	C41–C40–H33	121.(2)
C7–P1–Pd1	113.64(11)	C24–C19–P2	122.1(3)	C39–C40–H33	118.(2)
C1–P1–Pd1	117.81(11)	C20–C19–P2	118.7(3)	C42–C41–C40	119.3(3)
C31–P2–C19	103.29(15)	C21–C20–C19	120.4(3)	C42–C41–H34	120.(3)
C31–P2–C25	104.04(15)	C21–C20–H16	117.(2)	C40–C41–H34	121.(3)
C19–P2–C25	102.28(15)	C19–C20–H16	122.(2)	C41–C42–C43	121.9(3)
C31–P2–Pd1	115.38(11)	C22–C21–C20	120.2(4)	C41–C42–N1	119.3(3)
C19–P2–Pd1	118.86(10)	C22–C21–H17	121.(2)	C43–C42–N1	118.7(3)
C25–P2–Pd1	111.22(10)	C20–C21–H17	119.(2)	C44–C43–C42	118.6(3)
O1–N1–O2	123.4(3)	C21–C22–C23	120.3(3)	C44–C43–H35	121.(2)
O1–N1–C42	118.5(3)	C21–C22–H18	120.(3)	C42–C43–H35	120.(2)
O2–N1–C42	118.0(3)	C23–C22–H18	120.(3)	C43–C44–C39	121.1(3)
O6–N2–O5	124.2(3)	C22–C23–C24	119.6(4)	C43–C44–H36	118.(2)
O6–N2–C49	117.7(3)	C22–C23–H19	123.(2)	C39–C44–H36	121.(2)

continued on next page

Table A.54 – continued from previous page

O5–N2–C49	118.1(3)	C24–C23–H19	118.(2)	O3–C45–O4	121.3(3)
C46–O4–C45	118.2(3)	C19–C24–C23	120.4(3)	O3–C45–C38	128.2(3)
C6–C1–C2	118.7(3)	C19–C24–H20	120.(2)	O4–C45–C38	110.5(3)
C6–C1–P1	118.1(3)	C23–C24–H20	120.(2)	C51–C46–O4	116.8(3)
C2–C1–P1	123.2(3)	C30–C25–C26	119.4(3)	C51–C46–C47	121.2(3)
C3–C2–C1	119.9(3)	C30–C25–P2	122.7(3)	O4–C46–C47	121.8(3)
C3–C2–H1	120.(2)	C26–C25–P2	117.9(2)	C48–C47–C46	119.2(3)
C1–C2–H1	120.(2)	C27–C26–C25	120.9(3)	C48–C47–H37	121.(2)
C4–C3–C2	120.5(4)	C27–C26–H21	116.(2)	C46–C47–H37	120.(2)
C4–C3–H2	121.(3)	C25–C26–H21	123.(2)	C47–C48–C49	118.9(3)
C2–C3–H2	118.(3)	C26–C27–C28	119.5(4)	C47–C48–H38	122.(2)
C3–C4–C5	120.2(3)	C26–C27–H22	121.(2)	C49–C48–H38	120.(2)
C3–C4–H3	118.(2)	C28–C27–H22	120.(2)	C48–C49–C50	122.5(3)
C5–C4–H3	122.(2)	C29–C28–C27	119.5(4)	C48–C49–N2	119.2(3)
C4–C5–C6	119.6(4)	C29–C28–H23	120.(3)	C50–C49–N2	118.2(3)
C4–C5–H4	121.(2)	C27–C28–H23	120.(3)	C49–C50–C51	118.2(3)
C6–C5–H4	120.(2)	C28–C29–C30	120.8(4)	C49–C50–H39	121.(2)
C5–C6–C1	121.0(4)	C28–C29–H24	122.(3)	C51–C50–H39	121.(2)
C5–C6–H5	117.(3)	C30–C29–H24	117.(3)	C46–C51–C50	119.9(3)
C1–C6–H5	122.(3)	C25–C30–C29	119.7(4)	C46–C51–H40	120.(3)
C8–C7–C12	118.6(3)	C25–C30–H25	120.(2)	C50–C51–H40	120.(3)
C8–C7–P1	118.2(2)	C29–C30–H25	120.(2)	C53–C52–H52A	109.500
C12–C7–P1	123.2(3)	C36–C31–C32	119.1(3)	C53–C52–H52B	109.500
C9–C8–C7	120.4(3)	C36–C31–P2	118.4(3)	H52A–C52–H52B	109.500
C9–C8–H6	119.(2)	C32–C31–P2	122.4(3)	C53–C52–H52C	109.500
C7–C8–H6	121.(2)	C33–C32–C31	120.3(3)	H52A–C52–H52C	109.500
C10–C9–C8	120.5(4)	C33–C32–H26	119.(2)	H52B–C52–H52C	109.500
C10–C9–H7	122.(3)	C31–C32–H26	120.(2)	C58–C53–C54	117.8(4)

continued on next page

Table A.54 – continued from previous page

C8–C9–H7	118.(3)	C32–C33–C34	120.2(3)	C58–C53–C52	122.1(4)
C11–C10–C9	119.5(4)	C32–C33–H27	119.(2)	C54–C53–C52	120.0(4)
C11–C10–H8	120.(2)	C34–C33–H27	120.(2)	C55–C54–C53	121.2(4)
C9–C10–H8	121.(2)	C35–C34–C33	120.0(3)	C55–C54–H44	122.(3)
C10–C11–C12	120.7(4)	C35–C34–H28	120.(3)	C53–C54–H44	117.(3)
C10–C11–H9	120.(3)	C33–C34–H28	120.(2)	C56–C55–C54	119.7(4)
C12–C11–H9	120.(3)	C34–C35–C36	119.9(4)	C56–C55–H45	124.(3)
C11–C12–C7	120.4(4)	C34–C35–H29	120.(2)	C54–C55–H45	116.(3)
C11–C12–H10	118.(3)	C36–C35–H29	120.(2)	C55–C56–C57	119.6(5)
C7–C12–H10	122.(3)	C31–C36–C35	120.5(3)	C55–C56–H46	120.(3)
C18–C13–C14	118.7(3)	C31–C36–H30	120.(2)	C57–C56–H46	120.(3)
C18–C13–P1	123.9(3)	C35–C36–H30	120.(2)	C58–C57–C56	120.1(4)
C14–C13–P1	117.4(3)	C38–C37–C39	124.3(3)	C58–C57–H47	120.(3)
C15–C14–C13	120.7(4)	C38–C37–Pd1	72.01(19)	C56–C57–H47	120.(3)
C15–C14–H11	120.(3)	C39–C37–Pd1	111.8(2)	C53–C58–C57	121.4(4)
C13–C14–H11	120.(3)	C38–C37–H31	115.(2)	C53–C58–H48	122.(3)
C14–C15–C16	119.8(4)	C39–C37–H31	116.(2)	C57–C58–H48	116.(3)
C14–C15–H12	120.(3)	Pd1–C37–H31	106.(2)		
C16–C15–H12	120.(3)	C37–C38–C45	118.2(3)		

Pd(dppe)PhNO$_2$–^{i}Pr

Table A.55.: Structural parameters of **Pd(dppe)PhNO$_2$–^{i}Pr**:

molecular formula	C$_{38}$H$_{37}$NO$_4$P$_2$Pd
molecular weight [a.m.u.]	740.03
colour/habit	orange fragment
crystal system	triclinic
space group	P-1

continued on next page

Table A.55 – continued from previous page

unit cell dimensions		$a = 8.6025(2)$ Å	$\alpha = 102.603(2)°$
		$b = 9.8853(3)$ Å	$\beta = 93.391(2)°$
		$c = 22.1723(6)$ Å	$\gamma = 92.431(2)°$
cell volume [Å³]		1833.88(9)	
Z		2	
T [K]		123(2)	
calculated density [gcm^{-3}]		1.340	
absorption coefficient μ [mm^{-1}]		0.631	
F(000)		760	
Θ range [°]		2.83 - 26.02	
index ranges [h, k, l]		$\pm10, +11/-12, +27/0$	
no. of independent reflections / R_{int}		7123	
no. of absorbed reflections		5003	
no. of data		7123	
no. of restraints		0	
no. of parameters		417	
R_1 $(I > 2\sigma(I))$		0.0455	
wR_2 $(I > 2\sigma(I))$		0.1010	
R_1 (all data)		0.0673	
wR_2 (all data)		0.1055	
goodness of fit (on F^2)		0.943	

Table A.56.: Bond lengths of **Pd(dppe)PhNO$_2$–^{i}Pr** [Å]:

Pd1–C5	2.108(4)	C5–C6	1.411(5)	C21–C22	1.372(8)
Pd1–C6	2.131(4)	C6–C7	1.465(6)	C22–C23	1.365(8)
Pd1–P2	2.2926(11)	C7–C12	1.400(5)	C23–C24	1.375(7)
Pd1–P1	2.3082(10)	C7–C8	1.404(5)	C25–C26	1.538(5)
P1–C27	1.820(4)	C8–C9	1.377(6)	C27–C32	1.378(6)

continued on next page

Table A.56 – continued from previous page

P1–C33	1.823(4)	C9–C10	1.376(6)	C27–C28	1.387(6)
P1–C26	1.834(4)	C10–C11	1.383(6)	C28–C29	1.390(6)
P2–C18	1.811(4)	C10–N1	1.439(6)	C29–C30	1.362(7)
P2–C19	1.816(4)	C11–C12	1.366(6)	C30–C31	1.361(7)
P2–C25	1.835(4)	C13–C14	1.379(6)	C31–C32	1.391(6)
O1–C4	1.354(5)	C13–C18	1.380(6)	C33–C38	1.384(6)
O1–C3	1.455(6)	C14–C15	1.384(8)	C33–C34	1.391(5)
O2–C4	1.219(5)	C15–C16	1.367(8)	C34–C35	1.389(6)
O3–N1	1.228(5)	C16–C17	1.377(6)	C35–C36	1.364(6)
O4–N1	1.242(5)	C17–C18	1.381(6)	C36–C37	1.370(7)
C1–C3	1.442(9)	C19–C24	1.364(6)	C37–C38	1.369(6)
C2–C3	1.515(8)	C19–C20	1.386(6)		
C4–C5	1.452(6)	C20–C21	1.362(7)		

Table A.57.: Bond angles of **Pd(dppe)PhNO$_2$–^{i}Pr** [°]:

C5–Pd1–C6	38.87(15)	C6–C5–Pd1	71.5(2)	C21–C20–C19	121.1(6)
C5–Pd1–P2	155.96(11)	C4–C5–Pd1	112.1(3)	C20–C21–C22	120.4(6)
C6–Pd1–P2	117.49(11)	C5–C6–C7	125.4(4)	C23–C22–C21	119.0(5)
C5–Pd1–P1	118.18(11)	C5–C6–Pd1	69.7(2)	C22–C23–C24	120.5(6)
C6–Pd1–P1	156.04(11)	C7–C6–Pd1	114.2(3)	C19–C24–C23	121.1(5)
P2–Pd1–P1	85.82(4)	C12–C7–C8	117.1(4)	C26–C25–P2	110.0(3)
C27–P1–C33	103.31(17)	C12–C7–C6	119.8(4)	C25–C26–P1	108.3(3)
C27–P1–C26	104.21(18)	C8–C7–C6	123.1(4)	C32–C27–C28	119.1(4)
C33–P1–C26	102.83(18)	C9–C8–C7	121.3(4)	C32–C27–P1	117.4(3)
C27–P1–Pd1	126.49(13)	C10–C9–C8	119.6(4)	C28–C27–P1	123.5(3)
C33–P1–Pd1	110.59(13)	C9–C10–C11	120.5(4)	C27–C28–C29	119.9(4)
C26–P1–Pd1	106.97(13)	C9–C10–N1	119.8(4)	C30–C29–C28	120.5(5)
C18–P2–C19	102.87(18)	C11–C10–N1	119.6(4)	C31–C30–C29	119.9(4)

continued on next page

Table A.57 – continued from previous page

C18–P2–C25	103.63(18)	C12–C11–C10	119.7(4)	C30–C31–C32	120.6(4)
C19–P2–C25	105.39(19)	C11–C12–C7	121.7(4)	C27–C32–C31	119.9(4)
C18–P2–Pd1	123.16(14)	C14–C13–C18	120.8(5)	C38–C33–C34	118.4(4)
C19–P2–Pd1	113.37(13)	C13–C14–C15	119.1(5)	C38–C33–P1	119.5(3)
C25–P2–Pd1	106.84(13)	C16–C15–C14	120.7(5)	C34–C33–P1	122.1(3)
C4–O1–C3	119.6(4)	C15–C16–C17	119.5(5)	C35–C34–C33	120.2(4)
C1–C3–O1	105.7(5)	C16–C17–C18	120.9(5)	C36–C35–C34	120.2(4)
C1–C3–C2	115.1(5)	C13–C18–C17	118.9(4)	C35–C36–C37	119.8(4)
O1–C3–C2	108.1(5)	C13–C18–P2	119.1(3)	C38–C37–C36	120.7(5)
O2–C4–O1	122.2(4)	C17–C18–P2	121.9(3)	C37–C38–C33	120.7(4)
O2–C4–C5	127.0(4)	C24–C19–C20	117.9(4)	O3–N1–O4	122.0(4)
O1–C4–C5	110.8(4)	C24–C19–P2	126.0(3)	O3–N1–C10	119.1(4)
C6–C5–C4	119.6(4)	C20–C19–P2	116.1(4)	O4–N1–C10	118.8(4)

Pd(dppp)PhNO$_2$–Et

Table A.58.: Structural parameters of **Pd(dppp)PhNO$_2$–Et**:

molecular formula	$C_{76}H_{74}N_2O_8P_3Pd_2$	
molecular weight [a.m.u.]	1449.08	
colour/habit	orange block	
crystal system	triclinic	
space group	P-1	
unit cell dimensions	a = 12.190(4) Å	α = 67.930(5)°
	b = 12.590(2) Å	β = 76.261(12)°
	c = 13.333(3) Å	γ = 61.170(10)°
cell volume [Å^3]	1657.3(7)	
Z	1	
T [K]	123(2)	

continued on next page

Table A.58 – continued from previous page

calculated density [gcm^{-3}]	1.452
absorption coefficient μ [mm^{-1}]	0.674
F(000)	745
Θ range [°]	4.09 - 26.02
index ranges [h, k, l]	+14/-15, +14/-15, ±16
no. of independent reflections / R_{int}	6496
no. of absorbed reflections	5550
no. of data	6496
no. of restraints	0
no. of parameters	416
R_1 ($I > 2\sigma(I)$)	0.0269
wR_2 ($I > 2\sigma(I)$)	0.0627
R_1 (all data)	0.0369
wR_2 (all data)	0.0656
goodness of fit (on F^2)	1.016

Table A.59.: Bond lengths of **Pd(dppp)PhNO$_2$–Et** [Å]:

Pd1–C8	2.126(2)	C2–C3	1.375(3)	C21–C22	1.387(3)
Pd1–C7	2.148(2)	C3–C4	1.414(3)	C22–C23	1.394(3)
Pd1–P2	2.3022(8)	C4–C5	1.413(3)	C24–C25	1.532(3)
Pd1–P1	2.3117(9)	C4–C7	1.459(3)	C25–C26	1.538(3)
P1–C17	1.834(2)	C5–C6	1.379(3)	C27–C28	1.391(3)
P1–C23	1.835(2)	C7–C8	1.423(3)	C27–C32	1.403(3)
P1–C24	1.841(2)	C8–C9	1.470(3)	C28–C29	1.388(3)
P2–C33	1.836(2)	C10–C11	1.505(4)	C29–C30	1.383(3)
P2–C27	1.836(2)	C12–C17	1.388(3)	C30–C31	1.393(3)
P2–C26	1.837(2)	C12–C13	1.395(3)	C31–C32	1.387(3)
N1–O2	1.226(3)	C13–C14	1.390(3)	C33–C34	1.388(3)

continued on next page

Table A.59 – continued from previous page

N1–O1	1.228(3)	C14–C15	1.383(3)	C33–C38	1.399(3)
N1–C1	1.464(3)	C15–C16	1.387(3)	C34–C35	1.395(3)
O3–C9	1.215(3)	C16–C17	1.403(3)	C35–C36	1.385(3)
O4–C9	1.362(3)	C18–C23	1.387(3)	C36–C37	1.384(4)
O4–C10	1.447(3)	C18–C19	1.389(3)	C37–C38	1.387(3)
C1–C2	1.383(3)	C19–C20	1.383(4)		
C1–C6	1.393(3)	C20–C21	1.381(4)		

Table A.60.: Bond angles of **Pd(dppp)PhNO$_2$–Et** [°]:

C8–Pd1–C7	38.89(8)	C2–C3–C4	121.6(2)	C21–C20–C19	119.2(2)
C8–Pd1–P2	110.58(6)	C5–C4–C3	117.2(2)	C20–C21–C22	120.9(2)
C7–Pd1–P2	148.92(6)	C5–C4–C7	122.87(19)	C21–C22–C23	120.3(2)
C8–Pd1–P1	152.71(6)	C3–C4–C7	119.91(19)	C18–C23–C22	118.5(2)
C7–Pd1–P1	113.83(6)	C6–C5–C4	121.6(2)	C18–C23–P1	117.23(17)
P2–Pd1–P1	96.37(3)	C5–C6–C1	118.8(2)	C22–C23–P1	124.27(19)
C17–P1–C23	102.49(10)	C8–C7–C4	122.10(19)	C25–C24–P1	115.78(15)
C17–P1–C24	99.47(10)	C8–C7–Pd1	69.73(12)	C24–C25–C26	114.51(19)
C23–P1–C24	103.97(11)	C4–C7–Pd1	111.24(15)	C25–C26–P2	113.55(16)
C17–P1–Pd1	119.99(8)	C7–C8–C9	120.41(19)	C28–C27–C32	119.3(2)
C23–P1–Pd1	115.42(7)	C7–C8–Pd1	71.37(12)	C28–C27–P2	119.53(16)
C24–P1–Pd1	113.16(7)	C9–C8–Pd1	112.30(15)	C32–C27–P2	121.18(17)
C33–P2–C27	101.28(10)	O3–C9–O4	122.8(2)	C29–C28–C27	120.2(2)
C33–P2–C26	102.75(10)	O3–C9–C8	127.3(2)	C30–C29–C28	120.5(2)
C27–P2–C26	101.21(10)	O4–C9–C8	109.89(18)	C29–C30–C31	119.7(2)
C33–P2–Pd1	123.69(8)	O4–C10–C11	109.6(2)	C32–C31–C30	120.1(2)
C27–P2–Pd1	109.69(7)	C17–C12–C13	120.5(2)	C31–C32–C27	120.1(2)
C26–P2–Pd1	115.23(7)	C14–C13–C12	119.5(2)	C34–C33–C38	119.0(2)
O2–N1–O1	123.0(2)	C15–C14–C13	120.4(2)	C34–C33–P2	123.51(17)

continued on next page

Table A.60 – continued from previous page

O2–N1–C1	118.80(19)	C14–C15–C16	120.1(2)	C38–C33–P2	117.50(17)
O1–N1–C1	118.2(2)	C15–C16–C17	120.1(2)	C33–C34–C35	120.8(2)
C9–O4–C10	117.60(17)	C12–C17–C16	119.30(19)	C36–C35–C34	119.9(2)
C2–C1–C6	121.6(2)	C12–C17–P1	119.28(17)	C37–C36–C35	119.6(2)
C2–C1–N1	119.5(2)	C16–C17–P1	121.25(17)	C36–C37–C38	120.9(2)
C6–C1–N1	118.9(2)	C23–C18–C19	121.0(2)	C37–C38–C33	119.9(2)
C3–C2–C1	119.1(2)	C20–C19–C18	120.2(3)		

Pd(dppb)PhNO$_2$–PhOMe

Table A.61.: Structural parameters of **Pd(dppb)PhNO$_2$–PhOMe**:

molecular formula	$C_{56}H_{53}NO_4P_2Pd$
molecular weight [a.m.u.]	988.33
colour/habit	dark red block
crystal dimensions [mm^3]	$0.5 \times 0.5 \times 0.4$
crystal system	triclinic
space group	P-1
unit cell dimensions	a = 9.7516(2) Å α = 114.200(2)°
	b = 15.3223(2) Å β = 90.494(2)°
	c = 18.6487(4) Å γ = 106.366(2)°
cell volume [Å^3]	2413.93(8)
Z	2
T [K]	150(2)
calculated density [gcm^{-3}]	1.360
absorption coefficient μ [mm^{-1}]	0.500
F(000)	1024
Θ range [°]	2.72 - 25.50
index ranges [h, k, l]	±11, +18/-14, ±22

continued on next page

Table A.61 – continued from previous page

no. of independent reflections / R_{int}	8953
no. of absorbed reflections	7426
no. of data	8953
no. of restraints	0
no. of parameters	587
R_1 $(I > 2\sigma(I))$	0.0284
wR_2 $(I > 2\sigma(I))$	0.0720
R_1 (all data)	0.0373
wR_2 (all data)	0.0738
goodness of fit (on F^2)	1.048

Table A.62.: Bond lengths of **Pd(dppb)PhNO$_2$–PhOMe** [Å]:

Pd2–C7	2.124(2)	C16–H16A	0.9800	C36–H36	0.9500
Pd2–C8	2.128(2)	C16–H16B	0.9800	C37–C38	1.385(3)
Pd2–P2	2.2952(5)	C16–H16C	0.9800	C37–H37	0.9500
Pd2–P1	2.3062(6)	C17–C22	1.377(3)	C38–H38	0.9500
P1–C23	1.826(2)	C17–C18	1.395(3)	C39–C44	1.392(3)
P1–C17	1.827(2)	C18–C19	1.380(3)	C39–C40	1.393(3)
P1–C29	1.846(2)	C18–H18	0.9500	C40–C41	1.377(3)
P2–C33	1.823(2)	C19–C20	1.378(4)	C40–H40	0.9500
P2–C39	1.827(2)	C19–H19	0.9500	C41–C42	1.383(4)
P2–C32	1.835(2)	C20–C21	1.370(4)	C41–H41	0.9500
O1–N1	1.221(3)	C20–H20	0.9500	C42–C43	1.371(4)
O2–N1	1.229(3)	C21–C22	1.394(3)	C42–H42	0.9500
O3–C9	1.205(3)	C21–H21	0.9500	C43–C44	1.387(3)
O4–C9	1.381(3)	C22–H22	0.9500	C43–H43	0.9500
O4–C10	1.404(2)	C23–C28	1.385(3)	C44–H44	0.9500
O5–C13	1.369(3)	C23–C24	1.391(3)	C45–C46	1.357(5)

continued on next page

Table A.62 – continued from previous page

O5–C16	1.426(3)	C24–C25	1.390(3)	C45–C50	1.388(6)
N1–C1	1.459(3)	C24–H24	0.9500	C45–H45	0.9500
C1–C6	1.378(3)	C25–C26	1.368(4)	C46–C47	1.349(4)
C1–C2	1.382(3)	C25–H25	0.9500	C46–H46	0.9500
C2–C3	1.378(3)	C26–C27	1.368(4)	C47–C48	1.348(5)
C2–H2A	0.9500	C26–H26	0.9500	C47–H47	0.9500
C3–C4	1.400(3)	C27–C28	1.391(3)	C48–C49	1.343(5)
C3–H3	0.9500	C27–H27	0.9500	C48–H48	0.9500
C4–C5	1.400(3)	C28–H28	0.9500	C49–C50	1.383(6)
C4–C7	1.465(3)	C29–C30	1.528(3)	C49–H49	0.9500
C5–C6	1.373(3)	C29–H29A	0.9900	C50–H50	0.9500
C5–H5	0.9500	C29–H29B	0.9900	C51–C53	1.370(4)
C6–H6	0.9500	C30–C31	1.525(3)	C51–C52	1.378(4)
C7–C8	1.422(3)	C30–H30A	0.9900	C51–H51	0.9500
C7–H7	1.0000	C30–H30B	0.9900	C52–C53	1.380(4)
C8–C9	1.458(3)	C31–C32	1.534(3)	C52–H52	0.9500
C8–H8	1.0000	C31–H31A	0.9900	C53–C51	1.370(4)
C10–C11	1.371(3)	C31–H31B	0.9900	C53–H53	0.9500
C10–C15	1.383(3)	C32–H32A	0.9900	C54–C55	1.361(4)
C11–C12	1.385(3)	C32–H32B	0.9900	C54–C56	1.374(4)
C11–H11	0.9500	C33–C34	1.387(3)	C54–H54	0.9500
C12–C13	1.389(3)	C33–C38	1.392(3)	C55–C56	1.362(4)
C12–H12	0.9500	C34–C35	1.384(3)	C55–H55	0.9500
C13–C14	1.383(3)	C34–H34	0.9500	C56–C54	1.374(4)
C14–C15	1.372(3)	C35–C36	1.380(3)	C56–H56	0.9500
C14–H14	0.9500	C35–H35	0.9500		
C15–H15	0.9500	C36–C37	1.381(3)		

Table A.63.: Bond angles of **Pd(dppb)PhNO$_2$–PhOMe** [°]:

C7–Pd2–C8	39.08(8)	C14–C15–C10	119.8(2)	C38–C33–P2	123.12(16)
C7–Pd2–P2	147.84(6)	C14–C15–H15	120.100	C35–C34–C33	120.9(2)
C8–Pd2–P2	108.83(6)	C10–C15–H15	120.100	C35–C34–H34	119.600
C7–Pd2–P1	107.32(6)	O5–C16–H16A	109.500	C33–C34–H34	119.600
C8–Pd2–P1	146.35(6)	O5–C16–H16B	109.500	C36–C35–C34	120.2(2)
P2–Pd2–P1	104.63(2)	H16A–C16–H16B	109.500	C36–C35–H35	119.900
C23–P1–C17	103.84(10)	O5–C16–H16C	109.500	C34–C35–H35	119.900
C23–P1–C29	103.49(10)	H16A–C16–H16C	109.500	C35–C36–C37	119.6(2)
C17–P1–C29	100.79(10)	H16B–C16–H16C	109.500	C35–C36–H36	120.200
C23–P1–Pd2	116.75(7)	C22–C17–C18	118.5(2)	C37–C36–H36	120.200
C17–P1–Pd2	113.36(7)	C22–C17–P1	121.82(17)	C36–C37–C38	120.2(2)
C29–P1–Pd2	116.60(7)	C18–C17–P1	119.62(17)	C36–C37–H37	119.900
C33–P2–C39	103.2(1)	C19–C18–C17	120.8(2)	C38–C37–H37	119.900
C33–P2–C32	101.02(9)	C19–C18–H18	119.600	C37–C38–C33	120.6(2)
C39–P2–C32	103.91(10)	C17–C18–H18	119.600	C37–C38–H38	119.700
C33–P2–Pd2	114.00(7)	C20–C19–C18	120.0(2)	C33–C38–H38	119.700
C39–P2–Pd2	115.67(7)	C20–C19–H19	120.000	C44–C39–C40	118.9(2)
C32–P2–Pd2	117.06(7)	C18–C19–H19	120.000	C44–C39–P2	122.39(17)
C9–O4–C10	118.52(16)	C21–C20–C19	119.9(2)	C40–C39–P2	118.66(16)
C13–O5–C16	117.86(19)	C21–C20–H20	120.000	C41–C40–C39	120.7(2)
O1–N1–O2	122.8(2)	C19–C20–H20	120.000	C41–C40–H40	119.700
O1–N1–C1	118.8(2)	C20–C21–C22	120.2(2)	C39–C40–H40	119.700
O2–N1–C1	118.5(2)	C20–C21–H21	119.900	C40–C41–C42	119.9(2)
C6–C1–C2	121.3(2)	C22–C21–H21	119.900	C40–C41–H41	120.100
C6–C1–N1	118.8(2)	C17–C22–C21	120.5(2)	C42–C41–H41	120.100
C2–C1–N1	119.9(2)	C17–C22–H22	119.700	C43–C42–C41	120.1(2)
C3–C2–C1	118.9(2)	C21–C22–H22	119.700	C43–C42–H42	119.900

continued on next page

Table A.63 – continued from previous page

C3–C2–H2A	120.600	C28–C23–C24	118.9(2)	C41–C42–H42	119.900
C1–C2–H2A	120.600	C28–C23–P1	118.90(17)	C42–C43–C44	120.4(2)
C2–C3–C4	121.7(2)	C24–C23–P1	122.23(18)	C42–C43–H43	119.800
C2–C3–H3	119.100	C25–C24–C23	120.0(2)	C44–C43–H43	119.800
C4–C3–H3	119.100	C25–C24–H24	120.000	C43–C44–C39	119.9(2)
C5–C4–C3	117.2(2)	C23–C24–H24	120.000	C43–C44–H44	120.000
C5–C4–C7	118.7(2)	C26–C25–C24	120.3(3)	C39–C44–H44	120.000
C3–C4–C7	124.03(19)	C26–C25–H25	119.900	C46–C45–C50	119.9(3)
C6–C5–C4	121.7(2)	C24–C25–H25	119.900	C46–C45–H45	120.000
C6–C5–H5	119.100	C27–C26–C25	120.3(2)	C50–C45–H45	120.000
C4–C5–H5	119.100	C27–C26–H26	119.900	C47–C46–C45	120.2(3)
C5–C6–C1	119.2(2)	C25–C26–H26	119.900	C47–C46–H46	119.900
C5–C6–H6	120.400	C26–C27–C28	120.1(3)	C45–C46–H46	119.900
C1–C6–H6	120.400	C26–C27–H27	119.900	C48–C47–C46	120.8(3)
C8–C7–C4	124.2(2)	C28–C27–H27	119.900	C48–C47–H47	119.600
C8–C7–Pd2	70.64(11)	C23–C28–C27	120.4(2)	C46–C47–H47	119.600
C4–C7–Pd2	113.73(14)	C23–C28–H28	119.800	C49–C48–C47	120.5(3)
C8–C7–H7	113.700	C27–C28–H28	119.800	C49–C48–H48	119.800
C4–C7–H7	113.700	C30–C29–P1	115.68(16)	C47–C48–H48	119.800
Pd2–C7–H7	113.700	C30–C29–H29A	108.400	C48–C49–C50	120.5(3)
C7–C8–C9	118.25(19)	P1–C29–H29A	108.400	C48–C49–H49	119.800
C7–C8–Pd2	70.29(11)	C30–C29–H29B	108.400	C50–C49–H49	119.800
C9–C8–Pd2	109.61(14)	P1–C29–H29B	108.400	C49–C50–C45	118.2(3)
C7–C8–H8	116.600	H29A–C29–H29B	107.400	C49–C50–H50	120.900
C9–C8–H8	116.600	C31–C30–C29	116.19(18)	C45–C50–H50	120.900
Pd2–C8–H8	116.600	C31–C30–H30A	108.200	C53–C51–C52	119.7(2)
O3–C9–O4	121.8(2)	C29–C30–H30A	108.200	C53–C51–H51	120.200
O3–C9–C8	128.1(2)	C31–C30–H30B	108.200	C52–C51–H51	120.200

continued on next page

Table A.63 – continued from previous page

O4–C9–C8	110.11(19)	C29–C30–H30B	108.200	C51–C52–C53	119.9(3)
C11–C10–C15	120.2(2)	H30A–C30–H30B	107.400	C51–C52–H52	120.100
C11–C10–O4	118.75(19)	C30–C31–C32	113.58(19)	C53–C52–H52	120.100
C15–C10–O4	120.80(19)	C30–C31–H31A	108.900	C51–C53–C52	120.4(3)
C10–C11–C12	120.3(2)	C32–C31–H31A	108.900	C51–C53–H53	119.800
C10–C11–H11	119.900	C30–C31–H31B	108.900	C52–C53–H53	119.800
C12–C11–H11	119.900	C32–C31–H31B	108.900	C55–C54–C56	120.2(3)
C11–C12–C13	119.6(2)	H31A–C31–H31B	107.700	C55–C54–H54	119.900
C11–C12–H12	120.200	C31–C32–P2	114.53(14)	C56–C54–H54	119.900
C13–C12–H12	120.200	C31–C32–H32A	108.600	C54–C55–C56	120.1(3)
O5–C13–C14	115.24(19)	P2–C32–H32A	108.600	C54–C55–H55	120.000
O5–C13–C12	125.21(19)	C31–C32–H32B	108.600	C56–C55–H55	120.000
C14–C13–C12	119.5(2)	P2–C32–H32B	108.600	C55–C56–C54	119.7(3)
C15–C14–C13	120.5(2)	H32A–C32–H32B	107.600	C55–C56–H56	120.100
C15–C14–H14	119.700	C34–C33–C38	118.5(2)	C54–C56–H56	120.100
C13–C14–H14	119.700	C34–C33–P2	118.24(16)		

PdPhNO$_2$–^{i}PrOMe

Table A.64.: Structural parameters of **PdPhNO$_2$–^{i}PrOMe**:

molecular formula	C$_{49}$H$_{45}$NO$_5$P$_2$Pd
molecular weight [a.m.u.]	896.20
colour/habit	orange block
crystal system	monoclinic
space group	C1$_2$/c
unit cell dimensions	a = 48.4636(12) Å α = 90.00°
	b = 9.9207(2) Å β = 108.055(3)°
	c = 19.1960(6) Å γ = 90.00°

continued on next page

Table A.64 – continued from previous page

cell volume [Å^3]	8774.8(4)
Z	8
T [K]	123(2)
calculated density [gcm^{-3}]	1.357
absorption coefficient μ [mm^{-1}]	0.543
F(000)	3696
Θ range [°]	2.51 - 26.02
index ranges [h, k, l]	±59, +12/-10, +21/-23
no. of independent reflections / R$_{int}$	8643
no. of absorbed reflections	5010
no. of data	8643
no. of restraints	4
no. of parameters	506
R$_1$ ($I > 2\sigma(I)$)	0.0455
wR$_2$ ($I > 2\sigma(I)$)	0.0829
R$_1$ (all data)	0.0991
wR$_2$ (all data)	0.0918
goodness of fit (on F^2)	0.868

Table A.65.: Bond lengths of **PdPhNO$_2$–^{i}PrOMe** [Å]:

Pd1–C43	2.131(4)	C7–C12	1.378(5)	C28–C29	1.380(6)
Pd1–C44	2.138(4)	C7–C8	1.400(5)	C29–C30	1.390(6)
Pd1–P1	2.3095(11)	C8–C9	1.374(6)	C31–C32	1.378(5)
Pd1–P2	2.3192(11)	C9–C10	1.380(6)	C31–C36	1.401(6)
P2–C25	1.820(4)	C10–C11	1.385(5)	C32–C33	1.390(6)
P2–C31	1.822(4)	C11–C12	1.402(5)	C33–C34	1.390(7)
P2–C19	1.828(4)	C13–C14	1.383(5)	C34–C35	1.364(6)
P1–C12	1.819(4)	C13–C18	1.384(5)	C35–C36	1.382(6)

continued on next page

Table A.65 – continued from previous page

P1–C6	1.828(4)	C14–C15	1.376(6)	C37–C38	1.375(5)
P1–C18	1.831(4)	C15–C16	1.376(6)	C37–C42	1.381(6)
N1–O1	1.215(4)	C16–C17	1.382(6)	C38–C39	1.386(5)
N1–O2	1.221(5)	C17–C18	1.399(5)	C39–C40	1.392(5)
N1–C37	1.465(5)	C19–C20	1.380(5)	C40–C41	1.414(5)
O3–C45	1.205(4)	C19–C24	1.380(6)	C40–C43	1.464(5)
O4–C45	1.362(5)	C20–C21	1.381(6)	C41–C42	1.376(5)
O4–C46A	1.464(6)	C21–C22	1.361(6)	C43–C44	1.417(5)
C1–C2	1.382(5)	C22–C23	1.368(6)	C44–C45	1.462(5)
C1–C6	1.385(5)	C23–C24	1.378(6)	C46A–C47A	1.484(8)
C2–C3	1.384(6)	C25–C30	1.378(5)	C46A–C49A	1.505(8)
C3–C4	1.375(6)	C25–C26	1.381(6)	C47A–O5A	1.563(7)
C4–C5	1.385(6)	C26–C27	1.392(6)	C48A–O5A	1.424(9)
C5–C6	1.383(5)	C27–C28	1.359(6)	C48B–O5B	1.398(9)

Table A.66.: Bond angles of **PdPhNO$_2$–^{i}PrOMe** [°]:

C43–Pd1–C44	38.77(14)	C8–C9–C10	120.7(4)	C36–C31–P2	116.6(3)
C43–Pd1–P1	103.7(1)	C9–C10–C11	119.6(4)	C31–C32–C33	121.5(4)
C44–Pd1–P1	142.10(11)	C10–C11–C12	120.6(4)	C34–C33–C32	118.8(5)
C43–Pd1–P2	143.17(10)	C7–C12–C11	118.8(4)	C35–C34–C33	120.6(5)
C44–Pd1–P2	104.88(11)	C7–C12–P1	123.7(3)	C34–C35–C36	120.4(5)
P1–Pd1–P2	112.97(4)	C11–C12–P1	117.0(3)	C35–C36–C31	120.3(4)
C25–P2–C31	103.56(17)	C14–C13–C18	121.3(4)	C38–C37–C42	121.6(4)
C25–P2–C19	102.13(18)	C15–C14–C13	119.5(4)	C38–C37–N1	119.2(4)
C31–P2–C19	103.19(19)	C16–C15–C14	120.2(4)	C42–C37–N1	119.2(4)
C25–P2–Pd1	112.74(13)	C15–C16–C17	120.5(4)	C37–C38–C39	118.6(4)
C31–P2–Pd1	110.81(14)	C16–C17–C18	120.1(4)	C38–C39–C40	122.1(4)
C19–P2–Pd1	122.41(13)	C13–C18–C17	118.4(4)	C39–C40–C41	117.2(4)

continued on next page

Table A.66 – continued from previous page

C12–P1–C6	104.75(18)	C13–C18–P1	119.6(3)	C39–C40–C43	120.1(3)
C12–P1–C18	105.50(18)	C17–C18–P1	121.9(3)	C41–C40–C43	122.7(4)
C6–P1–C18	101.53(18)	C20–C19–C24	117.0(4)	C42–C41–C40	121.2(4)
C12–P1–Pd1	111.69(13)	C20–C19–P2	119.3(3)	C41–C42–C37	119.3(4)
C6–P1–Pd1	114.59(13)	C24–C19–P2	123.7(3)	C44–C43–C40	123.4(3)
C18–P1–Pd1	117.45(13)	C19–C20–C21	121.7(4)	C44–C43–Pd1	70.9(2)
O1–N1–O2	124.1(4)	C22–C21–C20	120.2(4)	C40–C43–Pd1	110.5(2)
O1–N1–C37	118.6(4)	C21–C22–C23	119.2(5)	C43–C44–C45	117.5(4)
O2–N1–C37	117.3(4)	C22–C23–C24	120.5(4)	C43–C44–Pd1	70.3(2)
C45–O4–C46A	116.1(3)	C23–C24–C19	121.3(4)	C45–C44–Pd1	108.4(3)
C2–C1–C6	120.5(4)	C30–C25–C26	118.2(4)	O3–C45–O4	122.3(4)
C1–C2–C3	120.4(4)	C30–C25–P2	119.1(3)	O3–C45–C44	126.3(4)
C4–C3–C2	119.4(4)	C26–C25–P2	122.6(3)	O4–C45–C44	111.4(4)
C3–C4–C5	120.2(4)	C25–C26–C27	120.3(4)	O4–C46A–C47A	106.6(4)
C6–C5–C4	120.8(4)	C28–C27–C26	120.8(4)	O4–C46A–C49A	107.6(5)
C5–C6–C1	118.8(4)	C27–C28–C29	119.9(4)	C47A–C46A–C49A	113.7(5)
C5–C6–P1	117.1(3)	C28–C29–C30	119.2(4)	C46A–C47A–O5A	91.6(5)
C1–C6–P1	124.1(3)	C25–C30–C29	121.6(4)	C48A–O5A–C47A	105.2(8)
C12–C7–C8	120.6(4)	C32–C31–C36	118.4(4)		
C9–C8–C7	119.6(4)	C32–C31–P2	125.0(3)		

6.3

Table A.67.: Structural parameters of **6.3**:

molecular formula	$C_{117}H_{99}P_6Pd$
molecular weight [a.m.u.]	1903.58
colour/habit	yellow block
crystal dimensions [mm^3]	$0.13 \times 0.11 \times 0.08$

continued on next page

Table A.67 – continued from previous page

crystal system	triclinic
space group	P-1
unit cell dimensions	a = 13.0295(5) Å α = 90.307(2)°
	b = 18.1620(13) Å β = 90.9250(10)°
	c = 20.7355(5) Å γ = 101.7400(10)°
cell volume [Å^3]	4803.4(4)
Z	2
T [K]	173(2)
calculated density [gcm^{-3}]	1.316
absorption coefficient μ [mm^{-1}]	0.524
F(000)	1966
Θ range [°]	0.98 - 25.43
index ranges [h, k, l]	±15, +21/-20, +22/-25
no. of independent reflections / R_{int}	17143
no. of absorbed reflections	14097
no. of data	17143
no. of restraints	9
no. of parameters	1082
R_1 ($I > 2\sigma(I)$)	0.0403
wR_2 ($I > 2\sigma(I)$)	0.0997
R_1 (all data)	0.0541
wR_2 (all data)	0.1069
goodness of fit (on F^2)	1.060
largest difference peak [eÅ^3]	1.240
largest difference hole [eÅ^3]	-0.632

Table A.68.: Bond lengths of **6.3** [Å]:

C21–H21	0.9500	C64–H64	0.9500	C107–H107	0.9500
C22–C23	1.381(5)	C65–C66	1.386(5)	C108–H108	0.9500
C22–H22	0.9500	C65–H65	0.9500	P1–Pd1	2.3298(9)
C23–C24	1.388(5)	C66–H66	0.9500	P2–Pd1	2.3232(9)
C23–H23	0.9500	C67–C72	1.384(5)	P3–Pd1	2.3275(9)
C24–H24	0.9500	C67–C68	1.388(5)	P4–Pd2	2.3045(8)
C25–C26	1.384(5)	C67–P5	1.840(3)	P5–Pd2	2.3241(8)
C25–C30	1.388(5)	C68–C69	1.385(5)	P6–Pd2	2.3165(9)
C25–P2	1.851(3)	C68–H68	0.9500	C110–C111	1.3900
C26–C27	1.392(5)	C69–C70	1.374(6)	C110–C115	1.3900
C26–H26	0.9500	C69–H69	0.9500	C110–H110	0.9500
C27–C28	1.375(6)	C70–C71	1.376(6)	C111–C112	1.3900
C27–H27	0.9500	C70–H70	0.9500	C111–H111	0.9500
C28–C29	1.371(6)	C71–C72	1.386(5)	C112–C113	1.3900
C28–H28	0.9500	C71–H71	0.9500	C112–H112	0.9500
C29–C30	1.389(6)	C72–H72	0.9500	C113–C114	1.3900
C29–H29	0.9500	C73–C78	1.388(5)	C113–H113	0.9500
C30–H30	0.9500	C73–C74	1.389(5)	C114–C115	1.3900
C31–C32	1.394(5)	C73–P6	1.845(3)	C114–H114	0.9500
C31–C36	1.396(5)	C74–C75	1.386(5)	C115–H115	0.9500
C31–P2	1.841(4)	C74–H74	0.9500	C117–C118	1.3900
C32–C33	1.394(5)	C75–C76	1.373(6)	C117–C122	1.3900
C32–H32	0.9500	C75–H75	0.9500	C117–H117	0.9500
C33–C34	1.374(6)	C76–C77	1.369(6)	C118–C119	1.3900
C33–H33	0.9500	C76–H76	0.9500	C118–H118	0.9500
C34–C35	1.378(6)	C77–C78	1.390(6)	C119–C120	1.3900
C34–H34	0.9500	C77–H77	0.9500	C119–H119	0.9500

continued on next page

Table A.68 – continued from previous page

C35–C36	1.382(5)	C78–H78	0.9500	C120–C121	1.3900
C35–H35	0.9500	C79–C84	1.389(5)	C120–H120	0.9500
C36–H36	0.9500	C79–C80	1.390(5)	C121–C122	1.3900
C37–C38	1.387(5)	C79–P6	1.846(3)	C121–H121	0.9500
C37–C42	1.395(5)	C80–C81	1.383(5)	C122–H122	0.9500
C37–P3	1.843(4)	C80–H80	0.9500	C124–C125	1.3900
C38–C39	1.385(6)	C81–C82	1.380(5)	C124–C129	1.3900
C38–H38	0.9500	C81–H81	0.9500	C124–H124	0.9500
C39–C40	1.375(6)	C82–C83	1.380(5)	C125–C126	1.3900
C39–H39	0.9500	C82–H82	0.9500	C125–H125	0.9500
C40–C41	1.362(7)	C83–C84	1.386(5)	C126–C127	1.3900
C40–H40	0.9500	C83–H83	0.9500	C126–H126	0.9500
C41–C42	1.392(6)	C84–H84	0.9500	C127–C128	1.3900
C41–H41	0.9500	C85–C90	1.384(5)	C127–H127	0.9500
C42–H42	0.9500	C85–C86	1.393(5)	C128–C129	1.3900
C43–C44	1.380(5)	C85–P6	1.848(4)	C128–H128	0.9500
C43–C48	1.395(5)	C86–C87	1.388(5)	C129–H129	0.9500

Table A.69.: Bond angles of **6.3** [°]:

C6–C1–C2	118.4(3)	C48–C47–H47	119.800	C95–C94–H94	120.100
C6–C1–P1	118.1(3)	C47–C48–C43	120.8(4)	C93–C94–H94	120.100
C2–C1–P1	123.4(3)	C47–C48–H48	119.600	C94–C95–C96	120.2(4)
C3–C2–C1	120.5(4)	C43–C48–H48	119.600	C94–C95–H95	119.900
C3–C2–H2	119.700	C54–C49–C50	118.7(3)	C96–C95–H95	119.900
C1–C2–H2	119.700	C54–C49–P3	124.0(3)	C95–C96–C91	120.7(4)
C4–C3–C2	120.1(4)	C50–C49–P3	117.4(3)	C95–C96–H96	119.700
C4–C3–H3	119.900	C51–C50–C49	120.2(4)	C91–C96–H96	119.700
C2–C3–H3	119.900	C51–C50–H50	119.900	C98–C97–C102	118.4(3)

continued on next page

Table A.69 – continued from previous page

C5–C4–C3	119.7(4)	C49–C50–H50	119.900	C98–C97–P4	117.5(2)
C5–C4–H4	120.100	C52–C51–C50	120.6(4)	C102–C97–P4	124.1(3)
C3–C4–H4	120.100	C52–C51–H51	119.700	C99–C98–C97	120.9(3)
C4–C5–C6	120.4(4)	C50–C51–H51	119.700	C99–C98–H98	119.500
C4–C5–H5	119.800	C53–C52–C51	120.0(4)	C97–C98–H98	119.500
C6–C5–H5	119.800	C53–C52–H52	120.000	C100–C99–C98	120.0(4)
C5–C6–C1	120.7(3)	C51–C52–H52	120.000	C100–C99–H99	120.000
C5–C6–H6	119.600	C52–C53–C54	120.1(4)	C98–C99–H99	120.000
C1–C6–H6	119.600	C52–C53–H53	119.900	C99–C100–C101	120.1(3)
C12–C7–C8	118.0(3)	C54–C53–H53	119.900	C99–C100–H100	120.000
C12–C7–P1	118.2(3)	C49–C54–C53	120.4(4)	C101–C100–H100	120.000
C8–C7–P1	123.8(3)	C49–C54–H54	119.800	C100–C101–C102	119.9(3)
C9–C8–C7	121.0(4)	C53–C54–H54	119.800	C100–C101–H101	120.000
C9–C8–H8	119.500	C56–C55–C60	118.5(4)	C102–C101–H101	120.000
C7–C8–H8	119.500	C56–C55–P5	123.8(3)	C101–C102–C97	120.7(3)
C10–C9–C8	120.0(4)	C60–C55–P5	117.7(3)	C101–C102–H102	119.700
C10–C9–H9	120.000	C57–C56–C55	120.6(4)	C97–C102–H102	119.700
C8–C9–H9	120.000	C57–C56–H56	119.700	C104–C103–C108	118.0(3)
C9–C10–C11	119.8(4)	C55–C56–H56	119.700	C104–C103–P4	118.6(2)
C9–C10–H10	120.100	C58–C57–C56	119.8(4)	C108–C103–P4	123.5(2)
C11–C10–H10	120.100	C58–C57–H57	120.100	C105–C104–C103	120.8(3)
C10–C11–C12	120.3(4)	C56–C57–H57	120.100	C105–C104–H104	119.600
C10–C11–H11	119.900	C59–C58–C57	120.3(4)	C103–C104–H104	119.600
C12–C11–H11	119.900	C59–C58–H58	119.800	C104–C105–C106	120.6(3)
C11–C12–C7	120.9(4)	C57–C58–H58	119.800	C104–C105–H105	119.700
C11–C12–H12	119.500	C58–C59–C60	120.6(4)	C106–C105–H105	119.700
C7–C12–H12	119.500	C58–C59–H59	119.700	C107–C106–C105	119.5(3)
C14–C13–C18	118.1(3)	C60–C59–H59	119.700	C107–C106–H106	120.300

continued on next page

Table A.69 – continued from previous page

C14–C13–P1	125.0(3)	C59–C60–C55	120.3(4)	C105–C106–H106	120.300
C18–C13–P1	116.9(3)	C59–C60–H60	119.900	C106–C107–C108	120.0(3)
C13–C14–C15	120.8(4)	C55–C60–H60	119.900	C106–C107–H107	120.000
C13–C14–H14	119.600	C62–C61–C66	118.5(3)	C108–C107–H107	120.000
C15–C14–H14	119.600	C62–C61–P5	123.0(3)	C107–C108–C103	121.1(3)
C16–C15–C14	120.1(4)	C66–C61–P5	117.9(3)	C107–C108–H108	119.400
C16–C15–H15	119.900	C61–C62–C63	120.5(4)	C103–C108–H108	119.400
C14–C15–H15	119.900	C61–C62–H62	119.800	C13–P1–C1	103.00(16)
C15–C16–C17	119.9(4)	C63–C62–H62	119.800	C13–P1–C7	98.88(15)
C15–C16–H16	120.000	C64–C63–C62	120.3(4)	C1–P1–C7	99.99(15)
C17–C16–H16	120.000	C64–C63–H63	119.800	C13–P1–Pd1	116.19(11)
C16–C17–C18	120.3(4)	C62–C63–H63	119.800	C1–P1–Pd1	111.20(11)
C16–C17–H17	119.800	C65–C64–C63	119.7(4)	C7–P1–Pd1	124.50(11)
C18–C17–H17	119.800	C65–C64–H64	120.100	C31–P2–C19	103.60(16)
C17–C18–C13	120.7(4)	C63–C64–H64	120.100	C31–P2–C25	103.06(16)
C17–C18–H18	119.700	C64–C65–C66	120.0(4)	C19–P2–C25	97.66(15)
C13–C18–H18	119.700	C64–C65–H65	120.000	C31–P2–Pd1	108.20(11)
C24–C19–C20	118.3(3)	C66–C65–H65	120.000	C19–P2–Pd1	122.66(11)
C24–C19–P2	118.8(3)	C65–C66–C61	120.9(4)	C25–P2–Pd1	119.12(11)
C20–C19–P2	122.8(3)	C65–C66–H66	119.600	C49–P3–C37	102.92(17)
C21–C20–C19	120.6(3)	C61–C66–H66	119.600	C49–P3–C43	98.67(16)
C21–C20–H20	119.700	C72–C67–C68	118.1(3)	C37–P3–C43	103.67(17)
C19–C20–H20	119.700	C72–C67–P5	116.7(3)	C49–P3–Pd1	120.31(12)
C22–C21–C20	120.4(3)	C68–C67–P5	125.2(3)	C37–P3–Pd1	105.23(11)
C22–C21–H21	119.800	C69–C68–C67	120.6(3)	C43–P3–Pd1	123.39(12)
C20–C21–H21	119.800	C69–C68–H68	119.700	C91–P4–C97	101.41(15)
C23–C22–C21	119.4(4)	C67–C68–H68	119.700	C91–P4–C103	102.52(15)
C23–C22–H22	120.300	C70–C69–C68	120.7(4)	C97–P4–C103	98.91(14)

continued on next page

Table A.69 – continued from previous page

C21–C22–H22	120.300	C70–C69–H69	119.600	C91–P4–Pd2	108.15(11)
C22–C23–C24	120.1(4)	C68–C69–H69	119.600	C97–P4–Pd2	120.73(10)
C22–C23–H23	119.900	C69–C70–C71	119.3(4)	C103–P4–Pd2	122.04(11)
C24–C23–H23	119.900	C69–C70–H70	120400	C55–P5–C61	102.34(16)
C19–C24–C23	121.1(3)	C71–C70–H70	120400	C55–P5–C67	104.12(15)
C19–C24–H24	119.400	C70–C71–C72	120.2(4)	C61–P5–C67	100.55(15)
C23–C24–H24	119.400	C70–C71–H71	119.900	C55–P5–Pd2	115.16(11)
C26–C25–C30	118.4(3)	C72–C71–H71	119.900	C61–P5–Pd2	115.51(11)
C26–C25–P2	124.9(3)	C67–C72–C71	121.1(3)	C67–P5–Pd2	117.02(11)
C30–C25–P2	116.6(3)	C67–C72–H72	119.400	C73–P6–C79	101.37(15)
C25–C26–C27	120.5(4)	C71–C72–H72	119.400	C73–P6–C85	101.54(15)
C25–C26–H26	119.800	C78–C73–C74	118.2(3)	C79–P6–C85	99.94(15)
C27–C26–H26	119.800	C78–C73–P6	116.9(3)	C73–P6–Pd2	113.44(11)
C28–C27–C26	120.5(4)	C74–C73–P6	124.9(3)	C79–P6–Pd2	121.90(11)
C28–C27–H27	119.700	C75–C74–C73	120.8(4)	C85–P6–Pd2	115.70(12)
C26–C27–H27	119.700	C75–C74–H74	119.600	P2–Pd1–P3	118.41(3)
C29–C28–C27	119.5(4)	C73–C74–H74	119.600	P2–Pd1–P1	120.20(3)
C29–C28–H28	120.300	C76–C75–C74	120.3(4)	P3–Pd1–P1	118.71(3)
C27–C28–H28	120.300	C76–C75–H75	119.900	P4–Pd2–P6	124.52(3)
C28–C29–C30	120.4(4)	C74–C75–H75	119.900	P4–Pd2–P5	120.05(3)
C28–C29–H29	119.800	C77–C76–C75	119.6(4)	P6–Pd2–P5	115.05(3)
C30–C29–H29	119.800	C77–C76–H76	120.200	C111–C110–C115	120.000
C25–C30–C29	120.7(4)	C75–C76–H76	120.200	C111–C110–H110	120.000
C25–C30–H30	119.600	C76–C77–C78	120.6(4)	C115–C110–H110	120.000
C29–C30–H30	119.600	C76–C77–H77	119.700	C112–C111–C110	120.000
C32–C31–C36	117.9(3)	C78–C77–H77	119.700	C112–C111–H111	120.000
C32–C31–P2	123.7(3)	C73–C78–C77	120.4(4)	C110–C111–H111	120.000
C36–C31–P2	118.2(3)	C73–C78–H78	119.800	C111–C112–C113	120.000

continued on next page

Table A.69 – continued from previous page

C33–C32–C31	120.6(4)	C77–C78–H78	119.800	C111–C112–H112	120.000
C33–C32–H32	119.700	C84–C79–C80	118.3(3)	C113–C112–H112	120.000
C31–C32–H32	119.700	C84–C79–P6	117.2(3)	C114–C113–C112	120.000
C34–C33–C32	120.3(4)	C80–C79–P6	124.5(3)	C114–C113–H113	120.000
C34–C33–H33	119.800	C81–C80–C79	120.4(3)	C112–C113–H113	120.000
C32–C33–H33	119.800	C81–C80–H80	119.800	C113–C114–C115	120.000
C33–C34–C35	119.7(4)	C79–C80–H80	119.800	C113–C114–H114	120.000
C33–C34–H34	120.100	C82–C81–C80	120.7(3)	C115–C114–H114	120.000
C35–C34–H34	120.100	C82–C81–H81	119.600	C114–C115–C110	120.000
C34–C35–C36	120.4(4)	C80–C81–H81	119.600	C114–C115–H115	120.000
C34–C35–H35	119.800	C83–C82–C81	119.5(3)	C110–C115–H115	120.000
C36–C35–H35	119.800	C83–C82–H82	120.200	C118–C117–C122	120.000
C35–C36–C31	121.0(4)	C81–C82–H82	120.200	C118–C117–H117	120.000
C35–C36–H36	119.500	C82–C83–C84	119.8(3)	C122–C117–H117	120.000
C31–C36–H36	119.500	C82–C83–H83	120.100	C117–C118–C119	120.000
C38–C37–C42	117.6(4)	C84–C83–H83	120.100	C117–C118–H118	120.000
C38–C37–P3	117.3(3)	C83–C84–C79	121.2(3)	C119–C118–H118	120.000
C42–C37–P3	124.8(3)	C83–C84–H84	119.400	C118–C119–C120	120.000
C39–C38–C37	121.7(4)	C79–C84–H84	119.400	C118–C119–H119	120.000
C39–C38–H38	119100	C90–C85–C86	118.4(3)	C120–C119–H119	120.000
C37–C38–H38	119100	C90–C85–P6	117.9(3)	C121–C120–C119	120.000
C40–C39–C38	119.7(4)	C86–C85–P6	123.5(3)	C121–C120–H120	120.000
C40–C39–H39	120.100	C87–C86–C85	120.8(4)	C119–C120–H120	120.000
C38–C39–H39	120.100	C87–C86–H86	119.600	C122–C121–C120	120.000
C41–C40–C39	119.7(4)	C85–C86–H86	119.600	C122–C121–H121	120.000
C41–C40–H40	120.200	C88–C87–C86	119.8(4)	C120–C121–H121	120.000
C39–C40–H40	120.200	C88–C87–H87	120.100	C121–C122–C117	120.000
C40–C41–C42	121.2(4)	C86–C87–H87	120.100	C121–C122–H122	120.000

continued on next page

Table A.69 – continued from previous page

C40–C41–H41	119.400	C89–C88–C87	120.0(4)	C117–C122–H122	120.000
C42–C41–H41	119.400	C89–C88–H88	120.000	C125–C124–C129	120.000
C41–C42–C37	120.0(4)	C87–C88–H88	120.000	C125–C124–H124	120.000
C41–C42–H42	120.000	C88–C89–C90	120.1(4)	C129–C124–H124	120.000
C37–C42–H42	120.000	C88–C89–H89	119.900	C126–C125–C124	120.000
C44–C43–C48	117.8(4)	C90–C89–H89	119.900	C126–C125–H125	120.000
C44–C43–P3	118.9(3)	C85–C90–C89	120.9(4)	C124–C125–H125	120.000
C48–C43–P3	123.3(3)	C85–C90–H90	119.600	C125–C126–C127	120.000
C43–C44–C45	121.4(4)	C89–C90–H90	119.600	C125–C126–H126	120.000
C43–C44–H44	119300	C92–C91–C96	118.3(3)	C127–C126–H126	120.000
C45–C44–H44	119300	C92–C91–P4	118.0(3)	C126–C127–C128	120.000
C46–C45–C44	120.0(4)	C96–C91–P4	123.7(3)	C126–C127–H127	120.000
C46–C45–H45	120.000	C91–C92–C93	120.5(4)	C128–C127–H127	120.000
C44–C45–H45	120.000	C91–C92–H92	119.700	C129–C128–C127	120.000
C47–C46–C45	119.6(4)	C93–C92–H92	119.700	C129–C128–H128	120.000
C47–C46–H46	120.200	C94–C93–C92	120.4(4)	C127–C128–H128	120.000
C45–C46–H46	120.200	C94–C93–H93	119.800	C128–C129–C124	120.000
C46–C47–C48	120.4(4)	C92–C93–H93	119.800	C128–C129–H129	120.000
C46–C47–H47	119.800	C95–C94–C93	119.8(4)	C124–C129–H129	120.000

6.9

Table A.70.: Structural parameters of **6.9**:

molecular formula	$C_{148}H_{124}P_8Pd_4Cl_4$
molecular weight [a.m.u.]	2717.89
crystal system	monoclinic
space group	$P2_1/n$
unit cell dimensions	a = 9.8643(5) Å $\alpha = 90.00°$

continued on next page

Table A.70 – continued from previous page

	b = 14.6865(8) Å	β = 96.910(2)°
	c = 20.9721(11) Å	γ = 90.00°
cell volume [Å³]	3016.20(27)	
Z	2	
calculated density [gcm⁻³]	1.496	
absorption coefficient μ [mm⁻¹]	0.74	
F(000)	1288	
no. of independent reflections / R$_{int}$	5530	
no. of absorbed reflections	4883	
no. of data	5530	
no. of restraints	0	
no. of parameters	370	
R$_1$ ($I > 2\sigma(I)$)	0.0540	
R$_1$ (all data)	0.0607	
largest difference peak [eÅ³]	1.593	
largest difference hole [eÅ³]	-1.044	

Table A.71.: Bond lengths of **6.9** [Å]:

PD1–CL1	2.3920(1)	C31–P2	1.8224(1)	C19–C20	1.3909(1)
P2–PD1	2.3253(1)	C31–C32	1.3983(1)	C18–C37	1.5153(1)
P1–PD1	2.2924(1)	C3–C4	1.3764(1)	C17–C18	1.3945(1)
C9–C10	1.3835(1)	C29–C30	1.3772(1)	C16–C17	1.3882(1)
C8–C9	1.3644(1)	C28–C29	1.3857(1)	C15–C16	1.3822(1)
C7–C12	1.3839(1)	C27–C28	1.3766(0)	C14–C15	1.3850(1)
C7–P1	1.8099(1)	C26–C27	1.3760(1)	C13–C14	1.3965(1)
C7–C8	1.4064(1)	C25–P2	1.8292(1)	C13–C18	1.3969(1)
C5–C6	1.3854(0)	C25–C30	1.3829(0)	C13–P1	1.8188(1)
C4–C5	1.3729(1)	C25–C26	1.4011(1)	C11–C12	1.4048(1)

continued on next page

Table A.71 – continued from previous page

C37–PD1	2.0630(1)	C23–C24	1.3787(1)	C10–C11	1.3907(1)
C35–C36	1.3802(1)	C22–C23	1.3811(1)	C1–P1	1.8177(1)
C34–C35	1.3772(1)	C21–C22	1.3796(1)	C1–C6	1.3788(1)
C33–C34	1.4004(1)	C20–C21	1.3882(1)	C1–C2	1.3960(1)
C32–C33	1.3647(0)	C2–C3	1.3938(0)		
C31–C36	1.3901(1)	C19–P2	1.8345(1)		

Table A.72.: Bond angles of **6.9** [°]:

C1–C2–C3	119.853(3)	C2–C3–C4	119.866(4)	C32–C31–P2	117.651(3)
C1–P1–PD1	118.814(2)	C20–C19–P2	118.324(3)	C32–C33–C34	119.735(3)
C1–P1–C7	103.216(2)	C20–C19–C24	118.850(3)	C32–C31–C36	118.568(3)
C1–C6–C5	121.227(4)	C20–C21–C22	120.658(4)	C33–C34–C35	119.542(3)
C10–C11–C12	118.877(3)	C21–C22–C23	119.673(3)	C34–C35–C36	120.606(3)
C12–C7–P1	122.441(3)	C22–C23–C24	120.033(3)	C36–C31–P2	123.761(3)
C13–C18–C17	118.296(3)	C24–C19–P2	122.448(3)	C37–PD1–P2	86.205(2)
C13–P1–PD1	104.050(2)	C25–C30–C29	121.036(3)	C37–PD1–P1	82.263(2)
C13–C18–C37	119.776(3)	C25–P2–PD1	109.661(2)	C4–C5–C6	119.395(3)
C14–C13–C18	121.034(4)	C25–C26–C27	120.745(4)	C6–C1–P1	121.537(3)
C14–C15–C16	120.594(3)	C26–C25–P2	120.730(3)	C7–P1–C13	105.735(3)
C15–C16–C17	119.873(4)	C26–C25–C30	118.080(4)	C7–C12–C11	120.917(3)
C16–C17–C18	120.918(4)	C26–C27–C28	120.377(3)	C7–C8–C9	120.782(3)
C18–C13–P1	112.302(3)	C27–C28–C29	119.413(4)	C8–C9–C10	120.559(3)
C18–C37–PD1	116.750(3)	C28–C29–C30	120.287(4)	C8–C7–C12	118.576(3)
C19–P2–PD1	123.380(3)	C3–C4–C5	120.736(4)	C8–C7–P1	118.748(3)
C19–C24–C23	120.847(4)	C30–C25–P2	121.183(3)	C9–C10–C11	120.273(3)
C19–C20–C21	119.841(3)	C31–C32–C33	121.194(3)	P1–PD1–CL1	95.164(2)
C2–C1–C6	118.917(3)	C31–P2–PD1	113.175(2)	P2–PD1–CL1	97.146(2)
C2–C1–P1	119.516(3)	C31–C36–C35	120.346(3)		

A.2. List of compounds

Organic compounds

Complexes

Pt(cod)Ph-bp (η^4-cycloocta-1,5-diene)-(η^2-2-benzoylphenyl cinnamate)platinum(0) 254

A.3. List of figures

A.4. List of graphs

A.5. List of schemes

A.6. List of tables